Lecture Notes in Computer Science 1351
Edited by G. Goos, J. Hartmanis and J. van Leeuwen

Springer
*Berlin
Heidelberg
New York
Barcelona
Budapest
Hong Kong
London
Milan
Paris
Santa Clara
Singapore
Tokyo*

Roland Chin Ting-Chuen Pong (Eds.)

Computer Vision – ACCV'98

Third Asian Conference on Computer Vision
Hong Kong, China, January 8-10, 1998
Proceedings, Volume I

Springer

Series Editors

Gerhard Goos, Karlsruhe University, Germany

Juris Hartmanis, Cornell University, NY, USA

Jan van Leeuwen, Utrecht University, The Netherlands

Volume Editors

Roland Chin
Ting-Chuen Pong
Hong Kong University of Science and Technology
Computer Science Department
Clear Water Bay, Kowloon, Hong Kong, China
E-mail:(roland,tcpong)@cs.ust.hk

Cataloging-in-Publication data applied for

Die Deutsche Bibliothek - CIP-Einheitsaufnahme

Computer vision : proceedings / ACCV '98, Third Asian Conference on Computer Vision, Hong Kong, China, January 8 - 10, 1998. Roland Chin ; Ting-Chuen Pong (ed.). - Berlin ; Heidelberg ; New York ; Barcelona ; Budapest ; Hong Kong ; London ; Milan ; Paris ; Santa Clara ; Singapore ; Tokyo : Springer

Vol. 1 (1998)
(Lecture notes in computer science ; Vol. 1351)
ISBN 3-540-63930-6

CR Subject Classification (1991): I.3.5, I.5, I.2.9-10, I.4

ISSN 0302-9743
ISBN 3-540-63930-6 Springer-Verlag Berlin Heidelberg New York

This work is subject to copyright. All rights are reserved, whether the whole or part of the material is concerned, specifically the rights of translation, reprinting, re-use of illustrations, recitation, broadcasting, reproduction on microfilms or in any other way, and storage in data banks. Duplication of this publication or parts thereof is permitted only under the provisions of the German Copyright Law of September 9, 1965, in its current version, and permission for use must always be obtained from Springer-Verlag. Violations are liable for prosecution under the German Copyright Law.

© Springer-Verlag Berlin Heidelberg 1997
Printed in Germany

Typesetting: Camera-ready by author
SPIN 10661298   06/3142 – 5 4 3 2 1 0   Printed on acid-free paper

Preface

We are very pleased to have the opportunity to organize the 3rd Asian Conference on Computer Vision (ACCV'98). The conference is sponsored by the IEEE Hong Kong Section, Computer Chapter, the Sino Software Research Institute and the Department of Computer Science of the Hong Kong University of Science and Technology, and the Hong Kong Industry Department.

We received over 300 submissions of full papers (not including the invited papers for the special sessions) from 30 countries in April 1997. In order to provide a quality conference and quality proceedings, each paper was reviewed by at least three members of the program committee. The program committee selected and accepted 58 papers for oral presentation and 112 papers for poster presentation after the review process. Some of these papers were jointly submitted to ACCV'98 and ICCV'98 (to be held in Bombay in January 4–7, 1998) and they were reviewed in a coordinated effort. We must add that the program committee and the reviewers have done an excellent job within a tight schedule and we are very pleased with the quality of the papers.

Four eminent invited speakers, Professors Brian Funt of Simon Fraser University, Krishna Nathan of IBM, Eric Grimson of MIT, and Shoji Tominaga of Osaka Electro-Communication University, have contributed to the conference. We are grateful to them. In addition, we wish to thank Professors Jake Aggarwal, Shashi Buluswar, Yi-Ping Hung, Anil Jain, and Sharatchandra Pankanti for organizing the very high-quality special sessions. Last but not least, we would like to express our gratitude to all the contributors, reviewers, program committee and organizing committee members, and sponsors, without whom the conference would not have been possible.

Finally, we hope that you will benefit from these proceedings.

Roland T. Chin
Ting-Chuen Pong

January 1998

Conference Chair:
Helen Shen (Hong Kong U. of Science & Technology)
Song De Ma (Inst. of Automation, Beijing)

Program Co-Chairs:
Roland Chin (Hong Kong U. of Science & Technology)
T.C. Pong (Hong Kong U. of Science & Technology)
Saburo Tsuji (Wakayama U.)

Program Committee:
Jake Aggarwal (U. of Texas)
Narendra Ahuja (U. of Illinois)
Carlo Arcelli (Institute for Cybernetics, Italy)
Terry Caelli (Curtin U. of Technology)
Larry Davis (U. of Maryland)
Xiaoqing Ding (Tsinghua U.)
Charles Dyer (U. of Wisconsin)
Olivier Faugeras (INRIA)
Jun-ichi Hasegawa (Chyukyo U.)
Thomas Huang (U. of Illinois)
Katsushi Ikeuchi (U. of Tokyo)
Horace Ip (City U. of Hong Kong)
Anil Jain (Michigan State U.)
Ramesh Jain (U. of California, San Diego)
Ben Jang (IBM)
Ray Jarvis (Monash U.)
Rangachar Kasturi (Penn State U.)
Kwang Ik Kim (POSTECH)
Les Kitchen (U. of Melbourne)
Josef Kittler (U. of Surrey)
Kok Fung Lai (ITI, Singapore)
Louisa Lam (Hong Kong Inst. of Education)
Chung-Nim Lee (POSTECH)
Hsi-Jian Lee (Chiao Tung U.)
Seong-Whan Lee (Korea U.)
Takashi Matsuyama (Kyoto U.)
Dinesh Mital (Nanyang Technological U.)
Shree Nayar (Columbia U.)
Ram Nevatia (U. of Southern California)
Yuichi Ohta (Tsukuba U.)
Shmuel Peleg (Hebrew U.)
Brent Seales (U. of Kentucky)
Yoshiaki Shirai (Osaka U.)
Arnold Smeulders (U. of Amsterdam)
Ching Y. Suen (Concordia U.)
Michael Swain (U. of Chicago)
Eam KhwangTeoh (Nanyang Technological U.)
Baba Vemuri (U. of Florida)
Kazuhiko Yamamoto (Gifu U.)
Naokazu Yokoya (AIST-Nara)

Organising Committee:
Oscar Au (Hong Kong U. of Science & Technology)
Ronald Chung (Chinese U. of Hong Kong)
Horace Ip (City U. of Hong Kong)
Tong Lee (Chinese U. of Hong Kong)
Chiew Lan Tai (Hong Kong U. of Science & Technology)
H.T. Tsui (Chinese U. of Hong Kong)
Christopher Yang (U. of Hong Kong)

Sponsored by
IEEE Hong Kong Section, Computer Chapter
Sino Software Research Institure, Hong Kong U. of Science and Technology
Department of Computer Science, Hong Kong U. of Science and Technology
Hong Kong Industry Department

Contents of Volume I

Invited Talk

Pen Computing - An Overview
   Krishna Nathan..I-1

Session T1A: Biometry I

Research Issues in Biometrics
   Ruud M. Bolle, Nalini K. Ratha, and S. Pankanti................................. I-2

Automatic On-line Signature Verification
   Vishvjit S. Nalwa...I-10

Integrating Faces and Fingerprints for Personal Identification
   Lin Hong and Anil Jain..I-16

Automated Fingerprint Pattern Classification Error Analysis
   Weicheng Shen... I-24

A High-Dimensional Indexing Scheme for Scalable Fingerprint-Based Identification
   Andrea Califano, Bob Germain, and Scott Colville............................. I-32

Session T1B: Physics-Based Vision

Sign of Surface Curvature from Shading Images Using Neural Network
   Yuji Iwahori, Masamitsu Murakami, Robert J. Woodham and
   Naohiro Ishii..I-40

On the Classification of Singular Points for the Global Shape from Shading Problem: A Study of the Constraints Imposed by Isophotes
   Takayuki Okatani and Koichiro Deguchi...I-48

Determination of Sign of Gaussian Curvature of Surface Having General Reflectance Property
   Takayuki Okatani and Koichiro Deguchi... I-56

Estimating Depth Through the Fusion of Photometric Stereo Images
   João L. Fernandes and José R. A. Torreão..I-64

Out of the Dark: Using Shadows to Reconstruct 3D Surfaces
  M. Daum and G. Dudek I-72

Session T2A: Color Vision I

Estimation of Reflection Parameters from a Color Image
  Shoji Tominaga I-80

A Natural Norm for Color Processing
  Ron Kimmel I-88

A Color Normalization Algorithm for Image Indexing
  In Kyu Park, Il Dong Yun and Sang Uk Lee I-96

Adaptive Color-Image Embeddings for Database Navigation
  Yossi Rubner, Carlo Tomasi and Leonidas J. Guibas I-104

A Large Capacity Steganography Using Color BMP Images
  Koichi Nozaki, Michiharu Niimi, Richard O. Eason and Eiji Kawaguchi I-112

Session T2B: Robot Vision and Navigation

Dynamic Calibration of an Active Vision System to Compute the Ground Plane Transformation
  Fuxing Li and Michael Brady I-120

Identification of 3D Reference Structures for Video-Based Localization
  Darius Burschka and Stefan A. Blum I-128

Directing Robots with Visual Primitives for Navigation and Micro-manipulation
  W. B. Tong, S.K. Tso, S. Lang, G.Z. Lu and S.D. Ma I-136

Combining Camera and Laser Radar for ALV Navigation
  Qi Zhang and Weikang Gu I-144

Stereo Vision-Based Obstacle Detection for Partially Sighted People
  Stephen Se and Michael Brady I-152

Session T3A: OCR and Applications

Evaluation and Application of Recognition Confidence in OCR
  Xiaofan Lin, Xiaoqing Ding, Youbin Chen, Jinhui Liu, and Youshou Wu I-160

A New Nonlinear Shape Normalization Method for Off-line Handwritten Chinese Character Recognition
    Youbin Chen, Xiaoqing Ding, Youshou Wu and Ming Chen I-168

A Novel Triangulation Procedure for Thinning Cursive Text
    Stanley S. Ipson, Muhammed Melhi and William Booth I-176

Digital Geometric Methods in Image Analysis and Compression
    Ari Gross and Longin Latecki I-184

Detection and Enhancement of Small Masses via Precision Multiscale Analysis
    Dongwei Chen, Chun-Ming Chang and Andrew Laine I-192

A Method of Industrial Parts Surface Inspection Based on an Optics Model
    Norifumi Katafuchi, Mutsuo Sano, Shuichi Ohara and Masashi Okudaira I-200

Poster Session I

Illumination Color from the Blurred Inter-reflection of a Reference Nose
    Mohamed Abdellatif, Yutaka Tanaka, Akio Gofuku and Isaku Nagai I-208

Shape Recovery from One Image under Multiple Light Sources
    Ying-li Tian, H.T. Tsui and S.Y. Yeung I-216

Spherical and Cylindrical Light Source Models for Shape Recovery
    Ying-li Tian, H.T. Tsui and S.Y. Yeung I-224

Polyhedral Shape Recovery Based on Interreflections
    Jun Yang, Dili Zhang, Noboru Ohnishi and Noboru Sugie I-232

Improved Supervised Color Constancy for Color Inspection
    Xuesheng Bai and Guangyou Xu I-240

Unsupervised Filtering of Munsell Spectra
    M. Hauta-Kasari, W. Wang, S. Toyooka, J. Parkkinen and R. Lenz I-248

Foveated Vision for Scene Exploration
    Naoki Oshiro, Atsushi Nishikawa, Noriaki Maru and Fumio Miyazaki I-256

Evolutionary Methods Applied to Binocular Disparity Estimation
    Carla L. Pagliari and Tim J. Dennis I-264

Robust Epipolar Geometry Estimation Using Genetic Algorithm
Jinxiang Chai and SongDe Ma I-272

New Development of Stereo Vision: A Solution of Motion Stereo Correspondence
M. Xie I-280

Acquisition of Three-Dimensional Information Using Omnidirectional Stereo Vision
Atsushi Chaen, Kazumasa Yamazawa, Naokazu Yokoya and Haruo Takemura I-288

Error Analysis in Stereo Vision
R.S. Ramakrishna and B. Vaidyanathan I-296

Detecting Targets in SAR Images: A Machine Learning Approach
Qi Zhang, Zoran Duric and Ryszard S. Michalski I-305

Precise Matching by Robust Estimation of Deformation and Local Coherence
Zhong-Dan Lan, Roger Mohr and Long Quan I-313

Active Viewpoint Control for Shape from Occluding Contours
Takashi Akutsu, Kenichi Arakawa and Hiroshi Murase I-321

Point Selection: A New Comparison Scheme for Size Functions (With an Application to Monogram Recognition)
Massimo Ferri, Patrizio Frosini, Alberto Lovato and Chiara Zambelli I-329

Sketch Up: Towards Qualitative Shape Data Management
Costantino Collina, Massimo Ferri, Patrizio Frosini and Eleonora Porcellini I-338

Robust Matching and Hierarchical Recognition of 2-D Shapes Using "Chain of Circles"
Jae-Moon Chung and Noboru Ohnishi I-346

Finding the Center of Rotational Symmetry from Noisy Forms
Hyoung Seop Kim, Nachi Motomura and Seiji Ishikawa I-354

Recognition in Wavelet-Compressed Imagery
Wei Hu and W. Brent Seales..I-362

Fast Image Template and Dictionary Matching Algorithms
Sung-Hyuk Cha I-370

Recognition of Planar Shapes Using Algebraic Invariants from Higher Degree Implicit Polynomials
    Satish Kaveti, Eam Khwang Teoh, and Han Wang I-378

Object Recognition and Orientation via Zernike Moments
    Samer M. Abdallah, Eduardo M. Nebot and David C. Rye I-386

A Study of Zernike Moment Computing
    Simon X. Liao and Miroslaw Pawlak I-394

Query Expansion by Raw Image Features and Text Annotations in Image Retrieval
    Kok F. Lai, Hong Zhou and Syin Chan I-402

Montage: An Image Database for the Fashion, Textile, and Clothing Industry in Hong Kong
    Tak Kan Lau and Irwin King I-410

Auto Cameraman Via Collaborative Sensing Agents
    Qian Huang, Yuntao Cui, Supun Samarasekera and
    Michael Greiffenhagen I-418

Dynamic Adaptive Data Structures for Semantic Analysis and Synthesis of Video Information
    V.V. Alexandrov, E.V. Laikov and B.E. Frenkel I-426

Recognition of Simple Curved Surfaces from 3D Surface Data
    Alan M. McIvor and Peter T. Waltenberg I-434

A Recursive Fitting-and-Splitting Algorithm for 3-D Object Modeling Based on Superquadrics
    Hongbin Zha, Tsuyoshi Hoshide and Tsutomu Hasegawa I-442

Learning and Recognizing 3D Objects by Using Partial Planar Curve Matching Method
    Jin Jia and Keiichi Abe I-450

Contour Matching Technique for 3D Object Recognition Using Kalman Filter
    M. Hanmandlu and V. Shantaram I-458

Kalman Filter Based Matching Technique for 3D Object Recognition
    M. Hanmandlu and V. Shantaram I-466

A Generating Method for 3-dimensional Knitting Cloth Shapes
    Tatsushi Funahashi, Tsuyoshi Miyazaki, Masashi Yamada,
    Hirohisa Seki and Hidenori Itoh I-474

A Fast Mesh Deformation Method to Build Spherical Representation Models of 3D Objects
  Antonio Adán, Carlos Cerrada and Vicente Feliu I-482

Semi-automatic 3D Object Digitizing System Using Range Images
  C. Schütz, T. Jost and H. Hügli I-490

Invited Talk

Image Guided Surgical Systems
  Eric Grimson I-498

Session F1A: Biometry II

Technical Evaluation of Biometric Systems
  Brigitte Wirtz I-499

Face Recognition from Sequences Using Models of Identity
  Stephen J. McKenna and Shaogang Gong I-507

Enhancing Human Face Detection Using Motion and Active Contours
  Kin Choong Yow and Roberto Cipolla I-515

Learning Identity and Behaviour with Neural Networks
  A. Jonathan Howell and Hilary Buxton I-523

Open Sesame! Speech, Password or Key to Secure Your Door?
  Stéphane H. Maes and Homayoon S.M. Beigi I-531

Session F1B: Low-Level Processing

A Unified Framework for Image-Derived Invariants
  Yuan-Fang Wang and Ronald-Bryan O. Alferez I-542

Stereo Correspondences in Scale Space
  Christian Menard I-550

Fast Stereo Matching in Compressed Video
  Michael S. Brown and W. Brent Seales I-558

Robust Total least Squares Based Optic Flow Computation
  Alireza Bab-Hadiashar and David Suter I-566

Image Processing via the Beltrami Operator
  R. Kimmel, R. Malladi and N. Sochen I-574

Session F2A: Color Vision II

Efficient Contour Extraction in Color Images
  Aldo Cumani I-582

Color Edge Detection Using Orthogonal Polynomials
  R. Krishnamoorthi and P. Bhattacharyya I-590

Fast and Robust Segmentation of Natural Color Scenes
  Volker Rehrmann and Lutz Priese I-598

Segmentation and Tracking Using Color Mixture Models
  Yogesh Raja, Stephen J. McKenna and Shaogang Gong I-607

Object Tracking Using Adaptive Color Mixture Models
  Stephen J. McKenna, Yogesh Raja and Shaogang Gong I-615

Session F2B: Active Vision

A Learning Approach to Fixating on 3D Targets with Active Cameras
  Narayan Srinivasa and Narendra Ahuja I-623

Automatic Detection and Tracking of Human Heads Using an Active
Stereo Vision System
  Cheng-Yuan Tang, Yi-Ping Hung, and Zen Chen I-632

Front Propagation and Level-Set Approach for Geodesic Active
Stereovision
  Rachid Deriche, Christophe Bouvin and Olivier Faugeras I-640

A Bayes Nets-Based Prediction/Verification Scheme for Active Visual
Reconstruction
  Éric Marchand and François Chaumette I-648

Actively Building Models with VIRTUE
  J. Lang and Michael R.M. Jenkin I-656

Session F3A: Face and Hand Posture Recognition

Using RBF Networks to Map GWT Ridge Images to Pose
  Alexandra Psarrou and Jonathan Tanner I-664

3-D Pose Estimation and Model Refinement of an Articulated Object from
a Monocular Image Sequence
    Nobutaka Shimada, Yoshiaki Shirai, Yoshinori Kuno and Jun Miura I-672

Face Synthesis with Arbitrary Pose and Expression from Several Images -
An Integration of Image-Based and Model-Based Approaches
    Yasuhiro Mukaigawa, Yuichi Nakamura and Yuichi Ohta I-680

Live Facial Expression Generation Based on Mixed Reality
    Hiromi T. Tanaka, Akira Ishizawa and Hiroaki Adachi I-688

Real-Time Tracking of Human Hands from a Sign-Language Image Sequence
    Kazuyuki Imagawa, Shan Lu and Seiji Igi I-698

The Model-Based Dynamic Hand Posture Identification Using Genetic
Algorithm
    Cheng-Chang Lien and Chung-Lin Huang I-706

Poster Session II

Parallel Implementation of Fractal Image Compression Using Multiple
Digital Signal Processors
    S.K. Chow, M. Gillies and S.L. Chan I-714

Comparison of Mean Field Annealing and Multiresolution Analysis in
Missing Data Estimation
    Hairong Qi, Wesley E. Snyder and Griff L. Bilbro I-722

Segmentation of MRF Based Image Using Hierarchical Genetic Algorithm
    Jin Wook Kim, Eun Yi Kim, Se Hyun Park and Hang Joon Kim I-730

Motion Compensated Color Video Classification Using Markov Random Fields
    Zoltan Kato, Ting-Chuen Pong and John Chung-Mong Lee I-738

Edge-Preserving Smoothing by Convex Minimization
    S.Z. Li, Y. H. Huang, J. S. Fu and K. L. Chan I-746

Author Index...I-755

Contents of Volume II

Poster Session II

On Typical Implementations of Hough Transform for Improving Its Performances
  Jun-ichiro Hayashi, Kunihito Kato, Toshio Endoh, Kazuhito Murakami, Takashi Toriu and Hiroyasu Koshimizu II-1

Hierarchical Segmentation and Representation with Dynamic Link Architecture Neural Network
  Yunqiang Chen and SongDe Ma II-9

Perceptually Consistent Segmentation of Texture Using Multiple Channel Filter
  Nan Zhang and Wee Kheng Leow II-17

Optimal Edge Detection under Difficult Imaging Conditions
  Md. Shoaib Bhuiyan, Yuji Iwahori and Akira Iwata II-25

Restoring Image Quality Through Structure Preserving De-noising
  Krishna Ratakonda and Narendra Ahuja II-33

Feature Saliency from Noise Variations in Invariants
  Mark Jenkinson and Michael Brady II-41

Multiscale Image Representation and Edge Detection
  Fang Chen and David Suter II-49

Rotation Invariant Texture Features from Gabor Filters
  S.R. Fountain and T.N. Tan II-57

Euclidean Invariants of Linear Scale-Spaces
  Alfons Salden II-65

Segmenting Objects at Multiple Scales : A Robust Approach
  Farzin Mokhtarian II-73

Multi-grid Edge Models for Magnifying Digital Images
  G. Qiu II-81

Scale and Rotation Invariant Recognition Method Using Higher-Order Local Autocorrelation Features of Log-Polar Image
  Takio Kurita, Kazuhiro Hotta and Taketoshi Mishima II-89

Script and Language Identification from Document Images
    G.S. Peake and T.N. Tan II-97

Document Categorization for Document Image Understanding
    Hiroyuki Masai and Toyohide Watanabe II-105

Recognition of Various Bar-graph Structures Based on Layout Model
    Naoko Yokokura and Toyohide Watanabe II-113

Word-Class Bigram Statistics Language Model for a Hand-Written
Chinese Character Recognizer
    Pak-Kwong Wong and Chorkin Chan..II-121

Log Classification by Single X-ray Scans Using Texture Features
from Growth Rings
    Xinli Wang II-129

Precise and Fast Form Identification Method by Using Adaptive Base
Lines for Matching
    Hiroaki Takebe, Yutaka Katsuyama and Satoshi Naoi II-137

Combinatorial Coarse Classification Method for OLCCR
    Jing Zheng, Xiaoqing Ding, Youshou Wu and Fanxia Guo II-145

Detecting Characters in Grey-Scale Scene Images
    Yongmei Liu, Tsuyoshi Yamamura, Noboru Ohnishi and Noboru Sugie........II-153

Conic Based Image Transfer for 2-D Objects: A Linear Algorithm
    Akihiro Sugimoto II-161

Minimal Conditions on Intrinsic Paramenters for Euclidean Reconstruction
    Anders Heyden and Kalle Åström II-169

Surface Based Hypothesis Verification in Intensity Images Using
Geometric and Appearance Data
    J.H.M. Byne and J.A.D.W. Anderson II-177

Next Best Viewpoint (NBV) Planning for Active Object Modeling Based
on a Learning-by-Showing Approach
    Hongbin Zha, Ken'ichi Morooka and Tsutomu Masegawa II-185

Object Recognition by Matching Symbolic Edge Graphs
    Tino Lourens and Rolf P. Würtz II-193

Interpretation of Complex Scenes Using Bayesian Networks
  Mark F. Westling and Larry S. Davis II-201

Recognition of Urban Scene Using Silhouette of Buildings and City Map Database
  Peilin Liu, Wei Wu, Katsushi Ikeuchi and Masao Sakauchi II-209

A Cooperative Inference Mechanism for Extracting Road Information Automatically
  Masakazu Nishijima and Toyohide Watanabe II-217

Model-Based Active Object Recognition Using MRF Matching and Sensor Planning
  Tianrong Liu, Kap Luk Chan and Stan Ziqing Li II-225

Improved Image Classification Using Morphing
  W. Brent Seales and Cheng Jiun Yuan II-233

Reconstruction of Non-manifold Objects from Two Orthographic Views
  Chang-Hun Kim and Tae-Jung Suh II-241

3D Object Recognition Using Segment-Based Stereo Vision
  Yasushi Sumi and Fumiaki Tomita II-249

Invited Talk

The State of Color Vision Research
  Brian Funt II-257

Color Vision and Color Media Processing Research in Asia
  Shoji Tominaga II-258

Session S1A: Recent Advances in Computer Vision

Recent Advances in Detection and Description of Buildings from Multiple Aerial Images
  Sanjay Noronha and Ram Nevatia II-259

Visual Surveillance of Human Activity
  Larry Davis, Sandor Fejes, David Harwood, Yaser Yacoob,
  Ismail Hariatoglu and Michael J. Black II-267

Bayesian Paradigm for Recognition of Objects - Innovative Applications
    J. K. Aggarwal and Shishir Shah II-275

Toward Motion Picture Grammars
    Ruud Bolle, Yiannis Aloimonos and Cornelia Fermüller II-283

Session S1B: Segmentation and Grouping

Hierarchical Texture Segmentation
    P. Bajcsy and N. Ahuja II-291

Range Image Segmentation: Adaptive Grouping of Edges into Regions
    Xiaoyi Jiang and Horst Bunke II-299

Optimising the Complete Image Feature Extraction Chain
    M. Mirmehdi, P. L. Palmer and J. Kittler II-307

A Unified Framework for Salient Curves, Regions, and Junctions Inference
    Mi-Suen Lee and Gérard Medioni II-315

Learning Multiscale Image Models of 2D Object Classes
    Benoit Perrin, Narendra Ahuja and Narayan Srinivasa II-323

Session S2A: Computer Vision & Virtual Reality

3D Model Centered Framework for CV and VR
    Michihiko Minoh II-332

Image-Based Geometrically-Correct Photorealistic Scene/Object
Modeling(IBPhM): A Review
    Zhengyou Zhang II-340

Measuring Object Surface Shape and Reflectance Properties
    Yoichi Sato, Mark D. Wheeler, and Katsushi Ikeuchi II-350

Robust Image Composition Algorithms for Augmented Reality
    Marie-Odile Berger and Gilles Simon II-360

Context-Based Recognition of Manipulative Hand Gestures for
Human Computer Interaction
    Kang-Hyun Jo, Yoshinori Kuno and Yoshiaki Shirai II-368

Session S2B: Motion Analysis

An Algorithm for Recursive Structure and Motion Recovery under Affine Projection
    Miroslav Trajković and Mark Hedley II-376

Relative Affine Depth: Structure from Motion by an Uncalibrated Camera
    Zhong-Ying Zhang and Hung-Tat Tsui II-384

The Eigenspace Method for Rigid Motion Recovery from less than Eight Point Correspondences
    Miroslav Trajković and Mark Hedley II-392

3D Shape and Motion Analysis from Image Blur and Smear: A Unified Approach
    Yuan-Fang Wang and Ping Liang II-400

3D Line's Extraction from 2D Spatio-temporal Image Created by Sine Slit
    Pingtao Wang, Katsushi Ikeuchi and Masao Sakauchi II-408

Toward Non-intrusive Motion Capture
    A. Bottino, A. Laurentini and P. Zuccone II-416

Session S3A: Object Recognition and Modeling

Appearance Based Visual Learning and Object Recognition with Illumination Invariance
    Kohtaro Ohba, Yoichi Sato and Katsushi Ikeuchi II-424

Evidence-Based Scene Interpretation Considering Subjective Certainty of Recognition
    Yasuhiro Taniguchi and Yoshiaki Shirai II-432

Robust Hypothesis Verification for Model Based Object Recognition Using Gaussian Error Model
    Frederic Jurie II-440

Shape Modeling from Multiple View Images Using GAs
    Satoshi Kirihara and Hideo Saito II-448

3-D Reconstruction of Multipart Self-Occluding Objects
    Nebojsa Jojic, Jin Gu, Helen C. Shen and Thomas S. Huang II-455

On Analysis of Cloth Drape Range Data
    Nebojsa Jojic and Thomas S. Huang II-463

Poster Session III

VR Models from Epipolar Images: An Approach to Minimize Errors in
Synthesized Images
    Mikio Shinya, Takafumi Saito, Takeaki Mori and Noriyoshi Osumi II-471

Shape and Pose Parameter Estimation of 3D Multi-Part Objects
    Satoshi Yonemoto, Naoyuki Tsuruta and Rin-ichiro Taniguchi II-479

Generating 3D Models of Objects Using Multiple Visual Cues in Image
Sequences
    Jiang Yu Zheng, Akio Murata and Norihiro Abe II-487

Strategical Tracking of Polyhedral Objects by Reactive Change of
Projection Pattern - Reactive Range Finder
    Takeshi Mita, Shinsaku Hiura, Hirokazu Kato and Seiji Inokuchi II-495

Autonomous Vision-Guided Robot Manipulation Control
    Wey-Shiuan Hwang and John (Juyang) Weng II-503

A New Adaptive Approach on Rapid Obstacle Detection in Range Image
    Qi Zhang, Weikang Gu and Xiuqing Ye II-511

Recognition of Shape Model for General Roads
    Keiichi Uchimura and Zhencheng Hu II-519

Visual Detection of Obstacles Assuming a Locally Planar Ground
    Manolis I.A. Lourakis and Stelios C. Orphanoudakis II-527

Potential-Based Modeling of 2D Regions Using Non-uniform Source
Distributions
    Jen-Hui Chuang, Chi-Hao Tsai, Wei-Hsin Tsai and Chuei-Yaw Yang II-535

A Linear Algorithm for Motion from Three Weak Perspective Images
Using Euler Angles
    Gang Xu and Noriko Sugimoto II-543

On Learning Spatio-Temporal Relational Structures in Two Different
Domains
    Adrian R. Pearce, Terry Caelli and Simon Goss.. II-551

An Efficient Iterative Pose Estimation Algirithm
    S.H. Or, W.S. Luk, K.H. Wong and I. King II-559

A New Multistage Approach to Motion and Structure Estimation by
Gradually Enforcing Geometric Constraints
    Zhengyou Zhang II-567

Tracking a Person with Pre-recorded Image Database and a Pan, Tilt,
and Zoom Camera
    Yimimg Ye, John K. Tsotsos, Karen Bennet and Eric Harley II-575

Recovery of Motion and Structure from Optical Flow under Perspective
Projection by Solving Linear Simultaneous Equations
    Toshiharu Mukai and Noboru Ohnishi II-583

Vector Coherence Mapping: A Parallelizable Approach to Image Flow
Computation
    Francis K.H. Quek and Robert K. Bryll II-591

Robust Motion Segmentation Using Rank Ordering Estimators
    Alireza Bab-Hadiashar and David Suter II-599

Optical Flow in the Scale Space
    Qing Yang and SongDe Ma II-607

Motion Detection in Temporal Clutter
    Phillip M. Ngan II-615

A Novel Fast Three-Step Search Algorithm for Block-Matching Motion
Estimation
    William Booth, James M. Noras and Donglai Xu II-623

Moving Vehicle Detection and Tracking in Image Sequences
    Yi Lu, Jason Miller and Tie Qi Chen II-631

Gesture Recognition from Image Motion Based on Subspace Method and
HMM
    Yoshio Iwai, Tadashi Hata and Masahiko Yachida II-639

Identifying Faces under Varying Pose Using a Single Example View
    Dadet Pramadihanto, Yoshio Iwai, Masahiko Yachida and Haiyuan Wu II-647

Multiple Camera Based Human Motion Estimation
    Akira Utsumi, Hiroki Mori, Jun Ohya and Masahiko Yachida II-655

An Autonomous Facial Caricaturing Based on a Model of Visual Illusion-
Experimental Modeling of Visual Illusion
    Kazuhito Murakami, Mikiko Takai and Hiroyasu Koshimizu II-663

3D Estimation of Facial Muscle Parameter from the 2D Marker Movement
Using Neural Network
  *Takahiro Ishikawa, Hajime Sera, Shigeo Morishima and
  Demetri Terzopoulos* II-671

Appearance-Based Face Recognition under Large Head Rotations in Depth
  Shaogang Gong, Eng-Jon Ong and Peter J. Loft II-679

Skin-Color Modeling and Adaptation
  Jie Yang, Weier Lu and Alex Waibel...II-687

Human Information Retrieval by Face Extraction and Recognition on TV
News Images Using Subspace Method
  Yasuo Ariki, Noriyuki Ishikawa and Yoshiaki Sugiyama................................II-695

Converting Facial Expressions Using Recognition-Based Analysis of
Image Sequences
  Takahiro Otsuka and Jun Ohya II-703

Muscle-Based Feature Models for Analyzing Facial Expressions
  Hiroshi Ohta, Hitoshi Saji and Hiromasa Nakatani...................................... II-711

A Morphological Method for Moving Object Segmentation and Posture
Recognition
  Yi Li, Songde Ma and Hanqing Lu II-719

Detection of Glasses in Facial Images
  Xiaoyi Jiang, M. Binkert, B. Achermann and H. Bunke................................II-726

Non-monotonic Continuous Dynamic Programming for Spotting Recognition
of Hesitated Gestures from Time-Varying Images
  T. Nishimura, T. Mukai and R. Oka..II-734

Face Recognition Using a Face-Only Database: A New Approach
  *Hong-Yuan Mark Liao, Chin-Chuan Han, Gwo-Jong Yu, Hsiao-Rong Tyan,
  Meng Chang Chen and Liang-Hua Chen*..II-742

Author Index..II-751

Pen Computing – An Overview

Krishna Nathan

IBM T.J. Watson Research Center
Yorktown Heights, NY
U.S.A.

ABSTRACT

Pen computing has had its share of ups and downs over the past few years - from the heady promises of handwriting recognition and Personal Digital Assistants (PDAs) to the disenchantment due to poor sales of pen based computers and the lack of credible applications for them. In my talk, I will discuss the issues that have led to this state of affairs, as well the current trends in the industry. The main challenges facing pen computing today will also be explored. These include improving algorithms for handwriting recognition, building better and more natural hardware platforms and designing user interfaces that take into account the demands multimodal input. The talk will also touch upon work in this area that is being carried out at the T.J. Watson Research Center at IBM.

Research Issues in Biometrics

Ruud M. Bolle, Nalini K. Ratha and S. Pankanti
{bolle,ratha,sharat}@watson.ibm.com

IBM Thomas J. Watson Research Center, Yorktown Heights, NY 10598

Abstract. Accurately determining the identity of a person is becoming critical to our increasingly interconnected information society. As increasing number of biometric-based identification systems are being deployed for many civilian and forensic applications, biometry and applications of biometric-based identification have evoked considerable interest. In this paper, we attempt to describe the key research issues involved in the design of a biometric system.

1 Introduction

Associating an identity with an individual is identification. The problem of resolving the identity of a person can be categorized into two fundamentally distinct types of problems with different inherent complexities: (i) Verification and (ii) Identification. Verification (authentication) refers to the problem of confirming or denying a person's claimed identity (Am I who I claim I am?). Identification (Who am I?) refers to the problem of establishing a subject's identity – from a set of already known identities. Positive personal identification term typically refers (in both verification as well as identification context) to verification or identification of a person with high certainty.

2 Identification Methods

The engineering approach to the (abstract) problem of authentication of identity of a person is to reduce this problem to the problem of authentication of a concrete entity related to the person. Typically, the problem of authenticating a person's identity is reduced (i) to that of a person's possession (*"something that you possess"*), e.g., permit physical access to building to all persons whose identity could be authenticated by possession of a key. (ii) to that of person's knowledge of a piece of information (*"something that you know"*), e.g., permit login access to a system to a person who knows existence of a user-id and a password associated with it. Some systems, e.g., ATMs, use a combination of "something that you have" (ATM card) and "something that you know" (PIN) to establish an identity.

Yet another approach to positive identification has been to reduce the problem of identification to the problem of identifying a person by their physical characteristics of the person. The characteristics could be either a person's physiological traits, e.g., fingerprints, hand geometry etc. or her behavioral characteristics, e.g., voice, signature. This method of identification of person based on his/her physiological/behavioral characteristics is called biometrics. The primary advantage of this identification method over the methods of identification utilizing "something that you possess" or "something that you know" approaches is that a biometrics cannot be misplaced or forgotten; it represents a tangible component of "something that you are". While biometrics are not an identification panacea, they, especially, when combined with the other methods of identification, could become very powerful tools in providing solutions requiring positive identification.

3 Issues

The general problem of identification raises a number of important research issues: What identification technologies are the most effective way to achieve accurate and reliable identification of individuals? In this paper, we attempt to describe some of the challenges in the biometrics research. Some of these issues are well-known open problems in the allied areas (e.g., pattern recognition and computer vision); while others need a systematic cross-disciplinary effort.

The design of a biometric-based identification system is essentially the design of a pattern recognition system. The conventional pattern recognition system designers have adopted a sequential phase-by-phase modular architecture. Although, it is generally known in the research community that a more integrated, parallel, active, system architecture involving feedback/feedforward control have a number of advantages, these concepts have not yet been fully exploited in commercial biometrics-based systems.

Given the performance specifications (cost, speed, and accuracy) of the end-to-end identification system, the design of the system involves resolving the following issues: (i) how to acquire the input data/measurements (biometric), (ii) what internal representation (features/patterns) of the input data is invariant and amenable for an automatic extraction process; (iii) given the input data, how to extract the internal representation from the input data (iv) given two input samples in the required internal representation, provide a matching metric definition that translates the intuition of "similarity" among the patterns (v) an effective method of implementing the matching metric. For identification systems, the design should also address the issues involving (vi) organization of a number of (representations) input samples into a database and (vii) effective methods of searching a given input sample representation in the database.

Researchers in pattern recognition have realized that effectively resolving these issues is very difficult and there is a need to engineer the solutions in absence of a definitive solution. The following sections delve into the specific problems facing each phase of the design of biometric-based identification system.

4 Acquisition

Acquiring relevant data for the biometric is one of the most critical and yet neglected processes in the design of pattern recognition system. The amount of care taken in acquiring the data determines the performance of the entire system. Further, it is also important to sense the measurements with maximal ease (ergonomics) and with least demands on the user. Two of the tasks associated with acquisition are

(a) Quality assessment: automatically assessing the suitability of the input data for automatic processing, and (b) Segmentation: separation of the input data into foreground (object of interest) and background (irrelevant information).

A number of opportunities exist for capturing (i) the context of the capture which may further help performance of the system (ii) avoiding capture of undesirable measurements and (subsequent) recapture of desirable. With increased availability of inexpensive computing and sensing resources, typically, the context of the capture could be made richer to improve the performance. For instance, a fingerprint is traditionally captured from its 2D projection on a flat surface. Why not a 3D capture? Why not a color capture? Why not active sensing? Such enhancement typically significantly improves the performance of systems. Similarly, utilizing a sequence of face images (rather than a single image) could result in the performance of face recognition system.

Although a number of existing identification systems routinely assess the quality of the input measurement indicating the desirability of the measurement for matching, the approach to such a quality assessment metric is subjective, debatable, and typically, inconsistent. A lot of research effort needs to be focussed in this area to systematize both (i) the rigorous and realistic models of the input measurements and (ii) metrics for assessment of desirability of a measurement. Similarly, the conventional foreground/background separation also typically relies on a very ad hoc empirical processing of input measurements and enhancing the information bandwidth of input channel often provides very effective avenues for the segmentation. Further, rigorous and realistic models of the object of interest often facilitate lead to a cleaner and better design of segmentation algorithms.

5 Representation

Which machine-readable representations completely capture the invariant and discriminatory information in a the input measurements? This representation issue constitutes the essence of system design and has far reaching implications on the design of the rest of the system. The unprocessed measurement values are typically not invariant over the time of capture and there is a need to determine salient features of the input measurement which both discriminate between the identities as well as remain invariant for a given individual. Thus, the problem of representation is to determine a representation (feature) space which is invariant (less variant) for the input signals belonging to the same identity and which differ maximally for those that belong to different identities. To systematically determine the discriminatory power of an information source and arrive at an effective feature space is a challenging problem. Additionally, when storage space is at a premium, the representation also needs to be parsimonious. The issues of most salient features of an information source also need to be investigated.

Representation issues cannot be completely resolved independent of a specific biometric domain and involve complex trade-offs. Take, for instance, the fingerprint domain. Representations based on the entire gray scale profile of a fingerprint image are prevalent among the verification systems using optical matching [1]. However, the utility of the systems using such representation schemes may be limited due to factors like brightness variations, image quality variations, scars, and large global distortions present in the fingerprint image because these systems are essentially resorting to template matching strategies for verification. Further, in many verification applications terser representations are desirable which preclude representations that involve the entire gray scale profile fingerprint images. Some system designers attempt to circumvent this problem by restricting that the representation is derived from a *small* (but consistent) part of the finger [2]. However, if this same representation is also being used for identification applications, then the resulting systems might stand at a risk of restricting the number of unique identities that could be handled, simply because of the fact that the number of distinguishable templates is limited. On the other hand, an image-based representation makes fewer assumptions about the application domain (fingerprints) and therefore, has the potential to be robust to wider varieties of fingerprint images. For instance, it is extremely difficult to extract a landmark-based representation from a (degenerate) finger devoid of any ridge structure.

6 Feature Extraction

A given arbitrarily complex representation scheme should be amenable to automation without any human intervention. For instance, the manual system of fingerprint identification uses more than a dozen number of features [4]. However, it is not feasible to incorporate these features into fully automatic fingerprint system because it not easy to reliably detect these features using automatic image processing system. Determining features that are amenable to automation has not received much attention for research and is especially important in the biometrics which are entrenched in the design philosophies of an associated mature manual system of identification.

Traditionally, the feature extraction system follow a staged sequential architecture which precludes effective integration of extracted information available from the measurements. As CPUs and sensors become cheaper, it is possible to use better architectures/methods for information processing to detect features reliably.

Once the features are determined, it is also a common practice to design the feature extraction process in a somewhat ad hoc manner. The efficacy of such methods is limited especially when input measurements are noisy. Rigorous models of the feature representations are helpful in reliable extraction of the features from the input measurements, especially, in the noisy situations. Determining terse and effective models for the features is a challenging research problem.

7 Matching

The crux of a matcher is a similarity function which quantifies the intuition of similarity between two representations of the biometric measurements. Determining a similarity metric is a very difficult problem since it should be able to discriminate between the representations of two different identities despite noise, structural and statistical variations in the input signals, aging, and artifacts of the feature extraction processing. In many biometrics, say signature verification, it is difficult to even define the ground truth [5]: do/should these two signatures belong to the same person or different person?

A representation scheme and a similarity metric determine the accuracy performance of the system for a given population of identities; hence the selection of appropriate similarity scheme and representation is critical.

Given a complex operating environment, it is critical to identify a set of valid assumptions upon which the matcher design could be based. Often, there is a choice between whether it is more effective to exert more constraints by incorporating more engineering or to build a more sophisticated similarity function for the given representation. For instance, in a fingerprint domain, it is known that the elastic distortion of a finger has a significant adverse effect on the matcher performance. In such a situation, the designer needs to characterize a realistic model of the variations among the representations of mated pairs. Is distortion significant for the given imaging? Is it easier to prevent distortion or is it more effective to take into account all the distortions possible and formulate a clever similarity function?

8 Search, Organization, and Scalability

Systems dealing with a large number of identities should be able to effectively operate as number of users in the system increases to its operational capacity and should only gracefully degrade as the system accommodates more users than envisaged at the time of its design. As civil applications (e.g., drivers and voter registration, National ID systems and IDs involving health, medical, banking, cellular, transportation, e-commerce) involving a very large number of identities (billions) are being designed and integrated, we are increasingly looking toward biometrics to solve our problems of authentication and identification.

Identification of an individual among a large number of identities becomes increasingly complex as the number of identities stored in the system increases [3, 8]. Many applications like National ID systems, passport and visa issuance further require a constant throughput and a very small turnaround time. Design of such systems need to adopt radically different strategies and mode of operation than those adopted by traditional forensic identification systems. This has a profound influence on every aspect of the system including the choice of biometric, features, metric of similarity, matching criteria, operating point, *etc.* None of these design issues have been rigorously studied either in biometrics or even in pattern recognition research.

All these criteria point to unique biometrics which remain invariant over a long period of time. Designing constant length, one dimensional, indexable features would become increasingly important for identification applications involving a large number of identities.

9 Evaluation

An end-user is interested in determining if the system can effectively perform for *his specific situation*: does this system identify sufficiently accurately? is the system sufficiently quick? how much would be the cost of the system? Among these issues, characterizing accuracy performance is the most difficult; we will only address accuracy performance issues here.

Unfortunately, the performance of a system in the context of a given application population is a random variable and strictly speaking cannot be computed but can only be estimated.

The performance metrics described in the literature [6] provide us means of estimating performance with respect to a specific representative database. For such empirical performance metrics to be able to precisely generalize to the entire population of interest, the test data should (i) be large enough to represent the population and (ii) contain enough samples from each category of the population [7]. To obtain fair and honest test results, enough samples should be available, the samples should be representative of the population, and adequately represent all the categories (impostors and genuine). In reality, especially for the large novel emerging applications, almost never do we have access to a sufficient number of test samples nor are the samples representative of the actual population. In such situations, there is a need to model, validate predictive models of performance in terms of controllable and measurable parameters of the available data. Such predictive models of performance may be useful both for bootstrapping a small number of available samples as well as obtaining realistic estimates of the utility of a given biometric technology to a given application.

Further, irrespective of the choice performance metric, error bounds that indicate the confidence of the estimates are valuable for understanding the significance of the test results. Estimating confidence measures without using unrealistic naive models of the hypothesized population distributions is challenging.

10 Integration

The accuracy of an identification system improves as we effectively utilize and integrate an increasing number of information sources related to an individual to confirm her identity; it becomes increasingly difficult to abuse the system privileges. However, the challenge of integration is to ascertain that the system performance degrades gracefully as some of the information sources become unavailable or unreliable. As better performance is demanded of identification systems and as a variety of different sensors become affordable, integration of different biometrics will become an important issue. Integration of different technologies is also becoming critical for imparting capabilities to the identification system. For instance, biometric sensor integrated smart cards could provide facilities for identity authentication without divulging any information about biometric measurements.

Deciding efficient architectures for integration is an open research problem and they are perhaps the single most important determinants of behavior of an integrated system. Decision level, feature level, and measurement level integration architectures have been studied in the literature. Determination of which integration strategies are appropriate for a given identification application needs more focussed research.

11 Circumvention

Some forms of fraud involve transcending the means and mechanisms of identification used by the system (extra-system) and hence, in principle, cannot be completely eliminated using any strategies embedded inside system (intra-system). Some problems of identification/security plague all identification technologies alike: collusion, coercion (means of access being used against a user's will), covert acquisition (means of access being stolen without a user's knowledge), and denial (users denying their legitimate access). Currently, the attempts to reduce fraud in the system are process-based and ad hoc. There is a need to focus research efforts on a systematic and technology-intensive approach to combat fraud in the system. This is especially true in terms of biometrics-based identification systems where the captured biometric measurements and context may have sufficient information to deter and detect some forms of fraud.

Some other problems related to identification are more specific to biometrics-based systems. For instance, skilled humans have an uncanny ability to disguise their identity and are able to assume (forge/mimic) a different specific identity. The "chameleon" identities pose an additional problem to the reliability of the identification systems based on some biometrics and warrants more research.

12 Privacy

- Privacy: As identification becomes more and more foolproof, the process of getting identified itself leaves trails of undeniable private information. In case of biometric-based identification, this problem is even more serious because the biometric features may additionally inform one about the medical history or susceptibilities of an individual. Consequently, there is a legitimate concern about privacy issues associated with the biometric-based identification.
- Proscription: This issue is somewhat related to the previous issue. When a biometric measurement is offered to a given system, the information contained in it should not be used for any other purpose than its intended use. In any (networked) information processing system, it is challenging to ensure that the biometric measurements captured will only be used for its intended purpose.

Wide-spread use of biometric-based identification systems should not only address the above mentioned issues from technical standpoint but also from the public perception point of view.

13 Summary

Biometrics is science of automatically identifying individuals based on their unique biological characteristics. As number of civil and commercial applications of biometric-based identification are emerging, several legitimate concerns are raised against use of biometrics for various applications; three of them appear to be the most significant: cost, privacy, and performance.

In order for the wide-spread use of the biometrics to happen, it is necessary to undertake systematic studies of fundamental research issues underlying the design of identification systems. Further, it is critical to engineer the match between the application needs and available technologies. We believe that biometrics technology alone may not be sufficient to resolve these issues effectively; the solutions to the outstanding open problems may lie in innovative engineering designs exploiting constraints otherwise unavailable to the applications and in harnessing the biometric technology in a combination with other allied technologies.

References

1. R. Bahuguna, "Fingerprint verification using hologram matched filterings," in *Proc. Biometric Consortium Eighth Meeting*, (San Jose, California), June 1996.
2. Mytec Technologies, "Access control applications using optical computing," (http://www.mytec.com/), 1997.
3. IBM Global Government Industry, "A Challenge to the Biometrics Industry: Technical Papers I,II, and III", (http://www.government.ibm.com/gov/ais.nsf), 1997.
4. Federal Bureau of Investigation, The Science of Fingerprints: Classification and Uses, U.S. Government Printing Office, Washington, D. C., 1984.
5. V. Nalwa, "Automatic on-line signature verification," *Proceedings of IEEE*, vol. 85, pp. 213–239, February 1997.
6. A. Jain, L. Hong, S. Pankanti, and R. Bolle, "On-line identity authentication using fingerprint verification," *Proceedings of IEEE (Special Issue on Automated Biometrics)*, vol. 85, No. 9, pp. 1365–1388, September 1997.
7. J. G. Daugman and G. O. Williams, "A proposed standard for biometric decidability," in *Proc. CardTech/SecureTech Conference*, (Atlanta, GA), pp. 223–234, 1996.
8. R. S. Germain, R. Bolle, A. Califano, S. Colville, S. Pankanti, N. Ratha, "Issues in large scale automatic biometric identification", in *Proc. Workshop on Automatic Identification Advanced Technologies* (Stony Brook, NY), November 6&7, 1997.

Automatic On-Line Signature Verification[†]

Vishvjit S. Nalwa
Bell Labs, Holmdel, NJ 07733, U.S.A.

Signature verification is an art: Whereas we may bring objective measures to bear on the problem, in the final analysis, the problem remains subjective. This art is both well studied and well documented as it applies to the verification by humans of signatures whose only records are visual—that is, as it applies to signatures during whose production no measurement is made of the pen trajectory or pen dynamics. Let us call such signatures, for which we have only a static visual record, **off-line**, and let us call signatures during whose production the pen trajectory or pen dynamics is captured, **on-line**. Whereas attempts to automate the verification of off-line signatures have fallen well short of human performance to this point, I demonstrate that automatic on-line signature verification is feasible.

In a break with tradition, I challenge the notion that the success of automatic on-line signature verification hinges on the capture of velocities or forces during signature production. Whereas velocities and forces can assist us in automatic on-line signature verification, I contend that we should not depend on them solely, or even primarily. If we were indeed unavoidably consistent over the dimensions of time and force when we signed, the use of pen dynamics during signature production—over and above that of signature shape—would be very useful in detecting forgeries, as dynamic information pertinent to a signature is not as readily available to a potential forger as is the shape of the signature given just the signature's off-line specimens. However, I have seen no substantive evidence to the effect that our pen dynamics is as

---

[†] This is an extented abstract of the following previously published paper: "Automatic On-Line Signature Verification," V. S. Nalwa, *Proceedings of the IEEE*, pp. 215–239, February 1997.

consistent as, or more consistent than, our final signature shape when we sign. My own informal experiments indicate that we typically exhibit similar temporal variations over the production of similar handwritten curves: In general, our speed along high-curvature curve segments is low relative to our speed along low-curvature curve segments, our average overall speed varying greatly from one instance of a pattern to another irrespective of whether we are producing our own pattern or forging someone else's. This observation suggests that at least the requirement of consistency over time during signature production is of limited value beyond that of consistency over shape. At any rate, irrespective of the velocities and forces generated during the production of a signature, for us to declare two signatures to be produced by the same individual, clearly, it is necessary that the shapes of the two signatures match closely.

Hence, I have based my signature-verification strategy primarily on the shapes of signatures; although, at this point, I do depend on time, this dependence is weak and could be removed. Thus, although my verification technique does require the capture of pen trajectories during signature production, unlike other reported on-line signature verification techniques, my technique can do without the explicit capture of any temporal, force, or pressure information during signature production. I propose that each handwritten on-line signature be represented by multiple models—local and global, shape-based and time-based, including a model that is local and purely shape-based. Whereas global models are easier to devise than local models—and, hence, global models are more widely used than local models—for signatures whose various instances are shaped consistently, global models are less discriminating than and less robust than local-shape-based models. My principal contributions to automatic on-line signature verification are two-fold. One, I suggest the weighted and biased harmonic mean as a graceful mechanism of combining errors from multiple models of which at least one model is applicable but not necessarily more than one model is applicable. Two, I devise a robust, reliable, and elastic local-shape-based model for handwritten on-line curves—this model generated by first parametrizing each on-line curve over its normalized arc-length, and then representing along the length of the curve, in a moving coordinate frame, measures of the curve within a sliding window that are analogous to the position of the center of mass, the torque exerted by a force, and the moments of inertia of a mass distribution about its center of mass. I have implemented and tested my signature-verification algorithm successfully both on databases and in live experiments.

Successful on-line signature verification based on comparing the varying local shapes of signatures offers several important advantages over alternative techniques, especially over those that are not shape-based:

- Local-shape–based signature verification is more likely than alternative techniques to reject only those genuine signatures that will be accepted by original signers as unrepresentative of their signatures, because such signers would typically base their judgment on the fidelity of their signatures on an a posteriori visual examination of the detailed shapes of their signatures, rather than on the velocities or forces generated during the production of these signatures. Such acceptance by nonfraudulent signers—of the inevitable rejection of some genuine signatures by a signature-verification system—is key to the acceptance of the signature-verification system by consumers in the marketplace.

- Local-shape–based comparisons of signatures, in contrast to global comparisons, avoid lumping together differences between signatures irrespective of their causes, which is important to us because we would like to distinguish between errors that are caused by isolated mistakes, such as inadvertent isolated gaps in writing, and errors caused by systematic deviations, such as those due to different writing styles.

- Local-shape–based signature verification can potentially highlight, for human consumption, local "nonobvious" similar- ities and discrepancies between the shapes of signatures—perhaps, so that a customer or a court of law can *see* why a particular signature was accepted or rejected.

- Shape-based signature verification does not require us to be consistent over the additional dimensions of time and force when we sign, a requirement that would alter the traditional expectation from us that we be consistent over only the shape of our signature when we sign. Alteration of this traditional expectation, it seems, would force many of us to change "the way in which we do business," weakening what is probably the strongest argument in favor of the continued use of handwritten signatures for transactions.

Of course, if we were unavoidably consistent over the dimensions of time and force when we signed, this last item would not be an issue. I would like to point out here, that because of their time independence, most of the tools I have developed for the elastic local comparison of handwritten shapes are immediately applicable both to on-line handwriting verification—which could be used to verify a user's identity by requesting the user to write something specific— and to on-line handwriting recognition. Note here, however, that my use of the pen trajectory to parametrize each on-line curve implies that visually identical curves that are traversed differently will be represented differently. Whereas this aspect of the representation I propose is advantageous to verification, it is disadvantageous to recognition. Further, note that signature verification is both easier than and more difficult than handwriting recognition. Verification is easier than recognition because, in verification, we know a priori what pattern to expect: All that successful verification entails is the comparison of an input pattern with a stored model. However, verification is more difficult than recognition because, unlike in recognition, where we are justified in assuming a cooperative human, in verification, we must allow for an adversary who is keenly intent on deceiving the system. Hence, whereas the answer in verification might simply be a *yes* or a *no*, successful verification requires the ability to detect subtle differences between patterns, an ability not required by recognition. More specific- ally, successful signature verification hinges on the ability to distinguish between inadv- ertent intrasigner variations on the one hand, and intersigner variations and advertent intrasigner variations on the other hand.

My presentation proceeds as follows, with the fundamental concepts underlying my approach fleshed out in Section 5:

1. Introduction
2. Alternatives to Signature Verification
3. Performance Evaluation
4. Prior Art
5. Key Concepts
    5.1 Harmonic Mean
    5.2 Jitter
    5.3 Aspect Normalization
    5.4 Parametrization over Normalized Length
    5.5 Sliding Computation Window
    5.6 Center of Mass
    5.7 Torque
    5.8 Moments of Inertia
    5.9 Moving Coordinate Frame and Saturation
    5.10 Weighted Cross-Correlation and Warping
6. Algorithm
7. Example
8. Database Results
9. A Signature-Verification System
10. Conclusion

In Section 2, I categorize the various techniques commonly used to verify the identities of individuals. In Section 3, I describe what constitutes successful signature verification. In Section 4, I summarize the state of the art of automatic on-line signature verification as recorded in the published literature. In Section 5, I highlight the key features of my approach, several of these features differentiating my approach from prior art. In Section 6, I outline my algorithm, which I have implemented and tested both on databases and in live experiments. In Section 7, I illustrate my algorithm with a detailed example. In Section 8, I describe the performance of my implementation on three databases created by Bell Laboratories. In Section 9, I describe a particular signature-verification system that runs my algorithm in real time on a notebook personal computer, this computer coupled to an electronic writing tablet for capturing on-line signatures. Finally, in Section 10, I list some of the outstanding issues in automatic on-line signature verification.

I refer you to the following references from my original paper for more on automatic on-line signature verification:

H. D. Crane and J. S. Ostrem, "Automatic Signature Verification Using a Three-Axis Force-Sensitive Pen," *IEEE Transactions on Systems, Man, and Cybernetics*, Vol. SMC-13, No. 3, pp. 329–337, May–June 1983.

T. Hastie, E. Kishon, M. Clark, and J. Fan, "A Model for Signature Verification," in *Proceedings of the 1991 IEEE International Conference on Systems, Man, and Cybernetics*, Vol. 1, Charlottesville, Virginia, pp. 191–196, October 1991.

F. Leclerc and R. Plamondon, "Automatic Signature Verification: The State of the Art—1989–1993," *International Journal of Pattern Recognition and Artificial Intelligence*, Vol. 8, No. 3, pp. 643–660, June 1994.

G. Lorette and R. Plamondon, "Dynamic Approaches to Handwritten Signature Verification," in *Computer Processing of Handwriting*, R. Plamondon and C. G. Leedham, Eds., World Scientific Publishing Co., Singapore, 1990, pp. 21–47.

W. Nelson and E. Kishon, "Use of Dynamic Features for Signature Verification," in *Proceedings of the 1991 IEEE International Conference on Systems, Man, and Cybernetics*, Vol. 1, Charlottesville, Virginia, pp. 201–205, October 1991.

R. Plamondon and G. Lorette, "Identity Verification from Automatic Processing of Signatures: Bibliography," in *Computer Processing of Handwriting*, R. Plamondon and C. G. Leedham, Eds., World Scientific Publishing Co., Singapore, 1990, pp. 65–85.

Y. Sato and K. Kogure, "Online Signature Verification Based on Shape, Motion, and Writing Pressure," in *Proceedings of the Sixth International Conference on Pattern Recognition*, Munich, Germany, pp. 823–826, October 1982.

T. K. Worthington, T. J. Chainer, J. D. Williford, and S. C. Gundersen, "IBM Dynamic Signature Verification," in *Computer Security: The Practical Issues in a Troubled World*, J. B. Grimson and H.-J. Kugler, Eds., North-Holland (Elsevier Science Publishers), Amsterdam, 1985, pp. 129–154.

P. Zhao, A. Higashi, and Y. Sato, "On-Line Signature Verification by Adaptively Weighted DP Matching," *IEICE Transactions on Information and Systems*, Vol. E79-D, No. 5, pp. 535–541, May 1996.

Integrating Faces and Fingerprints for Personal Identification

Lin Hong and Anil Jain

Dept. of Computer Science, Michigan State University, East Lansing, MI 48824

Abstract. An automatic personal identification system based solely on fingerprints or faces is often not able to meet the system performance requirements. Face recognition is fast but not reliable while fingerprint verification is reliable but inefficient in database retrieval. We have developed a prototype biometric system which integrates faces and fingerprints. The system overcomes the limitations of face recognition systems as well as fingerprint verification systems. The integrated prototype system operates in the identification mode with an admissible response time. The identity established by the system is more reliable than the identity established by a face recognition system. In addition, the proposed decision fusion schema enables performance improvement by integrating multiple cues with different confidence measures. Experimental results demonstrate that our system performs very well.

1 Introduction

Biometrics is a technology which identifies an individual based on his/her physiological or behavioral characteristics [7]. It is becoming more and more important to our increasingly inter-connected information society. A biometric system may operate in (*i*) the *verification mode* or (*ii*) the *identification mode* [7]. A verification system authenticates an individual's identity by comparing the individual with his/her own template(s). An identification system recognizes an individual by searching the entire template database for a match. With current technology, some biometric approaches are more suitable for operating in the identification mode than the others. For example, although significant progress has been made in fingerprint identification, an indexing schema which is both fast and efficient in guaranteeing that the correct query is in the top n matches, where n is a small number, say 10, is still very difficult to build even on a relatively small size fingerprint database (several thousand images). On the other hand, it is feasible to design a face recognition system operating in the identification mode, because (*i*) face comparison is a relatively less expensive operation, and (*ii*) efficient indexing techniques are available and the performance is admissible. In addition, different biometric characteristics possess different discrimination capability in terms of accuracy [7]. At the one extreme, we have biometric characteristics such as face and dynamic signature that are inherently better at accepting genuine individuals, but do not perform well in deterring impostors. At the other extreme, we have the biometric characteristics such as retinal scans, fingerprints, and iris that are better at preventing impostors but are less efficient in identifying genuine individuals.

In order to build a biometric system that is able to (i) operate efficiently in identification mode and (ii) achieve desirable accuracy, an integration schema which combines two or more different biometric approaches may be necessary. For example, a biometric approach that is suitable for operating in the identification mode may be used to index the template database and a biometric approach that is reliable in deterring impostors may be used to ensure the accuracy. Each biometric approach provides a certain confidence about the identity being established. A decision fusion schema which exploits all the information at the output of each approach can be used to make a more reliable decision. We will introduce a prototype integrated biometric system which makes personal identification by integrating both faces and fingerprints. The prototype integrated biometric system shown in Figure 1 operates in the identification mode. The proposed system integrates two different biometric approaches (face recognition and fingerprint verification) and incorporates a decision fusion module to improve the identification performance.

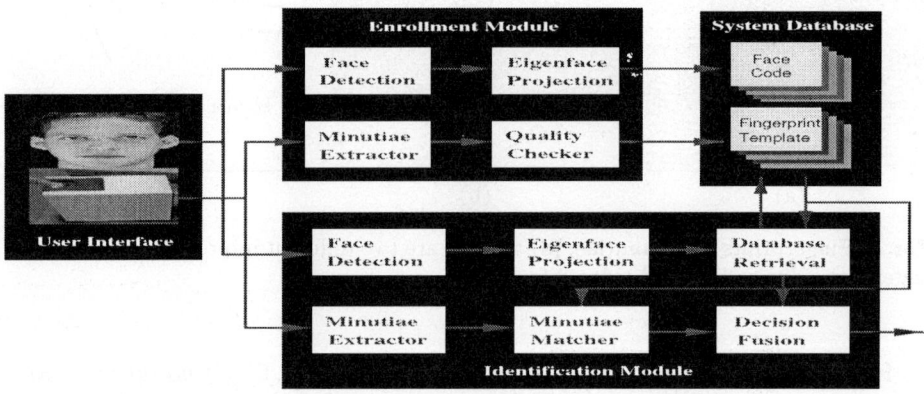

Fig. 1. System architecture of the prototype integrated biometric identification system.

2 Face Recognition and Fingerprint Verification

In our system, the eigenface approach is used to retrieve the top n faces, where n is a system parameter. The eigenface-based face recognition consists of the following two stages [8]: (i) training stage in which a set of training face images are collected; eigenfaces that correspond to the M highest eigenvalues are computed from the training set; and each face is represented as a point in the M-dimensional eigenspace, and (ii) operational stage in which each test image is first projected onto the M-dimensional eigenspace; the M-dimensional face representation is then deemed as a feature vector and fed to a classifier to establish the identity of the individual. In our system, a *k-nearest neighbor classifier* is used, in which the distance, d, called Distance From Feature Space (DFFS) [8] between a template, Φ, and a test pattern, Π, is defined as $\|\Phi - \Pi\|$, where $\|\bullet\|$ means L_2 *norm*.

The uniqueness of a fingerprint is exclusively determined by the local ridge characteristics and their relationships. The two most prominent ridge characteristics, called minutiae, are (*i*) *ridge ending* and (*ii*) *ridge bifurcation*. Fingerprint matching generally depends on the comparison of local ridge characteristics and their relationships [6, 5]. A fingerprint typically contains about 50 minutiae. Examples of minutiae are shown in Figure 2(c). For a given fingerprint, a minutiae can be characterized by its type, its x and y coordinates, and its direction, θ (Figure 2(c)).

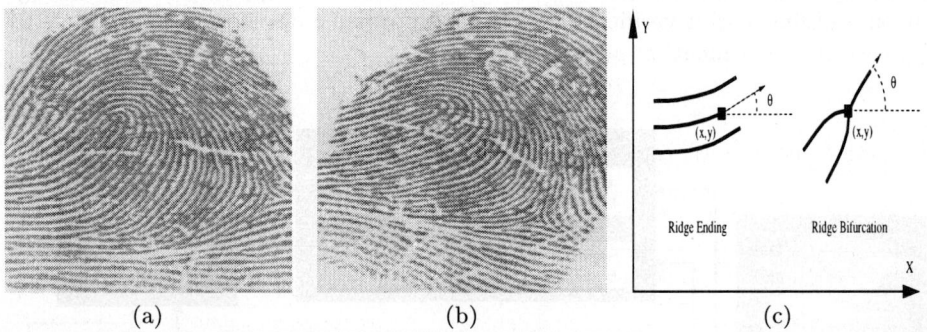

Fig. 2. Fingerprints and minutiae; (a) and (b) are two different impressions of the same finger; (c) ridge ending and ridge bifurcation.

Fingerprint verification consists of two main stages [5, 6]: (*i*) *minutiae extraction* and (*ii*) *minutiae matching*. Minutiae extraction is to extract minutiae from input fingerprint images. Minutiae matching determines whether two minutiae patterns are from the same finger or not. In our system, an alignment-based "elastic" matching algorithm [5] is used, which is capable of finding the correspondences between minutiae without resorting to an exhaustive search and has the ability to adaptively compensate for the nonlinear deformations and inexact transformations between different fingerprints. The alignment-based matching algorithm decomposes the minutiae matching into two stages: (*i*) *Alignment stage*, where transformations such as translation, rotation and scaling between an input and a template in the database are estimated and the input minutiae are aligned with the template minutiae according to the estimated parameters; and (*ii*) *Matching stage*, where both the input minutiae and the template minutiae are converted to "strings" in the polar coordinate system, an "elastic" string matching algorithm is used to match the resulting strings, and finally, the normalized number of corresponding minutiae pairs is reported.

3 Decision Fusion

Each of the top n possible identities established by the face recognition module needs to be verified by the fingerprint verification module. In order to carry out such a decision fusion schema, (i) measures that indicate the confidence of the decision criterion need to be defined and (ii) a decision fusion criterion needs to be formulated. The confidence of a given decision criterion may be characterized by its *false acceptance rate* (FAR), *i.e.* the rate of an impostor being accepted. In order to estimate FAR, the *impostor distribution*, which is defined as the distribution of incorrect matching scores, needs to be computed.

3.1 Imposter Distribution for Fingerprint Verification

Let us assume that the region of interest of all fingerprints is of the same size, a $W \times H$ (for example, 500×500) region. The $W \times H$ region is tessellated into small cells of size $w \times h$ which are assumed to be sufficiently large (for example, 50×50) such that possible deformations and transformation errors are within the bound specified by the cell size. Therefore, there are a total of $\frac{W}{w} \times \frac{H}{h} (= N_c)$ different cells in the region of interest of a fingerprint. Assume that each fingerprint has the same number of minutiae, N_m ($\leq N_c$), which are distributed randomly in different cells. Assume that each cell contains at most one minutiae. Each minutiae is directed towards one of the D (for example, 8) possible orientations with equal probability. Thus, for a given cell, the probability, P_{empty}, that the cell is empty is $\frac{N_m}{N_c}$ and the probability, P, that the cell has a minutiae directed in a specific direction is $\frac{1-P_{empty}}{D}$. A pair of corresponding minutiae between a template and an input is considered to be identical if and only if they are in the cells at the same position and directed in the same direction. With the above assumptions, the number of corresponding minutiae pairs between any two randomly selected minutiae patterns is a random variable, Y, which has a binomial distribution with parameters N_m and P:

$$g(Y) = \frac{N_m!}{Y!(N_m - Y)!} P^Y (1-P)^{(N_m - Y)}. \tag{1}$$

Thus, the probability that the number of corresponding minutiae pairs between any two minutiae patterns is less than a given threshold value, y, is

$$G(y) = g(Y < y) = \sum_{k=0}^{y-1} g(k). \tag{2}$$

3.2 Imposter Distribution for Face Recognition

The top n matches are obtained by searching through the entire database, in which N comparisons are conducted explicitly (in the linear search case) or implicitly (in organized search cases such as the tree search), where N is the total number of templates stored in the database. The impostor distribution for face recognition is a function of both the relative DFFS, Δ, and the rank order, i:

$$F_i(\Delta)P_{order}(i), \qquad (3)$$

where $F_i(\Delta)$ represents the probability that the consecutive DFFS between impostors and their claimed individuals at rank i are larger than Δ and $P_{order}(i)$ represents the probability that the retrieved match at rank i is an impostor.

Let $\Phi_1, \Phi_2, ...\Phi_N$ be the N face templates stored in the database. In order to simplify the analysis, we assume that each individual has only one face template in the database. Thus, there are a total of N individuals enrolled in the database and $I_1, I_2, ..., I_N$ are used as identity indicators. Let X^α denote the DFFS between an individual and his/her own template which is a random variable with density function $f^\alpha(X^\alpha)$ and let $X_1^\beta, X_2^\beta, ..., X_{N-1}^\beta$ denote the DFFS's between an individual and the templates of the other individuals in the database, which are random variables with density functions, $f_1^\beta(X_1^\beta), f_2^\beta(X_2^\beta), ..., f_{N-1}^\beta(X_{N-1}^\beta)$, respectively. Assume that X^α and $X_1^\beta, X_2^\beta, ..., X_{N-1}^\beta$ are statistically independent and $f_1^\beta(X_1^\beta) = f_2^\beta(X_2^\beta) = ...f_{N-1}^\beta(X_{N-1}^\beta) = f^\beta(X^\beta)$. For an individual, Π, which has a template stored in the database, $\{\Phi_1, \Phi_2, ...\Phi_N\}$, the rank, I, of X^α among $X_1^\beta, X_2^\beta, ..., X_{N-1}^\beta$ is a random variable with probability

$$P(I = i) = \frac{(N-1)!}{i!(N-1-i)!}p^i(1-p)^{(N-1-i)}, \qquad (4)$$

where

$$p = \int_{-\infty}^{\infty}\int_{-\infty}^{X_\alpha} f^\alpha(X^\alpha)f^\beta(X^\beta)dX^\beta dX^\alpha. \qquad (5)$$

Obviously, P(I=i) is exactly the probability that matches at rank i are genuine individuals. Therefore,

$$P_{order}(i) = 1 - P(I = i). \qquad (6)$$

Without any loss of generality, we assume that, for a given individual, Π, $X_1^\beta, X_2^\beta, ..., X_{N-1}^\beta$ are arranged in increasing order of values. Define the nonnegative distance between the $(i+1)th$ and ith DFFS values as the ith DFFS distance,

$$\Delta_i = X_{i+1}^\beta - X_i^\beta, \quad 1 \leq i \leq N-1. \qquad (7)$$

The distribution, $f_i(\Delta_i)$, of the ith distance, Δ_i, is obtained from the joint distribution $w_i(X_i^\beta, \Delta_i)$ of the ith value, X_i^β, and the ith distance, Δ_i,

$$f_i(\Delta_i) = \int_{-\infty}^{\infty} w_i(X_i^\beta, \Delta_i)dX_i^\beta, \qquad (8)$$

$$w_i(X_i^\beta, \Delta_i) = CF^\beta(X_i^\beta)^{i-1}[1 - F^\beta(X_i^\beta + \Delta_i)]^{N-i}f^\beta(X_i^\beta)f^\beta(X_i^\beta + \Delta_i), \qquad (9)$$

$$C = \frac{(N-1)!}{(i-1)!(N-2-i)!}, \qquad (10)$$

where $F^\beta(X^\beta) = \int_{-\infty}^{X^\beta} f^\beta(X^\beta)dX^\beta$ [3]. With the distribution, $f_i(\Delta_i)$, of the ith distance defined, the probability that the DFFS of the impostor at rank i is larger than a threshold value, Δ, is

$$F_i(\Delta) = \int_\Delta^\infty f_i(\Delta_i)d\Delta_i. \tag{11}$$

3.3 Decision Fusion

Since facial similarity between two individuals does not imply that they have similar fingerprints, and vice versa, it is safe to assume that the DFFS between two different individuals is statistically independent of the fingerprint matching score between them. Let $F_i(\Delta)P_{order}(i)$ and $G(Y)$ denote the impostor distribution at rank i for face recognition and fingerprint verification modules, respectively. The composite impostor distribution at rank i may be defined as

$$H_i(\Delta, Y) = F_i(\Delta)P_{order}(i)G(Y). \tag{12}$$

Let $I_1, I_2, ...I_n$ denote the n possible identities established by face recognition, $\{X_1, X_2, ...X_n\}$ denote the corresponding n DFFS's, $\{Y_1, Y_2, ...Y_n\}$ denote the corresponding n fingerprint matching scores, and FAR_o denote the specified value of FAR. The final decision, $E(\Pi)$, for a given individual Π is determined by the following criterion:

$$E(\Pi) = \begin{cases} I_k, & if \begin{cases} H_k(\Delta_k, Y_k) < FAR_o, \text{ and} \\ H_k(\Delta_k, Y_k) = \min\{H_1(\Delta_1, Y_1), ..., H_n(\Delta_n, Y_n)\} \end{cases} \\ imposter, otherwise. \end{cases} \tag{13}$$

where $\Delta_i = X_{i+1} - X_i$. Since $H_i(\Delta, Y)$ defines the probability that an impostor is accepted at rank i with consecutive relative DFFS, Δ, and fingerprint matching score, Y, the above decision criterion satisfies the FAR specification.

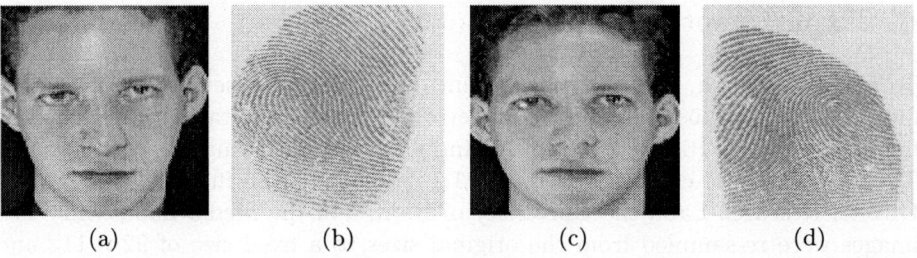

(a)　　　　　(b)　　　　　(c)　　　　　(d)

Fig. 3. Face and fingerprint pairs.

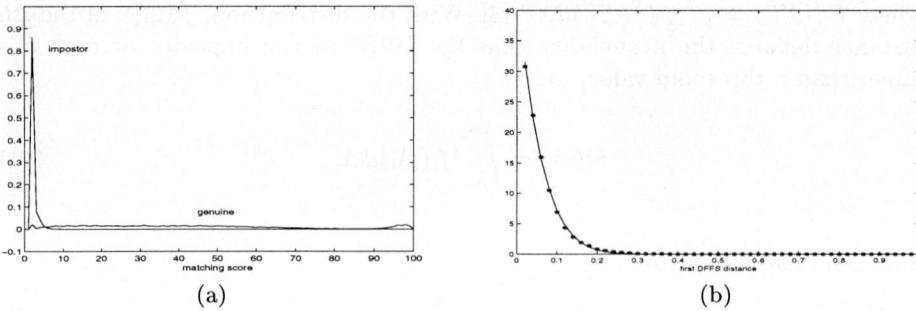

Fig. 4. Impostor distributions; (a) impostor distribution for fingerprint verification; (b) the impostor distribution for face recognition at rank No. 1, where the stars (*) represent empirical data and the solid curve represents the fitting curve; the mean square error between the empirical distribution and the fitted distribution is 0.0014.

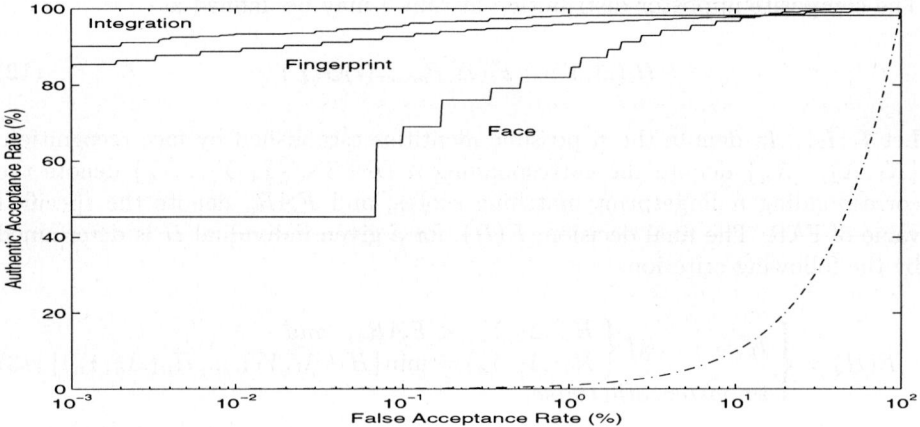

Fig. 5. Receiver Operating Curves; the vertical axis is (1-FRR); the dotted curve in the lower-right corner is the ROC corresponding to the situation when genuine distribution and the imposter distribution are identical.

4 Experimental Results

To test our system, the MSU fingerprint database and a set of public domain face databases are used. The MSU fingerprint database contains a total of 1,500 fingerprint images (640×480) from 150 individuals with 10 images per individual. The face database contains a total of 1,132 images of 86 individuals from the Olivetti Research Lab., the University of Bern, and the Media Lab., MIT. The images were re-sampled from the original sizes to a fixed size of 92 × 112 and normalized to zero mean.

In the test, we randomly selected 640 fingerprints of 64 individuals as the training set and the remaining as the test set. The mean and variance of the impostor distribution (Figure 4(a)) were estimated to be 0.70 and 0.64, respectively. A total of 542 face images were used as training samples. Eigenfaces were estimated from the 542 training samples and the first 64 eigenfaces were used. The top $n = 5$ impostor distributions were approximated. Figure 4(b) shows the impostor distribution at rank no. 1. Each of the remaining individuals in the MSU fingerprint database was randomly assigned to an individual in the face database (Figure 3). A total of 1000 different assignments were tried. A total of 590,000 (590 × 1000) face and fingerprint test pairs were generated and tested. The receiver operating curve (ROC), as well as the ROC's obtained by "all-to-all" verifications using only fingerprints (2,235,000 = 1500 × 1490 tests) or faces (342,750 = 350 × (590 − 5) + 240 × (590 − 15) tests) are plotted in Figure 5. We can see from the figure that the integration schema can greatly improve the identification accuracy. The average CPU time on a Sun SPARC 20 workstation is 4.1 seconds.

5 Summary

We have developed a prototype biometric system which integrates faces and fingerprints in authenticating a personal identification. The proposed system overcomes the limitations of both face recognition systems and fingerprint verification systems. The integrated system operates in the identification mode. The decision fusion schema formulated in the system enables performance improvement by integrating multiple cues with different confidence measures. Experimental results demonstrate that our system performs very well. It meets the response time as well as the accuracy requirements.

References

1. E. S. Bigün, J. Bigün, B. Duc, and S. Fischer, Expert Conciliation for Multi Modal Person Authentication Systems by Bayesian Statistics, *Proc. 1st Int. Conf. on AVBPA*, Crans-Montana, Switzerland, pp. 327-334, March., 1997.
2. R. Brunelli and D. Falavigna, Personal Identification Using Multiple Cues, *IEEE Trans. PAMI*, Vol. 17, No. 10, pp. 955-966, 1995.
3. E. J. Gumbel, Statistics of Extremes, Columbia University Press, New York, 1958.
4. Z. Hong, Algebraic Feature Extraction of Image for Recognition, *Pattern Recognition*, Vol. 24, No. 2, pp. 211-219, 1991.
5. A. Jain, L. Hong, and R. Bolle, On-line Fingerprint Verification, *IEEE Trans. PAMI*, Vol. 19, No. 4, pp. 302-314, 1997.
6. H. C. Lee and R. E. Gaensslen, editors, Advances in Fingerprint Technology, Elsevier, New York, 1991.
7. E. Newham, The Biometric Report, SJB Services, New York, 1995.
8. M. Turk and A. Pentland, Eigenfaces for Recognition, *Journal of Cognitive Neuroscience*, Vol. 3, No. 1, pp. 71-86, 1991.

Automated Fingerprint Pattern Classification Error Analysis

Weicheng Shen
Pacer Infotec Inc.
1420 Spring Hill Rd., Suite 205
McLean, VA 22102

1. Introduction

Automated Fingerprint Identification System (AFIS) has been widely deployed by law enforcement agencies, social welfare, national ID, voter registration, etc. in various countries in the world [1-3]. It is one of the most reliable and matured automated identification systems based on biometrics. In an identification system, the key functionality is to compare the biometrics of an individual of unknown identity (search subject) with the biometrics of the individuals (file subjects) stored in a biometrics repository to discover the identity of the individual. For a fingerprint identification system, the fingerprints collected either from an individual or lifted from some objects (latent prints) are compared with those stored in a fingerprint repository to either discover the identity of the individual or relate it to other unsolved latent prints. The fingerprints of a search subject are referred to as the search prints, while the fingerprints of file subjects are referred to as the file prints. The file print that belongs to the search subject is referred to as the "mate" of the search print.

Such automated identification systems normally demand great computing power since the number of comparisons to be made is fairly large. The cost of an automated identification system is normally proportional to the number of fingerprints comparisons performed in each second. In many cases, the performance requirements and budgetary constraints prevented the exhaustive searches of the data repository. The collection of fingerprints compared to the fingerprints of the search subject is often referred to as the search space. The complete search space is the entire fingerprint repository. Search space reduction techniques are often used to reduce the number of fingerprint comparisons. If the "mate" of the fingerprints of the search subject is somehow erroneously excluded from the reduced search space, the automated fingerprint matcher will never be able to find it. A "miss" in fingerprint matching is thus produced, which is caused due to the error in search space reduction. This paper addresses some of the issues related to errors introduced in the process of search space reductions.

2. Statement of Problems

For an AFIS that conducts fingerprint searches over a large-scale fingerprint repository, the number of pair wise fingerprint comparisons that the system must perform becomes critical to the throughput rate of the system. If the system has to compare any search print with the fingerprint of each subject in the fingerprint repository, the search space for a search print is the entire repository. To reduce the number of comparisons for each search, we need to somehow partition the search space such that for any given search print, the search space is only the fingerprints of a subset of subjects in the repository. There are many ways of partitioning a search space, such as fingerprint pattern classification, subordinate level classification, or hierarchical filtering. The basic idea is to partition the search space so that only those fingerprints in the repository that are "similar", in certain sense, to the search print will be compared with it. Since it is expected that only a subset of fingerprints of the entire fingerprint repository are similar to the fingerprints of the search subject, comparing the fingerprints of the search subject with these "similar" fingerprints would almost certainly reduce the total number of comparisons needed. The search space reduction can thus be achieved. Such partitions can either be achieved before the search print is available and is fixed (static or hard binning), or be achieved after the search print is available and is non-fixed (dynamic or soft binning). This paper focuses only on the error analysis of the hard (static) binning approach, *i.e.*, the search space is partitioned before the search print is available and is fixed. Furthermore, we assume that we use the fingerprint pattern classes as the basis for binning. In other words, we partition the fingerprints into subsets according to their respective pattern classes. For a 10-print card, if we use all ten rolled fingerprints of the card for pattern classification, the pattern class for this 10-print card is denoted as a vector of 10 components, where each component represents the pattern class of one fingerprint. We have partitioned 10-print cards into a ten-dimensional space. Each 10-print card is represented by one point in this space.

Fig. 1. depicts an AFIS with pattern classification capability. A fingerprint card is used as the search input to the fingerprint pattern classifier. The output of this pattern classifier points to the "electronic file cabinet" (pattern class bin) containing the all the fingerprint cards which have the same pattern classes as the fingerprint search card.

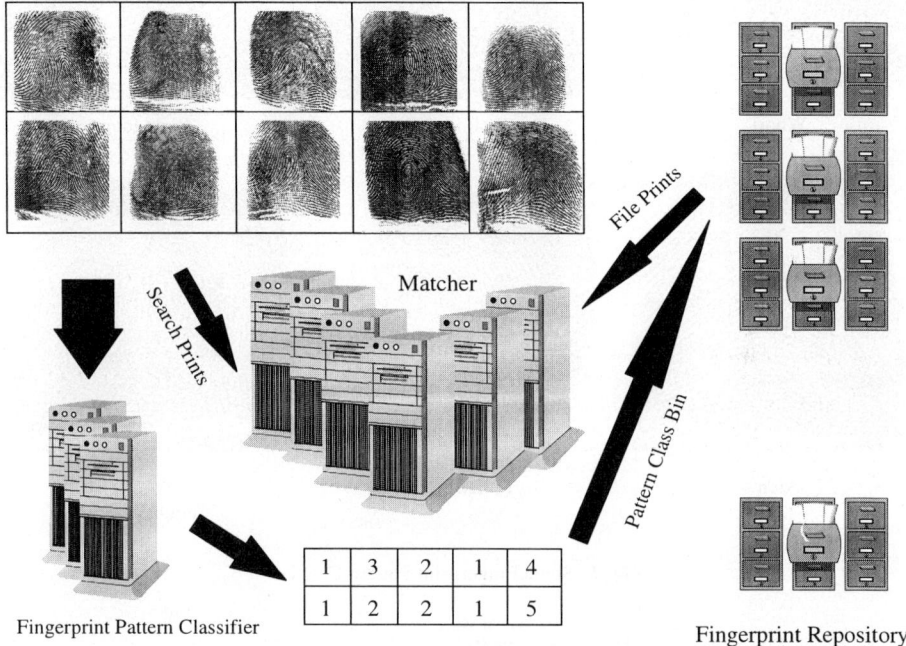

Fig. 1. Operation of An Automated Fingerprint Identification System with Pattern Classification

In Fig. 1., the input is a 10-print card. The fingerprint pattern classifier produces a pattern class key with 10 digits, each of which represents the pattern class of one fingerprint on the card. This pattern class key is then used to locate the pattern class bin (or bins) in the fingerprint repository that might contain the mate of the search print. All file prints in the located pattern class bin are compared with the search print by the matcher to discover the identity of the search subject.

Some of the fingerprint pattern classes are depicted in Fig. 2. These fingerprints are commonly known as the rolled impressions. As one may observe, each of these fingerprints has quite wide coverage in width. They normally can better capture the "deltas" in loops or whorls. The deltas are often very critical to correctly assigning pattern classes.

Since fingerprints have various qualities and the "deltas" and "cores" are not always captured for correct and precise classifications, it is inevitable to introduce fingerprint pattern classification errors. This paper attempts to answer the following questions: 1) How do we measure the fingerprint pattern classification error for a single finger? 2) How do we measure the fingerprint pattern classification error based on multiple fingers? 3) What can we do to reduce pattern classification errors? 4) What is the cost associated with the reduction in pattern classification errors?

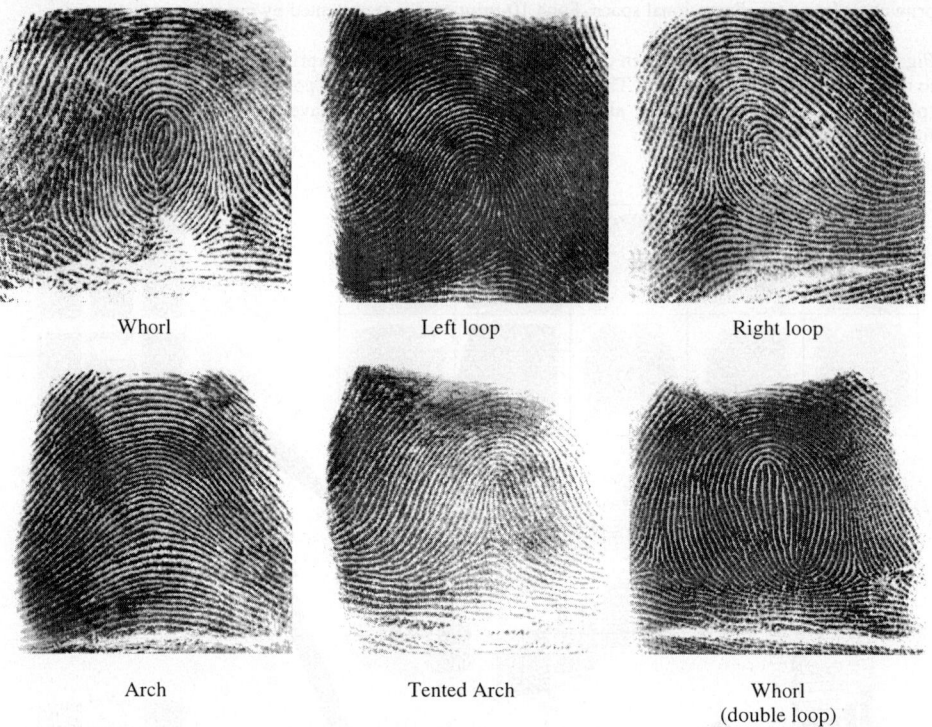

Fig. 2. Some Fingerprint Pattern Classes

3. Pattern Classification Error Analysis

As we discussed in the previous section, the pattern classification will introduce errors in the fingerprint search process. Since correct assignment of pattern classes to fingerprints is essential to achieving highly reliable fingerprint searches, it is important to understand the pattern classification error measures and their impact to the automated fingerprint identification systems.

In the following discussions, let the number of pattern classes of our fingerprint classifier be N, and assume that we have seven classes, $N = 7$. In other words, each fingerprint is assumed to be in one of the seven classes: whorl, left loop, right loop, arch, tented arch, scar, or amputated. A number from the set $\{1, 2, 3, ..., 7\}$ denotes each pattern class. In a fingerprint classification system based on the fingerprint patterns of 10 fingers, there are a total of $7^{10} \approx 282 \times 10^6$ different classes, or bins. Unfortunately, fingerprint patterns are not uniformly distributed over the entire population. There are more of certain patterns than others, which results in some unusually large bin sizes if a "real world" fingerprint classification system is built based on these pattern classes. Such observations lead to the development of subordinate level classification systems and others to further reduce the search space, which is beyond the scope of this paper.

Assume that the fingerprints in the repository are correctly classified. Each search print can be classified into each of the seven classes with certain probability. For example, a fingerprint of subject A had been classified as $j = 3$ in the repository, the search print of this subject can be classified as i with a probability of

$$P(i \mid j) = \Pr\{\text{search print assigned pattern class } i \mid \text{"true" class } j\}.$$

Now let's assume that the pattern class distribution for that finger is *a-priori*,

$$P(i) = \Pr\{\text{the probabilty that the search print has a "true" class } i\}.$$

It follows that for a given search print classified as i pattern class, the probability that it has a "true" class j, can be calculated using the Bayes formulation [4]:

$$P(j \mid i) = \frac{P(i \mid j)P(j)}{\sum_j P(i \mid j)P(j)}.$$

Although multiple fingers are normally used for pattern classification systems, we start our analysis with a hypothetical single fingerprint pattern classification system. The following confusion matrix illustrates the probabilistic relationships between the "true" pattern class of a fingerprint and it's assigned pattern class by an automated fingerprint classification system.

	1	2	3	4	5	6	7
1	$P(1\mid1)$	$P(1\mid2)$	$P(1\mid3)$				$P(1\mid7)$
2	$P(2\mid1)$	$P(2\mid2)$	$P(2\mid3)$				$P(2\mid7)$
3	$P(3\mid1)$	$P(3\mid2)$	$P(3\mid3)$				$P(3\mid7)$
4							
5							
6							
7	$P(7\mid1)$	$P(7\mid2)$	$P(7\mid3)$				$P(7\mid7)$

Each column of the confusion matrix represents an assigned pattern class for a fingerprint, while each row represents a "true" pattern class for a fingerprint. The entry at i-th row and j-th column, (i, j), contains the probability that a fingerprint of assigned pattern class i is indeed of the pattern class j by the fingerprint classifier under consideration. It is clear that the fingerprint classifier assigned a pattern class correctly to the fingerprint if $i = j$. In other words, the diagonal entries of the confusion matrix represents the probability of a fingerprint being classified correctly for each "true" pattern class, respectively. The off-diagonal entries of the matrix represents the probabilities that the fingerprint pattern classifier assigned incorrect class to a fingerprint. Consider the entries in one column of the confusion matrix, column 3, *i.e.*, a fingerprint is assigned with the pattern class 3. Only the entry (3,3) contains the correct classification probability, while the rest contains the probability of incorrect classifications given a fingerprint has the assigned pattern class 3. Let the error probability of a fingerprint of assigned pattern class i be denoted by E_i as expressed below:

$$E_i = \sum_{j \ne i} P(j \mid i),$$

where $P(j \mid i)$ is the probability that a fingerprint of assigned pattern class i is of "true" pattern class j.

It follows that the probability of a fingerprint being incorrectly classified can be expressed:

$$E = \sum_i E_i P(i) = \sum_i \sum_{j \neq i} P(j|i) P(i).$$

The probability that a fingerprint is correctly classified is thus equal to

$$C = 1 - E,$$

which is the weighted sum of the diagonal entries of the confusion matrix.

Next, consider a fingerprint pattern classifier based on two fingerprints of a subject. Denote the probabilities of incorrectly classifying the fingerprints of a subject as $E^{(1)}$ and $E^{(2)}$, respectively. The probabilities of correctly assigning pattern classes to these two fingerprints are denoted as $C^{(1)} = 1 - E^{(1)}$ and $C^{(2)} = 1 - E^{(2)}$, respectively. What is the probability that the pattern classes of both fingerprints are correctly assigned? If we neglect the correlation between the pattern classes of different fingers of a subject, the probability that pattern classes of both fingerprints are correctly assigned is

$$C = C^{(1)} \cdot C^{(2)} = \left(1 - E^{(1)}\right)\left(1 - E^{(2)}\right) \approx 1 - E^{(1)} - E^{(2)},$$

when both $E^{(1)}$ and $E^{(2)}$ are very small. The classification error is $E \approx E^{(1)} + E^{(2)}$. In other words, the classification error of a pattern classifier using two fingerprints is roughly equal to the sum of the classification errors of two fingerprints.

Similarly, for a fingerprint pattern classification system using M $(0 < M \leq 10)$ fingerprints of a subject, the probability that the pattern classes of all M fingerprints are correctly assigned is

$$C = \prod_{k=1}^{M} C^{(k)} = \left(1 - E^{(1)}\right)\left(1 - E^{(2)}\right) \cdots \left(1 - E^{(M)}\right) \approx 1 - \sum_{k=1}^{M} E^{(k)},$$

if $E^{(k)}$ is small for $k = 1, 2, \ldots, M$. The classification error is $E \approx \sum_{k=1}^{M} E^{(k)}$. This observation demonstrates that the classification error increases as the number of fingerprints used for pattern classification increases. On the other hand, it is observed that the increases in the number of fingerprints used for pattern classification leads to more search space reduction. It implies that more search space reduction (smaller search bins) inevitably results in pattern classification errors. Therefore, our capacity of reducing the search space has been limited by the bounds of the pattern classification error.

To overcome this limitation, the concept of pattern class referencing is introduced. When a fingerprint is assigned with a pattern class i, a probability of being of "true" pattern class j $(j \neq i)$ for a fingerprint with assigned pattern class i is calculated, i.e., $P(j|i)$. Note that the classification error probability $P(j|i)$ can be calculated using the Bayes formula as described before. The pattern class j_R that attains the highest value of $P(j|i)$ is called the reference class. This reference class j_R is indeed our "second guess" of the "true" pattern class given that the fingerprint is assigned to the primary class i. We observe that j_R is the second most likely pattern class for the fingerprint in question (the most likely pattern class for this fingerprint is i). Therefore, we do not only compare the search print with file prints of class i, but also j_R.

In this scenario, the error probability of a fingerprint of assigned pattern class i be denoted by $E_i(R)$ as expressed below:

$$E_i(R) = \sum_{\substack{j \neq i \\ j \neq j_R}} P(j|i),$$

where the argument R indicates that this is an error measure for fingerprint pattern classifiers with reference classes. For a fingerprint pattern classifier using at most one reference class, $R = 1$. Since $P(j|i)$ are non-negative functions, this error measure is less than or equal to the error measure previously derived for the fingerprint pattern classifier without reference class. It clearly demonstrates the accuracy advantage of introducing reference classes. The price paid for this gain is the increases in search space (the number of fingerprints to be compared), which now includes the fingerprints in both pattern class i and pattern class j_R.

How much do we gain by expanding the search space? In the worst scenario, $P(j|i) = 1/6[1 - P(i|i)]$ are all equal $\forall j \neq i$. This provides an error reduction of $(1/6)E_i$, i.e., $E_i(1) = (5/6)E_i$ for the fingerprint pattern classification of one fingerprint of assigned pattern class i. In the following discussions, we assume the worst scenario of the pattern classification errors. It follows that the pattern classification error for one fingerprint, over all pattern classes, is

$$E(R=1) = \sum_{i=1}^{N} E_i(R=1)P(i) = \sum_{i=1}^{N} \sum_{\substack{j \neq i \\ j \neq j_R}} P(j|i)P(i) = \frac{N-1}{N} \sum_{i=1}^{N} E_i P(i) = \frac{N-1}{N} E,$$

where E is the pattern classification error for one fingerprint without a reference class. This equation demonstrates the relationship between the pattern classification errors with and without a reference class when $R = 1$. Similarly, one can obtain the relationship between the pattern classification errors with and without reference classes when $1 < R \leq N$:

$$E(R) = \frac{N-R}{N} E,$$

where E is the pattern classification error for one fingerprint without a reference class. It follows that the classification error decreases as the number of reference classes increases. The probability that the pattern classifier assigned a correct class to the fingerprint in question is defined as

$$C(R) = 1 - E(R).$$

Now, let's consider the fingerprint pattern classification using multiple fingerprints. As in the case of pattern classification without reference class, we denote the probability that the fingerprint pattern classifier assigns a correct pattern class to the k-th fingerprint of the subject by $C^{(k)}(R) = 1 - E^{(k)}(R)$. For a fingerprint pattern classification system using M ($0 < M \leq 10$) fingerprints of a subject, the probability that the pattern classes of all M fingerprints are correctly assigned is

$$C(R) = \prod_{k=1}^{M} C^{(k)}(R) = \left(1 - E^{(1)}(R)\right)\left(1 - E^{(2)}(R)\right) \cdots \left(1 - E^{(M)}(R)\right) \approx 1 - \sum_{k=1}^{M} E^{(k)}(R),$$

where $E^{(k)}(R)$ is small for $k = 1, 2, \ldots, M$. The pattern classification error is consequently $E(R) \approx \sum_{k=1}^{M} E^{(k)}(R)$, which is the sum of classification error of each finger if the higher order items are ignored. This observation again demonstrates the fact that the classification error increases as the number of fingers involved increases, given that $E^{(k)}(R)$ is unchanged.

Furthermore, we can observe that error measure

$$E(R) \approx \sum_{k=1}^{M} E^{(k)}(R) = \frac{N-R}{N} \sum_{k=1}^{M} E^{(k)},$$

implies that the pattern classification error $E(R)$ decreases as the number of reference classes, R, increase. This is the performance gain achieved through reference classes.

On the other hand, consider the price that must be paid to achieve the aforementioned performance gain, *i.e.*, the increases in search space in comparison to a fingerprint pattern classifier without reference classes. For simplicity, assume that the reference class bins have about the same sizes for the primary class bins. Let the size of the search space of a subject, using M fingerprints for classification, be $S_M(R=0,m)$, where the first argument, R, represents the maximum number of reference classes used for each fingerprint. The second argument of $S_M(R=0,m)$, m, represents the number of fingerprints (out of M) that has the maximum number of reference classes. $S_M(R=0,m)$ represents the search space of a fingerprint pattern classifier without reference class. If only one fingerprint out of the M fingerprints has a reference class, then the $S_M(1,1) \approx 2 \times S_M(0,M)$. Similarly, one can obtain that $S_M(1,m) \approx 2^m \times S_M(0,M)$. As expected, the search space grows exponentially in case of equal size primary and reference pattern class bins. This estimation only illustrates the order at which the search space grows as the reference pattern class bins are introduced into the search space. Note that the assumption of equal size reference class bins is not always true. It was made for providing a cost growth "trend" estimation.

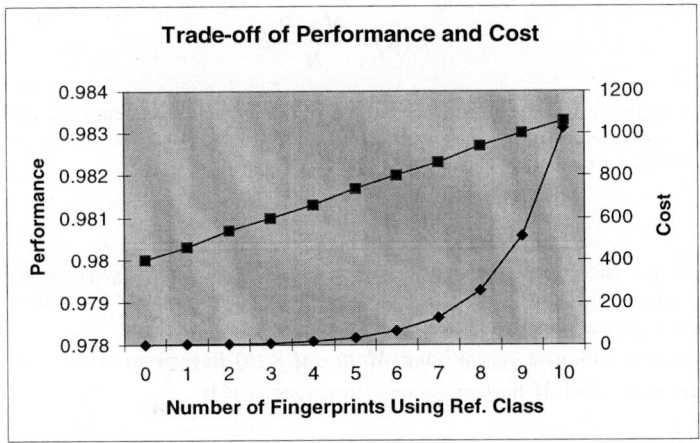

Fig. 3. Plot of Fingerprint Pattern Classifier Performance and Cost vs. The Number of Fingerprints Using Reference Classes

To demonstrate the trade-off between the performance and the cost of a fingerprint pattern classifier, according to the previous assumption, we calculate the correct pattern classification rate and the corresponding computing cost for various number of fingers using reference class. It is assumed that the correct pattern classification rate for a single fingerprint is 99.8% for any finger. Let the computing cost be unit when no reference class is used for pattern classification. The results is plotted in Fig. 3. The "almost linear" curve representing the improvement in rate of correct pattern classification. The "almost exponential" curve representing the computing cost increases. As the number of fingerprints using reference classes increases from zero (no reference class is used) to ten (each fingerprint has a primary and a reference class), the rate at which the pattern classifier correctly assigns pattern classes to the search prints increases from 98% to 98.33%. Meanwhile, the search space grows by over 1,000 times.

As a result, for a pattern classifier with a known single fingerprint classification error, $E^{(k)}$, it is quite expensive to reduce its multiple fingerprint pattern classification error by means of reference classes.

4. Conclusions

Fingerprint pattern classifiers are expensive means to reduce the search space of fingerprints identification systems. The more fingers of the search subject are used for pattern classification, the more search space reduction can be achieved. However, the more fingers of the search subject are used for pattern classification, the higher the classification errors will incur. The use of reference class is a practical way of improving the pattern classification accuracy. Its cost is an important factor in deciding whether the solution is cost effective. It is often the case that the number of comparisons made by the matcher is limited by hardware capacity. Hence, the search space size becomes constrained given the response time requirements for an AFIS. Thus, the utilization of reference classes might be limited.

5. Acknowledgement

The author wishes to thank Mr. Mike Mahoney of the Pacer Infotec Inc., for his support that makes this presentation possible. The author also thanks Mr. Federico Tirso Jr. of the Hughes Information Systems for his comments which improves the readability of the paper.

6. References

[1] *Proceedings of the 8-th Biometric Consortium Meeting*, June 1996, San Jose, CA.

[2] *Proceedings of the 9-th Biometric Consortium Meeting*, April 1997, Crystal City, VA

[3] *Proceedings of the IEEE*, Special Issue on Automated Biometrics, September, 1997.

[4] Rohatgi, V. K., An Introduction to Probability Theory and Mathematical Statistics, John Wiley & Sons, 1976, New York, NY

A High-Dimensional Indexing Scheme for Scalable Fingerprint-Based Identification

Andrea Califano
Bob Germain
Scott Colville[1]

IBM TJ Watson Research Center
PO Box 704, Yorktown Heights, NY 10598

Abstract

This paper describes an application of the Flash algorithm [3] to scalable fingerprint-based identification (one-to-many searches). Flash is a high-dimensional indexing algorithm akin to Geometric Hashing [11] that has been used for similarity searching in a number of other domains, including model based object recognition [3], genomic sequences homology detection [4], and 3D flexible molecular matching and docking [16].

A large number of independent properties of each model fingerprint that are invariant under Euclidean transformations are extracted and stored in a look-up table. Match candidates and their pose transformation parameters are then formed by indexing in this table, using identical invariants generated from a query fingerprint. This has the desirable properties of determining appropriate match candidates without having to compare the query to each individual model fingerprint in the database. Candidate hypothesis evidence is then gathered in a parameter space with an approach similar to the Generalized Hough Transform [2].

Results of this preliminary implementation on a database of 100,000 models show good scalability properties. Measured False Positive and False Negative Rates allow this approach to be extended to databases with tens of millions of fingerprints. Reported performance measurements show an equivalent 1 to 1 matching rate of about 150,000 prints/sec. on an 8-way SMP PowerPC workstation or, equivalently, on an 8-node SP/2 platform.

Introduction

Biometrics is the science of identifying individuals by a particular physical characteristic such as eye color, fingerprints, height, facial appearance, iris texture, or signature. Fingerprints are arguably the most popular biometric currently in use because of their long history in law enforcement applications. The pattern of ridges on each person's finger is uniquely characteristic of that individual and contains sufficient information to be used to distinguish that person from any other. A review of the history of fingerprint identification and some reviews of recent technology may be found in [12]. Descriptions of two fingerprint matching systems including a discussion of the kind of feature extraction used in this work can be found in [10, 15].

There are two general classes of problems which a fingerprint matcher is expected to address. The first class of problems involves situations for which it is necessary to verify or authenticate an indi-

---

1. Current adrress: Computer Science Dpt., University of Wisconsin-Madison, 1210 W. Dayton St. Madison WI 53706

vidual's identity. This is a one-to-one matching problem and is of interest here primarily as a conceptual basis for one-to-many matching.

The second, more challenging, problem occurs when it is important to ensure that a particular database contains only a single entry for any given individual. This occurs in the case of social services wherein one wishes to prevent individuals from collecting welfare under multiple aliases or in the case of identity card issuance. This identification problem requires that one search a large database of individuals and determine whether this person is already in the database.

Indexing Paradigms: FLASH

The standard approach in model-based object recognition, where both the model and the query are affected by several noise processes including sensor noise and distortions due to the sensing operation, is to directly compare the query to each model in the database [9]. More recently, however, indexing paradigms have shown that greater efficiency can be achieved, at the cost of increased storage, by performing comparisons in an invariant space rather than directly in the model database using a look-up table [11, 3]. The invariants choice depends critically on the assumptions about the class of allowed geometric transformations.

One of the earliest approaches, Geometric Hashing [11], used low-dimensional invariants to recognize three-dimensional objects from two-dimensional projections under affine transformation assumptions. Invariants, in this case, are computed directly from affine invariant properties of the geometric configuration of a chosen subset of feature points, usually a triplet.

Another approach, High-Dimensional Indexing [3], uses high-dimensional invariants to Euclidean transformations to search very large databases without saturating the look-up table, thereby providing increased false positive rejection and speed. This type of indexing is the basis for FLASH (Fast Look-up Algorithm for Structural Homology) in other related fields [4, 16]. It is also the basis for the scalable, index-based fingerprint searching framework described in this paper.

Application to Fingerprint Matching

Unlike the general shape recognition problem, fingerprints provide a convenient canonical representation for local shapes in the form of *minutiae*. Both automatic and manual fingerprint recognition schemes use these well-established features, determined by singularities in the finger ridge pattern. Minutiae are placed at points where a ridge either ends or splits into two ridges. No distinction is made between these two cases. Each minutia is characterized by its position and the local direction of the ridge with respect to a reference frame common to all of the minutiae in a given impression. Thus, each feature point is represented by three independent variables, (x, y, θ), respectively the coordinates of the minutia location and the angle of its corresponding ridge, as shown in Fig. 1.

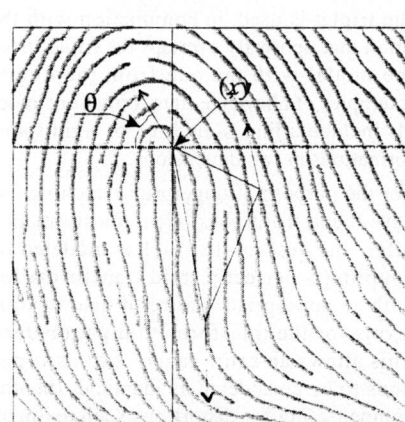

Fig. 1: Minutiae Triplet

A typical "dab" impression has approximately 40 minutiae which are recognized by the feature extraction software, but the number of minutiae can vary from zero to over one hundred depending on the finger morphology and imaging conditions. Not all of these minutiae will be reproducible from imprint to imprint and therefore the redundancy in the combinatorial index formation process described below is essential.

In the proposed approach, high-dimensional invariants are computed from combinations of 3 minutiae (triplets), using the length of each side of the triangle formed by the triplet, the angle of each minutia with respect to the longest side of the triangle, and the number of ridge crossings along each side, see Fig. 1. These values are appropriately quantized and bit-packed to fit in a 32 bit integer index. In Fig. 2, l_i is the length of the i-th side, θ_i are the minutiae angles encoded in a transformation invariant fashion, and rc_i are the number of ridges crossed along the i-th side. Side lengths are ordered so that the largest one is always encoded first in the index. Successive sides are enumerated proceeding always in the same predefined orientation (e.g., always in a clockwise fashion). Ridge counts and angles are also ordered in a deterministic fashion based on the side length ordering. For example, the first angle can be always taken to be the one of the most counter-clockwise minutia on the first side. This procedure is invariant under rotations and translations, but *not* under reflection. Triplets that do not allow deterministic ordering, i.e., those forming equilateral triangles, are not used to generate indices.

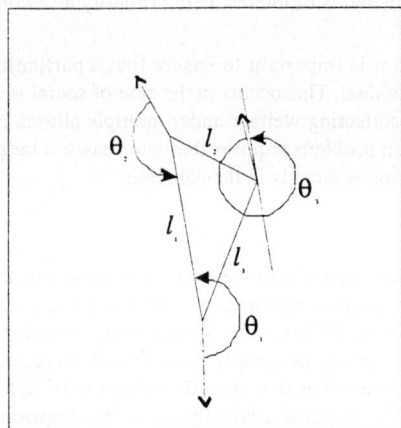

Fig. 2: Ordered lengths and angles

During model storage, triplets are assembled in a deterministic fashion from all available fingerprint minutiae. However, to provide some normalization of the potential support for matches, a hard upper limit is imposed on the total number of triplets generated by each fingerprint, as explained in the next subsection. Each triplet, through its corresponding 32 bit index, is used to uniquely identifies a bin in a large look-up table structure where an entry is appended. The entry consists of the model fingerprint Id and a set of parameters later required to recover pose. These could be, for instance, the $\{(x_i, y_i): i = 1, 2, 3\}$ coordinates of each minutiae in the triplet although a number of other choices are also possible.

During recognition, high-dimensional invariants are computed from query fingerprint triplets in identical fashion. Their 32 bit integer encoding is then used to locate a bin in the look-up table and to recover all the corresponding entries. These will identify all the model fingerprint that contain at lease one triplet identical to the one in the query fingerprint, within quantization limits.

For each of the retrieved entries, the geometry of the triplet is used, in conjunction with the retrieved shape parameters, to compute the transformation that optimally overlaps the triplet from the model with the one from the query. Examples of allowable transformations models include affine, similarity, or rigid transformations in two dimensions.

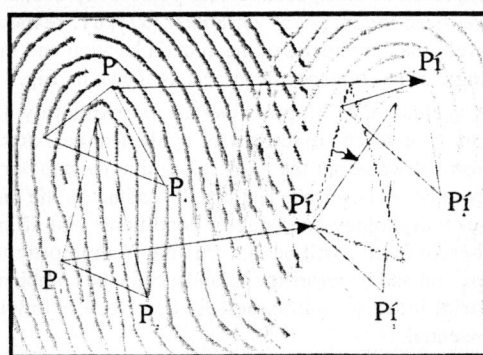

Fig. 3: Transformation recovery

For the purpose of fingerprint matching, we have found that a *quasi-rigid* transformation offers the best results. This is modeled by coupling a perfectly rigid transformation with an appropriate choice of quantization levels for the transformation parameters. This allows an entire range of transformations to be considered as identical, thereby compensating for distortions such as linear and radial shear. Under these assumptions, then, a *match hypothesis* $H_i(Id, \mathbf{T}, \mathbf{R})$ is uniquely identified by its Id, its quantized translation vector $\mathbf{T}$ and its quantized rotation matrix $\mathbf{R}$. Since the 3-point rigid transformation problem is overconstrained, we use a least squares pose estimation technique to compute $\mathbf{T}$ and $\mathbf{R}$ [8].

When a good match exists under the specific transformation assumptions, many of the triplets that generate identical indices will also produce the same transformation parameters. Fig. 3, for instance, shows a match with two triplets, and their relative transformations, for an identical fingerprint rotated and shifted to a different pose, where

$$\bar{P}'_2 - \bar{P}'_1 = \mathbf{T} + \mathbf{R}(\bar{P}_2 - \bar{P}_1)$$
$$\bar{P}'_4 - \bar{P}'_3 = \mathbf{T} + \mathbf{R}(\bar{P}_4 - \bar{P}_3)$$
(1)

Hypotheses H_i are therefore scored based on the number of triplets that independently generated them.

Both acquisition and recognition are efficient; specifically, recognition is not exponential in the problem size as it is the case for many systems [7] and grows very slowly with the number of models in the database. A full analysis of the approach is available in [3].

Index Generation

The proposed algorithm uses redundant, combinatorial index generation techniques to achieve some level of immunity against noise, either in the form of misidentified or mislabeled minutiae due to finger or imaging quality problems or in the form of image shear due to the coupling of the deformable surface of the finger with the sensing device. As shown in [3], combinatorial indices are also effective in producing more uniquely descriptive invariants than single feature point.

An exhaustive listing of all the possible combinations of n minutiae in groups of 3, would generate

$$\binom{n}{3} = \frac{n!}{(n-3)!3!}$$
(2)

Because this number can be very large, restrictions are placed on the "acceptable" combinations of minutiae used to form a triplet. The primary restriction used is a limitation on the distance between minutiae: only those separated by a distance in a pre-specified range are used to form a given triplet. This imposes a limit on the minimum and maximum length of the triangle sides of a triplet which can be used to effectively bit-pack the lengths.

Even the restriction on pairwise separations, however, leads to large variability in the total number of triplets, and hence of indices, generated by different fingerprints. In order to guarantee a relatively constant number of indices generated, a deterministic selection process is used to select a fixed sampling of all the allowed triplets.

Characterizing Accuracy for Verification and Identification

First consider the problem of determining whether or not two fingerprints were made by the same finger (verification). This problem amounts to assigning the pair to either of the mated or nonmated pair populations. The objective is to find a test that assigns a pair of prints to one of these two populations while making smallest number of mistakes in large number of trials. This problem in statistical decision making has a very long history [13]. The decision framework described here is similar in spirit to that used to describe recent work on iris identification [6]. In the case where one of two mutually exclusive hypotheses, H_0 and H_1, must be selected, two classes of errors can be made. Suppose that H_0 is the hypothesis that the pair of prints belongs to the non-mated population and that H_1 is the hypothesis that the pair of prints belongs to the mated population. Four scenarios are possible:

test says H_0 is true and H_0 is true	test says H_0 is true and H_0 is false
test says H_1 is true and H_1 is true	test says H_1 is true and H_1 is true

The test breaks down in two of the four scenarios; two distinct types of errors can be made:

- **False Negative or Miss**: incorrectly assigning a mated pair to the non-mated population
- **False Positive or False Alarm**: incorrectly assigning a non-mated pair to the mated population

The number of matching triangles, henceforth referred to as the score, that generate a consistent rigid transformation between two prints can be used as the basis for a test that assigns a pair of fingerprints to the mated pair population or to the non-mated pair population. Histograms of the scores achieved by the matcher on the two test populations can be used as estimates for the conditional probability densities of the score, $f_M(x) = f_{\text{Mated}}(x)$ and $f_N(x) = f_{\text{Non-mated}}(x)$. It is natural to use a threshold x_{th} to assign a pair of imprints to one of the two possible populations. Any pair whose score exceeds x_{th} will be assigned to the mated pair population and other pairs will be assigned to the non-mated pair population. Note that there is a tradeoff between the false positive error rate (FPR) and the false negative error rate (FNR). The FNR can be reduced to an arbitrarily small value by decreasing x_{th} sufficiently, but a large number of false alarms will result with a corresponding increase in the FPR. If this decision criterion is used, it is straightforward to compute the two error rates from the conditional probability densities computed from the test populations. The error rate for incorrectly assigning a mated pair to the non-mated population (false positive rate) is given by the distribution function defined below:

$$FPR = F_M(x_{th}) = \int_0^{x_{th}} f_M(t)(dt) \qquad (3)$$

Similarly, the error rate for incorrectly assigning a non-mated pair to the mated pair population (false negative rate) is given by the following function of the conditional probability distribution function:

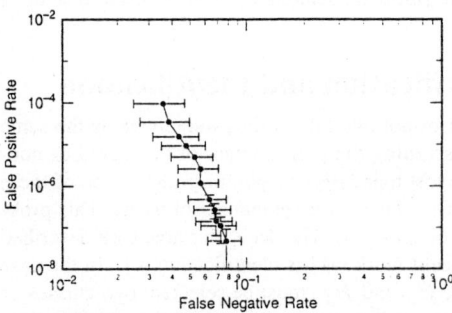

Fig. 4: Receiver Operating Curve

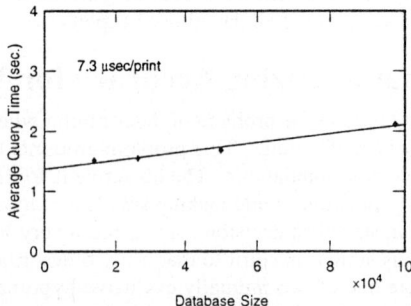

Fig. 5: Search Times

$$FNR = 1 - F_N(x_{th}) = 1 - \int_0^{x_{th}} f_N(t)dt \qquad (4)$$

Insofar as the mated and non-mated pair test populations form representative samples of the real populations, the estimates may be used to extrapolate to behavior on the real populations. The measured accuracy of a matcher is a strong function of the database from which estimates of the error rates are derived, most importantly, the variation of the false negative error rate with threshold depends on the care with which fingerprint images were acquired. Note that these estimates of the error rates are independent of the size of the test database used although the uncertainties in the estimates depend on the sizes of the sample pair populations. Now consider a one-to-many identification query, which may be viewed as a series of one-to-one verifications executed against every print in the database. With the assumption that at most one mate to the query is present, and the assumption that the candidate list of hypothesized matches is formed by taking all prints from the reference database whose verification matching scores with the query print exceed some fixed threshold, the false positive rate (FPR) and the false negative rate (FNR) for an identification search against a database of N individuals is as follows:

$$FNR(N) = d$$
$$FPR(N) = 1 - (1 - FPR(1))^N \qquad (5)$$
$$\approx N \times FPR(1), \text{ for } FPR(1) \ll 1/N$$

The false positive rate increases drastically with database size because each additional entry in the database provides another opportunity to randomly achieve a high score. A matcher operating at a point where its false positive verification rate is 1% may be satisfactory in a verification application, but in even a small scale identification application, the error rate will become unacceptable.

For example, when used on a ten person database, this matcher will generate false matches at a rate of $1 - 0.99^{10}$ or 9.5%. On a hundred person database this matcher's false positive error rate will be $1 - 0.99^{100}$ or 63%. Fig. 6 shows the extrapolated false positive rate versus population size for a variety of one-to-one error rates. In order to keep the false positive rate within reasonable bounds when operating on large population sizes, a matcher must be operating in a mode for which its false positive rate for verification is in the range of 10^{-6} to 10^{-9}. To make model-independent estimates of false positive rates in this range requires a correspondingly large sample population of mis-matched pairs of prints. Because of the tradeoff between FPR and FNR, the need to operate at very small values of the false positive rate in identification applications may lead to unacceptable miss rates (FNR) when using only a single finger. The system miss rate can be reduced dramatically by executing searches using two different query fingers and considering a match on either finger to be a hit while causing a modest increase in the false positive rate.

Results

There are two aspects of the system to be characterized, the accuracy and the matching speed. Results for these two system characteristics are presented in Fig. 4 and Fig. 5.

In order to characterize the accuracy of the system, a reference database of model prints was constructed from approximately 100,000 inked dab images (actually 97,492) acquired in 1995 which were processed by feature extraction code developed by the Exploratory Computer Vision Group at the IBM Thomas J. Watson Research Center. A description of this class of feature extraction algorithms can be found in [10]. A set of 657 queries were then executed against this database. The query set of prints were a subset of the 100,000 models. Conceptually, $657 \times 97,492$ comparisons of pairs took place. These pairs can be divided into three groups:

1. pairs consisting of identical fingerprints (657)
2. pairs consisting of different impressions of the same finger (768)
3. pairs consisting of impressions of different fingers (64,050,819)

The pairs in the first group are excluded from the analysis of results because they represent an experimental artifact. The reason that there are 768 pairs in the second group is because some query prints had a single mate while others had two or even three mates in the reference database.

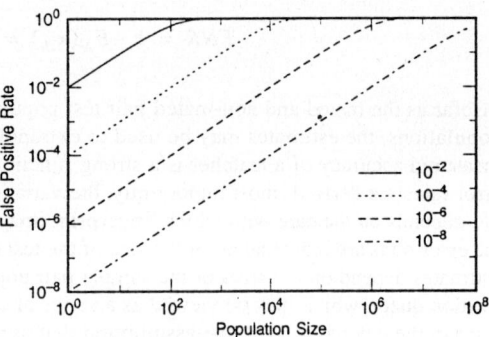

Fig. 6: Extrapolated FPR

Fig. 4 shows the experimental False Positive verification error Rate (FPR) plotted versus the False Negative verification error Rate(FNR) for a variety of operating thresholds on the number of index matches required to assess identity. The error bars represent the 90% confidence intervals for the estimates of the corresponding error rates obtained from this set of experiments. This presentation is similar to the Receiver Operating Curve (ROC) used to characterize the ability of a statistical test to distinguish between two alternative hypotheses [13]. Fig. 5 shows measured average query times for a series of database sizes: 13726, 26317, 50047, and 97492 fingerprints respectively. Each point represents the average over 657 queries and the interpolating line is obtained through a least squares fit to the data. No front-end filtering was used in these tests. That is, each data-point represents a search of the entire database. Both hardware and software configurations were identical through all the tests. Note that the non-zero intercept is expected as it is a direct consequence of the indexing approach. Since the look-up table is too large to fit in real memory, it is hosted on a series disks. Therefore even when the database is empty a certain number of disk seeks are necessary to retrieve the corresponding zero entries from the query generated indices. In this series of runs, 32 disks were used in parallel through independent I/O threads to spread out the I/O burden.

In this configuration, using an 8 node IBM SP2 system or an 8-way SMP workstation, the incremental time required to search the databases for each additional fingerprint is approximately 7 microseconds as shown in Fig. 5. Thus, the system configuration used for these trials could search a database of 10 million prints in approximately 70 seconds. Additional experiments indicate that the load balancing of the I/O is such that disk parallelism can be used effectively to reduce the I/O contribution to the query time. Identification searches of very large databases of fingerprints without pre-filtering are thus possible through indexing subsets of features on the fingerprints.

The work presented here comprises the preliminary findings of an IBM Research Division project initiated in early 1995. No inferences about the current state of technology resulting from continued Research and Development for commercial application should be drawn from this paper.

References

[1] ANSI/IAI, "Automated fingerprint identification systems-benchmark tests of relative performance." Technical Report ANSI/IAI 1-1988, American National Standards Institute, 1988.

[2] D.H. Ballard, "Generalizing the Hough transform to detect arbitrary shapes," Pattern Recognition, Vol. 13, No. 2, 1981, pp. 111-122.

[3] A. Califano and R. Mohan, "Multidimensional indexing for Recognizing Visual Shapes." IEEE Transactions on Pattern Analysis and Machine Intelligence, 16(4):373-392, April 1994. Also, U.S. Patent Number 5,351,310, September 1994

[4] A.Califano, I.Rigoutsos: "FLASH: A Fast Look-up Algorithm for String Homology," of IEEE Conf. on Computer Vision and Pattern Recognition 1993, New York.

[5] C.J. Date, "An Introduction to Database Systems." Addison-Wesley, 1995.

[6] J.G. Daugman, "High Confidence Visual Recognition of Persons by a Test of Statistical Independence." IEEE Transactions on Pattern Analysis and Machine Intelligence, 15(11):1148-1161, 1993.

[7] W.E.L. Grimson, "The combinatorics of heuristic search termination for object recognition in cluttered environments," MIT AI Memo 1111, MIT, May 1989.

[8] R.M. Haralick, H. Joo, C. Lee, X. Zhuang, V.G. Vaidya, and M.B. Kim, "Pose Estimation from Corresponding Point Data." IEEE Transactions on Systems, Man, and Cybernetics, 19(6):1426-1446, 1989.

[9] D.P. Huttenlocher and S. Ulmann, "Object Recognition Using Alignment." In Proceedings of the First International Conference on Computer Vision, pp. 102-111. IEEE Computer Society Press, 1987.

[10] A. Jain, L. Hong, S. Pankanti, and R. Bolle. An Identity Authentication System Using Fingerprints. Proceedings of the IEEE, 85(9):1365, September 1997.

[11] Y. Lamdan and H.J. Wolfson, "Geometric Hashing: A General and Efficient Model-based Recognition Scheme." In Proceedings of the Second International Conference on Computer Vision, pp. 238-249, 1988.

[12] H.C. Lee and R.E. Gaensslen, editors. Advances in Fingerprint Technology, CRC Press, 1994.

[13] E.L. Lehmann, "Testing Statistical Hypotheses." Second Edition. Springer-Verlag, 1986.

[14] J.Neyman and E.S. Pearson. "On the Problem of the Most Efficient Tests of Statistical Hypotheses." Philosopical Transactions of the Royal Society of London, 231A:289, 1933.

[15] N.K. Ratha, K. Karu, S. Chen, and A.K. Jain. "A Real-Time Matching System for Large Fingerprint Databases." IEEE Transactions on Pattern Analysis and Machine Intelligence, 18(8):799-813, August 1996.

[16] I.Rigoutsos, D.Platt, A.Califano: "Flexible 3D-Substructure Matching & Novel Conformer Derivation in Very Large Databases of 3D Molecular Information," to be published in Journal of Computer-Aided Molecular Design.

[17] M.K. Sparrow. "Measuring AFIS Matcher Accuracy." The Police Chief, pp. 147-151, April 1994.

[18] G. Stockman, "Object Recognition and Localization Via Pose Clustering." Computer Vision, Graphics, and Image Processing, 40:361-387, 1987.

[19] B. Stroustrup, "The C++ Programming Language, Third Edition." Addison-Wesley, 1997.

[20] J.D. Ullman and J. Widom. "A First Course in Database Systems." Prentice Hall, 1997.

Sign of Surface Curvature from Shading Images Using Neural Network

Yuji Iwahori[1], Masamitsu Murakami[1], Robert J. Woodham[2] and Naohiro Ishii[1]

[1] Faculty of Engineering, Nagoya Institute of Technology, Gokiso-cho, Showa-ku, Nagoya 466-8555, Japan
[2] Dept. of Computer Science, The University of British Columbia, Vancouver, B.C. Canada V6T 1Z4

Abstract. This paper proposes a new approach to recover the sign of local surface curvature of object from three shading images using neural network. The RBF (Radial Basis Function) neural network is used to learn the mapping of three image irradiances to the position on a sphere. Using the property that basic five kinds of surface curvature has the different relative locations of the local five points mapped on the sphere, not only the Gaussian curvature but also the kind of curvature is directly recovered locally from the relation of the locations on the mapped points on the sphere without knowing the values of surface gradient for each point. The entire approach is non-parametric, empirical in that no explicit assumptions are made about light source directions or surface reflectance. Results are demonstrated by the experiments for real images.

1 Introduction

Surface gradient and curvature are the essential information for the shape representation. Especially, the surface curvature is the invariant and effective feature for the viewing direction, and curvature feature can be used to many applications such as the shape recovery, shape modeling, segmentation, the object recognition and pose determination in the field of computer vision.

Recently, theoretical analysis of shape-from-shading and photometric stereo seem to dominate the literature at the expense of practice. Theoretical work often considers Lambertian reflectance alone. Regrettably, this convinces many potential implementors that the approach is of little practical value. In "physics-based vision," serious attention now is paid to other reflectance models [Healey et al., 1992]-[Healey and Jain, 1994]. Clearly, this is fundamental. But, it has not yet proven effective in practice.

Based on the physics based vision approach, to extract surface curvature from photometric stereo, Woodham [Woodham, 1994] recently developed a method to get surface curvature using the values of the surface gradients. This method obtains the local surface gradients by the empirical photometric stereo using a calibration sphere. Then, using the mathematical and geometrical equations, the method gets the local value of the surface curvature from the local value of surface gradients. In [Iwahori et al., 1995a], a neural network implementation was used for direct comparison with the lookup table (LUT) implementation described in [Woodham, 1994]. The comparison was favorable indicating that a neural network implementation is a viable alternative to an explicit LUT. Further extension to the case of a nearby moving light source photometric stereo, using principal components analysis (PCA) in conjunction with a neural network is reported in [Iwahori et al., 1995b].

While, Wolff et al. [Wolff and Fan, 1994] [Fan and Wolff, 1994] recently developed the method to extract the local Gaussian curvature sign without knowing the values of the surface gradients for the Lambertian model. The method also uses three light sources but needs one same light source direction as the viewing direction to eliminate the nonlinear terms from three equations for the Lambertian reflectance function.

Direct and robust determination of the sign of curvature provides very important information for both object recognition in automated vision tasks and manipulation of objects by robot vision. As with both previous non-parametric, empirical implementations [Woodham, 1994] [Iwahori et al. 1995a], no explicit assumptions need to be made either about light source directions or about the functional form of surface reflectance. Under these conditions, the method can recover not only the sign of Gaussian curvature but also the local basic five surface curvature directly from three shading images. Experiments on real data are demonstrated.

2 Problem Formulation
2.1 Empirical Photometric Constraint

Let the image irradiances at (x_{obj}, y_{obj}) on the test object be $(E_{1obj}, E_{2obj}, E_{3obj})$ and that at (x_{sph}, y_{sph}) on the sphere be $(E_{1sph}, E_{2sph}, E_{3sph})$. Under the condition that the surface material is the same for both of a test object and a sphere, if the following constraint

$$\begin{aligned} E_{1obj}(x_{obj}, y_{obj}) &= E_{1sph}(x_{sph}, y_{sph}) \\ E_{2obj}(x_{obj}, y_{obj}) &= E_{2sph}(x_{sph}, y_{sph}) \\ E_{3obj}(x_{obj}, y_{obj}) &= E_{3sph}(x_{sph}, y_{sph}) \end{aligned} \quad (1)$$

is satisfied, the corresponding surface normal vector should be the same between a test object and a sphere. This constraint is used to determine the curvature sign of a test object using neural network.

2.2 Surface curvature and sign

Five kinds of surface curvature exist. Let k_1 and k_2 be the maximum curvature and the minimum curvature, respectively. Based on the sign of k_1 and k_2. Five kinds of surface curvature are shown in Table 1. Any surface curvature is one of the convex surface, the concave surface, the hyperbolic surface, the cylindrical surface, the plane as shown in Table1. The Gaussian curvature G is defined as $k_1 \cdot k_2$ and the mean curvature M is defined as $(k_1 + k_2)/2$. Determining five kinds of surface curvature leads to the sign of the Gaussian curvature G, further for the case of $G > 0$, it leads the sign of the mean curvature M, i.e., it can classify the convex or concave curved surface uniquely.

Table 1. Principal curvature k_1 and k_2

	$k_2 > 0$	$k_2 = 0$	$k_2 < 0$
$k_1 > 0$	convex	cylindrical	hyperbolic
$k_1 = 0$	—	plane	cylindrical
$k_1 < 0$	—	—	concave

2.3 RBF Networks and OLS Learning

Neural networks are attractive for non-parametric functional approximation. A radial basis function (RBF) neural network [Chen et al., 1991] is one choice suitable for many applications. In particular, it has been widely used for strict interpolation in multidimensional spaces. It is argued that RBF neural networks often can be designed in a fraction of the time it takes to train standard feed-forward networks. They are expected to work well when many training vectors are available. The architecture is shown in Figure 1. The learning procedure adopted here is based on the orthogonal least squares (OLS) method. The OLS method can be employed as a forward regression procedure to select a suitable set of centers (regressors) from a large set of candidates. At each step of the regression, the increment to the explained variance of the desired output is maximized. It chooses radial basis function centers one by one in a systematic way until an adequate network has been constructed. The algorithm has the property that each selected center maximizes the increment to the explained variance of the desired output while remaining numerically well-conditioned.

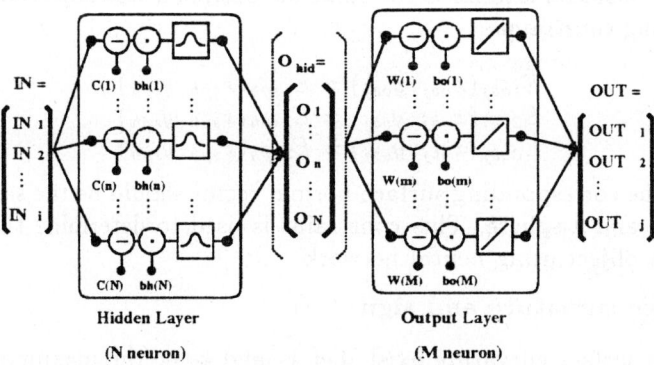

Fig. 1. Architecture of RBF Neural Network

3 Curvature Sign from Shading Images

The detailed strategy is given by the following steps.

[step1] Mapping onto the sphere by neural network. Here, a neural network to do the mapping of $(E_1(x_{sph}, y_{sph}), E_2(x_{sph}, y_{sph}), E_3(x_{sph}, y_{sph}))$ to (x_{sph}, y_{sph}) is constructed for the sphere.

With this learning procedure, two RBF networks are trained using input / output data from a sphere. Many training vectors are available since data from the calibration sphere are dense and include all possible visible surface normal. Each RBF network used consists of two layers, (i.e., a hidden layer of N neurons and an output layer of 2 neurons), as shown in Figure 1.

The resulting network generalizes in that it predicts a position (x_{sph}, y_{sph}) on the sphere, given any triple of input values, $[E_1, E_2, E_3]$ of the test object. The resulting network can then be used to estimate the corresponding coordinate (x, y) onto the sphere for each point of other test objects. Actually, this neural network outputs the real range values for locations on the sphere as the cue for the surface curvature at any point (x_{obj}, y_{obj}) on the test object.

[step2] Local confidence estimate by two step RBF neural networks. We use the simultaneous training of a second neural network for the same sphere, to inversely predict the input, $[E_1, E_2, E_3]$ from the estimated output, (x, y). Comparison between the actual input and the inversely predicted input then serves as a suitable confidence estimate.

If the three image irradiances from the sphere is input to this two step network, the output of the second step, which we call the resynthesized input space, should be very similar to the original input. However, if an impossible triple such as cast shadows and interreflection (i.e., one that could not have arisen from the sphere) is input, we expect the resynthesised image irradiances to be quite different since the resynthesized values necessarily correspond to points on the sphere.

For the evaluation of the local confidence for the surface curvature, image irradiances at five local points are input to this two step network one by one, then we can calculate the local confidence $C(x_{obj}, y_{obj})$ as

$$C(x_{obj}, y_{obj}) = \frac{\sqrt{\sum_{l=1}^{3} |E_l(x_{obj}, y_{obj}) - E'_l(x_{obj}, y_{obj})|^2}}{3} \quad (2)$$

The final local confidence for any object point (x_{obj}, y_{obj}) is defined as the mean value $C_m(x_{obj}, y_{obj})$ of local five points around any point.

$$C_m(x_{obj}, y_{obj}) = \{C(x_{obj}, y_{obj}) + C(x_{obj}+1, y_{obj}) + C(x_{obj}-1, y_{obj})$$
$$+C(x_{obj}, y_{obj}+1) + C(x_{obj}, y_{obj}-1) \} / 5 \quad (3)$$

The larger the value of $C_m(x_{obj}, y_{obj})$, the larger is the deviation of the test point from a point that could have arisen on the sphere.

Thus, once two neural networks for both of $(E_1, E_2, E_3) - (x_{sph}, y_{sph})$ and the inverse $(x_{sph}, y_{sph}) - (E_1, E_2, E_3)$ are learned for the sphere, the two step neural network of $(E_1, E_2, E_3) - (x_{sph}, y_{sph}) - (E'_1, E'_2, E'_3)$ can be used to get the local confidence estimate.

[step3] Determination of curvature sign. Let the local five points on the test object be labeled as ⓪ for the center point, ① for its upper point, ②, ③, ④ for clockwise, respectively.

The feature of the surface curvature at these local points is appeared after the mapping for the location on the sphere, i.e., the location on the sphere mapped by neural network holds the relative relation of the orientations of the local points around any point on the test object. The method classifies the kind of surface curvature from this relation of the location of the mapped points onto the sphere. In the following, it is shown that how the mapping of the neighbor five local points is done onto the sphere for each of five kinds of curvatures.

Convex surface: Since a sphere is also the convex surface, the relative locations for the labeled number holds the same numbering relations as the local points on the test object after mapping onto the sphere. The geometrical relations for mapping of ⓪ to ④ are located in clockwise for both of the local test points and the sphere.

Concave surface: The geometrical relations for mapping of ⓪ to ④ are located in clockwise for the sphere after the mapping, but the mapped ① to ④ have the phase difference of around 180 degrees for each of ① to ④ in comparison with the case for the convex surface.

Hyperbolic surface: Let the maximum curvature and the minimum curvature be k_1 and k_2, then $k_1 > 0$ and $k_2 < 0$ hold. The mapping of five local points of ⓪ to ④ onto the sphere gives the inverse rotation of ① to ④ for the center point ⓪.

Plane: Definitely, the plane has the same surface normal for the local points. Therefore, the mapping of five local points of ⓪ to ④ onto the sphere gives the same one point on the sphere.

Cylindrical surface: The sign of Gaussian curvature G is 0 for cylindrical surface, i.e., $k_1 = 0$ or $k_2 = 0$ holds. The mapping of five local points of ⓪ to ④ onto the sphere arises on the same line.

4 Experiments on Real Data

Two objects are used in the experiments reported. One is a pottery sphere, and the other is a pottery doll-face. In this case, both objects are made of the same material with the same reflectance properties. No particular assumptions for the surface reflectance or light source directions are used or needed for the experiments. The pottery doll-face was mounted on a rotational motion stage of a calibrated imaging facility (CIF) developed at the UBC Laboratory for Computational Intelligence (LCI) to control both scene parameters and conditions of imaging. Three shading images of a sphere were used to the learning and generalization of neural network for the doll-face test object.

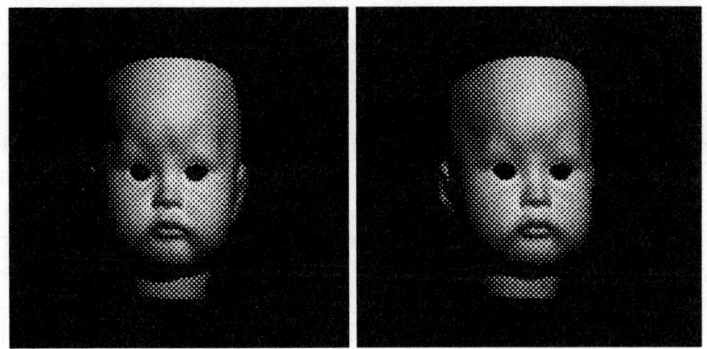

(a) doll-face1　　　　　　　(b)doll-face2
Fig. 2. Shading image set (one among three images)

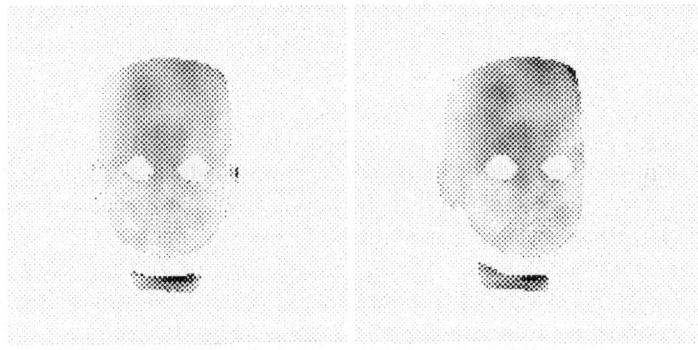

(a) doll-face1　　　　　　　(b)doll-face2
Fig. 3. Local confidence for curvature

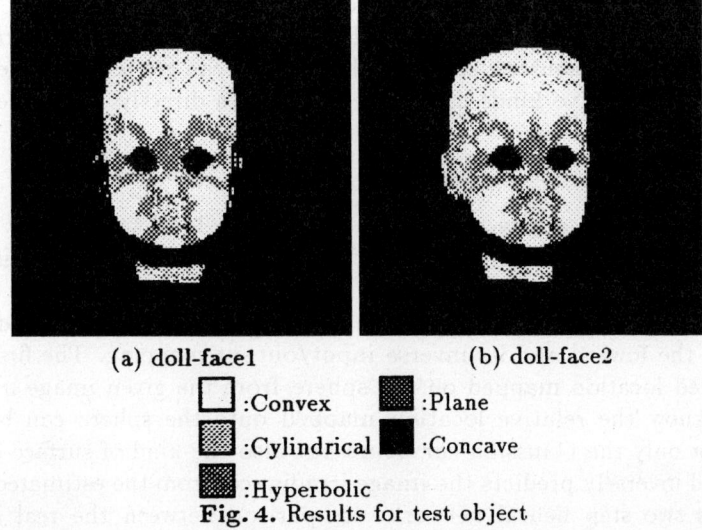

(a) doll-face1　　　　　　　(b) doll-face2

☐ :Convex　　　▨ :Plane
▨ :Cylindrical　■ :Concave
▨ :Hyperbolic
Fig. 4. Results for test object

These two different cases for doll-face are shown in Figure 2, one is the face image set of face (doll-face1) towards the viewing direction, the other is another image set (doll-face2) rotated with 12 degrees. Condition of three light sources is the same for both cases.

654 points which were taken 8 dots apart from the sphere and these points were used for the neural network learning, while 3955 points were sampled taken 4 dots apart from the doll-face1, whose 3230 points have the four neighbor points and those points were tested for the classification of surface curvature. For the doll-face2, 3358 points were used for the test among 4015 sampling points taken 4 dots apart. In the learning of the first RBF neural network of (E_1, E_2, E_3) — (x, y), the spread constant was set to be 50 and the number of the learning epochs was set to be 50. While, for the inverse neural network of (x, y) — (E_1, E_2, E_3), the spread constant was set to be 10 and the number of the learning epochs was set to be 50. Learned two step neural networks were used to estimate the local confidence for the curvature sign. The final results of this experiment are shown in Figure 4. It is shown that five kinds of surface curvature are extracted from this method.

Based on the original local geometrical shape of this test object, we can see some points with random curvature sign on around the forehead for the results. However, the method gets almost reasonable and robust results, as the human can guess from the image set, and it is shown that the method is applicable as one of the physics based vision approaches. From the results, it is also shown that the curvature is invariant feature from the position of viewing direction for the movement of object between two image sets. Figure 3 shows the encoded image of the local confidence estimate for the results. The darker point indicates the lower confidence, while the brighter point indicates the higher confidence. Neck regions are relatively darker. This suggests that the neck regions are apparently effected by the cast shadow. Interreflection and cast shadow region has the lower local confidence than other normal regions.

It is also an advantage that the method need not the explicit information for the direction of light sources, and that the method can get not only the Gaussian curvature but also five kinds of surface curvatures directly from three shading images.

5 Conclusion

This paper proposed a new method to recover the local surface curvature sign and determine the local surface curvature directly from three shading images using neural network.

For the implementation, RBF neural network are effectively learned using the sphere for the foward and its inverse input/output mapping. The first predicts the intended location mapped on the sphere from the given image irradiances. One can know the relative location mapped onto the sphere can be used to recover not only the Gaussian curvature but also the kind of surface curvature. The second inversely predicts the image irradiances from the estimated location. Using this two step neural network. comparison between the real input and the inversely predicted input for the local five points is used as the confidence estimate for the curvature sign. The local confidence estimate helps to detect regions of cast shadow and interreflection.

It is valuable to stress that the entire approach is non-parametric, empirical in that no explicit assumptions are made about light source directions or surface reflectance. It is sufficient that the sphere used in training and the subsequent test objects be viewed under the same pattern of illumination and be made of the same material (i.e, have the same reflectance properties.) Further, the method need not the calibration for the empirical photometric stereo because we need not to know the value of the surface normal for the calibration as the standard photometric stereo.

Acknowledgements

Support for the work described in this paper was provided by the HORI INFORMATION SCIENCE PROMOTION FOUNDATION, Japan. Support for Woodham's research is provided by the Institute for Robotics and Intelligent Systems (IRIS) and by the Natural Sciences and Engineering Research Council of Canada (NSERC).

References

[Woodham, 1994] R. J. Woodham, "Gradient and curvature from the photometric stereo method, including local confidence estimation," *Journal of the Optical Society of America, A*, vol. 11, pp. 3050–3068, 1994.

[Iwahori et al., 1994] Y. Iwahori, H. Tanaka, R. J. Woodham, and N. Ishii, "Photometric stereo for specular surface shape based on neural network," *IEICE Transactions on Information and Systems*, vol. E77-D, no. 4, pp. 498–506, 1994. (Special issue on neurocomputing).

[Iwahori et al., 1995a] Y. Iwahori, A. Bagheri, and R. J. Woodham, "Neural network implementation of photometric stereo," *Proc. of Vision Interface 95*, (Quebec City, Canada), pp. 81–88, May 1995.

[Iwahori et al., 1995b] Y. Iwahori, R. J. Woodham, and A. Bagheri, "Principal components analysis and neural network implementation of photometric stereo," in *Proc. IEEE Workshop on Physics-Based Modeling in Computer Vision*, pp. 117–125, June 1995.

[Healey et al., 1992] G. Healey, S. Shafer, and L. Wolff, eds., *Physics-Based Vision: Principles and Practice (Vol. 1 Radiometry, Vol. 2 Color and Vol. 3 Shape Recovery)*. Boston, MA: Jones and Bartlett Publishers, Inc., 1992.

[Healey and Jain, 1994] G. Healey and R. Jain, "Physics-based machine vision," *Journal of the Optical Society of America, A*, vol. 11, p. 2922, 1994. (Introduction to special issue).

[Wolff and Fan, 1994] L. B. Wolff and J. Fan, "Segmentation of surface curvature using a photometric invariant," *Proc. of IEEE-CVPR 1994*, pp. 23-30, 1994.

[Fan and Wolff, 1994] J. Fan and L. B. Wolff, "Surface Curvature from Integrability," *Proc. of IEEE-CVPR 1994*, pp 520-525, 1994.

[Chen et al., 1991] S. Chen, C. F. N. Cowan, and P. M. Grant, "Orthogonal least squares learning algorithm for radial basis function networks," *IEEE Transactions on Neural Networks*, vol. 2, no. 2, pp. 302–309, 1991.

On the Classification of Singular Points for the Global Shape from Shading Problem: A Study of the Constraints Imposed by Isophotes

Takayuki Okatani and Koichiro Deguchi

Faculty of Engineering, University of Tokyo, Bunkyo-ku, Tokyo 113, Japan

Abstract. This paper concerns with the global surface reconstruction from a shaded image. It is known that the key to the global reconstruction lies in classifying singular points in the image into three categories—convex, concave or saddle classes. Several methods have been proposed for this problem. All of them require explicit solution of the shape from shading equation. Therefore the results may suffer from image noise and modeling errors of the surface reflecting properties and illumination. If the classification of singular points without such an explicit solution of the equation would have been possible, then the global reconstruction could have been made much more precise. The present work aims towards developing a classification technique where such a explicit solution is not required. Our first step is to show that isophotes (lines of equal brightness) provide some information on the types of singular points; if two isophotes exist which evolve from two singular points and meet each other, then the one of the two singular points is elliptic and the other is hyperbolic. This is derived from intrinsic properties of a smooth (twice differentiable) surface.

1 Introduction

Shape from shading is one of the most widely studied problems in computer vision. Horn [1] formulated the problem as a first-order PDE and proposed a method for solving the equation based on the characteristic strip expansion. The method draws characteristic curves in the image and determines the surface height along them from a starting point. As the starting point, the singular point, i.e., maximally bright point in the image, is used. Since that method suffered from numerical instability, iterative algorithms based on the variational methods were proposed (the first one is in [2] and some successors are in [3]). In all these works the primary considerations were stability and convergence of the algorithm.

Apart from the development of these algorithms, attention has also been paid to the uniqueness of the solutions [4-7]. It has been shown that singular points well limit the number of possible solutions. The result is summarized as follows: if the type of surface shape at each singular point in the image is known as convex, concave, or saddle, then the surface can be uniquely reconstructed within some image neighborhood of the singular points [6, 7]. The image region where the solution is uniquely determined is referred to as the domain of attraction of the singular points, which usually covers most part of the image. From this viewpoint, shape from shading becomes a well-constrained problem. Thus, the essential problem is to identify the types of singular points from the image itself. This is often referred to as the problem of *global* shape from shading [8], in order to distinguish it from *local* problems of reconstructing the shape around singular points.

Several methods for this global shape from shading problem have been proposed. Oliensis and Dupuis [8] proposed a method using the nature of characteristic strips. They used the result of [7] that characteristic strips on the imaged object are curves of steepest ascent in the direction away from the light source. From this, the behavior of characteristics around a singular point can be used to identify the types of the singular point; a convex type singular point corresponds to source of the characteristic curve, and so forth. In their method, at least one singular point which is known not to be a saddle point must be chosen beforehand. Kimmel and Bruckstein [9] proposed another algorithm which used global topological properties of the surface. This algorithm computes the local surface around singular points by tracking iso-height contours of the object surface. The classification of the singular points is based on the fact that the first singular point which a propagating iso-height contour encounters must be a saddle point. This fact is owing to the mountaineers theorem. The theorem states that the number of extrema located inside a closed iso-height contour of a smooth surface is one more than the number of saddle points inside the contour. Their algorithm carry out the trials for all the combinations of types of singular points in the image and propagate contours from the singular points. When two propagating contours meet at the same singular point, this singular point can be identified as the saddle type. The two local surfaces are merged at this saddle point. The global shape is obtained by iterating this process.

To classify singular points, both of the above methods require computing the local surface by solving the shape from shading equation using image gray level. Of course, to obtain the total surface shape or, to obtain a dense depth map, the explicit solution of the equation is necessary. However, if classification of singular points is possible prior to such an explicit solution, we can reconstruct the global shape more stably regardless of image noise and errors of modeling the surface reflecting property. This will certainly enable more accurate reconstruction of the total shape. Moreover, we believe that a classification result of singular points is valuable in itself. Such information about the object surface shape has rather applications for practical problems, since, as was argued by Forsyth and Zisserman [10], accurate surface reconstruction using only shading information was impossible because of interreflections.

This paper describes a constraint on the classification of singular points provided automatically by the image. In this paper attention is restricted to classifying singular points as elliptic or hyperbolic points. (Singular points at which the surface is elliptic correspond to the convex and concave type singular points, while singular points at which the surface is hyperbolic correspond to the saddle type singular points.) Our result is that the isophotes (contours of equal brightness) provide some information on the types of singular points; if two isophotes exist which evolve from two singular points and meet each other, then the one of the two singular points is elliptic and the other is hyperbolic. This constraint along with those imposed by occluding contours and so forth [6] well limit the number of possible classifications.

2 Problem Formulation and Background

2.1 The Image Irradiance Equation

To simplify our discussion we assume orthographic projection. We represent an object surface by a height function z with respect to image coordinates (x, y). We assume that the surface is smooth, by which we mean that the function $z(x, y)$ is twice differentiable. The local surface normal $\hat{n}$ is parallel to $(p, q, -1)$, where the variables p and q are the gradients of the surface. They are given by the first-order derivatives of z as $p = \partial z/\partial x$ and $q = \partial z/\partial y$.

We assume that the illumination is a distant point source. Let $l = (\alpha, \beta, \gamma)$ be a vector representing the product of the light source strength with the unit vector parallel to the lighting direction. We assume that there are no interreflections in the image. The image $E(x, y)$ is determined from the reflectance function or *reflectance map* $R(\hat{n})$ by the image irradiance equation [11]. For the surface obeying the Lambert's law, the image intensity $E(x, y)$ is written using the albedo and a inner product of $\hat{n}$ and l as

$$E(x, y) = \rho \hat{n}^\top l = \rho \frac{\alpha p + \beta q - \gamma}{\sqrt{p^2 + q^2 + 1}}. \tag{1}$$

2.2 Assumptions About Types of Singular Point

For the Lambertian case, a singular point is generated at an image point whose corresponding surface normal $\hat{n}$ is parallel to the lighting direction l. Thus, a singular point is detected as a maximally bright point in the image. Note that the positions of singular points depend only on the lighting direction for a given surface.

In this paper we restrict our attention to the class of surfaces on which parabolic points do not form any area but at most a curve. (This restriction rules out, for example, a cylindrical surface.) For such surfaces we assume that each of singular points lies on either elliptic or hyperbolic point. In other words, we assume that there is no singular point of parabolic type in the image. In the following sections we will refer to a singular point as *elliptic singular point* or *hyperbolic singular point* according as the surface on which it lies is elliptic or hyperbolic.

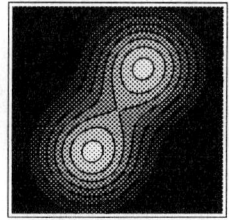

Fig. 1. Starting from two different singular point, two isophotes evolve and encounter a critical point (a saddle point of the image) from the opposite sides. In this case the type (i.e., elliptic or hyperbolic) of each singular point is different from the other's.

We further assume the Hessian matrix **H** of $z(x,y)$ is not zero matrix. This rules out surfaces having pessimistic singularities such as, for example, the monkey saddle ($z = x^3 - 3x^2y$). Also in [7, 8] as well as [12], assumptions of similar kind were made.

3 Isophotes and the Types of Singular Points

3.1 Isophotes Around Singular Point

Isophotes are defined as contours of equal brightness and they form a family of curves in the image. For each singular point in the image, isophotes near the point are closed curves; they contain the singular point and do not contain another one. As we track such an isophote outward starting from a singular point in the direction that their brightness decreases, we encounter a critical point ($E_x = E_y = 0$) that is a saddle point of the image $E(x,y)$. There exists another isophote which reaches the same critical point from the opposite side. Hereafter the two isophotes are merged into one isophote as shown in Fig. 1.

We consider the case where, for such two evolving isophotes in the image merged into one at a critical point, it is the first time for both evolving isophotes to encounter a critical point starting from their respective singular points. Then we can know that these two singular points have different sign from each other, i.e., the one of the two singular points is elliptic and the other is hyperbolic. We will show this in the rest of this section.

3.2 Mapping onto the Gradient Space

Consider a mapping which maps each image point (x,y) on its gradients (p,q): $p = p(x,y)$ and $q = q(x,y)$. Let T denote this mapping. The Jacobi matrix of T is nothing but the Hessian matrix **H** of $z(x,y)$. As described earier, we assume $\mathbf{H} \neq \mathbf{O}$ at every image point. Since the Gaussian curvature at a surface point has the same sign as $\det \mathbf{H}$, if the Gaussian curvature isn't zero at a point, then $\det \mathbf{H} \neq 0$ and thus **H** is nonsingular there.

An isophote is a curve of $E(x,y) =$ constant. Thus, from Eq. (1), isophotes are mapped on ellipses in the gradient space.

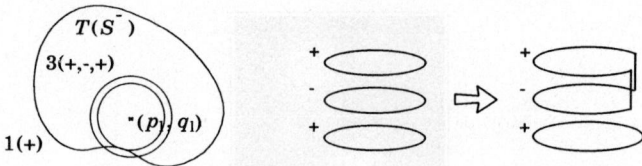

Fig. 2. LEFT: The multiplicity of T^{-1} and isophotes as ellipses in the gradient space. For the ellipses completely contained in $T(S^-)$, the original isophotes are seperated from each other. For the ellipses even small part of which lies outside $T(S^-)$, the original isophotes can not be seperated. RIGHT: Merged ellipses. When isophotes go outside S^-, two of the three are merged. It can be known that the merged isophotes are around different types of singular points.

3.3 The Case of Single Hyperbolic Region

First we consider an ovoid shape (i.e., a shape such that the Gaussian curvature is positive everywhere) having single hyperbolic region S^-. We assume the region S^- to be simply connected (i.e., S^- does not have any hole). We denote the imaged region of the whole shape by S and the elliptic region by S^+. Since we assumed that parabolic points did not form any area, $S^+ = S \setminus S^-$.

We define the regions $T(S^-)$, $T(S^+)$, and $T(S)$ on the gradient space as,

$$T(S^-) = \{(p,q)|(p,q) = T(x,y), (x,y) \in S^-\}.$$

and so forth. Since $\det \mathbf{H} \neq 0$ inside the region S^- and S^+ (except on the boundary), there exists the reverse mapping T^{-1}. The forward mapping T is single-valued, while this reverse mapping T^{-1} may be multivalued. The maximal multiplicity of T^{-1} is three in this case of single hyperbolic region. More definitely, for every point (p,q) inside $T(S^-)$ in the gradient space, there exist exactly three image points (x,y) which are mapped on the same point (p,q) by T. Moreover, the two of the three points are inside S^+ and the other one is inside S^-. For every point outside $T(S^-)$, i.e., every point inside $T(S) \setminus T(S^-)$, exactly one corresponding image point exists and it is inside S^+. These are derived from the global topological properties of a smooth surface, if its Euler characteristic is two (i.e., the surface doesn't have any hole). We use these without proofs.

The multiplicity of T is directly related to the number of singular points on the surface. For a Lambertian surface, which we now consider, the singular point is a point where the surface normal is parallel to the light source orientation. We may represent the light source orientation as (p_l, q_l) in the gradient space. If (p_l, q_l) is inside $T(S^-)$, then the number of singular points equals the multiplicity of T^{-1} there, that is, three and two of them are elliptic and one is hyperbolic. If (p_l, q_l) is outside $T(S^-)$, then only one singular point exists and it is elliptic.

Now we suppose that three singular points are on the surface. For each singular point, small isophotes exist around the point and, as described earlier, they are mapped on ellipses around (p_l, q_l) in the gradient space. When two isophotes are merged into one, the two corresponding ellipses are merged into one in the

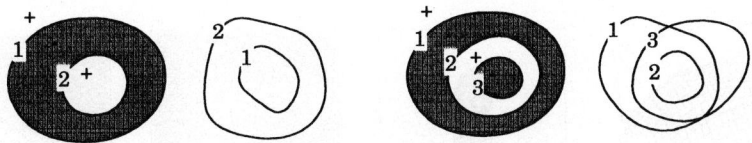

Fig. 3. LEFT: A hyperbolic region containing simply connected elliptic region and the distribution of parabolic curves in the gradient space. RIGHT: A hyperbolic region containing not simply connected elliptic region and the distribution of parabolic curves in the gradient space.

gradient space, as shown in Fig. 2. In the gradient space, this merge occurs when the ellipses touch the boundary of $T(S^-)$ as shown in Fig. 2.

Of the two merged isophotes, one is around the hyperbolic singular point and the other is around the elliptic singular point. This can be roughly proven as follows. The third isophote, which is still not merged and isolated in the image, must be mapped on a complete ellipse in the gradient space. It is not possible that this isolated isophote is around the hyperbolic singular point (i.e., it is not possible that both of the merged two are around elliptic singular points). When the merge occurs, the isolated isophote goes outside S^-. Recall the boundary of S^- is a parabolic curve. For any point on one side of a parabolic curve, there exists another point at the other side whose normal is parallel to that of the first point. This means that the isolated isophote can not form a complete ellipse in the gradient space when it just goes outside S^-. Thus, it is shown that the isolated isophote is the one around the elliptic singular point, and the merged two isophotes are ones around singular points of different types.

3.4 The Case of Multiple Hyperbolic Regions

Next we consider the case where multiple hyperbolic regions are on an ovoid shape. We still assume each hyperbolic region to be simply connected. When the hyperbolic regions are mapped onto the gradient space, if they overlap with each other there, then the mutiplicity of T^{-1} may be more than three. From the topological properties of a smooth surface, it will be odd numbers such as five, seven, and so forth. It increases or decreases by two, where one is for the elliptic region and the other for the hyperbolic region. Thus, based on the above discussions, every two isophotes which are merged first are around singular points of different types.

3.5 More Complicated Shapes

We have described the case where the hyperbolic regions are simply connected. We further consider the case of multiply connected hyperbolic regions.

First we consider the case where a single elliptic region which is simply connected lies in a hyperbolic region (See Fig. 3). In this case, when the inner and

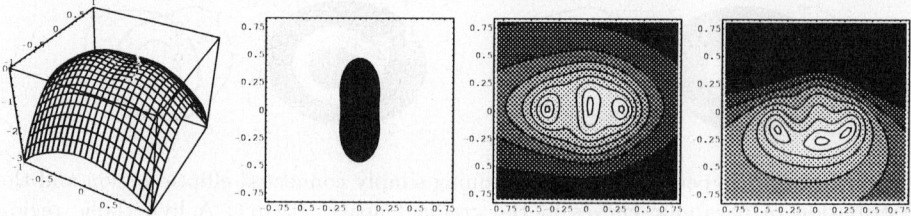

Fig. 4. An example for the behavior of isophotes under different lighting directions; the shape of the surface, the distribution of the sign of the Gaussian curvature, isophotes when the lighting direction is $(0.1, -0.06, -1)$, and isophotes when $(0.1, 0.5, -1)$.

the outer parabolic curve are mapped in the gradient space, they does not cross each other. The distribution of the multiplicity of T^{-1} is still orderd in the gradient space (i.e., $3, 5, \ldots$). The ellipses, which are the image of isophotes, are merged on the boundary. The centers of such two ellipses belong to the regions of different types. Thus, the two isophotes which are merged first are ones around singular points of different types. Therefore, our claim is true for this case.

Next we consider the case where, inside a hyperbolic region, there is an elliptic region, and it further contains a hyperbolic region (See Fig. 3). There are three parabolic curves. The most outer parabolic curve does not cross the second outer one in the gradient space. Further, the second outer one does not cross the most inner one. However, it is not guaranteed that none of the three curves crosses each other in the gradient space. If they cross each other, although the ellipses are merged on the boundary, it can not be said which types of regions the centers of the ellipses belong to. Therefore, it can not be said that our claim is always true for this case.

4 Examples

Here we show examples in Fig. 4 and Fig. 5. For both figures, the first 3D images are shapes of surfaces which are represented as the depth from the image plane. The second black and white images show the distributions of the sign of the Gaussian curvature of the surfaces. The black regions indicate hyperbolic ones and the white regions indicate elliptic ones. The third and fourth images show synthesized images of the surfaces for different lighting directions and they are displayed with isophotes. We can see that two isophotes which are merged first are always isophotes around different types of singular points. Note that this is true regardless of the lighting direction.

5 Conclusion

We have described a constraint on the classification of singular points imposed by the topology of isophotes. This is useful for the classification without explicit solution of the shape from shading equation.

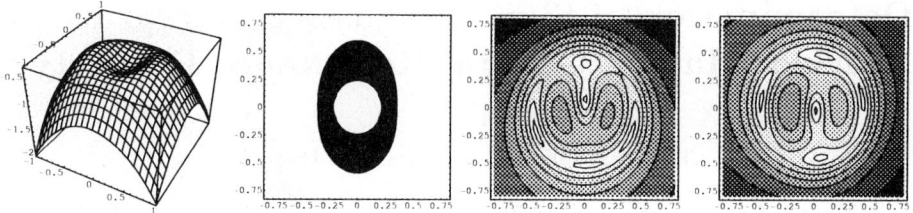

Fig. 5. Another example of the behavior of isophotes; the shape of the surface, the distribution of the sign of the Gaussian curvature, isophotes when the lighting direction is $(0.02, 0.2, -1)$, and isophotes when $(0.1, -0.06, -1)$.

It is usually not possible that the classification is completed only by this constraint. For elliptic singular points, it is necesarry to identify them as convex or concave. Furthermore, this constraint gives limited information on a few singular points in the image. However, along with the constraints imposed by occluding contours and shadowed edges [6], it well restrict the number of possible classifications. It should be noted that knowledge of illumination and albedo is not necessary.

References

1. B. K. P. Horn, "Shape from shading: a method for obtaining the shape of a smooth opaque object from one view," Tech. Rep. MAC-TR-79, MIT, Cambridge, MA, 1970.
2. K. Ikeuchi and B. K. P. Horn, "Numerical shape from shading and occluding boundaries," *Artificial Intelligence*, vol. 17, no. 3, pp. 141–184, 1981.
3. B. K. P. Horn and M. J. Brooks, *Shape from shading*. Cambridge, MA: MIT Press, 1989.
4. A. Blake, A. Zisserman, and G. Knowles, "Surface descriptions from stereo and shading," *Image and Vision Computing*, vol. 3, no. 4, pp. 183–191, 1985.
5. B. V. H. Saxberg, "A modern differential geometric approach to shape from shading," Tech. Rep. 1117, MIT AI Laboratory, 1989.
6. J. Oliensis, "Uniqueness in shape from shading," *IJCV*, vol. 6, no. 2, pp. 75–104, 1991.
7. J. Oliensis, "Shape from shading as a partially well-constructed problem," *CVGIP*, vol. 54, pp. 163–183, September 1991.
8. J. Oliensis and P. Dupuis, "A global algorithm for shape from shading," in *4th ICCV Conference, Berlin*, pp. 692–701, 1993.
9. R. Kimmel and A. M. Bruckstein, "Global shape from shading," *Computer Vision and Image Understanding*, vol. 62, no. 3, pp. 360–369, 1995.
10. D. Forsyth and A. Zisserman, "Reflections on Shading," *IEEE Trans. Pattern Anal. Mach. Intell.*, vol. 13, no. 7, pp. 671–679, 1991.
11. B. K. P. Horn, *Robot Vision*. MIT Press: Cambridge, MA; and McGraw-Hill: New York, 1986.
12. R. Kimmel and A. M. Bruckstein, "Tracking level sets by level sets: A method for solving the shape from shading problem," *Computer Vision and Image Understanding*, vol. 62, pp. 47–58, July 1995.

Determination of Sign of Gaussian Curvature of Surface Having General Reflectance Property

Takayuki Okatani and Koichiro Deguchi

Faculty of Engineering, University of Tokyo, Bunkyo-ku, Tokyo 113, Japan

Abstract. A new method for determining sign of the Gaussian curvature at points on the surface of an object using images taken under three different lighting directions is presented. A smooth surface is segmented into two regions based on the sign of Gaussian curvature. The boundaries of these two regions are independent of the orientation of the object in 3D space and the position of the viewer. Thus they are useful for various applications such as object recognition and pose estimation. The present method does not require knowledge of the directions and strengths of the light sources except the rotation orientation of the directions about the viewing direction if they are linearly independent. It is applicable for almost any kind of surface reflecting properties.

1 Introduction

The image irradiance provides information about surface shape of the imaged object. The problem of reconstructing the shape of the object surface from single irradiance pattern, i.e., *shape from shading*, has been widely studied in computer vision [1]. To solve this problem, precise knowledge of lighting direction and surface reflectance property are required. On the other hand, much more practically, using multiple images of an object from a fixed viewpoint with different lighting directions, the object shape can be reconstructed. This notion of *photometric stereo* was introduced by Woodham [2]. According to this notion local surface orientation can be determined from three shaded images if the lighting directions are precisely known. Furthermore, a method determining local surface curvature directly from the three images was also proposed [3].

Recently, Wolff [4] proposed a method for calculating sign of the Gaussian curvature on the surface of an object from three images with different lighting directions which are *unknown*.

The Gaussian curvature at a point on a smooth surface is defined as the product of two primary curvatures. It is a local intrinsic property of the surface, and its sign well describes the local geometric shapes of the surface as shown in Fig. 1. The surface points having positive, zero and negative Gaussian curvature are called elliptic (Fig. 1a), parabolic (Fig. 1b), and hyperbolic points (Fig. 1c), respectively. By this sign of the Gaussian curvature, a smooth surface is segmented into two types of regions; one is a region consisting only of elliptic points and the other is a region of hyperbolic points. Examples of such a segmentation

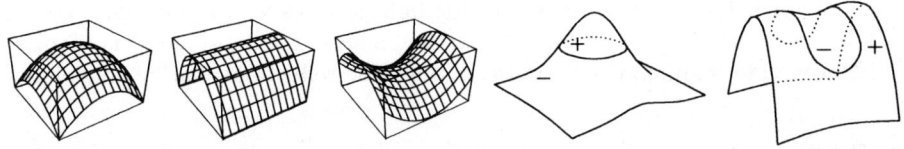

Fig. 1. The shapes of local surface patch and the signs of the Gaussian curvature; the first, the second, and the third show patches having positive, zero and negative Gaussian curvature, respectively. The rest are examples of surface segmentation by the sign of the Gaussan curvature.

are shown in Fig. 1d and e. Since their boundaries across the object surface are independent of the viewing direction and the orientation of the object in 3D space, they are useful for many vision tasks such as object recognition and pose estimation.

This paper presents a method for accurate determination of the sign of the Gaussian curvature and segmentation of the surface. The present method is applicable to wider range of objects than the previous ones especially in terms of surface reflectance. This is due to the new notion of the *pseudo Gaussian image* which is made from photometric data.

Prior to Wolff's work[4], Koenderink and van Doorn [5] showed that, for the Lambertian case, the orientation of the image intensity gradient at a parabolic point always remained the same for different lighting directions. Then, Blake, Zisserman and Knowles [6] extended this theory to the non-Lambertian cases.

Here let us summarize these researches. Under the assumption of orthographic projection, an object surface can be represented by a height function z with respect to image coordinates (x, y). The local surface normal is parallel to $(p, q, -1)$, where p and q are the gradients of the surface. They are given by the first-order derivatives of z as $p = \partial z/\partial x$ and $q = \partial z/\partial y$. Then, it is assumed that the image $E(x, y)$ is determined from the reflectance function or *reflectance map* $R(p, q)$ and the surface albedo $\rho(x, y)$ by the image irradiance equation [7] of the form,

$$E(x, y) = \rho(x, y) R(p, q). \tag{1}$$

In [5] the case of the Lambertian surface with constant albedo $\rho(x, y)$ was considered. In [6] general reflectance property was considered but $\rho(x, y)$ was still assumed to be constant. In [4] non-constant albedo was considered but the surface was assumed to be Lambertian. In contrast, the present paper considers non-constant albedo and also general reflectance. Furthermore, in [4], one of the three lighting directions must be chosen to be parallel to the viewing direction in order to free the calculation from the albedo. In our method such a constraint is not imposed on the three lighing directions.

To sum up, our method has the following properties.

1. The lighting directions can be chosen arbitrally unless they are linearly dependent. The knowledge of their directions is not necessary except for their rotation orientation about the viewing direction.

2. The knowledge of the strengths of the light sources is not necessary.
3. For the reflecting property of the surface, a sort of monotonicity is required for complete segmentation. However, specular reflections can be considered.

2 Surface Normals and Triples of Image Grey Levels

The image $E(x,y)$ is given by Eq. (1), where the gradients (p,q) are equivalent to the surface normal $\hat{n}$. Thus we rewrite Eq. (1) as $E(x,y) = \rho(x,y)R(\hat{n})$. For the Lambertian case, the above equation becomes $E(x,y) = \rho(x,y)\,\hat{n}^\top l$, where $l = (\alpha, \beta, \gamma)$ is a vector representing the product of the light source strength with the unit vector parallel to the lighting direction. In the rest of this paper we first treat the case of Lambertian surface and next the general case.

Here we represent the image coordinates by discrete arrays with pixel sizes Δx and Δy as $(x_i, y_j) \equiv (i\Delta x, j\Delta y)$. That is, $E_{i,j} \equiv E(x_i, y_j)$, $\rho_{i,j} \equiv \rho(x_i, y_j)$, and $\hat{n}_{i,j} \equiv \hat{n}(x_i, y_j)$. Let us suppose that we take three images $E_{i,j}^1$, $E_{i,j}^2$, and $E_{i,j}^3$ by changing lighting directions l^1, l^2, and l^3. For the Lambertian case, each image can be written as $E_{i,j}^k = \rho_{i,j}\,(\hat{n}_{i,j} \cdot l^k)$, ($k = 1, 2, 3$). For each pixel (i,j), we define 3-dimensional vector consisting of the triple of image grey levels as $\mathbf{e}_{i,j} \equiv [E_{i,j}^1, E_{i,j}^2, E_{i,j}^3]^\top$. We also define 3×3 matrix $\mathbf{L} \equiv [l^1, l^2, l^3]^\top$ for the lighting directions. Then, the vector $\mathbf{e}_{i,j}$ is written as

$$\mathbf{e}_{i,j} = \rho_{i,j}\,\mathbf{L}\,\hat{n}_{i,j}. \quad (2)$$

Therefore, $\mathbf{e}_{i,j}$ is related to $\hat{n}_{i,j}$ by an affine transformation defined by $\mathbf{L}$.

For the non-Lambertian case, the discrete form can be written as $E_{i,j}^k = \rho_{i,j}\,R^k(\hat{n}_{i,j})$, ($k = 1, 2, 3$), since the change of the lighting direction results in the change of reflectance function $R^k(\hat{n}_{i,j})$, ($k = 1, 2, 3$). Although there is no simple relationship between $\mathbf{e}_{i,j}$ and $\hat{n}_{i,j}$ like Eq. (2) in this case, we define a function $\mathbf{f}$ ($\mathbb{R}^3 \to \mathbb{R}^3$) and write $\mathbf{e}_{i,j}$ as

$$\mathbf{e}_{i,j} = \rho_{i,j}\,\mathbf{f}(\hat{n}_{i,j}). \quad (3)$$

3 Determination of the Sign of the Gaussian Curvature

3.1 The Gaussian Image of a Surface

The Gaussian image of a surface is an image obtained by mapping each surface point on its normal or, equivalently, on a point of a unit sphere (often called Gaussian sphere). In the discrete domain, this becomes mapping each point (i,j) on its normal $\hat{n}_{i,j}$. We refer to the result of this mapping as the discrete Gaussian image.

This notion of the Gauss mapping is closely related to the Gaussian curvature on the surface. That is, the Gaussian curvature equals *ratio of solid angle on the Gaussian sphere to surface area of the corresponding surface patch* (See, for example, [8] and [9]). From this geometric interpretation of the Gaussian

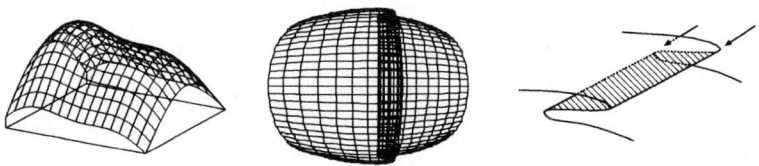

Fig. 2. LEFT: Original surface. MIDDLE: Its Gaussian image. RIGHT: An illustration of folds of Gaussian image. They are the images of parabolic curves (curves consisting of points of zero Gaussian curvature) on the original surface.

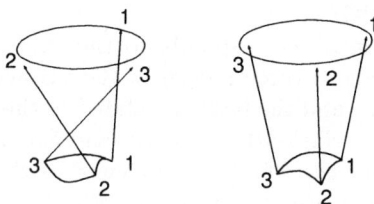

Fig. 3. The geometric relationships between the sign of the Gaussian curvature and local surface shape. LEFT: For the negative sign. RIGHT: For the positive sign.

curvature, the changes of its sign across the surface makes a *fold* of the Gaussian image as shown in Fig. 2. (Such topological structures of the Gaussian image for various types of surface are intensively described in [10].)

Based on this geometric interpretation, the sign of the Gaussian curvature at a point on the surface is determined by the rotation orientation of the surface normals at neighboring three points around the point. This is illustrated in Fig. 3. In the discrete domain, the sign at a point (i,j) depends on whether the three normals $\hat{n}_{i,j}$, $\hat{n}_{i+1,j}$, and $\hat{n}_{i,j+1}$ form a left-handed system or a right-handed system. Thus, the sign equals the following determinant of 3×3 matrix:

$$\det[\hat{n}_{i,j}, \hat{n}_{i+1,j}, \hat{n}_{i,j+1}]. \tag{4}$$

3.2 The Pseudo Gaussian Image

Our method need not know the illumination conditions. We compute the sign of the Gaussian curvature only from the set of three images.

In the same manner as in the discrete Gaussian image, a discrete image is constructed from the image intensity vectors $e_{i,j}$ for three lighting conditions. We call this image *pseudo Gaussian image* (PGI). More definitely, it is *an image created by mapping each surface point or, equivalently, each image point, on the orientation of* $e_{i,j}$. This orientation is represented as a point on a unit sphere.

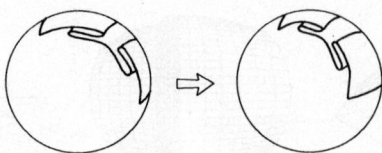

Fig. 4. The PGI can be viewed as the result of transforming the Gaussian image by Eq. (3). The structural property of the true Gaussian image is almost preserved after the transformation.

Note that the norm of the vector $\mathbf{e}_{i,j}$ is only dependent on the albedo and does not affect the resulting PGI.

The true Gaussian image corresponds to the shape of the object surface, while the PGI corresponds to three images of the surface. The former is related to the surface normals $\hat{\mathbf{n}}_{i,j}$ and the latter is related to the image intensity triplets $\mathbf{e}_{i,j}$. We already have the relationships Eq. (2) and Eq. (3) between $\hat{\mathbf{n}}_{i,j}$ and $\mathbf{e}_{i,j}$ at each image pixel. Thus, the PGI can be viewed as the result of transforming the true Gaussian image by Eq. (2) or Eq. (3) (Fig. 4).

As described earlier, the Gaussian image has topological structures reflecting the sign of the Gaussian curvature. In the rest of this section, we show that the PGI preserves the structures and that the sign of the Gaussian curvature can be determined only from the PGI.

3.3 The Lambertian Case

We discuss first the Lambertian case, where Eq. (2) is to be considered. Let us consider $\det[\mathbf{e}_{i,j}, \mathbf{e}_{i+1,j}, \mathbf{e}_{i,j+1}]$, which corresponds to a substitution of $\mathbf{e}_{i,j}$ for $\hat{\mathbf{n}}_{i,j}$ in Eq. (4). Since $\det AB = \det A \det B$, using Eq. (2),

$$\det[\mathbf{e}_{i,j}, \mathbf{e}_{i+1,j}, \mathbf{e}_{i,j+1}] = \rho_{i,j}\, \rho_{i+1,j}\, \rho_{i,j+1} \det \mathbf{L} \det[\hat{\mathbf{n}}_{i,j}, \hat{\mathbf{n}}_{i+1,j}, \hat{\mathbf{n}}_{i,j+1}]. \quad (5)$$

The last determinant is Eq. (4) itself which is the sign of the Gaussian curvature. The albedos $\rho_{i,j}$, $\rho_{i+1,j}$, and $\rho_{i,j+1}$ are always positive. Further, the matrix $\mathbf{L}$ contains the vector $\mathbf{l}^1$, $\mathbf{l}^2$, and $\mathbf{l}^3$ in its columns and thus the sign of $\det \mathbf{L}$ is determined by the relative rotation orientation of the lighting directions about the viewing direction. Therefore, if this rotation orientation is known, the sign of the Gaussian curvature can be determined by computing the sign of $\det[\mathbf{e}_{i,j}, \mathbf{e}_{i+1,j}, \mathbf{e}_{i,j+1}]$.

3.4 More General Case

We extend the discussions in the last subsection to the case of general reflectance. In this case Eq. (3) is to be considered. Recall that $\mathbf{f}$ is a function $\mathbb{R}^3 \to \mathbb{R}^3$ dependent on the surface reflectance and the illumination conditions. This function may be nonlinear. However, the proposed method still works in many cases without any modification. That is, the sign of the Gaussian curvature can be computed by $\det[\mathbf{e}_{i,j}, \mathbf{e}_{i+1,j}, \mathbf{e}_{i,j+1}]$.

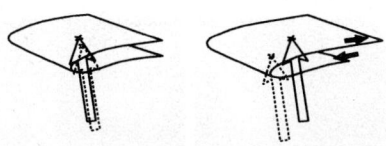

Fig. 5. LEFT: The folds of the Gaussian image are arisen from two surface points having the same normal. Since such two points are always mapped on the same position of the PGI, the folds of the PGI are occurred at the same position as the Gaussian image. RIGHT: An impossible case.

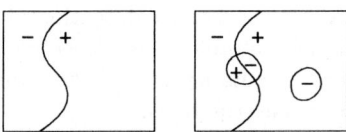

Fig. 6. A typical example of errors of segmentation. LEFT: True segmentation. RIGHT: Segmentation with errors. There may appear small wrong regions. Nevertheless, the original boundary is still extracted correctly.

It is difficult to prove our claim strictly for all types of function **f**. We show this by considering the topological structure of the PGI. Here, attention is payed to the following underlying principle of the image formation model; in the case of the constant albedo, *two surface points whose normals are the same appear equally bright*. For the PGI, this means that *the two surface points mapped onto the same position in the Gaussian image are also mapped onto the same position in the PGI*. This remains true even for the case where the albedo varies across the surface, since the albedo doesn't affect the resulting PGI as described earlier. Recall that the folds of the Gaussian image correspond to parabolic curves on the surface. These folds are arisen from two surface points having the same surface normal (See Fig. 5). Therefore, it is not difficult to see that the PGI has the folds at the same position as the Gaussian image has (See Fig. 5). The PGI preserves the topological structure of the Gaussian image at least concerning the folds. From this, we can conclude that the segmentation of the surface by the sign of $\det[\mathbf{e}_{i,j}, \mathbf{e}_{i+1,j}, \mathbf{e}_{i,j+1}]$ is *almost* independent of the surface reflecting properties. More definitely, the segmentation boundaries are precisely detected with respect to their position.

Other structures than folds of the Gaussian image are not necessarily preserved for the PGI. This means that the segmentation is not completely independent of the reflecting properties. In fact, the segmentation results obtained by the above determinant at times may be generate extra regions. Figure 6 shows typical examples of such errors. There may appear incorrect segmented regions. Such extra regions often appear around the specular reflection. (Such case will be demonstrated in the experimental results.) It is worth noting that, even in the reflecting properties having a large component of specular reflection, the component of diffuse reflection usually dominates. Also, the *true* segmentation boundaries are always successfully detected in the true position. The specular

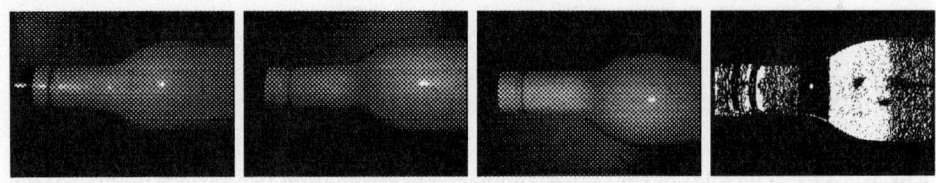

Fig. 7. A result for a white bottle. The first thee images are taken under three lighting directions. The last image is segmentation result.

reflection either does not affect the precise detection of a true boundary or make the boundary disappear and be unable to be detected at all. Even if the surface reflectance is different from the Lambertian reflectance, the boundaries never move to different positions from true ones.

4 Experimental Results

We present experimental results for three objects, a white glass bottle, a figure of duck made of rubber, and a plaster work of a face. The original images for them were 640×480 pixels taken by an ordinary CCD camera. We scaled the images to 320×240 pixels to smooth out the image quantization error and noise. Then we computed the sign of $\det[\mathbf{e}_{i,j}, \mathbf{e}_{i+1,j}, \mathbf{e}_{i,j+1}]$ for these images.

Figure 7 shows the result for the bottle. Hyperbolic points are displayed in black and elliptic points are in white. There are noisy regions consisting of many random white and black dots around the base and the head of the bottle. These parts of the bottle surface are composed of parabolic points. Since we only computed the sign, these regions suffered from the image noise. However, around the neck of the bottle, elliptic points and hyperbolic points are identified correctly. The vertical boundary line at the neck corresponding to a parabolic curve appears distinctly. Note that there exist circular segmentation errors due to the specular reflections. However, it is easy to identify the error regions in the segmentation by referring the specular reflections occurred in the original images. Thus, we may distinguish true segmentations from such wrong segmentations.

Figure 8 shows the result for the figure of a duck. As shown in Fig. 8d, except for some noisy region due to the zero Gaussian curvature, the segmentations are almost correct. Hyperbolic points around the neck and the beak of the duck are distinguished from elliptic points at the body and the head. In this experiment we varied the strength of the light source intentionally. This resulted in the relative different brightness of the three images but it did not affect the results.

Figure 9 shows the result for the plaster work. Although the shape of this object is more complicated than the above two objects, the segmentation results shown in Fig. 9 looks good. The nose, cheek and chin are correctly identified as elliptic points and the regions below the eyes and the mouth are correctly identified as hyperbolic points. The undulation running horizontally on the forehead, which was hardly seen in the original three images, could be extracted.

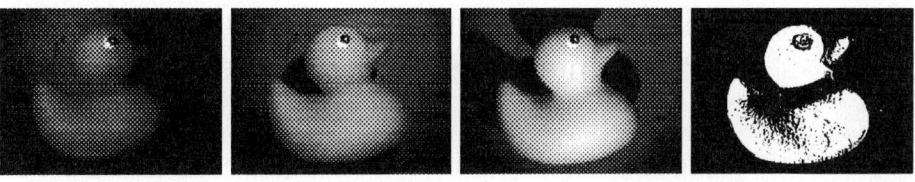

Fig. 8. Another result for a duck figure.

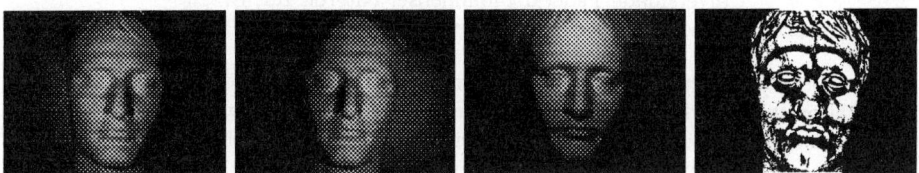

Fig. 9. Another Result for a plaster figure.

5 Conclusion

We have presented a new method for determining the sign of the Gaussian curvature at points on the surface of an object using three images taken from a fixed viewpoint under different unknown lighting directions. Neither knowledge of surface reflectance nor that of light sources is necessary. We have shown that the position of the segmentation boundary computed by our method is not affected at all by the surface reflectance. This feature enables applications of our method to wide variety of objects.

References

1. B. K. P. Horn and M. J. Brooks, *Shape from shading*. Cambridge, MA: MIT Press, 1989.
2. R. J. Woodham, "Photomtric method for determining surface orientation from multiple images," *Optical Engineering*, vol. 19, no. 1, pp. 139–144, 1980.
3. R. J. Woodham, "Determining surface curvature with photometric stereo," in *Proceedings of IEEE International Conference on Robotics and Automation*, pp. 36–42, 1989.
4. L. B. Wolff and J. Fan, "Segmentation of surface curvature using a photometric invariant," *J. Opt. Soc. Am. A*, vol. 11, no. 11, pp. 3090–3100, 1994.
5. J. J. Koenderink and A. J. van Doorn, "Photometric invariants related to solid shape," *Optica Acta*, vol. 27, no. 7, pp. 981–996, 1980.
6. A. Blake, A. Zisserman, and G. Knowles, "Surface descriptions from stereo and shading," *Image and Vision Computing*, vol. 3, no. 4, pp. 183–191, 1985.
7. B. K. P. Horn, *Robot Vision*. MIT Press: Cambridge, MA; and McGraw-Hill: New York, 1986.
8. M. P. do Carmo, *Differential geometry of curves and surfaces*. New Jersey: Prentice-Hall, Inc., 1976.
9. B. K. P. Horn and M. J. Brooks, "The variational approach to shape from shading," *Comput. Vision. Graph. Image Process*, vol. 2, no. 33, pp. 174–208, 1986.
10. J. J. Koenderink, *Solid Shape*. Cambridge, MA: MIT Press, 1988.

Estimating Depth Through the Fusion of Photometric Stereo Images

João L. Fernandes and José R.A. Torreão

Computação Aplicada e Automação/TEE
Universidade Federal Fluminense, Niterói, RJ, Brazil

Abstract. Here we revisit a recently introduced process of shape estimation through the matching of photometric stereo images, which are monocular images obtained under different illuminations. By considering the general solution of the differential equation which relates surface depth to the disparity map produced by the matching process, we are able to obtain a more consistent formulation than previously for such disparity-based approach to photometric stereo. We also employ a simple least-squares regression in a calibration strategy for estimating the parameters required by this approach. Finally, we introduce a multiscale matching procedure, based on a new stochastic metaheuristic for combinatorial optimization, which yields more reliable disparity maps in shorter processing times.

1 A Disparity-Based Photometric Stereo (DBPS)

We consider two photometric stereo images, $I_1(s)$ and $I_2(s)$, where $s = (x, y)$ denotes a general point on the image plane. If such images correspond to illumination directions which are not far apart, and if the underlying surface is smooth, we may attempt to match them to obtain a disparity field similar to the ones resulting in stereoscopy. Calling $\mathbf{D}(s) = (D_x(s), D_y(s))$ the disparity field, we would have

$$I_1(x, y) \approx I_2(x + D_x(s), y + D_y(s)), \tag{1}$$

from which we obtain, through a Taylor-series expansion,

$$\Delta I(s) \equiv I_1(x, y) - I_2(x, y) \approx D_x(s)\frac{\partial I_2}{\partial x} + D_y(s)\frac{\partial I_2}{\partial y} . \tag{2}$$

Now, if we assume that a linear approximation of the reflectance map function is applicable [1], we may rewrite $\Delta I(s)$ as

$$\Delta I(s) = k_0 + k_1 p + k_2 q, \tag{3}$$

where $k_i = k_{i1} - k_{i2}, i = 0, 1, 2$, with $k_{0i} = \overline{k}_{0i} - k_{1i}p_0 - k_{2i}q_0, i = 1, 2$, and

$$\overline{k}_{0i} = R_i(p_0, q_0), \; k_{1i} = \frac{\partial R_i}{\partial p}(p_0, q_0) \text{ and } k_{2i} = \frac{\partial R_i}{\partial q}(p_0, q_0), \tag{4}$$

where $R_i(p, q)$, for i=1,2, denote the reflectance maps associated with the two images, which are functions of the surface gradient components, $p = \partial z/\partial x$

and $q = \partial z/\partial y$, with (p_0, q_0) denoting the orientation around which the linear expansion is taken.

Equation (3) should give an accurate approximation for ΔI, provided that the image region considered is small enough so that it contains only a restricted range of (p, q) values around (p_0, q_0). From (3) and (2), we thus get

$$k_1 p + k_2 q \approx D_x(s)\frac{\partial I_2}{\partial x} + D_y(s)\frac{\partial I_2}{\partial y} - k_0, \qquad (5)$$

which is the differential equation relating the photometric disparity field to surface depth.

Now, for two sufficiently close illumination directions, it is possible to obtain another relation between $\mathbf{D}(s)$ and the depth function, $z(s)$, by requiring that the displacement of a given irradiance patch over the imaged surface be approximately perpendicular to the local normal vector, which is given by

$$\hat{n} = \frac{(-p, -q, 1)}{\sqrt{p^2 + q^2 + 1}} \; . \qquad (6)$$

Since, for an orthographic projection geometry, any such displacement can be denoted by $(D_x(s), D_y(s), v(s))$, where $v(s)$ is the unobservable displacement component along the optical-axis direction (direction z), we thus have

$$D_x(s)p + D_y(s)q \approx v(s) \; . \qquad (7)$$

In order to obtain $z(s)$ in terms of the disparity field, we must therefore find a solution to (5) which is also consistent with equation (7). Employing Lagrange's method [3] for those two equations, we find that the following relations must hold

$$\frac{dx}{k_1} = \frac{dy}{k_2} = \frac{dz}{D_x(s)\frac{\partial I_2}{\partial x} + D_y(s)\frac{\partial I_2}{\partial y} - k_0}, \qquad (8)$$

and

$$\frac{dx}{D_x(s)} = \frac{dy}{D_y(s)} = \frac{dz}{v(s)} \; . \qquad (9)$$

From these, it then follows that, by matching the image pair along the straight line given by

$$\frac{D_y(s)}{D_x(s)} = \frac{k_2}{k_1}, \qquad (10)$$

we may obtain a depth map which is a solution to equation (5), and also satisfies equation (6) for $v(s) = k_1 p + k_2 q \equiv \Delta I(s) - k_0$.

Proceeding with the solution to (5), we must now solve the ordinary differential equation

$$\frac{dz}{D_x(s)\frac{\partial I_2}{\partial x} + D_y(s)\frac{\partial I_2}{\partial y} - k_0} = \frac{k_1 dx + k_2 dy}{k_1^2 + k_2^2}, \qquad (11)$$

which is equivalent to

$$(\mathbf{k} \cdot \mathbf{ds})(\mathbf{D}(s) \cdot \nabla I_2 - k_0) = (k_1^2 + k_2^2) dz, \qquad (12)$$

for $\mathbf{k} = (k_1, k_2)$, $\nabla I_2 = (\partial I_2/\partial x, \partial I_2/\partial y)$, and $\mathbf{ds} = (dx, dy)$.

Using the property of the double vector product, the factor $(\mathbf{k} \cdot \mathbf{ds})\mathbf{D}(s)$ can be rewritten as $[(\mathbf{k} \cdot \mathbf{D}(s))\mathbf{ds} + \mathbf{k} \times (\mathbf{D}(s) \times \mathbf{ds})]$, and the second term inside the square brackets is found to vanish, due to (9). Thus, (12) becomes

$$(\mathbf{k} \cdot \mathbf{D}(s))\nabla I_2 \cdot \mathbf{ds} - k_0(\mathbf{k} \cdot \mathbf{ds}) = (k_1^2 + k_2^2)dz \ . \tag{13}$$

The term on the left-hand side of the above equation becomes a complete differential, df, if we assume, as was done in the original formulation of DBPS [1], that $\mathbf{D}(s)$ varies slowly with position across the image plane, when compared to $I_2(s)$. In such case, we will have

$$f = (\mathbf{k} \cdot \mathbf{D}(s))I_2(s) - k_0(k_1 x + k_2 y), \tag{14}$$

and the general solution to (5) can be given as [3]

$$z(x, y) = \frac{(\mathbf{k} \cdot \mathbf{D}(s))I_2(s) - k_0(k_1 x + k_2 y) + F(k_2 x - k_1 y)}{k_1^2 + k_2^2}, \tag{15}$$

where F is an arbitrary function of its argument, which comes from the first relation in (8). It is interesting to remark that, if we employ relation (10) in the above equation, we obtain an expression for $z(x, y)$ in terms of only one of the components of the disparity map, similarly to what was suggested in [1].

2 Estimating the Coefficients of the Reflectance Map

Apart from the measured values of $\mathbf{D}(s)$ and $I_2(s)$, the estimation of $z(x, y)$ through (15) depends only on the parameters of the linear approximation to the reflectance function, which can be easily estimated through the calibration strategy described below.

Let us assume that a pair of photometric stereo images of a known surface, as for instance a sphere, is available. If such a calibration surface has the same reflectance properties as the surface to be reconstructed, and if its images, $C_1(s)$ and $C_2(s)$, are obtained under the same pair of illuminations as $I_1(s)$ and $I_2(s)$, we have that the relation

$$\Delta C(s) \equiv C_1(s) - C_2(s) = k_0 + k_1 p + k_2 q \tag{16}$$

will hold for the same linear expansion coefficients as in (3).

Now, let us consider square regions of n^2 pixels (with $n \geq 2$) over the difference image ΔC. Labeling the pixels of the i-th such window by the index j, with $j=1,2,...,n^2$, we obtain from (16) the set of equations

$$\Delta C_i(j) = k_0^{(i)} + k_1^{(i)} p_i(j) + k_2^{(i)} q_i(j), \ j = 1, 2, ..., n^2, \tag{17}$$

from which a least-squares regression yields

$$\mathbf{K}(i) = (\beta_i^T \beta_i)^{-1} \beta_i^T \Delta \mathbf{C}_i, \tag{18}$$

where $\mathbf{K}(i) = [k_0^{(i)}, k_1^{(i)}, k_2^{(i)}]^T$ is the vector of the reflectance map coefficients in the i-th window; $\mathbf{\Delta C}_i = [\Delta C_i(1), \Delta C_i(2), ..., \Delta C_i(n^2)]^T$ is the vector of the difference image intensities there, and β_i is an $n^2 \times 3$ matrix whose j-th line is $(1, p_i(j), q_i(j))$.

Such calibration procedure thus yields the coefficients of the linear expansion of the reflectance map function about the mean surface orientation in the window considered.

3 Estimating the Photometric Disparities

With the reflectance map coefficients obtained as above, there only remains the estimation of the disparity field, $\mathbf{D}(s)$, for the implementation of DBPS. Here we will consider, for this purpose, a multiscale version of a recently introduced stochastic metaheuristic called the microcanonical optimization algorithm (μO).

μO is an optimization strategy derived from statistical physics, which has proven more efficient than alternative approachs - such as annealing, tabu search, and genetic algorithms - in applications of combinatorial optimization [2, 4]. The algorithm consists of two procedures which are alternately applied: initialization and sampling. In the initialization phase, starting from an arbitrary solution, the goal is to reach a local minimum of the cost function (also called energy function) associated with the optimization problem. The procedure thus implements a local search, randomly generating new solutions and accepting only those which lead to lower costs. Once the local minimum has been approached, the proposed solutions will start to be consistently rejected, and this will trigger the sampling phase.

In the sampling, the algorithm tries to free itself from the local minimum, by moving to another solution of similar cost. One may therefore picture the microcanonical heuristic, once stuck in a local minimum at the end of the initialization phase, as trying to break loose by moving around the hills in the solution space, instead of by trying to climb them, as does, for instance, simulated annealing [2]. With this purpose, the so-called Creutz algorithm of statistical physics is implemented [5]: an extra degree of freedom, called the *demon*, is introduced which generates controlled perturbations on the current solution. The demon is defined by its initial energy (D_i) and its capacity ($D_{\max}$), which are much smaller than the energies (costs) associated with the solution. In the sampling phase, changes in the current solution are proposed which are accepted only if the demon is capable of accomodating or supplying the cost difference entailed. Calling E_S the final cost obtained in the initialization phase, the sampling thus generates solutions whose costs E lie in the range $E_S - D_{\max} + D_i < E < E_S + D_i$.

After the sampling, which is implemented for a few iterations, a new initialization is run, and the algorithm thus proceeds, alternating the two phases, until a stopping condition is reached - which can be signalled by the occurence of a number of initialization/sampling cycles without decrease of the solution cost.

Apart from the stopping criteria for the whole procedure and for each of its phases, the only free parameters of μO are the demon values, D_i and $D_{\max}$. In our work, we took $D_i = D_{\max}$, and employed the following strategy for determining such value: during the initialization, a list is compiled of the cost changes associated to those moves (proposed solution changes) which are rejected for leading to higher costs, and the demon parameter is taken as one of the lower entries in this sorted list [4].

In our application of μO to the matching of PS images, we employed the cost function introduced in [6] for stereoscopy, which incorporates a photometric criterium and a smoothness criterium to be satisfied by the disparity map. Such criteria are represented, respectively, by the first and the second terms in the functional

$$E(\{\mathbf{D}(s)\}) = \sum_s |I_1(s) - I_2(s + \mathbf{D}(s))| + \lambda \sum_s S(s), \qquad (19)$$

where the smoothness factor is taken, over a neighborhood of $(2a+1) \times (2a+1)$ pixels, as $S(s) = \sum_{k=-a}^{k=a} \sum_{l=-a}^{l=a} ||\mathbf{D}(x,y) - \mathbf{D}(x+k, y+l)||$, for an euclidean norm. The positive constant λ in (19) represents a weighting factor which must be empirically determined.

In our work, the above cost function was used in a multiscale implementation of μO for a sequence of images, I_{i_k} ($k = 0, 1, 2, 3$), of dimension $2^{n-k} \times 2^{n-k}$, created by the subdivision of the original image, I_i (of size $N = 2^n \times 2^n$), in blocks of $2^k \times 2^k$ pixels, each pixel in I_{i_k} taking the average of the intensities in the corresponding block in I_i.

Our algorithm starts with the images of lowest resolution (corresponding to $k = 3$), from which an initial disparity map is obtained. Such initial estimate will then guide the matching of the images at the finer level $k = 2$, and thus consecutively: each disparity d found at level k is taken as $2d$ to the upper level, $k-1$, and the matching at this level is then restricted to the interval $[2d-2, 2d+2]$ about the disparity previously found.

Such simple multiscale strategy has proven very efficient, not only in terms of processing speed, but also in what concerns the robustness and the quality of the matching. For comparison, we show in Table 1 the final costs and the running times required in two implementations (described below) of the standard and the multiscale versions of the μO algorithm.

Table 1. Matching of photometric stereo images. Final costs and running times for two implementations (see Experiments) of the standard and the multiscale versions of the microcanonical optimization algorithm (Pentium 133MHz processor)

Experiments	μO		Multiscale μO	
	Time (s)	Final Cost	Time (s)	Final Cost
1	122	52524	28	51241
2	119	59706	29	58344

4 Experiments

In Figs. 1 and 2, we illustrate a couple of reconstructions yielded by the DBPS strategy described above. The calibration surfaces employed for the estimation of the reflectance map coefficients were spheres of approximately the same reflectance as the test objects, and, for the illumination directions chosen ($\pm 0.36, 0, 1$), yielded $k_2 \approx 0$. For each experiment, we show the input image pair, the disparity map for matching along the x direction (see (10)), and a view of the reconstructed surface, with the intensities of one of the input images mapped onto it. Equation (15) was employed for depth estimation, with the function F taken equal to zero.

5 Concluding Remarks

The human visual system has the ability of obtaining 3-D shape cues through the binocular fusion of monocular images captured under different illuminations, as first noted in [1]. This has led to the proposition of a disparity-based approach to the computational process of photometric stereo, where the input images are matched to yield a disparity map from which relative depth information can be recovered. Here, we have presented what we believe to be the definitive formulation of this process, showing that, under the general assumptions that the illumination directions considered are not too far apart, and that the disparity map varies slowly with position as compared to the image intensities, a 3-D reconstruction of the imaged surfaces can be performed. We have also introduced a new multiscale stochastic matching algorithm, and a calibration procedure for the estimation of the parameters required by our reconstruction strategy.

References

1. Torreão, J.R.A., Carvalho, B.M., Mattos, G.M.: Binocular fusion of photometric stereo images. Procs. 2nd. Asian Conference on Computer Vision, vol. 2, Singapore (1995) 336-340
2. Torreão, J.R.A., Roe, E.: Microcanonical optimization applied to visual processing. Physics Letters **A 205** (1995) 377-382
3. Hildebrand, F.B.: Advanced calculus for applications. Prentice-Hall, New Jersey (1962)
4. Linhares, A.: Microcanonical optimization applied to the traveling salesman problem. Mater's Thesis (in Portuguese), CAA-UFF, Niterói, Brazil (1996)
5. Creutz, M.: Microcanonical Monte Carlo simulation. Physical Rev. Letts. **50** (1983) 1411-1414
6. Barnard, S.T.: A stochastic approach to stereo vision. Procs. 5th National Conference on AI (1986) 676-680

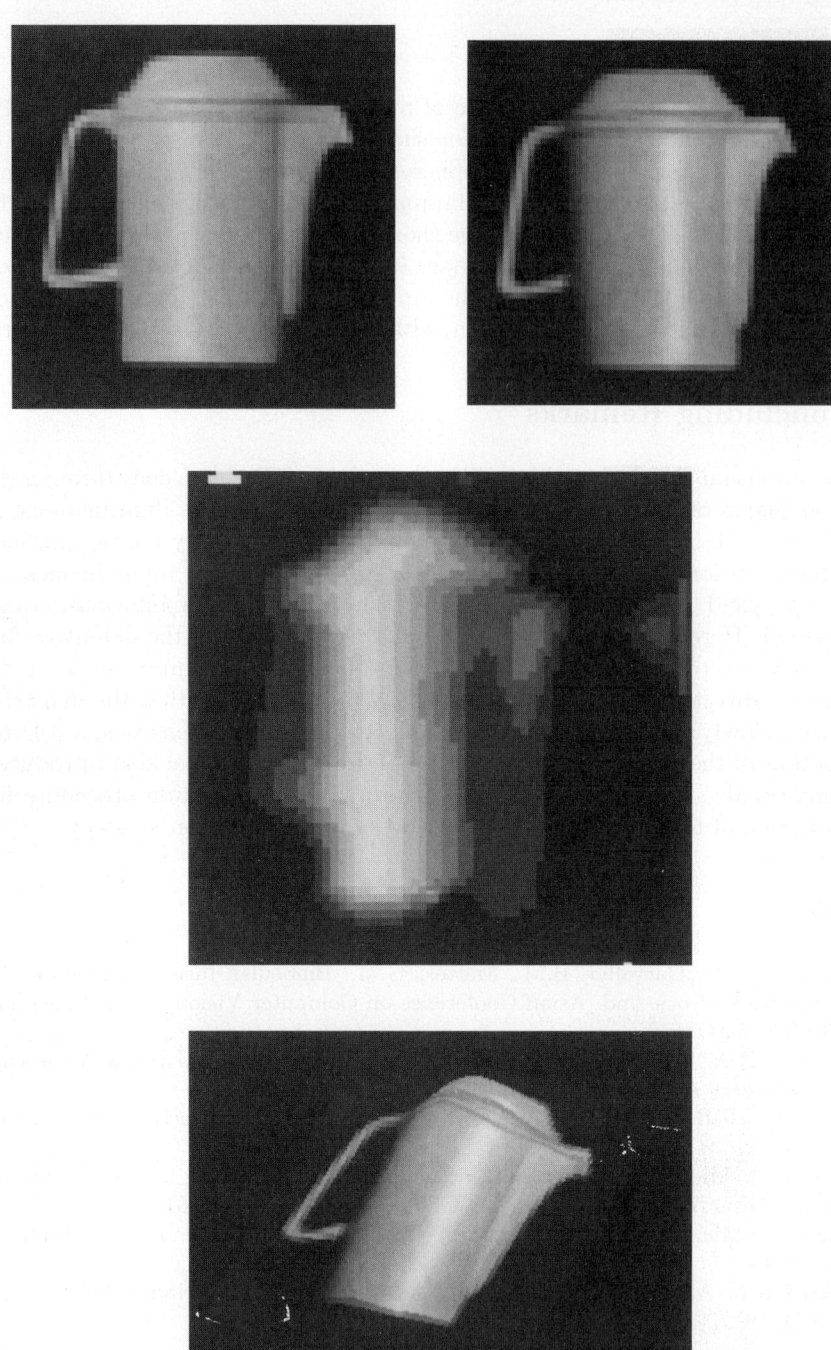

Fig. 1. Depth estimation through DBPS. From top to bottom: input image pair, estimated photometric disparity map, and estimated depth map

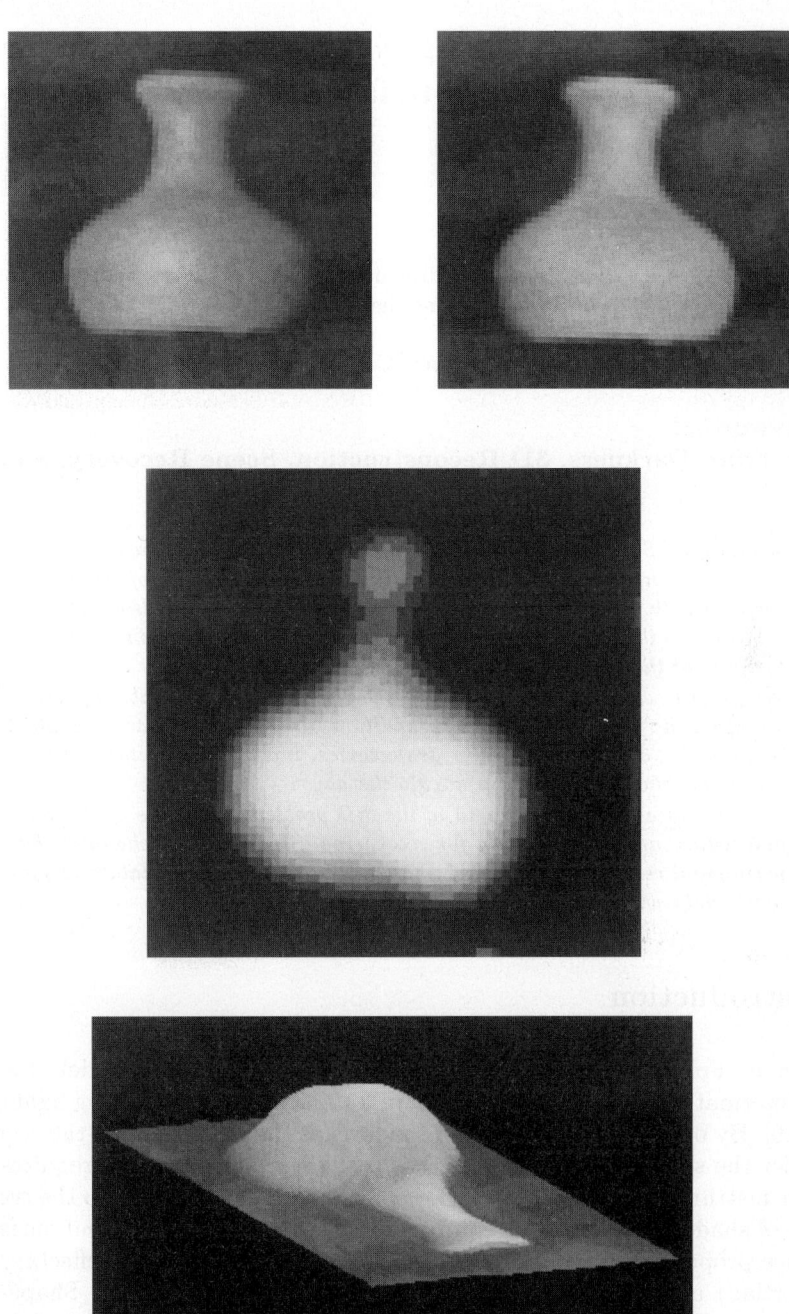

Fig. 2. Depth estimation through DBPS. From top to bottom: input image pair, estimated photometric disparity map, and estimated depth map

Out of the Dark: Using Shadows to Reconstruct 3D Surfaces *

M. Daum
mdaum@cim.mcgill.ca

G. Dudek
dudek@cim.mcgill.ca

McGill University
Centre For Intelligent Machines
School of Computer Science
Montreal, Quebec, Canada H3A 2A7

Keywords:
Shape From Darkness, 3D Reconstruction, Scene Recovery, Shadows

Abstract. *Shape From Darkness refers to using the shadows cast by a scene to reconstruct the structure of the scene. A collection of images associated with different light source positions is used. Previously published solutions to this problem have performed the reconstruction only for cross sections of the scene.*

We propose a variant of Shape From Darkness which is capable of reconstructing the entire 3-D scene. In addition, this algorithm can be applied to a broader class of light source trajectories, including trajectories which mimic the motion of the sun during the day.

We present a formal statement of the 3-D problem and some of its characteristics, and an algorithm for recovering a surface from shadows. Experimental results are presented and discussed for both real data and synthetic data with associated ground truth.

1 Introduction

The Shape From Darkness method allows one to construct a model of a scene using information on cast shadows under illumination from a moving light source [9, 1, 10]. By observing the shapes of shadows as they move across the scene, we can infer the shapes of the surfaces that cast them. This method requires inexpensive instrumentation and allows for efficient computation due to the compact nature of shadow data. It also requires only weak assumptions about surface reflectance properties, as opposed to shape-from-shading's strong reflectance (eg. Lambertian) assumptions. Furthermore, it has been shown that Shape From Darkness can be used to infer the shapes of surfaces in the scene even if they are not directly visible to the camera Pragmatically, the technique may be useful in contexts where traditional range sensors may not be suitable (eg. Martian Exploration).

---
* This work was supported by the Canadian Centres of Excellence IRIS Project IS-5

Shape from darkness is superficially related to the shape-from-shading problem. Shape from shading has been examined in a variety of contexts, using methods including relaxation labelling and regularization [7, 8, 12, 5, 3, 4, 11]. Shape from shading also appears to be a psychophysically relevant process [14]. Note, however, that shape from shading is a fundamentally local phenomenon (assuming the standard assumptions including that of a single distant light source and an absence of mutual illumination are made).

Much of the prior work on shape-from-darkness deals with two-dimensional instances of the problem, where the light source and the surface to be recovered all lie in the same plane [9, 1].

Typical existing approaches to shape-from-darkness make three critical assumptions regarding the problem [9, 6, 2, 10].

- The world, including the camera and light source is two-dimensional (i.e the light source and the surface to be recovered all lie in the same plane)
- The light source and camera geometry can both be modeled using orthographic projection (i.e. they are extremely distant)
- The surface to be reconstructed is a *terrain* described by a function $z(x)$ (i.e. a graph surface)

Our approach to the problem allows us to relax all three of the assumptions, although in this paper we will focus only on the first (a 3-D instead of a 2-D world).

We present a method by which scene reconstruction can be performed given a light source moving through an arbitrary set of three dimensional positions. During the reconstruction, the scene is modelled as a region lying between two bounding surfaces. As more shadow information is integrated into the estimate, the bounding surfaces move closer together until an exact reconstruction is achieved.

2 Problem Definition

2.1 Two Dimensional Problem

Previous work on Shape From Darkness has focused on the solution of the two dimensional version of the surface reconstruction problem. In this version of the problem a surface is defined as a function $z = f(x)$. If a surface $f(x)$ assigns a single value to each x in a given *range*, then this surface is *terrain-like*. If the surface is more complex, then it is *non-terrain-like*.

To reconstruct a surface, a light source must be moved through a trajectory of angles above the surface (in practice, these can be arbitrary discrete sample locations.)

A stationary camera records a series of images of the surface as the light source moves overhead. Both the light source and the camera are considered to be an "infinite" distance away from the scene. This has the effect of creating a camera with orthographic projection and a light source which casts rays which are parallel to one another. Thus, in the 2D formulation a single angular parameter θ suffices to describe the position of the light source. An important effect of this positioning is that every pixel in the image is guaranteed to be lit when the light source is directly overhead (in the "noon" position).

Shadow information can be described using an intermediate representation known as a *Shadowgram*.[9] As shown in Figure 1, the shadowgram is a binary

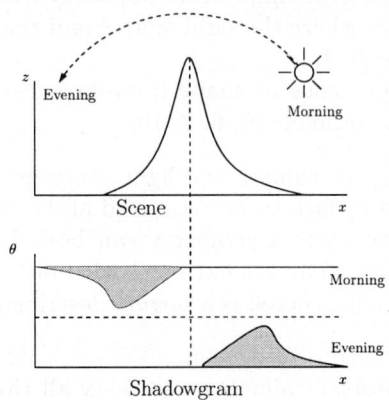

Fig. 1. Two Dimensional Shadowgram

function $s(x, \theta)$ on the angle θ and a spatial dimension x. A white entry in the shadowgram indicates that image pixel x was lit when the light source was at angle θ, while a black entry indicates that it was shadowed. It was shown in [9] that the shadowgram of a terrain-like surface can be completely described by two curves: θ^+ and θ^-, representing the first light in the "morning" and the last in the "evening" respectively. It is possible to reconstruct the surface by integrating θ^+ and θ^- [9].

When the surface is non terrain-like, the shadowgram possesses not only two curves, but also some white *holes* where one would expect darkness if the surface were a terrain. Here θ^+ and θ^- are not defined as first and last lighting curves but rather as the envelope of the shadowgram which lies closest to the noon position. It is shown in [10] that using these two curves to reconstruct the surface will in fact produce a terrain-like superset of the surface. Furthermore, the holes in the shadowgram may then be used to carve pieces out of this terrain, allowing one to reconstruct some or all of the hidden surfaces in the scene.

2.2 Three Dimensional Problem

The surface for reconstruction in the 3D problem is a function $z = f(x, y)$. As before, we say that the scene is terrain-like if this function is single valued over the ranges of x and y which are presented in the image. Because the scene no longer lies within a single vertical cross section, the light source is allowed to point freely, and must be described by two angles, ϕ and θ.

One may view the original two dimensional problem as a special case of the larger 3D formulation. One in which the light source travels through a trajectory through a series of θ angles while keeping its ϕ fixed at noon. In fact, given this type of trajectory one can, in fact, reconstruct each scan-line of the image individually using the 2D algorithm. Because the process underlying the reconstruction is integration, however, these internally consistent scan-lines can not be combined into a whole as they are each reconstructed to within an unknown additive constant.

2.3 Constraining Shadower and Exact Reconstruction

Consider a shadowed point in an image representing a single light source position. If a ray is cast from this point in the direction of the light source (the point's *light seeking ray*), any surface point lying above this ray is a potential shadower of this point. Of these possible shadowing points, the point which lies highest above this ray (and furthest from the casting point, if this height is not unique) has special significance, and is called the point's *constraining shadower*.

If a point is a constraining shadower of another point, we have the guarantee that this point lies along a shadow boundary in the image in question. We know that this is a contact shadow boundary on the surface. As a result, we can identify the constraining shadower as the first shadow boundary point encountered in image space along the image projection of a point's light seeking ray.

3 Approach to Reconstruction

The reconstruction problem lends itself naturally to a solution through the iterative relaxation of constraints. It is natural that two types of constraints exist: *expect light* for pixels and source directions resulting in light, and *expect darkness* for pixels in shadow.

Consider the constraints in terms of the behaviour of a light seeking ray cast from a point on the surface. In the case of expect light, such a ray is expected to pass freely out of the scope of the image without intersecting the working surface. On the other hand, a ray expecting darkness must certainly intersect the surface in at least one place in order to shadow the pixel. It is assumed that all shadowers lie within the image.

During the reconstruction, we model the scene as an upper and a lower bounding surface which are incrementally brought together. The two types of constraints (expect light and darkness) and the two surfaces (upper and lower bounds) yield four rules for extraction of shadow information:

- **Expect Light**
  - The **Upper Bound** of any pixel lying in the image projection of a light seeking ray cast from the lit point's upper bound is lowered to the level

of the ray (if it was previously above). These points cannot be higher than this as they would shadow the point.
- The lit point's **Lower Bound** is raised until a light seeking ray cast from said lower bound will be pass above (or just touch) all lower bounds along the ray. This bound cannot be lower as the point would then be shadowed.
- **Expect Darkness**
  - The **Upper Bound** of the shadowed pixel is lowered until the light seeking ray which it casts intersects the upper bound at some other point. If the bound were higher, then the point could not be shadowed.
  - The **Lower Bound** of the shadowed point's *constraining shadower* is raised to the level of the light seeking ray cast from the lower bound of the shadowed point. The shadower must be at least this high in order to shadow the point at its lower bound.

From these rules, one can see that the upper and lower bound surfaces are not directly coupled to one another. Points on the upper bound will only effect other upper bound points, while lower bounds only effect other lower bounds. Thus, the two surfaces are related only through the shadow information and may be computed separately.

It also follows that points on the upper bound are only ever lowered, while points on the lower bound are raised. This shows that at worst, the application of new shadow data will leave the bounds unchanged, and will never degrade the estimate. The distance between corresponding upper and lower bound points cannot increase, guaranteeing termination.

The global character of the discretized shape from darkness problem suggests that it is an ill-posed problem in the sense of Tikhonov [13]. That is, small changes in one part of the surface or shadowgram can have very large repercussions for the solution. As a result, it is necessary when solving the discrete version of the problem to apply a stringent confidence-based threshold on the shadow information. Only constraints which pass this test are allowed to contribute to the reconstruction. In practice, it is possible to base such a filter on the immediate neighborhood in shadow-space of the constraint in question, avoiding the complicating effects of non-local shadowing relationships.

4 Implementation and Results

The algorithm used to reconstruct the bounding surfaces involves the iterative application of constraints . The work surfaces are initially set to be flat. For each pixel in the work surfaces a constraint is enforced for each light source direction in the trajectory. The complexity of a single iteration is thus $\mathbf{O}(n \times m \times t \times r)$ for an image of width n and height m, a trajectory of t source directions, and an average ray length of r. If we assume a square image and an equal number of source directions then this reduces to $\mathbf{O}(n^4)$.

Each iteration involves two waves of reconstruction. In the first wave all of the expect darkness constraints are applied, followed by expect light in the second. Each image row's constraints are applied in parallel, with changes being written back between successive rows. Rows are processed either front-to-back or vice-versa in order to minimize repetitive work. To avoid sensitivity to errors introduced by discretization, all constraints in a neighborhood containing both shadow and light are discarded.

4.1 Experimental Data

Generated Our implementation of shape from darkness accomplishes surface recovery on 64x64 pixel images with 64 shadow images in roughly 5 minutes on a sparc-20 workstation.

Figure (2) presents the input surfaces and the associated reconstructed upper

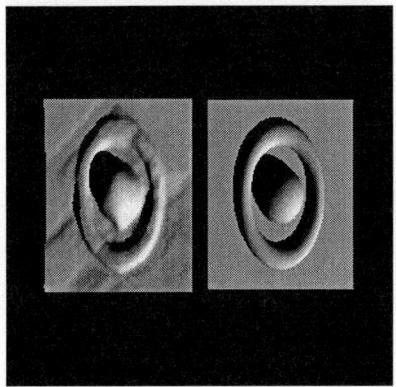

Fig. 2. Surface Reconstruction from Shadow Data for a Complex Terrain. Original surface is shown on the right

bound for a surface of moderate complexity. The shadow information for this scene was extracted from a series of artificial images rendered from a CAD model of the scene. The average error in this reconstruction was approximately 2.5% of the total scene height.

For portions of the surface that are not shadowed, or that are always in shadow, the absence of sufficient constraints on the surface geometry can sometimes lead to significant artifacts in the upper bound surface. These artifacts reflect the dearth of information about these points present in the shadow set. As expected, these effects are most often seen in the extreme fronts and backs of both images and the individual objects within them.

Observe that the surface is accurately reconstructed both on the front surface (facing the light source) as well as along the back surface (away from the light source). This is possible since information on surface geometry is obtained both based on shadows cast *by* a surface as well as by the behaviour of shadows that are cast *upon* a surface.

Real As a demonstration of the applicability of this algorithm, reconstruction was performed on a simple scene containing a four-sided pyramid with a flat top. The camera and scene were both mounted on a platform, which was then rotated under constant lighting by a single source. The light-source "trajectory" generated was that of a cone of directions whose axis lay in the image plane. The base angle of the cone was approximately $70°$. Photographs of the scene were taken in 64 light source positions (figure 3). The resulting images were then cropped and thresholded, as depicted in figure (4). The resulting reconstruction

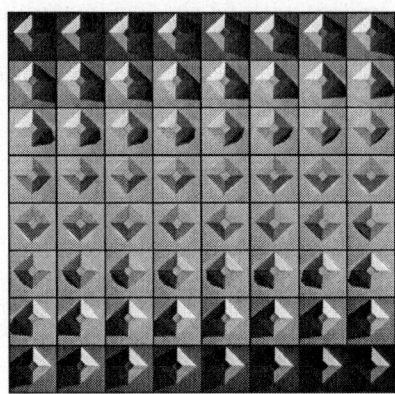

Fig. 3. Shadow Information Used in Pyramid Reconstruction

Fig. 4. Image Preparation for Reconstruction

was performed in 33 iterations, taking 347 seconds on a sparc 20 workstation.

5 Summary and Conclusion

Initial results show that the 3D Shape From Darkness algorithm is capable of estimating the structure of a scene using only the shadows within the scene. Furthermore, this reconstruction can be performed in a manner which uses far less data than the theoretically complete set, and which generates a full, three dimensional description of the surface.

This algorithm is well suited to terrain reconstruction from images taken of the earth from a geosynchronous satellite throughout a single day. It can also be applied to the problem of environment mapping by allowing for two explorer robots, one of which holds a camera while the other moves with a light source. Questions for future and ongoing research are those of the stability of uniqueness properties of reconstructed surface and its envelope. Our results show that the robustness of the solution is acceptable.

The use of shadow data provides for a robust reconstruction which uses extremely compact data due to the boolean nature of shadow information, which makes this method a computationally efficient tool for terrain reconstruction.

The use of shadow data permits a new form of three-dimensional surface reconstruction that exploits the rich geometric information contained in cast shadows. This appears to permit computational surface reconstruction in new contexts. Several interesting problems remain to be resolved, including uniqueness of the computed solutions and smoothness constraints.

References

1. K.A. Loparo D. Raviv, Y. Pao. Reconstruction of three-dimensional surfaces from two-dimensional binary images. *IEEE Transactions on Robotics and Automation*, 5(5):701–710, 1989.
2. J. Kender D. Yang. Shape from shadows under error. In *Image Understanding Workshop 1993*, pages 1083–1090, Washington, D.C., August 1993.
3. P. Dupuis and J. Oliensis. Direct method for reconstructing shape from shading. pages 453–458.
4. P. Dupuis and J. Oliensis. Shape from shading: Provably convergent algorithms and uniqueness results. volume 2, pages 259–268.
5. Robert T. Frankot and Rama Chellappa. A method for enforcing integrability in shape from shading algorithms. *IEEE Trans. Pattern Analysis and Machine Intelligence*, 10(4):439–451, July 1988.
6. M. Hatzitheodorou and J.R. Kender. An optimal algorithm for the derivation of shape from shadows.
7. Berthold Horn. *Robot Vision*. The MIT Press, Cambridge, Massachusetts, 1986.
8. Katsushi Ikeuchi and Berthold K. P. Horn. Numerical shape from shading and occluding boundaries. In Michael Brady, editor, *Computer Vision*, pages 141–184. Elsevier Science Publishing Company, New York, NY, August 1981.
9. E.M. Smith J.R. Kender. Shape from darkness: Deriving surface information from dynamic shadows. In *AIII*, pages 539–546, 1987.
10. Michael Langer, Gregory Dudek, and Steven W. Zucker. Space occupancy using multiple shadowimages. In *Proceedings IEEE/RSJ International Conference on Intelligent Robots and Systems (IROS)*, pages 390–396, Pittsburgh, PA, August 1995. IEEE Press.
11. M.S. Langer and S. W. Zucker. Shape-from-shading on a cloudy day. *Journal of the Optical Society of America A*, 11(2):467–478, 1994.
12. Alex P. Pentland. Local shading analysis. *IEEE Trans. Pattern Analysis and Machine Intelligence*, 6(2):170–187, March 1984.
13. Andrei Nikolaevich Tikhonov and Vasilii Iakovlevich Arsenin. *Solutions of ill-posed problems [Metody resheniia nekorrektnykh zadach]*. Halsted Press, New York, 1977.
14. James T. Todd and Ennio Mingolla. Perception of surface curvature and direction of illumination from patterns of shading. *Journal of Experimental Psychology: Human Perception and Performance*, 9(4):583–595, 1983.

Estimation of Reflection Parameters from a Color Image

Shoji Tominaga

Osaka Electro-Communication University
Neyagawa, Osaka 572, Japan

Abstract. The present paper proposes a method for estimating various parameters of a reflection model from a single color image measured with a camera. The dichromatic reflection model of the Phong type is used for an inhomogeneous dielectric object. The object's shape is assumed to be cylindrical. First we analyze a relationship between the reflection properties of a cylinder object and the color histogram. Second, Algorithms are developed for estimating several parameters of (1) object color vector, (2) illumination direction, (3) illumination color vector, (4) surface roughness, and (5) ratio of interface to body intensity. Finally the feasibility of the proposed method is shown in an experiment.

1 Introduction

Modeling the reflection properties of material surfaces is needed in many fields including computer graphics and computer vision. In computer graphics, reflection models have been used for generating realistic images [1-3]. Machine vision applications often require reflection models for image analysis, object recognition [4-5], and realizing color constancy [6-7]. The dichromatic reflection model for inhomogeneous dielectric materials like plastics suggests that under all illumination and viewing geometries the spectral reflectance function can be expressed as the weighted sum of two functions: the constant interface function and the body reflectance function [4, 8]. Tominaga and Wandell [9] and Lee et al. [10] showed the adequacy of this standard dichromatic reflection model for most materials, and Tominaga [11] proposed an extension of the dichromatic reflection model
 Generally a reflection model consists of three essential properties of the spectral factor on interface and body colors, the geometric factor on illumination and viewing, and the surface factor on surface roughness. The dichromatic reflection model assumes that the spectral factor and the nonspectral factors of geometry and roughness are separable. This separability and the constant interface reflection property make it possible to estimate the illumination color and the object color [12]. In this case the nonspectral factors are treated as being included in the weighting coefficients of the reflection functions.
 Novak and Shafer [13] proposed an algorithm for analyzing color histograms that yields estimates of various parameters including surface roughness and illumination intensity. This estimation was based on interpolation between histograms that came from many computer graphics images simulated with the known parameters. Tominaga [14] proposed a method for determining the magnitude terms to complete the dichromatic reflection models for a variety of materials.
 This paper proposes a method for estimating various parameters of a reflection model from a single color image. The dichromatic reflection model of the Phong type is used. The object's shape is limited to cylinders, and the object's material is assumed to be inhomogeneous dielectrics including plastics and paints. Sheets made of these materials are also available as our estimation objects by putting

the sheets on a cylinder. Moreover the reflection properties of different paints can be determined from the painted surfaces on cylinders. Therefore this study is applicable to determination of the reflection properties for a wide range of object materials.

2 Reflection Model and Color Histogram

2.1 Color reflection

Fig. 1 shows a simple description of measuring a cylinder object with a camera. The light source emits parallel beams perpendicularly to the cylinder axis. It is assumed that the angle between viewing and illumination directions is less than 90 degree.

Suppose that the object surface is composed of an inhomogeneous dielectric material. Then light reflection from the surface is described by the dichromatic reflection model. This model suggests that the reflected light is composed of two additive components, the body reflection and the interface reflection. The interface reflection is assumed to have the same spectral composition as the illumination.

The observed color vector $\mathbf{c}(x)$ at pixel x can be described as follows:
$$\mathbf{c}(x) = w_b(x)\, \mathbf{c}_b + w_i(x)\, \mathbf{e}, \qquad (1)$$
where $\mathbf{c}_b$ and $\mathbf{e}$ are, respectively, the object color vector by the body reflection and the illumination color vector. These color vectors are three-dimensional and normalized to unit length. The scalars $w_b(x)$ and $w_i(x)$ are weighting coefficients for the body reflection and the interface reflection, respectively.

The distribution of the reflected light intensity by two components is depicted on the cylinder surface in Fig. 1. The body reflection component takes the maximum intensity at a location of the surface which are perpendicular to the illumination. The interface reflection component takes the maximum at a location of the surface which bisects the angle between viewing and illumination.

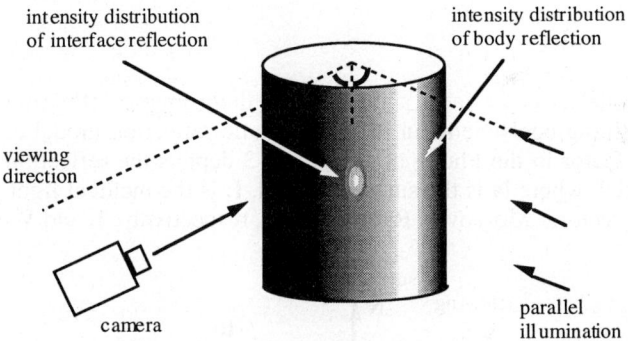

Fig. 1. Intensity distribution on a cylinder surface.

2.2 Relationship between reflection model and color histogram

We utilize some features of the color histogram of a single color image to estimate various parameters of the reflection model. Fig. 2 demonstrates the color histogram of the measured image for a real object in a RGB space. The histogram is composed of two linear cluster: the body-reflection cluster and the highlight cluster. The body-reflection cluster from the origin corresponds to the matte part of a surface without specular highlight, which is based on the body reflection only. The direction of this cluster coincides with the direction of the object color vector. The highlight cluster, which is based on both the interface and body reflections, starts from a location close to the end of the body-reflection cluster. Klinker et al. [12] showed that, if an object is convex and smooth, the pixels in the color space will be distributed in the shape of a skewed T as shown in Fig. 2. If the phase angle between viewing and illumination is small, the histogram shape becomes a skewed L or a dogleg.

According to the reflection model of Eq. (1), the observed color vectors from a cylinder are two-dimensional and fall in a two-dimensional subspace spanned by two vectors c_b and e. This subspace is called the color signal plane. Therefore the color histogram in a three-dimensional RGB space is projected onto the color signal plane without loss of information. The color signal plane can be computed as a plane spanned by two principal components for the distribution of color pixels.

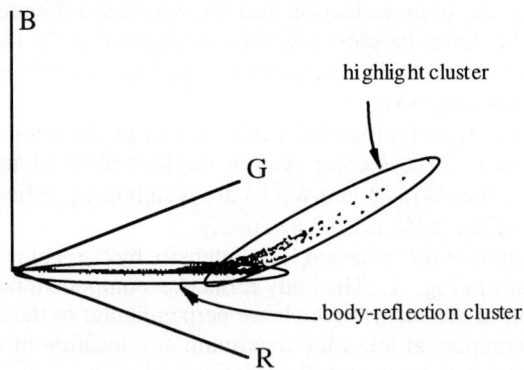

Fig. 2. Color histogram in a RGB space.

2.3 The Phong model

The Phong model is essentially a type of the dichromatic reflection model. The unknown weighting coefficients in the dichromatic reflection model are specified as the geometric factor in the Phong model. Fig. 3 depicts the reflection geometry for the Phong model, where **N** is the surface normal, **L** is the incident light vector, and **V** is the viewing vector. Moreover, R_l and R_v are, respectively, **L** and **V** mirrored

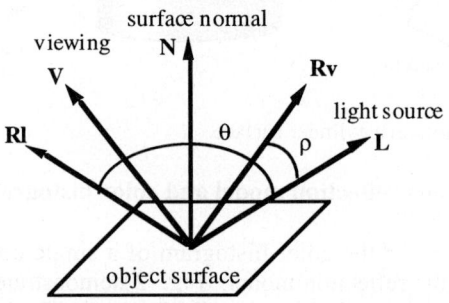

Fig. 3. Reflection geometry for the Phong model.

about **N**. We use the symbol n as a measure of surface roughness. Then the Phong model is described as

$$\begin{aligned} \mathbf{c}(x) &= w_b(x)\, \mathbf{c}_b + w_i(x)\, \mathbf{e} \\ &= \alpha\, (\mathbf{N} \cdot \mathbf{L})\, \mathbf{c}_b + \beta\, (\mathbf{R}_v \cdot \mathbf{L})^n\, \mathbf{e} \quad , \\ &= \alpha\, (\cos \theta)\, \mathbf{c}_b + \beta\, (\cos^n \rho)\, \mathbf{e} \end{aligned} \quad (2)$$

where θ is the angle of incidence and ρ is the angle between $\mathbf{R}_v$ and $\mathbf{L}$. Moreover the scalars α and β are the weighting coefficients for representing the relative intensity of the body and interface reflection components. Therefore, $\mathbf{c}_b$, $\mathbf{e}$, n, and α/β are the parameters to be estimated in this study.

3 Estimation Method

3.1 Object color vector

The extension of the linear cluster of the body-reflection cluster passes through the origin of the color signal plane, because this origin corresponds to the origin in the RGB space. The estimation problem of the object color vector $\mathbf{c}_b$ is reduced to detection of the line passing through the origin. The algorithm based on the principal component analysis can be used for detecting a linear cluster [11].

Next, the point P_b, where the intensity of body reflection is maximized, is determined on the body-reflection cluster. Fig. 4 is a sketch of the color histogram on the color signal plane, where the point P_b indicates the largest color vector of the body reflection component. That is, this point corresponds to the most saturated object color on the cylinder surface. Moreover note that the distance between the origin and P_b corresponds to the weighting coefficient α in Eq. (2).

Thus, the estimate of the object color $\mathbf{c}_b$ is obtained by transforming the vector of $\overline{OP_b}$ on the color signal plane into the RGB space.

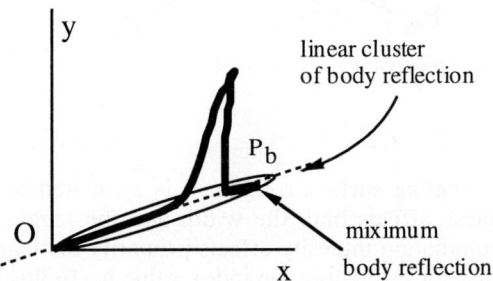

Fig. 4. Estimation of object color.

3.2 Illumination direction

Let us estimate the illumination direction from a highlight peak in the measured image. We first determine the peak of the highlight cluster in the color histogram, and then detect the corresponding pixel on the measured image. Fig. 5 shows the relationship between the highlight peak on the histogram and the strongest highlight point on the object surface.

3.3 Illumination color vector

The highlight peak is also available for estimating the illumination color vector **e**. This vector is determined from the principal component line passing through the point H_m in the highlight cluster (see Fig. 6). For this purpose, we determine the intensity of the body reflection component at the location Z_m on the surface. The point P_m in Fig. 6 indicates the corresponding coordinates on the x-axis in the color signal plane. Both points P_m and H_m are located at the same spatial location Z_m, where P_m means no illumination component, whereas H_m means the maximum illumination component. Therefore the segment $\overline{P_m H_m}$ suggests the direction vector of illumination color. The coordinates of P_m are computed with only the body reflection term in Eq.(2) as follows: Let P_{bx} be the x-coordinate value of P_b. Then $\alpha = P_{bx}$. The angle of incidence θ is specified as $\theta = \theta_m$ at the highlight peak. Hence the x-coordinate value of P_m is given by

$$P_{mx} = P_{bx} \cos(\theta_m), \qquad (5)$$

where the estimates of P_{bx} and θ_m are given in Sections **3.1** and **3.2**. Finally the directional vector of $\overline{P_m H_m}$ is transformed into the RGB space to obtain the estimate of an illumination color vector **e**.

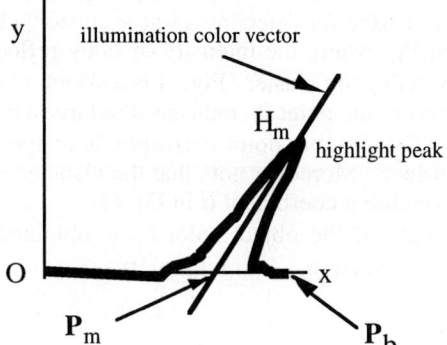

Fig. 6. Estimation of illumination color.

3.4 Surface roughness

The index value n representing surface roughness is estimated from the measured image. Surface roughness affects both the width and the length of the highlight cluster, although the illumination intensity affects primarily the length [13]. By this property the width is used for estimating the index value n. In this paper we use the full-width at half-maximum (fwhm) of the highlight peak for making the estimation process easy. The following relation should hold for the interface reflection term at half maximum of the peak.

$$(\mathbf{R_v} \cdot \mathbf{L})^n = 0.5. \qquad (6)$$

There are two solutions of the vector $\mathbf{R_v}$ satisfying Eq.(6) which correspond to two edge points h_1 and h_2 on the highlight cluster. First, the corresponding spatial points to h_1 and h_2 are found on the cylinder surface. Second, the vectors $\mathbf{R_{v1}}$ and $\mathbf{R_v}_2$ are determined at the respective spatial points. Two possible values of n are obtained as

$$n_1 = -\frac{\log 2}{\log(\mathbf{R_{v1}} \cdot \mathbf{L})}, \qquad n_2 = -\frac{\log 2}{\log(\mathbf{R_{v2}} \cdot \mathbf{L})}. \qquad (7)$$

The index of surface roughness n is estimated as the average value of n_1 and n_2.

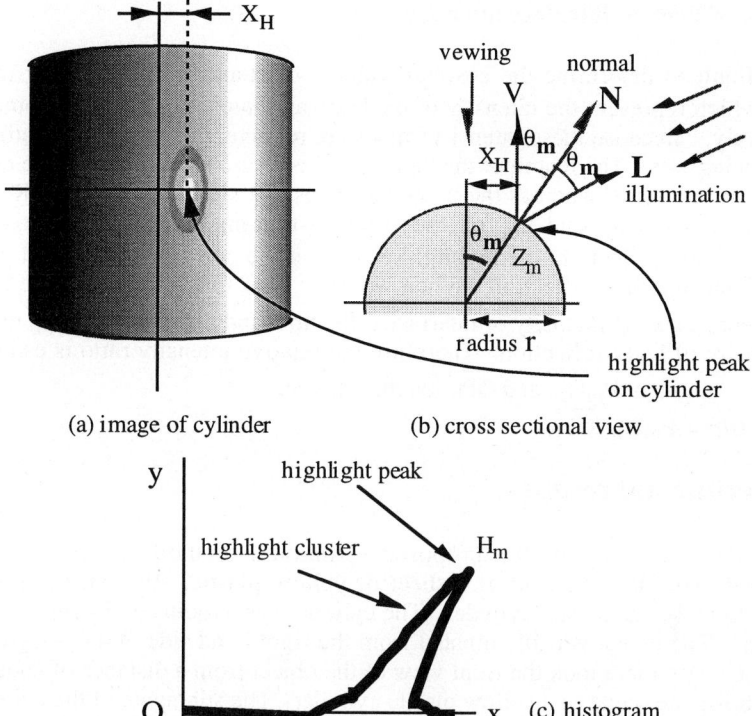

Fig. 5. Highlight peak on histogram and object surface.

To determine a highlight peak, the original color histogram is rotated on the color signal plane so that the segment $\overline{OP_b}$ of the object color vector coincides with the x-axis as shown in Fig. 5 (c). The highlight peak is then determined as the point H_m with a maximum y-value in the transformed histogram.
Next the corresponding pixel to the peak H_m is determined on the image plane. Moreover we determine the corresponding spatial location on the cylinder surface.

For simplicity, suppose that the viewing direction is vertical to the cylinder as shown in Fig. 5 (b). At the location Z_m of the highlight peak, the interface reflection term in Eq. (2) becomes maximum. Note this case that the incident light vector **L** coincides with the mirrored viewing vector $\mathbf{R_v}$ at the location . That is, $\angle NV = \angle NL \equiv \theta_m$. Hence the angle of illumination direction $\angle LV$ is given as
$$\angle LV = 2\,\theta_m. \tag{3}$$
Finally the angel is determined as follows: Let X_H be the distance between the two points of the cylinder center and the highlight peak on the image plane. The radius r of the cylinder may be estimated by detecting the edges of an object in the measured image. With these values, the angle θ_m is then computed as
$$\theta_m = \sin^{-1}(X_H/\,r). \tag{4}$$

3.5 Ratio of body to interface intensity

It is difficult to determine the absolute values of α and β in the Phong model of Eq.(7), which represent the intensity of the body and interface reflection components, respectively. Because these intensity values are relative, the ratio α/β is estimated in the following way. The length of the highlight cluster grows as the interface reflection component increases, whereas the length shortens as the ratio decreases. Recall that three points P_b, H_m, and P_m on the color histogram indicate, respectively, the maximum body reflection, the highlight peak, and the corresponding point on the x-axis without interface reflection. When we draw lines on these points, the length of $\overline{P_m H_m}$ represents the intensity of interface reflection, and the length of $\overline{OP_b}$ represents the intensity of body reflection. Therefore the relative intensity ratio is estimated as the ratio of segment $\overline{P_m H_m}$ and $\overline{OP_b}$ lengths, that is,

$$\alpha/\beta = \overline{P_m H_m}/\overline{OP_b}. \tag{8}$$

4 Experimental results

To test the feasibility of the proposed estimation method, an experiment was performed using the material of a sheet of yellow plastic. We put the sheet on a cylinder to make the plastic cylinder. The cylinder has a radius of 15 cm and a height of 40 cm. The object was illuminated from the right-hand side of a 45-degree angle. A color CCD camera took the front view of the object from a distance of about 1.8 m. Fig. 7 shows the scene of a yellow plastic cylinder. The silhouette of the cylinder was first extracted from the measured image. Next the image region surrounded with white lines was cut out from the cylinder as shown in Fig. 7, and only this narrow region was used for the parameter estimation. Fig. 8 shows the histogram for the cut-out region. To investigate the estimation accuracy, the light source and object color were measured in a separate way. Table 1 shows a comparison between the estimation results and the direct measurement results for the color vectors of illumination and object. The angle $\angle \mathbf{LV}$ of the illumination direction was estimated as 41 degree, and the index value n of surface roughness was estimated as 71.

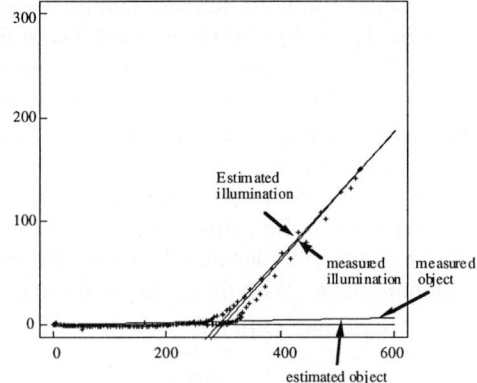

Fig. 7. Scene of a yellow plastic cylinder.   **Fig. 8.** Histogram and estimation results

	illumination color vector	object color vector
measurement	[0.608, 0.569, 0.553]	[0.823, 0,565, 0.058]
estimate	[0.620, 0.566, 0.542]	[0.842, 0.535, 0.057]

Table 1. Measurement and estimation results.

5 Conclusion

This paper has proposed a method for estimating various reflection parameters of the Phong reflection model for an object surface from the measured image with a camera. These parameters are object color, illumination color, illumination direction, surface roughness, and ratio of interface to body intensity. It is assumed that the object's material is inhomogeneous dielectric, and the shape is cylindrical. The same method is applicable to spherical objects.

The proposed method has the advantage that all the parameters of the reflection method can be estimated from a single color image for an object. The light source is assumed to emit parallel beams perpendicularly to the cylinder axis. Unless this condition is satisfied, some errors will be contained in estimating illumination direction and surface roughness. Moreover the angel between viewing and illumination directions is restricted to less than 90 degree. This observation condition is needed because the point with maximum body reflection on the surface must be included in the measured image.

References

1. B.T. Phong, Illumination for computer-generated pictures, *Comm. ACM*, **18**: 311 (1975).
2. J.F. Blinn, Model of light reflection for computer synthesized pictures, *Computer Graphics*, **11**: 192 (1977).
3. R.L. Cook and K.E. Torrance, A reflection model for computer graphics, *Computer Graphics*, **15**: 307 (1981).
4. G. J. Klinker, S. A. Shafer, and T. Kanade, The measurement of highlights in color images, *Int. J. of Computer Vision*, **2**: 87 (1988).
5. G. Healey, Using color for geometry-insensitive segmentation, *J. Opt. Soc. Am. A*, **6**: 920 (1989).
6. M.S. Drew, Optimization approach to dichromatic images, *J. Math. Imaging and Vision*, **3**: 187 (1993).
7. S. Tominaga, Realization of color constancy using the dichromatic reflection model, *2nd Color Imaging Conf.*, 37 (1994).
8. S. Shafer, Using color to separate reflection components, *Color Res. Appl.*, **10**: 210 (1985).
9. S. Tominaga and B. A. Wandell, The standard surface reflectance model and illuminant estimation, *J. Opt. Soc. Am. A*, **6**: 576 (1989).
10. H. C. Lee, E. J. Breneman, and C. Schulte, Modeling light reflection for computer color vision, *IEEE Trans. Patt. Anal. Mach. Intell*, **12**: 402 (1990).
11. S. Tominaga, Dichromatic reflection models for a variety of materials, *Color Research and Application*, **19**: 277 (1994).
12. G.J. Klinker, S.A. Shafer, and T. Kanade, A physical approach to color image understanding, *Int. J. Computer Vision*, **4**: 7 (1990).
13. C. L. Novak and S.A. Shafer, Method for estimating scene parameters from color images, *J. Opt. Soc. Am. A*, **11**: 3020 (1994).
14. S. Tominaga, Dichromatic reflection models for rendering object surfaces, *J. Imaging Science and Technology*, **40**: 549 (1996).

A Natural Norm for Color Processing

Ron Kimmel[*]

Lawrence Berkeley National Laboratory and Dept. of Mathematics
University of California, Berkeley, CA 94720

Abstract. We show that the geometrical framework, in which color images are considered as surfaces, results in a meaningful operator for enhancing color images. The area functional, or "norm", captures the way we would like the smoothing process to act on the different color channels while exploring the coupling between them. Next, the steepest descent flow associated with the first variation of this functional is shown to be a natural selective smoothing filter for the color case. Here we justify the usage of the area norm and the Beltrami steepest descent flow in the color case. We list the requirements, compare to other recent norms, relate to line element methods in color, and conclude with simulation results.

1 Introduction

In a recent work [15, 4, 3], a geometrical framework for image diffusion was introduced. Minimizing the area of the image surface was claimed to yield a natural filter for color image enhancement. The area norm may serve for intermediate asymptotic analysis in low level vision, that is referred to as 'scale space' in the computer vision community [9]. The norm may be coupled with variance constraints that are implemented via projection methods that were used for convergence based denoising [10] for image processing. Another popular option is to combine the norm with lower dimensional measures to create variational segmentation procedures, like the Mumford-Shah [7]. In this note we justify the usage of the area norm obtained by the geometric framework and the Beltrami flow as its natural scale-space.

We limit our comparisons to variational methods in non linear scale space image processing, and to Euclidean color space. Given other significant group of transformations in color, one could design the invariant flow with respect to that group based on the philosophy of images as surfaces in the hybrid space (x, y, R, G, B) through an arclength definition.

The structure of this paper goes as follows: In Section 2 we briefly review the geometric framework and the Beltrami flow and explore its relation to line element theory in color. Section 3 lists the coupling requirements for the color

---

[*] This work is supported in part by the Applied Mathematics Subprogram of the Office of Energy Research under DE-AC03-76SFOOO98, and ONR grant under NOOO14-96-1-0381.

case. A simple 'color image formation' model defines a 'natural' order of events for color image enhancement. It is shown that this sequence of events is captured by the area norm, We conclude with some experimental results and a short comparison review on previous norms in color.

2 The Geometric Framework and Beltrami Flow

Recently [15, 4, 3], a new geometric framework for image processing was introduced. For the gray level case, based on the geometry of the image and its interpretation as a surface, the geometric framework finds a seamless link between the L_1 ($\int |\nabla I|$) and the L_2 ($\int |\nabla I|^2$) norms that are often used in image processing.

According to the geometric framework, images are considered as surfaces rather than functions. The area of the image surface minimized in a special way yields filters for texture, volume, movie, and color image enhancement;

Usually, a color image is considered as 3 images Red, Green, and Blue, that are composed into one. How should we treat such a composition? To answer this question, we view color images as *embedding maps*, that flow towards *minimal surfaces*. See [22] for a non variational related effort.

At this point we would like to go back more than a hundred years, when physicists started to describe the human color perception as simple geometric space. Helmholtz [18] was the first to define a 'line element' (arclength) in color space. He used a Euclidean R, G, B defined by the arclength $ds^2 = dR^2 + dG^2 + dB^2$. His model failed to represent empirical data of human color perception. Schrödinger [13] fixed Helmholtz model by introducing the arclength

$$ds^2 = \frac{1}{l_R R + l_G G + l_B B} \left(\frac{l_R (dR)^2}{R} + \frac{l_G (dG)^2}{G} + \frac{l_B (dB)^2}{B} \right), \qquad (1)$$

where l_R, l_G, l_B are constants. Schrödinger's model was later found to be inconsistent with findings on threshold data of color discrimination.

If we summarize the existing models for color space, we have two main cases: 1. The *inductive* line elements that derive the arclength by simple assumptions on the visual response mechanisms. For example, we can assume that the color space can be simplified and represented as a Riemannian space with zero Gaussian curvature, e.g. Helmholtz [18] or Stiles [17, 21] models. Another possibility for inductive line elements is to consider color arclengths like Schrödinger or Vos-Walraven [19]. These models define color spaces with non zero curvature ('effective' arclength). 2. The *empirical* line elements, in which the metric coefficients are determined to fit empirical data. Some of these models describe a Euclidean space like the CIELAB (CIE 1976 ($L^*a^*b^*$)) [21], recently used in [12]. Others, like MacAdam [5, 6], are based on an effective arclength.

The geometric framework is not limited to zero curvature spaces, and can incorporate any inductive or empirical color line element. See for example [16].

In case we want to perform any meaningful processing operation on a given image, we need to define a spatial relation between the points in the image plane **x**. As a first step define the image plane to be Euclidean, which is a straightforward assumption for 2D images, that is: $ds_{\mathbf{x}}^2 = dx^2 + dy^2$.

In order to construct a valuable geometric measure for color images we need to combine the spatial and color measures. The simplest combination of this hybrid spatial-color space is given by:

$$ds^2 = ds_{\mathbf{x}}^2 + \beta^2 ds_c^2. \tag{2}$$

For a large β it defines the natural regularization of the color space.

Given the above arclength for color images, we pose the following question: How should a given image be simplified? In other words: What is the measure/norm/functional that is meaningful? What kind of variational method should be applied in this case?

The next geometrical measure after arclength is area. Minimization of area is a well known and studied physical phenomena. Once the area is defined as a meaningful measure, one still needs to determine the parametrization for the steepest decent flow. The geometric flow for area minimization, that preserves edges the most is given by the Beltrami flow.

Let x and y be the *spatial* coordinates and the intensity R, G, B the *feature* coordinates, and describe color images as 2D surfaces in the 5D (x, y, R, G, B) space. The arclength is given by

$$ds^2 = dx^2 + dy^2 + dR^2 + dG^2 + dB^2. \tag{3}$$

Next, we *pull back* the image surface *induced metric* from the arclength definition. By applying the chain rule $dR = R_x dx + R_y dy$, and rearranging terms, we obtain a distance measure on the surface defined via

$$ds^2 = g_{11}dx^2 + 2g_{12}dxdy + g_{22}dy^2,$$

where $g_{\mu\nu} = \delta_{\mu\nu} + \sum_i I_\mu^i I_\nu^i$ are the induced metric coefficients, $i \in \{1, 2, 3\}$ indicates the different color channels: $I^1 = R$, $I^2 = G$ and $I^3 = B$.

We plug the induced metric $g_{\mu\nu}$ into an action known as 'Polyakov action' [8] which is a general form for measuring area: Denote by (Σ, g) the image manifold and its metric and by (M, h) the space-feature manifold and its metric, then the map $\mathbf{X} : \Sigma \to M$ has the following weight

$$S[X^i, g_{\mu\nu}, h_{ij}] = \int d^m \sigma \sqrt{g} g^{\mu\nu} \partial_\mu X^i \partial_\nu X^j h_{ij}(\mathbf{X}), \tag{4}$$

where m is the dimension of Σ, g is the determinant of the image metric, $g^{\mu\nu}$ is the inverse of the image metric, the range of indices is $\mu, \nu = 1, \ldots, \dim \Sigma$, and $i, j = 1, \ldots, \dim M$, and h_{ij} is the metric of the embedding space. For more details see [15].

The minimization of the area action yields the steepest decent direction. If we vary with respect to the feature coordinate (fixing the x and y coordinates), we obtain the area minimization direction given by applying the *second order differential operator of Beltrami* on the feature coordinates. Filtering the image based on this result, yields an efficient geometric flow for smoothing the image while preserving the edges. It is written as

$$\mathbf{I}_t = \Delta_g \mathbf{I}, \tag{5}$$

where, for color $\mathbf{I} = (R, G, B)$. Beltrami operator, denoted by Δ_g, that is acting on $\mathbf{I}$ is a generalization of the Laplacian from flat spaces. It is defined by

$$\Delta_g \mathbf{I} \equiv \frac{1}{\sqrt{g}} \partial_\mu (\sqrt{g} g^{\mu\nu} \partial_\nu \mathbf{I}). \tag{6}$$

For the color $2D$ surfaces in $5D$, the flow is given by

$$I^i_t = \frac{1}{g}\left(p^i_x + q^i_y\right) - \frac{1}{2g^2}\left(g_x p^i + g_y q^i\right) \tag{7}$$

where $g_x = \partial_x g$ ($g_y = \partial_y g$), $g_{\mu\nu} = \delta_{\mu\nu} + \sum_i I^i_\mu I^i_\nu$, $g = g_{11} g_{22} - g_{12}^2$, and

$$p^i = g_{22} I^i_x - g_{12} I^i_y, \qquad \text{and} \qquad q^i = -g_{12} I^i_x + g_{11} I^i_y. \tag{8}$$

For the gray level case, the above evolution equation is the mean curvature flow of the image surface divided by the induced metric $g = \det(g_{\mu\nu})$. It is the evolution via the $\mathbf{I}$ components of the mean curvature vector $\mathbf{H}$. I.e. for the surface $(\mathbf{x}(\sigma_1, \sigma_2), \mathbf{I}(\sigma_1, \sigma_2))$ in the Euclidean space $(\mathbf{x}, \mathbf{I})$, the curvature vector is given by $\mathbf{H} = \Delta_g(\mathbf{x}(\sigma_1, \sigma_2), \mathbf{I}(\sigma_1, \sigma_2))$. If we identify $\mathbf{x}$ with σ then $\Delta_g I^i(\mathbf{x}) = \mathbf{H} \cdot \hat{I}^i$. Where, this direct computation applies for co-dimensions > 1. The determinant of the induced metric matrix $g = \det(g_{ij})$ may be considered as a generalized form of an edge indicator. Therefore, the flow (5) is a selective smoothing mechanism that preserves edges and can be generalized to any dimension. In [15, 3], methods for constraining the evolution and the construction of convergent schemes based on the knowledge of the noise variance, were reported.

For the Euclidean color case, the norm is $\int \sqrt{g}$. Here g is the determinant of the metric matrix $g = \det(g_{ij}) = g_{11}g_{22} - g_{12}^2$ given by its components $g_{\mu\nu} = \delta_{\mu\nu} + \sum_i I^i_\mu I^i_\nu$. If we multiply the intensities by a constant β, this action functional is given explicitly by

$$S = \int \sqrt{1 + \beta^2 \sum_i |\nabla I^i|^2 + \beta^4 \frac{1}{2} \sum_{ij} (\nabla I^i, \nabla I^j)^2} \, dx dy. \tag{9}$$

where $(\nabla R, \nabla G) \equiv R_x G_y - R_y G_x$ is the magnitude of the cross product of the vectors ∇R and ∇G. The action in Eq. (9) is the area of the image as a surface.

The steepest descent flow for this functional depends on the scalar β. For $\beta \gg \sup_{i,\mathbf{x}} |\nabla I^i|$ it practically means mapping the intensity values that usually range between 0 and 255 to, let say, [0, 1000]. Roughly speaking, for this limit of β, the order of events along the scale of the flow is as follows: First the different colors align together, then starts the selective smoothing geometric flow (similar to the single channel TV-L_1). On the other limit, where $\beta \ll \sup_{i,\mathbf{x}} |\nabla I^i|$, the smoothing tends to occur uniformly as a multi channel heat equation (L_2).

3 From Color Model to Coupling Requirements

Let us elaborate on the selection of area as a proper measure for color images. The question we try to answer is how should we link between the different spectral channels. Let us assume that each color is 'equally important' and thus the measure we define should be symmetric. Within the scale space philosophy, we want the different spectral channels to get smoother in scale. This requirement leads to the minimization of the different color channels gradient magnitudes combined in one why or another.

An important demand for color image processing is the alignment requirement of the different color channels. That is, we want the color channels to align together as they become smoother in scale. Figure 1 shows one level set of the Red and Green colors and their corresponding gradient vectors at one point along the level set. The requirement that the color channels align together as they evolve, amounts to minimizing the cross products between their gradient vectors.

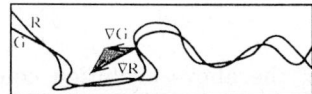

Fig. 1. The cross product between ∇R and ∇G, $\frac{(\nabla G, \nabla R)}{2}$ displayed as the area of the gray triangle, measures the alignment between them.

A simplified color image formation model is a result of viewing Lambertian surface patches (not necessarily flat). Such a scene is a generalization of a 'Mondriaan world'. Each channel is considered as the projection of the real 3D world surface normal $\hat{\mathbf{N}}(\mathbf{x})$ onto the light source direction $\mathbf{l}$, multiplied by the albedo $\rho(x, y)$. The albedo captures the characteristics of the 3D object's material, and is different for each spectral channel. The 3 color channels may then be written as

$$I^i(\mathbf{x}) = \rho_i(\mathbf{x})\hat{\mathbf{N}}(\mathbf{x}) \cdot \mathbf{l}. \tag{10}$$

Which means that the different colors capture the change in material via the albedo that multiplies the normalized shading image $\tilde{I}(\mathbf{x}) = \hat{\mathbf{N}}(\mathbf{x}) \cdot \mathbf{l}$. Let us also assume that the material, and therefore the albedo, are the same within a given object in the image, e.g. $\rho_i(\mathbf{x}) = c_i$, where c_i is a given constant. The intensity gradient for each channel within a given object is then given by

$$\nabla I^i(\mathbf{x}) = \tilde{I}(\mathbf{x})\nabla \rho_i(\mathbf{x}) + \rho_i(\mathbf{x})\nabla \tilde{I}(\mathbf{x})$$
$$= \tilde{I}(\mathbf{x})\nabla c_i + c_i \nabla \tilde{I}(\mathbf{x}) = c_i \nabla \tilde{I}(\mathbf{x}). \tag{11}$$

Under the above assumptions, all color channels should have the same gradient direction within a given object. Moreover, the gradient direction should be orthogonal to the boundary for each color, since both the normalized shading image $\tilde{I}$ and the albedo ρ_i change across the boundaries.

As a conclusion, the first step in color processing is the alignment of the colors so that their gradients agree. Only next comes the diffusion of all the colors simultaneously. For a large enough β, Eq. (9) follows exactly these requirements and the area norm is a regularization form of $\int \sqrt{\sum_i |\nabla I^i|^2 + \beta^2 \sum_{ij}(\nabla I^i, \nabla I^j)^2} dxdy$, that captures the order of events described above. For an even larger β it can be considered as a regularization of the *affine invariant* norm $\int \sqrt{\sum_{ij}(\nabla I^i, \nabla I^j)^2} dxdy$. If we also add the demand that edges should be preserved and search for the simplest geometric parametrization for the flow, we end up with the Beltrami flow as a natural selection.

Experimental Results: Figure 3 shows 3 snapshots from the Beltrami scale space in color. Next, the flow is used to selectively smooth the JPEG compression distortions in Figure 2. Observe how the color perturbation are smoothed: The cross correlation between the colors holds the edges while selectively smoothing the non correlated data.

Previous norms for color Images: Let us review the previous norms suggested for color processing. We start with two non-variational methods that will lead us to the variational norms: Chambolle [2], suggested a flow by the second derivative in the direction of minimal change with respect to the spectral channel with the largest gradient. Sapiro and Ringach [12] considered a different evolution, see [20] for a previous related effort. They used the eigenvalues of the matrix (though not a metric!) $g_{\mu\nu} = \sum_i I^i_\mu I^i_\nu$ as a generalized edge detector to preserve edges.

These eigenvalues may be written as

$$\lambda_\pm = \sum_i |\nabla I^i|^2 \pm \sqrt{\sum_i |\nabla I^i|^4 + \sum_{ij, i \neq j} (\nabla I^i \cdot \nabla I^j)^2 - \sum_{ij}(\nabla I^i, \nabla I^j)^2}. \quad (12)$$

Observe that the square root includes cross (vector products) and gradient magnitude in different signs. We have shown that this combination is not natural for color image processing.

In [11] Sapiro suggested to consider the variational method of the general form $\int f(\lambda_-, \lambda_+)$. As we have just argued, the terms that appear in the square root results in a 'weekly coupled definition' for the arclength in color space. This observation was made from a different perspective by Blomgren and Chan in [1]. They also claimed that from the class of all possible norms of the form $f(\lambda_+, \lambda_-)$, the $f(\lambda_+ + \lambda_-)$ is the most natural one. This brings us to Shah's multi channel model [14], that is based on the norm $\int \sqrt{\sum_{i=1} |\nabla I^i|^2}$ as part of a generalized Mumford-Shah functional.

Blomgren and Chan [1] defined a different color TV norm:

$\text{TV}_m = \sqrt{\sum_{i=1}^m \left(\int |\nabla I^i|\right)^2}$, with a constraint. In this case the coupling between the colors is only by the constraint. Actually, without the constraint the minimization yields a channel by channel curvature flow.

In order to preserve the edge and resolve color fluctuations one needs to use the cross alignment within the definition of the norm. While non of the previous norms included the cross-alignment terms in a proper way, the geometric framework of images as surfaces lead us to the norm that resolves the twist (torsion) between the channels via the cross-alignment term. We have thereby shown that the geometric framework yields a natural norm with respect to all previous existing norms, and with respect to a list of objective requirements and considerations of color image formation.[2]

References

1. P Blomgren and T F Chan. Color TV: Total variation methods for restoration of vector valued images. cam TR, UCLA, 1996.
2. A Chambolle. Partial differential equations and image processing. In *Proceedings IEEE ICIP*, Austin, Texas, November 1994.
3. R Kimmel, N Sochen, and R Malladi. From high energy physics to low level vision. In *Lecture Notes In Computer Science: First International Conference on Scale-Space Theory in Computer Vision*, volume 1252, pages 236–247. Springer-Verlag, 1997.
4. R Kimmel, N Sochen, and R Malladi. Images as embedding maps and minimal surfaces: Movies, color, and volumetric medical images. In *Proc. of IEEE CVPR'97*, pages 350–355, Puerto Rico, June 1997.
5. D L MacAdam. Visual sensitivity to color differences in daylight. *J. Opt. Soc. Am.*, 32:247, 1942.
6. D L MacAdam. Specification of small chromaticity differences. *J. Opt. Soc. Am.*, 33:18, 1943.
7. D Mumford and J Shah. Boundary detection by minimizing functionals. In *Proceedings of CVPR, Computer Vision and Pattern Recognition*, San Francisco, 1985.
8. A M Polyakov. *Physics Letters*, 103B:207, 1981.
9. In B M ter Haar Romeny, editor, *Geometric–Driven Diffusion in Computer Vision*. Kluwer Academic Publishers, The Netherlands, 1994.
10. L Rudin, S Osher, and E Fatemi. Nonlinear total variation based noise removal algorithms. *Physica D*, 60:259–268, 1992.
11. G Sapiro. Vector-valued active contours. In *Proceedings IEEE CVPR'96*, pages 680–685, 1996.
12. G Sapiro and D L Ringach. Anisotropic diffusion of multivalued images with applications to color filtering. *IEEE Trans. Image Proc.*, 5:1582–1586, 1996.
13. E Schrödinger. Grundlinien einer theorie der farbenmetrik in tagessehen. *Ann. Physik*, 63:481, 1920.
14. J Shah. Curve evolution and segmentation functionals: Application to color images. In *Proceedings IEEE ICIP'96*, pages 461–464, 1996.
15. N Sochen, R Kimmel, and R Malladi. A general framework for low level vision. *IEEE Trans. on Image Processing*, to appear, 1997.
16. N Sochen and Y Y Zeevi. Using Vos-Walraven line element for Beltrami flow in color images. EE-Technion and TAU HEP report, Technion and Tel-Aviv University, March 1997.
17. W S Stiles. A modified Helmholtz line element in brightness-colour space. *Proc. Phys. Soc. (London)*, 58:41, 1946.
18. H Helmholtz von. *Handbuch der Psychologishen Optik*. Voss, Hamburg, 1896.
19. J J Vos and P L Walraven. An analytical desription of the line element in the zone-fluctuation model of colour vision II. The derivative of the line element. *Vision Research*, 12:1345–1365, 1972.

---

[2] **Acknowledgments:** I thank Dr. Nir Sochen, Dr. Ravi Malladi, and Dr. Sherif Makram-Ebeid for interesting discussions, and Dr. Yacov Hel-Or for the sails benchmark image.

20. J Weickert. Scale-space properties of nonlinear diffusion filtering with diffusion tensor. Report no. 110, laboratory of technomathematics. University of Kaiserslautern, P.O. Box 3049, 67653 Kaiserslautern, Germany, 1994.
21. G Wyszecki and W S Stiles. *Color Science: Concepts and Methods, Qualitative Data and Formulae, (2nd edition)*. Jhon Wiley & Sons, 1982.
22. A. Yezzi. Modified curvature motion for image smoothing and enhancement. *IEEE Trans. IP,* 1997.

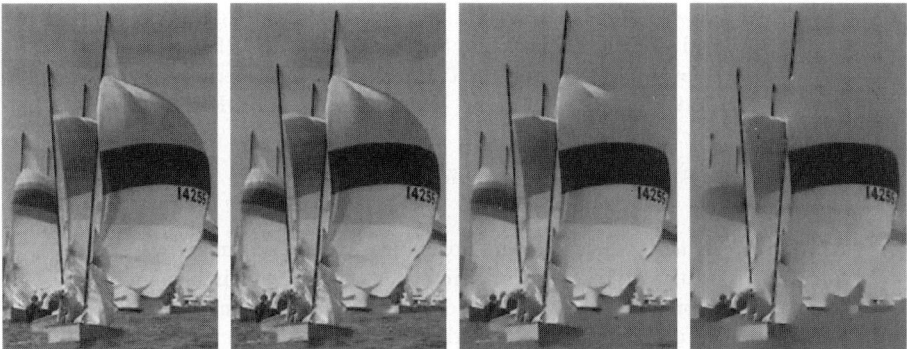

Fig. 2. Three snapshots along the scale space (left most is the original image). [This is a color figure]

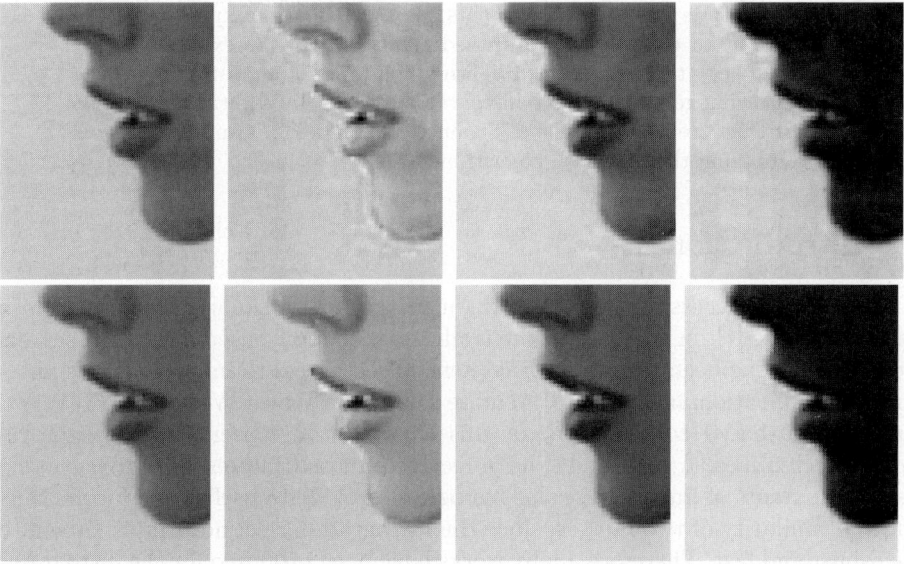

Fig. 3. Denoising JEPG lossy compression perturbations. The original image and the three colors (R,G,B). Before and after the flow. (70 numerical iterations. $\Delta t = 0.21, \Delta x = 1$). [This is a color figure]

A Color Normalization Algorithm for Image Indexing

In Kyu Park[1], Il Dong Yun[2], and Sang Uk Lee[1]

[1] Real Time Vision Laboratory, School of Electrical Engineering
Seoul National University, Seoul, 151-742, KOREA
[2] Dept. of Control and Instrumentation Engineering
Hankuk University of F. S., Yongin, 449-791, KOREA

E-mail: {pik, yun, sanguk}@sting.snu.ac.kr

Abstract. In this paper, a color normalization algorithm is proposed to compensate the difference of illumination between two images, which could be used for pre-processing, *i.e.*, color constancy step in a histogram-based indexing algorithm. Unlike traditional color constancy algorithms, we attempt to transform the query image, so that the lighting condition is adjusted to be same with the reference image. The proposed algorithm assumes the Maloney and Wandel's reflectance model [6], and normalizes the magnitude of color components of input image. Experiments are carried out to evaluate the proposed algorithm. In the experiments, it is shown that the transformed lighting condition is almost same as the reference image in the color histogram domain. In addition, it is also shown that the performance of Swain's color indexing can be enhanced by combining the proposed algorithm.

1 Introduction

Content-based access is a new paradigm in the field of image retrieval from a large database [1]. In contrast to the traditional context-based approach, several advantages have been reported for content-based access to image, such as automatic classification and retrieval, remote search via World Wide Web (WWW), and content-based compression. In this context, the content-based image retrieval techniques [1, 2, 3, 4, 11] have received a great interest recently. Among several contents of image, the color histogram provides a useful clue for measuring the similarity of two images, since the histogram-based indexing is known to be robust and fast. Therefore, many works have been proposed for the histogram-based color image retrieval techniques [2, 3, 11].

However, the histogram can be distorted severely, when the illumination varies or the characteristics of the frame grabber or camera are not time-invariant. Indeed, many of the previous algorithms assume that the input image should be preprocessed with a certain color constancy algorithm [2].

The color constancy is another important issue in color vision, and several algorithms have been already proposed [3, 4, 5, 6]. Funt and Finlayson [3] proposed an algorithm, called color constant color indexing, that recognizes object, by matching the distributions of color ratio. But [3] is observed to be sensitive to the input noise because of the employment of the differentiation. Healey and Slater [5] proposed an efficient algorithm for extracting common invariants between two object with different illumination. In [5], affine algebraic moment invariants are computed for two histogram. Then, the moment-matrix eigenvalues are chosen for the invariants for two object. Despite significant progress in color constancy, automated color constancy algorithms have not yet reported [5]. It is also difficult to predict the inaccuracies, caused by color constancy algorithms, that might affect the color indexing process. Moreover, the color constancy algorithms usually require side information on the sensor or lighting *etc.*

As an alternative to the color constancy algorithm, we propose a color normalization algorithm, which is specially designed for color histogram indexing. In our approach, the input image is transformed to another image, in which the illumination condition is approximately equivalent to the corresponding model. In this stage, brightness distortion, due to the scaling of the object, is also taken into account. It is also worth to note that the proposed algorithm is independent to the translation and rotation of object. Our approach differs from the previous techniques in that it does not estimate the spectral reflectance coefficients and does not require any side information. Since the main application area of the proposed algorithm is image retrieval, we aim to achieve a high successful matching rate. Note that the proposed algorithm can be applicable to any histogram indexing techniques.

This paper is organized as follows. In Section 2, the proposed algorithm is presented in detail. Section 3 provides the experimental results of color normalization and its application to the image indexing. Thereafter, Section 4 draws conclusions from our work.

2 The Proposed Algorithms

In this section, the proposed algorithm is described in detail. First, we introduce the color reflectance model, then describe the color normalization algorithm. Finally, it is shown that the proposed algorithm is very relevant to the color indexing.

2.1 Color Reflectance Model

First, let us introduce the Maloney and Wandel's color reflectance model [6]. The input to the kth sensor on the position (x, y) is determined by a linear combination of ambient light L, its albedo $S^{(x,y)}$ and the input response of the kth sensor R_k, given by

$$\rho_k^{(x,y)} = \int_\lambda L(\lambda) S^{(x,y)}(\lambda) R_k(\lambda) d\lambda \tag{1}$$

where λ is the spectral wave length. In our approach, it is assumed that the variation in the ambient light in the spatial domain is negligible. As in [3], we also assume the narrow-band sensitivity of the input sensor, so that (1) can be approximated to the coefficient model, yielding

$$\rho_k^{(x,y)} = L(\lambda_s)S^{(x,y)}(\lambda_s)R_k(\lambda_s) \qquad (2)$$

where λ_s is the dominant wavelength of the input characteristics of the sensor.

2.2 Normalization Algorithm

In general, the purpose of the color constancy is to estimate the spectral reflectance coefficients of each point of the object [6]. Thus, if ambient light is known, then the actual appearance of the object in real world can be modeled. However, it would be very difficult to obtain the information on the illumination condition and the input characteristics of color sensors to acquire the input image. Even if that is possible, a real time implementation of the color constancy might be very difficult. Instead of direct estimation of the spectral reflectance coefficient, in our approach, the illumination condition of the input image is transformed to that of the reference image.

In order to examine the effects of sensor characteristics and ambient light, the spatial averaging is applied to the color reflectance model (2), yielding

$$E\{\rho_k^{(x,y)}\} = E\{L(\lambda_s)S^{(x,y)}(\lambda_s)R_k(\lambda_s)\} \qquad (3)$$

where $E\{\cdot\}$ denotes the averaging over (x,y). Since we assume L and R_k are independent of position, we can obtain

$$E\{\rho_k^{(x,y)}\} = L(\lambda_s)R_k(\lambda_s)E\{S^{(x,y)}(\lambda_s)\}$$
$$\bar{\rho}_k = a_k\bar{S}(\lambda_s). \qquad (4)$$

From (4) it is found that the spatial averaging of the kth input does not depend on the ambient light and the input sensor characteristics, but only on the albedo. Based on this observation, by adjusting the mean of each sensor input, the lighting condition can be changed, although it is not known exactly. Thus, attempts are made to normalize the pixel value using the means of input and reference image.

Let $^i\mu_k$ and $^r\mu_k$ be the means of the input and reference image, respectively. Then, the normalized magnitude, $^i\tilde{\rho}_k^{(x,y)}$, is obtained by

$$^i\tilde{\rho}_k^{(x,y)} = \left(\frac{^r\mu_k}{^i\mu_k}\right){^i\rho_k^{(x,y)}}. \qquad (5)$$

As a result, the mean of $^i\tilde{\rho}_k^{(x,y)}$ is almost equal to that of the reference image. Actually, by (4), the ratio of two means is

$$\frac{^r\mu_k}{^i\mu_k} \simeq \frac{L^r(\lambda_s)R_k^r(\lambda_s)}{L^i(\lambda_s)R_k^i(\lambda_s)} \qquad (6)$$

where $L(\lambda_s)$ and $R_k(\lambda_s)$ are the magnitude of the ambient light and the input response of the kth sensor, respectively. Thus, the normalized image (5) is proved to be

$$i\tilde{\rho}_k^{(x,y)} \simeq \left(\frac{L^r(\lambda_s)R_k^r(\lambda_s)}{L^i(\lambda_s)R_k^i(\lambda_s)}\right) L^i(\lambda_s)S^{(x,y)}(\lambda_s)R_k(\lambda_s) \qquad (7)$$

$$= L^r(\lambda_s)S^{(x,y)}(\lambda_s)R_k^r(\lambda_s). \qquad (8)$$

It is shown in (7) that the ambient light and sensor input response input image is transformed to those of reference image. Note that the ambient light is not considered as a point light source, that is to say, it is assumed to be uniform over the view plane of the image.

2.3 Relevance to the Color Histogram Indexing

In the histogram-based image indexing, the most similar image is retrieved from the database by applying some similarity metric [2, 11]. However, if the illumination in the input image is quite different from the model to be retrieved, the retrieval result would not be convincing. To avoid this kind of risk, the input is preprocessed using the normalization technique. That is, during the sequential process of measuring the similarity between the input and the selected model, the input is transformed before applying metric. The overall procedure is shown in Figure 1. Consequently, the input could be now considered to have the same illumination with the model. Although it seems to be simple, this procedure reduces the rate of the retrieval failure due to the illumination difference significantly.

Also, since the proposed color normalization algorithm is independent of the rotation, translation and scaling of the object, *i.e.*, it does not affect the desirable features of the histogram indexing technique, it could be used with any histogram-based color indexing techniques.

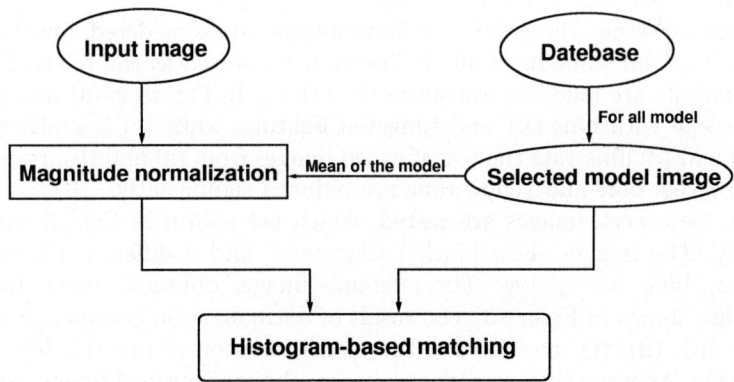

Fig. 1. The overview of color normalized image indexing

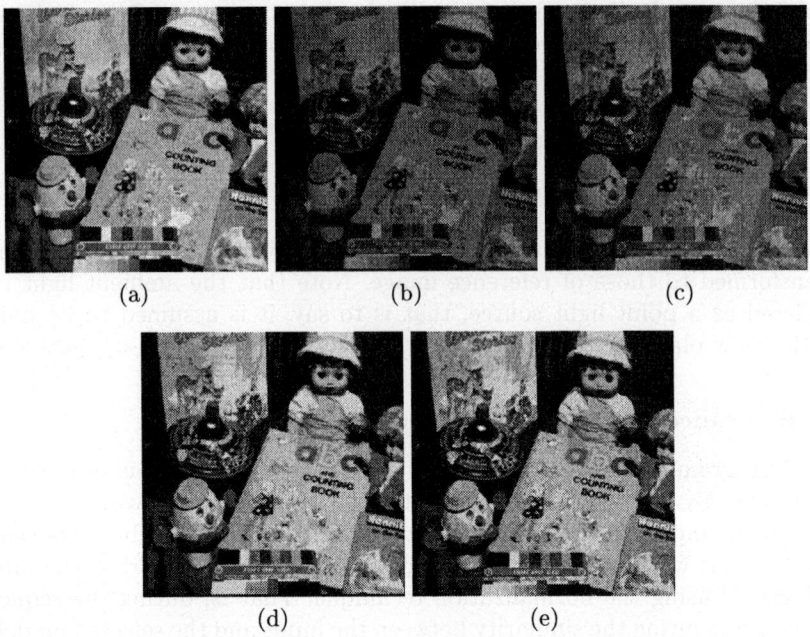

Fig. 2. Results of toy scene (a) Reference image under sunlight (b) Image under blue sky (c) Image under tungsten light (d) Normalized image of (b) (e) Normalized image of (c)

3 Experimental Results

In this section, we present several experimental results to demonstrate the performance of the proposed algorithm. Experiments are carried out on 2 sets of images: toy scene and Newsweek images.

For the toy scene, three different illuminations are considered: blue sky, tungsten lighting, and sunlight. Sunlight image is employed as the reference, thereafter, attempts are made to transform the others. In Figure 2, (a) and (b) show the toy scene with blue sky and tungsten lighting, while (c) is under sunlight. Figs. (d) and (e) illustrate the transformed images from (a) and (b), respectively. As is observed, blue and yellow tone are reduced significantly.

Next, Newsweek images are tested, which are shown in Figs. 3 and 4, respectively. The images show black background, and 4 different illuminations: red, green, blue, and yellow. The reference images obtained under fluorescent lighting are shown in Figure 3. The result of normalization is shown in Figure 4, in which (b), (d), (f), and (h) are the transformation of (a), (c), (e), and (g), respectively. As was the case with toy scene, the transformed images appear to be almost same as the model images.

As was described previously, the proposed algorithm is relevantly applicable to image indexing based on color histogram. In order to validate the performance, a small database is constructed with 100 Newsweek images. For the query image, the models are distorted with rotation, translation, scaling, and changing illumination as shown in Figure 5. Using histogram intersection method [2], all, except one, are retrieved perfectly. The result is shown in Table 1, in which it is shown that the performance is enhanced significantly after the normalization process.

To evaluate the retrieval performance for a more large database, Virage database was tested. There are about 1000 images in the database, and the retrieval results are compared depending on that the normalization is applied or not. Some of the results are shown in Figure 6. It is shown that the retrieval performance is improved better in terms of the rank and similarity index on the window title bar.

Fig. 3. Model images under fluorescent lighting

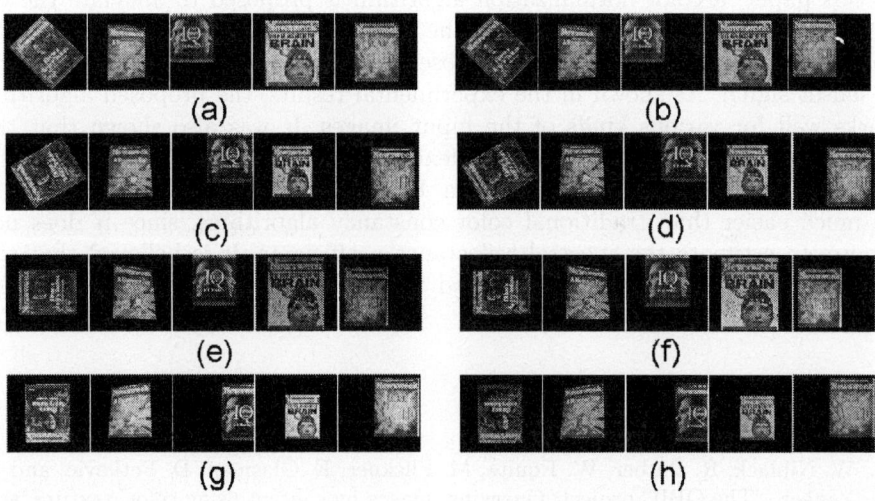

Fig. 4. Color normalization is applied to Newsweek images with several illuminations (a)(c)(e)(g) before normalization (b)(d)(f)(h) after normalization

Table 1. Matching results for Fig. 5 (18 inputs, 100 models)

Normalization	Rank 1	Rank 2	Rank 3	≥ Rank 4	Ratio
With	17	1	0	0	0.9444
Without	2	3	0	13	0.1111

Fig. 5. Query images with motion, scaling, and change of lighting condition

4 Conclusions

In this paper, a color normalization algorithm is proposed to alleviate the effect of variation in illumination. On the assumption that the ambient light is uniform on the planar object, the proposed algorithm normalizes the magnitude of sensor signal. As shown in the experimental results, the proposed algorithm works well for various kinds of the input images. It was also shown that the retrieval performance of histogram indexing is much enhanced by applying the proposed color normalization algorithm. In addition, the implementation would be much easier than traditional color constancy algorithms, since it does not require to estimate the spectral reflectance coefficients. It is believed that the proposed algorithm could be also used very efficiently as a preprocessing for other color-related algorithms [9, 10].

References

1. W. Niblack, R. Berber, W. Equitz, M. Flickner, E. Glasman, D. Petkovic, and P. Yanker, "The QBIC project: Querying images by content using color, texture, and shape," *SPIE 1908, Storage and Retrieval for Images and Video Dbases*, Feb. 1993
2. M. J. Swain and D. H. Ballard, "Color indexing," *International Journal of Computer Vision,* vol. 7, no. 1, pp. 11-32, Nov. 1991.
3. B. Funt and G. Finlayson, "Color constant color indexing," *IEEE Trans. on Pattern Analysis and Machine Intelligence,* vol. 17, no. 5, pp. 522-529, May 1995.
4. G. Finlayson, M. Drew, and B. Funt, "Spectral sharpening: sensor transformations for improved color constancy," *Journal of Optical Society of America A,* vol. 11, no. 5, pp. 1553-1563, May 1994.

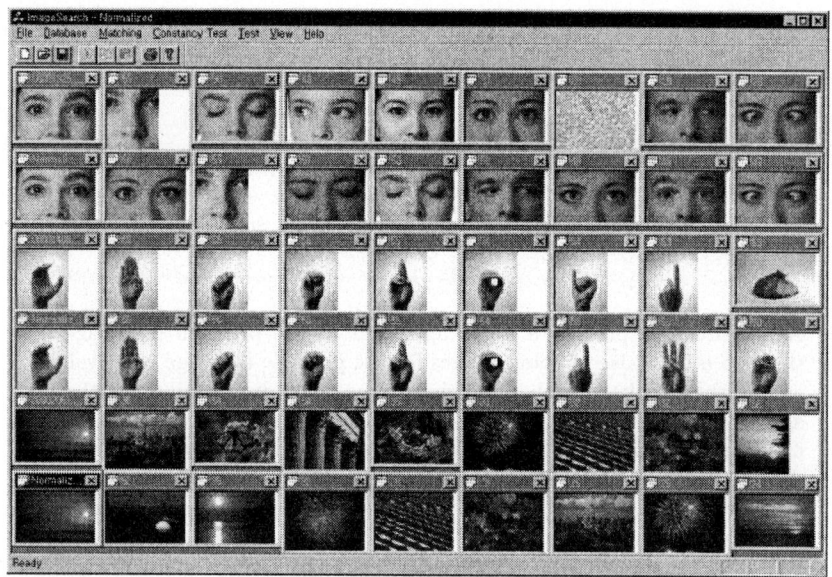

Fig. 6. Test on Virage database: Some of the results without normalization

5. G. Healey and D. Slater, "Global color constancy: recognition of objects by use of illumination-invariant properties of color distributions," *Journal of Optical Society of America A*, vol. 11, no. 11, pp 3003-3010, Nov. 1994.
6. L. T. Maloney and B. A. Wandell, "Color constancy: a method for recovering surface spectral reflectance," *Journal of Optical Society of America A*, vol. 3, no. 1, pp 29-33, Jan. 1986.
7. B. A. Wandell, "The synthesis and analysis of color images," *IEEE Trans. on Pattern Analysis and Machine Intelligence*, vol. 9, no. 1, pp. 2-13, Jan. 1987.
8. D. Slater and G. Healey, "The illumination-invariant recognition of 3D objects using local color invariants," *IEEE Trans. on Pattern Analysis and Machine Intelligence*, vol. 18, no. 2, pp206-210, Feb. 1996.
9. G. Healey and D. Slater, "Computing Illumination-invariant Descriptors of Spatially Filtered Color Image Regions," *IEEE Tran. on Image Processing*, vol. 6, no. 7, pp. 1002-1013, July 1997.
10. D. J. Jobson, Z. Rahman, and G. A. Woodell, "A Multiscale Retinex for Bridging the Gab Between Color Images and the Human Observation of Scenes," *IEEE Tran. on Image Processing*, vol. 6, no. 7, pp. 965-976, July 1997.
11. I. K. Park, I. D. Yun, and S. U. Lee, "Models and algorithms for efficient color image indexing," *Proceedings of IEEE Workshop on Content-based Access of Image and Video Libraries*, pp. 36-41, San Juan, PR, June 1997.
12. W. K. Pratt, *Digital Image Processing*, John Wiley & Sons, 1991.

Adaptive Color-Image Embeddings for Database Navigation *

Yossi Rubner, Carlo Tomasi, and Leonidas J. Guibas

Computer Science Department, Stanford University, Stanford, CA 94305, USA

Abstract. We present a novel approach to the problem of navigating through a database of color images for the purpose of image retrieval. We endow the database with a metric for the color distributions of the images. We then use multi-dimensional scaling techniques to embed a group of images as points in a two-dimensional Euclidean space so that their distances reflect image dissimilarities as well as possible. Such geometric embeddings allow the user to perceive the dominant axes of variation in the displayed image group, and form a mental picture of the database contents. Furthermore, since these embeddings group similar images together, away from dissimilar ones, the user can refine the query in a perceptually intuitive way. By iterating this process, the user can quickly navigate to the portion of the image space of interest.

1 Introduction

The user of an image retrieval system would typically like to specify queries in semantic terms (e.g. "children playing in a park"). Unfortunately, the state-of-art in computer vision does not yet allow for such queries. Instead, systems use simpler syntactic image features such as color, texture and shape [2, 1, 3, 6, 7], in the hope that these correlate well with semantic features. This discrepancy between syntactic and semantic queries causes a basic problem with the traditional query/response style of interaction. An overly generic query yields a large jumble of images, which are hard to examine, while an excessively specific query may cause many good images to be overlooked by the system. This is the traditional trade-off between good precision (few false positives) and good recall (few false negatives). Striving for both good precision and good recall may pose an excessive burden on the definition of a "correct" measure of image similarity. While most image retrieval systems, including the ones above, recognize this and allow for an iterative refinement of queries, the number of images returned for each query is usually kept low so that the user can examine them one at a time.

In contrast, we propose that with an appropriate display technique, which is the main point of this paper, many more images can be returned without overloading the user's attention. Specifically, if images can be arranged on the screen so as to reflect similarities and differences between their color distributions, the initial queries can be very generic, and return a large number of images. The

---

* Research supported by grants DARPA DAAH04-94-G-0284 and NSF IRI-9712833.

consequent low initial precision is an advantage rather than a weakness. In fact, the user can see large portions of the database at a glance, and form a global mental model of what is in it. Rather than following a thin path of images from query to query, as in the traditional approach, the user now *zooms in* to the images of interest. Precision is added incrementally in subsequent query refinements, and fewer and fewer images are displayed as the desired images are approached.

In our system, we use the distributions of colors in images as our retrieval features. These have been shown [12, 2, 11, 1, 3, 6, 7] to be useful retrieval cues. When a (usually vague) query is specified or drawn by the user, we locate and display a large number of neighboring images in the database. Since queries in our system are image-like, neighborhood can be defined in terms of the distance between images. The resulting images are then used for more focused queries that return fewer and fewer images. At every step, query results are embedded in two-dimensional space by using *multi-dimensional scaling (MDS)* [10, 4], by which we place picture thumbnails on the screen so that screen distances reflect as closely as possible the distances between the images. While more traditional displays list images in order of similarity to the query, thereby representing n distances if n images are returned, our display conveys information about all the $\binom{n}{2}$ distances between images. As shown by the examples in this paper, this display makes it easy for the user to grasp the entire set of returned images at a glance, understand how the query actually performed, and decide where to go next. In fact, such geometric embeddings allow the user to perceive the dominant axes of variation in the displayed image group. When the user selects a region of interest on the display, a new, more specific query is automatically generated, and returns a smaller set of images. These are again displayed by a new MDS, which now reflects the new dominant axes of variation. Thus, the embeddings are *adaptive*, in the sense that they use the screen's real estate to emphasize whatever happen to be the main differences and similarities among the particular images at hand. By iterating this process, the user is able to quickly navigate to the portion of the image space of interest, typically in very few mouse clicks.

In the next section, we introduce the data structures we use to summarize the color content of images, and a distance measure between them. Section 3 then describes the visualization technique, and Section 4 shows its use for database navigation. Section 5 argues that MDS image embeddings can be usefully applied to other modalities besides color and discusses topics for future work.

2 A Metric for Color Images

This section describes the color signature as our basic representation for color distributions, and introduces the earth mover's distance as a measure of distance between signatures. More details on these concepts can be found in [8].

2.1 Color Signatures

The color information of each image is reduced to a compact representation that we call the *color signature* of the image. A color signature contains a varying number of points, each representing a cluster of similar colors in the CIE-Lab color space [14]. The number of points in a signature varies with the color complexity of the image. A weight describing the fraction of the image area with that color is attached to each point.

To compute the signature of a color image, we first slightly smooth each band of the image's RGB representation in order to reduce possible color quantization and dithering artifacts. We then transform the image into the CIE-Lab color space. This nonlinear transformation deforms the RGB color space so that Euclidean distance between nearby color coordinates approximates how well colors are discriminated by humans. We coalesce the three-dimensional CIE-Lab distribution of colors in the image into clusters by the algorithm described in [8]. In our database[1], a signature contains eight color clusters on average.

2.2 Distance Between Color Signatures

In order to define a similarity measure between two color signatures, we introduce the notion of the *Earth Mover's Distance (EMD)*. This is the minimal amount of 'work' needed to transform one signature into the other, in the following sense. The work needed to move a point, or a fraction of a point, to a new location is the portion of the weight being moved, multiplied by the Euclidean distance between the two locations. When changing one signature to another, the work is the sum of the work done by moving the weights of the individual points of the source signature to those of the destination signature. We allow the weight of a single source signature point to be partitioned among several destination signature points, and vice versa. Although we do not claim that the EMD is a perceptual distance, it is an extension of distances of single colors in the CIE-Lab color space, which are perceptual distances, to distances between sets of colors. In practice, as we show in the following sections, the EMD leads to good results. It also allows for partial matches like "give me images with 20% orange and I don't care about the remaining 80%". The ability to use partial matches is important for navigation as we shall see in section 4. See [8] for more details about the EMD properties, and for efficient algorithms for its computation.

3 Database Visualization

While the EMD is indeed at the core of our image retrieval system, and has proven very effective, in this paper we want to emphasize a related but distinct use of this metric. Once image retrieval systems find the best matches for a given query, they usually display them in a list, sorted by their similarity to the query. While this might suffice if the image we are after is in that list, this is usually not the case, especially when we have only a vague idea of what we are looking for. At this point it is desirable to display a coherent view of the query results where the returned

---

[1] Our database contains a rather diverse set of 20,000 color images from the Corel Stock Photo Library.

images should be displayed not only in order of their distance from the query, but also arranged according to their mutual distances. With such view, the user can see the relations between the images, better understand how the query performed, and be guided to successive queries. How can such a global picture of part of an image database be created? Our EMD approximates the perceptual difference that separates two signatures. Consequently, each signature can be represented by a single point in a suitably high-dimensional space, such that distances between these points are equal to the EMDs between the corresponding signatures. The computation of the coordinates of these high-dimensional points is called an *embedding*. However, humans can only visualize low-dimensional spaces, typically in two or three dimensions. We then look for an approximate embedding, rather than an exact one, in two dimensions. Such approximate embeddings are an instance of the problem of multi-dimensional scaling (MDS).

Given a set of n objects together with the dissimilarities δ_{ij} between them, the MDS technique [10,4] computes a configuration of points $\{p_i\}$ in a low-dimensional Euclidean space $\mathbf{R}^d$, (we use $d = 2$) so that the Euclidean distances $d_{ij} = \|p_i - p_j\|$ between the points in $\mathbf{R}^d$ match as well as possible the original dissimilarities δ_{ij} between the corresponding objects. Kruskal's [4] formulation of this problem requires minimizing the following quantity

$$\text{STRESS} = \left[\frac{\sum_{i,j}(d_{ij} - \delta_{ij})^2}{\sum_{i,j} \delta_{ij}^2} \right]^{1/2}$$

with the additional constraint that the d_{ij}s are in the same rank ordering as the corresponding δ_{ij}s. STRESS is a nonnegative number that indicates how well distances are preserved in the embedding. Zero STRESS indicates a perfect fit. Rigid transformations and reflections can be applied to the MDS result without changing the STRESS. Embedding methods such as SVD and PCA are not appropriate here because our signatures do not form a linear space, and we do not have the actual points, only the non-Euclidean distances between them. In our system we used the ALSCAL MDS program [13].

An example is shown in Figure 1. Suppose that we are looking for images of skiers. These images can be characterized by blue skies and white snow, so we use as our query "find images with 20% blue, 20% white and 60% don't care". Part (a) shows the eight best matches out of 20,000 pictures, sorted by their similarities to the query. Notice that the best match has nothing to do with skiing (although its color signature matches the query well), and that consecutive images in the list can be very different from each other. Part (b) displays the MDS embedding of the best twenty matches where images of skiers are placed on the right: images with more snow at the bottom-right, and images with more sky at the top-right. The bottom-left holds images which contain also green, and the top-left holds images which contain also darker colors. Notice that image thumbnails placed at the coordinates returned by the MDS algorithm might occlude other thumbnails. Up to a point, this is not really a problem since these images are likely to be similar, and are therefore represented well by the topmost thumbnail.

Fig. 1. Looking for ski images; (a) Traditional display. Only the eight best matches are shown; (b) MDS display of the best twenty images (STRESS=0.148); A color version of this figure can be found at http://vision.stanford.edu/~rubner.

4 Navigation

Using the MDS embedding can assist navigation in the space of images, as we now illustrate. In our system the user starts a query by specifying the color content of the requested image. Using "don't care" is encouraged when the user is not confident about certain colors. A large set of images is returned and embedded in 2D space using MDS. Once the user selects an area of interest, an appropriate new query is generated and submitted. Now a smaller number of images is returned by the system. The new set of images is not necessarily a subset of the previous set, so images which were not returned by the previous query can still be retrieved at later stages, as the query becomes more precise. A new MDS is computed with new axes of variation which are based on the new image set.

To this end, in order to generate the new query, our user interface allows the user to draw a circle in the MDS map around images that are to be used to form the next query (In general, it would be better to let the user draw any closed region). The color signatures of these k images are used to generate a color signature that will be used for the next query. We want the new color signature to reflect the common clusters in the k color signatures, and ignore the others by using "don't care". This is done by representing the clusters as points in the CIE-Lab color space, and clustering these points using a similar algorithm to the one used in Section 2.1 to generate the color signatures. We reject clusters that are not represented by points from at least 50% of the k images. Each cluster is assigned a weight which is the median of the weights of the points in that cluster.

An example of navigation is given in Figure 2, where we are looking for images of deserts. The initial generic query was 20% blue for the sky and 80% "don't care". The MDS embedding of 400 returned images is shown in Part (a) of the figure. By glancing at the results, we see immediately the organization of the returned images: images of airplanes are in the bottom-left, diving images are in the top-left, images with green trees are in the mid-right, and so forth. Our desert images are in the bottom-right, so we select them (indicated in the figure by a black circle) and ask for 80 images. In the new MDS map shown in Part (b), most of the images are desert-like, with some buildings on the left. Although we could probably stop here and pick our favorite desert image, one more iteration is shown in Part (c) where we asked for only 20 images. Now all the images are deserts. Notice the cacti at the top-right, and the cougars at the mid-left.

Although MDS embeddings can be computed quickly for small sets of images (about 2 seconds for 80 images, and 0.15 seconds for 20 images on an SGI Indigo 2 with a 250 MHz processor), the computation time grows rapidly as the number of images increases (about a minute for 500 images). This is mostly because of the full distance matrix computation. The MDS technique, however, can tolerate missing data [4], and we are currently investigating ways to decrease computation times by computing only sparse distance matrices.

5 Conclusions

The methods presented in this paper open a novel set of tools and possibilities for image database navigation and visualization. Our color signatures and the EMD between them seem to approximate well the perceptual similarity or dissimilarity of images in terms of their color content. Furthermore, the low-dimensional embeddings we compute by MDS provide an intuitive way for the user to refine a query and to continue exploring interesting neighborhoods of the image space — or to see large portions of it all at once.

The idea of an adaptive image embedding can be applied to other modalities besides color, as long as some notion of similarity, metric or not, continuous or discrete, can be defined. For instance, for texture, shape, eigenimage similarity, or any other image features [1-3, 5-7, 9, 11] . In this context, the key question, which we leave for future work, is then to determine whether the main axes of variation

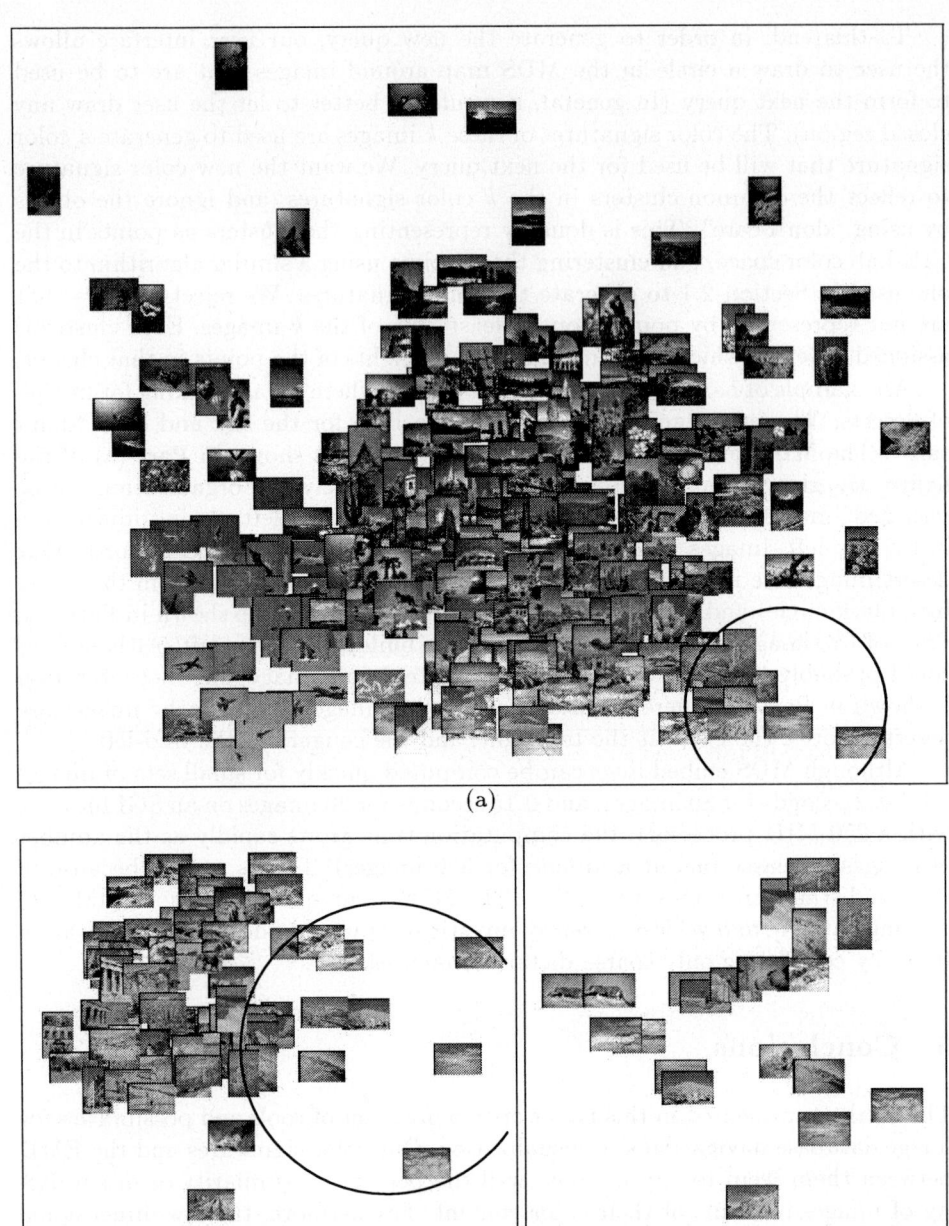

Fig. 2. Looking for a desert landscape; (a) 400 images; (b) 80 images; (c) 20 images. The black circles shows the user selection. A color version of this figure can be found at http://vision.stanford.edu/~rubner.

"discovered" by MDS for each of these distances and for various types of image distributions is perceptually meaningful. We believe that since the MDS groups together similar images — away from dissimilar ones — this is often the case. For instance, in [8] we defined a metric for textures, based on their spectral contents. We used this metric to compute a two-dimensional MDS on texture patches. The results were perceptually meaningful and agreed with psychophysics results. We also plan to study more the relations between the axes chosen by MDS for related or overlapping image sets. Knowing the correspondence between these 'local charts' (in the sense of topology) of the image space may help in providing a globally consistent sense of navigation.

All image query systems are ultimately based on computational approximations to perceptual image distance — approximations whose quality we are often asked to take for granted. Our approach appears to be the first one to allow the user to explore, in an intuitive way, the area of the image space beyond what the system considers the neighborhood of the query. Such an exploration can provide increased confidence that what is wanted will not be missed.

References

1. J. R. Bach, C. Fuller, A. Gupta, A. Hampapur, B. Horowitz, R. Humphrey, R. Jain, and C. Shu. Virage image search engine: an open framework for image management. *SPIE*, 2670:76–87, 1996.
2. C. Faloutsos, R. Barber, M. Flickner, J. Hafner, W. Niblack, D. Petkovic, and W. Equitz. Efficient and effective querying by image content. *Journal of Intelligent Information Systems*, 3:231–262, 1994.
3. D. Forsyth, J. Malik, M. Fleck, H. Greenspan, and T. Leung. Finding pictures of objects in large collections of images. *International Workshop on Object Recognition for Computer Vision*, 1996.
4. J. B. Kruskal. Multi-dimensional scaling by optimizing goodness-of-fit to a nonmetric hypothesis. *Psychometrika*, 29:1–27, 1964.
5. W. Y. Ma and B. S. Manjunath. Texture features and learning similarity. *CVPR*, 425–430, 1996.
6. G. Pass and R. Zabih. Histogram refinement for content-based image retrieval. *IEEE Workshop on Applications of Computer Vision*, 1996.
7. A. Pentland, R. W. Picard, and S. Sclaroff. Photobook: content-based manipulation of image databases. *IJCV*, 18(3):233–254, 1996.
8. Y. Rubner, C. Tomasi, and L. J. Guibas. A metric for distributions with applications to image databases. *IEEE ICCV*, 1998.
9. S. Santini and R. Jain. Similarity queries in image databases. *CVPR*, 646–651, 1996.
10. R. N. Shepard. The analysis of proximities: Multidimensional scaling with an unknown distance function, i and ii. *Psychometrika*, 27:125–140,219–246, 1962.
11. M. Stricker and M. Orengo. Similarity of color images. *SPIE*, 2420:381–392, 1995.
12. M. J. Swain and D. H. Ballard. Color indexing. *IJCV*, 7(1):11–32, 1991.
13. Y. Takane, F. W. Young, and J. Leeuw. Nonmetric individual differences multidimensional scaling: an alternating least squares method with optimal scaling features. *Psychometrika*, 42:7–67, 1977.
14. G. Wyszecki and W. S. Stiles. *Color Science: Concepts and Methods, Quantitative Data and Formulae*. Wiley, 1982.

A Large Capacity Steganography Using Color BMP Images

Koichi Nozaki*, Michiharu Niimi**, Richard O. Eason***
and Eiji Kawaguchi**
*Nagasaki University
**Kyushu Institute of Technology
***University of Maine

Abstract

A new steganography (information hiding technique) is proposed. It uses a color image as the information hiding dummy image, i.e., the container, or carrier of the secret information. This new technique is not based on a programming technique, but is based on a property of human vision system such that human eyes are blind to very complex binary patterns. In other word, human can not see the effect of the data change, even if the "noise-like" portions in the bit-planes of a multi-valued image are all changed to other noise-like patterns. In order to assure this property, we made a replacement experiment of noise-like portions of a color photo with random binary patterns, and it turned out in a surprising result. This human vision property is the key to the large capacity steganography which uses a color image in a BMP file format. This new technique may open a new step to an internet communication age.

1. Introduction

Digital images are getting more and more commonly used in internet communications. Almost all Web pages today are decorated with color images. People enjoy looking at such colorful information on computer. So, sending and receiving color images through internet are not any more specially paid attention to.

In the meantime, security about internet communication becomes a large concern for all network users. For example, e-mail is not reliable to send secret messages without being tapped. Of course, we can encrypt our messages when we send them. Such encryption software is available on a commercial base. However, the problem is, there are some cases when we want to send a mail to a person without being noticed by anyone else. Encrypting a message, even if it is easy, can not solve the problem. Encryption can hide the content of the message, but can not hide the message data itself. Anyone can see the "unreadable message." This does not meet our requirement.

In recent years, several steganographic programs are posted on internet home pages [*]. Most of them use image data for the container (or, carrier) of the secret information. Some of them use the least significant bits of the image data to hide secrets. Other program embeds the secret information in a specific band of the spatial frequency component of the carrier. Some other program makes use of the sampling error in image digitization. However, all those steganographies are insufficient in terms of information hiding capacity. They can embed only 5-15 % of the carrier image at the best. Therefore, current steganography is more oriented to water marking of computer data than to secret human-human communication applications.

We have invented a new technique to hide secret information in a color image. This is not based on a programming technique, but is based on the property of human vision system. Its information hiding capacity is as large as 50% of the original image data. This could open a new step for a steganography toward a secure internet communication age.

Digital images are categorized in either binary (black-and-white) or multi-valued pictures despite their actual color. We can decompose an n-bit image into a set of n binary images by bit-slicing operations[1][2]. Therefore, binary image analysis is essential to all digital image processings. Bit-slicing is not necessarily the best in the Pure-Binary Coding system (PBC), but in some case the Canonical Gray Coding system (CGC) is much better [3].

2. The complexity of binary images

There is no standard definition of image complexity. Kawaguchi discussed this problem in connection with the image thresholding problem, and proposed three types of complexity measures[4][5][6].

In the present paper we adopted a black-and-white border image complexity.

The definition of image complexity

The length of black-and-white border in a binary image is a good measure for an image complexity. If the border is long, the image is complex, otherwise it is simple. The total length of black-and-white border equals to the summation of the number of color-changes along the rows and columns in an image. For example, a single black pixel surrounded by white background pixels has the boarder length of 4.

We will define the image complexity α by the following.

$$\alpha = \frac{k}{\text{The max. possible B - W changes in the image}} \quad (2.1)$$

Where, k is the total length of black-and-white border in the image. So, the value ranges over:

$$0 \leq \alpha \leq 1 \quad (2.2)$$

(2.1) is defined globally, i.e., α is calculated over the whole image area. It gives us the global complexity of a binary image. However, we can also use α for a local image (e.g., 8×8 pixel size area) complexity.

Our principle to see a binary image is that the informative regions are simple, while the noise-like regions are complex. We will use α as our complexity measure in this paper.

3. Analysis of Informative and Noise-Like regions

Informative images are simple, while noise-like images are complex. However, this is only true in case such binary images are part of natural image. In this section we will discuss how many image patterns are informative and how many patterns are noise-like. First, we will introduce a "conjugation" operation of a binary image.

3.1 Conjugation of binary image

Let P be a $2^N \times 2^N$ size black-and-white image with black area as the foreground and white area as the background. W and B denote all white and all black patterns, respectively. We introduce two checkerboard patterns Wc and Bc, where Wc has a white pixel at the upper-left position, and Bc is its complement, i.e., the upper-left pixel is black (See Fig. 1). We regard black and white pixels have logical value "1" and "0", respectively.

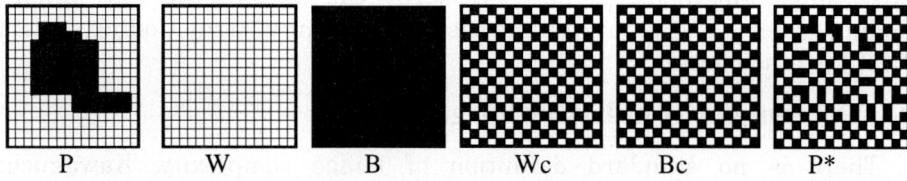

P   W   B   Wc   Bc   P*

Fig. 1 Illustration of each binary pattern (N=4)

P is interpreted as follows. Pixels in the foreground area have B pattern, while pixels in the background area have W pattern.

Now we define P* as the conjugate of P which satisfies:
1) The foreground area shape is the same as P.
2) The foreground area has the Bc pattern.
3) The background area has the Wc pattern.

Correspondence between P and P* is one-to-one, onto. The following properties hold true for such conjugation operation. "$\oplus$" designates the exclusive OR operation.

A) $P^* = P \oplus Wc$ (3.1)
B) $(P^*)^* = P$ (3.2)
C) $P^* \neq P$ (3.3)

We can easily prove those properties.

The most important property about conjugation is the following.

D) Let α (P) be the complexity of a given image P, then we have,

$$\alpha(P^*) = 1 - \alpha(P). \qquad (3.4)$$

It is evident that the combination of each local conjugation (e.g., 8×8 area) makes an overall conjugation (e.g., 512×512 area).

(3.4) says that every binary image pattern P has its counterpart P*. The complexity value of P* is always symmetrical against P regarding $\alpha = 0.5$.

3.2 Criterion to segment a bit-plane into informative and noise-like regions

We are interested in how many binary image patterns are informative and how many patterns are noise-like respecting α.

Firstly, as we think 8×8 is a good size for local area, we want to know the total number of 8×8 binary patterns in relation to α value. This means we must check all 2^{64} different 8×8 patterns. However, 2^{64} is too huge to make an exhaustive check by any means.

Our practical approach is as follows. We first generate random 8×8 binary patterns as many as possible, where each pixel value is set random, but has equal black-and-white probability. Then we make a histogram of all generated patterns in terms of α. This simulates the distribution of 2^{64} binary patterns.

Fig.3 shows the histogram for 4,096,000 8×8 patterns generated by our computer. This histogram shape almost exactly fit the normal distribution function as shown in the figure. The average value of the complexity α was exactly 0.5. The standard deviation was 0.047 in α. We denote this deviation by σ ("sigma" in Fig. 2)

Secondly, our next question is how much image data we can discard without deteriorating image quality, or, how small informative data are indispensable to keep image information.

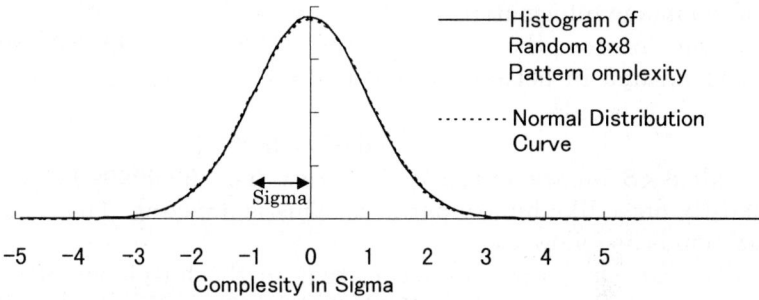

Fig. 2 Histogram of randomly generated 8×8 binary patterns

To discard data means to replace local image areas in a bit-plane with random noise patterns. If we replace all the local areas having complexity value $\alpha_L \leq \alpha$, yet it still maintains a good quality, then we can discard more. If it is not any more a good quality, we can not discard that much. If $\alpha = \alpha_L$ is the minimum complexity value to be good, such α_L is the threshold value.

To be indispensable for the image to be "informative" means the following. If the image data is still "picture-like" after we have discarded (randomized) certain amount of image data for such an α that $\alpha \leq \alpha_U$, and if we discard more, then it becomes only noise-like. Then, that α_U is regarded as the limit of the informative image complexity.

If α_L and α_U coincide($\alpha_0 = \alpha_L = \alpha_U$), we can conclude α_0 is the complexity threshold to divide informative and noise-like regions in a bit-plane.

We made a "random pattern replacing" experiment on a bit-plane of a color image. Fig. 3 illustrates the result.

 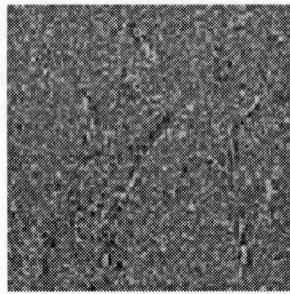 

A) Original image   B) Randomization (simple side)   C) Randomization (complex side)

Fig. 3 Randomization of the less and the more complex than $\alpha = 0.5\text{-}8\sigma$.

Fig. 3 shows that if we randomize the less complex regions in each bit-plane than $0.5\text{-}8\sigma$, the image can not be image-like any more. While, we can randomize the more complex regions than $0.5\text{-}8\sigma$ without losing much of the image information.

This means the most of the informative image information is concentrated in between 0 and $0.5\text{-}8\sigma$ in complexity scale. Surprising enough, it is only

$$6.67 \times 10^{-14} \%$$

of all 8×8 binary patterns. The rest (i.e., 99.9999999999999333%) are mostly noise-like binary patterns. This is amazing. The conclusion of this section is as follows.

We can categorize the local areas in the bit-planes of a multi-valued image into three portions (1) Natural informative portions (2) Artificial informative portions (3) Noise-like portions.

The reason we categorize the excessively complicated patterns as "informative" is base on our experiments[7].

4. The BPCS-Steganography

The Bit-Plane Complexity Segmentation Steganography is our new steganographic technique which has a large information hiding capacity.

As we have studied in Section 3 that replacing the complex regions in each bit-plane of a color image with random binary patterns is invisible to human eye. We can use this property for information hiding (embedding) strategy. Our practical method is as follows.

In our method we call a carrier image a "dummy" image. It is a color image in BMP file format, which hides (or, embeds) the secret information (files in any format). We segment each secret file into a series of blocks having 8 bytes data each. They are regarded as a 8×8 image patterns. We

call such blocks the secret blocks. The file embedding algorithm into a dummy image takes the following steps.
1) Transform the dummy image from PBC to CGC system.
2) Segment each bit-plane of the dummy image into informative and noise-like regions by using a threshold value (α_0). A typical value is $\alpha_0 = 0.3$.
3) Segment the secret file into the series of secret blocks.
4) If a block (S) is less complex than the threshold (α_0), then conjugate it to make it a more complex block (S*). It becomes more complex than α_0
5) Embed each secret block into the noise-like regions of the bit-planes (or, replace all the noise-like regions with series of secret blocks). If the block is conjugated, then record it in a "conjugation map."
6) After the secret file embedding, embed the conjugation map, too.
7) Convert the embedded dummy image from CGC to PBC.

The Decoding algorithm (i.e., the extracting operation of the secret information from an embedded dummy image) is just the reverse procedure of the embedding steps.

The novelty in BPCS-Steganography is itemized in the following.
A) Segmentation of each bit-plane of a color image into "Informative" and "Noise-like" regions.
B) Introduction of the B-W boarder based complexity measure (α) for region segmentation
C) Introduction of the Conjugation operation to convert simple secret blocks to complex blocks.
D) Using CGC image plane instead of PBC plane

5. Experiments

We have developed a PBCS-Steganography program for both Windows and Unix. They are rather compact programs at the moment. In each program, we took a 8×8 square as the local image size. Fig. 6 (A) is an example of the original dummy image (512x512, full color). (B) is the recursively embedded image of four other images shown in Fig. 5

(A) Original dummy image　　　(B) Embedded dummy image
Fig.4　Example of the dummy image

(A) 380KB   (B) 167KB   (C) 79KB   (D) 37KB

Fig.5   Embedded files (four color image files)

Through our file embedding experiments, we found that most of the color images taken by a digital camera are used as dummy images. In almost all cases, the information hiding capacity was around 50% of each dummy image. This score is 4 - 5 times as large as currently known steganographies.

6. Customization of the program

The BPCS-Steganography algorithm has several embedding parameters for a practical program implementation. Some of them are:
(1) The embedding location of the header(s) of the secret file(s)
(2) Encryption parameters of the secret file(s)
(3) The compression parameters of the secret file(s)

Each program user can give his/her special parameters when installing. So, a single BPCS-Steganography program can be customized into many ways. A customized program is not compatible to any other program. It is very easy for this program to have such optional parameters.

7. Possible applications

Each PBCS-Steganography program consists of the encoder component and the decoder component. It can be used either "**Encoder-and-Decoder**", or the "**Encoder-and-Decoder**" and the "**Decoder.**"
- Private use of a customized Encoder-and-Decoder
- Sharing a customized Encoder-and-Decoder within a group
- The case the leader retains each Encoder-and-Decoder of the group member.

8. Conclusions and future study

The objective of this paper was to demonstrate our PBCS-Steganography which is based on the property of human visual system. The most important point in this technique is that human can not see any

information in the bit-planes of a color image if it is complex. We have discussed the following points and showed our experiments.
1. We can categorize the bit-planes of a natural image as informative areas and noise-like areas by the complexity thresholding.
2. Human see informative information only in a very simple binary pattern.
3. We can replace complex regions with secret information in the bit-planes of a natural image without changing image quality. This leads to our PBCS-Steganograpy.
4. A BPCS-Steganography program can be customized for each user. Thus it guarantees a secret internet communication.

We are very convinced that this steganography is a very strong information security technique. Our next research should be directed to making a more complete steganography program. Its application programs should be also developed.

Acknowledgement

This project is partly funded by the Advanced Information Technology Program (AITP) of Information-technology Promotion Agency (IPA), Japan.

References

[1] Hall, Ernest L., Computer Image Processing and Recognition, Academic Press, New York, 1979.
[2] Jain, Anil K., Fundamentals of Digital Image Processing, Prentice Hall, Englewood Cliffs, NJ, 1989.
[3] Kawaguchi, E., Endo, T. and Matsunaga, J., "Depth First picture expression viewed from digital picture processing", IEEE Trans. on PAMI, vol.5, no.4, pp.373-384, 1988.
[4] Kawaguchi, E. and Taniguchi, R., "Complexity of binary pictures and image thresholding - An application of DF-Expression to the thresholding problem", Proceedings of 8th ICPR, vol.2, pp.1221-1225, 1986.
[5] Kawaguchi, E. and Taniguchi, R., "The DF-Expression as an image thresholding strategy", IEEE Trans. on SMC, vol.19, no.5, pp.1321-1328, 1989.
[6] Kamata, S, Eason, R. O., and Kawaguchi, E., "Depth-First Coding for multi-valued pictures using bit-plane decomposition", IEEE Trans. on Comm., vo.43, no.5, pp.1961-1969, 1995.
[7] Kawaguchi, E. and Niimi M, "Modeling Digital Image into Informative and Noise-Like Regions by Complexity Measure", Preprint of the 7th European-Japanese Conference on Information Modeling and Knowledge Basses, pp.268-278, May, Toulouse, 1997.
[*] http://patriot.net/~johnson/html/neil/sec/steg.htm
http://members.iquest.net/ ~mrmil/stego.html

Dynamic Calibration of an Active Vision System to Compute the Ground Plane Transformation

Fuxing Li, Michael Brady

Department of Engineering Science, University of Oxford, Oxford OX1 3PJ

Abstract. Calibration is fundamental, but difficult problem in computer vision. It is even difficult to calibrate an active vision system where the head geometry and/or camera intrinsic parameters change. Conventionally, calibration aims to calibrate a global model for various tasks, and this is difficult, if not impossible, for an active vision system. We propose that calibration be task-related with different models for different tasks. This simplifies the calibration process yet yields sufficient accuracy for the related task. A general model for computing the ground plane transformation for a common elevation active stereo head is derived and an approach to identifying the related parameters using real image data is given. We demonstrate that after compensating for manufacturing errors, this general ground plane transformation model can be used in a real system to compute the ground plane transformation in real-time from according to the head feedback state.

1 Introduction

Accurate camera calibration can greatly simplify solutions to many important vision problems such as stereo vision, 3D visual tracking, mobile robot visual guidance, and 3D reconstruction. Therefore, extensive work has been devoted to developing camera calibration techniques [12] [13], though accuracy is always a problem, especially for 3D reconstruction. With the development of active vision, a number of vision problems have become well-defined and easier to solve. However, the changeable geometry of active camera platforms makes calibration complex and difficult. For this reason, most existing active vision systems are not calibrated accurately. It has been found that, due to the changeable geometry and the changeable camera static parameters, it is difficult, if not impossible, to find a global model to calibrate an active vision system, then to use this global model to perform various tasks with the active vision system.

Nevertheless, an accurate model of an active stereo vision system remains extremely useful and so that a number of efforts have been made towards improving the calibration accuracy of active binocular heads [8] [10].

But we propose that, rather than calibrating a global model for use in all the tasks that the vision system will perform, we would calibrate the system with different models for different tasks so that each model is sufficient for that task but not necessarily for others. This leads us to propose task-related calibration.

Previous work that can be viewed as related to our approach, but which were not explicitly regarded as such includes efforts to develop visual modules which rely either on rough camera models [5] or self-calibration techniques [1] [3] [9]. In this paper, we illustrate our approach with respect to ground plane obstacle detection, based on a model of the ground plane transformation. Another example of calibrating the fundamental matrix model for the epipolar geometry computation can be found in [7].

2 Model derivation and analysis

An active stereo vision head with four degrees of freedom and a common elevation platform has been developed by Du & Brady [2] for visual navigation of mobile robots. Currently the system has fixed lenses and apertures.

It has been realised that for a stereo vision system with fixed geometry, the ground plane transformation between the two images can be constructed *a priori* and used for ground plane obstacle detection [14] [4]. If, in practice, we could compute the ground plane transformation when the head changes its geometry, then a similar idea could be used for ground plane obstacle detection with an active stereo vision system.

Previously, an on-line active stereo vision head initialisation system has been developed that attains an accuracy of half a degree [11]. Also, the stereo head encoders can dynamically and accurately provide the camera orientations relative to its initial position. The camera intrinsic parameters are not altered by changes of orientation of the cameras. These observations encourage us to calibrate a model of the ground plane transformation using changeable but known head geometry hence to detect ground plane obstacles.

2.1 The ground plane transformation matrix

Figure 1 shows the coordinate systems and their relationships for a common elevation stereo head mounted on an AGV moving on the ground plane. $(OXYZ)_h$ are the head coordinates, $(OXYZ)_{hg}$ the ground-head coordinates, $(OXYZ)_{cl}$ and $(OXYZ)_{cr}$ the left and right camera coordinates respectively. In the figure, B is the baseline, h the height of the head above the ground plane. θ_t is the head tilt angle, and θ_l and θ_r are the left and right camera angles. Because the head pan motion does not affect the ground plane transformation matrix, we do not consider it here.

The ground plane transformation $\mathbf{H_{rl}}$ has following analytical form:

$$\mathbf{H_{rl}} = \begin{pmatrix} \frac{\alpha_r}{\alpha_l}c^v - \frac{u_{0r}}{\alpha_l}s^v & 0 & (u_{0r} - \frac{\alpha_r}{\alpha_l}u_{0l})c^v + (\frac{u_{0l}u_{0r}}{\alpha_l} + \alpha_r)s^v \\ -\frac{v_{0r}}{\alpha_l}s^v & \frac{\beta_r}{\beta_l} & \frac{u_{0l}v_{0r}}{\alpha_l}s^v + v_{0r}c^v - \frac{v_{0l}\beta_r}{\beta_l} \\ -\frac{1}{\alpha_l}s^v & 0 & \frac{u_{0l}}{\alpha_l}s^v + c^v \end{pmatrix}$$

$$+ \frac{B}{h} \begin{pmatrix} -(\frac{\alpha_r}{\alpha_l}c^l s^r + \frac{u_{0r}}{\alpha_l}c^l c^r)s^t & -(\frac{\alpha_r}{\beta_l}s^r + \frac{u_{0r}}{\beta_l}c^r)c^t \\ -\frac{v_{0r}}{\alpha_l}c^l c^r s^t & -\frac{v_{0r}}{\beta_l}c^r c^t \\ -\frac{1}{\alpha_l}c^l c^r s^t & -\frac{1}{\beta_l}c^r c^t \end{pmatrix}$$

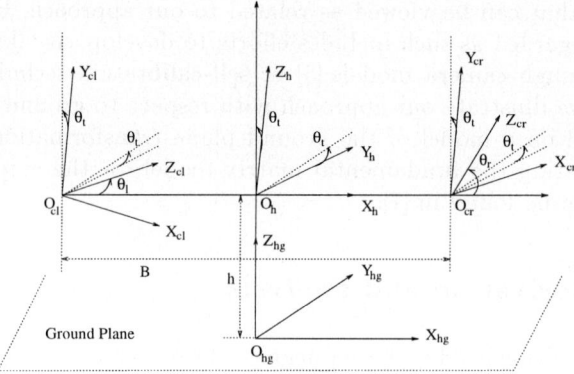

Fig. 1. *The coordinate system for a common elevation stereo head mounted on an AGV moving on the ground plane.*

$$\begin{pmatrix} \frac{\alpha_r}{\alpha_l} u_{0l} C^l S^r S^t + \frac{u_{0l} u_{0r}}{\alpha_l} C^l C^r S^t + \frac{\alpha_r}{\beta_l} v_{0l} S^r C^t + \frac{u_{0r} v_{0l}}{\beta_l} C^r C^t + \alpha_r S^l S^r S^t + u_{0r} S^l C^r S^t \\ \frac{u_{0l}}{\alpha_l} v_{0r} C^l C^r S^t + \frac{v_{0l} v_{0r}}{\beta_l} C^r C^t + v_{0r} S^l C^r S^t \\ \frac{u_{0l}}{\alpha_l} C^l C^r S^t + \frac{v_{0l}}{\beta_l} C^r C^t + S^l C^r S^t \end{pmatrix}, \quad (1)$$

where $S^l = sin\theta_l$, $C^l = cos\theta_l$, $S^r = sin\theta_r$, $C^r = cos\theta_r$, $S^t = sin\theta_t$, and $C^t = cos\theta_t$, $S^v = sin\theta_v = sin(\theta_r - \theta_l)$, $C^v = cos\theta_v$. Here, $\theta_v = \theta_r - \theta_l$ is the camera vergence angle, α_l and α_r are the left and the right camera focal lengths (in pixel units) in the horizontal direction, β_l and β_r are the left and the right camera focal lengths in the vertical direction, and $(u_{0l}, v_{0l})^\top$ and $(u_{0r}, v_{0r})^\top$ are the left and right image centres.

Corresponding points from the ground plane in the left image $(u_l, v_l)^\top$ and in the right image $(u_r, v_r)^\top$ are related by $\mathbf{H_{rl}}$ by:

$$\begin{pmatrix} u_r \\ v_r \\ 1 \end{pmatrix} = \mathbf{H_{rl}} \begin{pmatrix} u_l \\ v_l \\ 1 \end{pmatrix}. \quad (2)$$

2.2 Sensitivity analysis

In this section, we present a typical test result using synthetic data to analyse the sensitivity of the ground plane transformation to variations in the parameters. Figure 2 shows the mean (E) and standard deviation (σ) of the distance (d), the horizontal coordinate difference (u), and the vertical coordinate difference (v) between the corresponding point and that transferred using the ground plane transformation computed with different parameters (typically, one degree variation for the camera angles and 10% variation for all other parameters). The head states 0 to 11 correspond to symetrically verging two cameras with vergence angles $0°$, $10°$, $20°$, and $30°$ for each of the following tilt angles $0°$, $-5°$ and $-10°$, respectively. In Figures 2(1), 2(3) and 2(5), the solid lines correspond to u_0; the dotted lines to v_0; the dashdot lines to α_c; and the dashed lines to β_c. In Figures 2(2), 2(4) and 2(6), the solid lines correspond to θ_l; the dotted lines to θ_r; the dashdot lines to θ_t; and the dashed lines to the head height. We suppose that the two cameras have the same camera intrinsic parameters, ie. $\alpha_l = \alpha_r = \alpha_c$, $\beta_l = \beta_r = \beta_c$, $u_{0l} = u_{0r} = u_0$, and $v_{0l} = v_{0r} = v_0$. Several conclusions can be drawn from these and other experiments:

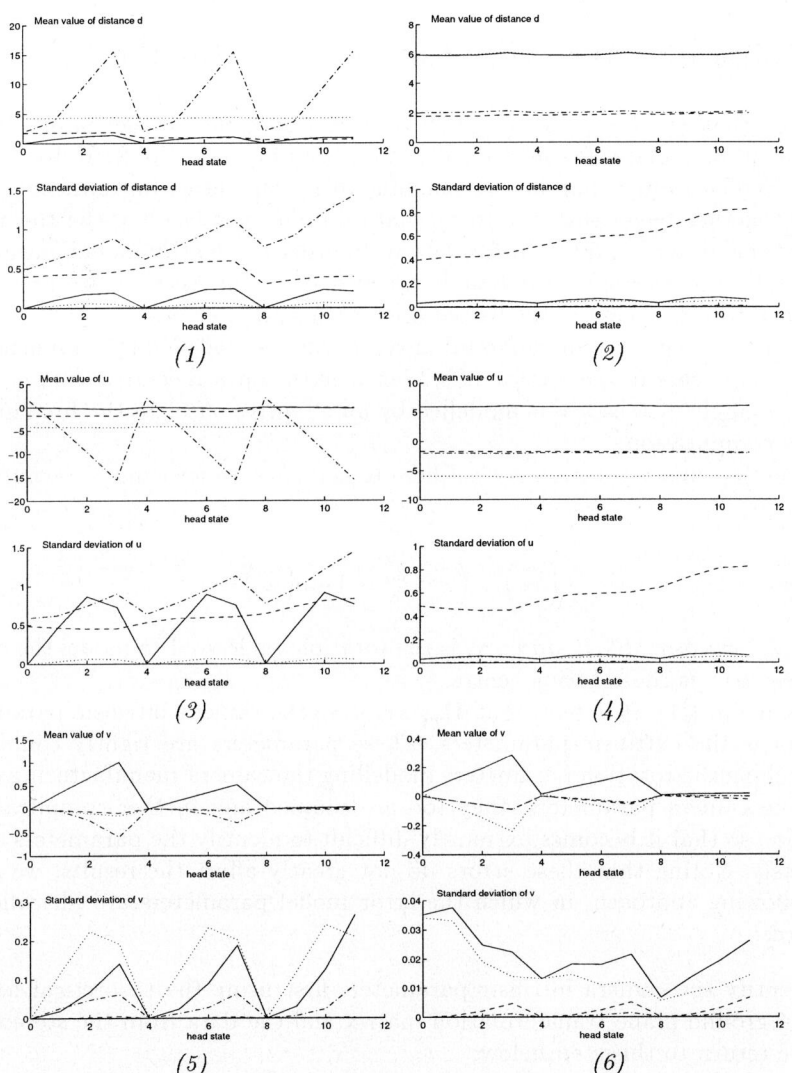

Fig. 2. *Sensitivity Analysis. See text for interpretation.*

- The vertical coordinate v is not sensitive to parametric variation (see Figures 2(5) and 2(6));
- Apart from α_c, the parameters have an approximately constant effect on the horizontal coordinate u (see Figures 2(3) and 2(4)). This suggests the possiblity of modelling these errors with constant values as described below;
- The effect of α_c varies when the head state changes, in particular, the effect is large when the vergence angle increases.

3 Calibrating the ground plane transformation

Suppose that $(u_l, v_l)^T$, and $(u_r, v_r)^T$ are corresponding image points arising from a ground plane feature, ideally, they should obey Eq. 2. However, real images suffer from a number of manufacturing and head initialisation errors which together mean that the image data do not exactly obey the theoretical form of the ground plane transformation. In order to identify model parameters accurately, these errors must first be identified. In this section, we present an approach to identifying both the parameters and their errors.

As shown in [6], the manufacturing errors can be modelled by an image rotation. The increase in the vertical offset between the optical centres as the camera vergence angle increases was modelled by an adaptive offset in the fundamental matrix computation.

A similar image rotation is used here to compensate for camera manufacturing errors:

$$\begin{pmatrix} u_r \\ v_r \\ 1 \end{pmatrix} = \begin{pmatrix} C^e & S^e & x_0 \\ -S^e & C^e & y_0 \\ 0 & 0 & 1 \end{pmatrix} \mathbf{H_{rl}} \begin{pmatrix} u_l \\ v_l \\ 1 \end{pmatrix}. \tag{3}$$

where $C^e = cos\gamma_e$, $S^e = sin\gamma_e$; γ_e is the rotation angle used to model the errors, and $(x_0, y_0)^T$ is the rotation centre.

From Eq. (1), it is seen that $\mathbf{H_{rl}}$ involves the camera intrinsic parameters as well as the extrinsic parameters. These parameters are tightly coupled. If we combine the rotation parameters modelling the camera manufacturing errors with the camera parameters, they too are coupled and further complicate the situation so that it becomes extremely difficult to identify the parameters simultaneously. Noting that these errors do not greatly affect the results, we adopt the following approach, in which the error model parameters are identified afterwards:

1. Identify the camera intrinsic parameters first using the theoretical form of the ground plane transformation matrix and the data from the stereo pair. We return to this step below;
2. Compute the transferred image data $(u'_r, v'_r)^T$ from the resulting $\mathbf{H_{rl}}$ as follows:

$$\begin{pmatrix} u'_r \\ v'_r \\ 1 \end{pmatrix} = \mathbf{H_{rl}} \begin{pmatrix} u_l \\ v_l \\ 1 \end{pmatrix}. \tag{4}$$

Because the parameters for $\mathbf{H_{rl}}$ are computed by a least squares process, $(u'_r, v'_r)^\top$ is in general different from $(u_r, v_r)^\top$;

3. Identify the error parameters (γ_e, x_0, y_0) using the real image data and the transferred image data (a simple SVD suffices) :

$$\begin{pmatrix} u_r \\ v_r \\ 1 \end{pmatrix} = \begin{pmatrix} c^e & s^e & x_0 \\ -s^e & c^e & y_0 \\ 0 & 0 & 1 \end{pmatrix} \begin{pmatrix} u'_r \\ v'_r \\ 1 \end{pmatrix} = \begin{pmatrix} c^e & s^e & x_0 \\ -s^e & c^e & y_0 \\ 0 & 0 & 1 \end{pmatrix} \mathbf{H_{rl}} \begin{pmatrix} u_l \\ v_l \\ 1 \end{pmatrix}. \tag{5}$$

This process could be repeated several times. In practice, we find that a single iteration produces sufficiently good results for our AGV application.

The remaining problem is how to identify the camera parameters in step 1 above. $\mathbf{H_{rl}}$ is sufficiently complicated that it is impossible in practice to identify these coupled parameters without simplification. Based on the experience of [6], we make the following assumptions:

- the intrinsic parameters of the two cameras are identical, except that v_{0l} is not assumed equal to v_{0r}. This latter is quite apparent in the real stereo pairs;
- the camera initialisation errors (that is, errors in $\theta_l, \theta_r, \theta_t$) can individually be ignored but the error in the camera vergence angle can not. This is based on the sensitivity analysis that θ_t variation has a negligible effect on the ground plane transformation while θ_l and θ_r do not. But θ_l and θ_r separately are multiplied by small scalars while their difference (the vergence angle) is not.

Based on the above assumptions, and the magnitude of each entry of $\mathbf{H_{rl}}$, which determines whether or not the camera initialisation error has a significant effect, $\mathbf{H_{rl}}$ becomes $\mathbf{H_{rl}^c}$:

$$\mathbf{H_{rl}^c} = \begin{pmatrix} c^v - \frac{u_0}{\alpha_c} + e_{11} c^v s^v & 0 & (\frac{u_0 u_0}{\alpha_c} + \alpha_c) s^v \\ -\frac{v_{0r}}{\alpha_c} s^v + e_{21} c^v & 1 & \frac{u_0 v_{0r}}{\alpha_c} s^v + v_{0r} c^v - v_{0l} \\ -\frac{1}{\alpha_c} s^v & 0 & \frac{u_0}{\alpha_c} s^v + c^v + e_{33} c^v \end{pmatrix}$$

$$+ \frac{B}{h} \begin{pmatrix} -(c^l s^r + \frac{u_0}{\alpha_c} c^l c^r) s^t & -(\frac{\alpha_c}{\beta_c} s^r + \frac{u_0}{\beta_c} c^r) c^t \\ -\frac{v_{0r}}{\alpha_c} c^l c^r s^t & -\frac{v_{0r}}{\beta_c} c^r c^t \\ -\frac{1}{\alpha_c} c^l c^r s^t & -\frac{1}{\beta_c} c^r c^t \end{pmatrix}$$

$$\begin{matrix} u_0 c^l s^r s^t + \frac{u_0 u_0}{\alpha_c} c^l c^r s^t + \frac{\alpha_c}{\beta_c} v_{0l} s^r c^t + \frac{u_0 v_{0l}}{\beta_c} c^r c^t + \alpha_c s^l s^r s^t + u_0 s^l c^r s^t \\ \frac{u_0}{\alpha_c} v_{0r} c^l c^r s^t + \frac{v_{0l} v_{0r}}{\beta_c} c^r c^t + v_{0r} s^l c^r s^t \\ \frac{u_0}{\alpha_c} c^l c^r s^t + \frac{v_{0l}}{\beta_c} c^r c^t + s^l c^r s^t \end{matrix} \Bigg), \tag{6}$$

where $v_{0r} = v_{0l} + p_{23} + K_v \theta_v$. $\alpha_c = \alpha_l = \alpha_r$, $\beta_c = \beta_l = \beta_r$, $u_0 = u_{0l} = u_{0r}$. Here e_{11}, e_{21} and e_{33} are the camera vergence angle initialisation errors caused by $\theta_{v0} \neq 0$. $p_{23} = v_{0r} - v_{0l}$ when vergence angle is zero. K_v is a proportional cofficient used to approximate the effect of camera vergence angle to v_{0r}.

The above parameters can be identified by capturing image pairs with particular head states and using these images to compute the simplified ground plane transformation matrix. Three steps are involved in the parameter identification process [6]:

1. Capture image pairs with parallel cameras ($\theta_v = 0$) but different tilt angles. In this case, $\mathbf{H_{rl}^c}$ can be further simplified and used to identify α_c, β_c, v_{0l}, and p_{23} given that $\frac{B}{h}$ is measurable.
2. The remaining unknown parameter is u_0 and K_v. u_0 has to be identified using a nonlinear method since it appears quadratically. Notice from the form of $\mathbf{H_{rl}}$ that v_{0r} is only related to the vertical coordinate of the ground plane transformation. This property can be used to solve for v_{0r} by SVD with different vergence angles. Images with different vergence angles but the same zero tilt angle, are used to compute u_0 and different v_{0r} according to a further simplified form of $\mathbf{H_{rl}^c}$. u_0 is computed by minimising the sum of distance between real correspondence points and the transformed points using $\mathbf{H_{rl}^c}$ for which v_{0r} can be computed for different vergence angles given a value of u_0. From the results it is clear that as the camera vergence angle increases v_{0r} increases by a small amount. The increase is nearly linear and can be represented by K_v.
3. Based on the above linear estimations, the parameters (α_c, β_c, e_{11}, e_{21}, e_{33}, u_0, v_{0l}, p_{23}, K_v) are refined using nonlinear optimisation to minimise the image distance between the transformed point and the real corresponding point over frames with different head geometries.

Figure 3 shows a typical example of ground plane transformation computed using the calibration results. The biggest distance error between the real correspondence points and the transformed points over frames is less than 4 pixels which, in our case, represents $2cm$ to $8cm$ over the ground plane corresponding to the range from $1m$ to $5m$. The obstacles to be detected in our case are usually higher than $10cm$ so it is practical to accommodate this error. It is observed that after identifying the parameters for the ground plane transformation model and error compensation, the results computed directly from the combined ground plane transformation model are sufficient for ground plane obstacle detection.

References

1. R. Deriche, Z. Zhang, Q. T. Luong, and O. Faugeras. Robust recovery of the epipolar geometry for an uncalibrated stereo rig. In *Proc. 3rd European Conference on Computer Vision*, pages 567–576. Springer-Verlag, 1994.
2. F. Du and J. M. Brady. A four degree-of-freedom robot head for active vision. *International Journal of Pattern Recognition and Artificial Intelligence*, 8(6):1439–1469, 1994.
3. F. Du and M. Brady. Self-calibration of the intrinsic parameters of cameras for active vision systems. *IEEE Conf. on Computer Vision and Pattern Recognition*, pages 477–482, 1993.
4. F. Ferrari, E. Grosso, G. Sandini, and M. Magrassi. A stereo vision system for real time obstacle avoidance in unknown environment. *IEEE International Workshop on Intelligent Robots and Systems*, 2:703–8, 1990.
5. E. Krotkov. Active computer vision by cooperative focus and stereo. *Springer Series in Perception Engineering*, 1989.

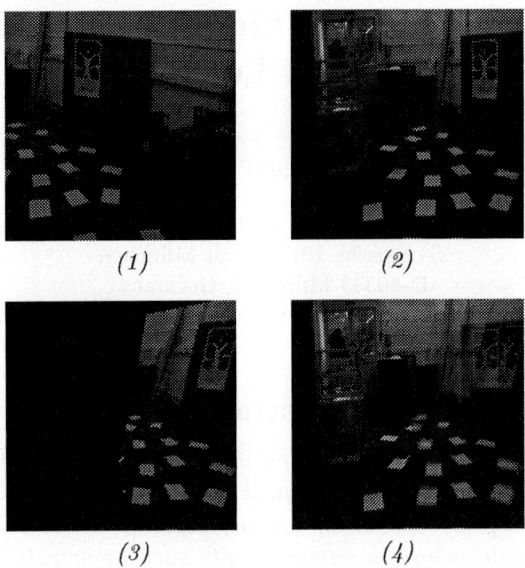

Fig. 3. Ground plane transformation for images captured with $\theta_l = 75°$, $\theta_r = 105°$ and $\theta_t = -10°$. (1) the left image; (2) the right image; (3) the transformed right image; (4) the superimposed image of the transformed right image onto the left image.

6. F. Li. *Active Stereo for AGV Navigation*. PhD thesis, University of Oxford, 1996.
7. F. Li, J. M. Brady, and C. Wiles. Fast computation of the fundamental matrix for an active stereo vision system. In *Proc. of ECCV'96*, pages (I) 157–166, Cambridge, England, 1996. Springer-Verlag.
8. M. Li. Camera calibration of the KTH head/eye system. Technical report, CVAP147, Computational Vision and Active Perception Laboratory, Royal Institute of Technology, S-100 44,Stockholm, Sweden, February, 1994.
9. S. D. Ma. A self-calibration of technique for active vision systems. *IEEE Transactions on Robotics and Automation*, 12(1):114–120, 1996.
10. P. F. McLauchlan and D. W. Murray. Active camera calibration for a head-eye platform using the variable state-dimension filter. *IEEE Transactions on Pattern Analysis and Machine Intelligence*, 18(1):15–22, 1996.
11. I.D. Reid and P. A. Beardsley. Self-alignment of a binocular robot. In *Proc. 6th British Machine Vision Conference*, Birmingham, 1995.
12. R. Y. Tsai. An efficient and accurate camera calibration technique for 3d machine vision. In *Proc. of IEEE International Conference on Computer Vision and Pattern Recognition*, pages 364–374, Miami, FL, June 22-26, 1986.
13. G-Q. Wei and S-D. Ma. Implicit and explicit camera calibration: Theory and experiments. *IEEE Transactions on Pattern Analysis and Machine Intelligence*, 16(5):469–480, 1994.
14. M. Xie. Ground plane obstacle detection from stereo pairs of images without matching. In *Proc. ACCV'95*, pages II–280, Singapore, 1995.

Identification of 3D Reference Structures for Video–Based Localization*

Darius Burschka and Stefan A. Blum

Laboratory for Process Control and Real-Time Systems
Technische Universität München
D-80333 München, Germany
e-mail: {burschka|blum}@lpr.e-technik.tu-muenchen.de

Abstract

The bootstrap–problem of self–localization in indoor environments is a demanding task for initial localization and topological navigation. The ability to determine the position in an a priori known or already explored environment allows unsupervised use of mobile robots in environments such as private households. This paper presents our approach to identify a given set of known reference structures in a three–dimensional map of the local environment. This map is constructed from the data extracted by a line-based stereo camera system mounted on a mobile vehicle. We present the method used to identify objects and to compute the vehicle's position in a world frame.

1 Motivation

Navigation in indoor environments requires dependable knowledge about the robot's position and a precise model of the environment. Dynamic changes in the environment caused by human influence or the operation of other autonomous mobile vehicles increase the differences between the sensor data and the model prediction that is based on an *a priori* model of the environment. The *a priori* model becomes gradually useless [4, 5] as the environment changes.

The explored information is often incomplete and uncertain. The significance of the correct interpretation decreases compared to other possible matches. Therefore, continuous exploration of the changes in the environment is necessary. In [2], a three–dimensional model capable of storing the changing sensor data during an exploration of the environment was presented. The three–dimensional lines are stored at their geometrical position and are be used for prediction of future sensor readings or for obstacle avoidance. All objects are referenced to a local structure visible in this region. The size of this local region is restricted by the range of the applied sensor or the structure of the environment. For example, the walls of a room can reduce the possible size of the local region defined by the sensor range.

---

* The work presented in this paper was supported by the Deutsche Forschungsgemeinschaft as a part of an interdisciplinary research project on "Information Processing in Autonomous Mobile Robots"(SFB331).

2 Sensor Feature Extraction

2.1 Geometrical Constraints

The three–dimensional information can be retrieved from a pair of images if the correspondence between extracted elements (e.g., lines) is established and the exact position of the cameras is known. The number of possible matching candidates can be reduced by applying constraints such as the epipolar constraint, uniqueness and continuity [3]. Our sensor system computes the 3D information for the endpoints of the detected lines [1].

It is almost impossible to align the two optical systems precisely and stably to a parallel arrangement. This orientation error must be taken into account to get accurate results. Most algorithms compute the geometry of the epipolar lines for a given situation. We propose an alternative approach for transforming the original data into the parallel case with minimum computation. For example, a rotation φ around the y–axis results in the following transformation of the image coordinates:

$$\begin{aligned} x'_p &= \frac{x'}{z'} = \frac{\frac{x}{z}\cdot\cos\varphi+\sin\varphi}{-\frac{x}{z}\cdot\sin\varphi+\cos\varphi} = \frac{x_p\cdot\cos\varphi+\sin\varphi}{-x_p\cdot\sin\varphi+\cos\varphi} \\ y'_p &= \frac{y'}{z'} = \frac{\frac{y}{z}}{-\frac{y}{z}\cdot\sin\varphi+\cos\varphi} = \frac{y_p}{-y_p\cdot\sin\varphi+\cos\varphi}. \end{aligned} \quad (1)$$

The rotation θ around the z–axis results in a simple rotation of the image coordinates. The advantage of these equations is that the world coordinates of the projected obstacles may remain unknown. We use these equations (1) to transform the camera system to the parallel cameras case. The initial orientation errors are estimated in an off-line calibration of extrinsic and intrinsic camera parameters [7]. The orientation of the cameras may change during a mission due to vibrations and camera repositioning errors. It is corrected in an on-line re-calibration method based on the explored information [2].

2.2 Virtual 3D Sensor

In our system we use a model of a "virtual 3D sensor" situated between the two cameras. The properties of this sensor can be achieved by different real sensors. It is possible to replace the stereo system by any other sensor capable of reconstruction of three–dimensional lines.

3 Dynamic Local Feature Map

3.1 Possible Input Sources

The Dynamic Local Map (DLM) stores a local region at the abstraction level of the applied sensor system to support topological modeling. The data stored in the DLM come from different sources [2]. The most important source is the sensor system, which registers the recent changes in the environment. Another source is a global model of the environment that stores reliable information generated from CAD–models or previously-verified lines. The map also stores hypothetical features of the environment generated by the Predictive Spatial Completion (PSC) module. These hypothetical features are based on statistics and are used to control the path planning and to support the sensor system.

3.2 Data Formats

Currently, we store only line shaped features described by their three-dimensional endpoints and orientation. This information is refined in consecutive steps. In addition, each feature in the map is described by its confidence and accuracy. This information is used in navigation tasks to decide, which features should be employed for localization to reduce the errors caused by 3D lines resulting from false correspondences in the stereo system.

3.3 Internal Structure

Flaws in the sensor feature extraction can be reduced by comparing new explored data with the former sensor readings stored in the DLM. The DLM stores a local region of the environment described by the three–dimensional lines. Multiple and partially contradictory requirements forced us to develop a multi–level indexing structure. The upper layers consist of a two–dimensional grid containing octrees to store the explored features efficiently and to minimize time-consuming memory transfers for feature updates or localization changes [1, 2]. This structure adapts to the features' distribution in the local environment.

3.4 Update of the Stored Information

Our fast reconstruction of the 3D information is based on fast interaction with the DLM. The possible match candidates for a given feature exceeding a specified quality value are stored in the DLM. Therefore, the DLM also stores false information. In consecutive sensor readings this information is verified from different positions. False features are deleted if they cannot be verified. Therefore, it is important that our sensor data processing not be free-running, but be triggered from the planning instance. In this way, sensor readings from different positions are guaranteed. Multiple readings from the same position would also stabilize the false features. This procedure permits the handling of moving objects in the environment.

Each endpoint is described by its precision. The precision p of a detected endpoint depends on its distance from the camera. The precision of an endpoint describes its maximum position error and is adjusted each time the feature can be verified. The feature is specified by a confidence value c for its existence. The confidence value c is modified with an exponential function $c(f) = 1 - e^{-g(f_{old}+f)} = 1 - (1 - c_{old})e^{-gf}$, $0 \leq f \leq 1$ each time it can be matched. The value f describes the confidence in the current step. It may vary from 1 (only one matching candidate) to 0 (useless information). The value g describes the speed at which this value is changed. It must be adapted to the applied matching algorithm.

4 Object Recognition

Our approach is based on a hypothesize&test strategy underlying an interpretation tree. Because of the high computational cost of these algorithms, several tests for truncations are added, including rigidity and visibility constraints. Therefore, a relevant-first search is implied. In our application, a set $\mathcal{O} = \{O_1, O_2, \ldots, O_r, \ldots, O_R\}$ of relevant objects that are described with 3D line segments[1] is presumed. Therefore, an object O_r consists of a set of segment lines $\mathcal{S}_r = \{S_1, S_2, \ldots, S_j, \ldots, S_m\}$[2]. Our approach to identify those *a priori* known reference structures from a list of line features $\mathcal{S}' = \{S'_1, S'_2, \ldots, S'_i, \ldots, S'_n\}$ delivered from the DLM is composed of two parts: an off-line object preprocessing to gain a minimal set of relevant line segment triplets and on-line identification, triggered by these triplets.

4.1 Object Preprocessing

Our approach to preprocess the reference objects is to find some of their properties that are not explicitly derivable from their geometric description. We propose to use only the most relevant features for generation of an object's hypotheses. Grouping of three linearly independent line features at a time seems to be suitable for this application (section 4.2). The visibility of the groups is decided from a set of synthetic projections $\mathcal{P} = \{P_1, P_2, \ldots, P_p, \ldots, P_Z\}$ of the object's CAD-model generated for different aspects.

A virtual camera is positioned at different locations on a spheric surface enclosing the model. A facility to exclude regions on the sphere's surface is provided. The visibility of each single line segment is determined for each projection with a Z-buffering algorithm. Because only discrete points on the surrounding sphere are selected, a binary criterion v for the visibility of a line feature is sufficient:

$$v_{pjr} = \begin{cases} 1, \text{ if 70\% of the object's } r \text{ segment } j \text{ are visible in projection } p \\ 0, \text{ otherwise.} \end{cases} \quad (2)$$

We created a simple heuristic calculation of a rating $0 \leq \mathcal{R}_j \leq 1$ for each line segment S_j consisting of a rating r_l of the segment length l_j, an evaluation of the information content that is derived from a term $0 \leq \mathcal{A}_{jj'} \leq 1$ [3], and the number of aspects in which a segment is visible in all projections.

$$\mathcal{R}_j = r_l(l_j) \cdot (\text{ld} \sum_{j'=1}^{m} \mathcal{A}_{jj'} + 1)^{-1} \cdot \frac{1}{Z} \sum_{p=1}^{Z} v_{pj}. \quad (3)$$

The obtained visibility information is used for the generation of triple line groups for each object. At least one visible segment triplet must exist in each projection P_p. The number T of chosen triplets is considered to be as small as possible[4]. A generated triplet is supposed to describe the sight of an object uniquely. The distances and angles between line segments should be as large as possible combined with a high rating $\mathcal{R}_j$. If possible, a segment line should only be used once for triplet generation because even high rated segment lines may not be detected, due to poor illumination for example. This leads to another heuristic rating for each triplet combination $\hat{\mathcal{T}}_u$:

$$\mathcal{U}_u = (\mathcal{R}_{u1} + \mathcal{R}_{u2} + \mathcal{R}_{u3}) \cdot |det(\mathbf{M}_u)| \cdot (d_{u1,u2} + d_{u2,u3} + d_{u3,u1}), \quad (4)$$

---
[1] For occlusion check (described below), a plane-based description is also necessary.
[2] In case of uniqueness, the index r will be left out in the following.
[4] The cost of the online identification is proportional to T.

where $\mathbf{M}_u$ is a matrix of the unit direction vectors, $d_{uk,uk'}$ is the distance between the two line segments $\hat{S}_{uk}$ and $\hat{S}_{uk'}$.[5] An iterative strategy is applied to extract the best triplet set reducing stepwise the number of projections and resulting in a number of 3 to 9 relevant triplets.

4.2 Object Identification

Clustering of Image Features The lines delivered from the DLM are clustered first. The implied clustering algorithm groups neighboring, connected segments, tolerating small errors. Clusters consisting of only a few line segments are deleted. The remaining position clusters are processed in another clustering step to build groups of approximately parallel segments called *direction clusters*. At least three direction clusters must emerge from each position cluster, otherwise the corresponding position cluster is deleted. This condition is necessary for unique determination of the object's pose.

Initial Guess for Object's Pose All emerged segment triplets are searched, primarily to build a correspondence tree for each position cluster without pose restrictions. Neighborhood relationships such as the angle and distance between two segment lines [6] are criteria for establishing correspondences between image and reference segments. Matching candidates for the second and third feature of a group must be searched only in certain direction clusters, reducing the number of required comparisons.

A *prehypothesis* is established if all three reference segments of a triplet could be associated. Based on the prehypothesis we are able to guess the pose and orientation of the reference object. We define two 3 × 3-matrices $\mathbf{K}$ and $\mathbf{K}'$, that store the unit direction vectors of the image and the reference triplets. A transformation between the reference frame and the image frame can be expressed as a matrix $\mathbf{R} = \mathbf{K}' \cdot \mathbf{K}^{-1}$, which is, in the best case, an orthogonal rotation matrix, in which small errors are tolerated. $\mathbf{R}$ is normalized in order to compute the three rotational degrees of freedom (DOF). The remaining three translational DOF are calculated by simply using a reference point for each reference triplet and associated image triplet. The translation can be determined by $\boldsymbol{T} = \boldsymbol{a} + \boldsymbol{S} - \boldsymbol{b}$, where the offset between the segment triplet's center and the object's center in the object frame is denoted as $\boldsymbol{a}$. $\boldsymbol{S}$ is the offset between the position of center C_T of model segment triplet in the object frame and the center C_T''' of the associated triplet in the image space. Because $\boldsymbol{b} = \mathbf{R} \cdot \boldsymbol{a}$, the translation can be denoted as $\boldsymbol{T} = \boldsymbol{S} + (\mathbf{I} - \mathbf{R}) \cdot \boldsymbol{a}$, where $\mathbf{I}$ is identity matrix.

Hypothesis Generation The probability of the existence of a certain object depends on the existence and, especially, on the accuracy of the associations of the predicted features.

A segment wise linear accuracy function $0 \leq a(e) \leq 1$ is defined by orientation accuracy $a_\delta(\varphi)$, longitudinal accuracy $a_l(u)$ and parallel accuracy $a_p(d)$.

---

[5] Note that the number of possible combinations is about $0.125 \cdot m^3$, assuming that the probability that a single line feature is visible in a single projection is about 0.5.

A total accuracy $\mathcal{A}_{ji}$ of a pairing of predicted line segment $\tilde{S}_j$ and image line segment S'_i is obtained by

$$\mathcal{A}_{ji} = a_\delta(\varphi) \cdot a_l(u) \cdot a_p(d). \tag{5}$$

Any non-zero accuracy value effects an association between S_j and S'_i and is kept. The obtained accuracy of a model line segment S_j is denoted as $\mathcal{A}_j$.

A probability $\mathcal{P}'$ of the object's prehypothesis is calculated by

$$\mathcal{P}' = \frac{\sum_{j=1}^m \mathcal{R}_j \cdot \mathcal{A}_j \cdot v_{p^*j}}{\sum_{j=1}^m \mathcal{R}_j \cdot v_{p^*j}}. \tag{6}$$

A probability of the hypothesis $\mathcal{P}$ is created with $v_{p^*j} := 1.0 \quad \forall_j$ by using (6). Only if $\mathcal{P}'$ and $\mathcal{P}$ exceed certain thresholds will a hypothesis be generated.

Hypothesis Pruning Hypotheses belonging to the same object are fused if they are consistent with respect to their poses. Then their probability is recalculated.

Often, two or more hypotheses are generated for different, but similar objects at similar pose. Even if an object has moderate symmetries, different hypotheses for its orientation may be created. In our approach, the object's hypothesis with the highest probability survives. The threshold for recognition $\mathcal{P}_{reg}$ has to be chosen high in order to keep the probability for misdecisions low.

Correction of Recognized Object's Pose The method of gradient descent is applied to minimize a squared error sum $S(t_x, t_y, t_z, \alpha, \beta, \gamma)$, which is defined by

$$\begin{aligned} S &= \frac{1}{m'} \cdot \sum_{i=1}^m |M_{i1} B'_{i1}|^2 + |M_{i2} B'_{i2}|^2 \\ &= \frac{1}{m'} \cdot \sum_{i=1}^m d_{i1}^2 + d_{i2}^2, \end{aligned} \tag{7}$$

where m' is the number of predicted model segment lines to which any image line can be associated[6].

4.3 Results of Object Recognition

The object recognition algorithm was tested on several objects by measuring the hypothesis' probability with different degrees of distortion. Here, a parallel shift error distorting image features within a range of $[0; p_{max}]$ and an orientation error of their direction within a range of $[0; \delta_{max}]$ were simulated. The maximal errors were varied between 0cm and 50cm in 5cm increments and 0° and 20° in 2° increments, respectively. In order to normalize and keep the results independent from a certain aspect, all features of a object's model were registrated into the DLM. The hypothesis' probability was averaged on 25 measurements at a time, the bounds for parallel accuracy were selected to $10cm \leq d_{min} \leq 15cm$, and orientation accuracy to $5° \leq \varphi_{min} \leq 10°$. The threshold for

---

[6] m' is used only as a scale factor and has no influence on the result of the gradient descent method.

the prehypothesis' probability $\mathcal{P}'$ was chosen to be zero to enable the desired behavior as stated above. Fig. 1 shows the testing results of the objects "quader" (a) and the object "cubicle" (c). The influence of a threshold was also examined, whereby the value 0.75 has to be exceeded for a hypothesis' probability to count as a recognized object.

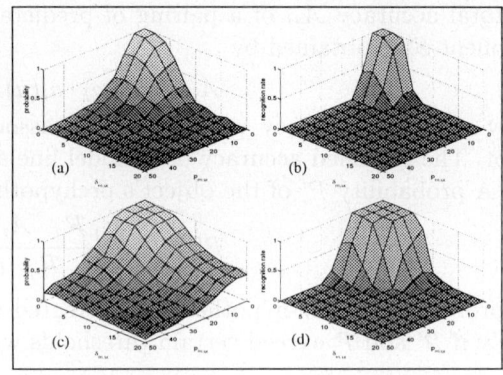

Fig. 1. Probability of object's hypotheses

5 Localization

The pose estimation of the mobile robot is performed by comparing object's pose referred to the exploration frame to its reference pose in the landmark description. A recognized object $\hat{O}_b$ consistent with its stored position and orientation ($|z_r - Z_l| \leq z_{max}, |\beta_r - \beta_l| \leq \beta_{max}$ and $|\gamma_r - \gamma_l| \leq \gamma_{max}$[7] generates a *local hypothesis* $\mathcal{L}_b$.

Fig. 2 shows the relationships between the frame of a local map and the exploration frame for a single unique recognized object. The transformation parameters (X,Y,φ) stored in the local hypothesis are calculated as:

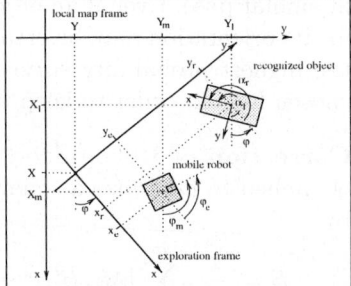

Fig. 2. Affine transformations between frames

$$\begin{pmatrix} X \\ Y \end{pmatrix} = \begin{pmatrix} X_l \\ Y_l \end{pmatrix} - \begin{pmatrix} \cos\varphi & -\sin\varphi \\ \sin\varphi & \cos\varphi \end{pmatrix} \cdot \begin{pmatrix} x_r \\ y_r \end{pmatrix}, \quad \varphi = \alpha_l - \alpha_r. \tag{8}$$

Hence, if possible local hypotheses can be generated for similar transformation parameters, an algorithm to fuse those candidates will be applied. If both of the following conditions are true, two local hypotheses $\mathcal{L}_i$ and $\mathcal{L}_j$ will be combined into a new local hypothesis denoted as $\mathcal{L}_{ij}$:

$$\mathcal{D}_{ij} = |(X_i, Y_i)^T - (X_j, Y_j)^T| \leq \mathcal{D}_{max} \quad \text{and} \quad \varphi_{ij} = |\varphi_i - \varphi_j| \leq \varphi_{max}. \tag{9}$$

The thresholds $\mathcal{D}_{max}$ and φ_{max} can be chosen depending on the required accuracy of the localization. The new generated combined local hypothesis $\mathcal{L}_{ij}$ is provided with weighted mean values. Its quality is defined by

$$\mathcal{Q}_{ij} = (\mathcal{Q}_i + \mathcal{Q}_j) \cdot (1 - \frac{\mathcal{D}_{ij}}{4 \cdot \mathcal{D}_{max}} - \frac{\varphi_{ij}}{4 \cdot \varphi_{max}}). \tag{10}$$

---

[7] r-indexed parameters are recognized parameters and l-indexed are stored landmarks parameters

This fusing procedure has to be repeated until all local hypotheses are combined. As a result, a set of combined and not combined local hypotheses $\mathcal{W} = \{\mathcal{L}_1, \mathcal{L}_2, \ldots, \mathcal{L}_b, \ldots, \mathcal{L}_B\}$ emerges.

The highest qualified local hypothesis is denoted as $\mathcal{L}_{b_\star} = (X_\star, Y_\star, \varphi_\star)$. The resulting transformation parameters $(X_\star, Y_\star, \varphi_\star)$ are used to transform the stored DLM content. To determine the 3 DOFs (X_m, Y_m, φ_m) of the mobile robot referring to the local map frame from $(X_\star, Y_\star, \varphi_\star)$, a similar affine transformation as in (8) is applied.

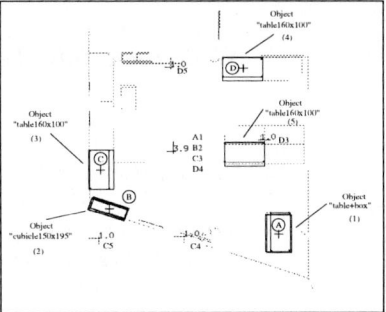

Fig. 3. The local hypotheses in an exemplary scenario

This strategy was tested in a project room consisting of various table groups, walls, cubicles and other items (fig. 3). The four landmarks $A-D$ were a priori stored in the representation of the robot. The object recognition algorithm delivered five identified objects (1) – (5), whereby object (5) was a misinterpretation[8]. As a result, eight local hypotheses emerged. Landmark A and B produced correct local hypotheses at each case. Because landmark C and D are related to the same object and three interpretations were created, six further local hypotheses were generated.

6 Future Work

We plan to enhance particularly the precision of the explored information by fusing camera data with data retrieved from the laser range finder. A mission expert for topological navigation will be developed to use the presented tools.

References

1. D. Burschka, C. Eberst, and C. Robl. Vision Based Model Generation for Indoor Environments. In *ICRA97*, pages 1940–1945, 1997.
2. D. Burschka and G. Färber. Active Controlled Exploration of 3D Environmental Models Based on a Binocular Stereo System. In *ICAR97*, pages 971–977, Monterey, California, USA, July 1997.
3. Oliver Faugeras. *Three-Dimensional Computer Vision*. Massachusetts Institute of Technology, The MIT Press, Cambridge, Massachusetts London, England, 1993.
4. A. Hauck and N. O. Stöffler. A hierarchic world model supporting video-based localisation, exploration and object identification. In *2. Asian Conference on Computer Vision, Singapore, 5. – 8. Dec.*, pages (III) 176–180, 1995.
5. Gunter Magin, Achim Ruß, Darius Burschka, and Georg Färber. A dynamic 3D environmental model with real-time access functions for use in autonomous mobile robots. *Robotics and Autonomous Systems*, 14:119 – 131, 1995.
6. G. Stockman. Object recognition and localization via pose clustering. *Computer Vision*, 40:361–387, 1987.
7. Roger Y. Tsai. A Versatile Camera Calibration Technique for High Accuracy 3D Machine Vision Metrology Using Off-the-Shelf TV Cameras and Lenses. *IEEE Transactions of Robotics and Automation*, RA-3(4):323–344, August 1987.

---

[8] The recognized object was a smaller table that did not belong to the a priori knowledge of the scenario.

Directing Robots with Visual Primitives for Navigation and Micro-manipulation

W.B. Tong[1], S.K. Tso[1], S. Lang[1], G.Z. Lu[2] and S.D. Ma[3]

[1] Centre for Intelligent Design, Automation and Manufacturing
City University of Hong Kong, Tat Chee Avenue, Kowloon, Hong Kong
[2] AI and Robotics Lab. Dept.of Computer and System Science
Nankai University, Tianjing, P.R.China, 300071
[3] National Lab. of Pattern Recognition, Institute of Automation
Chinese Academy of Sciences, Beijing, 100080, P.R. China

Abstract. For most present-day industrial robots, the user programs the robot using a robot language or teaches the robot using a pendant and/or joystick. In telerobots, the user controls the mobile robot or the robot arm by manually operating a joystick where the user himself acts as a cognition and decision making unit in the control loop. One of the important sensed information given to the user is visual feedback. Inspired by behavioral control of mobile robots, we propose a novel interface between the user and the robot to improve the performance of the robot system. In the proposed paradigm, the visual feedback exists in two control loops: one for the human supervisory control and the other for the robot itself for visual servoing. The human operator only provides the visual primitives to the robot and tells the robot its current situation which will activate a certain action, as in situation-to-action reactive/reflex control of mobile robots.

1 Introduction

Scientific and technological advances have reduced the human workload and improved mass production through the development of industrial robots. To extend human scientific explorations into remote and hostile environments, telerobots are being developed to conduct some tasks in place of human beings.

Currently, there are two approaches in pursuing this goal in robotics. One is the development of fully autonomous robots which are capable of acting without human intervention, but for complex applications, it is still difficult to achieve successful systems with high intelligence and full automation. The other approach adopted is via developing teleoperated robots called teleoperators which require human intervention at different levels in the perception-decision-control loop, or advanced human robot interfaces to let the end user operate the robots more intuitively.

To circumvent shortcomings of traditional AI methods for developing autonomous robot systems in the real world, some researchers have proposed some alternative control architectures[2, 3], such as the subsumption approach and the behavioral schema. Many researchers have investigated vision servoing control in the camera space[1, 5, 6, 9].

1.1 Visual feedback control in telerobots

In telerobots, visual feedback[7, 8] is an important source of information for the human operator to control the remote robot. It is difficult for the human operator to perceive depth cues by monoscopic video cameras, or to control the robot precisely by directly viewing the robot. Besides teleoperation from a distant site, teleoperation can act in a scale space such as micro-manipulation(for example, for cutting chromosomes, and injecting genes into cells under a microscope). Although researchers have developed stereoscopic video cameras recently to improve the human operator's performance, it is still difficult for teleoperation to carry out accurate positioning.

1.2 Providing visual primitives to the robot

We investigate a novel human-robot interface named interactive computer vision for robot navigation and manipulation. In this paradigm, the human operator provides some visual primitives to the robot, tells the robot of its situation and then the robot will automatically select an action and execute this action by visual servoing.

2 Visual primitives

Visual primitives can be considered as abstractions of those informative subsets of an image which are of interest in a given vision task. Features can be considered as basic elements in an image, and a pattern can be referred to as a feature vector which characterizes an object. What features to be selected depends on the given task. Boundaries or edges of objects are most frequently used features. A primal sketch is a data structure for representing gray level intensity changes in an image which includes zero crossings, blobs, terminations and discontinuities, edge segments, and boundaries. Here, a visual primitive is taken as a set of primal sketches around a fixation point which can be extracted by the vision system and is a function of the robot task and can act as visual constraints for guiding the robot. In this context, a visual primitive may be a function of a set of image features, such as textures, colors, convex edges, line segments and corners, etc., or a projection of an object model.

3 Directing a mobile robot with visual primitives

Figure 1 shows a sketch map of the paradigm of interactive teaching by showing.

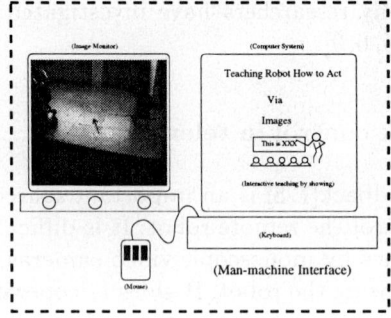

Fig. 1. Interactively driving a mobile vehicle via images

3.1 Telling the robot how to walk

To drive a mobile robot, the user can interact with the robot and the environment via images, i.e. visual feedback. In traditional methods, a human operator manually controls the robot via images while he/she perceives the remote site. The interactive vision method not only depends on a human operator's feeling of the remote site, but also depends on constraints from the images which the interactive vision system itself can understand or learn from the human operator. The human operator does not necessarily control the robot at all times via a pendant or a joystick, but just needs to give primitives to tell the robot how to act and show the robot where to go by indicating the goal point or drawing a path in the image, i.e. teaching by showing in images.

In this paradigm, the operator manipulates the mouse (ideally a 3-D mouse), moving the cursor on the image monitor screen. The mouse acts as an indicator or a pointer, as well as a commander.

The most frequently used mouse has three buttons and functional definitions are attached to each of the three buttons as follows:

- **Left button**: Drive the robot to rotate left or right so that the heading of the robot is directed toward the position which the mouse cursor points to. If the mouse cursor is placed at the left of the image center, the robot rotates to left, if it is at the right of the image center, the robot rotates to the right.
- **Right button**: Draw a trajectory in an image and drive the robot to "walk along" it.
- **Middle button**: Drive the robot to move so that the center of the image reaches the position which the cursor points to.

For example, let a camera tilt downward so that it can see the floor of the office or the hallway. Move the mouse and the cursor also moves in the image monitor screen. With the cursor located at the point A, press the left button, then the position of the point A in the image plane can be found via reading the mouse position. The position difference between the point A and the center point of the image can be computed, referring to the camera parameters(intrinsic and extrinsic), such as focal length, scale factors and height of the camera. Then the angle of the robot rotation

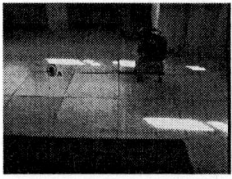  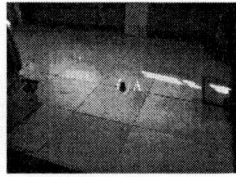

Fig. 2. Rotating left and driving forward the mobile vehicle via images and a mouse

can be computed and the visual servoing control strategies as described in section 2 can be applied. If there are some prespecified features with the point A, the vision system can automatically extract these features as tokens based on the intensity information or low-level features around the point A, and the robot automatically tracks and maintains these tokens. The human operator shows the object of interest to the robot. The robot extracts the descriptions of the object and then maintains these descriptions while it is interacting with the environment. Of course, these descriptions need to be defined and given to the robot by a human operator in advance. Figure 2 shows how to direct the robot to turn left and drive forward via images and a mouse.

3.2 The mobile robot tracking the cursor

To realize this simple tracking problem, we can get the cursor position and compute the difference between the current cursor position and the center position of the image, using visual servoing strategies to drive the robot head or the turret. In this paradigm, the cursor can only locate the object in the image and cannot locate the object in the 3D physical space. Drascic and Milgram of the University of Toronto proposed an interface called ARGOS (Augmented Reality through Graphic Overlays on Stereovideo) for the human operator to improve understanding of spatial relationships at a remote site. By applying stereo-video and stereoscopic graphics technology to teleoperation [4], the human operator can control the virtual graphic cursor to move in the camera space, through fusion of the human operator's left and right eye's images(making use of the human capability to solve the correspondence problem), and to locate the object in the 3D physical space.

3.3 Aligning with a line segment

In this section, we will explain in detail the implementation of a robot task: going through a door. Selected features in the camera space are projections of features of an object lying on the ground, (for example, a line segment of a door frame intersecting the floor, or a line segment of a wall intersecting the floor). Assuming that the visual primitives lie on the ground, we can compute these visual primitives in the robot motion space by the relationship between the image plane and the ground plane. When the human operator shows the door to the robot, the robot tracks the door and aligns itself with the door in order to get through.

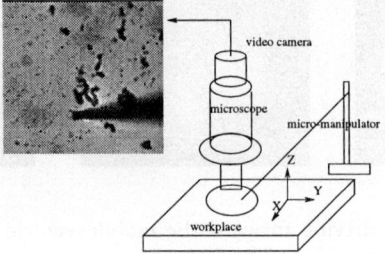

Fig. 3. Micro-vision and micro-manipulator

4 Directing a micro-manipulator under microscope

4.1 Teleoperation in a scale space

In genetic engineering, manipulations of cells, chromosomes, and small plant seeds are required, such as cutting a pair of chromosomes or transplanting genes. A special instance is transplanting genes into wheat seeds or animal cells. It is a tedious and difficult task for the human operator to manually control the gene gun(usually a slender glass tube) to inject the genes into the seed at the right location. Some injections require the gene gun be inserted into the cell nucleus along a specific axis to avoid damaging the cell. Thus how to make these operations automatic or semi-automatic and reduce the human operator's workload under the microscope deserves to be investigated. Here we will consider semi-automation of these operations. The human operator shows *where* and *what* to be manipulated in images, then the micro-mainipulator will finish the corresponding operations automatically.

Teleoperation or telemanipulation not only means "operation at a remote distance site far away from the human operator in the physical space", but also means "operation at a scale space which the human operator can observe with some special-purpose instruments", such as micro-manipulation under a microscope. which has many potential applications in bio-engineering.

4.2 Micro-vision and Micro-manipulation under microscope

Typical micro-manipulations under a microscope include "cutting" and "injecting". Here we will discuss how to implement these two manipulations based on visual guidance. As we do in guiding the mobile robot to walk by showing visual primitives according to a given task, micro-manipulation under the microscope can be carried out in the same way.

Recently, a microscope setup has been built with a micro-manipulator and video cameras, shown in Figure 3. The workplace of the microscope has three degrees of freedom(X, Y, Z). The micro-manuipulator also has three degrees of freedom. They can be driven by stepper motors with nanometer precision or manually driven by the human operator's hand. The video cameras can be monoscopic or stereoscopic, take images of the workplace through the microscope optical path, display images in an image monitor or send stereoscopic images to a special pair of glasses which the human operator wears. If the camera system is called a micro-vision system, it has many characteristics distinct from the present vision system in robotics and computer vision. Because the depth of view of the microscope is very narrow (image formation is almost for a object plane), the traditional (pin-hole) camera model cannot be applied to the micro-vision system. This feature may be useful to reconstruct 3D structures of an object using "focus".

Under manual control, the human operator controls the micro-manipulator by turning the handwheels to transform macro-motion of the human hand to micro-motion of the micro-manipulator or micro-motion of the workplace through mechanical transferring gears. But a human operator becomes fatigued easily when he repeatedly does the same thing such as injecting genes into cells, cutting chromosomes or cells, etc.

Some micro-mainipulations need the micro-manipulator's translational movements, such as cutting chromosomes, and some actions need the micro-manipulator to move along a constant direction, such as injecting genes into the nucleus of cells. To automatically control the micro manipulator to finish these operations, we can "point out", *where* and *what* (cells, seeds, chromosomes) to manipulate in the image, i.e. provide some visual primitives to the micro-vision system which will constrain and guide the micro-manipulator.

4.3 Considerations of visual guidance under microscope

The micro-manipulator under a microscope mainly does 2D motions in the workplace of the microscope. When it approaches the manipulation point, then it is controlled to move into that point or along a certain direction. These visual primitives can be very simple. For operations such as injecting into a cell or a seed once the human operator to indicates a "point" feature, the micro-manipulator is driven using this feature as a visual primitive by visual servoing techniques.

Focus on "point feature" or "tip" of the micro-manipulator. The micro-vision system working under the microscope has a very narrow depth of view and when the workplace or the micro-manipulator is a small distance away from the object plane, their images will blur seriously. The human operator can control his mouse to show the object of interest in the image which is dim, then the micro-vision system will bring that object into the focus plane by controlling the workplace to move along the Z axis.

A clear image has abundant details, i.e. it has many components with high frequencies in its Fourier transform. Assuming that the focal length of the microscope is unchanged, to get a clear image, adjusting the motion of the workplace along the Z axis is required.

Assume that the microscope's composite focal length is unchanged, and the image plane is also unchanged. To get a clear image, we must change the object distance d_1, i.e. move the workplace along the Z axis to let the object get nearer of farther from the lens. Thus we can use the sharpness of the image to control the workplace of the microscope or the micro-manipulator.

Define a quantity named discernment around the "point feature" which is indicated by the human operator. A small circular area W around this point is used to compute a statistical variable:

$$SMD = \sum_{(i,j) \in W} (|I(i,j) - I(i,j-1)| + |I(i,j) - I(i-1,j)|)$$

Fig. 4. Micro-manipulation under micro-vision using visual servoing

We can control the workplace to move so as to maximize the SMD. Because blurring of the image will take place on both sides of the object plane, if the SMD is decreasing continuously, the workplace must move back, until it reaches the maximum value of SMD.

Driving the micro-manipulator to approach "point feature". If the point feature and the 'tip' of the microscope are all clear, then we can drive the micro manipulator to approach the point feature using visual servoing.

We compute the difference between the point feature and the tip of the micro manipulator:

$$X_e = X_o - X_m, Y_e = Y_o - Y_m$$

(X_o, Y_o) stands for the feature point, and (X_m, Y_m) stands for the tip of the micro-manipulator. Using the orientation matrix from the micro-vision system to the micro-manipulator system, the differential motion of the micro-manipulator is given by:

$$\begin{bmatrix} \Delta x_m \\ \Delta x_m \\ 0 \end{bmatrix} = \begin{bmatrix} a_{11} & a_{21} & a_{31} \\ a_{12} & a_{22} & a_{32} \\ a_{13} & a_{23} & a_{33} \end{bmatrix} \cdot \begin{bmatrix} X_e \\ Y_e \\ 0 \end{bmatrix} = A \cdot R \cdot \begin{bmatrix} X_e \\ Y_e \\ 0 \end{bmatrix}$$

where R is an orientation transform matrix from the micro-vision system to the micro-manipulator, and A is a matrix that contains the microscope intrinsic parameters.

Cutting operation by visual guidance of the micro-manipulator. We can draw a line which defines the cutting path of the micro-manipulator in the image. The task is to control the micro-manipulator to track the defined path. Because this path is virtually defined by the human operator, the micro-vision system can not extract the real path of cutting, so the micro-vision system can at first extract the real feature trajectory according the path which the human operator indicates, then the micro-manipulator will follow the real feature trajectory. During the cutting operation, the human can monitor this procedure in progress.

Figure 4 shows an image taken from the micro-manipulatior system by the micro-vision system.

5 Future considerations for interactive computer vision

Visual tasks can be divided into two classes, recognition(identification) and manipulation (navigation), i.e. the *what* and *where* problems. Given an image, a human operator can quickly recognize what is in the image, but how does he/she tells that to the machine (vision system)? The vision system can focus on the *"where"* problem, if it knows what the object of interest is via the human operator in advance; or conversely, if human operator tells the vision system where the object of interest is in the image, then implicitly the vision system can focus on the *"what"* problem. In the former case, the vision system only needs to precisely locate the object position or analyze some features of it, In the latter case, the vision system can focus attention on operation in that area. The most important and difficult problem in the interactive computer vision paradigm is how to design a middle description where a human operator and the vision system can share the information. Generally, one person can easily and intuitively understand what other people do, because they have the same description of the world in their 'thinking' brain. How does the human operator interact with the vision system in the environment? The interactive vision is not a full autonomous image understanding system which the traditional computer vision community is tackling. It is also not an information transferring tool for human perception similar to traditional teleoperation. It is directed by the human operator with high-level perception. It has its own automatic image understanding abilities to make decisions, report results and act on the environment based on sensory information received at the low-level perception. The paper describes an approach to interactive vision useful for robot navigation and micro-manipulation.

References

1. Special issue on vision-based control of robot manipulators, IEEE Trans. on Robotics and Automation, Vol.12, No. 5, October, 1996.
2. R. C. Arkin,"Motor Schema based mobile robot navigation," Int. J. of Rob. Res., Vol.8, No. 4, 1989, pp.92-112.
3. R.A. Brooks,"A Robust layered control system for a mobile robot", IEEE Journal of Robotics and Automation, Vol.RA-2, No.1, March 1986, pp.14-23.
4. D. Drascic and P. Milgram, "Positioning accuracy of a virtual stereographic pointer in a real stereoscopic video world", SPIE Vol.1457: Stereoscopic Displays and applications II, 58-69, San Jose, California, September 1991.
5. J.J. Feddema, C.S.G. Lee, and O.R. Mitchell, "Weighted selection of Image features for resolved rate visual feedback control", IEEE Trans. on Robotics and Automation, Vol. 7, No. 1, February 1991, pp.31-47.
6. K. Hashimoto, (Ed.), Visual servoing : real-time control of robot manipulators based on visual sensory feedback, World Scientific, 1993.
7. N.P. Papanikolopoulos and P.K. Khosla, "Shared and traded telerobotic visual control", In Proc. IEEE Int. Conf. on Robotics and Automation, 1992, pp. 878-885.
8. F. Tendick, J. Voichick, G. Tharp and L. Stark, "A supervisory telerobotic control system using model-based vision feedback", In Proc. IEEE Int. Conf. on Robotics and Automation, 1991, pp.2280-2285.
9. L.E. Weiss, et al., "Dynamic sensor-based control of robots with visual feedback", IEEE J. Robotics and Automation, Vol. RA-3, Oct. 1987, pp. 404-417.

Combining Camera and Laser Radar for ALV Navigation

Zhang Qi

Post & Telecommunications Project Institute of Zhejiang Province,
8 Liu-Yuan-Qian-Dao-Lu Road, Zhaohui 2nd District, Hangzhou, 310014, P. R. China

Gu Weikang

Institute of Information and Intelligent System, Department of Information and Electronics,
Zhejiang University, Hangzhou, 310027, P. R. China

Abstract — Efficient approaches to environment modeling and obstacle detection for the navigation of Autonomous Land Vehicle(ALV) based on multisensor data fusion are presented. The fusion algorithms are based on the new generalized Dempster-Shafer's theory of evidence. Experiments on real outdoor environments have proved the presented algorithms are effective.

1 Introduction

Recently there has been an increasing interest in the development of autonomous systems capable of using many different sources of sensory information. This interest arises from a realization that there are fundamental limitations on any attempt at modeling the environment based on a single source of information: A single source of sensory information can only provide partial and imprecise information. Diverse information from different sources can be used to overcome the limitations inherent in the use of single sensors [1], thus make possible robust results of detection, modeling and recognition etc.

This paper describes an integrated vision system in ALV on which a CCD color camera and a LIRS are coupled. The main purpose of this system is to provide knowledge about the environment, especially the obstacle information, for the path planner when ALV is navigating on outdoor world, by integrating the pre-processed 2D color image obtained by CCD color camera and the 3D range image sensed by LIRS. Camera is faster in image acquisition and the color image is higher in resolution but it is difficult to recover 3D information from it. In contrast, range image provides 3D information about the environment but usually noisy and with lower resolution. So the fusion of both would provide more accurate, complete and reliable information that are less variant to scene conditions, occlusion and viewing position than single sensor systems.

As the D-S approach for evidential reasoning explicitly models "uncertainty" and "ignorance" and its formulation for computation is simple, it is more suited to the task of environment modeling and obstacle detection, which has been proven with experiments on real outdoor images. The D-S theory evidential reasoning has been successfully utilized in a variety of applications of multisensor fusion [2]. In this paper, it is used to provide environment descriptions and obstacle information for path planning in ALV.

Dempster's rule of combination is the most important tool for belief value updating after the acquisition of new evidence. However, the rule has some limitations. The first problem is the evidence independence requirement [3], which is seldom met in ALV while it is navigating. In addition, Zadeh has pointed out situations in which the form of normalization used in the rule leads to some counterintuitive results [4]. In order to solve the two problems, a new generalized fusion algorithm which not only relaxes the requirement of evidence independence but also avoids the problem pointed out by Zadeh is presented.

2 The Integrated Vision System and Environmental Model

The integrated vision system of our ALV includes CCD color camera and LIRS.

LIRS is an amplitude-modulated continuous-wave laser radar that sensed the range by measuring the shift in phase between an emitted beam and its echo. Since the range to a target is proportional to the difference of phase and the phase is defined modulo 2π, LIRS only senses the relative range, not absolute one.

For the purpose of navigation, it is usually sufficient to know where the ALV can go through and where can not. Based on this reason, we adopt a terrain representation structure known as "Global Common Grid"(GCG) map to model the environment.

On this structure GCG map with the size of $m \times n$ is established on the ground in front of ALV. The size of each grid depends on the resolution of LIRS, and usually is the multiple of the minimum distinguishable size of LIRS; therefore the information acquired by two sensors can be projected onto the same representation model, i.e. the GCG map. A multisensor data fusion process is implemented in each grid and as a result the ground map is labeled with "Free" or "Obstacle" at each grid. Our experiments show that this representation model of GCG map not only simplifies the fusion process and is convenient for path planning but also can obtain a good trade-off between speed of fusion and efficiency of environment description.

3 D-S Theory of Evidence and Its New Generalized Form
3.1 The Dempster-Shafer's Theory of Evidence

Let Θ be a set of mutually exclusive and exhaustive hypotheses about some problem domains. Relevant propositions are presented as subsets of this set Θ which is called *the frame of discernment*.

Definition 1 Let Θ be a frame of discernment(Θ will always be assumed to be finite).

(i) A *basic probability assignment function*(BPAF) on Θ is a function m from 2^{Θ} to $[0,1]$ such that
$$m(\emptyset) = 0 \quad \text{and} \quad \sum_i m(A_i) = 1$$
where the non-null subset A_i of Θ is called *focal element*.

(ii) The *belief function* of B, denoted as *bel(B)*, induced by the BPAF m on Θ is defined as:
$$bel(B) = \sum_{A \subseteq B} m(A)$$

(iii) The *plausibility function* of B, denoted as *pls(B)*, induced by the BPAF m on Θ is defined as:
$$pls(B) = \sum_{A \cap B \neq \emptyset} m(A)$$

An important issue in the D-S theory is the procedure for aggregating multiple belief structures on the same variable. This procedure of combination is based upon the Dempster's rule for aggregation and be seen as a kind of conjunction of the belief structures. The combination is performed by the orthogonal sum of Dempster, expressed for n sources as

$$\oplus_{i=1}^{n} m_i(A) = \frac{1}{1-k_{conflict}} \sum_{\cap_{i=1}^{n} B_{ij}=A} \prod_{i=1}^{n} m_i(B_{ij}) \qquad (1)$$

where $A, B_{1j}, B_{2j}, ..., B_{nj}$ are subsets of Θ, and

$$k_{conflict} = \sum_{\cap_{i=1}^{n} B_{ij}=\varnothing} \prod_{i=1}^{n} m_i(B_{ij})$$

is a measure of conflict between the n belief structures.

3.2 Generalization of the D-S Theory of Evidence

Dempster's rule of combination is the most important tool for belief value updating. However, this rule requires that the bodies of the two pieces of evidence must be independent, which is considered to be a very strong constraint. Besides, the combination of highly conflicting structures is in a dubious fashion and will produce some counterintuitive results.

3.2.1 The Dependency Degree of the Two Evidences

When a new evidence source is joined, several new focal elements are produced by the source. It is easy to see that the capability of one evidence to provide information is related to the cardinal number of its focal element set, which can be described by the function $Eng(EVI)$ defined as follows.

Definition 2 The function $Eng(EVI)$ is defined as

$$Eng(EVI) = \sum_{i=1}^{|\{A_i\}|} \frac{m(A_i)}{|A_i|} \quad A_i \neq \Theta$$

where A_i is the focal element, $|A_i|$ is the cardinal number of focal elements A_i, and $|\{A_i\}|$ is the number of focal element sets.

Suppose the BPAF of evidence EVI_1 and EVI_2 are m_1 and m_2 respectively, and their focal element are A_i and B_j. It is possible that some focal elements of EVI_1 and EVI_2 will be dependent, and the dependence degree is related to the number of the dependent focal elements and the sum of their BPAF values. Now we define the dependence degree as follows.

Definition 3 The dependency coefficient of EVI_1 to EVI_2 and EVI_2 to EVI_1 is defined respectively as

$$R_{12} = \frac{Eng(EVI_1, EVI_2)}{Eng(EVI_1)+Eng(EVI_2)} \frac{Eng(EVI_2)}{Eng(EVI_1)} \qquad (2)$$

$$R_{21} = \frac{Eng(EVI_1, EVI_2)}{Eng(EVI_1) + Eng(EVI_2)} \frac{Eng(EVI_1)}{Eng(EVI_2)} \qquad (3)$$

3.2.2 Normalization and Conflict in the Dempster's Rule of Combination

In the Dempster's rule the quantity k is a measure of the degree to which the combining structures disagree with each other; therefore if $k=1$, we can not use Dempster's rule to combine the structures. This is a problem in many case. Zadeh has shown the use of this normalization factor in highly conflicting evidence can lead to some counterintuitive results. So highly conflicting structures combine in a dubious fashion.

3.2.3 Modification of the Dempster's Rule of Combination

As we have shown in a previous section, the Dempster's rule of combination needs the independence assumption and in the meantime can lead to some counterintuitive results sometimes when there is conflict between two pieces of evidence. In this section we shall suggest a modification of the Dempster's rule in order to solve the two problems.

Definition 4 Let m_1 and m_2 are two evidence structures on the set Θ, and $\{A_i\}$ and $\{B_j\}$ be their sets of focal elements. We shall define a modified combination rule $\hat{m} = m_1 \hat{\oplus} m_2$ as follows:

$$\begin{cases} \hat{m}(A) = \sum_{A_i \cap B_j = A} m_1'(A_i) m_2'(B_j) & A \neq \emptyset, \Theta \\ \hat{m}(\emptyset) = 0 \\ \hat{m}(\Theta) = \left(\sum_{A_i \cap B_j = \Theta} m_1'(A_i) m_2'(B_j) \right) + k_{conflict} \end{cases} \qquad (4)$$

where

$$m_1'(A_i) = \begin{cases} m_1(A_i)(1 - R_{12}) & A_i \neq \Theta \\ 1 - \sum_{A_i \subset \Theta} m_1'(A_i) & A_i = \Theta \end{cases}$$

$$m_2'(B_j) = \begin{cases} m_2(B_j)(1 - R_{21}) & B_j \neq \Theta \\ 1 - \sum_{B_j \subset \Theta} m_2'(B_j) & B_j = \Theta \end{cases}$$

$$k_{conflict} = \sum_{A_i \cap B_j = \emptyset} m_1'(A_i) m_2'(B_j)$$

in which R_{12} and R_{21} are defined according to formula (2) and (3) respectively.

The fundamental distinction between this modified combination rule and the Dempster's rule is: (1)In essence, the above formula first transforms the dependent evidences to independent ones then combines them. (2) With the use of this new rule the conflicted portion of the basic belief is put back into the set Θ. In this case we are saying that since we don't really know anything about the conflicted portion, we let it be distributed among all the elements rather than just those in the focal sets. As the new rule combines dependent bodies of evidence and can obtain more intuitive results, this generalization is reasonable.

4 Implementation of the Algorithm in ALV

The fusion level in the system was determined by the system's tasks and the data to be processed in the system: as what this system will do is to detect obstacle and provide environment modeling for path planning in ALV and both of the precision and speed of fusion must be considered simultaneously, the feature level fusion is suitable to the fusion, namely, the fusion between the pre-processed 2D and 3D information.

4.1 The Pre-processing of Information for Fusion

As the range image contains strong noise, a 3×3 median filtering operator is first used. Then the original relative range is converted into an absolute one. The next step of pre-processing is to convert the range image from the spherical coordinate system into a sensor-centered Cartesian coordinate $\{x, y, z\}$. The converted image whose pixel values reflect the height of the points above ground is called elevation map. The last step is to threshold the elevation map. Points whose elevation is higher than threshold are determined as obstacle.

At the stage of color image pre-processing, a 3×3 median filter is also first used. Because on the flat road environment the color spectrum can be classified in finite classes. The source color image can be compressed by clustering, and the color number can be reduced. So each pixel in the color image representing road, edge or obstacle can be determined.

In a multisensor fusion system, because there are differences of sensor system in geography or geometry as well as differences of information in resolution and types, one of the important problem is to calibrate each sensor and determine the transformations from the coordinate system of one sensor to the others. In this paper, the sensors are calibrated by the technique presented by Ke Liu etc. [5], and the transformation matrices are T_c and T_r, which are used to project the pre-processed color image and range image onto the GCG map.

4.2 Selection of The Evidence

In implementing the D-S theory, the selection of the discernment frame Θ and the construction of BPAFs are the key problems. For our system, the Θ is defined as $\{F, O\}$ to represent two types of regions, i.e. the free region on the road(denoted as F), the obstacle region(denoted as O). For simplicity, we assume(these assumption are satisfied in many cases):

(1)There is no obstacle at the edges of the image.
(2)Regions except the obstacle region are all free.

With the two assumption, we can use the information of edge to extract features of free regions.

Because the range images are lower in resolution and badly spoiled by stronger noise, we use two kinds of features extracted from the range image to classify the two region classes to get more accurate fusion results and make range information sufficiently redundant. The two kinds of features are: (1) the average height of each grid $\bar{h}$; (2) the height distribution variance in a grid σ_h^2. The $\bar{h}$ and the σ_h^2 are defined as:

$$\bar{h} = \sum_{i=1}^{N} \frac{z_i}{N} \qquad \sigma_h^2 = \frac{1}{N}\sum_{i=1}^{N}(z_i - \bar{h})^2$$

where N is the total numbers of obstacle points in the grid.

The two features are easy to compute from the elevation map.

Now we use a $n \times n$ window moving in the image. Assume the mean values of a feature on the free and the window region are μ_F and μ_W respectively, and the probability distribution are $p_F(x)$ and $p_W(x)$, and A is the area between μ_F and μ_W under curve $p_F(x)$. Thus a distance measure d_{WF} can be defined as

$$d_{WF} = 2\left|\int_{\mu_W}^{\mu_F} p_F(x)dx\right| = 2A$$

So if the mean values of $\bar{h}$ and σ_h^2 in free region are $\mu_{\bar{h}}^F$ and $\mu_{\sigma_h^2}^F$, and the probability distributions are $p_{\bar{h}}^F(x)$ and $p_{\sigma_h^2}^F(x)$ respectively, and the mean values of $\bar{h}$ and σ_h^2 in the $n \times n$ window region are $\mu_{\bar{h}}^W$ and $\mu_{\sigma_h^2}^W$ respectively, then we define:

$$m_{\bar{h}}(\{F\}) = 1 - 2\left|\int_{\mu_{\bar{h}}^W}^{\mu_{\bar{h}}^F} p_{\bar{h}}^F(x)dx\right|, \quad m_{\bar{h}}(\Theta) = 1 - m_{\bar{h}}(\{F\}), \quad m_{\bar{h}}(A_i) = 0$$

$$m_{\sigma}(\{F\}) = 1 - 2\left|\int_{\mu_{\sigma}^W}^{\mu_{\sigma}^F} p_{\sigma}^F(x)dx\right|, \quad m_{\sigma}(\Theta) = 1 - m_{\sigma}(\{F\}), \quad m_{\sigma}(A_i) = 0$$

where A_i are other subsets of Θ except $\{F\}$ and Θ.

The $p_{\bar{h}}^F(x)$ and $p_{\sigma_h^2}^F(x)$ can be obtained by sample training.

As to color image, the prior distribution of colors in an obstacle region, denoted as C_0, can be found by a statistical method applied to the compressed color images. Assume C_0 is steady(which is true in many cases). If the color set extracted from a grid is C, we use the value of function

$$f_c = \|C \cap C_0\| / \|C_0\|$$

as the criterion to determine whether the grid is obstacle or free region, using color information. If the color distribution in a grid C is similar to the distribution in the free region C_0, then f_c tend to be 1 and the grid is classified as free grid in high probability.

We use the above three parameters as grid classification evidence. In addition, because of the obvious dependence and sometimes conflict, the fusion algorithm is based on the new generalized Dempster-Shafer's theory of evidence.

4.3 The Process of Fusion

Applying the above three evidence, we can use Formula (4) to calculate the combined BPAF. For every grid in one level, after combining the BPAFs of all features, the next stage is to classify the grid as one of the two classes (F, O) based on the fused BPAFs or belief intervals. The criterion and method for classification based on BPAFs should be selected according to the problem to be solved. In this system, a rule-base method is adopted. Through analyzing the probability sense of BPAF, the following rules are formed:

$$\begin{cases} m_{object} > m_{other} \\ m_{object} - m_{other} > T_{belief} \\ m_{object} > m(\Theta) \\ m(\Theta) < T_{uncertainty} \end{cases}$$

where m_{object} and m_{other} are the BPAFs of object class and any other class respectively, and $m(\Theta)$ is the BPAF for uncertainty, and T_{belief} and $T_{uncertainty}$ are thresholds.

5 Conclusion

The efficiency of the presented method has been proven by experiments on the real outdoor environment. Fig. 1b shows one of the experimental environment, an outdoor image with a road flanked by the neatly pruned shrubs. Fig. 1a is the range image of the same scene. The pre-processed results of source color and range image are shown in Fig. 1c and Fig. 1d respectively.

From Fig. 1e we can see, the results of fusion is satisfying. The locations of roads from the 2D and 3D image are almost coincide in the fused resultant image, and the error is less than *10cm* at the distance *15m* from the radar. The validity of the generalized Dempster-Shafer's theory of evidence has been verified by the experiments. Because several features used here may be dependent on each other and sometimes be conflicting, the generalization of Dempster's rule of combination is reasonable for ALV navigation.

In addition, the time cost is also decreased, partly because fusion is achieved on feature level and partly because the GCG map for environment modeling which has been proven to be convenient for multisensor fusion and path planning is unitized.

In short, experiments show our algorithm is simple, robust and efficient. It can satisfy the requirement of ALV navigation in a relatively flat road environment.

Reference

1 Durrant-Whyte H. F. Sensor models and multi-sensor integration. Int. J. of Robotics Research, 1988, 7(6):97 ~ 113.
2 Lowrance J. D. and Garvey T. D. Evidential reasoning: A developing concept. In: Proc. of Int. Conf. on Cybernetics and Society, Oct., 1982: 6 ~ 9.
3 Shafer G. A mathematical theory of evidence. Princeton Univ. Press, Princeton, NJ, 1976.
4 L. A. Zadeh. A simple view on the Dempster-Shafer theory of evidence and its implication for the rule of combination. AI Mag. 7, 1986.
5 Ke Liu, Ren Jiang, Qian Zhang, Yong-Qing Cheng and Jing-Yu Yang. Calibration of multiple sensors by a planar calibration object. In: SPIE Vol. 1828 Sensor Fusion V, 1992: 499 ~ 509.

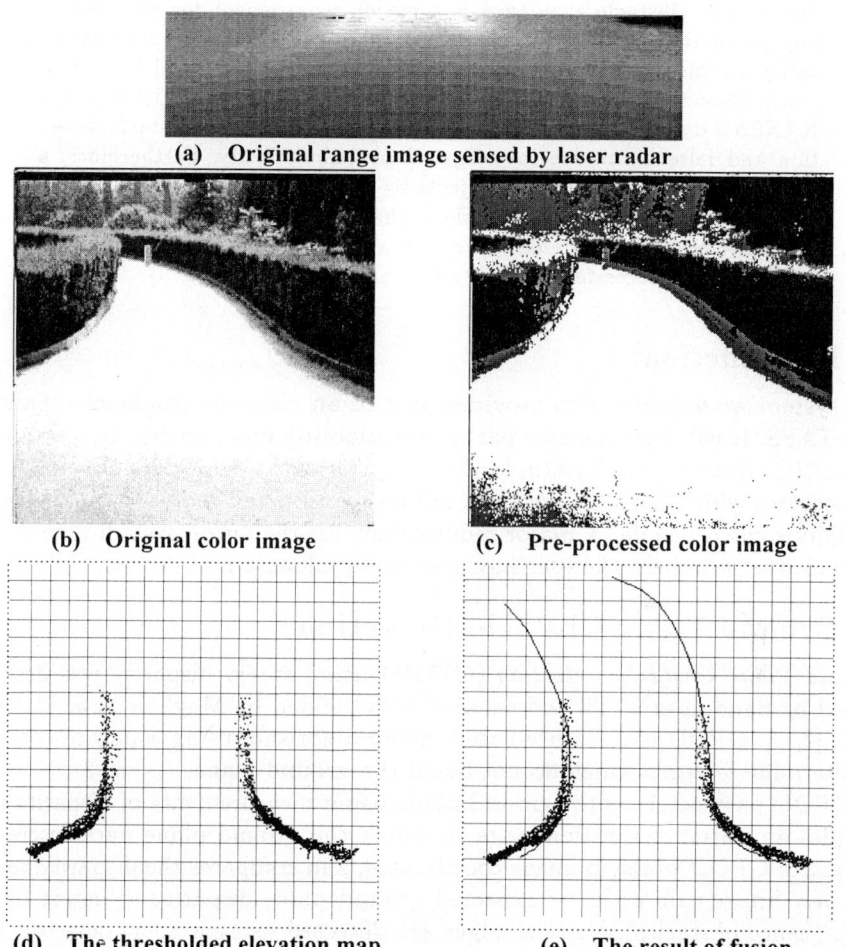

(a) Original range image sensed by laser radar

(b) Original color image (c) Pre-processed color image

(d) The thresholded elevation map (e) The result of fusion

Fig. 1 The experimental result of the modified D-S theory in multisensor fusion

Stereo Vision-Based Obstacle Detection for Partially Sighted People

Stephen Se and Michael Brady

Department of Engineering Science, University of Oxford, Oxford OX1 3PJ, U.K.
{syss,jmb}@robots.ox.ac.uk

Abstract. Obstacle avoidance is a major requirement for any technological aid aimed at helping partially sighted (TAPS) people to navigate safely. In this paper, a stereo vision-based algorithm (Ground Plane Obstacle Detection) is extended to detect small obstacles for TAPS using RANSAC dynamic recalibration and Kalman Filtering. Obstacle detection and false alarm are investigated probabilistically. Furthermore, a technique is developed to find objects by matching their edges with some heuristic criteria. Experiments show that obstacle edges are extracted much better with our dynamic recalibration approach and that objects can be found successfully by the edge matching technique.

1 Introduction

The system we describe here provides part of an obstacle avoidance capability for a TAPS. It will form a major part of the mobility function of a larger project, ASMONC (Autonomous System for Mobility, Orientation, Navigation and Communication) which aims to provide a full navigation and mobility capability for partially sighted people. A major requirement for the vision system is to detect small obstacles to help the user navigate safely along a path.

2 Ground Plane Obstacle Detection

Ground Plane Obstacle Detection (GPOD) using stereo disparity was first reported by Sandini *et al.* [3] and subsequently refined by Mayhew *et al.* [10] and by Li [7, 8]. GPOD is a feature-based stereo algorithm using a pair of cameras to determine features which do not lie on the ground plane.

GPOD parameterises the ground plane using measurements of disparity and includes an initial calibration stage in which the ground plane parameters are extracted. GPOD works in image coordinates, and compares the disparity values in a new image pair with the expected ground plane disparity to detect differences (hence obstacles). Vertical edges are detected using a Sobel detector, and stereo matching uses the PMF algorithm [13, 14]. Images of the ground with line features but no obstacles are used to initialise the ground plane estimate.

The ground plane disparity d varies linearly with cyclopean image plane coordinates (u, v) [8], that is

$$d = au + bv + c \qquad (1)$$

where (a, b, c) are the *ground plane parameters*.

3 Probability and False Alarms

In purely geometric terms, we can easily derive how many pixels are subtended by an obstacle, however, noise makes obstacle detection a stochastic process and geometry alone does not capture the stochastic element. Therefore, reliability is inevitably expressed probabilistically [9].

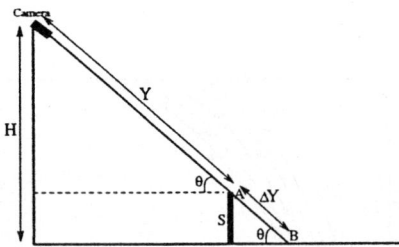

Fig. 1. *The camera geometry showing the various parameters.*

Disparity is given by $d = \frac{fI}{Y}$, where f is related to the camera intrinsic parameters, I is the interocular distance, and Y is the distance to the object or ground [5]. Referring to Figure 1, the quantity used in GPOD for checking obstacles is the difference between the measured disparity at A and the predicted disparity at B, which is

$$\Delta d = \frac{fI}{Y} - \frac{fI}{Y + \Delta Y} = \frac{fI}{Y}\left[\frac{\frac{\Delta Y}{Y}}{1 + \frac{\Delta Y}{Y}}\right] = \frac{fI}{Y}\left(\frac{S}{H}\right) \quad (2)$$

The probability of detecting an obstacle of size S is the probability of its disparity difference being larger than the threshold. Assuming that the disparity difference is normally distributed, from Equation 2, we have

$$P(\Delta d > threshold|S) = \frac{1}{\sqrt{2\pi}\sigma_{\Delta d}} \int_{threshold}^{\infty} e^{-\frac{1}{2}\left(\frac{\Delta d - \frac{fIS}{YH}}{\sigma_{\Delta d}}\right)^2} d(\Delta d) \quad (3)$$

We can illustrate this by analysing the case in which we want to detect an object of height 10cm (e.g. a small step, sufficiently high to inconvenience a partially sighted person) at a distance of 5m. With our known camera configuration and parameters, this corresponds to a disparity difference of 1.7. In Equation 3, $\sigma^2_{\Delta d}$ is the variance of the disparity difference given by

$$\sigma^2_{\Delta d} = \sigma^2_{d_{measured}} + \sigma^2_{d_{expected}}$$

where $\sigma^2_{d_{measured}} = 1$ is assumed and $\sigma^2_{d_{expected}}$ is given by the ground plane parameters uncertainty analysis [16] assuming the image coordinates variances are unity.

A vertical edge string of length m is regarded as an obstacle if at least one pixel of the edge string is detected as an obstacle. Therefore, if p_i is the probability obtained in Equation 3 for the ith pixel, then assuming independence, the probability of obstacle detection is $1 - \prod_{i=1}^{m}(1 - p_i)$.

Figure 2 shows the probability of obstacle detection for different obstacle sizes at various distances. We see that with our current configuration, to obtain 90% detection rate, the minimum obstacle size increases from 6cm to 11cm as the distance increases from 2.5m to 5m.

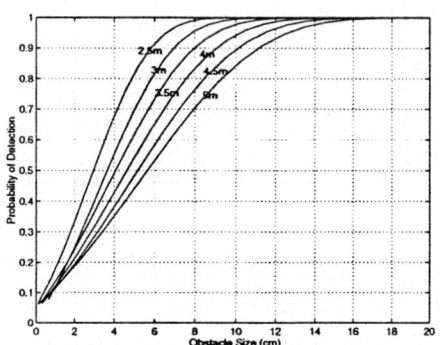

Fig. 2. *Obstacle detection probability at various distances (threshold=1.7).*

False alarm probability is the likelihood that an obstacle is detected when nothing is actually there:

$$P(False\ Alarm) = \frac{1}{\sqrt{2\pi}\sigma_{\Delta d}} \int_{threshold}^{\infty} e^{-\frac{(\Delta d - LSFE)^2}{2\sigma_{\Delta d}^2}} d(\Delta d)$$

where $LSFE$ is the least-squares fitting error, i.e. $d_{(u,v)} - (au + bv + c)$, in which $d_{(u,v)}$ is the measured disparity for a ground plane point (u, v). Figure 3(a) shows the false alarm rate for a typical obstacle-free scene when $threshold$ is 1.7. It indicates quite a high false alarm rate.

However, we can discard any single-pixel obstacles detected and not regard them as false alarms, since it is highly likely that they are due to noise. Let pixel $(i-1)$, pixel i and pixel $(i+1)$ be three consecutive pixels on a vertical edge, and let P_{i-1} be the probability of a false alarm for pixel $(i-1)$, P_i that for pixel i, and P_{i+1} that for pixel $(i+1)$. Then, the false alarm probability (excluding single-pixel obstacles) is given by $P_i - (1 - P_{i-1})P_i(1 - P_{i+1})$.

Figure 3(b) shows the false alarm rate in this case which is reduced substantially compared to Figure 3(a), as the chance of two consecutive points being affected by noise in the same way is much lower. A real obstacle can reasonably be expected to occupy at least a few contiguous pixels, so this simple heuristic does not affect obstacle detection much.

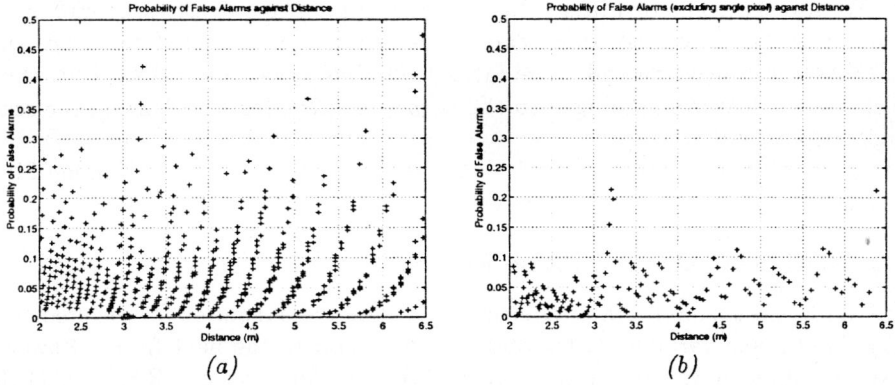

Fig. 3. *False alarm probability for real ground plane data. (a) With threshold being 1.7. (b) With exclusion of single-pixel obstacles and threshold being 1.7.*

4 RANSAC Dynamic Ground Plane Recalibration

For wheeled mobile robots moving over flat ground, there is no relative change in position of the ground plane from the cameras, hence the cyclopean ground-plane disparity function is fixed. However, cameras attached to the shoulders of a partially sighted person in a TAPS move up and down while he/she is walking around. The cameras move with six degrees of freedom. Therefore, a one-time, fixed ground plane calibration cannot be used to detect small obstacles.

4.1 Dynamic Ground Plane Recalibration (DGPR)

We propose dynamic recalibration of the ground plane to prevent human movement affecting obstacle detection and to obtain a better estimate of the ground plane for slopes, hills or non-flat ground.

DGPR [15, 12] recalibrates the ground plane parameters at each step. Iteratively, it uses step k's ground plane parameters for obstacle detection, partitions the features found into ground plane features and obstacle features. The ground plane features and the estimated camera movement between steps are then used to obtain step k+1's ground plane parameters.

4.2 Kalman Filter Tracking

In a multiple target tracking system, we confront the data association problem, which addresses how to associate predictions of target positions with actual measurements. The Mahalanobis distance [17] quantifies the likelihood of a measurement originating from a specific geometric feature.

The nearest-neighbour approach uses this distance metric to associate the measurements to their closest geometric features. It may perform badly as the closest measurements are not always correct [2]. Nevertheless, it is both computationally and conceptually simple, and is employed in our current work.

We use the Kalman Filter [1] to track ground plane features as well as obstacle features. Their positions can be determined more accurately and, with suitable track initiation and termination techniques, we can deal with situations such as new features coming into the scene, existing features leaving the scene and temporary occlusion.

In addition to tracking ground plane *features*, we use a further Kalman Filter to track the ground plane *parameters* in order to better estimate them.

4.3 RANSAC Ground Plane Fitting

In DGPR, the six d.o.f. camera motion parameters are required for the prediction of ground plane parameters for obstacle detection in the next frame. However, they are difficult to obtain accurately [11], therefore, we use RANSAC [4] for ground plane fitting in each frame.

RANSAC takes all the image features (provided that there are sufficiently many ground plane features, and not all obstacle features lie on the same plane), fits the ground plane features and discards the obstacle features as outliers.

A sequence of stereo images of a real outdoor scene was captured at 128x128 resolution. The environment is a tiled pavement with various obstacles. There was some camera motion between the images with translations up to 20cm and rotations up to 5 degrees, which cover the extreme case for human movement [6].

Figure 4(a) shows an image of the sequence where the white rectangle indicates the window of interest. Results from GPOD and RANSAC-DGPR are shown in Figures 4(b) and 4(c) respectively, where detected obstacle edges are marked. It can be seen that obstacles are missed by GPOD but are detected by RANSAC-DGPR, showing that using only the initial ground plane parameters is insufficient to detect obstacles in the presence of camera motion, and that the RANSAC-DGPR approach gives promising results. Detected obstacles in the scene include a 10cm-high brick at 1.5m and a 15cm-high box at 3m.

5 Objects

So far, only obstacle edges have been detected, as edges are sometimes weak, we cannot tell which pairs of vertical edges correspond to the sides of an object. This makes it difficult to advise the partially sighted person how to avoid the obstacles. An edge matching technique is developed using notions similar to disparity constraints, intensity correlations and mutual admiration in stereo.

5.1 The Algorithm

Among the edges in the candidate edge list, for each pair of edges i and j, compute a 'score' indicating the likelihood of them being the two sides of an object (see the next section).

For edge i, find its partner by choosing an edge which gives a *lowest* score among all the other edges. If there exist edges i and j ($i \neq j$) such that edge i chooses edge j and edge j chooses edge i (mutual admiration), then edges i and j are declared as a pair and removed from the candidate edge list.

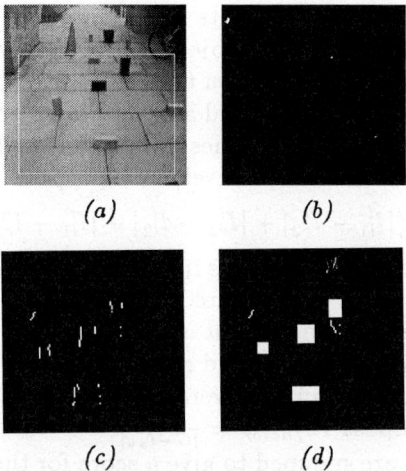

Fig. 4. *Obstacle detection. (a) The original image. (b) Result from GPOD. (c) Result from RANSAC-DGPR. (d) Edge matching on RANSAC-DGPR result.*

Any score needs to be below a threshold, otherwise, the pair is not declared. This is to allow cases where there does not exist a match for a certain edge.

Repeat until all possible pairs are picked out.

5.2 Scoring

For each prospective edge pair, we compute a score to determine how likely it is that they are the two sides of a single object. This score consists of three factors: disparity, position and intensity.

Normally, the two sides of an object should be at about the same distance from the camera, therefore they should have similar disparities. Let the average disparity of the two edges be $d1$ and $d2$ respectively, then we have $D_{factor} = k_1|d1 - d2|$.

There are three components to the position factor: u-direction, v-direction and length.

Let $u1$ and $u2$ be the horizontal coordinates of the two edges under consideration. We would like to eliminate the case of them being too close, in particular, to avoid $u1 = u2$ as they could not be a pair in that case: we have a term inversely proportional to $|u1 - u2|$. However, by itself this would bias towards picking two edges as far away as possible which is contrary to the usual situation. Therefore, we propose $Pu_{factor} = \frac{k_2}{|u1-u2|} + k_3|u1 - u2|$.

The midpoint of the two edges forming a pair should be at a similar distance. Let $v1$ and $v2$ be the vertical coordinates of the midpoint of the two edges, we propose $Pv_{factor} = k_4|v1 - v2|$.

The two edges forming a pair should be of similar length. Let $l1$ and $l2$ be the lengths of the two edges, then we propose $Pl_{factor} = k_5|l1 - l2|$.

Similarly, there are two components to the intensity factor.

If two edges are the sides of an object, the intensity to the left of one edge (I_{l1}) is in many cases similar to that on the right of the other (I_{r2}); the intensity to the right of the one edge (I_{r1}) should also be similar to the intensity to the left of the other edge (I_{l2}). The same applies to both left and right images, indicated by superscribing with l and r respectively. We propose

$$I1_{factor} = k_6(|I_{l1}^l - I_{r2}^l| + |I_{r1}^l - I_{l2}^l| + |I_{l1}^r - I_{r2}^r| + |I_{r1}^r - I_{l2}^r|)$$

Moreover, we also need to check the intensity of a patch between the prospective pair against the average background intensity. Let G be the average intensity of the whole image, assuming most of it is ground region with relatively little obstacle clutter, then G is close to ground region intensity. We can use a 3x3 patch between the two edges and find the average intensity I_{av} to compare against G. For this reason, we propose $I2_{factor} = \frac{k_7}{|G - I_{av}|}$.

Finally, the factors are summed to give a score for the prospective pair under consideration. The constants k_is appearing in the factors are partly experimental and are individually weighed according to their importance and stability under noise. For example, since the D_{factor} is important and stable to noise, it is given a larger weight. On the other hand, we find that the image coordinates and pixel length are susceptible to noise, hence have smaller weights. We also need to set the threshold t above which a match is not declared.

5.3 Results

Sensitivity analysis shows that the following chosen parameters ($k_1 = 100$, $k_2 = 500$, $k_3 = 10$, $k_4 = 10$, $k_5 = 10$, $k_6 = 1$, $k_7 = 5000$, $t = 1000$) give stable results. Applying this to the result from Figure 4(c), we obtain Figure 4(d). Each object block, whose two sides are paired, is indicated.

The result shows that the score function does provide sufficient discriminating power for pairs to be matched in scenes without too much clutter. We can locate the objects now which will enable effective navigation for the partially sighted.

6 Conclusions

We have extended the GPOD algorithm developed originally for mobile robots to include dynamic recalibration of the ground plane, Kalman Filter tracking of features and RANSAC ground plane fitting in successive images. The experimental results from our implementation show that RANSAC-DGPR can detect small obstacles much better than GPOD in the presence of camera motion. Moreover, we have investigated obstacle detection and false alarm in probabilistic terms and demonstrated that objects can be found from matching their edges with criteria based on heuristics.

There are, however, some assumptions made which require further investigation and justification, such as the variances of the image coordinates. The current DGPR implementation takes 1.5 seconds on the average to process a pair of 128x128 images on an Ultra-Sparc machine. So before it can be actually used by the partially sighted, we will need to achieve at least near real-time speed by parallelisation and other optimisations.

Acknowledgements

We thank Fuxing Li for many useful discussions about GPOD, and David Lee, Nick Molton & Penny Probert for collaborations on ASMONC. JMB thanks the EPSRC for support during his Senior Fellowship. SS thanks EPSRC for his Graduate Studentship support.

References

1. Y. Bar-Shalom and T.E. Fortmann. *Tracking and data association*. Academic Press, Boston, London, 1988.
2. I.J. Cox. A review of statistical data association techniques for motion correspondence. *International Journal of Computer Vision*, 10(1):53–66, February 1993.
3. F. Ferrari, E. Grosso, G. Sandini, and M. Magrassi. A stereo vision system for real time obstacle avoidance in unknown environment. In *Proceedings of IEEE International Workshop on Intelligent Robots and Systems IROS '90*, pages 703–708, 1990.
4. M.A. Fischler and R.C. Bolles. Random sample consensus: a paradigm for model fitting with application to image analysis and automated cartography. *Commun. Assoc. Comp. Mach.*, 24:381–395, 1981.
5. B.K.P. Horn. *Robot Vision*. The MIT Press, 1986.
6. D. Lee. The movement of sensors carried on the trunk of a walking person. Oxford University, January 1996.
7. F. Li. Visual control of AGV obstacle avoidance. DPhil First Year Report, Department of Engineering Science, University of Oxford, 1994.
8. F. Li, J.M. Brady, I. Reid, and H. Hu. Parallel image processing for object tracking using disparity information. In *Second Asian Conference on Computer Vision ACCV '95*, pages 762–766, Singapore, December 1995.
9. L. Matthies and P. Grandjean. Stochastic performance modeling and evaluation of obstacle detectability with imaging range sensors. *IEEE Trans. on Robotics and Automation*, 10(6):783–792, 1994.
10. J.E.W. Mayhew, Y. Zheng, and S. Cornell. The adaptive control of a four-degrees-of-freedom stereo camera head. In H.B. Barlow, J.P. Frisby, A. Horridge, and M.A. Jeeves, editors, *Natural and Artificial Low-level Seeing Systems*, pages 63–74. The Royal Society, London, 1992.
11. N. Molton. Egomotion recovery from stereo. DPhil First Year Report, Department of Engineering Science, University of Oxford, 1996.
12. N. Molton, S. Se, J.M. Brady, D. Lee, and P. Probert. A stereo vision-based aid for the visually impaired. *Image and Vision Computing*, 1997. to appear.
13. S.B. Pollard, J.E.W. Mayhew, and J.P. Frisby. Implementation details of the pmf stereo algorithm. In J.E.W. Mayhew and J.P. Frisby, editors, *3D Model Recognition From Stereoscopic Cues*, pages 33–39. MIT, 1991.
14. S.B. Pollard, J. Porrill, J.E.W. Mayhew, and J.P. Frisby. Disparity gradient, lipschitz continuity, and computing binouclar correspondences. In J.E.W. Mayhew and J.P. Frisby, editors, *3D Model Recognition From Stereoscopic Cues*, pages 25–32. MIT, 1991.
15. S. Se. Visual aids for the blind. DPhil First Year Report, Department of Engineering Science, University of Oxford, 1996.
16. S. Se and M. Brady. Vision-based detection of kerbs and steps. In *Eighth British Machine Vision Conference BMVC '97*, pages 410–419, September 1997.
17. Z. Zhang and O. Faugeras. *3D Dynamic Scene Analysis*. Springer-Verlag, 1992.

Evaluation and Application of Recognition Confidence in OCR

Xiaofan Lin, Xiaoqing Ding, Youbin Chen, Jinhui Liu, Youshou Wu

Image Processing Division, Department of Electronic Engineering
Tsinghua University, Beijing 100084, P.R. China
E-mail: lxf@ocrserv.ee.tsinghua.edu.cn

Abstract

Recognition confidence plays an important role in the selection of rejection threshold and the combination of multiple classifiers. In this paper, we first present a systematic theory on classifier's confidence, which includes the definition, the concept of generalized confidence, optimal rejection theorem and the relationship between confidence value and recognition rate. Then we propose a method for the evaluation of recognition confidence. The theory and method are strongly supported by the practice in handwritten numeral recognition and off-line handwritten Chinese character recognition.

KEY WORDS: confidence, optimal rejection, handwritten numeral recognition, off-line handwritten Chinese character recognition, classifier combination

1.Introduction

For a pattern classifier, it's often desirable to know its reliability on a certain input sample, namely, the confidence. Confidence can be used for various purposes. First, it can provide basis for the selection of rejection region. In many applications, we hope to drop the substitution rate to a certain acceptable level. For a given classifier, we must reject some samples in order to achieve this goal. Then comes the question as to which part of the samples should be rejected. As shown in Section 2, if the rejected samples are the ones that the classifier has the lowest confidence in, the substitution rate can drop to the minimum under a given rejection rate. Second, confidence plays an critical role in the combination of multiple experts(CME). Today CME is a focus in the field of pattern recognition and researchers are not satisfied with simple schemes such as majority voting. Instead, they seek to take full advantage of the information that individual classifiers provide. Confidence value is a kind of important information that can be employed in CME[1]. It's easy to understand the concept: when the decisions of individual classifiers disagree, we should choose the decision with relatively high confidence, or at least assign more weight to it.

Due to the significance of confidence, many pattern classifiers estimate the confidence in some ways[6][8]. However, systematic theory is seldom seen. In this paper we attempt to build a theoretical framework of pattern classifier's confidence and propose its various applications in OCR.

2. Confidence and generalized confidence for pattern classifiers

Definition 1(Classification Confidence):

Given a pattern classifier S, suppose that x is the feature vector extracted from an input pattern, the classification result is $e_s(x)$ (one of the M possible classes) and the class x belongs to is $\omega(x)$. Then the probability of the event that $e_s(x)$ is correct can be denoted as:

$$c_s(x) = P(e_s(x) = \omega(x)) \qquad (1)$$

$c_s(x)$ is defined as S's classification confidence at point at x. In this paper it's often called confidence for short.

The relation between confidence defined here and recognition rate is as follows: confidence reflects the classifier's reliability at a certain point where recognition rate is the average confidence across the entire region of the feature space. Moreover, the conditional error probability P(e|x) frequently referred in literature[3] is the probability that the classifier makes a mistake at point x. It's closely related to the confidence defined above:

$$P(e|x) + c_s(x) = 1 \qquad (2)$$

Definition 2(Generalized Confidence):

If there exists function $f_s(x)$ and a monotonous function g(.) which satisfy:

$$f_s(x) = g(c_s(x)) \qquad (3)$$

$f_s(x)$ is called generalized confidence.

Of course, $c_s(x)$ is a special case of generalized confidence. We can say that confidence is an absolute metric while generalized confidence is a relative metric. In many situations, what we can directly observe just reflects the relative reliability. That's why we propose the concept of generalized confidence

Theorem 1: For a pattern classifier S, given rejection rate Pr, when the selected rejection region R={x|$c_s(x)$<TH(Pr)}, the substitution rate can be lowered to minimum. TH(Pr) is a threshold decided by Pr.(proof ommited)

Corollary 1: For a pattern classifier S, given rejection rate Pr, when the selected rejection area R={x|$f_s(x)$<TH(Pr)}, the substitution rate can be lowered to minimum. $f_s(x)$ is generalized confidence and TH(Pr) is a threshold decided by Pr.

Theorem 2: A classifier's expected recognition rate on a given sample set is equal to its average recognition confidence on that sample set:(proof ommited)

$$E\{P_a\} = \frac{\sum_{i=1}^{N} c_s(x_i)}{N} \qquad (4)$$

Theorem 1 and its corollary show that when confidence or generalized confidence is chosen as the rejection threshold, optimal rejected region can be obtained. From Theorem 2 we know that the average confidence is an unbiased estimate of the classifier's recognition rate on a given sample set. These conclusions lay the theoretical foundation for using confidence in the practice of handwritten character recognition.

3. Estimation of classifier's confidence

We must find ways to estimate confidence before we can use it. This can be done in two steps. The first step is to estimate generalized confidence. This step is decided by the types of classifiers. In this paper we'll focus on two widely used classifiers in character recognition. Then we can map generalized confidence to confidence through a universal approach.

3.1 Estimate the generalized confidence for distance metric based classifiers:

Distance metric based pattern classifiers such as nearest neighbor classifier[3] and SOFM(Self Organized Feature Mapping)[10] classifier are widely used in character recognition. For this category of classifiers, the decision rule is quite straightforward: the unknown sample should have the same label as its nearest neighbor in the representative sample set. Intuitively, confidence should be related to distances between unknown sample and representative samples. Some possible estimation formulae are listed below[6][10]:

$$\hat{f}_1(x) = -d_1(x) \tag{5}$$

$$\hat{f}_2(x) = d_2(x) - d_1(x) \tag{6}$$

$$\hat{f}_3(x) = 1 - d_1(x) / d_2(x) \tag{7}$$

where $d_1(x)$ is the distance between unknown sample x and its nearest neighbor x_m and $d_2(x)$ is the distance between x and its nearest neighbor among those representative samples whose label is different from that of x_m.

At first glance, these formulae are all plausible estimations. Which one should we adopt? The authors have done some theoretical analysis for nearest neighbor classifier and have proved that for 1-dimensional 2-class problem, if there are enough training samples, the expectation of $\hat{f}_3(x)$ in (7) is a kind of generalized confidence while $\hat{f}_1(x)$ and $\hat{f}_2(x)$ have no such characteristics. In [6] numerical optimization is employed to find the expression. The result is just $\hat{f}_3(x)$. In [10] experimental result shows $\hat{f}_3(x)$ is superior to other expressions. Both examples are related to handwritten numeral recognition. In [6] a nearest neighbor classifier is used and in [10] an SOFM classifier is adopted. Considering above theoretical analysis and practice, we believe that for distance metric based classifier $\hat{f}_3(x)$ is a reasonable choice for the estimation of generalized confidence.

3.2 Estimate the generalized confidence for feed-forward neural network classifiers:

In [2] it's proved that when mean square error(MSE) or cross-entropy is used as the cost function, the expected outputs feed-forward neural networks such as BP networks, RBF networks and high-order polynomial networks correspond to the posteriori probabilities. If o_i is the output corresponding to class ω_i, then:

$$E\{o_i\}=P(\omega(x)=\omega_i)$$

Because we'll choose the label corresponding to the largest output as the recognition result, c(x) is equal to $E\{\max o_i\}$. So we can use the maximum output to estimate the confidence.

3.3 Map generalized confidence to confidence

Although generalized confidence is competent for tasks such as optimal rejection, in cases such as combination of different kinds of classifiers absolute confidence is more desirable. Because a monotonous function g(.) must have an inverse function $g^{-1}(.)$, if $g^{-1}(.)$ can be derived we can get $c_s(x)=g^{-1}(f_s(x))$. For this purpose, statistics can be used:

Assume the domain of generalized confidence $f_s(x)$ is T. For $\forall y \in T$, we choose a small close set $[y-\delta, y+\delta]$. For a large testing set S_t, we can calculate $g^{-1}(y)$ using:

$$g^{-1}(y) = \frac{count(\{x|x \in S_t \text{ and } f_s(x) \in [y-\delta, y+\delta] \text{ and } e_s(x)=\omega(x)\})}{count(\{x|x \in S_t \text{ and } f_s(x) \in [y-\delta, y+\delta]\})} \quad (8)$$

where function count(.) can count the number of elements in a set.
The meaning of (8) is as follows:
The numerator is the number of correctly recognized samples whose generalized confidence fall in $[y-\delta, y+\delta]$ and denominator is the total number of samples whose generalized confidence fall in $[y-\delta, y+\delta]$. The ratio is the recognition rate when the generalized confidence is y, i.e., $g^{-1}(y)$.

4. Applications in handwritten Character recognition

In this section we demonstrate the application of the theory and method proposed in this paper through handwritten numeral recognition and handwritten Chinese character recognition. Both the training set and testing set of handwritten numeral recognition system are provided by NIST(National Institute of Standard and

Technology)[4]. The training set consists of 12,000 samples while the testing set has 3,000 samples, and 288 features are extracted[1]. The classifier is based on SOFM. With regard to handwritten Chinese character recognition, the samples are collected by ourselves, which include 3,755*50 training samples and 3755*3 testing samples (because there are 3,755 classes of frequently used characters in Chinese character set). Two approaches are employed to recognized handwritten Chinese characters. One is directional element based statistical approach[5] and the other is contour model guided structural approach[9]. The decisions of both methods are based on distance metric.

4.1 Selection of rejection region based on generalized confidence

In handwritten numeral recognition, we must reject part of the samples in order to obtain desirable accuracy. Testing sample set St consists of N handwritten numerals. Both rejection rate and substitution rate are functions of confidence threshold TH:

$$\Pr(TH) = \frac{count(\{x|x \in S_t, c_s(x) < TH\})}{N} \times 100\%$$

$$\Pe(TH) = \frac{count(\{x|x \in S_t, c_s(x) < TH, \omega(x) \neq e_s(x)\})}{N} \times 100\%$$

When TH change from 0. i.e. no rejection, to a sufficiently large value, a curve on the Pe-Pr plane can be obtained. Fig 1. is such a curve for our handwritten numeral recognition system. Solid line A is obtained if $\hat{f}_3(x)$ is used to estimate the generalized confidence while dotted line B and C are based on $\hat{f}_1(x)$ and $\hat{f}_2(x)$ respectively. It's obvious that under the same rejection rate, A has the lowest error rate. This result confirms our conclusion that (7) is an good estimation of generalized confidence for distance metric based classifiers.

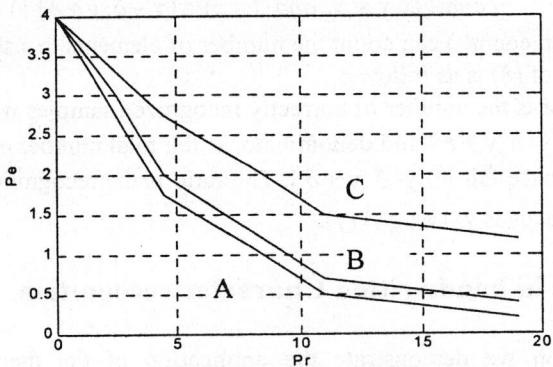

Fig.1: Relationship between Pe and Pr

4.2 Estimate recognition rate using confidence

Recognition rate is the most important index in pattern recognition. How to increase it is always the central task in research. However, we must conduct supervised testing in order to get the recognition rate. Now based on Theorem 2, by using the concept of confidence we can estimate recognition rate even **without** knowing the correct label of each testing sample. This can be done in several steps:

(1). Estimate generalized confidence according to the types of classifiers;
(2). Find the mapping from generalized confidence to confidence $g^{-1}(.)$, using (8);
(3). Estimate the recognition rate using (4).

Table 1 shows the result for handwritten numeral recognition. The differences between estimated recognition rates and tested recognition rates are all less than 2%, which prove the effectiveness of our theory.

Table 1 : Estimation of Recognition Rate for Handwritten Numeral Recognition

No	Source of Data	Number of Samples	Estimated Recog. Rate(%)	Observed Recog. Rate(%)
1	NIST	2969	95.25	95.25
2	ETL	6804	97.18	97.95

4.3 Multiexpert combination

Recently Multiexpert combination is a trend in character recognition[12][13]. By far researchers have proposed many methods of combining multiple classifiers. These methods can be divided into three categories according to the information they are based on[12]. Among the three levels of information mentioned, measurement level provides the most information. But there is one problem: the comparability of measurement information provided by different classifiers. In fact, the concept of confidence can be used to handle the problem. We'll give an example in handwritten Chinese character recognition:

Off-line Chinese character recognition is an extremely difficult pattern recognition task. Generally speaking, statistical method is the mainstream because of its high recognition rate and ease of machine learning. However, it is very difficult for statistical classifiers to distinguish between similar patterns. So it's widely accepted that in order to improve the recognition performance we must use structural method in addition to statistical method. But there is a disadvantage: the recognition rate of statistical method is much higher than that of structural method(see Table 2). Here confidence analysis plays an important role. The mappings from generalized confidence to confidence for the two methods are given in Fig. 2. We can draw two conclusions from it. On the one hand, the overall confidence of statistical method is much higher than that of structural method. On the other hand, for similar patterns the confidence of statistical method is about 0.5 while structural method performs better with a confidence of about 0.6. This is in accordance with the common sense that structural method is superior to statistical method in distinguishing similar patterns.

Based on the above idea, we employ a dynamic classifier selection strategy based on confidence. When two classifiers disagree we choose the decision with higher confidence. Classifier S_1 and S_2's decisions on x are $e_1(x)$ and $e_2(x)$, with the recognition confidence of $c_1(x)$ and $c_2(x)$ respectively. Then the final decision is given by:

$$e(x) = \begin{cases} e_1(x), \text{when } c_1(x) > c_2(x) \\ e_2(x), \text{when } c_2(x) > c_1(x) \end{cases}$$

As can be seen in Table 2, the recognition rate of integrated system is higher than that of statistical method. Through combination the recognition rate can be raised by 0.35% even when the statistical method has already achieved a very high recognition rate of about 98%.

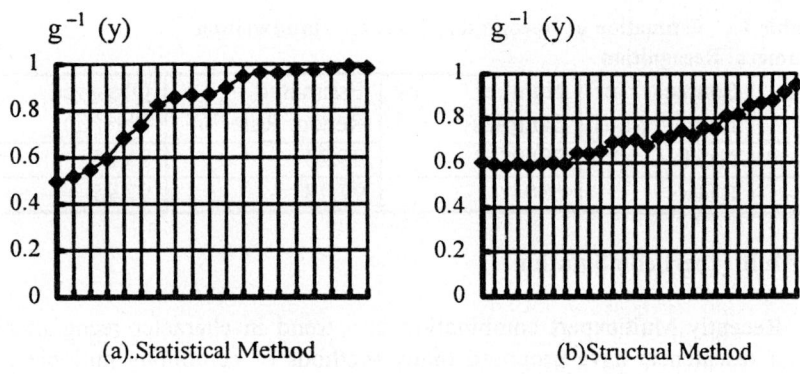

(a).Statistical Method   (b)Structual Method

Fig 2 : Mapping from Generalized Confidence to Confidence

Table 2 : Comparison of Several Approaches for Handwritten Chinese Character Recognition

Methods \ Test Set	No.1	No. 2	No. 3
Statistical Method	86.34%	95.89%	98.48%
Structural Method	70.09%	84.13%	94.73%
Combination	87.70%	96.35%	98.83%
Improvement of Recognition Rate	1.36%	0.46%	0.35%

5. Conclusion

In this paper we discuss classifier's confidence systematically. Explicate definition of classification confidence is given and the concept of generalized confidence is introduced. The optimal selection of rejection region based on confidence is proved and the relationship between confidence value and recognition rate is presented. In addition, the estimation of several widely used classifiers' generalized confidence is discussed and a universal method is proposed to map generalized confidence to confidence. The theory and method are applied in handwritten character recognition with promising result.

References

[1]   Lin Xiaofan, et al, "Handwritten Numeral Recognition Using MFNN Based Adaptive Multiexpert Combination ", ICDAR'97, 1997.8
[2]   Michael D. Richard, et al, "Neural Network Classifiers Estimate Bayesian a posteriori Probabilities", Neural Computation, pp461-483, 1991.3
[3]   T. M. Cover and P.E. Hart, "Nearest Neighbor Pattern Classification," IEEE Tans. Inform. Theory, vol.IT-13, pp. 21-27, 1967
[4]   Michael D. Garris, et. al, "NIST Form-Based Handprint Recognition System",National Institute of Standards and Technology, 1994.6
[5]   Youbin Chen, et al , " New Method for the Extraction of Handwritten Chinese Characters " , submitted to Signal Processing
[6]   Stephen J. Smith, et al, "Handwritten Character Classification Using Nearest Neighbor in Large Database", IEEE. Trans. Pattern Anal. Machine Intell, vol.16, no.9, pp915-919, 1994.9
[7]   Guo Hong,et al, "Comprehensive Recognition Method for Improving the Robustness of Recognition of Printed Chinese Characters", Proc. of the 1st Conf. on Multimodal Interface, pp.221-226, 1996.10
[8]   F.F Soulie, E. Viennet, "Multi-Modular Neural Network Architectures: Applications in Optical Character and Human Face Recognition", Advances in Pattern Recognition Systems Using Neural Network Technologies, pp77-111, 1993
[9]   Jinhui Liu, " Research on Contour Model Based Handwritten Chinese Character Recognition " , Ph.D Thesis, Tsinghua University, 1997
[10] Teuvo Kohonen, "The Self-Organizing Map", Proc. of IEEE, vol 78, no 9,1990.9
[11] Xiaofan Lin, et al, " Theoretical Analysis of the Confidence Estimation for Nearest Neighbor Classfier " , Chinese Science Bulletin
[12] Lei Xu, et al, "Methods of Combining Multiple Classifiers and Their Applications to Handwritten Recognition", IEEE Trans. System, Man and Cybernetics, vol. 22, no 3, 1992.6
[13] C.Y.Suen, et. Al, "Computer Recognition of Unconstrained Handwritten Numerals", IEEE Trans. on Pattern Anal. And Machine Intell., vol.80, no.7, pp1162-1180
[14] Dar-Shyang Lee, "A Theory of Classifier Combination: The Neural Network Approach", Ph.D Thesis, SUNY at Buffalo, 1995.4

A New Nonlinear Shape Normalization Method for Off-line Handwritten Chinese Character Recognition[*]

Youbin Chen, Xiaoqing Ding, Youshou Wu and Ming Chen

*Image Processing Division, Dept. of Electronic Engineering
Tsinghua University, Beijing 100084, P.R.China
Email: chenyo@cs.colostate.edu, cm@ocrserv.ee.tsinghua.edu.cn*

Abstract

How to correct the shape variations of handwritten Chinese characters is very important in off-line handwritten Chinese character recognition. In order to get rid of shape variations and reduce within-category variances directly from handwritten Chinese character image, a new nonlinear shape normalization method is proposed in this paper. In this new method, both of the stroke pixels and the background pixels are assigned with different feature densities. In addition, the feature density is local and two-dimensional. It is more reasonable and more effective than other nonlinear shape normalization methods. Its effectiveness has been demonstrated by our experiments.

Key words: off-line handwritten Chinese character recognition, preprocessing, feature density, feature density equalization, nonlinear shape normalization

1. Introduction

Off-line handwritten Chinese character recognition is one of the most difficult problems in the area of character recognition. These difficulties are: (1)A large number of Chinese character categories. (2)Very complicated shape structures of most Chinese characters. (3)Existence of many similar Chinese characters. (4)Existence of shape variations of handwritten Chinese characters. Among the four kinds of difficulties, the shape variation is the most difficult problem to be solved. In order to get rid of the shape variations and reduce the within-category variances directly from off-line handwritten Chinese character image, some researchers have proposed some nonlinear shape normalization(NSN) methods[1-7]. In this paper, a new nonlinear shape normalization method named nonlinear shape normalization based on different feature densities for both the stroke pixels and the background pixels is proposed. Our new method has the following characteristics: (1) Different pixels of input image are assigned with different feature densities. (2) Since the shape variations of off-line handwritten Chinese characters are local and two-dimensional, the feature density for each pixel is also local and two-dimensional in our method. From the following analysis, we can see that our definition of feature density for each

---

[*] supported by China National Natural Science Foundation

pixel is more reasonable than other nonlinear shape normalization methods. From our experimental results, it can be seen that our new method has the best recognition performance compared with other nonlinear shape normalization methods.

This paper is organized as follows. In section 2, some nonlinear shape normalization methods are reviewed. In section 3, our new nonlinear shape normalization method is described. Our experimental system and results are introduced in section 4. Finally, the summary is given in section 5.

2. Review of Some Nonlinear Shape Normalization Methods

Because of the localities and irregularities of the shape variations of handwritten Chinese characters, it can be easily seen that in the input image some parts are crowded and some parts are very sparse. It means that strokes are not well-separated. In order to make the strokes uniformly distributed, some researchers have proposed some nonlinear shape normalization methods[1-7]. All these methods are based on feature density projection and feature density equalization. The former makes feature density projection histogram by projecting a certain feature at each point onto horizontal- or vertical-axis and the latter equalizes feature densities of input image by re-sampling the feature density projection histogram. The difference between these methods is that different method has different feature density definition. In this section, all these methods are reviewed from four points: feature density definition, feature density projection, feature density equalization and performance analysis.

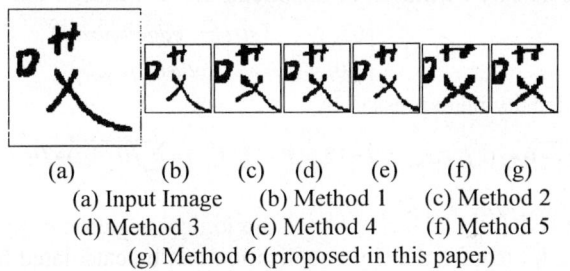

(a)　　　(b)　　(c)　　(d)　　(e)　　(f)　　(g)
(a) Input Image　　(b) Method 1　　(c) Method 2
(d) Method 3　　(e) Method 4　　(f) Method 5
(g) Method 6 (proposed in this paper)

Fig. 1　some examples of shape normalization methods

Let f(i, j) be the input image (i=1, 2, ..., I;　j=1, 2, ..., J), G(m, n) be the normalized image (m=1, 2, ..., M; n=1,2,..., N), d(i, j) be the feature density of each point f(i, j). Let H(i) and V(j) represent the feature density projection functions at each point onto horizontal-and vertical-axis respectively, i.e.:

$$H(i)=\sum_{j=1}^{J}[d(i,j)+\alpha_H(i,j)] \quad i=1,2,...,I \quad V(j)=\sum_{i=1}^{I}[d(i,j)+\alpha_V(i,j)] \quad j=1,2,...,J \quad (2\text{-}1)$$

Usually α_H(i, j) and α_V(i, j) are constants. Feature density equalization transforms input image into normalized image by equalizing the feature densities of the input image. It calculates the new position (m, n) of G by re-sampling the feature density projection histogram as follows:

$$m = \sum_{k=1}^{i} H(k) \times \frac{M}{\sum_{k=1}^{I} H(k)} \quad n = \sum_{l=1}^{j} V(l) \times \frac{N}{\sum_{l=1}^{J} V(l)} \quad (2\text{-}2)$$

where: i =1, 2, ..., I;   j =1, 2, ..., J;   m = 1, 2, ..., M;   n = 1, 2, ..., N

2.1 Linear Shape Normalization (Method 1)

LSN can only concerned with the linear shape variations such as size, position, rotation, etc. It can not resolved the nonlinear shape variations such as distortion which is local and irregular. Obviously, LSN is a special instance of NSN.

2.2 NSN based on Dot Density (Method 2) [1][5]

This is the simplest NSN method. Its feature density is defined as follows:

$$d(i,j) = f(i,j) = \begin{cases} 1, & black \ pixels \ (stroke \ pixels) \\ 0, & white \ pixels \ (background \ pixels) \end{cases}$$

Its feature density projection and feature density equalization can be calculated by equations (2-1) and (2-2) respectively. From the above definition, it can be seen that only the stroke pixels of input image have feature density. The merits of this method are that character shape unbalance can be roughly corrected and the calculation is simple and fast. The demerits of this method are that stroke thickness can not be homogenous and the localities and irregularities of shape distortion can not be well corrected as shown in Fig.1(c).

2.3 NSN based on Line Density by Crossing Lines (Method 3) [2][5]

The feature density definition of this method is as follows:

$$d(i,j) = \begin{cases} 0.5, & (stroke \ edge \ pixels) \\ 0, & (other \ pixels) \end{cases}$$

Its feature density be expressed by:

$$H(i) = \sum_{j=1}^{J} \overline{f(i,j-1)} \bullet f(i,j) + \alpha_H \quad i=1,2,...,I \quad V(j) = \sum_{i=1}^{I} \overline{f(i-1,j)} \bullet f(i,j) + \alpha_V \quad j=1,2,...,J$$

where, $f(i,0) = f(0,j) = 0$, $\overline{f(i,j)}$ is a logical negation of image f(i, j), α_H and α_V are constants. Its feature density equalization can be calculated by equation (2-2). In this method, only the stroke edge pixels of input image are given feature densities. As show in Fig.1 (d), a local part of high line density by crossing lines is expanded and that of low line density is shrunk. The merits of this method are that character shape unbalance can be roughly corrected, the stroke thickness is almost the same and the calculation is simple and fast. The demerits of this method are that the uniform relocation of inner strokes is not possible and the localities and irregularities of shape distortion can not be well corrected because the definition of feature density is not local nor two-dimensional.

2.4 NSN based on Line Density by Line Interval (Method 4) [3][5]

Unlike the above two NSN methods, this method assigns both the stroke pixels and background pixels with different feature density definitions. For each point of input image, it has horizontal feature density $F_H(.)$ and vertical feature density $F_V(.)$ as as described in literature [3]. Its feature density projection functions are:

$$H(i) = \sum_{j=1}^{J} F_H(i,j) + \alpha_H \qquad i=1,2,\ldots,I \qquad\qquad V(j) = \sum_{i=1}^{I} F_V(i,j) + \alpha_V \qquad j=1,2,\ldots,J$$

where, α_H, α_V are constants.

Compared with Method 2 and Method 3, this method gives feature density definitions to all the pixels of input image. The definition of feature density is local but one-dimensional. As shown in Fig.1(e), the dense parts are expanded and the sparse parts are shrunk. Its merits are that the stroke thickness is almost the same, the character shape unbalance can be roughly corrected and inner strokes can be uniformly relocated. But all the stroke pixels have the same feature density and all the background pixels in a scanning segment have the same feature density. This is not reasonable. Its calculation is relatively slow compared with method 2 and method 3. Moreover, it is not very effective to correct the nonlinear shape variations such as distortions because its feature density is only one-dimensional. However, as we know, the Chinese character structure is two-dimensional.

2.5 NSN based on Line Density by Virtual Inscribed Circle(Method 5) [4][5]

In this method, the following four kinds of edges are defined as described in literature[4].

$$L_1 = \max\{i' \mid i' < i,\ f(i',j) \bullet \overline{f(i'+1,j)} = 1\} \qquad L_2 = \min\{i' \mid i' \geq i,\ f(i',j) \bullet \overline{f(i'+1,j)} = 1\}$$

$$L_3 = \max\{i' \mid i' < i,\ \overline{f(i'-1,j)} \bullet f(i',j) = 1\} \qquad L_4 = \min\{i' \mid i' \geq i,\ \overline{f(i'-1,j)} \bullet f(i',j) = 1\}$$

Assume that the size of input image is W × W. At each point (i, j) the horizontal line interval L_X is defined as:

$$L_X = \begin{cases} 4W, & (c)\ L_1, L_2, L_3, L_4\ \text{are undefined} \\ 2W, & (b)\ \text{Only}\ L_1, L_3\ \text{are undefined} \\ 2W, & (b)\ \text{Only}\ L_2, L_4\ \text{are undefined} \\ 2W, & (f)\ \text{Only}\ L_1, L_4\ \text{are undefined} \\ L_4 - L_3, & (e)\ \text{Only}\ L_1\ \text{is undefined} \\ L_2 - L_1, & (e)\ \text{Only}\ L_1\ \text{is undefined} \\ (L_2 - L_1 + L_4 - L_3)/2 & (a)(d) \qquad\qquad \text{otherwise} \end{cases}$$

The vertical line interval L_Y can be defined similar to L_X. At each point (i, j) its feature density is defined as:

$$d(i,j) = \begin{cases} \max\left(W/L_X,\ W/L_Y\right) & \text{if}\ L_X + L_Y < 6W \\ 0 & \text{if}\ L_X + L_Y \geq 6W \end{cases}$$

Similar to Method 4, all the pixels of input image have feature density definition. From the above definition, it can be seen that the line interval L_X and L_Y can be imagined as the width and height of a virtual circumscribed rectangle and the feature density is a function of the diameter of the virtual inscribed circle of this rectangle. Obviously, the feature density is local and two-dimensional. As shown in Fig.1(f), it is effective to correct the nonlinear shape variations which is local and two-dimensional. The local part with a small inscribed circle is expanded and that with a large inscribed circle is shrunk. The stroke thickness is almost the same. But the algorithm is complex and the calculation is time-consuming.

2.6 NSN based on Line Density by Real Inscribed Circle[6-7]

This is an improved NSN method of Method 5. It uses the real inscribed circle to replace the above virtual inscribed circle described in method 5. But it is very time-consuming to find the center and diameter of the real inscribed circle at each point of input image. So it is very difficult to use this method in a practical system. The more detailed information about this method can be found in literatures [6-7].

3. NSN based on Different Feature Densities for Both the Stroke Pixels and the Background Pixels(Method 6)

As we know, the structure of Chinese character is two-dimensional and the nonlinear shape variations are local. In order to correct the shape variations more effectively, the feature densities should be local and two-dimensional. Among the above five NSN methods, method 5 has the best performance and method 4 has the better performance as shown in Fig.1. But in method 4 and method 5, pixels on the same line segment with the same line interval have the same feature density. It is not reasonable. For example, as shown in Fig. 2 (a), point O is the center point of the background line segment, point A and point B are the stroke edge pixels. In method 4 and method 5, all the pixels on line segment AB have the same feature density. Obviously, point O is the most sparse point, point A and point B are denser points. Point C and point D are the densest points. So different pixels on line segment CD should have different feature densities.

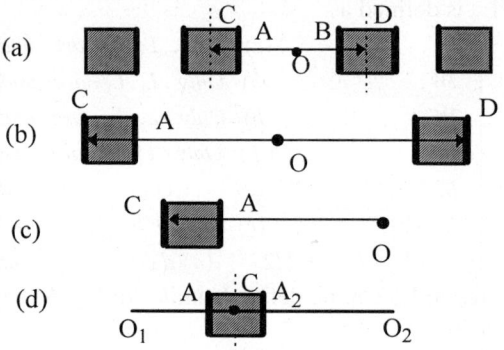

Fig. 2 Definition of Line Interval of Method 6

On the other hand, the four peripheral areas of Chinese characters have important information for recognition. Though shape variations exist in off-line handwritten Chinese characters, strokes located in the four peripheral areas are relatively stable according to our inspection. In order to keep the information of the four peripheral areas, the feature densities of pixels distributed in the four peripheral areas should be enhanced.

In this section, a new nonlinear shape normalization method (Method 6) is proposed according to the above analysis.

3.1 Feature Density Definition
As shown in Fig. 4, the horizontal feature density is defined as follows:
(1) Fig.2 (a)(b)

Let (i, j) be the coordinate of point O, (i_c, j_c) be the coordinate of point C, the distance between point C and point D is $|CD|$, then we can define the horizontal feature density as:

$$d_X(i,j) = \frac{1}{\frac{1}{2}|CD|+1} + \alpha_{BFH}(i,j) \qquad d_X(i',j) = \frac{1}{\frac{1}{2}|CD|-|i'-i|+1} + \alpha_{BFH}(i',j) \qquad \begin{pmatrix} i \le i' \le i_D \\ i_C \le i \le i \end{pmatrix}$$

where:
$$\alpha_{BFH}(i,j) = \begin{cases} \alpha_{H_1}, & \text{if } (i,j) \text{ is a background pixel} \\ \alpha_{H_2}, & \text{if } (i,j) \text{ is a stroke pixel} \\ \alpha_{H_3}, & \text{if } (i,j) \text{ is a peripheral pixel} \end{cases}$$

Usually, $\alpha_{H_3} > \alpha_{H_2} > \alpha_{H_1}$ and they are constants.

(2) Fig.2 (c)

$$d_X(i,j) = \frac{1}{|OC|+1} + \alpha_{BFH}(i,j) \qquad d_X(i',j) = \frac{1}{|OC|-|i'-i|+1} + \alpha_{BFH}(i',j) \qquad (i_C \le i' \le i)$$

(3) Fig.2 (d)

Let (i_1, j_1) be the coordinate of point O_1, (i_2, j_2) be the coordinate of point O_2, (i_c, j_c) be the coordinate of point C, where $j_1 = j_2 = j_c$. Then the horizontal feature density can be defined as:

$$d_X(i_1,j_1) = \frac{1}{|O_1C|+1} + \alpha_{BFH}(i_1,j_1) \qquad d_X(i_1',j_1) = \frac{1}{|O_1C|-|i_1'-i_1|+1} + \alpha_{BFH}(i_1',j_1) \qquad (i_1 \le i_1' \le i_C)$$

$$d_X(i_2,j_2) = \frac{1}{|O_2C|+1} + \alpha_{BFH}(i_2,j_2) \qquad d_X(i_2',j_2) = \frac{1}{|O_2C|-|i_2'-i_2|+1} + \alpha_{BFH}(i_2',j_2) \qquad (i_C \le i_2' \le i_2)$$

Similarly, we can define the vertical feature density $d_Y(i, j)$. Then at each point (i, j) the feature density can be defined as:

$$d(i,j) = \max[d_X(i,j), d_Y(i,j)]$$

3.2 Feature Density Projection
Its feature density projection and feature density equalization can be calculated by equation (2-1) and equation (2-2) respectively.

3.3 Performance Analysis
From the above feature density definition, it can be seen that all the pixels of input image have feature densities. Different pixels have different feature densities with different density enhancements. Pixels distributed in the four peripheral areas have the largest density enhancement. Stroke pixels have larger density enhancement. Background pixels have the smallest density enhancement. Using different density enhancement will make it possible to keep the four peripheral areas during normalization. Moreover, the feature density is local and two-dimensional. So it is very effective to correct the nonlinear shape variations which are local and irregular. Compared with the other five methods, our new method proposed in this paper has

the best corrected image as shown in Fig.1(g). Table 1 shows the performance comparison of these methods.

method	local global	1D 2D	Feature Density		stroke thickness	four peripheral areas	speed
			pixels	reasonable			
2	global	1D	stroke pixels	not reasonable	not even	not keep	quick
3	global	1D	stroke pixels	not reasonable	the same	not keep	quick
4	local	1D	all pixels	somewhat reasonable	the same	not keep	slow
5	local	2D	all pixels	somewhat reasonable	the same	not keep	slow
6	local	2D	all pixels	reasonable	the same	keep	slow

Table 1 Performance Comparison of NSN Methods

4. Experimental System and Results

The following is the recognition system used in our experiments as shown in Fig.3.

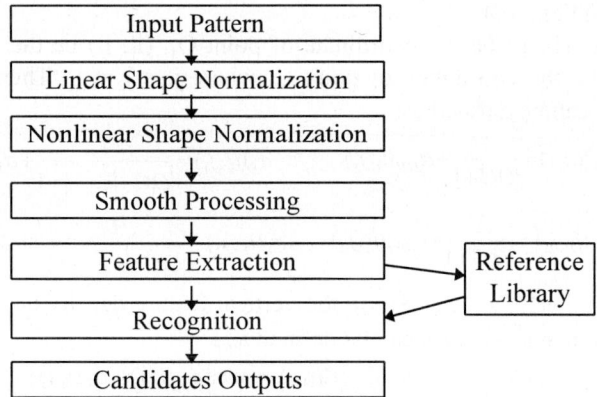

Fig.3 a recognition system used in our experiments

In our experiments, we used the off-line handwritten Chinese character sample library collected by OCR Lab, Tsinghua University, Beijing. We randomly selected 100 sets totally having 3755 × 100 handwritten Chinese characters. Half of these samples were used for training, the other half were used for testing. We used the improved directional element feature [8] and designed a tree classifier in which the weighted Euclidean distance was used. According to the accumulative classification rate, the performance order from the best to the worst of these nonlinear shape normalization methods is:

M6>M5>M4>M3>M2>M1

The comparison of accumulative classification rate of these methods is shown in table 2.

Candidate(s) / Method(s)	1	2	3	4	5	8	9	10
M1	84.38	89.57	91.06	92.27	93.41	95.98	96.31	96.76
M1+M2	88.76	93.02	95.15	96.24	96.95	98.22	98.54	98.57
M1+M3	89.91	94.78	95.34	96.89	97.21	98.86	98.96	99.01
M1+M4	92.24	96.43	97.54	98.04	98.53	99.36	99.38	99.41
M1+M5	92.96	97.08	97.76	98.12	98.67	99.45	99.51	99.53
M1+M6	95.02	98.24	98.56	98.89	98.96	99.52	99.61	99.68

Table 2 comparison of accumulative classification rate of NSN methods

5. Summary

Shape variation is the most difficult problem needed to be solved in off-line handwritten Chinese character recognition. Therefore, some researchers have proposed some nonlinear shape normalization methods as reviewed in section 2. In this paper, a new nonlinear shape normalization method is proposed. In this new method, all the pixels of input image have their own feature densities. Moreover, the feature density is local and two-dimensional. It is more effective than other methods to solve the nonlinear shape variations. Our experiments have demonstrated the effectiveness of our new method proposed in this paper.

References

[1] Yoshiyuki Yamashita et al, "Classification of Handprinted Kanji Characters by the Structured Segment Matching Method", Pattern Recognition Letters 1(1983) 475-479

[2] H.Yamada, T.Saito and K.Yamamoto, "Line Density Equalization—A Nonlinear Normalization for Correlation Method", Trans. IECE, Vol.J67-D, No.11, pp.1379-1383, 1984

[3] Jun Tsukumo, Haruhiko Tanaka, "Classification of Handprinted Chinese Characters Using Nonlinear Normalization and Correlation Methods", Proc. of 9th ICPR, Rome, Itlay, Nov.1988

[4] H.Yamada et al, "A Nonlinear Normalization Method for Handprinted Kanji Character Recognition—Line Density Equalization", Pattern Recognition, Vol. 23, No.9, 1990

[5] S.W.Lee and J.S.Park, "Nonlinear Shape Normalization Methods for the Recognition of Large-Set Handwritten Characters", Pattern Recognition, Vol.27, No.7, pp.895-902, 1994

[6] Dayin Gou, "Research on Subject-Free Handprinted Chinese Character Recognition", Ph.D. dissertation, Dept. of Electronic Engineering, Tsinghua University, Beijing, May 1995

[7] Dayin Gou, Xiaoqing Ding and Youshou Wu, A Handwritten Chinese Character Recognition Method Based on Image Shape Correction, Proc. of 1st National Conf. on Multimedia and Infomation Networks(CMIN'95), Beijing, Mar.1995, pp.254-259

[8] N.Sun et al, "A Handwritten Character Recognition System by Using Improved Directional Element Feature and Subspace Method", IEICE Vol. J78-D-II, No.6, pp.922-930, June 1995

A Novel Triangulation Procedure for Thinning Cursive Text

Stanley S Ipson, Muhammed Melhi and William Booth
Department of Electronic and Electrical Engineering
University of Bradford
Bradford, BD7 1DP, UK

Abstract
Thinning or skeletonization can contribute tremendously to efficient feature extraction and classification. This paper describes a novel thinning procedure which decomposes a polygonal approximation of the inner and outer boundaries of a word image into a set of triangles. These triangles are classified into three types and replaced either by three lines at branches or by single lines to form a completely connected skeleton which is one pixel wide. The procedure also automatically generates a graph representation of the skeleton for use by subsequent analysis steps. The procedure has been tested on a variety of Arabic and English handwritten words and found to produce very few spurious features. The speed of this method does not depend strongly on the resolution of the digital image unlike commonly used thinning methods which iteratively remove layers from the boundary of words. In addition, it can be used for thinning any line-like patterns such as fingerprints, chromosomes and line drawings.

Keywords
Thinning, skeleton, triangulation, polygonal approximation, handwriting, cursive Arabic text.

1 Introduction
Thinning plays a central role in a broad range of recognition systems including the automatic recognition of text [1], the classification of fingerprints [2], analysis of chromosomes [3], and automatic inspection of printed circuit boards [4]. This paper is primarily concerned with the application to handwritten Arabic text which is fundamentally cursive in nature and poses different problems to the recognition of most European texts. Most approaches to character recognition utilize a thinning (also called skeletonization) algorithm to obtain skeletons of printed or handwritten text by the process of reducing their stroke width to just a single pixel. Because such skeletons represent the structural shape of textual images they reduce the amount of information inherent in the image to the minimum necessary for recognition. This provision of a compressed format for images of text minimizes the computer memory needed to store the essential structural information and also simplifies the shape analysis and shortens processing time. The skeleton of an object region is formally defined by the medial axis transformation (MAT) method proposed by Blum [5] in 1967.

Thinning algorithms [6] can be classified into iterative and non-iterative procedures. The former involve the deletion of successive layers of pixels on the boundary of the pattern based on either the four or eight neighbours of each pixel, depending on the definition of connectivity employed, until only a skeleton of one pixel thickness is reached. In general these methods tend to be slow, although some work [7,8] on parallel iterative algorithms has been done. Non-iterative thinning produces the medial axis of shapes without iterative removal of contour pixels. Two important examples [9,10] of this approach are based on the medial axis transform which may be redefined as the set of local maxima in the distance transform. The main advantage of the latter methods is that skeleton generation is not dependent on the pattern sizes and hence is produced in a fixed number of passes. The main problems with these algorithms are that the skeletons produced sometimes have spurious tails and that they are not always one pixel in width.

The two most important characteristics of skeletonization algorithms are reliability and speed. Because it is sometimes necessary to capture images with high resolution, it is desirable to have a skeletonization procedure whose speed is not strongly dependent on the resolution. A method which guarantees a one pixel width is also essential to avoid further unnecessary processing steps. This paper describes a novel approach to skeletonization which attempts to achieve these characteristics by decomposing a polygonal approximation of a character image (with or without holes) into triangles. Triangulation techniques reported in the literature include one [11] which is O(n log log n). Details of the algorithms developed are provided below and are illustrated using mainly Arabic words as examples because this work was performed as part of a programme of work to recognize handwritten Arabic text.

2 The Triangulation Algorithm
2.1 Overview
A typical example of a handwritten Arabic word digitized by scanning is shown after thresholding in a magnified form in Figure 1. This particular word consists of four separated segments and contains three holes. Boundary pixels on each segment may lie either on external or internal boundaries as indicated in the figure.

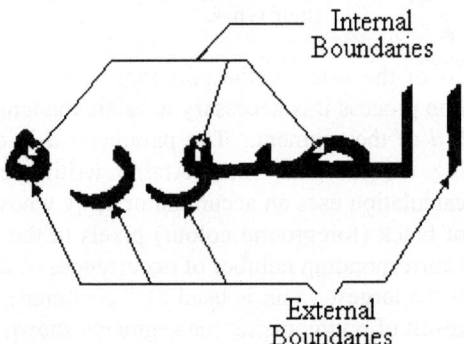

Fig. 1. A typical Arabic word indicating internal and external boundaries.

The first step in thinning such a word is to estimate the width of the pen stroke prior to obtaining a polygonal approximation of each type of boundary for each segment. Each separate region enclosed by polygons is then decomposed into neighboring non-overlapping triangles whose vertices are the vertices of the polygons. Each triangle is then reduced to either one or three lines, depending on the topology, and these lines are finally combined without discontinuities to form the required skeleton. In order to illustrate the approach, a sample segment without interior holes will be considered first and the method will then be extended to deal with the general case of segments possessing holes.

2.2 Skeletonization of a Word Segment Without Holes
A fast algorithm of O(n) for defining line segments from a sequence of n chain codes has recently been reported [12] and a version of this has been implemented in the current algorithm. The two polygons obtained by applying this method to the Arabic word "unn" shown in Figure 2a are displayed in Figure 2b.

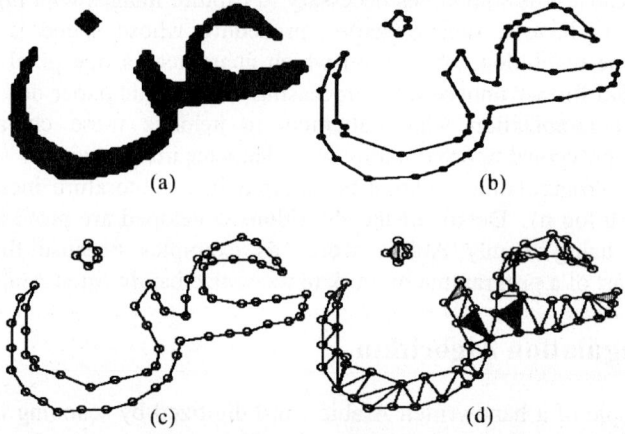

Fig. 2. The Arabic word "unn" is shown in (a) together with a polygonal approximation of its boundary in (b). The polygon resulting from taking account of stroke width is shown in (c) and the result of triangulation is shown in (d) with some triangles shaded according to their types.

It is clear that the lengths of the sides of the polygons vary widely but in order to facilitate the skeletonization process it is necessary to relate the lengths of the polygon sides to the stroke width d of the segment. The parameter d is estimated from the digitized text by utilizing the fact that most Arabic writing consists largely of horizontal strokes. The calculation uses an accumulator array whose indices equal the lengths of vertical runs of black (foreground colour) pixels in the image of the word and whose values are the corresponding number of occurrences of the runs. The index of the array element with the largest value is used as the estimate of d for the word. Figure 2c illustrates the result of polygonizing the segments shown in Figure 2a using a value of d estimated in this way. Essentially this stage of the processing increases the number of sides defining the polygon by creating additional points whose positions are judicially chosen to avoid segments with overlong sides.

2.2.1 Triangulating a Polygon Without Holes
The aim of the new algorithm presented here is to generate triangles which are directly useful to the task of representing skeletons by non-directed graphs. These graphs are normally used in a later stage of cursive handwriting recognition systems to break a

word into its constituent characters. The triangulation algorithm is recursive and is based on testing the diagonality of a line joining two vertices (P and Q) of the polygon. To be a diagonal, a candidate line must be wholly within the interior of the polygon and not intersect any sides of the polygon except at the vertices P and Q. The diagonal line testing algorithm developed by O'Rourke [13] has been used for this purpose because its avoidance of floating point operations results in greater execution speed. The triangulation algorithm uses the list of polygon vertices and commences with the line segment formed by joining the first point P and the last point V in the list. The algorithm then runs through the intermediate vertices Q ($P < Q < V$) to find the one which makes the sum of the distances QP and QV a minimum. If more than one point satisfies this condition, a vertex forming a triangle QPV with all angles acute is chosen, if it exists. The triangle QPV becomes the first entry in a list of triangles and the algorithm is called recursively using the new diagonal line as its starting segment until the polygon has been fully triangulated. In this process the sides of the triangles are tagged as external lines if they belong to the original polygon or diagonal if they link polygon vertices internally. The triangles formed in this way fall into the following three types:

    Type 1: a triangle with only one internal diagonal side;
    Type 2: a triangle with two internal diagonal sides;
    Type 3: a triangle with three internal diagonal sides.

In Figure 2d, which demonstrates the triangulation algorithm applied to the original polygon shown in Figure 2c, the three types of triangle are indicated by filling in grey, white and black respectively.

2.2.2 Skeleton Construction of a Polygon Without Holes

In the previous sub-section, the method of dividing the polygons making up a word into triangles of either types 1, 2, or 3 was described. These three types are associated with the major structural features of the words, namely ends of strokes, middles of strokes and branching of strokes. The manner in which triangles of each type are replaced by lines representing sections of the skeleton of the word is shown in Figure 3.

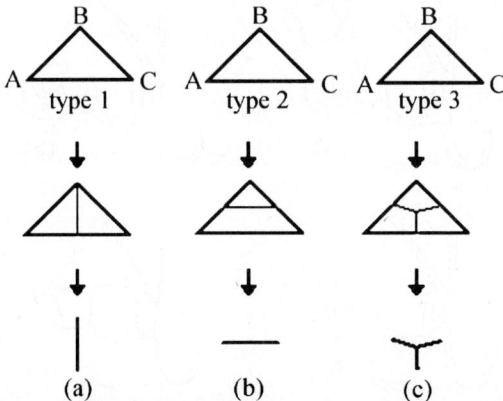

Fig. 3. Thinning the three types of triangles resulting from the triangularisation of polygons

A triangle of type 1 is thinned into a line with one end at the mid-point of the internal side and the other at the opposite vertex. A triangle of type 2 is thinned into a line connecting the mid points of its two internal sides. A triangle of type 3 is thinned into three lines which meet at the center of gravity of the triangle and have their other ends at the mid-points of the triangle's sides. After this triangle thinning procedure is applied to the triangles in Figure 2d, the Arabic word "unn" transforms into the skeleton shown in Figure 4a. Because neighbouring triangles share a side, the final skeleton is fully connected and is clearly one pixel in width. Furthermore, no tracing is required to build the final representation of the word in the form of a graph which is shown in Figure 4b.

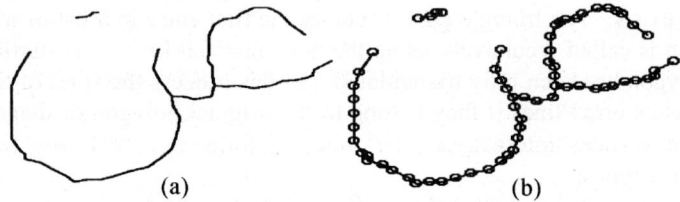

Fig. 4. The skeleton (a) and graph (b) obtained for the Arabic word "unn".

2.3 Skeletonization of a Word Segment with Holes

This section deals with the general case of a word containing one or more holes which, after polygonization, become separate internal polygons embedded in an external polygon. For each internal polygon, the approach is to make a suitable break in the stroke so that it becomes part of its surrounding external polygon. The previous method of triangularisation is then applied to the modified polygon. The application of the approach is illustrated in more detail using the word whose boundaries are shown in Figure 5a.

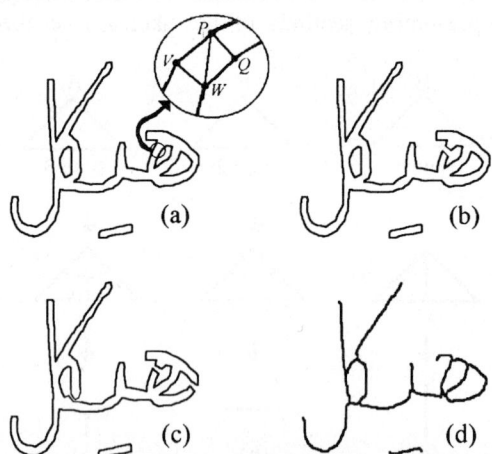

Fig. 5. The polygonal form of an Arabic word with three holes is shown in (a). In (b) it is shown after the removal of one hole and in (c) it is shown after the removal of all holes to produce one external polygon. The final skeleton of the example word is shown in (d).

After polygonization of this word, two external polygons and three internal polygons are obtained. For the upper right internal polygon shown in Figure 5a, the narrowest part of the stroke, shown magnified in the inset, is selected by finding the side *WQ* that is closest to the side *VP* of the surrounding external polygon such that the quadrilateral *VPQW* is completely inside the region of the stroke. The quadrilateral region is then detached and decomposed into two type 2 triangles *VPW* and *PWQ* which are added to the list of triangles used later to construct the skeleton. Following this detachment, all the internal polygon vertices are inserted in the list of vertices of the external polygon between *V* and *P* forming a longer external polygonal boundary and reducing the number of internal polygons by one. To facilitate this insertion process, the order of the vertices of the external polygon (clockwise or anti-clockwise) should be different from that of the internal polygon. In this case the insertion is accomplished by inserting after vertex *V* all vertices of the original internal polygon starting at *W* and ending with *Q*. The boundary of the resulting polygon after removing one hole is shown in Fig. 5b. The two sides linking the external and internal polygons, *VW* and *PQ* should be marked as internal although they will appear to be external in the new external polygon.

This procedure of quadrilateral detachment is repeated for all remaining internal polygons and the final result for the example word is shown in Figure 5c. Then, the resulting polygonized word without holes is triangulated using the algorithm of subsection 2.2.1 and the triangles so generated added to those resulting from the quadrilateral detachment stages. Finally, the skeletonization method described in subsection 2.2.2 is used to construct the skeleton of the pattern from the complete list of triangles. The skeleton produced for the word of Figure 5a is shown in Figure 5d.

3 Software Testing and Modifications

The algorithms described in section 2 were coded in C and tested on images of 228 words written by four writers taken from a much larger database constructed by the authors [14]. It was found that a few spurious branches were produced as a result of triangles of type 3 appearing in wrong places. These errors were either branches falsely generating short tails or branches where more than three lines should join at a point but the current form of the algoritm is limited to three. An attempt was made to eliminate branch errors of the first kind from the skeleton using the following two-step process to minimize the number of triangles of types 1 and type 3:

- Delete all type 1 triangles with areas less than $0.25\ d^2$;
- Change two neighboring type 1 and 3 triangles into two type 2 triangles by rearranging their vertices.

By adding both of these steps to the software, it was found that most of the unwanted branches were eliminated. Errors of the second kind are not dealt with in this paper although they could be eliminated by removing segments of length less than a suitably chosen threshold distance related to the stroke width. In particular, if a short segment joining points having triple branches is eliminated then a single point having four branches results. The authors have delayed introducing this modification because they want to minimize the number of nodes in the resulting graph first and then perform the removal.

4 Experimental Results

The results produced by the modified algorithm for a few Arabic and English words are shown in Figures 6 and 7 respectively. The quality of the skeletons obtained are acceptable for the intended purpose of character recognition since they contain the essential structure of the words. The branching problems referred to in section 3 can be seen in the skeletonized Arabic words shown in Figures 6b and 6c and also the skeletonized English words shown in Figures 7b and 7c.

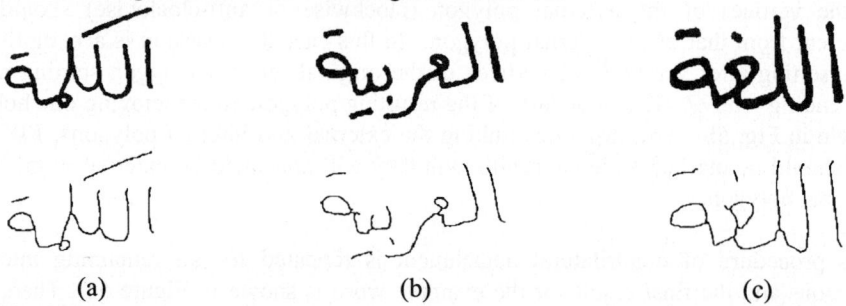

Fig. 6. Images of three Arabic words and their skeletons produced by the algorithm.

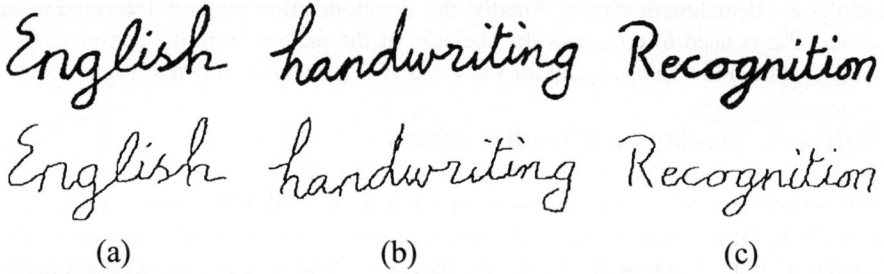

Fig. 7. Images of three English words and their skeletons produced by the algorithm.

5 Conclusions

Thinning is an important step in most machines designed to recognise characters. Accurate skeletons help to eliminate many unnecessary decisions in the recognition of characters, thereby improving the recognition rate and increasing speed. This paper has described an efficient method, based on converting the boundaries of words into polygons and triangulating these polygons, that offers several advantages compared with alternative techniques. The time taken by the authors' method does not depend strongly on the resolution of the digitized images unlike the commonly used methods that iteratively remove layers from the pattern. In addition, the skeletons obtained are guaranteed to be one pixel in width and have fewer spurious tails compared to those generated by methods based on the distance transform. Furthermore, this triangulation approach provides adequate decomposition for thinning unlike other methods based on polygonal approximation currently found in the literature. Another advantage of the algorithm is that the parameters used by it are calculated from the image being thinned unlike some of the the non iterative techniques described in the literature. Finally, it is not necessary to follow the thinning algorithm with a separate processing stage to represent words by graphs, needed for subsequent analysis, because this type of output is inherently generated by the algorithm. Although the authors' method was created specifically for application to the problem of character recognition, it could be used to skeletonize any line-like patterns such as are found in chromosomes, fingerprints, and line drawings.

References
1. Amin A., Al-Sadoun H., and Fischer S. Hand-printed Arabic character recognition system using an artificial network, Pattern Recognition **29**(4), 663-675 (1996).
2. Fitz A. and Green R. Fingerprint pre-processing on a hexagonal grid, Proceedings of the 1995 European Convention on Security and Detection , pp. 257-260, Brighton, UK (May 1995).
3. DiezHiguera J., DiazPernas F. and LopezCoronado J. Neural network architecture for automatic chromosome analysis, Proceedings of SPIE - Applications of Artificial Neural Networks in Image Processing, pp. 85-94, CA, USA (February 1996).
4. Ye Q. and Danielsson P. Inspection of printed circuits boards by connectivity preserving shrinking, IEEE Trans. Pattern Anal. and Mach. Intell. **10** (5), 737-743 (1988).
5. Blum H. "A transformation for extracting new descriptors of shape" in Models for the perception of speech and visual form, Wathen-Dunn, W., ed., MIT Press, Cambridge, Mass (1967).
6. Lam L., Lee S., and Suen C.Y. Thinning methodologies - a comprehensive Survey, IEEE Trans. Pattern Anal. and Mach. Intell. **14**, 869-885 (1992).
7. Datta A. and Parui S. A robust parallel thinning algorithm for binary images, Pattern Recognition **27** (9), 1181-1192 (1994).
8. Altwaijri M., Ayoubi R. and Bayoumi M. Skeletonization of Arabic characters using a neural network mapped on Maspar, World Congress on Neural Networks - International Neural Network Society Annual Meeting, **2**, pp. 653-658 (1994).
9. Niblack C.W., Gibbons P.B. and Capson D.W. Generating skeletons and centerlines from the distance transform, CVGIP: Graphical Models and Image Processing **54** (5), 420-437 (1992).
10. Shih F.Y. and Pu C.C. A skeletonization algorithm by maxima tracking on Euclidean distance transform, Pattern Recognition **28** (3), 331-341 (1995).
11. Tarjan R. and Wyk V. An O(n log log n)-time algorithm for triangulating a simple polygon, SIAM J. Comput. **17**, 143-178 (Feb. 1988).
12. Yuan J. and Suen C.Y. An optimal O(n) algorithm for identifying line segments from a sequence of chain codes, Pattern Recognition **28** (5), 635-646 (1995).
13. O'Rourke J. "Computational geometry in C", Cambridge University Press.
14. Melhi M.H., Ipson S.S. and Booth W. A database for research in Arabic handwriting recognition, Dept. of E & EE, Internal Report 595, University of Bradford, UK (1997)

Digital Geometric Methods in Image Analysis and Compression

Ari Gross[1] and Longin Latecki[2]

[1] University Graduate Center and Queens College,
City University of New York, Flushing NY 11367, USA
[2] University of Hamburg, Vogt-Kölln-Str. 30,
22527 Hamburg, Germany

Abstract. One of the important problems related to image analysis and compression is finding repeated structure. Although the focus of this paper is developing digital geometric models and methods for finding regular structure in digital document images, the applicability of the digital geometric approach is also demonstrated on images taken under orthographic and perspective projection. First, a fast linear-time algorithm is given to compute the static threshold that minimizes the non-well-composedness or weak connectivity of the document image. Next, a new digital similarity measure is introduced that outperforms the standard similarity measures, including the Hausdorff distance, with respect to determining if two discrete objects in the image are digitizations of the same prototype. This measure is then used in a model-based compression algorithm, and a variation of the algorithm is developed for finding structure in images taken under affine and perspective transformations.

1 Introduction

In previous work [3] [4] [8], the authors have focused on mathematically modeling the digitization process and on developing digitization rules and related algorithms that guarantee that a digitization is topology preserving. In this paper, we focus more on the application side, demonstrating how discrete spatial models and related digital similarity measures can be used in model-based compression of digital documents. Part of this paper focuses on the need for correct digital similarity measures. In particular, we will focus on the Hausdorff distance and its variations. There has been considerable important research on the Hausdorff distance [5] [10]. In this paper, however, we will focus on the relationship between the Hausdorff distance and the digitization process. It is shown that the Hausdorff distance is interesting exactly because it is closer to a digital similarity measure than the standard measures such as Hamming distance, weighted Hamming distance [11], residual entropy [6], and template distance [7].

The paper is structured as follows: First, a fast topology-preserving thresholding algorithm is presented. Next, we derive a new Hausdorff-based digital similarity measure and demonstrate its effectiveness with respect to document image compression. This measure is compared to the standard bidirectional Hausdorff distance used in document image analysis. Finally, a variation on the algorithm for finding repeated image structure is shown to be applicable to images taken under orthographic and perspective projection.

2 Computing a Topology Preserving Threshold

In previous work [2], the authors considered under what conditions a digital image is topology preserving. It was proven that for any **r parallel regular** set, i.e., any set that supports an inner and outer osculating ball of radius r at every point on its boundary (see [8]), a digitization resolution of r for the diameter of a grid square always guarantees that topology is preserved under monotonic digitization and that the resulting discrete set is well-composed, i.e., it has no checkerboard patterns. Well-composed sets have many desirable properties that make them particularly amenable to image processing algorithms, see [9]. It was shown in [2] that selecting a gray-level threshold that minimized the number of checkerboard neighborhoods also seemed to approximately minimize the sum of false topological connections and disconnections. This threshold also gave us very high recognition rates when applying subsequent OCR (using Omni-Page) to the thresholded binary document.

We now present an efficient algorithm for computing the threshold that minimizes the non-well-composedness of a gray-level image:

Starting at the top of the gray-level image, every 2 x 2 neighborhood is visited. For each such neighborhood, we consider the gray-level values for the 2 pairs of diagonal points a_1, a_2 and b_1, b_2. Let $a_{min} = \min(a_1, a_2)$, $a_{max} = \max(a_1, a_2)$, and similarly for b_{min} and b_{max}. Next, we consider the two closed intervals $[a_{min}, a_{max}]$ and $[b_{min}, b_{max}]$. If these two intervals intersect, then no threshold will cause this local 2 x 2 neighborhood to become non-well-composed and we need not consider it further. If, however, the two intervals are disjoint then assume, without loss of generality, that $a_{max} < b_{min}$. Assume further that when the threshold is set to some value t, $0 \leq t \leq 255$, that all pixels with gray-level values $\leq t$ are set to 0 (black), while all pixels with gray-level $> t$ are set to 255 (white). As we traverse the 2 x 2 neighborhoods of the image, two arrays plus_chk[] and minus_chk[] are maintained. These, respectively, keep track of the number of checked neighborhoods that are added or subtracted at each gray-level value. For each neighborhood with disjoint intervals, we increment the two arrays: ++plus_chk[a_{max}], ++minus_chk[b_{min}]. After the entire image has been traversed, these two arrays are used to compute the values for a third array chk[]. Initially, all the values in chk[] are set to zero. Then we compute the values of chk[] iteratively: chk[-1] = 0; chk[i]=chk[$i-1$]+plus_chk[i]-minus_chk[i].

The threshold value selected using this algorithm is appealing for several reasons. One important reason is that the algorithm minimizing the weak connectivity of the document consistently yields a value very close to the threshold value that minimizes the number of models required to match the connected components using the Hausdorff distance metric or a variation introduced in the next section.

3 Digitization Invariant Similarity Measures

In symbolic compression and model-based image coding [1] [11], there is a need for both fast and accurate techniques for comparing the similarity of two discrete planar shapes. A review of several of the similarity measures frequently used is given in [1] and includes the Hamming distance, the weighted Hamming distance, sum of weighted AND-NOTs, residual entropy, and degradation probability. These measures do not model the digitization process and are not invariant with respect to digitization. As a result, the problem of finding the correct number of models or prototypes is often considered a clustering problem, and it is accepted that a certain degree of mislabeling will occur. This mislabeling of connected components necessitates a correction phase, which takes the form of a residual map. This residual map ensures that the compression is lossless but is very expensive. As sensors improve, it is very desirable to put this model matching process into a more rigorous mathematical framework and discard the need for a corrective residual map. This is also desirable from the perspective of constructing an image compiler that converts a digital document back into MS Word or LaTeX format since a residual map is really not applicable in this context.

First, we define the Hausdorff distance and its relationship to monotonic topology-preserving digitization. The Hausdorff distance between two sets A and B is defined

as
$$H(A, B) = \max(h(A, B), h(B, A)), \text{ where}$$
$$h(A, B) = \sup_{a \in A} \inf_{b \in B} \|b - a\|,$$

and $\|\cdot\|$ is some norm.

In [8], we proved the following theorem:

Theorem 1 *Let A be a par(r)-regular set. Then A and $Dig(A, r)$ are homotopy equivalent for every digitization $Dig(A, r)$, and $H(A, Dig(A, r)) \leq r$, where H is the Hausdorff distance.*

Using Theorem 1, we have $H(A, Dig(A, r)) \leq 1$, where the diameter of a grid square is presumed equal to 1. Moreover, since H is a distance metric satisfying the transitivity property, then for any two digitizations $Dig^1(A, r)$ and $Dig^2(A, r)$ we have the constraint $H(Dig^1(A, r), Dig^2(A, r)) \leq 2$. We have used this constraint effectively to find supersets of a given connected component (i.e., digital instance of a model), where the superset consists of all the connected components on the document that could conceivably be digitizations of the same underlying prototype. An example of this is shown in Fig 1.c, where all the letter "b"s are detected using Hausdorff distance 2, in addition to some extraneous letters. The "b"s that are undetected are either connected to adjacent letters or are substantially corrupted by noise. In all the examples that we considered, this constraint was always successful at finding a superset that included all unperturbed homeomorphic digital instances of the underlying model corresponding to a given connected component. Thus, the Hausdorff distance is useful as a necessary condition for two instances to belong to the same prototype.

On the other hand, using the Hausdorff distance as a sufficient condition for class membership does not work effectively nor does it correctly model the digitization process. As an example, consider the digital document shown in Fig 1.d. In this case, a Hausdorff distance of 1 was used as a criterion for grouping connected components together, where the model selected was an instance of the letter "b". As can be seen, this criterion is neither necessary nor sufficient. Some of the "b" instances are missing while other extraneous components have been included. Instead, we introduce a new variation of the Hausdorff distance that corresponds more closely to the digitization process and, in the many documents we considered, admitted no false positive matches. As a result, there may be some degree of model fragmentation but there is no need for maintaining a residual map.

Define $Q_I, Q_{II}, Q_{III}, Q_{IV}$ to be the closed first, second, third and forth quadrants of R^2. For example,

$$Q_I = \{(x, y) \in R^2 \mid x \geq 0 \text{ and } y \geq 0\}.$$

Then we can define the first quadrant directional Hausdorff distance between two sets in R^2 as follows:

$$h_I(A, B) = \max_{a \in A} \min_{\substack{b \in B \\ b-a \in Q_I}} \|b - a\|, \tag{1}$$

and similarly for other quadrants.

Next we define
$$H_{i,j}(A, B) = \max(h_i(A, B), h_j(B, A)),$$

where i, j are quadrant numbers and $|i - j| = 2$.

Finally, the quadrant bidirectional Hausdorff is defined as

$$H_Q(A,B) = \min\{\ H_{I,III}(A,B), H_{III,I}(A,B), \\ H_{II,IV}(A,B), H_{IV,II}(A,B)\ \}\,. \qquad (2)$$

This measure is a distance similarity measure that models the fact that, at the very least, a model and its digitization can vary by a quadrant Hausdorff distance equal to 1. To see this, consider the fact that even if a connected component A was arbitrarily close to the original model, another digitization of this "model" could have its centroid vary by a translation of up to 1 grid square diameter. Thus, any digitization resolution where $H_Q(A = Dig(M_1,r), B = Dig(M_2,r)) \leq 1$, for distinct models M_1 and M_2, cannot be model preserving. Assuming we originally had a model-preserving digitization, $H_Q(A,B) \leq 1$ can be used effectively as a sufficient condition for two instances A and B to belong to the same model class.

An example is shown in Fig 2.a, where the initial "model" was an instance of the letter "b". Grouping based on this sufficient condition, no mismatches occur although some instances are missed. This matching algorithm can be improved upon by iterating the matching algorithm on the recomputed model. As shown in the example, initially a single instance of the letter "b" was taken to be the model. After the first iteration, additional "b" connected components were matched, as shown in Fig 2.a. Once these instances of the model are found, the model is recomputed using the dilated binary connected components as masks onto the corresponding gray-level components. This process is repeated until convergence of the model. In the example shown, the final set of matched "b" connected components is shown in Fig 2.b, and the final reconstructed gray-level model is shown in Fig 2.c. The thresholded version of the reconstructed image, as shown in Fig 2.d, is closer to a monotonic digitization than the thresholded original image shown in Fig 1.b, as evidenced by the fact that linear segments remain digitally linear (see [4]). The converged set of gray-level models is shown in Fig 2.e.

4 Discrete Sets of Discrete Spatial Objects

Consider once again the document image shown in Fig 1.a. In the algorithm given in the previous section, the model evolved by starting with a single connected component. Every other connected component that matched this initial model with quadrant Hausdorff ≤ 1 was added to the list. After all the connected components that could be matched were added to the list, the model was recomputed by averaging all the matched connected components and rethresholding. This process was iterated until no new matches were found. Using the quadrant Hausdorff as a sufficient condition for two instances to belong to the same object class, we can alternatively keep the evolving model as the set of instances that have been matched so far. Instead of "learning" more about the model by averaging the instances together, we take their union as the representation of the current underlying model that is evolving. Every new instance of the model that is added to the set must eventually be expanded so that every connected component that it matches in a quadrant Hausdorff sense is also added to the set. This process continues until convergence. For the document shown in Fig 1.a, this algorithm generated the reconstructed document image shown in Fig 3.a.

This method of representing the model as a discrete set of discrete spatial objects is also very useful in finding patterns in images and can be used as part of an image search engine to classify images based on symmetric structure. Consider the window tiles shown in Fig 3.b. These tiles are not all scaled versions of each other since the image was taken under perspective projection. Rather than solve for the transformation parameters, the algorithm simply uses a version of the rank quadrant Hausdorff distance to compare connected components to each other. When the algorithm finally converges, nearly all of the window instances are recognized. For the flag image shown in Fig 3.c, the tesselating elements we want to match (i.e., the stars) are mapped onto a non-planar surface under perspective projection. The algorithm described recognized almost all the stars, as shown in Fig 3.c. For the tire tread, the tread elements are mapped to an approximately cylindrical surface and have considerable variation. Nevertheless, the tread elements are matched quite well, as can be seen in Fig 3.d. This method allows us to find shapes that deform smoothly over time. Consequently, it can serve as a useful tool in finding regular structure in images without regard to the underlying surface or the projective transformation.

5 Conclusion

In this paper, a fast linear-time algorithm was presented to compute the static threshold that minimizes the non-well-composedness or weak connectivity of the document image. Next, a new digital similarity measure was introduced that outperforms the standard similarity measures, including the Hausdorff distance, with respect to determining if two discrete objects in the image are digitizations of the same prototype. This similarity measure was then used as the basis for a model-based compression algorithm. Finally, we demonstrated that a variation on the method can be extended to finding structure in images taken under affine and perspective transformations.

Acknowledgements: *The authors would like to acknowledge support for this research under NSF grant IRI-9707090 and the QC/CUNY Presidential Research Award. They would also like to acknowledge the very constructive support and assistance of Ruben Lusinyants, Ilya Dondoshansky, Elena Oranskaya, and Navdeep Tinna in this work.*

References

1. D. Doermann, *Document Image Understanding: Integrating Recovery and Interpretation*, PhD thesis, Univ of MD, College Park, 1993.
2. A. Gross and L. Latecki. Homeomorphic Digitization, Correction, and Compression of Digital Documents. *IEEE Workshop on Document Image Analysis*, Puerto Rico, June 1997.
3. A. Gross and L. Latecki. Digitizations Preserving Topological and Differential Geometric Properties. *Computer Vision and Image Understanding*, 62:370-381, Nov. 1995.
4. A. Gross and L. Latecki. A Realistic Digitization Model of Straight Lines. *Computer Vision and Image Understanding*, Vol. 67, No. 2, pp. 131-142, 1997.
5. D.P. Huttenlocher and W.J. Rucklidge, A multi-resolution technique for comparing images using the HHausdorff distance, In *Proceedings Computer Vision and Pattern Recognition*, pp. 705-706, NYC, NY, 1993.
6. S. Inglis and I. Witten. Compression-based template matching. In *Proceedings of the IEEE Data Compression Conference*, 1994.
7. T. Kanungo, R.M. Haralick, and I.T. Phillips. Global and local document degradation models. In *Proceedings of the International Conference on Document Analysis and Recognition*, pp. 730-734, 1993.
8. L. Latecki, C. Conrad, and A. Gross, Preserving Topology by a Digitization Process, to appear in *Journal of Mathematical Imaging and Vision*, 1997.
9. L. Latecki, U. Eckhardt, and A. Rosenfeld, Well-Composed Sets. *Computer Vision and Image Understanding*, 61:70–83, 1995.
10. W. Rucklidge. *Efficient Visual Recognition Using the Hausdorff Distance*. Number 1173 in Lecture Notes in computer Science. Springer-Verlag, 1996.
11. I. Witten, A. Moffat, and T. Bell. *Managing Gigabytes: Compressing and Indexing Documents and Images*. Van Nostrand Reinhold, 1994.

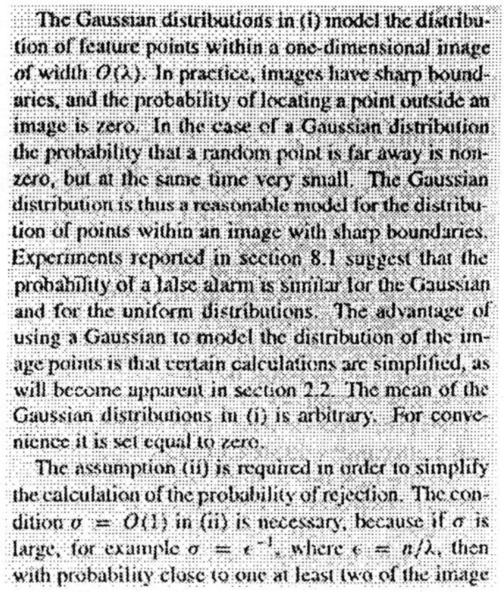

Fig 1a

The Gaussian distributions in (i) model the distribution of feature points within a one-dimensional image of width $O(\lambda)$. In practice, images have sharp boundaries, and the probability of locating a point outside an image is zero. In the case of a Gaussian distribution the probability that a random point is far away is nonzero, but at the same time very small. The Gaussian distribution is thus a reasonable model for the distribution of points within an image with sharp boundaries. Experiments reported in section 8.1 suggest that the probability of a false alarm is similar for the Gaussian and for the uniform distributions. The advantage of using a Gaussian to model the distribution of the image points is that certain calculations are simplified, as will become apparent in section 2.2. The mean of the Gaussian distributions in (i) is arbitrary. For convenience it is set equal to zero.

The assumption (ii) is required in order to simplify the calculation of the probability of rejection. The condition $\sigma = O(1)$ in (ii) is necessary, because if σ is large, for example $\sigma = \epsilon^{-1}$, where $\epsilon = n/\lambda$, then with probability close to one at least two of the image

Fig 1b

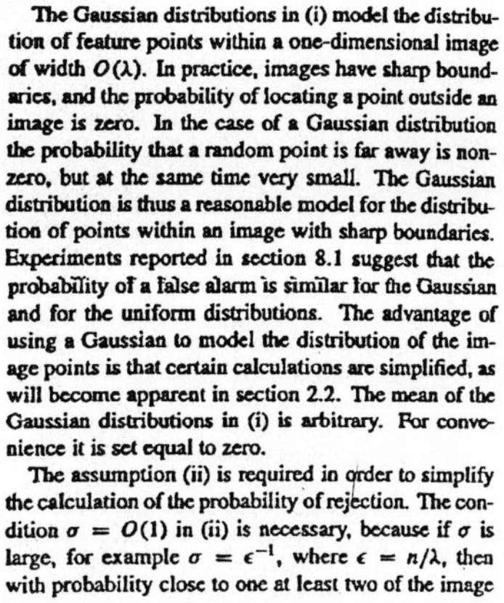

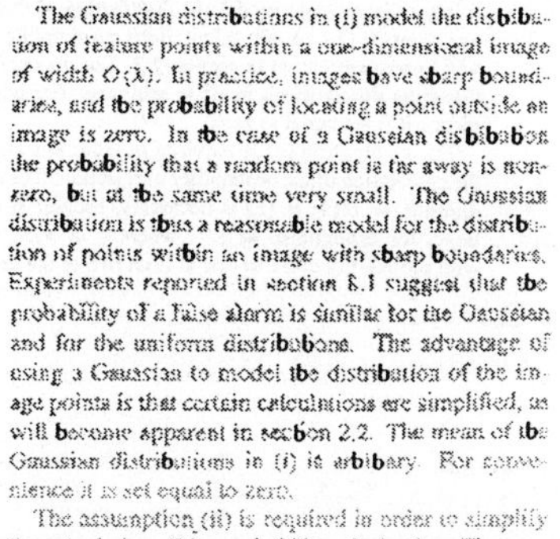

The Gaussian distributions in (i) model the distribution of feature points within a one-dimensional image of width $O(\lambda)$. In practice, images have sharp boundaries, and the probability of locating a point outside an image is zero. In the case of a Gaussian distribution the probability that a random point is far away is nonzero, but at the same time very small. The Gaussian distribution is thus a reasonable model for the distribution of points within an image with sharp boundaries. Experiments reported in section 8.1 suggest that the probability of a false alarm is similar for the Gaussian and for the uniform distributions. The advantage of using a Gaussian to model the distribution of the image points is that certain calculations are simplified, as will become apparent in section 2.2. The mean of the Gaussian distributions in (i) is arbitrary. For convenience it is set equal to zero.

The assumption (ii) is required in order to simplify the calculation of the probability of rejection. The condition $\sigma = O(1)$ in (ii) is necessary, because if σ is large, for example $\sigma = \epsilon^{-1}$, where $\epsilon = n/\lambda$, then with probability close to one at least two of the image

Fig 1c

The Gaussian distributions in (i) model the distribution of feature points within a one-dimensional image of width $O(\lambda)$. In practice, images have sharp boundaries, and the probability of locating a point outside an image is zero. In the case of a Gaussian distribution the probability that a random point is far away is nonzero, but at the same time very small. The Gaussian distribution is thus a reasonable model for the distribution of points within an image with sharp boundaries. Experiments reported in section 8.1 suggest that the probability of a false alarm is similar for the Gaussian and for the uniform distributions. The advantage of using a Gaussian to model the distribution of the image points is that certain calculations are simplified, as will become apparent in section 2.2. The mean of the Gaussian distributions in (i) is arbitrary. For convenience it is set equal to zero.

The assumption (ii) is required in order to simplify the calculation of the probability of rejection. The condition $\sigma = O(1)$ in (ii) is necessary, because if σ is large, for example $\sigma = \epsilon^{-1}$, where $\epsilon = n/\lambda$, then with probability close to one at least two of the image

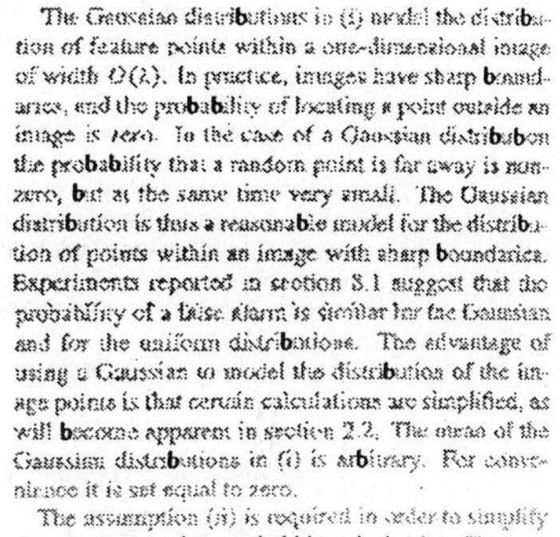

Fig 1d

The Gaussian distributions in (i) model the distribution of feature points within a one-dimensional image of width $O(\lambda)$. In practice, images have sharp boundaries, and the probability of locating a point outside an image is zero. In the case of a Gaussian distribution the probability that a random point is far away is nonzero, but at the same time very small. The Gaussian distribution is thus a reasonable model for the distribution of points within an image with sharp boundaries. Experiments reported in section 8.1 suggest that the probability of a false alarm is similar for the Gaussian and for the uniform distributions. The advantage of using a Gaussian to model the distribution of the image points is that certain calculations are simplified, as will become apparent in section 2.2. The mean of the Gaussian distributions in (i) is arbitrary. For convenience it is set equal to zero.

The assumption (ii) is required in order to simplify the calculation of the probability of rejection. The condition $\sigma = O(1)$ in (ii) is necessary, because if σ is large, for example $\sigma = \epsilon^{-1}$, where $\epsilon = n/\lambda$, then with probability close to one at least two of the image

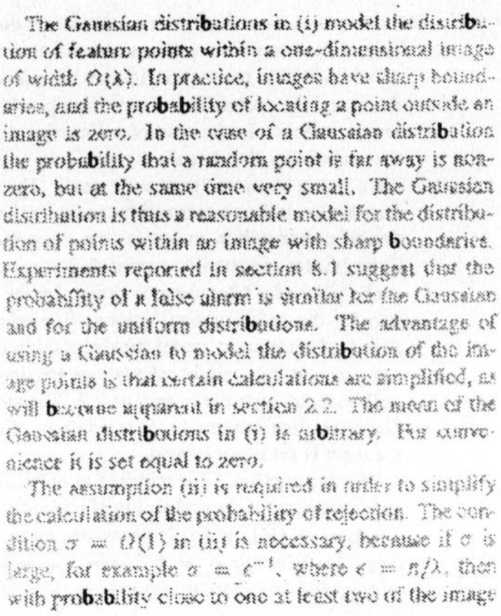

Fig 2a

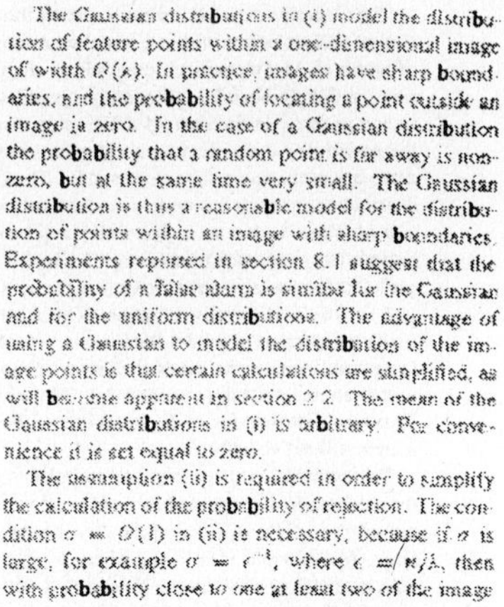

The Gaussian distributions in (i) model the distribution of feature points within a one-dimensional image of width $O(\lambda)$. In practice, images have sharp boundaries, and the probability of locating a point outside an image is zero. In the case of a Gaussian distribution the probability that a random point is far away is non-zero, but at the same time very small. The Gaussian distribution is thus a reasonable model for the distribution of points within an image with sharp boundaries. Experiments reported in section 8.1 suggest that the probability of a false alarm is similar for the Gaussian and for the uniform distributions. The advantage of using a Gaussian to model the distribution of the image points is that certain calculations are simplified, as will become apparent in section 2.2. The mean of the Gaussian distributions in (i) is arbitrary. For convenience it is set equal to zero.

The assumption (ii) is required in order to simplify the calculation of the probability of rejection. The condition $\sigma = O(1)$ in (ii) is necessary, because if σ is large, for example $\sigma = \epsilon^{-1}$, where $\epsilon = n/\lambda$, then with probability close to one at least two of the image

Fig 2b

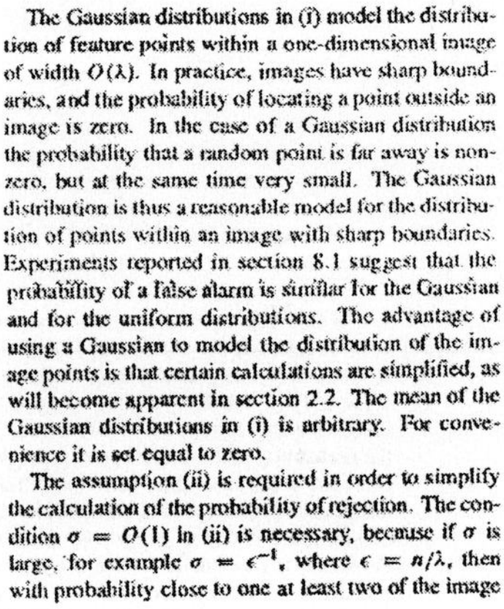

The Gaussian distributions in (i) model the distribution of feature points within a one-dimensional image of width $O(\lambda)$. In practice, images have sharp boundaries, and the probability of locating a point outside an image is zero. In the case of a Gaussian distribution the probability that a random point is far away is non-zero, but at the same time very small. The Gaussian distribution is thus a reasonable model for the distribution of points within an image with sharp boundaries. Experiments reported in section 8.1 suggest that the probability of a false alarm is similar for the Gaussian and for the uniform distributions. The advantage of using a Gaussian to model the distribution of the image points is that certain calculations are simplified, as will become apparent in section 2.2. The mean of the Gaussian distributions in (i) is arbitrary. For convenience it is set equal to zero.

The assumption (ii) is required in order to simplify the calculation of the probability of rejection. The condition $\sigma = O(1)$ in (ii) is necessary, because if σ is large, for example $\sigma = \epsilon^{-1}$, where $\epsilon = n/\lambda$, then with probability close to one at least two of the image

Fig 2c

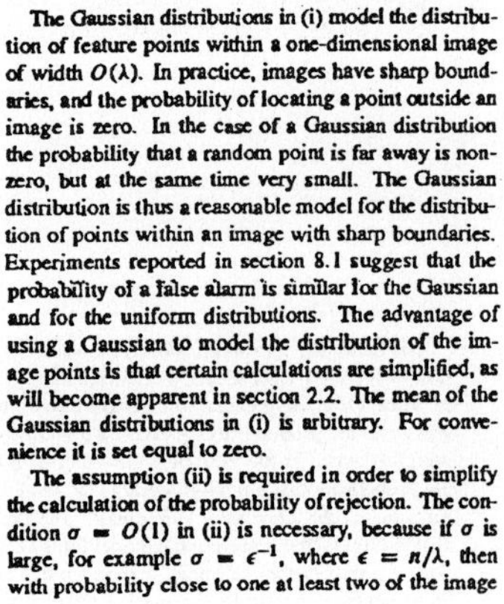

The Gaussian distributions in (i) model the distribution of feature points within a one-dimensional image of width $O(\lambda)$. In practice, images have sharp boundaries, and the probability of locating a point outside an image is zero. In the case of a Gaussian distribution the probability that a random point is far away is non-zero, but at the same time very small. The Gaussian distribution is thus a reasonable model for the distribution of points within an image with sharp boundaries. Experiments reported in section 8.1 suggest that the probability of a false alarm is similar for the Gaussian and for the uniform distributions. The advantage of using a Gaussian to model the distribution of the image points is that certain calculations are simplified, as will become apparent in section 2.2. The mean of the Gaussian distributions in (i) is arbitrary. For convenience it is set equal to zero.

The assumption (ii) is required in order to simplify the calculation of the probability of rejection. The condition $\sigma = O(1)$ in (ii) is necessary, because if σ is large, for example $\sigma = \epsilon^{-1}$, where $\epsilon = n/\lambda$, then with probability close to one at least two of the image

Fig 2d

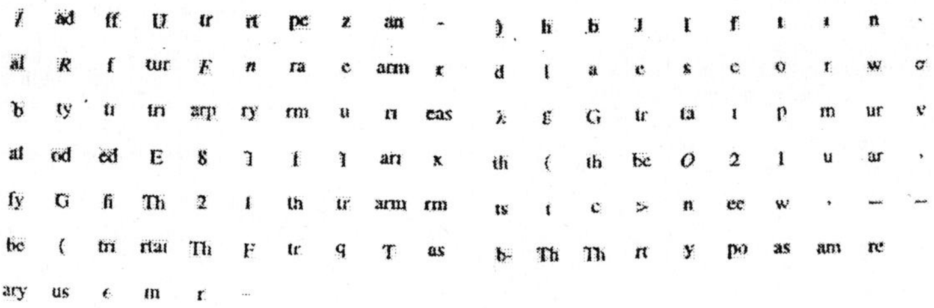

The Gaussian distributions in (i) model the distribution of feature points within a one-dimensional image of width $O(\lambda)$. In practice, images have sharp boundaries, and the probability of locating a point outside an image is zero. In the case of a Gaussian distribution the probability that a random point is far away is non-zero, but at the same time very small. The Gaussian distribution is thus a reasonable model for the distribution of points within an image with sharp boundaries. Experiments reported in section 8.1 suggest that the probability of a false alarm is similar for the Gaussian and for the uniform distributions. The advantage of using a Gaussian to model the distribution of the image points is that certain calculations are simplified, as will become apparent in section 2.2. The mean of the Gaussian distributions in (i) is arbitrary. For convenience it is set equal to zero.

The assumption (ii) is required in order to simplify the calculation of the probability of rejection. The condition $\sigma = O(1)$ in (ii) is necessary, because if σ is large, for example $\sigma = \epsilon^{-1}$, where $\epsilon = n/\lambda$, then with probability close to one at least two of the image

/	ad	ff	U	tr	rt	pe	z	an	-	)	h	b	J	I	f	t	s	n	.
al	R	f	tur	F	n	ra	e	arm	r	d	l	a	e	s	c	o	r	w	σ
b	ty	ti	tri	arp	ry	rm	u	n	eas	λ	g	G	tr	ta	t	p	m	ur	v
at	cd	ed	E	8	]	f	l	ari	x	th	(	th	be	O	2	1	u	ar	,
fy	G	fi	Th	2	1	th	tr	arm	rm	ts	t	c	>	n	ee	w	.	—	—
be	(	tri	rtat	Th	F	tr	q	T	as	b	Th	Th	rt	y	po	as	am	re	
ary	us	e	m	r															

Fig 2e

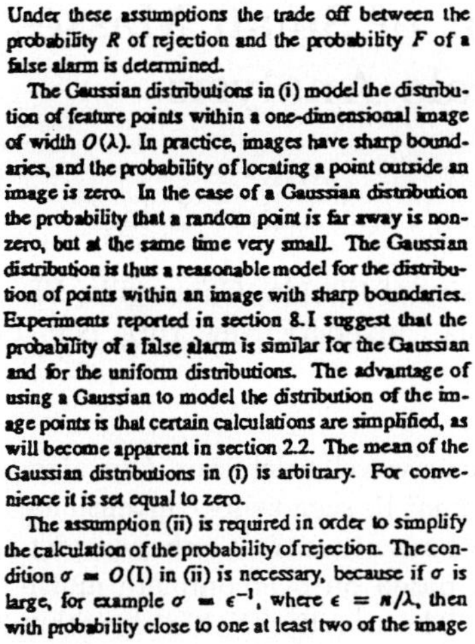

Under these assumptions the trade off between the probability R of rejection and the probability F of a false alarm is determined.

The Gaussian distributions in (i) model the distribution of feature points within a one-dimensional image of width $O(\lambda)$. In practice, images have sharp boundaries, and the probability of locating a point outside an image is zero. In the case of a Gaussian distribution the probability that a random point is far away is non-zero, but at the same time very small. The Gaussian distribution is thus a reasonable model for the distribution of points within an image with sharp boundaries. Experiments reported in section 8.1 suggest that the probability of a false alarm is similar for the Gaussian and for the uniform distributions. The advantage of using a Gaussian to model the distribution of the image points is that certain calculations are simplified, as will become apparent in section 2.2. The mean of the Gaussian distributions in (i) is arbitrary. For convenience it is set equal to zero.

The assumption (ii) is required in order to simplify the calculation of the probability of rejection. The condition $\sigma = O(1)$ in (ii) is necessary, because if σ is large, for example $\sigma = \epsilon^{-1}$, where $\epsilon = \pi/\lambda$, then with probability close to one at least two of the image

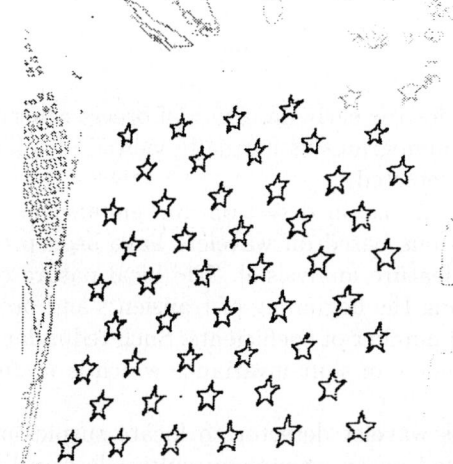

Fig 3a   Fig 3b

Fig 3c 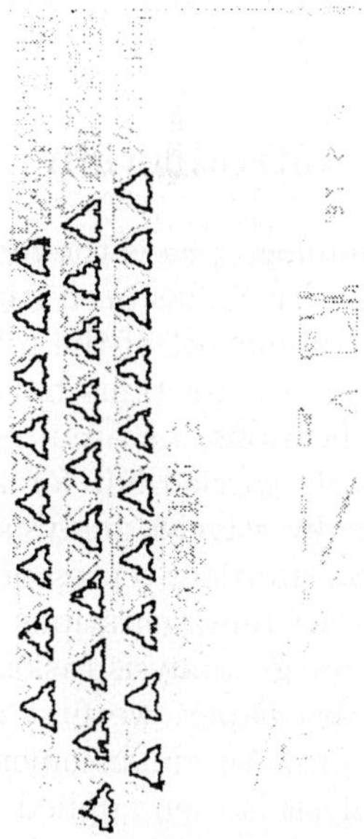   Fig 3d

Detection and Enhancement of Small Masses via Precision Multiscale Analysis

Dongwei Chen[1], Chun-Ming Chang[1], and Andrew Laine[2]

[1] Electrical and Computer Engineering, University of Florida, Gainesville, FL 32611
[2] Center for Biomedical Engineering, Columbia University, New York, NY 10027

Abstract. We introduce a continuous scale wavelet detector. Our algorithm was able to detect a mass that could not be seen using conventional windowing and leveling or traditional methods of contrast enhancement. An artifact free enhancement algorithm based on overcomplete multiscale wavelet analysis is then presented. The novelty of this algorithm lies in its detection of directional features and removal of unwanted perturbations.

1 Introduction

Mammography is the most effective tool for the early detection of breast cancer. However, it has been suggested that mammograms as normally viewed, display only about 3% of the total information detected.
Wavelet algorithms have found widespread use in detection and enhancement applications, *i.e.*, multiscale representations based on wavelets have been previously carried out for mammographic feature analysis[3]. The local nature of time by wavelet analysis enables capturing the beginning of transients and provides transient representation by a small number of coefficients. Such redundant wavelet representations provide the property of shift invariance which is useful for image analysis tasks.
In this paper, we first use a continuous wavelet detector to locate suspicious regions. An enhancement algorithm based on overcomplete multiscale wavelet analysis is then applied. The novelty and advantage of this algorithm compared to existing techniques lies in its detection of directional features and suppression of local perturbations (artifacts).

2 Continuous Scale Discrete Wavelet Detector

Let's first consider the hypotheses $H_1 : y(x) = s(x) + n(x)$ and $H_2 : y(x) = n(x)$, where x is a location, y is an observed signal, s is an underlying signal to be detected, and n is additive noise. Based on an observation y, a wavelet detection algorithm is used to choose between H_0 and H_1 by comparing the detector's output with a threshold determined by a statistical measure[5]. In particular, we shall develop a continuous scale wavelet detector compared to a M-Band wavelet detector[5].

2.1 A Continuous Scale Discrete Wavelet Transform

The Continuous Scale Discrete Wavelet Transform (CSDWT) is defined as the CWT sampled along the shifting parameter,

$$\mathbf{W}f(a,n) = \int_{-\infty}^{\infty} f(t)\psi\left(\frac{n-t}{a}\right)dt. \quad (1)$$

Scaling Space First, let's introduce scaling space to describe signals of distinct resolution. A compactly supported scaling function $\phi(t)$ satisfies the following conditions [10]:

1. Riesz basis condition: $A \leq \sum_{k \in \mathcal{Z}} |\Phi(\omega + 2k\pi)|^2 \leq B$, where A and B are strictly positive constants;
2. Order property: $\Phi(0) = 1, \Phi^{(m)}(2k\pi) = 0, k \in \mathcal{Z}, k \neq 0$, for $m = 0, ..., N-1$;
3. Two-scale relation: $\phi\left(\frac{t}{2}\right) = \sum_{k \in \mathcal{Z}} h(k)\phi(t-k)$;
4. Frequency constraint: $\Phi(\omega) \neq 0$, for $|\omega| \leq \pi$, and $\Phi(\omega)$ should be very small for $|\omega| > \pi$.

Initial Condition In the traditional DWT analysis literature, we assume that $x(n) = \int_{-\infty}^{\infty} f(t)\phi(n-t)dt$. However, in most cases, this assumption is not true. When the impulse response of an acquisition device is an ideal lowpass filter $\xi(t)$, $x(n) = \int_{-\infty}^{\infty} f(t)\xi(n-t)dt$. If the scaling function is a spline $\beta^n(t)$ of order $n-1$, $\xi(t) \approx \sum_k (b^n)^{-1}(k)\beta^n(t-k)$, $x'(n) = \sum_k x(k)(b^n)^{-1}(n-k) = \int_{-\infty}^{\infty} f(t)\phi(n-t)dt$. Therefore, we need to initialize data $x(n)$ with a prefilter $(b^n)^{-1}(k)$.

CSDWT A discrete approximation of $f(t)$ at scale a is defined as

$$\mathbf{S}_a^d f(n) = \int_{-\infty}^{\infty} f(t)\phi\left(\frac{n-t}{a}\right)dt. \quad (2)$$

It can be shown from the two-scale relation and Eq(2) that

$$\mathcal{F}\{\mathbf{S}_{2^{j+1}}^d f(n)\} = H(2^j\omega)\mathcal{F}\{\mathbf{S}_{2^j}^d f(n)\}. \quad (3)$$

From the initial condition, we have $\mathbf{S}_1 f(n) = x'(n)$, where $a = 1$ is the finest scale of analysis. As a increases, $\mathbf{S}_a f(n)$ becomes a coarser representation of $f(t)$.

Theorem 1. *Given $x(n)$, for $a > s_0$, if $\tilde{\Psi}_a(\omega)$ is band-limited between $[-\pi, \pi]$, the CSDWT can be calculated by $\mathbf{W}_a f(a,n) = \sum_{k \in \mathcal{Z}} p(a,k)x(n-k)$, where $P^c(a,\omega) = \frac{\tilde{\Psi}_a(\omega)}{\Phi(\omega)}$, $P^c(a,\omega) = \int_{-\infty}^{\infty} p^c(a,t)e^{-j\omega f_s t}dt$, $p(a,k) = p^c(a,t)|_{t=k}$.*

Proof. Note that $\tilde{\psi}_a(t)$ and $p^c(a,t)$ are band-limited at $[-\pi, \pi]$. We see that

$$\tilde{\psi}_a(t) = \int_{-\infty}^{\infty} p^c(a,u)\phi(t-u)du. \tag{4}$$

If we substitute Eq(4) into Eq(1) and exchange the sequency of the integral, we obtain

$$\mathbf{W}_a f(a,n) = \int_{-\infty}^{\infty} x^c(n-u)p^c(a,u)du. \tag{5}$$

It is straightforward to show that if $x^c(t)$ and $p^c(a,t)$ are band-limited to $[-\pi, \pi]$, $\int_{-\infty}^{\infty} x^c(n-u)p^c(a,u)du = \sum_{k \in \mathcal{Z}} p(a,k)x(n-k)$. Therefore,

$$\mathbf{W}_a f(a,n) = \sum_{k \in \mathcal{Z}} p(a,k)x(n-k). \tag{6}$$

If a wavelet $\psi(t)$ is local in the frequency domain, we need to select an s_0 such that $\tilde{\Psi}_{s_0}(\omega) = \Psi_{s_0}(\omega) Rect(\omega)$ has an acceptable approximation error onto the scaling space. The approximation error is defined as the ratio of the energy of $\Psi(a\omega)$ outside $[-\pi, \pi]$ to the total energy and decreases as s_0 increases. It is straightforward to show that $\tilde{\psi}_{s_0}(t)$ is also a mother wavelet.

When little is known about the best scale to detect a certain feature, it is desirable to first search the scale dyadically, then carry out a finer search. When an s_0 is selected, it may be considered as a "voice" and $P(2^L s_0, \omega)$ can be calculated as [9]

$$P(2^L s_0, \omega) = P(s_0, 2^L \omega)(\prod_{k=0}^{L-1} H(2^k \omega)). \tag{7}$$

Thus, Eq(3), (6) and (7) yield

$$\mathcal{F}\{\mathbf{W}_{2^j s_0} f\} = P(s_0, 2^j \omega) \mathcal{F}\{\mathbf{S}_{2^j} f\}. \tag{8}$$

Let $K(\omega)$ be a 2π periodic function and suppose $k(\omega)$ satisfies

$$K(\omega) P(s_0, \omega) + |H(\omega)|^2 = 1 \tag{9}$$

It is straightforward to derive from Eq(9), (3) and (8) that

$$\mathcal{F}\{\mathbf{S}_{2^j} f\} = H^*(2^j \omega) \mathcal{F}\{\mathbf{S}_{2^{j+1}} f\} + K(2^j \omega) \mathcal{F}\{\mathbf{W}_{2^j} f\}. \tag{10}$$

Thus, Eq(3), (8) and (10) show that a signal can be decomposed and reconstructed at an arbitrary scale $2^{J-1} s_0$.

2.2 Application of CSDWT detector

In this study, the wavelet applied was a second derivative of a spline of order 5 (n=4, d=2). The scaling function and mother wavelet is given by

$$\Phi(\omega) = \left(\frac{sin(\frac{\omega}{2})}{\frac{\omega}{2}}\right)^n, \quad \Psi(\omega) = (j\omega)^d \left(\frac{sin(\frac{\omega}{4})}{\frac{\omega}{4}}\right)^{n+d}$$

and

$$H(\omega) = e^{jt_0\omega}\cos^n\left(\frac{\omega}{2}\right), \quad P(s_0,\omega) = e^{jt_1\omega}\frac{j^d 2^{n+2d}}{s_0^n}\frac{\sin^{n+d}\left(\frac{s_0\omega}{4}\right)}{\sin^n\left(\frac{\omega}{2}\right)}.$$

$t_0 = \frac{1}{2}$ when n is odd and level $L = 0$. Otherwise, $t_0 = 0$. $t_1 = \frac{1}{2}$ when d is odd and level $L = 0$, $t_1 = 0$.

We first applied our detector to radiographic images of an RMI phantom, captured with technique factors of 22 kVp and 112 mAs. The smallest mass in the insert of the phantom cannot be seen using conventional window and leveling at 12-bit contrast resolution. In Figure 1(b), all masses within the phantom were detected correctly. Note that the scale used was not dyadic. Detectors searching at dyadic scales alone would miss the smallest mass, located in the upper right corner.

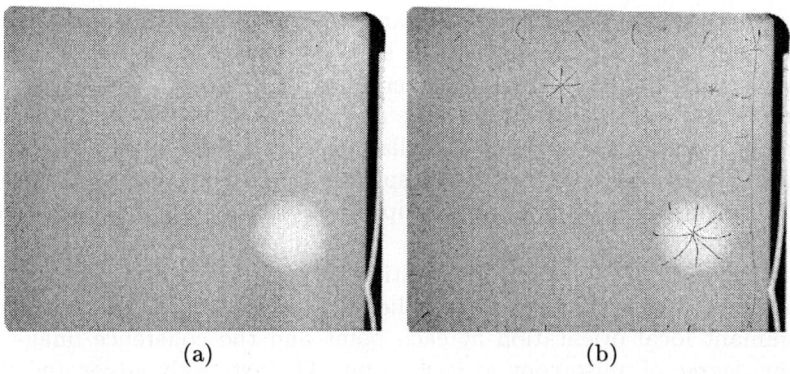

Fig. 1. (a) Digital X-ray image of the RMI156 phantom. (b) Maxima from precision scale detector obtained at scale 53.8.

We applied our detector to 6 cases of real mammograms with dense tissue. All the masses were detected correctly. Figure 2(b) shows a sample result.

3 Enhancement of Mammograms from Oriented Information

Reliable diagnosis by radiographs of malignant breast disease depends on observing local and distant changes in tissues produced by the disease. Of the visual signs of cancer found by radiologists, spiculated features are of great importance. Unfortunately, at the early stages of breast cancer, these signs are very subtle and varied in appearance, making diagnosis difficult and challenging even to specialists in mammography. In this section, an enhancement algorithm based on multiscale wavelet analysis is described. The novelty and advantage of this algorithm lies in its detection of spiculated features and removal of unwanted pertabations (artifacts).

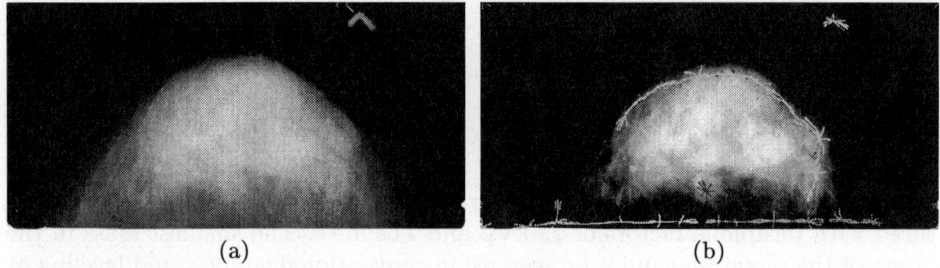

Fig. 2. (a) Mammogram lcc004. (b) Precision maxima superimposed on lcc004 enhanced by window and leveling.

3.1 Measure of Coherence

Texture plays an important role in many machine vision and image processing tasks including surface inspection, scene classification, surface orientation and shape determination. Texture patterns may be characterized by extracting measurements that quantify the nature and directions of pattern. Most breast carcinomas have the appearance of stellate lesions consisting of a central mass surrounded by radiating spicules. The spicules radiate outward in all directions and vary in length. These provide an important cue for early detection of breast cancer.

Rao and Schunck [8] defined the orientation field of a texture image to consist of two images — an angle image and a coherence image. The angle image denotes the dominant local orientation at each point and the coherence image represents the degree of anisotropy at each point. They strongly advocated the use of angle and coherence images as intrinsic images. In this paper, we investigated the efficiency of these two representations to capture and enhance features of importance to mammography.

3.2 Methodology

Our algorithm consists of the following four steps.
(1) Multiscale Wavelet Analysis: Wavelet transforms, owing to their localization characteristics, are powerful tools of analysis for many signal and image processing applications. Through multiscale analysis we can extract features at distinct scales and provide local information often hidden in an original (single resolution) mammogram. One major drawback of traditional wavelet transforms is their lack of translation invariance, making the content of wavelet subbands unstable under the translations of an input signal. In our algorithm, a digitized mammogram was decomposed using a fast wavelet transform algorithm (FWT) [4]. In order to obtain wavelet coefficients at each level without downsampling, a undecimated *"algorithme à trous"* (algorithm with holes) [2] was implemented. In the spatial domain, this redundancy corresponds to a representation without aliasing.

(2) Separable Steerable Filters: A filter is called "steerable" if the filter at an arbitrary orientation can be expressed as a linear combination of a set of basis filters, generated from rotations of a single kernel [1]. Steerable filters [1], which can be adaptively adjusted to arbitrary orientation, are used to detect stellate patterns of spicules and locate feature orientations more precisely. As pointed out by [7], the separability property of the filters speeds up computations considerably when convolved with large image matrices. In our algorithm, we used three basis functions as steerable filters. The x-y separable steerable approximations of filter kernels were generated by Singular Value Decomposition (SVD) [1, 7]. Using a set of separable steerable filters, the magnitude (M^i) and associated dominant directions (A^i) of local energy were determined by the basis functions of the constituent filter and its quadrature pair [6].

(3) Coherence Maps: A coherence map is an image showing a local measure of the degree of anisotropy of flow [8]. If the orientations of a texture pattern at any point (x_i, y_i) are coherent, then magnitude and phase information are important and should be emphasized. Conversely, if the orientations are not coherent, the magnitude and phase information can be neglected or attenuated. The measure of coherence proposed by Rao and Schunck [8] was obtained by weighting the energy with the normalized projection of energy within a specified window ($\mathcal{W}$) onto the central point (j, k) of each window. This coherence measure incorporates the gradient magnitude and hence places more weight on regions that have higher visual contrast. The combination of coherence and orientation structure was able to extract the more salient features of spiculated lesions. In overview, a schematic diagram of processing Steps 1–3 is shown in Figure 3.

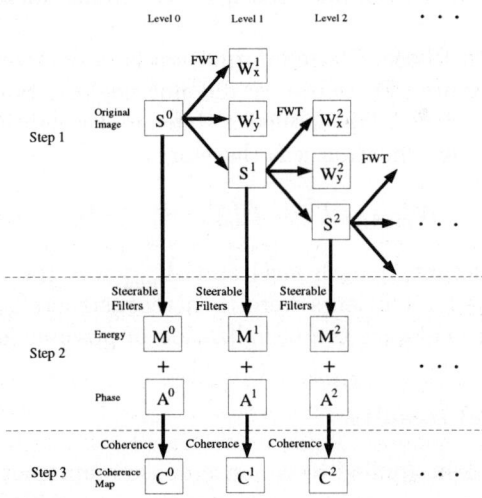

Fig. 3. Overview of processing for Steps 1–3.

(4) Nonlinear Operators: So far, we have precomputed all the information needed in our algorithm. A nonlinear operation was then applied within each

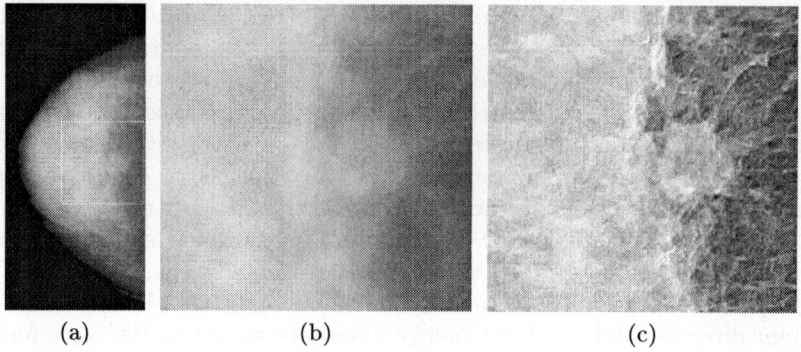

Fig. 4. Mammogram with a mass: (a) Original mammogram. (b) ROI image. (c) enhanced ROI image.

level to precisely modify transform coefficients. This operation integrates both coherence map and phase information.

A. Modification from Coherence Map: Let $C^i(j,k)$ denote the coherence measure of point (j,k) at some level i. Modifications of coherence computed in place were obtained by a nonlinear function expressed as

$$C^i_{mod}(j,k) = \begin{cases} \log\sqrt{C^i(j,k)}, & \text{if } C^i(j,k) > 1, \\ 0, & otherwise, \end{cases}$$

such that a coefficient was emphasized if its coherence measure was large and attenuated if small.

B. Modification from Phase: Phase information is important to distinctly characterize oriented texture. Therefore we did not neglect its contribution in the modification of coefficients. We applied a sinusoidal weighting to the phase information. The final modification was therefore

$$W^i_{mod} = T^{i-1} \cdot C^{i-1}_{mod} \cdot A^{i-1}_{mod} \cdot W^i,$$

where T^i was a constant at each level and A^i_{mod} was the modified gain from phase. These modified coefficients were then reconstructed, via an inverse fast wavelet transform, to enhance the visualization of possible lesions.

3.3 Experimental Results

Our algorithm was then applied to enhance dense mammograms using oriented information. In order to capture distinct directions of subtle features, steerable filters were used. Due to the limit of space we only show one example. A mammogram (mam004lcc) with a mass tumor is shown in Figure 4. The craniocaudal view of the left breast shown in Figure 4(b) shows an irregular spiculated mass in retroglandular fat. Note how the enhanced version shown in Figure 4(c) clearly delineates the margins of the mass.

4 Conclusion

The primary radiographic signs of breast cancer are related to tumor mass, its density, size, shape, borders and calcification content. Extraction of these features and enhancement may assist general radiologists to locate suspicious areas more reliably.

We first provided a more reasonable initialization procedure for wavelet processing. We showed that a continuous scale wavelet detector can provide a more precise matched basis than detectors with a limited number of scales. The importance of employing continuous scale was demonstrated by digital radiographs of a mammography phantom and digitized mammograms. Our investigation showed that the algorithm was able to detect very subtle masses, which were rated to be almost invisible by radiologist specializing in mammography.

Existing and previous multiscale enhancement approaches [3] attempted to enhance an image by detecting edges. Unfortunately, most edge detection algorithms can not distinguish between "authentic" edges and phantom edges. In contrast, we presented an algorithm which relied upon a coherence map and phase information. This resulted in an enhancement naturally close to the original image. Such artifact free processed images provide more familiar visual cues for radiologists.

References

1. W. T. Freeman and E. H. Adelson. The design and use of steerable filters. *IEEE Transactions on Pattern Analysis and Machine Intelligence*, 13(9):891–906, 1991.
2. M. Holschneider, R. Kronland-Martinet, J. Morlet, and Ph. Tchamitchian. A realtime algorithm for signal analysis with the help of the wavelet transform. In J. M. Combes, A. Grossmann, and Ph. Tchamitchian, editors, *Wavelets: Time-frequency Methods and Phase Space*, pages 286–304, Springer-Verlag, Berlin, Germany, 1990.
3. A. F. Laine, S. Schuler, J. Fan, and W. Huda. Mammographic feature enhancement by multiscale analysis. *IEEE Transactions on Medical Imaging*, 13(4):725–740, 1994.
4. S. Mallat and S. Zhong. Characterization of signals from multiscale edges. *IEEE Transactions on Pattern Analysis and Machine Intelligence*, 14(7):710–732, 1992.
5. S. D. Marco and J. Weiss. M-band wavepacket-based transient signal detector using a translation-invariant wavelet. *Optical Engineering*, 33(7):2175–2182, 1994.
6. M. C. Morrone and R. A. Owens. Feature detection from local energy. *Pattern Recognition Letters*, 6(5):303–313, 1987.
7. P. Perona. Deformable kernels for early vision. *IEEE Transactions on Pattern Analysis and Machine Intelligence*, 17(5):488–499, 1995.
8. A. R. Rao and B. G. Schunck. Computing oriented texture fields. In *Proceedings of the IEEE Computer Society Conference on Computer Vision and Pattern Recognition*, pages 61–68, San Diego, CA, 1989.
9. O. Rioul and P. Duhamel. Fast algorithms for discrete and continuous wavelet transforms. *IEEE Transactions on Information Theory*, 38(2):569–586, 1992.
10. M. Vrhel, C. Lee, and M. Unser. Fast computation of the continuous wavelet transform through oblique projections. In *Proceedings of the IEEE International Conference on Acoustics, Speech, and Signal Processing*, volume 3, pages 1459–1462, Atlanta, GA, 1996.

A Method of Industrial Parts Surface Inspection Based on an Optics Model

Norifumi Katafuchi, Mutsuo Sano, Shuichi Ohara, and Masashi Okudaira

NTT Human Interface Laboratories, Musashino-shi, Tokyo, 180 Japan.

Abstract. *This paper proposes a new approach based on an optics model for highly reliable surface inspection of industrial parts. This method uses multiple images taken under different camera conditions. Phong's model is employed for surface reflection and then the albedo and the reflection model parameters are estimated by the least squares method. Experimental results show the proposed method's advantages over the conventional binarization method for easily determining a threshold of product acceptability and coping with changes in light intensity when detecting defects.*

1 Introduction

The inspection process for checking the surface finish and detecting defects in industrial parts generally has to be a 100% inspection in which each product is tested. However, the level of automation used in factories has not keep up with the need in recent years; the inspection of complicated objects and metallic parts, excepting printed circuit boards, is often performed by human inspectors. Moreover, problems exist with a parts-inspection apparatus that uses image processing techniques such as binarization and spatial filters:

- The defect-detection reliability is low.
- The apparatus cannot accommodate changes in image intensity caused by the variety of normal products, peculiar patterns and noise.
- The diversion of instruments to other applications is difficult, because the instruments strongly depend on the shape of the inspected object.

A clear need exists for high reliability in parts-inspection systems. However, though instruments for limited applications have been developed *individually*[1], few systems that work accurately have actually been introduced at production sites.

In this study, as a step towards developing a highly reliable algorithm for surface inspection, we deal with the problem of detecting defects such as separation and destruction in objects such as products made of rubber. Two methods[2] often used for such defect detection are the hybrid method which combines binarization and 2D spatial filters, and the matching method which is based on normalized correlation using reference templates. However, these methods do not solve the problems described above. Moreover, by local processing, excessive mis-identification of non-defective area is not unusual. As an alternative approach, we propose a method[3], [4] based on an optics model that uses a set of images taken under different camera conditions.

Related work concerning shape recovery from multiple images includes Ref. [5] which used a set of gray-scale images and estimated the surface normal and reflectance of the object based on the reflection model that expresses the components of both Lambertian and specular reflection. Reference [6] estimated the surface normal, reflectance and surface-roughness parameters of an object from multiple gray-scale images based on the reflection model; this *tempered* the former model with the specular diffuse component. However, the purpose of these studies was not inspection, but rather 3D shape reconstruction or reflectance estimation. Also, the photometric sampling approach is complicated, requiring a large-scale device that includes several light-sources, and parameter tuning to ensure a high level of performance is difficult. Consequently, these approaches may not be suitable for practical use and are limited in their applicability to product inspection. An additional problem is recovering the 3D shape of an unfolded book surface from the shading information in a scanner image. This is examined by T. Matsuyama *et al.*[7], [8]. They solved the problem of dealing with book-like objects by limiting the environmental conditions.

Inspection does not require the 3D shape recovery, and very few papers have been presented on application of photometric methods to inspection. This study presents a new approach toward handling the surface condition of the object. Its main goal is automated visual inspection with high reliability. The key idea of this work is that a highly reliable judgement on product acceptability can be made based on the albedo and reflection properties. This is because these properties, as estimated from the obtained multiple images, seem to be unaffected by changes in the inspection environments.

In the following sections, we begin by discussing real-world defect detection based on the estimation of optics-model parameters. The effectiveness of the proposed method is then compared with that of a conventional method through several experiments, using images of real samples.

2 Image Acquisition

2.1 Target Object and Items to be Inspected

We used a conductive rubber pad as the object to be inspected (see Fig. 1). Conductive rubber pads are used in products such as TV remote controllers and pocket calculators. In the pad shown in Fig. 1, the bonding of a piece of conductive material in each circular region must be checked. The defects to be identified are:

– separation or destruction (a portion of the pill has peeled off or collapsed),
– distension (a piece of some other material is attached to the pill),
– cracks of hundred-micron-order width.

Here, we handle planar defects larger than about 20% of the pill area, such as separation and destruction but excluding cracks.

2.2 Multiple Images of an Object

Figure 2 shows the configuration of our image-input system. In this system, multiple object images are taken by changing the light-source direction; that is,

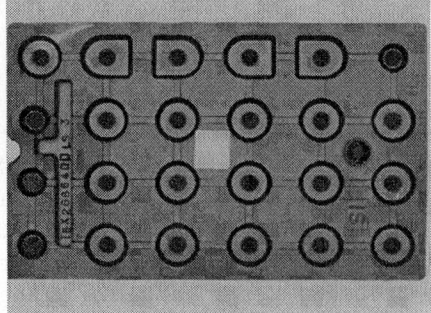
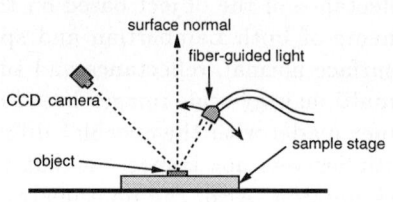

Fig. 1. Conductive rubber pad

Fig. 2. Configuration of the image-input system

by varying the incident angle i from 0 to 40° at intervals of 5°, with a fixed relative configuration between the camera and the object. This affords a set of images where each viewing direction is the same but each illumination direction is different can be obtained.

3 Problem Formulation

3.1 Assumptions in Solving Problem

In the development of our defect-detection algorithm, we assume several physical conditions:

1. The light-source direction and camera direction are known, and the location of the light source is fixed.
2. The light source is modeled as a point light source.
3. The object surface is not Lambertian and the albedo distribution over the sample surface is not uniform.

Such conditions are applicable to an object like the conductive rubber pad. Thus, under these conditions, the reflected-light intensity observed at a 2D point in the image $L_k(\boldsymbol{x})$ is formulated as follows:

$$L_k(\boldsymbol{x}) = r(\boldsymbol{p}(\boldsymbol{x})) \cdot I_k\big(d(\boldsymbol{p}(\boldsymbol{x}))\big) \cdot f(\boldsymbol{n}, \boldsymbol{l}_k, \boldsymbol{v}), \qquad (1)$$

where $\boldsymbol{p}(\boldsymbol{x})$ denotes the 3D point on the object surface corresponding to $\boldsymbol{x}$, $r(\boldsymbol{p}(\boldsymbol{x}))$ is the albedo at point $\boldsymbol{p}(\boldsymbol{x})$, $I_k\big(d(\boldsymbol{p}(\boldsymbol{x}))\big)$ is the illumination intensity at point $\boldsymbol{p}(\boldsymbol{x})$, $d(\boldsymbol{p}(\boldsymbol{x}))$ is the distance between the illumination and $\boldsymbol{p}(\boldsymbol{x})$, and $f(\boldsymbol{n}, \boldsymbol{l}_k, \boldsymbol{v})$ denotes the reflectance properties at point $\boldsymbol{p}(\boldsymbol{x})$. The suffix k expresses each direction of the illumination.

3.2 Optics Model of Surface Reflection

In this study, Phong's model[9] is employed. Hence, the reflectance properties f at point $\boldsymbol{p}(\boldsymbol{x})$ on the object to be inspected can be represented as:

$$f(\boldsymbol{n}, \boldsymbol{l}_k, \boldsymbol{v}) = \Big(1 - s(\boldsymbol{p}(\boldsymbol{x}))\Big)\cos\psi(\boldsymbol{l}_k, \boldsymbol{n}) + s(\boldsymbol{p}(\boldsymbol{x}))\cos^{\lambda(\boldsymbol{p}(\boldsymbol{x}))}\phi(\boldsymbol{l}'_k, \boldsymbol{v}), \qquad (2)$$

where n denotes the surface normal vector, l the direction of the illumination, and v the view point direction. Here, $l'_k(n, l_k, v)$ denotes the direction of specular reflection l_k to n, $\psi(l_k, n)$ denotes the angle between l_k and n, and $\phi(l'_k, v)$ denotes the angle between l'_k and v. Also, s and λ are the parameters specifying the reflectance properties.

In this formulation, finding the albedo distribution and reflectance properties of the object's surface requires computing the 3D absolute position. This is impossible to solve generally without some additional conditions concerning the shape of the target object. However, for an object with an even surface, such as the conductive rubber pad, the problem can be reduced to that of a 2D-shaped object to find the parameters that minimize the square error of the image intensity.

The significant characteristics are the more or less flat shape and the small inspection region—a circle with a radius of about 4 mm. Thus, the approximations $\cos\psi(l_k, n) = A_k$, $\cos\phi(l'_k, v) = B_k$, $I_k(d(p(x))) \simeq I_k$ seem to be acceptable. Note that A_k, B_k, and I_k are constant according to k. Furthermore, if we take into account the photoelectric transformation in the CCD camera, Eq. (1) can be transformed as follows:

$$P_k(x) = aI_k \cdot r(p(x))\left\{\left(1 - s(p(x))\right)A_k + s(p(x))B_k^{\lambda(p(x))}\right\} + \delta. \quad (3)$$

where $P_k(x)$ represents the image intensity at point x, and a and δ are parameters of the photoelectric transformation in the sensor. In practice, these parameters need to be estimated a priori using images from a flat white plate with known reflectance properties.

4 Surface-inspection Algorithm

4.1 Estimation of Albedo and Reflection-model Parameters

The cosine of the angle between the direction of specular reflection and the viewpoint direction is denoted as B_k. When the maximum of the angle is rather small, B_k is close to unity. Let B_k equal $1 - \Delta_k$, with $\Delta_k \ll 1$. Then B_k^λ can be expressed as:

$$B_k^\lambda = (1 - \Delta_k)^\lambda \simeq 1 - \lambda\Delta_k + \frac{\lambda(\lambda-1)}{2}\Delta_k^2. \quad (4)$$

For inspection, we believe that it is important to consider product acceptability as relative rather than absolute. On those grounds, we deem it satisfactory to consider simplicity and processing speed to be more important than accuracy. Ignoring the square term of Δ_k in Eq. (4), Eq. (3) can be transformed as follows:

$$\frac{P_k - \delta}{aI_k} \cdot \frac{1}{r} + \Delta_k s \cdot \lambda = A_k + (1 - A_k)s. \quad (k = 1, \cdots, K) \quad (5)$$

In Eq. (5), when s is fixed, a pair of parameters $(\hat{r}, \hat{\lambda})$ can theoretically be found that minimize the following function of the square error:

$$E(r, \hat{s}, \lambda) = \sum_{k=1}^{K}\left\{\frac{P_k - \delta}{aI_k} \cdot \frac{1}{r} + \Delta_k \hat{s} \cdot \lambda - A_k - (1 - A_k)\hat{s}\right\}^2. \quad (6)$$

Thus, if we change the value of parameter s in a range $[0, 1]$, the parameters (r, s, λ) can be calculated by numerical computation. In surface inspection, a product can be judged as either good or no good by using the difference in albedo between the normal and the defective area.

4.2 Tessellation

We have described pointwise processing at the pixel level. However, this approach is not appropriate from the technical point of view because noise often degrades the accuracy of the estimation. Hence, tessellation of the image is used to improve the computational efficiency and the stability. Tessellation is applied to approximate the 3D conductive rubber surface by means of planar rectangles with constant albedo and to estimate the parameters for each rectangle.

5 Experiments

5.1 Albedo Estimation

Experimental Conditions Ten pieces with good pills and four pieces with defective pills were taken as samples for evaluation from three sheets of conductive rubber pads. The input images were acquired by the method described in Sect. 2.2, and the partial images of rectangular regions externally tangent to them were cut out. Each of the four pieces of conductive pill had a planar defect of a certain size: *defect* A—a large separation; *defect* B—a small separation; *defect* C—a crescent-shaped destruction; *defect* D—a distension (i.e., a piece of some other material attached to the pill). Nine images of each sample were taken. The albedos r of each sample were estimated using the five images having the incident angles 0, 10, 20, 30, and 40°.

Results and Discussion The estimation results for *defect* B and *defect* D are shown in Figs. 3 and 4, respectively. Figure 5(a) shows the 3D gray-level distribution and Fig. 5(b) shows the estimated 3D albedo distribution for the sample having *defect* B with a surface intensity of 12900[lx]. This sample has a small separation on the pill slightly left of the center. Detecting a defect using only the gray-level data is difficult in a case like this. However, the albedo in the defective area is larger than that in the normal area, as can be seen in Fig. 5. Adding this fact makes it possible to judge whether a sample is good by selecting an appropriate threshold.

Tables 1 and 2 show the average value of r in the normal area (α), the maximum value of r in the defective area (β) and the SN ratio defined as $\frac{\beta-\alpha}{\beta}$ for each sample. This definition of the SN ratio represents the difference in the albedo between the normal area and the defective area divided by the maximum albedo in the defective area. So defined, it can be used as criteria to measure the ease of selecting a threshold. As listed in Tables 1 and 2, the SN ratio ranges from 50 to 70% (except for *defect* D), affording a margin for threshold selection. Good results shown above were obtained by using a linear approximation of Eq. (4) for an actual rubber product (except for *defect* D). No significant difference was detected between the good and the defective area for *defect* D. We think that this is because of the 3D shape and *difference in material*.

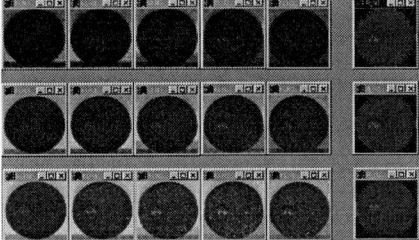
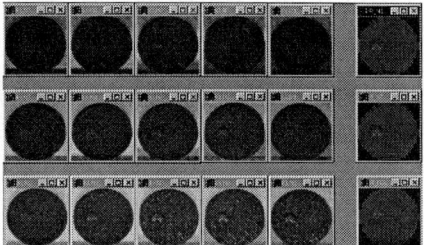

Fig. 3. *Defect* B: a small separation **Fig. 4.** *Defect* D: a distension

Multiple images and estimated albedo distributions

In both figures, the upper row was taken at a surface intensity of 12900[lx], the middle row at 22600[lx], and the lower row at 33000[lx]. The size of each image is 110×105 pixels. The five images on the left were used for estimation; their incident angle increase in 10° increments from $i = 0$ to 40°. The gray-scale images in the right-hand column are the albedo distributions (256 levels; the gray-level is 255 when $r = 1$). The size of the tessellation is 5×5 pixels.

Table 1. Estimated albedo with a surface intensity of 12900[lx]

defect type	average (α)	maximum (β)	ratio $(\beta - \alpha)/\beta$
defect A	0.25258	0.81008	**0.68821**
defect B	0.27195	0.59913	**0.54609**
defect C	0.25638	0.67016	**0.777274**
defect D	0.25464	0.35082	**0.27418**

Table 2. Estimated albedo with a surface intensity of 22600[lx]

defect type	average (α)	maximum (β)	ratio $(\beta - \alpha)/\beta$
defect A	0.26104	0.87828	**0.70278**
defect B	0.27957	0.63639	**0.56069**
defect C	0.25679	0.82004	**0.68686**
defect D	0.25670	0.36439	**0.29553**

In Tables 1 and 2, the average values in the normal area (α) are about the same, despite the differing surface-light intensities. This indicates that the proposed method is robust against changes in illumination intensity.

5.2 Comparison of the Proposed Method and Binarization

In this section, we compare the proposed method with binarization in terms of defect-detection performance and stability. The evaluation criteria are the probability of a false negative or a false positive[10], defined as follows:

- probability of a false negative: $Pr^N = 1 - \frac{R_S}{R_D}$,
- probability of a false positive: $Pr^P = \frac{R_E - R_S}{R_P - R_D}$.

Note that R_p denotes the area of the pill—the number of pixels in the inspection region (see Fig. 6).

Experimental Conditions In our experiments, we used as input the images with incident angles of 0, 10, 20, 30, or 40°. The binarization was done by 5×5 tessellation of each rectangular region and linear transformation of the gray-scale value to permit its distribution to range from 0 to 255. Our method of estimating the albedo was to use multiple images (five images per sample) and linear transformation of the albedo in a similar way.

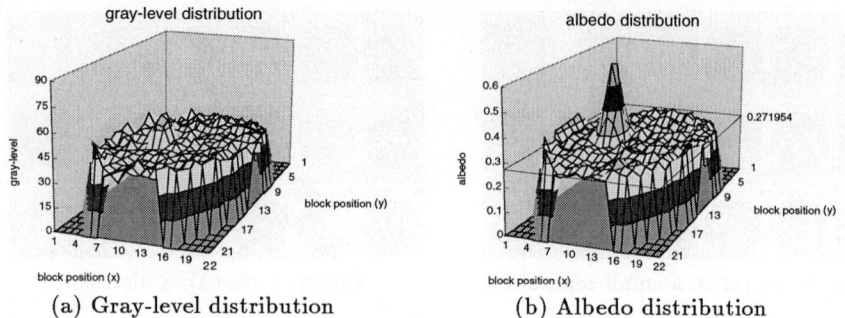

(a) Gray-level distribution (b) Albedo distribution

Fig. 5. Gray-level and albedo distribution for *defect* B with a surface intensity of 12900[lx]. For graph (a) $i = 40°$.

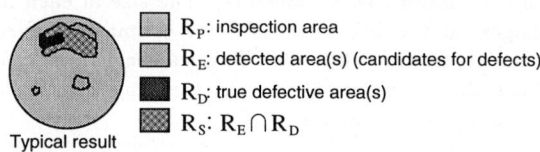

R_P: inspection area
R_E: detected area(s) (candidates for defects)
R_D: true defective area(s)
R_S: $R_E \cap R_D$

Fig. 6. Calculation of the probability of a false negative or a false positive

In this performance evaluation, the binarization method detected defects by thresholding of the gray-scale value, while the proposed method did this by selecting a threshold T of the albedo. The two types of probability can be calculated according to a threshold T. We obtained the reliability of confidence (ROC) curves by plotting each false probability value with T chosen as a parameter.

Results and Discussion The ROC curves obtained for *defect* B and *defect* D are Figs. 7 and 8, respectively. For *defect* B, for instance, the 20° curve follows the curve of our method most closely, and for *defect* D it is the 10° curve. The 20° images from *defect* B are better and have a higher contrast than those of the other angles; thus, it would be possible to detect this defect by simple binarization. However, generally speaking, we do not know beforehand whether good-quality, high-contrast images can be obtained or what configuration of the optical system is most suitable for image acquisition. That is a problem, especially when an inspection is performed by a single image, because the choice of the angle essentially depends on the object surface or the type of defects. Nevertheless, the proposed technique provides easy defect-detection that is negligibly affected by the angle choice problem.

6 Conclusion

We discussed the use of an optics model to detect defects on the surface of industrial products. We found that such defects on a conductive rubber pad could be detected by solving a square error minimization problem. This allows us to estimate the albedo and the parameters of the reflection model by the least-squares method. To improve the efficiency and reliability of this defect-detection, we used tessellation of the albedo distribution and the reflectance properties. Experimental results show that the proposed algorithm is capable

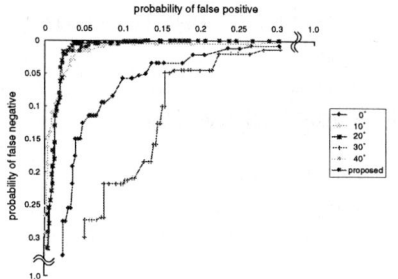
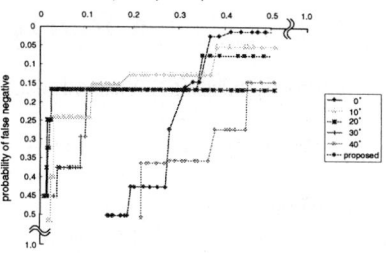

Fig. 7. ROC curve for *defect* B, with a surface intensity of 22600[lx]

Fig. 8. ROC curve for *defect* D, with a surface intensity of 22600[lx]

of judging correctly and reliably test sample acceptability despite changes in illumination intensity.

Acknowledgements

The authors would like to thank Mr. Takashi Sakai and Dr. Kenji Kogure, former director and director of our laboratory, respectively, and our colleagues for their guidance and helpful advice.

References

1. A. Ishii et al., "Technical Trend Research on Visual Inspection Technology," JSPE Technical Committee on Applied Image Processing, 1992.
2. T. S. Newman, "A survey of automated visual inspection," *Computer Vision and Image Understanding*, pp. 231–262, Vol. 61, No. 2, 1995.
3. N. Katafuchi, M. Sano and M. Okudaira, "A method of industrial parts surface inspection based on an optics model," IPSJ Technical Report, 96-CVIM-100, pp. 77–84, 1996.
4. N. Katafuchi, M. Sano and M. Okudaira, "A method of industrial parts surface inspection based on an optics model," *Trans. of IEICE*, Vol. J80-DII, No. 7, pp. 1802–1809, 1997.
5. S. K. Nayar, K. Ikeuchi and T. Kanade, "Determining shape and reflectance of hybrid surfaces by photometric sampling," *IEEE Trans. on Robotics and Automation*, Vol. 6, No. 4, pp. 418–431, 1990.
6. T. Kiuchi and K. Ikeuchi, "Determining surface roughness and shape of specular diffuse lobe objects using photometric sampling device,"*IAPR Workshop on Machine Vision Applications*, pp. 175–178, 1992.
7. T. Wada, H. Ukita and T. Matsuyama, "Recovering 3D shape of unfolded book surface from a scanner image (I)," *Trans. of IEICE*, Vol. J77-DII, No. 6, pp. 1059–1067, 1994.
8. T. Matsuyama, H. Ukita and T. Wada, "Shape from shading with interreflections under proximal light source," *Proc. of the Fifth International Conference on Computer Vision*, pp. 66–71, 1995.
9. D. H. Ballard and C. M. Brown, *"Computer Vision,"* Prentice-Hall, Inc., pp. 93–102, 1982.
10. A. Shimizu, J. Hasegawa and J. Toriwaki, "Minimum directional difference filter for extraction of circumscribed shadows in chest X-ray images and its characteristics," *Trans. of IEICE* , Vol. J76-DII, No. 2, pp. 241–249, 1993.

Illumination Color from the Blurred Inter-Reflection of a Reference Nose

Mohamed Abdellatif *, Yutaka Tanaka**, Akio Gofuku** and Isaku Nagai**

*Graduate School of Science and Technology, Okayama University, Japan.
E-mail: latif@computer.org
** Faculty of Engineering, Okayama University, 3-1-1,Tsushima- Naka, Okayama 700, Japan.
E-mail: field@apollo2.mech.okayama-u.ac.jp

Abstract. This paper presents a novel technique to estimate illumination color in a scene. The key principle is that the blurred inter-reflection of a camera-mounted " nose" represents the illumination color directly under weak scene assumptions. The nose surface profile is designed to reflect a blurred scene version into a small image area. The nose image is then spatially mapped to the scene image to correct its colors. Experiments showed that a nose surface represents illumination color robustly and as effective as a scene white patch with the merit of detecting smooth spatial changes of illumination color. The color constancy performance on real images is presented and compared with the retinex method [1].

1 Introduction

Color is an important surface property that can be used for several vision tasks. However, the color response in an image is illumination dependent, which is the well-known problem of color constancy. This problem may be defined as the derivation of surface color descriptors *despite* changes of illumination color and changes of scene color distribution.

To achieve color constancy, the illumination color should be known. Estimation of lighting color in the literature may be classified into spatial and global search methods. The main spatial method is the retinex theory [1]. This method relies on a key assumption that the average color in a local scene area is gray (Gray World Assumption, GWA). This assumption was criticized in several studies as it does not always hold in real scenes.

The second approach of global search can be further classified into scene-inserted patches and the use of specular highlights. Scene-inserted reference patches had been used in several studies to constrain a single illuminant color [2,3,4]. The drawback with scene patches is the difficulty to apply in real scenes. The use of highlights from dielectric surfaces can relax the need for reference patches and it was used in [5,6,7,8]. The problem in these methods is the measurement accuracy due to the limited dynamic range of the sensors.

In light of this discussion, we suggest that the inter-reflection from a reference surface " nose" can represent illumination color under weak assumptions. The specular reflection is weakened through the dispersion on nose and thus, the inter-reflection may fit in the sensor range. When the nose is designed to reflect the entire scene image, the spatial lighting color can be detected as well. The nose is attached to the camera as an integral device and applicable for all scenes.

In this paper, we study the feasibility of using the nose inter-reflection to represent scene illumination and use it to achieve color constancy. The paper is arranged as follows, Sec.2 presents the camera and nose models. The nose design is discussed in Sec.3. Experiments on real scenes are described in Sec.4 and then we conclude in Sec.5.

2 Nose-Camera Model

2.1 Camera Model

When a surface is illuminated by a single color light, it reflects light in the form

$$I(\lambda) = E(\lambda)S(\lambda) \tag{1}$$

where $S(\lambda)$ is the Spectral Power Distribution (SPD) of surface reflection and $E(\lambda)$ is the SPD of illumination and λ is the wavelength of light. The sensor integrates the irradiance over its area and deliver the following response

$$k_i(x,y) = \iiint R_i(\lambda) I(X,Y,\lambda) d\lambda. dX. dY \tag{2}$$

where (x,y) are the image coordinates, (X,Y) are the world coordinates referred to the camera, $R_i(\lambda)$ is the spectral response function of the i-th camera receptor, i=1,2,3 (red, green and blue). When the illumination color is known, the surface true color can be described by a vector q evaluated as

$$q_i = \left(\frac{k_i}{k_{in}}\right) L \tag{3}$$

where, k_i; is the color response, k_{in}; is the illumination color (the nose color response as illustrated in the next subsection), subscript i; is the number of the spectral channel, in; is the number of spectral channel in nose image, and L is a fixed scale factor.

2.2 Nose Inter-Reflection Model

The following assumptions are needed to recover surface colors in our model:
1. The structure of the scene in a local area *must* either:
   a) Reflect some *specularities* which follow the Neutral Interface Reflection model (NIR).
   b) Have an average chromaticity of *gray* (GWA).
2. Illumination color is uniform or changes *smoothly* in the scene.
3. Scene surfaces reflect light *measurable* with moderate signal to noise ratio in both main and nose images in all spectral channels (except for the highlights in the main image).

The first two assumptions ensure a reliable illumination measure while the third is needed for the validity of correction model in Eq.(3). The parallel dependence on either the existence of NIR specularities or satisfaction of the GWA increases the set of feasible scenes. Figure 1 shows the nose inter-reflection and a protection hood shielding the nose surface from lateral reflections. Figure 2.a shows the dispersed specular reflection pattern on the smooth nose surface. The ray emerging from a nose surface point can be described as

$$C(X_n,Y_n,\lambda) = \iint B_1(X,Y) S_n(X,Y,\lambda) S(X,Y,\lambda) E(X,Y,\lambda) dX dY \tag{4}$$

where $B_1(X,Y)$ is a blurring function, $S_n(X,Y,\lambda)$ is the spectral reflectance of nose

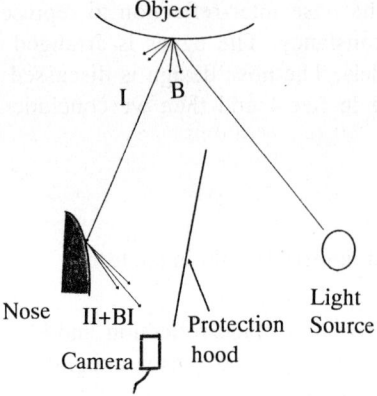

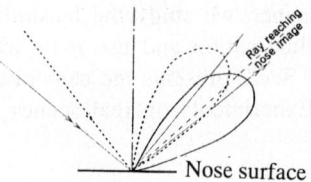

Fig.2. a First blurring layer on nose surface due to reflection pattern.

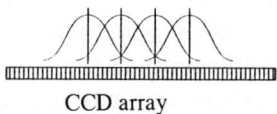

CCD array

Fig.1. The nose surface inter-reflection.

Fig.2. b Second blurring layer on CCD array due to defocusing.

surface in world coordinates (provided that mapping between nose and world coordinates is known in advance as will be shown in Sec.3.)

The nose image is seen defocused because it is near to lens. Figure 2.b shows the defocus blurring and the reflected spectrum may be described as

$$C_i^n(x_{ni}, y_{ni}, \lambda) = \iint B_2(X_n, Y_n) C(X_n, Y_n, \lambda) dX_n dY_n \tag{5}$$

where subscript *ni* refers to nose image pixel. The spatial blurring scale depends on the width of the specular reflection lobe and the defocus circle diameter.

The resulting inter-reflection color can be explained according to the dichromatic reflection model [9] as shown in Fig.1. When the nose is approximated as a smooth surface, the inter-reflection is composed of two colors; Interface-Interface II and Body-Interface BI. (The first letter refers to object while the second refers to nose reflections). The II component carries the illumination color, while BI reflection represents the illuminant color modified by surface color. Although, the scene spectral composition can affect the inter-reflection, the nose color is mostly dominated by the II component and this claim is confirmed in Sec.4. The following two subsections may explain why the nose inter-reflection is dominated by the light color.

2.2.1 Specular Dominance

The optical blurring for the nose intr-reflection has the advantage that strong specular reflections dominate the output color as their brightness is high. While blurring in the image plane using main image colors does not benefit from specularities due to pixel clipping.

2.2.2 Change of the Blurring Spatial Scale

The spatial blurring scale increases when object depth increases for the two blurring layers. The scene area enclosed by a single nose ray increases when object depth increases. Also, defocus blurring scale increases when the camera focus to a distant scene. This behavior is useful for illumination color representation as near scenes probably contain several color patches, while distant scenes (in particular outdoor scenes such as a forest) contain large areas of a *single color*. This behavior tries to enforce our model assumptions by including larger scene area for distant scenes.

3 The Nose Design

The derivation of the surface color vector in Eq.(3) requires knowledge of the nose pixel corresponding to a main image pixel. To achieve this, first the nose profile must be designed and second, the nose image must be mapped to the main image. Figure 3 shows the geometry of the nose and the CCD plane for one scan line consisting of main and nose image sections separated by the vertical nose edge. We consider the specular reflection as the nose surface is smooth. A general object point P reflects light across the nose at point N, and the following relation can be derived from simple geometry

$$\left(\frac{Z - Z_n}{X - X_n}\right) = \left(\frac{f - x_{ni}.\tan 2\alpha}{x_{ni} + f.\tan 2\alpha}\right) \quad (6)$$

where f is the lens focal length, subscript n refers to nose points, and α is the nose angle.

If we replace $\left(\frac{X}{Z}\right) = a$, and $\left(\frac{x_{ni}}{f}\right) = m$, then

$$m = \left(\frac{(a - \tan 2\alpha) - (X_n/Z) + (Z_n/Z)\tan 2\alpha}{1 + a.\tan 2\alpha - (X_n/Z)\tan 2\alpha - (Z_n/Z)}\right) \quad (7)$$

This equation describes the mapping between nose and main image in one horizontal line. The scene depth affects the mapping, but this effect can be reduced by keeping the nose profile near to the lens (smaller X_n and Z_n). The nose is designed for a nominal scene depth and its fixed mapping is used during operation. Figure 4 shows the profile of our prototype nose designed for a nominal depth of 2 meters. The mapping deviations do not exceed 4% of the nose image width for object depth of 5 meters and do not increase significantly for more distant scenes. Due to the lack of space, further design details can not be included here and will appear in a forthcoming paper[12].

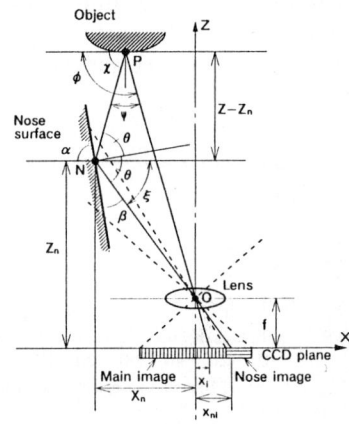

Fig.3. The Nose-Camera geometry.

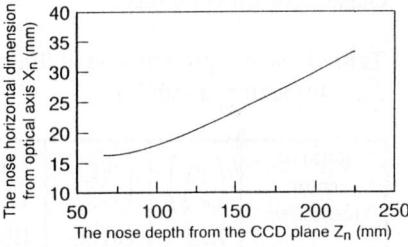

Fig.4. The nose profile for nominal scene depth of 2 meters.

4 Experiments

Color images are obtained using a linear CCD camera fitted with a 12.5 mm lens. The spectrum is sampled using Kodak Wratten filters (No.25,58,47). The nose is machined from steel and an Aluminum foil is pasted onto the profile and polished with varnish. The intensity across the nose image decreases toward the inner edge and this was experimentally corrected to make almost uniform nose brightness when viewing a uniformly illuminated scene. We have conducted a series of experiments to be described in the following subsections.

4.1 Representation of Single Lighting Color in Texture Scenes

We arranged 5 textured papers of smooth surface each fitted with a small white patch. The light color were changed 10 times using color filters, and both white patch and nose image responses were recorded. Figure 5 shows the white patch versus nose responses for the red channel in a texture. The relation is linear for the red and other color channels as well. Tables 1 and 2 shows the ratio of white patch to nose responses in color channels for two textures. The small variance shows good linearity implying that the nose reflection can represent the illumination color as effective as the scene white patch.

4.2 Representation of Illumination Spatial Changes

Figure 6 shows the spatial representation, where a white light spot is projected on white wall with dark surround and the circle is reflected as an ellipse in the nose image.

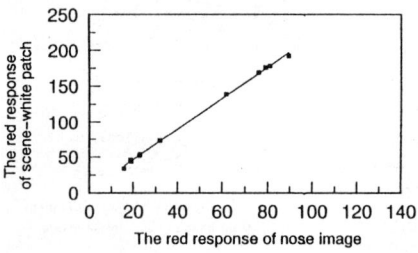

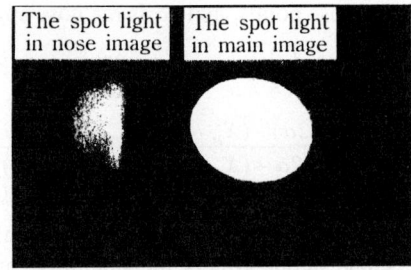

Fig.5. Relation between the white patch and nose responses in the red color channel for a texture scene.

Fig.6. Image of a spotlight on a white board and its nose reflection.

Table.1 Nose representation of illuminant color for texture number 1.

Ratio of color responses	$\left(\dfrac{k_1}{k_{1n}}\right)$ Red	$\left(\dfrac{k_2}{k_{2n}}\right)$ Green	$\left(\dfrac{k_3}{k_{3n}}\right)$ Blue
Mean, M	2.02	1.87	2.11
Standard deviation, S	.064	.052	.048
(S/M) (%)	3.17	2.81	2.27

Table.2 Nose representation of illuminant color for texture number 2.

Ratio of color responses	$\left(\dfrac{k_1}{k_{1n}}\right)$ Red	$\left(\dfrac{k_2}{k_{2n}}\right)$ Green	$\left(\dfrac{k_3}{k_{3n}}\right)$ Blue
Mean, M	1.92	2.09	2.16
Standard deviation, S	.063	.078	.068
(S/M) (%)	3.29	3.74	3.18

4.3 Representation for Scenes with Strong Specularities

We made an experiment using a rough red sheet occupying the entire image. As expected, the nose reflection is affected by the red color because assumption number 1 does not hold in the scene. When the scene was planted with small patches reflecting specularities, the nose estimate was quite improved, while the retinex estimate was still affected by the red scene color. This confirms our discussion in subsection 2.2.1.

4.4 Representation of Single Illumination Color in Mondrian Scenes

We arranged four color Mondrians, the first contains diverse and distributed 30 color patches, the others contain near red, near green and near blue color patches. The Mondrians were lighted by 10 colored lights and corrected by the nose, retinex and the white patch method. The images of nose and white patch correction show constant and comparable appearance, while the retinex method provides constant colors that appear desaturated. A statistical analysis was made with the constancy statistic being defined as the norm of the standard deviations of corrected color vector divided by the norm of the mean color vector. The scene-white patch, retinex and nose methods show low variance compared to the original color variance as shown in Fig.7, for the first and second Mondrians. The scene-white patch and the nose show comparable performance while the retinex provides more constant colors in all test Mondrians. However, this error measure does not encapsulate the performance as when the GWA is violated, the color is divided by itself yielding a *very* constant *gray* color.

The white patch indeed shows the limiting constancy beyond which the constancy is increased by loosing the color quality. In Table 3, it is shown that while the retinex method produces more constant colors, the variance of color chromaticity is also reduced and the retinex-corrected colors are scattered nearer to the gray color chromaticity denoting poor color maintenance.

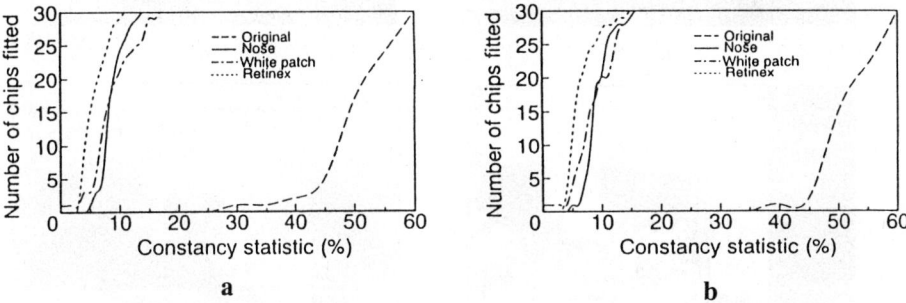

Fig.7. Cumulative histogram for constancy statistic of the color chips for
a) Diverse color Mondrian, the first, and
b) Near-red color Mondrian, the second.

Table.3 Comparison of color constancy and maintenance performance between the nose and retinex methods in the case of Mondrian scenes.

	Average constancy statistic	(S/M) for red chromaticity	(S/M) for green chromaticity
First Mondrian using nose	9.1%	50%	38%
First Mondrian using retinex	6%	48%	30%
Second Mondrian using nose	9.6%	27%	31%
Second Mondrian using retinex	6.4%	24%	26%

4.5 Correction of Real Indoor and Outdoor Scenes

Figure 8.a, b and c show an indoor image, the nose- and retinex-corrected images respectively. The nose correction performs well even in single color areas. The colors for the retinex-corrected image suffers gray color induction (center areas of the red and blue boxes) and the induction of complementary colors around the single color areas (yellow flare around the blue, and cyan flare around the red). These are the typical effects for the GWA violation. Figure 9.a, b and c show an outdoor image at daylight, the nose- and retinex-corrected images respectively. The green color of the trees in the nose correction is recovered and compares well to human observation while the retinex method fails to maintain the green color of the trees as the surround is also green and the output color is gray. The nose provides *very* fast color correction as one image is corrected in less than a second while it takes 200 minutes for the retinex-correction.

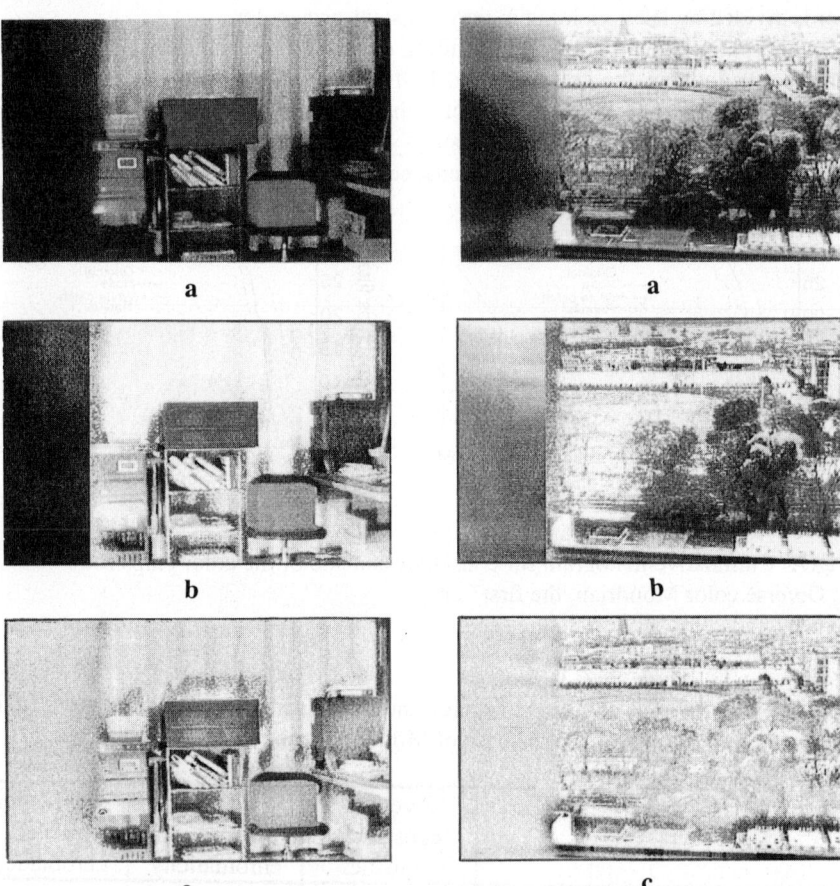

Fig.8. Real indoor image,
    a) Original image.
    b) Nose-corrected image.
    c) Retinex-corrected image.

Fig.9. Real outdoor image,
    a) Original image.
    b) Nose-corrected image.
    c) Retinex-corrected image.

5 Conclusion

We proposed a simple and effective method of color constancy on real images. The blurred inter-reflection from a nose surface is used as an illumination measure in the scene. The blurring layers have favorable features to robustify the illumination measure under weak scene assumptions. The nose is shown to represent the illumination color as effective as a scene-inserted white patch with the advantage of detecting smooth changes of illumination in the scene. For single illumination color scenes, the color constancy performance of the nose method compares favorably with the scene white patch method while the retinex method produces more constant colors at the expense of loosing coloration. The application of nose method to correct real images confirms the robustness of the nose illumination measure.

Acknowledgment

M. Abdellatif acknowledges the financial support of the Japanese Ministry of Education, Sports and Culture.

References

1. Land, E. 1987. An alternative technique for the computation of the designator in the retinex theory of color vision, In Proc. Nat. Acad. of Sci. USA, 83, pp.3078-3080.
2. Bajcsy, R., Lee, S.W., and Leonardis, A., 1996. Detection of diffuse and specular interface reflections and inter-reflections by color image segmentation, IJCV,17(3), pp.241-272.
3. Novak, C.L., and Shafer, S.A., 1992. Supervised color constancy, Color, Jones and Bartlett, pp.284-299.
4. Brill, M.H. 1979. Computer simulation of object color recognisers, J.Opt.Soc. Am.69.
5. Lee, H.-C., 1986. Method for computing the scene-illuminant chromaticity from specular highlights, J. Opt. Soc. Am. A 3(10):pp.1694-1699.
6. Klinker, G.I., Shafer, S.A., and Kanade, T., 1988. The measurement of highlights in color images, IJCV., 2(1):pp.7-32.
7. Tominaga, S. and Wandell, B., 1990. Component estimation of surface spectral reflectance, J. Opt. Soc. Am., 7.
8. Healey, G.1991, Estimating spectral reflectance using highlights, Image and vision computing, 9(5):pp.333-337.
9. Shafer, S.A., 1985. Using color to separate reflection components, Color Res. & Applications, 10:pp. 210-218.
10. Wandell, B.A. 1987. The synthesis and analysis of color images, IEEE Trans. PAMI, 9(1), pp.2-13.
11. Maloney, L.T. 1986. Evaluation of linear models of surface reflectance with small number of parameters, J. Opt. Soc. Am., 3:pp.29-33.
12. Abdellatif, M, Tanaka, Y, Gofuku, A, and Nagai,I. Color constancy using the inter-reflection from a reference nose, (submitted for the IJCV).

Shape Recovery from One Image under Multiple Light Sources

Ying-li Tian[1], H. T. Tsui[2] and S.Y.Yeung[2]

[1] National Laboratory of Pattern Recognition, Chinese Academy of Sciences, Beijing, China
[2] Department of Electronic Engineering The Chinese University of Hong Kong, Hong Kong

Abstract. In this paper, we propose a rectangular light source model and a novel method for recovering surface shape from one image under multiple planar light sources using shape-from-shading(SFS). This method can be extended to the case of multiple sources of different shapes [16, 15] . In indoor environments, the exact 3D positions (usually on the ceiling or walls) of the light sources are known. Assuming that the target object is small relative to the distances from the sources, we have derived a reflectance map for the Lambertian surface of an object. Hence, the shape recovery can be performed using SFS technique. This is a significant step towards the application of SFS in uncontrolled practical environments. Our method is verified by many examples of simulations and real experiments and the results are good.

1 Introduction

The shape-from-shading(SFS) technique for recovering the shape of an object with Lambertian surface from a single intensity image was first proposed by Horn in 1970[9]. Since then, many techniques on the same line have been developed. For example, shape from shading(SFS)[8, 10], shape from photometric stereo[18, 19] and shape from color[5, 11, 13]. In the past, many reflectance models have been used in computer graphics, computer vision and image analysis. They were also used in physics based techniques to improve shape recovery. In the 1960s, two accurate reflectance models were proposed by Beckmann and Spizzichino[1] and by Torrance and Sparrow [17]. The Torrance-Sparrow model[17] is the first theoretical reflectance model in the literature that takes specular reflection as well as diffusion into consideration[7]. Blinn[3] rewrote the Torrance-Sparrow model using vector notation. Cook[6] extended the model by including ambient light and spectral dependencies. Shafer[14] had developed a dichromatic reflectance model for color computer vision. Recently, Oren and Nayar [12] had proposed a general Lambertian Model. However, all the above methods, including the SFS and photometric stereo, have to work under an ideal parallel light source or a point light source at a great distance from the object. None of these methods is suitable for applications outside of special laboratories. Extending this approach to work in common lighting environments of laboratories, factories and offices is extremely useful and significant.

In this paper, we propose a novel rectangular light source model for shape recovery using SFS. We proved that a model with multiple rectangle light sources can be obtained by superposition of the individual sources. It is easy to see that this can be extended to cover the case of planar light sources of any shape because any planar shape can be approximated by a mosaic of rectangles. We had also developed a cylindrical light source model and a spherical light source model [16]. We further discovered that our method can be extended for shape recovery under a mixture of planar, spherical and cylindrical shapes[15]. This is of extreme practical importance because many ceiling lights and wall lights are of these shapes. We verify this method by simulations as well as real experiments and the results are quite good. However, the shape recovered in real experiments suffer a bit more distortion due to interreflection and non-uniform brightness of the extended sources.

The rest of this paper is organized as follows. In section 2, the rectangular light source model is derived. Based on this result, a model for multiple rectangular light sources for SFS of Lambertian surfaces is obtained. Section 3 is on the recovering surface shape under multiple rectangular light sources by SFS. Experimental results and discussions are given in Section 4. Section 5 is the conclusions.

2 Rectangular Light Source Model

2.1 Radiance of a Rectangular Light Source

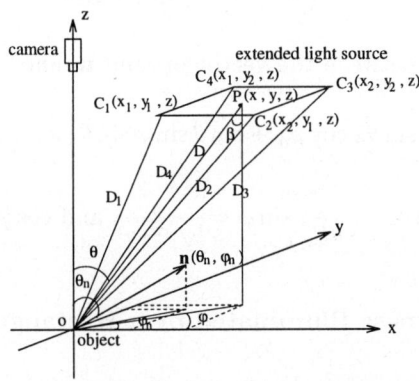

Fig. 1. Geometry of rectangular light in 3D coordinate system.

Consider a rectangular light source with uniform radiance (E), the size and position of the source are assumed known. Its geometry is as shown in Fig. 1.

The surface normal of a point on the object surface is denoted by $\mathbf{n}(\theta_n, \varphi_n)$. The camera is in the direction of the Z axis. θ_n is the polar angle between the surface normal and the Z axis, and φ_n is the azimuth angle. We use as reference the *camera-object* coordinate system where the original is at the centroid of the object and the optical axis of the camera is aligned with the Z axis. In this reference coordinate system, the rectangular light source is parallel to the XOY plane. The coordinates of the four corners are $C_1(x_1, y_1, z)$, $C_2(x_2, y_1, z)$, $C_3(x_2, y_2, z)$, $C_4(x_1, y_2, z)$ respectively and D_1, D_2, D_3, D_4 are the distances between the object centroid and the corners. Consider an arbitrary point of the rectangular light source $P(x, y, z)$. Its radiance in the direction of the object is[4]:

$$L(\theta) = E \cos \beta = \frac{Ez}{D}, \tag{1}$$

where E is the radiance of the rectangular light source, $D = \sqrt{x^2 + y^2 + z^2}$ is the distance between P and the object.

2.2 Surface Brightness Illuminated by One Point of a Rectangular Light Source

The source point $P(x, y, z)$ illuminates a point of the object having the surface normal $\mathbf{n}(\theta_n, \varphi_n)$. The brightness of the surface point P' is:

$$B = \begin{cases} \frac{L(\theta) \cos \Delta\theta}{D^2}, & \Delta\theta < 90° \\ 0, & \text{otherwise} \end{cases} \tag{2}$$

where $\Delta\theta$ is the angle between the vector $\mathbf{op}$ and $\mathbf{n}$, and

$$\cos \Delta\theta = \sin\theta \cos\varphi \sin\theta_n \cos\varphi_n + \sin\theta \sin\varphi \sin\theta_n \sin\varphi_n + \cos\theta \cos\theta_n, \tag{3}$$

where $\sin\theta = \frac{\sqrt{x^2+y^2}}{D}, \cos\theta = \frac{z}{D}, \sin\varphi = \frac{y}{\sqrt{x^2+y^2}}$ and $\cos\varphi = \frac{x}{\sqrt{x^2+y^2}}$.

2.3 Surface Brightness Illuminated by a Rectangular Light Source

An extended source can be thought of as a collection of many point light sources. The brightness of the object surface can be estimated up to a constant by integrating the light energy reflected from all points on the extended source. So we have:

$$B' = \int_{x_1}^{x_2} \int_{y_1}^{y_2} B \, dx \, dy = \int_{x_1}^{x_2} \int_{y_1}^{y_2} \frac{L(\theta) \cos \Delta\theta}{D^2} dx \, dy$$
$$= A \sin\theta_n \cos\varphi_n + B \sin\theta_n \sin\varphi_n + C \cos\theta_n, \tag{4}$$

where

$$A = Ez\{\frac{\gamma_4 - \gamma_1}{2\sqrt{x_1^2 + z^2}} + \frac{\gamma_2 - \gamma_3}{2\sqrt{x_2^2 + z^2}}\},$$

$$B = Ez\{\frac{\delta_2 - \delta_1}{2\sqrt{y_1^2 + z^2}} + \frac{\delta_4 - \delta_3}{2\sqrt{y_2^2 + z^2}}\},$$

$$C = E\{\frac{y_2(\delta_3 - \delta_4)}{2\sqrt{y_2^2 + z^2}} + \frac{y_1(\delta_1 - \delta_2)}{2\sqrt{y_1^2 + z^2}} + \frac{x_2(\gamma_3 - \gamma_2)}{2\sqrt{x_2^2 + z^2}} + \frac{x_1(\gamma_1 - \gamma_4)}{2\sqrt{x_1^2 + z^2}}\}. \qquad (5)$$

For the Lambertian surface, the brightness is equal in every direction. So the brightness of a surface point P' in the camera direction is proportional to B'. Consider the orthographic projection model, the intensity of the image point P'' corresponding to the object point P' can be determined by:

$$I' = S(A \sin\theta_n \cos\varphi_n + B \sin\theta_n \sin\varphi_n + C \cos\theta_n) \qquad (6)$$

For the multiple light sources I'_1, I'_2 and I'_i, the intensity of a pixel illuminated by all the light sources is:

$$I'_i = I'_1 + I'_2 + ... + I'_i = S[(A_1 + A_2 + ... + A_i) \sin\theta_n \cos\varphi_n$$
$$+ (B_1 + B_2 + ... + B_i) \sin\theta_n \sin\varphi_n + (C_1 + C_2 + ... + C_i) \cos\theta_n]$$
$$= S(A^* \sin\theta_n \cos\varphi_n + B^* \sin\theta_n \sin\varphi_n + C^* \cos\theta_n) \qquad (7)$$

2.4 Reflectance Map for Lambertian Surface Under Multiple Rectangular Light Sources

From Equation (6), the reflectance map for the Lambertian surface under one rectangular light source is:

$$R(x,y) = \begin{cases} S[An_x(x,y) + Bn_y(x,y) + Cn_z(x,y)], \\ \qquad \text{if } An_x(x,y) + Bn_y(x,y) + Cn_z(x,y) > 0 \\ 0, \qquad \text{otherwise} \end{cases} \qquad (8)$$

In the same way, the reflectance map for the Lambertian surface under multiple rectangular light sources is:

$$R'(x,y) = \begin{cases} S(A^*n_x(x,y) + B^*n_y(x,y) + C^*n_z(x,y)], \\ \qquad \text{if } A^*n_x(x,y) + B^*n_y(x,y) + C^*n_z(x,y) > 0 \\ 0, \qquad \text{otherwise} \end{cases} \qquad (9)$$

Using the surface gradients $p = dz/dx$ and $q = dz/dy$ to express the surface normal, we have $\mathbf{n} = (-p, -q, 1)$. So the reflectance maps for the rectangular light sources are:

$$R(x,y) = S\frac{(-Ap - Bq + C)}{\sqrt{1 + p^2 + q^2}}, \qquad (10)$$

$$R'(x,y) = S\frac{[-A^*p - B^*q + C^*]}{\sqrt{1 + p^2 + q^2}}, \qquad (11)$$

where p and q are the rates of change of depth in x and y directions respectively, $A^* = (A_1 + A_2 + ... + A_i)$, $B^* = (B_1 + B_2 + ... + B_i)$ and $C^* = (C_1 + C_2 + ... + C_i)$.

3 Recovering Surface Shape Under Multiple Light Sources

We had shown in [16] that we can obtain a similar equation as (7) for a cylindrical or a spherical light source. It is obvious that we can perform SFS using multiple lights of mixed shapes.

We use a global propagation approach for SFS proposed by Bichsel and Pentland[2]. Depth are recovered directly and the resulting surface is guaranteed to be continuous. Given the initial values at the singular points(brightest points), the algorithm looks in eight discrete directions in the image and propagates the depth information away from the singular points to ensure the proper termination of the process. Since slopes at the surface points in low brightness regions are close to zero for most directions, the image is initially rotated to align the light source direction with one of the eight directions. The inverse rotation is performed on the resulting depth map in order to restore to the original orientation.

The normalized surface normal can be expressed as a function of the partial derivatives (p, q):

$$\mathbf{n}(n_x, n_y, n_z) = \left(\frac{-p}{\sqrt{p^2 + q^2 + 1}}, \frac{-q}{\sqrt{p^2 + q^2 + 1}}, \frac{1}{\sqrt{p^2 + q^2 + 1}} \right). \qquad (12)$$

From the reflectance map given by Equations (10) and (11), the surface gradients (p, q), are precomputed by taking the derivative of Equations (10) and (11) with respect to q in the rotated coordinate system, setting it to zero, and then solving for p and q. The detail derivation is described in [2]. The solutions for p and q are given by:

$$p = \frac{-A^*C^* \pm \sqrt{[1 - R^2][R^2 - (B^*)^2]}}{R^2 - (A^*)^2 - (B^*)^2}, \qquad (13)$$

$$q = \frac{A^*B^*p - B^*C^*}{R^2 - (B^*)^2}. \qquad (14)$$

where A^*, B^* and C^* are the geometric parameters of the light sources and R is the reflectance map.

4 Experimental Results and Discussions

4.1 Synthetic Image Results

To verify our rectangular light source model, we use a raytracing package to generate some images of objects under several rectangular light sources. Then we recover the object shapes by our reflection models using SFS. The results show that our method works well. We use four rectangular light sources. Two light sources are in a plane parallel to X-Y plane. Other two light sources are in a plane parallel to Z-Y plane. Fig. 2(a) shows the image of a sphere generated by Raytracing under 4 rectangular light sources. Fig. 2(b) shows its depth map recovered by shape from shading using our rectangular light source model. Fig. 2(c) shows its depth maps viewed in another direction. Fig. 2(d), (e) and (f) show the image and depth maps of a frustum of a cone under four rectangular light sources.

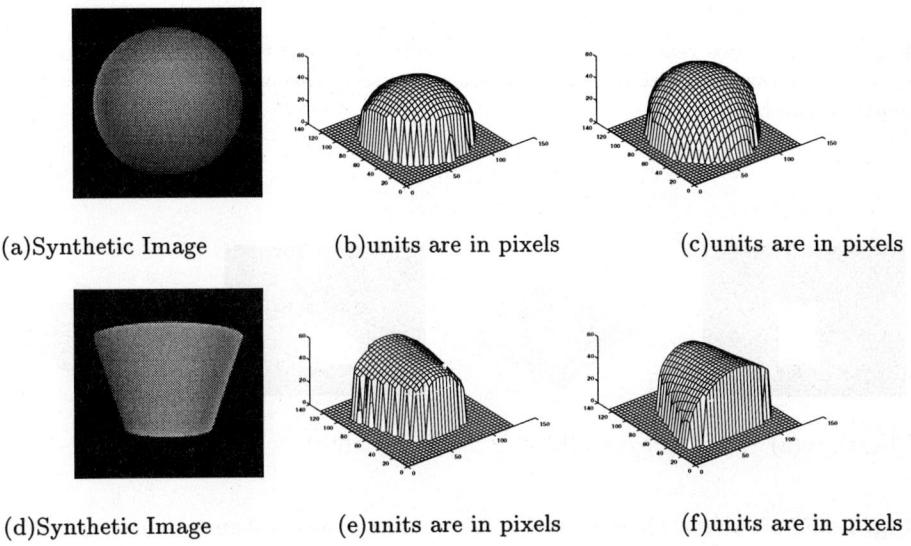

Fig. 2. SFS of two objects under four rectangular light sources(Simulation).

4.2 Real Image Results

In the real experiments, we use two rectangular light sources. Each of these is $10cm$ wide and $20cm$ long and is $100cm$ away from the object surface. Monochromatic images are obtained by a PULNIX TM-6 CCD camera with a $16mm$ lens. The distance between the camera and the object is $90cm$. The two rectangular lights are placed in position 1($x_{min} = 6.5cm$, $x_{max} = 16.5cm$, $y_{min} = -15.5cm$, $y_{max} = 4.5cm$) and position 2($x_{min} = -16.5cm$, $x_{max} = -6.5cm$, $y_{min} = -15.5cm$, $y_{max} = 4.5cm$) respectively. A mouse and a bowl are used in our experiments for shape recovery. Fig. 3(a) and Fig. 3(b) show the image of the mouse and its depth map recovered by SFS under the planar rectangular light sources. Similarly, Fig. 3(c) and (d) show the real image and the corresponding depth map of a bowl under the rectangular light sources. The results show that our method works well for the real images also.

To verify the flexibility of our method, we use simultaneously the spherical and cylinderical lights [16] together with the rectangular light for SFS. The real images and the recovered depth maps are shown in Fig. 4.

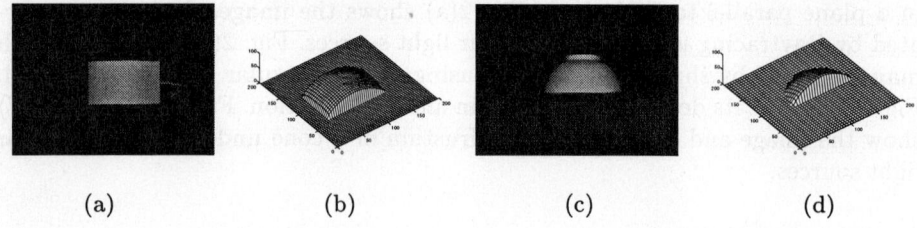

(a) (b) (c) (d)

Fig. 3. SFS of two real objects under two rectangular light sources(real images). (a) Image of a mouse. (b) Recovered depth of the mouse. (c) Image of a bowl. (d) Recovered depth of the bowl.

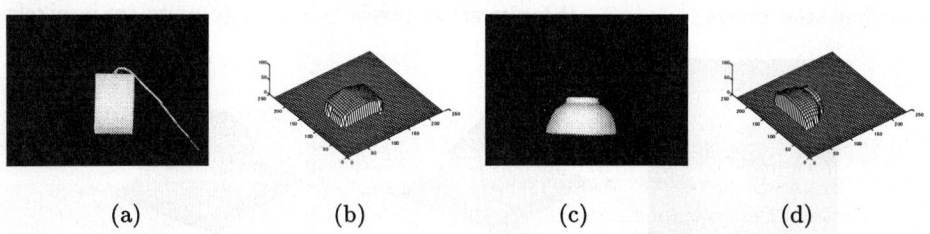

(a) (b) (c) (d)

Fig. 4. SFS of two real objects under 3 light sources of mixed shapes: planar, cylindrical and spherical lights. (a) Image of a mouse. (b) Recovered depth of the mouse in (a). (c) Image of a bowl. (d) Recovered depth of the bowl in (c).

5 Conclusions

In this paper, we proposed a novel method for recovering surface shape from one image under multiple light sources. We give the derivation of the rectangular light source model for a Lambertian surface. If the positions and sizes of the light sources are known *a priori*, by using our reflectance map derived, we can use a shape-from-shading technique to obtain the surface shape. We had released the restriction of a point light source at a great distance for the conventional SFS. However, there is a mild restriction that the locations of the lights and the target object must be known. This is not a serious limitation as the indoor lights on the ceilings and walls are usually fixed. The location of target object is often known or can be estimates using laser or ultrasound means. The efficiency of our method was verified by real and synthetic images. Real experiments of SFS using multiple sources of mixed shapes were also performed. In situations where there are strong interreflections, our method cannot get good results. However, our method can be used by an intelligent robot for shape estimation or object recognition in an environment where interreflection is low. It is obvious that this is a significant step towards SFS in uncontrolled indoor environments.

Acknowledgement

This project is supported by the RGC Earmarked research grant. RGC Ref. No. CUHK 4116/97E.

References

1. P. B. ans A. Spizzichino. *"The Scattering of Electromagnetic Waves from Rough Surfaces"*. Pergamon Press, Oxford, 1963.
2. M. Bichsel and A. Pentland. "A Simple Algorithm for Shape from Shading". In *Proc. of CVPR*, pages 459–465, 1992.
3. J. F. Blinn. "Models of Light Reflection for Computer Synthesized Pictures". In *Proc. of Computer Graphics, Vol. 11*, pages 192–198, 1977.
4. M. Born and E. Wolf. *"Principles of Optics"*. Pergamon Press, 1959.
5. D. Brainard and B. A. Wandell. "Asymmetric Color Matching: How Color Appearance Depends on the Illuminant". *Journal of the Optical Society of America A 9(9)*, page 1992, 1433–1448.
6. R. L. Cook and K. E. Torrance. "A Reflectance Model for Computer Graphics". *Computer Graphics 15(3)*, pages 307–316, 1981.
7. R. M. Haralick and L. G. Shapiro. *"Computer and Robot Vision, Volume 1, 2"*. Addison-Wesley Publishing Company, 1993.
8. K. Hartt and M. Carlotto. "A Method for Shape-from-Shading Using Multiple Images Acquired under Different Viewing and Lighting Conditions". In *Proc. of CVPR*, pages 53–60, 1989.
9. B. K. P. Horn. *"Robot Vision"*. The MIT Press, 1986.
10. B. K. P. Horn and M. J. Brooks. *"Shape from Shading"*. The MIT Press, 1989.
11. G. J. Klinker, S. A. Shafer, and T. Kanade. "The Measurement of Highlights in Color Images". *International Journal of Computer Vision*, pages 7–32, 1988.
12. M. Oren and S. Nayar. "Generalization of the Lambertian Model and Implications for Machine Vision". *International Journal of Computer Vision, 14*, page 1995, 227–251.
13. Y. Sato and K. Ikeuchi. "Temporal-Color Space Analysis of Reflection". In *Proc. of the Conf. of CVPR*, pages 570–576, 1993.
14. S. A. Shafer. "Using Color to Separate Reflection Components". *Color Research and Application*, pages 210–218, 1985.
15. Y. Tian, H. T. Tsui, and S. Y. Yeung. "Shape from shading using multiple light sources of mixed shapes". Technical Report, Dept. of Electronic Engineering, the Chinese University of Hong Kong., 1997.
16. Y. Tian, H. T. Tsui, and S. Y. Yeung. "Spherical and Cylindrical Light Source Models for Shape Recovery". In *ACCV98*, 1998.
17. K. E. Torrance and E. Sparrow. "Theory for off-Specular Reflection from Roughened Surface". *J. Opt. Soc. Amer. 57*, pages 1105–1114, 1967.
18. L. Wolff. "On Diffuse Reflection and Photometric Stereo". In *Proc. of Image Understanding Workshop*, 1992.
19. R. J. Woodham. "Photometric Stereo: A Reflectance Map Technique for Determining Surface Orientation from Image Intensity". In *Proc. of SPIE*, pages 136–143, 1978.

Spherical and Cylindrical Light Source Models for Shape Recovery

Ying-li Tian[1], H. T. Tsui[2] and S.Y.Yeung[2]

[1] National Laboratory of Pattern Recognition, Chinese Academy of Sciences, Beijing, China
[2] Department of Electronic Engineering, The Chinese University of Hong Kong, Shatin, Hong Kong

Abstract. Parallel rays of light coming from the same direction is the basic assumption for the lighting condition of the conventional shape from shading(SFS) methods. This is at best approximately true in controlled environments. To eliminate this limitation, we propose a spherical light model and a cylindrical light model to model practical light sources in common indoor environments. Each of these lights can be used in a SFS technique or a photometric stereo technique for shape recovery of objects with Lambertian surface. The validity of this method is verified by many examples of simulation and real experiments. Good results had been obtained in those experiments with little inter-reflection. This paper forms an important step towards SFS under ordinary indoor lighting environments.

1 Introduction

Since Horn [4] first proposed a SFS technique in 1970, many papers had been published on various versions of SFS under different constraints and boundary/initial conditions [1, 2, 3, 5, 6, 7]. In spite of these efforts, SFS are often inaccurate and impractical. This is because the two basic assumptions for SFS are only partially true in the majority of cases. First, the assumption that an object surface is Lambertian is only approximately true for most cases. Second, the assumption that the light rays are parallel and coming in from a single direction are often not strictly true except in well controlled lighting environments. Common ceiling lights and wall lights in indoor environments are often spherical, cylindrical or flat. Thus, the classical SFS techniques are not very practical in these environments. In this paper, we focus on tackling the problem caused by the second assumption and the first problem is tackled in another paper by Tian and Tsui [9]

A SFS technique must be able to work under a real practical light source for it to be useful for many applications. Inspired by [8], we had developed a spherical light model and a cylindrical light model. A model of flat light is described in [10]. Some researchers had reported the use of extended light sources recently [8], but did not use them explicitly for SFS.

The organization of this paper is as follows. Section 2 gives a spherical light source model. A cylindrical light source model is introduced in Section 3. Some real experiment results are given in Section 4. Section 5 gives the discussions and conclusions.

2 A Spherical Light Model in 3D Coordinate System

In this section, we propose a spherical light source model in 3D coordinate system and derive the relationship between the light source and the surface normal of a point on an object directly. The *source-object coordinate system* is one with the centroid of the object at the original and the center of the spherical light on the Z axis (Fig1(a)). The *camera-object coordinate system* is one where the object centroid is at the original point and the optical axis of the camera is aligned with the Z axis (Fig.1(b)).

2.1 Radiance of the Spherical Light Source

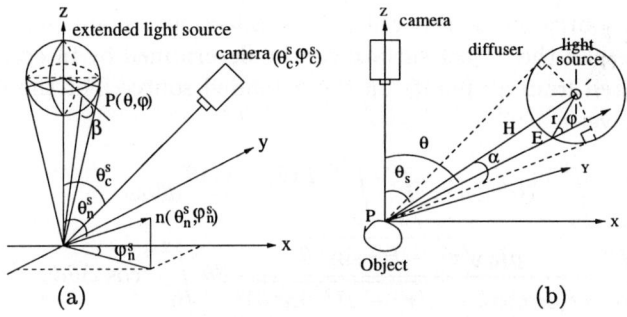

Fig. 1. (a)Geometry of the spherical light source in 3D coordinate system. (b) A cross-section of the spherical light.

In Fig. 1(a), the surface normal of one point on the object is in the direction $\mathbf{n}(\theta_n^s, \varphi_n^s)$. The target object is assumed small relative to the distance of the source from the object. The camera is in the direction $(\theta_c^s, \varphi_c^s)$. θ is the polar angle and φ is the azimuth angle where the superscript s mean in the source-object coordinate system. Let r be the radius of the spherical light source, H be the distance between the object and the center of the spherical light source, and $\alpha = \sin^{-1}(r/H)$ be the interval angle for the area of the spherical light source which illuminates the object (Fig. 1(b)). We assume the radiance intensity of the point light source is I_0, and the irradiance of a point of the surface of the spherical light source is $E = (\rho I_0)/r^2$, where ρ is the translucent parameter, and r is the radius of the spherical light source. The radiance of one point $p(\theta, \varphi)$ on the surface of the spherical light source in the object direction is:

$$L(\theta) = E \cos \beta = \frac{\rho I_0 \sqrt{r^2 - H^2 \sin^2 \theta}}{r^3}.$$

2.2 Surface Brightness Illuminated by One Point of the Spherical Light Source

The light source point $p(\theta,\varphi)$ illuminates a point of the object with the surface normal $\mathbf{n}(\theta_n^s,\varphi_n^s)$ in the source-object coordinate system. The brightness of the point on the object is:

$$B = \begin{cases} \frac{L(\theta)\cos\Delta\theta}{D^2}, & \Delta\theta < 90° \\ 0, & \text{otherwise} \end{cases} \quad (1)$$

where $D = H\cos\theta - \sqrt{r^2 - H^2\sin^2\theta}$ is the distance between the object and the source point $p(\theta,\varphi)$, $\Delta\theta$ is the angle between the vector **op** and the surface normal **n**, and

$$\cos\Delta\theta = \sin\theta\cos\varphi\sin\theta_n^s\cos\varphi_n^s + \sin\theta\sin\varphi\sin\theta_n^s\sin\varphi_n^s + \cos\theta\cos\theta_n^s. \quad (2)$$

2.3 Surface Brightness Illuminated by the Spherical Light Source

An extended source can be thought of as a collection of many point light sources. The brightness of the object surface can be determined by integrating the light energy reflected from all points on the extended source. For the spherical light source, we have:

$$B' = \int_0^\alpha \int_0^{2\pi} B d\theta d\varphi = \int_0^\alpha \int_0^{2\pi} \frac{L(\theta)\cos\Delta\theta}{D^2} d\theta d\varphi$$

$$= \int_0^\alpha \frac{\rho I_0 \sqrt{r^2 - H^2\sin^2\theta}}{r^3(H\cos\theta - \sqrt{r^2 - H^2\sin^2\theta})^2} d\theta \int_0^{2\pi} \cos\Delta\theta d\varphi = k\cos\theta_n^s, \quad (3)$$

$$k = \frac{\pi\rho I_0\{2H^3r^2\pi + Hr^4\pi + (4Hr^3 - 2H^3r - 4r)\sqrt{H^2 - r^2} + (2H^5 - 4H^2)\alpha\}}{4H^2r^3(H^2 - r^2)^2}.$$

2.4 Rotating the Source-Object Coordinate to the Camera-Object Coordinate

In our approach, we have to transfer Equation (3) from the source-object coordinate system to the camera-object coordinate system. In the camera-object coordinate system, the surface normal of the same point $\mathbf{n}(\theta_n^s,\varphi_n^s)$ in the source-object coordinate system will be presented as $\mathbf{n}(\theta_n,\varphi_n)$. The relationship between the two coordinate systems is:

$$\begin{bmatrix} X^s \\ Y^s \\ Z^s \end{bmatrix} = Q \begin{bmatrix} X^c \\ Y^c \\ Z^c \end{bmatrix} \quad (4)$$

where Q is the rotation matrix, $[X^s, Y^s, Z^s]$ and $[X^c, Y^c, Z^c]$ are the coordinates in the source-object coordinate system and the camera-object coordinate system respectively. In this case, the resultant or composite rotation matrix may be obtained from the following simple rules: (1) rotate the source-object coordinate system by an angle u about the OX_s axis, then rotate the source-object coordinate system by an angle v about the OY_s axis, where u is the angle between the component of the camera direction vector in $Y_s Z_s$ plane and the OZ_s axis, and v is the angle between the camera direction vector and its component in the plane $Y_s Z_s$; (2) The angle is positive if the rotation is clockwise. Therefore,

$$Q = \begin{bmatrix} \cos v & 0 & \sin v \\ \sin u \sin v & \cos u & -\sin u \cos v \\ -\cos u \sin v & \sin u & \cos u \cos v \end{bmatrix}. \quad (5)$$

The relationship between the surface normal $\mathbf{n}(X_n^s, Y_n^s, Z_n^s)$ and the surface normal $\mathbf{n}(X_n^c, Y_n^c, Z_n^c)$ is: $Z_n^s = -X_n^c \cos u \sin v + Y_n^c \sin u + Z_n^c \cos u \cos v$,

$$\cos \theta_n^s = \sin u \sin \theta_n \sin \varphi_n - \cos u \sin v \sin \theta_n \cos \varphi_n + \cos u \cos v \cos \theta_n$$

$$= \left(\frac{\sin \theta_c^s \sin \varphi_c^s}{\sqrt{1 - \sin^2 \theta_c^s \cos^2 \varphi_c^s}} \right) \sin \theta_n \sin \varphi_n$$

$$- \left(\frac{\sin \theta_c^s \cos \theta_c^s \cos \varphi_c^s}{\sqrt{1 - \sin^2 \theta_c^s \cos^2 \varphi_c^s}} \right) \sin \theta_n \cos \varphi_n + \cos \theta_c^s \cos \theta_n$$

$$= A \sin \theta_n \sin \varphi_n + B \sin \theta_n \cos \varphi_n + C \cos \theta_n \quad (6)$$

Since θ_c^s, φ_c^s are fixed because the light location is fixed, A, B, C are constant for a given light source. The brightness of the object point with surface normal when the light source in position i is: $B_i' = k \cos \theta_{ni}^s$. where θ_n and φ_n define the surface normal in the camera-object coordinate system.

Considering the orthography projection camera model, the intensity of the point P' on the image can be determined by:

$$I_i' = Sk \cos \theta_{ni}^s = Sk(A \sin \theta_n \cos \varphi_n + B \sin \theta_n \sin \varphi_n + C \cos \theta_n). \quad (7)$$

where S is an constant for the orthography projection camera model.

3 Cylindrical Light Source Model

3.1 Radiance of a Cylindrical Light Source

Let E be the radiance of a point on the cylindrical light source, The geometry of the light source is shown in Fig. 2(a). The surface normal at the point is $\mathbf{n}(n_x, n_y, n_z)$. The cylindrical light is parallel to the y-axis. The coordinates of the points at the two ends os the central axis of the cylinder fixed it's location. Assume that the radius of the light source is d, the coordinates of the two points are (x_0, y_1, z_0) and (x_0, y_2, z_0) respectively. The radiance of a point of the light source $P(x, y, z)$ in the direction of the object is $L(\theta) = E \cos \beta$, where β is the angle between the direction of $\mathbf{r}$ and the position vector $\mathbf{op}$. In Fig. 1(b), $\mathbf{r}$ is a vector normal to the central axis of the light. $\mathbf{r}(r_x, r_y, r_z) = (\cos\alpha, 0, \sin\alpha)$ and $\cos\beta = \frac{1}{D}\mathbf{r} \cdot \mathbf{op} = \frac{1}{D}(x\cos\alpha + z\sin\alpha)$ where $D = \sqrt{x^2 + y^2 + z^2}$ is the distance between the origin and the point P. Therefore, $L(\theta) = E\cos\beta = \frac{E}{D}(x\cos\alpha + z\sin\alpha)$

3.2 Surface Brightness Illuminated by One Point of the Cylindrical Light Source

The brightness of a point on the object surface with a normal $\mathbf{n}(n_x, n_y, n_z)$ illuminated by one point $P(x, y, z)$ of the light source is given by Equation (1). where θ is the angle between the vector $\mathbf{op}$ and $\mathbf{n}$, and $\cos\theta = \mathbf{op} \cdot \mathbf{n} = \frac{1}{D}(x \cdot n_x + y \cdot n_y + z \cdot n_z)$ From Fig.2(a), we can see that for simplicity, we can use the vector $\mathbf{op'}$ to approximate the vector $\mathbf{op}$ while the error is very small because the radius of the cylindrical light source is usually small with respect to the distance to the object. The vector $\mathbf{op}$ has three components: $x = x_0$ and $z = z_0$, and y varies from y_1 to y_2.

3.3 Surface Brightness Illuminated by the Cylindrical Light Source

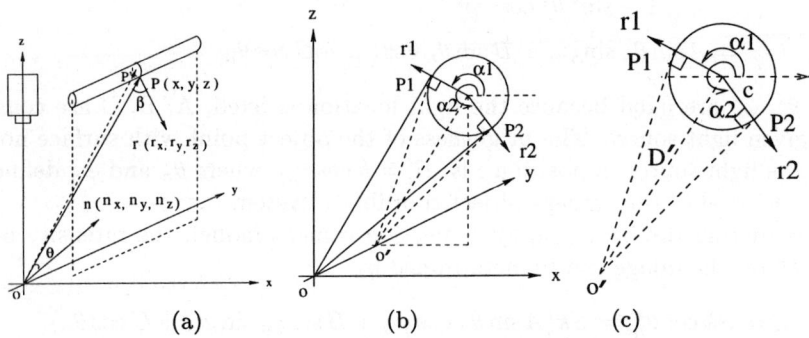

Fig. 2. (a) Geometry of cylindrical light source in 3D coordinate system. (b)(c) A cross-section of the cylindrical light source.

An enlarged cross-section of the cylindrical light source is shown in Fig.2(b) and (c). Consider a cross-section of the cylindrical light. All source points will have the same y-coordinate: y'. The circle can be seen as the intersection line of the cylindrical light source and a plane $y = y'$, and O' is the point where the y axis and the plane intersect. From Fig.2 we can see that the object can only be illuminated by the source points from P_1 to P_2. Vector $\mathbf{op_1}$ is perpendicular to the radiance vector $\mathbf{r_1}$, vector $\mathbf{op_2}$ is perpendicular to the radiance vector $\mathbf{r_2}$. To get the brightness of the object illuminated by this piece of light source, we may integrate the light source points along the circle from P_1 to P_2, i.e. from angle α_1 to α_2.

The radiance vector $\mathbf{r_1}$ is perpendicular to y-axis, vector $\mathbf{op_1}$ and vector $\mathbf{o'p_1}$. Let D' be the distance between point O' and the center point C of the light source, from Fig.2 we can see that: $\alpha_1 = \pi + \arctan \frac{z_0}{x_0} - \arccos \frac{d}{D'}$, and $\alpha_2 = \pi + \arctan \frac{z_0}{x_0} + \arccos \frac{d}{D'}$. Since $D' = \sqrt{x_0^2 + z_0^2}$ and the radius d are constant, both α_1 and α_2 are constants. We integrate the brightness illuminated by each point on the cylindrical light source. First, we integrate along the curve where all points have the same y coordinate (see Fig2(b)), and then we integrate along the y direction from y_1 to y_2.

$$B' = \int_{y_1}^{y_2} \int_{\alpha_1}^{\alpha_2} B d\alpha dy = \int_{y_1}^{y_2} \int_{\alpha_1}^{\alpha_2} \frac{L(\theta) \cos \theta}{D^2(y)} d\alpha dy = kE(a \cdot n_x + b \cdot n_y + c \cdot n_z)$$

where $k = \int_{\alpha_1}^{\alpha_2} (x_0 \cos \alpha + z_0 \sin \alpha) d\alpha = constant$ and D is a function of y only as $x = x_0$ and $z = z_0$ are both constants.

$$a = \int_{y_1}^{y_2} \frac{x_0}{D^4(y)} dy$$

$$= \frac{x_0}{2(\sqrt{x_0^2 + z_0^2})} \left(\frac{y_2 - y_1}{x_0^2 + (y_2 - y_1)^2 + z_0^2} + \frac{1}{\sqrt{x_0^2 + z_0^2}} \arctan \frac{y_2 - y_1}{\sqrt{x_0^2 + z_0^2}} \right),$$

$$b = \int_{y_1}^{y_2} \frac{y}{D^4(y)} dy = \frac{1}{4} \left(\frac{1}{x_0^2 + y_2^2 + z_0^2} - \frac{1}{x_0^2 + y_1^2 + z_0^2} \right),$$

$$c = \int_{y_1}^{y_2} \frac{z_0}{D^4(y)} dy$$

$$= \frac{z_0}{2(\sqrt{x_0^2 + z_0^2})} \left(\frac{y_2 - y_1}{x_0^2 + (y_2 - y_1)^2 + z_0^2} + \frac{1}{\sqrt{x_0^2 + z_0^2}} \arctan \frac{y_2 - y_1}{\sqrt{x_0^2 + z_0^2}} \right).$$

The intensity of the point P' on the image can be determined by:

$$I' = S(A \sin \theta_n \cos \varphi_n + B \sin \theta_n \sin \varphi_n + C \cos \theta_n). \tag{8}$$

where S is a constant, and $(n_x, n_y, n_z) = (\sin \theta_n \cos \varphi_n, \sin \theta_n \sin \varphi_n, \cos \theta_n)$.

4 Experimental Results for Shape Recovery

Real and simulated experiments were performed using our light source models for an algorithm due to Bichsel and Pentland [1]. Monochromatic images are grabbed by a PULNIX TM-6 CCD camera with a 16mm lens. Both the spherical and cylindrical light sources used in these experiments are common indoor lights. The centre of the spherical light source is located at the position (30cm, 15cm, 170cm). The real objects used in our experiments include a mouse and a bowl. Real images of the objects under the spherical light source and the corresponding recovered depth maps are shown in Fig.4. Fig.5 shows the real images and the recovered depth maps by SFS using the above set of real objects. The above experiments show that the results on using the spherical light and the cylindrical lights for SFS is quite reasonable. The method is robust in a practical environment. Results of simulated experiments are even better as there is no inter-reflection.

5 Discussion and Conclusion

5.1 Error analysis

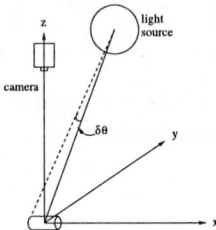

Fig. 3. The setup of a spherical light in a real experiment

In our method, an extended light source with light rays coming from every point seen by the object are represented by an equivalent virtual light ray coming from a given direction. For example, consider an object with its length less than 10cm and the light source is at a distance of 200cm from it (The minimum typical distance of an object on a workbench from the ceiling lights) as shown in Fig. 3. The maximum error, $\delta\theta$ of the virtual light ray direction at the outer rim of the object is: $\delta\theta = arctan(5/200) = 1.43 degrees$. The mean error is about 0.7 degree. These errors are quite acceptable for shape recovery using SFS.

5.2 Conclusion

Very good results had been obtained in most experiments for SFS under a light of cylindrical or spherical shape. It is obvious that models of other light shapes can be developed if needed using the same approach. We had proposed a flat light source model in another ACCV'98 paper [10]. From the light Equations (7) and (8) of the two extended light models proposed here, it is obvious that a multiple light source model of mixed shapes can be built by superposition. SFS using multiple rectangular light sources is being demonstrated in [10].

References

1. M. Bichsel and A. Pentland. "A Simple Algorithm for Shape from Shading". In *Proc. of CVPR*, pages 459–465, 1992.

2. P. Dupuis and J. Oliensis. "Direct Method for Reconstructing Shape from Shading". In *Proc. of Image Understanding Workshop, DARPA*, pages 563–571, 1992.
3. K. Hartt and M. Carlotto. "A Method for Shape-from-Shading Using Multiple Images Acquired under Different Viewing and Lighting Conditions". In *Proc. of CVPR*, pages 53–60, 1989.
4. B. K. P. Horn. "Shape from Shading: A method for Obtaining the Shape of a Smooth Opaque Object from One View". Technical Report 232, MIT Project MAC Internal Report TR79 and MIT AI Laboratoru Technical Report, November 1970.
5. B. K. P. Horn and M. J. Brooks. *"Shape from Shading"*. The MIT Press, 1989.
6. J. Oliensis. " Shape from Shading as a Partially Well-Constrained Problem". *CVGIP: Image Understanding, Vol. 54*, page 1993, 163–183.
7. J. Oliensis and P. Dupuis. "A Global Algorithm for Shape from Shading". In *Proc. of the 4th ICCV*, pages 692–701, 1993.
8. Y. Sato and K. Ikeuchi. "Temporal-Color Space Analysis of Reflection". In *Proc. of the Conf. of CVPR*, pages 570–576, 1993.
9. Y. Tian and H. Tsui. "Shape from Shading for Non-Lambertian Surfaces from One Color Image". In *Proceedings of the ICPR*, 1996.
10. Y. Tian, H. T. Tsui, and S. Y. Yeung. "Shape Recovery from One Image under Multiple Light Sources". In *ACCV98*, 1998.

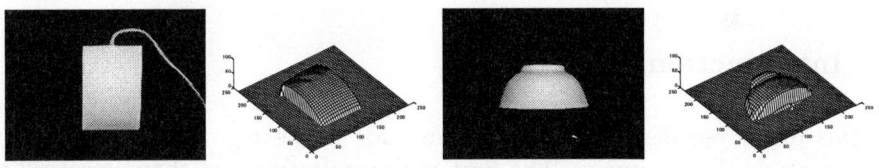

(a)Real Image (b) units are in pixels (c)Real Image(d) units are in pixels

Fig. 4. SFS of objects under a spherical light sources in real experiments. (a) Image of a mouse. (b) Recovered depth of the mouse in (a). (c) Image of a bowl. (d) Recovered depth of the bowl in (c).

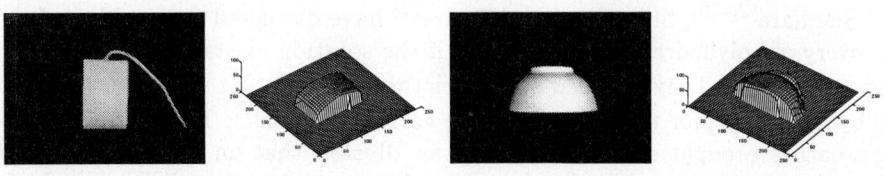

(a)Real Image (b) units are in pixels (c)Real Image (d) units are in pixels

Fig. 5. SFS of objects under a cylindrical light source in a real experiment. (a) Image of a mouse. (b) Recovered depth of the mouse in (a). (c) Image of a bowl. (d) Recovered depth of the bowl in (c).

Polyhedral Shape Recovery Based on Interreflections

Jun Yang[1], Dili Zhang[2], Noboru Ohnishi[1] and Noboru Sugie[3]

[1] Bio-Mimetic Control Research Center, RIKEN, Nagoya, 463 Japan
[2] Simutech Co., Inc., Nagoya, Japan
[3] Faculty of Science and Technology, Meijo University Nagoya, 468 Japan

Abstract. We discuss the uniqueness of 3-D shape recovery of a polyhedron from a single shaded image. First, we analytically show that multiple convex (and concave) shape solutions usually exist for a simple polyhedron if interreflections are not considered. Then we propose a new approach to uniquely determine the shape solution using interreflections as a constraint. Interreflections, which were considered to be deleterious in shape-from-shading, are used as a constraint to determine the shape solution in our approach.

1 Introduction

In this paper, we describe two problems for shape recovery of a polyhedron from a single shaded image. The first is the uniqueness of shape recovery of a polyhedron of which facets meet at an apex (like a pyramid). The second is how to uniquely determine the shape using interreflections as a constraint.

Horn has shown a numerical example in which two convex shape solutions and two concave shape solutions exist for a trihedral corner[1]. However, it is not described that how many solutions exist for a general polyhedron because two solutions may (or may not) exist for a quadratic equation (of image intensity constraint), and it must be discussed case by case. We will discuss the uniqueness using a reflectance map, and analytically show that multiple shape solutions usually exist for a pyramid.

Sugihara[2]-[6], Shimshoni and Ponce[7] have discussed the problem of shape recovery of polyhedra. They described if the solution exists under a condition of superstrictness. They avoided superstrictness by deleting some constraints[5][6] or by accounting for uncertainty in the vertex position[7]. The successes of these approaches brought some researchers an illusion that an arbitrary polyhedral shape can be always uniquely determined from a single shaded image. In fact, the shape solution is usually not unique for a simple polyhedron, although the shape solution can be uniquely determined for a complex polyhedron.

We propose a new approach to uniquely determine polyhedral shape solution using interreflections as a constraint that limits the shape solution. Because different interreflection distributions are caused by different shapes, the shape solution can be uniquely determined by using the interreflection distribution in the input image. We shows that interreflection distribution is one-to-one correspondence with polyhedral shape. Nayar et al. focused on eliminating interreflections to discuss shape recovery from interreflections[8], where interreflections were considered to be deleterious. However, our approach uses interreflections as a constraint to limit the shape solution.

2 Assumptions

We make the following assumptions: the object is a polyhedron with a given reflectance and the reflection of the object facets is Lambertian. An image is obtained in orthographical projection, and the illumination light consists of parallel rays with a given direction and unit irradiance. Hence we obtain the relation between the facet gradient and image intensity due to light source illumination:

$$I = \rho(\mathbf{n} \cdot \mathbf{s})/(|\mathbf{n}||\mathbf{s}|), \tag{1}$$

where I is the image intensity of a facet, ρ is the reflectance of the facet, $\mathbf{n} = (p, q, 1)^T$ is the surface normal of the facet, and $\mathbf{s} = (p_s, q_s, 1)^T$ is the light source vector. We define $p = -\partial z(x, y)/\partial x$ and $q = -\partial z(x, y)/\partial y$, where $z(x, y)$ is a facet depth function of x and y.

3 Uniqueness of Shape Recovery

To simplify the discussion of the uniqueness of 3D shape recovery, we assume that the light source vector is parallel to the viewing direction (camera optical axis), i.e., $\mathbf{s} = (0, 0, 1)^T$. Note that the results obtained for this special illumination direction can be easily extended to an arbitrary illumination direction by coordinate rotation because coordinate rotation does not affect the uniqueness of solutions.

Consider a simple polyhedron of which only three facets meeting at an apex are visible (see Fig.1). We obtain the line drawing shown in Fig.2 from an input image. There are three image intensity constraint conditions for the three facets and three orthogonality constraints (an edge intersected by two facets in

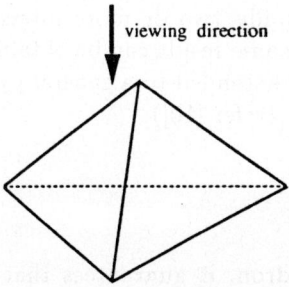
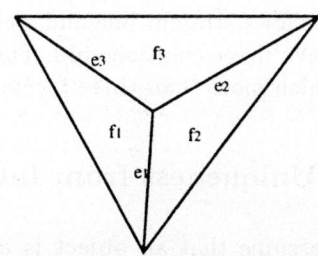

Fig. 1. A polyhedron in 3D space.   **Fig.2** A line drawing of the polyhedron.

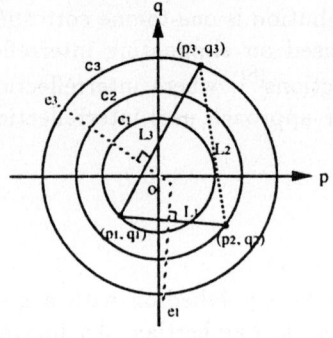
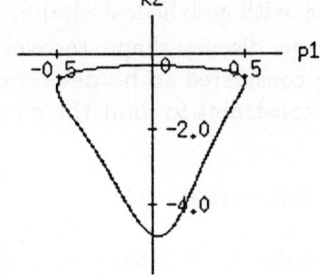

Fig. 3. Determining slope k_2 from (p_1, q_1). **Fig. 4** Locus of slope k_2 of line L_2.

an image must be orthogonal to the line linking the two facets in the gradient space[11]) for the three edges. But there are six variables in the six equations. Thereby, there are $2^3 = 8$ solutions. However, it is not guaranteed that all solutions are real (some of them may be imaginary). Hence, the number of possible solutions is an even number between 0 and 8 (half of the possible solutions are for convex shapes, and the other half are for concave shapes[12]), and the number depends on the values of the image intensities and the slopes of edges. It must be discussed case by case.

Next, we discuss the uniqueness of shape recovery using a reflectance map (refer [9] and [10]). Three image intensities I_i ($i = 1$ to 3) correspond to three circles c_i of radii r_i (see Fig.3). We assume $I_1 > I_2 > I_3$. To find a possible solution for a polyhedron with convex shape, first, we select a point (p_1, q_1) on circle c_1. Then we obtain the gradients (p_2, q_2) for f_2 and (p_3, q_3) for f_3 according to the orthogonality constraint. Thus, a line L_2 is obtained by linking of (p_2, q_2) to (p_3, q_3). If the linking-line L_2 is orthogonal to the edge e_2, it is one solution for this polyhedral surface gradients.

When we move the point (p_1, q_1) around the circle c_1, the slope k_2 of the line L_2 changes and returns to the starting value. Hence the locus of the slope k_2 must be a closed curve (see Fig.4). A straight line $k_2 = -1/e_2$ is drawn in Fig.4 (which corresponds to the orthogonality constraint $e_2 \perp L_2$) and its intersection points with the locus determine the possible solutions for convex shapes.

It is obvious that more than one possible solution usually exists for convex shapes from a real scene, because there are usually two or more intersection points for a straight line and a closed curve. The same result can be obtained for concave shape solutions too. This result has been extended to a general pyramid of which more than three facets meet at an apex (refer [10]).

4 Uniqueness from Interreflections

We assume that an object is a concave polyhedron, it guarantees that interreflections exist among object facets. Our basic idea is, first, to reduce the shape recovery problem to an optimization under image intensity and orthogonality constraints, so that multiple concave shapes are obtained. Then to use an interreflection constraint determines the correct shape. Here the interreflection constraint is that the interreflection distribution computed from the recovered shape must be identical to that in the input image.

4.1 Shape Recovery

We formulate 3D shape recovery as an optimization problem. We define an error function as

$$E = \sum_{i=1}^{m}(I'_i - I_i)^2 + \sum_{i=1}^{n}(1 + k_{ei}k_i)^2, \qquad (2)$$

where m is the number of facets, I'_i is the input image intensity of facet i, I_i is the value calculated using expression (1) for facet i; n is the number of edges intersected by the facets, k_{ei} is the slope of edge e_i obtained from the input image, and k_i is the slope of line L_i in gradient space. Surface gradients are obtained by optimizing the error function. Then its surface shapes are obtained.

4.2 Interreflections

The interreflection model proposed by Koenderink and van Doorn [13] is used in this research. As shown in Fig.5, consider two surface patches x and x' with unit area on image plane. The interreflection at patch x caused by patch x' is[8]

$$dI_{int}(x, x') = \frac{\rho}{\pi} \frac{1}{\mathbf{r}^2} \frac{(\mathbf{n'} \cdot \mathbf{r})}{|\mathbf{n'}||\mathbf{r}|} \frac{(\mathbf{n} \cdot -\mathbf{r})}{|\mathbf{n}||-\mathbf{r}|} \frac{|\mathbf{v}||\mathbf{n'}|}{(\mathbf{v} \cdot \mathbf{n'})} L(x') view(x, x'), \qquad (3)$$

where $\mathbf{r}$ is the vector from x' to x, $\mathbf{n}$ and $\mathbf{n'}$ are the surface normals of the patches x and x', $L(x')$ is the image intensity at the patch x', $\mathbf{v}$ is the view direction, and function $view(x, x')$ indicates the visibility of the two patches, it is unit when x and x' can view each other, and zero otherwise. The function is defined as[8]:

$$view(x, x') = \frac{(\mathbf{n} \cdot -\mathbf{r}) + |(\mathbf{n} \cdot -\mathbf{r})|}{2|(\mathbf{n} \cdot -\mathbf{r})|} \frac{(\mathbf{n'} \cdot \mathbf{r}) + |(\mathbf{n'} \cdot \mathbf{r})|}{2|(\mathbf{n'} \cdot \mathbf{r})|}. \qquad (4)$$

Hence, the total of interreflections at x is

$$I_{int}(x) = \int_{x' \in image} dI_{int}(x, x'). \qquad (5)$$

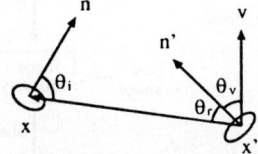

Fig. 5. Interreflection between two facets.

4.3 Uniqueness from Interreflections

The interreflections, as above description of expressions (3) and (5), are a non-linear function of position and shape. The non-linear function determines the uniqueness of shape solution. Because the direct reflection (caused by illumination of light source) is uniform on every facet, the variety of image intensities is decided by interreflections. Since the non-linear interreflection function, different shape will cause different interreflection distribution, also different image intensity distribution. It is a one-to-one correspondence between shape and interreflection distribution. This implies the image intensity distribution of an input image only corresponds to one specified shape. Therefore, interreflection distribution determines the uniqueness of shape recovery of a polyhedron.

Nayar et al. discussed the convergence of shape recovery[8]. They showed some cases which recovered facets converge on other pseudo facets different from actual facets. Their discussion considered the image intensities of pixels independently. However, our discussion considers the image intensity distribution globally, the shape recovery is unique determined by the interreflection distribution. This can be understood like that, it is easy to find multiple functions to fit one point, but it is difficult to find two functions to fit all points of a non-linear surface (the interreflection distribution on object surface).

4.4 Algorithm

As described before, optimization function (2) derives multiple shape solutions. However, the correct shape is determined by checking whether the interreflection distribution computed using recovered shape is identical to that in the input image. This implies, if a recovered shape is the actual one, the computed interreflection distribution is identical to that in the input image, so the image intensities with interreflections deleted are uniform for every facet. This is called interreflection constraint.

Figure 6 illustrates the flow of the algorithm. Initial interreflections are 0. The new image is obtained by subtracting interreflections from the input image. If the iterative number is greater than a limit, the process is stopped, and the recovered shape is considered to be an incorrect solution. The processes are implemented for every possible solution, respectively.

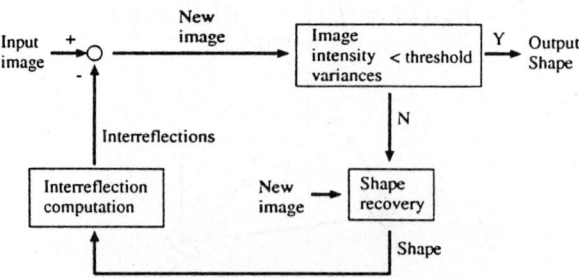

Fig. 6. Shape recovery algorithm.

5 Experimental Results

We have implemented our algorithm to test its validity and feasibility, using synthetic and real images. A experimental result is described next.

In a dark room, the position of the object was set as the origin, a camera (Victor KY-F30, CCD) was fixed at (0, 0, 169) (cm) and the light source, a halogen lamp with radius of about 1(cm) (LA-150S, Hayashi Tokei Kougyou Co. Ltd.), was set at (69, 2, 175) (cm). The object had a concave shape, likes an upside-down pyramid, made of paper. The paper (Japan Color Research Institute) had a Lambertian reflection property, and its reflectance was 0.69. The object size was about 6×6 (cm^2), and the depth was about 3 (cm).

The image size was 256×240 in our experiments, and there were 256 grey levels (8 bits). We extract edges from an input image using the DOG algorithm, then interactively select the edges which are used in the optimization for shape recovery. We select the conjugate gradient method [14] to solve the optimization problem. To avoid the optimization falls into a local minimum in the first iteration of shape recovery, random values were used as the initial facet gradients to find a possible solution, and we also set a threshold of $\epsilon = 10^{-10}$ for the error function E to check whether an solution is a optimal solution or a local minimum solution. We set T_v, the threshold of the variance of image intensity distribution, to be 1.0. The maximum number of iterations is 15. Our iteration algorithm was repeated until found all possible solutions.

Fig.7 shows a real image. Based on the real size of the object and the vertex coordinates in the input image, the vertex coordinates converted to the image pixel scale are [(41, 32, 76), (78, 218, 76), (222, 88, 76), (116, 113, 0)].

Two concave shape solutions which satisfy $E < \epsilon$ were obtained. The variance of image intensity distribution, after eliminating interreflections, was greater than the threshold T_v and that in other experiments using a synthetic image. This is due to such influences as those of noise and nonuniform illumination. Therefore the iteration was stopped after the maximum number of iterations for both shapes.

Fig. 7. A real image.

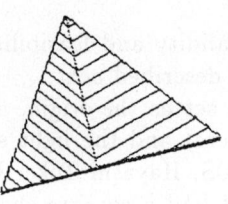 

(a). Recovered shape. (b). The image with interreflections deleted

Fig. 8. Recovered correct shape and the image with interreflections deleted.

Fig.8(a) shows the correct recovered shape. Its vertex coordinates are [(41, 32, 73.6), (78, 218, 72.5), (222, 88, 75.4), (116, 113, 0)]. The orientation errors of the three facets are 0.6, 0.9 and 1.1 degrees. Fig.8(b) shows the image obtained after eliminating interreflections. The variances of image intensity distributions in three facets are 7.6, 8.5 and 4.2.

The other recovered concave shape is shown in Fig.9(a). Its vertex coordinates are [(41, 32, -13.3), (78, 218, 9.8), (222, 88, 81.6), (116, 113, 0)]. Fig.9(b) shows the image obtained after eliminating interreflections. The variances of image intensity distributions in three facets are 95.3, 91.8 and 124.6.

The experimental results show that the proposed approach is valid and feasible. Only simple concave polyhedral were used in our experiments, because we focus on describing that multiple shape solutions exist for a simple polyhedron and that interreflections are an important constraint to uniquely determine the shape solution. The algorithm also works for a general convex or concave polyhedron.

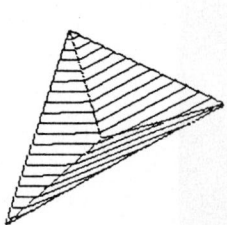 

(a). Recovered shape. (b). The image with interreflections deleted

Fig. 9. Recovered incorrect shape and the image with interreflections deleted.

6 Conclusion

In this paper, we discussed the uniqueness of shape recovery of a simple polyhedron from a single shaded image. We showed that multiple solutions usually exist for a convex shape with three or more facets intersecting at an apex (as in a pyramid), and there are the same number of solutions for a concave shape. We have discussed that interreflection distribution uniquely determines the shape solution, and described a new approach to determining the correct solution from the multiple solutions using the interreflection constraint. Our experimental results also show that our approach is valid and feasible. This work helps to clarify the conditions under which shape recovery is possible for shape-from-shading.

References

1. B.K.P. Horn, "Understanding Image Intensities", Artificial Intelligence, Vol.8, pp.201-231, 1977.
2. K. Sugihara, "Mathematical Structures of Line Drawings of Polyhedra - Toward Man-Machine Communication by Means of Line Drawings", IEEE Trans. PAMI, Vol.4, pp.458-469, 1982.
3. K. Sugihara, "Classification of Impossible Objects", Perception, Vol.11, pp.65-74, 1982.
4. K. Sugihara, "A Necessary and Sufficient Condition for a Picture to Represent a Polyhedral Scene", IEEE Trans. on PAMI, Vol.6, No.5, pp.578-586, 1984.9.
5. K. Sugihara, "An Algebraic Approach to Shape-from-Image Problems", Artificial Intelligence, Vol.23, No.1, pp.59-95, 1984.5.
6. K. Sugihara, "Algebraic Approach to the Recovery of Three-Dimensional Shapes from Single Images", Trans. IEICE Japan, J66-D, No.5, pp.541-548, 1983 (in Japanese).
7. I. Shimshoni and J. Ponce, "Recovering the Shape of Polyhedra Using Line-Drawing Analysis and Complex Reflectance Models", Proc. of CVPR'94, pp.514-519.
8. S. K. Nayar, K. Ikeuchi and T. Kanade, "Shape from Interreflections", Proc. of ICCV'90, pp.1-11, 1990.
9. J. Yang, D. Zhang, N. Ohnishi and N. Sugie, "Uniqueness of Shape Recovery Based on Interreflections", Proc. of Intelligent Robots and Computer Vision XV, SPIE's Photonics East '96 Symposium, Vol. 2904, pp.148-158, 1996.
10. J. Yang, D. Zhang, N. Ohnishi and N. Sugie, "Determining a Polyhedral Shape Using Interreflections", Proc. of CVPR'97, pp.110-115, 1997.
11. D.A. Huffman, "A Duality Concept for the Analysis of Polyhedral Scenes", Machine Intelligence 8, Elcock, E.W. and Michie, D.(eds.), Ellis Horwood, pp.475-492, 1977.
12. M.J. Brooks, "Two Results Concerning Ambiguity in Shape from Shading", Proc. of the National Conference on Artificial Intelligence, pp.36-39, 1983.
13. J.J. Koenderink and A.J. van Doorn, "Geometrical Modes as a General Method to Treat Diffuse Interreflections in Radiometry", Journal of Optical Society of America, Vol.73, No.6, pp.843-850, 1983.6.
14. W.H. Press, S.A. Teukolsky, W.T. Vetterling and B.P. Flannery, "Numerical Recipes in C", Cambridge University Press, 1988.

Improved Supervised Color Constancy for Color Inspection

Bai Xuesheng   Xu Guangyou
Computer Vision Laboratory,   Information Research Group
Department of Computer Science, Tsinghua University, Beijing 100084, P.R.C

Abstract

In industrial applications, color inspection is difficult to be implemented because the environment illumination may vary unexpectedly, thus color constancy algorithm must be applied. Novak proposed the supervised color constancy thoughts, in which a few color chips of known spectral reflectance are placed in the scene to correct illumination changes. But his algorithm requires camera sensitivity functions which are usually difficult to obtain, and only numerical simulations is presented. In this paper, we proposed an improved supervised color constancy algorithm and applied it to color inspection. The algorithm need not to know the reflectance functions of color chips and the camera sensitivity functions, thus is more suitable for industrial uses. To ensure that this algorithm work well in industrial applications, imaging process of real system is studied and incremental-linear imaging model is adopted. Combined with this model, we gave the algorithm implementation on real systems, which shows satisfactory performance. Main thoughts and implementation details of the algorithm are presented in this paper, with experiment results and analysis.

Key Words
color constancy, supervised color constancy, canonical illumination, canonical color, incremental-linear model, zero-point, illumination spatial distribution

1. Introduction

Color inspection has potentially broad applications in industrial manufacturing. In many occasions, the colors of the same kind of products may vary because of uncontrollable reasons. In such cases, a color inspection and classification process is needed where products are classified ensuring that the color difference between products in a same class are indiscernible to human eyes. But since the color response is relevant to environment illumination, and usually the illumination is difficult to be controlled at a constant level because of various reasons (one main reason is the fluctuation of electric voltage), color constancy algorithm which seeks illumination-invariant features must be applied.

From the middle of 1980s, many color constancy algorithms have been proposed, yet most of them are based on highlight image analysis [1][2] and can not be applied to color inspection in real environment effectively. In 1991 Novak proposed the supervised color constancy thoughts [3] in which a few patches of known spectral reflectance are placed in the scene to correct illumination changes. The algorithm is simple and efficient, yet it need to know the camera spectral sensitivity, which may be difficult to obtain in many applications, and only numerical simulations are given.

In this paper, we improved Novak's algorithm and apply it to color inspection. The algorithm differs from Novak's original method in that it does not need to know the spectral reflectance functions of color chips and the camera spectral sensitivity, thus loosen the requirements greatly. To implement the algorithm in real applications, we studied the imaging process of real color imaging system and adopted the incremental-linear imaging model. Combined with this model, we developed an algorithm for real color imaging system. The algorithm show satisfactory performance on our experimental system, which is designed for color inspection and classification for decorating color bricks. Main thoughts and implementation details of the algorithm is presented in this paper, also with experiment results and analysis.

The sections below are arranged as follows: in section 2, we briefly introduce the theory basis and basic thoughts of our algorithm. Implementation details on real systems are discussed in section 3. Section 4 focuses on the requirement on illumination spatial distribution for implementing the algorithm, while section 5 gives our experiment results and analysis.

2. Basic Thoughts

First, we introduce a theorem which forms the basis of our algorithm.

2.1 Relation between color responses under different illuminations

For ideal color imaging systems, we have a theorem on relation between color responses under different illuminations as below:

Theorem 1: For an ideal color imaging system, the color responses of color chips under two arbitrary illumination $e'(\lambda)$ and $e''(\lambda)$, c' and c'', are related by a matrix $\mathbf{M}$ which only depend on the system channel sensitivity functions and the two illuminations:

$$\mathbf{c'} = \mathbf{M}\mathbf{c''} \quad (1)$$

Proof: According to finite linear model [4], the reflectance functions of objects can be expressed to a fairly precise extent by the linear combination of the first three Cohen base functions, that is $s(\lambda) = \sum_{i=1}^{3} s_i S_i(\lambda)$, where $S_i(\lambda)$ (i = 1, 2, 3) are Cohen base functions. Then from the imaging model of ideal color imaging system, we have

$$\mathbf{c} = \begin{pmatrix} R \\ G \\ B \end{pmatrix} = \begin{pmatrix} \int_{\lambda_1}^{\lambda_2} e(\lambda)s(\lambda)C_1(\lambda)d\lambda \\ \int_{\lambda_1}^{\lambda_2} e(\lambda)s(\lambda)C_2(\lambda)d\lambda \\ \int_{\lambda_1}^{\lambda_2} e(\lambda)s(\lambda)C_3(\lambda)d\lambda \end{pmatrix} = \begin{pmatrix} \sum_{i=1}^{3} s_i \cdot \int_{\lambda_1}^{\lambda_2} e(\lambda)S_i(\lambda)C_1(\lambda)d\lambda \\ \sum_{i=1}^{3} s_i \cdot \int_{\lambda_1}^{\lambda_2} e(\lambda)S_i(\lambda)C_2(\lambda)d\lambda \\ \sum_{i=1}^{3} s_i \cdot \int_{\lambda_1}^{\lambda_2} e(\lambda)S_i(\lambda)C_3(\lambda)d\lambda \end{pmatrix} \quad (2)$$

Define 3*3 matrix $\mathbf{A} = (a_{k,j})_{3 \times 3}$ with entry $a_{k,j}$ defined as $a_{k,j} = \int e(\lambda)S_j(\lambda)C_k(\lambda)d\lambda$. At the same time, denote the reflectance vector $\mathbf{s} = (s_1\ s_2\ s_3)^T$, then

$$\mathbf{c} = \mathbf{A}\begin{pmatrix} s_1 \\ s_2 \\ s_3 \end{pmatrix} = \mathbf{A} \cdot \mathbf{s} \quad (3)$$

From the definition of **A**, we know that **A** is only relevant with illumination $e(\lambda)$ and system spectral sensitivity functions $C_1(\lambda), C_2(\lambda), C_3(\lambda)$. Then under different illuminations $e'(\lambda)$ and $e''(\lambda)$, the color responses of any color chip can be expressed as $\mathbf{c'} = \mathbf{A'} \cdot \mathbf{s}$ and $\mathbf{c''} = \mathbf{A''} \cdot \mathbf{s}$.

Under the condition that $\mathbf{A''}$ is not singular, we can define illumination transform matrix $\mathbf{M} = \mathbf{A'}(\mathbf{A''})^{-1}$, which leads to the conclusion

$$\mathbf{c'} = \mathbf{A'} \cdot \mathbf{s} = \mathbf{A'}(\mathbf{A''})^{-1} \cdot \mathbf{c''} = \mathbf{M} \cdot \mathbf{c''}$$

2.2 Basic thoughts of our algorithm

From theorem 1, we can see that given two illuminations the illumination change transform matrix **M** is the same for all color objects. If we select an illumination $e'(\lambda)$ as a standard illumination, any color response under unknown illuminations $e''(\lambda)$ can be transformed to its corresponding color response under this standard illumination. We call this standard illumination the canonical illumination, and color under this illumination the canonical color. Denote canonical color as $\mathbf{c}^s$, color response under unknown illumination as $\mathbf{c}$, then

$$\mathbf{c}^s = \mathbf{M} \cdot \mathbf{c} \tag{4}$$

Using this method, colors under different illumination can be transformed to its corresponding canonical color and compared, as long as we known matrix **M**. Since **M** is the same for all color objects, we can use responses of color chips under canonical and unknown illumination to calculate matrix **M**, then calculate the corresponding canonical color of a unknown object under this unknown illumination. The algorithm is as follows:

Algorithm 1 (Calculation of canonical color on ideal imaging system)
Input:　　Responses of m color chips under canonical illumination $\{\mathbf{c}_i^s = (R_i^s, G_i^s, B_i^s)^T, i = 1, \cdots, m\}$,

　　　　Responses of m color chips under unknown illumination $\{\mathbf{c}_i = (R_i, G_i, B_i)^T, i = 1, \cdots, m\}$,

　　　　Response of unknown color sample under unknown illumination $\mathbf{c} = (R, G, B)^T$.

output:　　canonical color of unknown color sample $\mathbf{c}^s = (R^s, G^s, B^s)^T$.

Algorithm
1) Let 3*m matrix $\mathbf{C} = (\mathbf{c}_1\ \mathbf{c}_2\ \cdots \mathbf{c}_m)$, $\mathbf{C}_s = (\mathbf{c}_1^s\ \mathbf{c}_2^s\ \cdots \mathbf{c}_m^s)$, then from (4)

$$\mathbf{MC} = \mathbf{C}_s$$
$$\mathbf{MCC}^T = \mathbf{C}_s\mathbf{C}^T$$

when $m \geq 3$, we can solve **M** using the LMS method:
$$\mathbf{M} = \mathbf{C}_s\mathbf{C}^T(\mathbf{CC}^T)^{-1}$$

2) The canonical color of unknown sample can be calculated as

$$\mathbf{c}^s = \begin{pmatrix} R^s \\ G^s \\ B^s \end{pmatrix} = \mathbf{M}\mathbf{c} = \mathbf{M} \cdot \begin{pmatrix} R \\ G \\ B \end{pmatrix}$$

3. Implementation on real imaging systems

The algorithms discussed above is based on ideal color imaging systems. In real imaging systems, the above algorithm need to be revised.

3.1 Imaging model of real imaging system

From the definition of ideal imaging model, we known it is totally linear which means it has the two properties below:

$$\mathbf{C}(a \cdot L(\lambda)) = a \cdot \mathbf{C}(L(\lambda)) \tag{5}$$

$$\mathbf{C}(L_1(\lambda) + L_2(\lambda)) = \mathbf{C}(L_1(\lambda)) + \mathbf{C}(L_2(\lambda)) \tag{6}$$

where $\mathbf{C}(L(\lambda))$ represents the color response vector of imaging system when the illumination spectrum of incident light into the system is $L(\lambda)$.

In most cases, real color imaging systems do not have these two properties. A typical color imaging system is usually composed of a color CCD camera and a image grabber. The CCD camera transforms illumination intensity into voltage signal, which is digitized by ADC in the grabber. The photo-electronic transform of the CCD camera is thought to bear high linearity [5], that is, the output voltage signal is proportional to the illumination intensity. Although there are γ – corrections in CCD cameras, most CCD camera now used for image processing provides γ – OFF function which preserves the linearity. In this paper, we regard the CCD camera as a linear device and neglect the influence of γ – correction.

In image grabber, the voltage signal is evenly digitized. It should be noted this procedure usually no longer bears the linear relation between output and input. Because of the reference voltage of ADC, the digitized output is not linear to the input voltage signal and the illumination intensity, but incremental-linear. As shown in Figure 1, where V represents the voltage signal of CCD which is proportional to the illumination intensity, N the digitized output, ADC converts a V_{min} to number 0, and in most cases this V_{min} does not correspond to the situation when illumination intensity is zero. So, the output of the whole imaging system is incremental-linear to the illumination-intensity, not perfect linear as in the ideal imaging system case.

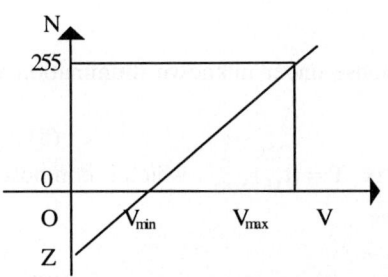

Figure 1 Incremental Linearity caused by ADC

To apply supervised color constancy algorithm to a real imaging system, we need to linearize the imaging system first. Considering the point Z in Figure 1, which corresponds to the situation when illumination intensity is zero, then the difference of digitized output N and Z, N-Z , is proportional to the illumination intensity. For a real color imaging system, there are Z_R, Z_G and Z_B for channels R, G, B. We define color vector $\mathbf{Z}=(Z_R \ Z_G \ Z_B)^T$ as the zero-point of the color imaging system.

3.2 Revised algorithm on real imaging systems

From the analysis above, a method for real systems can be derived which first linearize the system then process the problem in the ideal system way. But linearizing the system need calculating the zero-point, which may still be difficult in some occasions. We combined the original algorithm with incremental model, which gives a revised algorithm for real system. The algorithm need not calculating zero-point first, it can calculate the zero-point value itself from the correspondence of color chip values. Below we will give the algorithm. But before going any further, we first give a theorem for real system corresponding to theorem 1:

Theorem 2: For a real color imaging system, the color responses of color chips under two arbitrary illumination $e'(\lambda)$ and $e''(\lambda)$, $\mathbf{c}'$ and $\mathbf{c}''$, are related by a matrix $\mathbf{M}$ which only depend on the system channel sensitivity functions and the two illuminations:

$$\mathbf{c}' = \mathbf{M}\,\mathbf{c}'' + (\mathbf{I} - \mathbf{M})\,\mathbf{z} \qquad (7)$$

where $\mathbf{I}$ is the 3*3 identity matrix, and $\mathbf{z}$ is the zero-point of the system。

Proof: Select $\mathbf{c}-\mathbf{z}$ as output, then this new imaging system is an ideal color imaging system. From theorem 1, we known that the color responses of color chips under two arbitrary illumination $e'(\lambda)$ and $e''(\lambda)$ are related by a matrix $\mathbf{M}$ which only depend on the system channel sensitivity functions and the two illuminations, that is:

$$\mathbf{c}' - \mathbf{z} = \mathbf{M}(\mathbf{c}'' - \mathbf{z})$$

rearrange, gives

$$\mathbf{c}' = \mathbf{M}\,\mathbf{c}'' + (\mathbf{I} - \mathbf{M})\,\mathbf{z}$$

Denoting canonical color as $\mathbf{c}^s$, color response under unknown illumination as $\mathbf{c}$, then

$$\mathbf{c}^s = \mathbf{M}\,\mathbf{c} + (\mathbf{I} - \mathbf{M})\,\mathbf{z} \qquad (8)$$

Define 3*4 extended illumination change matrix $\mathbf{T} = (t_{ij})_{3 \times 4}$, which is composed by $\mathbf{T} = (\mathbf{M} \ (\mathbf{I}-\mathbf{M})\mathbf{z})$, then (8) can be written as

$$\mathbf{c}^s = (\mathbf{M} \ (\mathbf{I}-\mathbf{M})\mathbf{z}) \begin{pmatrix} \mathbf{c} \\ 1 \end{pmatrix} = \mathbf{T} \begin{pmatrix} \mathbf{c} \\ 1 \end{pmatrix} \qquad (9)$$

From (9), we can first use responses of color chips to calculate matrix T of a unknown illumination, then calculate the corresponding canonical color of a unknown object under this unknown illumination. The algorithm is as follows:

Algorithm 2 (Calculation of canonical color on real imaging system)
Input: Responses   of   m   color   chips   under   canonical   illumination

$\{c_i^s = (R_i^s, G_i^s, B_i^s)^T, i = 1, \cdots, m\}$,

Responses of m color chips under unknown illumination
$\{c_i = (R_i, G_i, B_i)^T, i = 1, \cdots, m\}$,

Response of unknown color sample under unknown illumination
$c = (R, G, B)^T$.

output: canonical color of unknown color sample $c^s = (R^s, G^s, B^s)^T$ 。
zero-point of imaging system z.

Algorithm:

1) Let 3*m matrix $C_s = (c_1^s \ c_2^s \ \cdots c_m^s)$, 4*m matrix

$$C = \begin{pmatrix} c_1 & c_2 & \cdots & c_m \\ 1 & 1 & \cdots & 1 \end{pmatrix} = \begin{pmatrix} R_1 & R_2 & \cdots & R_m \\ G_1 & G_2 & \cdots & G_m \\ B_1 & B_2 & \cdots & B_m \\ 1 & 1 & \cdots & 1 \end{pmatrix}$$

then
$$TC = C_s$$
$$TCC^T = C_s C^T$$

when $m \geq 4$, we can solve **T** using the LMS method:
$$T = C_s C^T (CC^T)^{-1}$$

2) The canonical color of unknown sample can be calculated as

$$c^s = \begin{pmatrix} R^s \\ G^s \\ B^s \end{pmatrix} = T \begin{pmatrix} c \\ 1 \end{pmatrix} = T \cdot \begin{pmatrix} R \\ G \\ B \\ 1 \end{pmatrix}$$

3) From the definition of **T**,and
$$M = T \cdot (I \ \ 0)^T$$
$$(I - M) z = T \cdot (0 \ \ 0 \ \ 0 \ \ 1)^T$$

then the system zero-point:

$$z = \left(I - T \cdot \begin{pmatrix} I \\ 0 \end{pmatrix} \right)^{-1} \cdot T \cdot \begin{pmatrix} 0 \\ 0 \\ 0 \\ 1 \end{pmatrix}$$

4. Requirement on Illumination Spatial Distribution

In supervised color constancy algorithm, since the color chips and unknown sample are distributed in the scene, we implicitly hypnotize that the illumination change matrix is the same everywhere in the scene, which means $M(x) = M(y)$. Below we will discuss the requirement of this condition on illumination spatial distribution.

Considering illumination changes with time, supposing that we observe the scene at time t_1 and t_2. Then from the definition of **M**, $M(x) = M(y)$ means $A(x,t_1)[A(x,t_2)]^{-1} = A(y,t_1)[A(y,t_2)]^{-1}$, gives

$$[A(y,t_1)]^{-1}A(x,t_1) = [A(y,t_2)]^{-1}A(x,t_2) \qquad (10)$$

which requires that $[A(y,t_1)]^{-1}A(x,t_1)$ be a matrix only depends on **x** and **y**. Below we give a sufficient condition of this requirement:

Theorem 3: if the illumination spatial distribution satisfies that the illumination intensity $e(u,\lambda,t)$ at position u time t and wavelength λ can be decomposed into the multiplex of position factor $\alpha(u)$ and position-irrelevant factor $\beta(\lambda,t)$

$$e(u,\lambda,t) = \alpha(u)\cdot\beta(\lambda,t) \qquad (11)$$

then $[A(y,t_1)]^{-1}A(x,t_1)$ is a matrix only depends on **x** and **y**.

Proof: when the illumination spatial distribution satisfies (11), we have
$$e(x,\lambda,t) = \alpha(x)\cdot\beta(\lambda,t)$$
$$e(y,\lambda,t) = \alpha(y)\cdot\beta(\lambda,t)$$

According to the definition of **A**, any entry of matrix $A(x,t_1)$, $a_{k,j}(x,t_1)$ is defined as

$$\begin{aligned}
\mathbf{a}_{k,j}(x,t_1) &= \int e(x,\lambda,t_1)S_j(\lambda)C_k(\lambda)d\lambda \\
&= \int \alpha(x)\beta(\lambda,t_1)S_j(\lambda)C_k(\lambda)d\lambda \\
&= \alpha(x)\int \beta(\lambda,t_1)S_j(\lambda)C_k(\lambda)d\lambda
\end{aligned}$$

let **S** be a 3*3 matrix with entry

$$S_{k,j} = \int \beta(\lambda,t_1)S_j(\lambda)C_k(\lambda)d\lambda$$

then $A(x,t_1) = \alpha(x)S$; Similarly, $A(y,t_1) = \alpha(y)S$ which gives

$$[A(y,t_1)]^{-1}A(x,t_1) = \alpha(x)/\alpha(y)\cdot I \qquad (12)$$

means that $[A(y,t_1)]^{-1}A(x,t_1)$ is a matrix only depends on **x** and **y**.

This theorem gives a sufficient condition on illumination spatial distribution for supervised color constancy algorithm to be applied. Generally speaking, the condition holds in most cases, especially when illumination is caused by single illuminant. It will be broken only when the illumination is caused by multiple illuminants and the illuminating structure changes (which means part of the illuminants change individually, such as a illuminant goes out).

5. Experimental Results

We test our algorithm on a real color imaging system which is designed for inspection and classification of decorating color bricks. The system we use for experiments is composed of a CCD camera and a ASPEX-PIPE image processing system. The camera is a XC-711 color CCD camera (γ – OFF), output signal of which is sent to ASPEX-PIPE to be sampled and digitized. The whole system is encapsulated to reducing the influence of ambient illumination.

Five squared color chips are used to calculate illumination changes and a illumination is selected as the canonical illumination. Besides illumination changes caused by electric voltage fluctuation, a group of filters (including intensity and light color filters)are used to change the illumination manually. Various color samples are tested, and the algorithm shows satisfactory color constancy against illumination change — the canonical color values calculated from same color sample under different illuminations vary within 2 graylevel range. This is sufficient for our application and many other applications.

It should be noted that the above results are achieved when system works in its linear range. Since color response values are restricted to [0, 255], when illumination is too bright response values will clipped to 255 and when illumination is too dim color clipping to 0 is also possible on systems whose zero-points have negative coordinates(such as our imaging system). If clipping occurs, canonical color values calculated from samples under different illuminations may be very different, and the system can not give satisfactory results.

To avoid this, careful selection of imaging system and color chips should be made. A good imaging system should has a positive zero-point which is approximately zero, which can avoid 0-clipping under dim illuminations. By carefully adjusting the camera, clipping to 255 can also be avoided.

As to color chip selection, high saturation colors should avoided. Such color has not enough responses in all RGB channels, and may clip to zero under dim illumination and clip to 255 under bright illumination. But on the other hand, from the view of reducing noise influence, the colors of chips should departs each other as far as possible. This two color selection criteria can not be satisfied to the maximum extent simultaneously. In real implementations, trade-off should be made based on possible changes of illumination, channel response of imaging system and the value of zero-point.

References

[1] S.A.Shafer, "Optical phenomena in computer vision", Proc. Canadian Soc. Comp. Studies of Intel., 1984, 572-577
[2] G.J.Klinder, S.A,Shafer and T.Kanade. "Using a color reflection model to separate highlights from object color" . Int. conf. on Computer vision, 1987, PP.145-150. 1987 IEEE.
[3] C.L.Novak, S.A.Shafer, "Supervised color constancy for machine vision", SPIE Vol.1453, Human Vision, Visual Processing, and Digital Display II, 1991, pp.353-368.
[4] L. T. Maloney. " Evaluation of linear models of surface spectral reflectance with small number of parameters ". J.Opt.Soc.Am.A Vol.3, No.10 1986. pp.1673-1682.
[5] G.E.Healey, R. Kondepudy, "Radiometric CCD Camera Calibration and Noise Estimation", IEEE. Trans. PAMI, Vol.16, No.3, March 1994, pp.267-276.

Unsupervised Filtering of Munsell Spectra

M. Hauta-Kasari[1], W. Wang[1], S. Toyooka[1], J. Parkkinen[2], and R. Lenz[3]

[1] Department of Environmental Science and Human Engineering,
Graduate School of Science and Engineering, Saitama University,
255 Shimo-okubo, Urawa, Saitama, 338 JAPAN
Email: mhk@mickey.mech.saitama-u.ac.jp

[2] Department of Information Technology, Lappeenranta University of Technology,
P.O.Box 20, FIN-53851 Lappeenranta, FINLAND

[3] Image Processing Laboratory, Department of Electrical Engineering,
Linköping University, S-58183 Linköping, SWEDEN

Abstract. We present a new method for producing color filters with positive coefficients to represent color reflectance spectra. The subspace method which is based on the KL-expansion can be used to define a basis to describe the spectral data accurately. However, due the orthogonality of the eigenvectors, the corresponding color filters usually contain negative coefficients and cannot be used in optical components directly. Our method finds the set of vectors which span a very similar color space as the subspace method does. These color filters contain only positive coefficients and can be directly used in optical implementations. We used an unsupervised competitive neural network (Instar) to find a set of positive color filters. The experiments with the Munsell spectra show that the filters produced by the neural network span a color space very similar to the color space spanned by the eigenvectors of the subspace method.

1 Introduction

Color is an important factor in many computer vision-, pattern recognition-, and industrial quality control applications. Usually the color analysis is based on three-dimensional color coordinate systems, like CIE xyY-, Lab- and Luv-color spaces. These three-dimensional color spaces are related to human color vision system, in which there are three different types of photoreceptors [1, 2].

The color of an object is a sensation, which is produced in the brain [1] and it is thus hard to define color. Color can however be defined indirectly through the cause of the sensation. This is called color spectrum, which can be measured physically: the electromagnetic spectrum in the wavelength range from 380 nm to 780 nm. Using the spectrum itself as color representation avoids problems such as metamerism [3], where several spectra have the same three-dimensional color coordinates.

In the subspace method the color coordinates are projections, i.e. the values of inner products between the color spectrum and the basis. This avoids the problem of metamerism and the accuracy is high [4, 5]. The Munsell [6] color spectra database can be described by a few basis vectors, and this basis can be also used for describing natural colors [5]. Usui et al. [7] described a multilayer perceptron based system where the weights in the hidden layer were used to reconstruct Munsell data from the three-dimensional color space representation.

The optical implementation of the subspace method gives the possibility of faster calculation, which is needed in many industrial applications. The basis produced by the subspace method is orthogonal and therefore usually contains negative coefficients. These cannot be directly implemented in optical components. For example, the liquid crystal spatial light modulator (LCSLM), which has been used to calculate the optical inner product [8, 9], takes only filters with positive coefficients. In Refs. [8, 9], the basis vector set produced by the subspace method was biased and multiplied to make suitable filters for LCSLM. It is also possible to divide the basis vectors to the positive and negative parts, and handle these parts separately, but this leads to more complicated optical systems. The aim of this study is to produce a vector set with positive coefficients, which can be directly used in optical pattern recognition. This problem was also addressed in [10]. There the positive color filters are found by optimizing an energy function based on second- and fourth-order statistical moments [11].

In this paper we present an unsupervised neural network based method to find filters with positive coefficients. The competitive neural network finds the centers of color clusters in the color space. After learning, the weight vectors of the neural network are used as filters, which span a color space very similar to the color space spanned by the eigenvectors of the subspace. Our experiments show that the Munsell color spectra can be described very accurately by these filters. The obtained filter systems are compared with the results reported in [10].

2 Subspace Method

In [4, 5] it is shown that the color space for the color spectra can be described accurately by the subspace method. A measured spectrum $s(\lambda)$ can be represented as a column vector $s(\lambda) = [s(\lambda_1), s(\lambda_2), \ldots, s(\lambda_n)]^T$, where λ is the wavelength and T denotes the transpose. Next we compute the eigenvectors of the correlation matrix $R = \sum_{i=1}^{N} s_i(\lambda) s_i(\lambda)^T$, where the index i indicates the ith spectrum in the set of N measured spectra. The eigenvectors ϕ are the solutions of the equation $R\phi = \sigma\phi$, where σ is an eigenvalue of R. The first n eigenvectors form a basis for the subspace. The subspaces can be designed for several different color regions in parallel [12]. In this study we formed only one subspace representing the whole Munsell color space.

The information content, i.e. the fidelity value k of the first n eigenvectors can be defined as $k = \sum_{i=1}^{n} \sigma_i / \sum_{i=1}^{N} \sigma_i$, where k is ratio of the information on the first n largest eigenvalues over the information on all N eigenvalues. In Ref. [4], eight basis vectors were needed for describing the Munsell database accurately.

The basis vector set for the Munsell spectra is orthogonal and the vectors contain negative coefficients. For optical pattern recognition, a small vector set with only positive coefficients spanning the color space as accurately as possible should be found. Since the subspace method produces systems with minimal squared error, it can be used as a reference.

3 Unsupervised Neural Network

Color filters suitable for optical pattern recognition should fulfill the following requirements: *1) the filters should contain only positive coefficients, 2) the filters should span the color space as accurately as possible, 3) the filters should be separated from each other.* To fulfill these conditions, we investigated the clustering properties of competitive learning and self-organization methods [13-15]. The main problem was if it is possible to cluster the input color spectra and use the centers of these clusters for a representation of the whole color space. The method we decided to use is based on competitive learning, which clusters the input data without any knowledge of the right cluster distributions.

In our experiments we used an unsupervised competitive neural network with a learning algorithm based on the Instar-algorithm by Grossberg [13]. We also incorporated Kohonen's [14] self-organizing map with a winner take all layer (WTA). The competitive neural network clusters the input data so that the weight vectors are the centers of these clusters [15]. In our study the input data are the measured color spectra containing only positive coefficients. Thus the weight vectors, i.e. the color filters are also positive. These weight vectors should span the color space like the eigenvectors of the subspace method. In the competitive neural network only the winner neuron learns during each learning cycle, which means that the filters are separated from each other.

The winner w_c is the weight vector, which has the closest euclidean distance to the input vector x. Next, the updating process of the weight vectors is defined as follows

$$w_i(t+1) = \begin{cases} w_i(t) + \alpha(t)[x(t) - w_i(t)] & \text{, if } i = c, \\ w_i(t) & \text{, otherwise,} \end{cases} \quad (1)$$

where t is the iteration parameter and $\alpha(t)$ is a learning rate. The learning rate is decreasing exponentially during the learning. In each cycle of the learning process the training sample is taken randomly from the input data. The weights are initialized by the average vector (≥ 0) of the input data. Equation 1 can be written as $w_i(t+1) = w_i(t)[1 - \alpha(t)] + \alpha(t)x(t)$, where $0 \leq \alpha(t) < 1$, $w_i(t) \geq 0$, $x(t) \geq 0$ and therefore the weights w are always positive. A detailed description of the competitive learning and self-organization can be found in Ref. [14].

4 Filtering the Spectral Database

To measure the accuracy of the produced color filters we reconstructed the Munsell database and compared the results to the subspace method. The reconstruction procedure is as follows:

If the basis vector set is orthonormal, the reconstructed spectrum s' can be calculated from the equation $s' = BB^T s$, where s is the original spectrum and B is the basis vector set.

In the nonorthonormal case, as in the case of filters with positive coefficients, the reconstruction is obtained by the generalized inverse matrix:

$s' = W(W^TW)^{-1}W^Ts$, where W is the filter set. In the optical implementation $W(W^TW)^{-1}$ is known and W^Ts is determined experimentally.

We calculate the relative error as the norm of spectrum's reconstruction error divided by the norm of original spectrum: error $= 100 \times \|s' - s\|/\|s\|$ (%).

5 Experiments

In our experiments the Munsell database was used. It consists of 1269 color spectra measured by a spectrophotometer from the *Munsell Book of Color* [6]. We used the wavelength range from 400 nm to 700 nm, at 5nm intervals, i.e. each spectrum contained 61 components. Fig. 1 shows the first eight basis vectors of the subspace method. The fidelity value for the first eight eigenvalues is 99.99%.

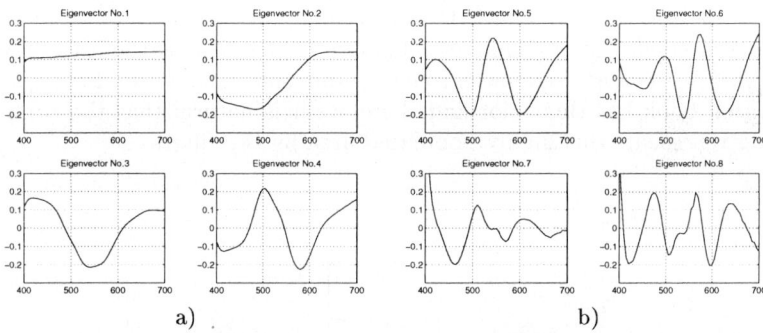

Fig. 1. Eigenvectors of the subspace method a) No.1-4, b) No.5-8.

5.1 Filter Design

The input data, i.e. the whole Munsell data was processed by the competitive neural network, with an included WTA-layer. The number of neurons was 8 and the total number of learning cycles was 50000. The learning data was first normalized to unit norm, the weights were initialized by the average vector of the input data and the learning rate was decreasing exponentially from 0.9 towards 0 during learning. The learned filter functions are shown in Fig. 2. The filters produced by the competitive neural network are the centers of color clusters in the color space. Fig. 2 shows that they represent regularly the spectral range from 400 nm to 700 nm. Fig. 3 visualizes the Munsell data and the filters in two-dimensional CIE 1931 xy-diagram. CIE 1931 xy-coordinates were calculated under daylight illumination D65. The xy-coordinates are only used for visualization, but not used in the experiments. Fig. 3 a) shows the xy-color coordinates of Munsell data and Fig. 3 b) shows the xy-coordinates of filters. The numbers in Fig. 3 b) correspond to the filter numbers shown in Fig. 2. It can

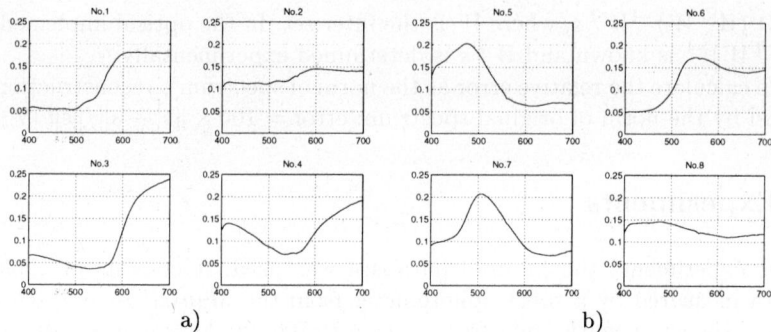

Fig. 2. Learned filter functions a) No.1-4, b) No.5-8.

be seen that the CIE xy-coordinates of the filter set have approximately equal distances between each other. These filters are the centers of eight color clusters of the Munsell database. The CIE 1931 xy-diagram shows that two filters, no.2 and no.8 are near the xy-values $x = 0.33$, $y = 0.33$. This is the area of Munsell colors with high value- and low chroma-components, near white. Each color page in Munsell book has this color and therefore it is natural that the competitive learning represents this highly populated area by two filters.

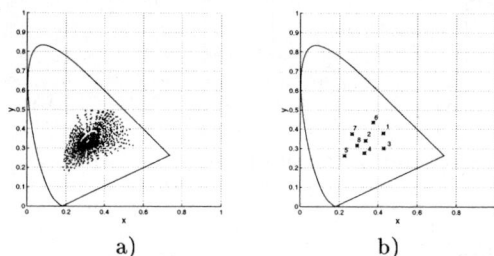

Fig. 3. CIE 1931 xy-diagram of a) Munsell-data, b) filters.

Next the learned filter set was orthogonalized by the singular value decomposition (SVD) method. The orthogonalized filters are highly correlated to the basis vectors of the subspace method shown in Fig. 1. The numerical correlations between the eigenvectors of subspace method and orthogonalized filters are 0.999, −0.974, −0.966, 0.984, −0.901, −0.551, −0.361, and −0.096, respectively. The anticorrelation in the 2nd, 3rd and 5th correlation values is not problem, since if ϕ is an eigenvector of R then also $-\phi$ is an eigenvector of R [4]. In the orthogonalized filter set, the last three filters are less correlated to the eigenvectors and they also contained some noise. The information content of these last three vectors is very small, since the fidelity value for the first five eigenvalues is 99.93%. The orthogonalized filter set was only used for showing the similarity to the eigenvectors of the subspace method and it was not used in filtering.

5.2 Filtering the Munsell Database

To obtain a detailed error analysis the Munsell database was first reconstructed by the subspace method using the basis vectors shown in Fig. 1. The reconstruction error of each Munsell spectrum is shown in Fig. 4 a). Next the database was reconstructed by the learned filters shown in Fig. 2. The reconstruction was done using generalized inverse matrix method and Fig. 4 b) shows the reconstruction error obtained. In Table 1, the correspondence between sample number and Munsell color category is tabulated.

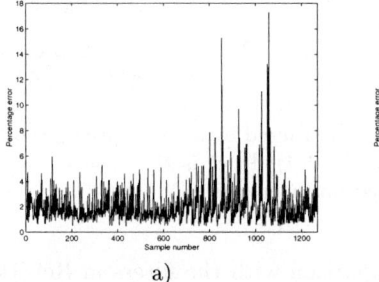
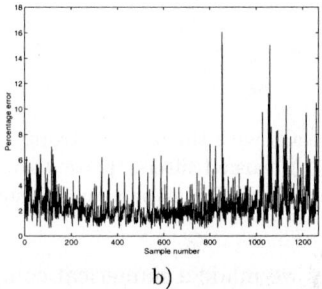

a)   b)

Fig. 4. The Munsell database reconstruction errors, using a) subspace method, b) learned filters.

Table 1. Correspondence between sample number and Munsell color category.

Sample number range	Munsell color category
1-139	Red (R)
140-261	Yellow-Red (YR)
262-404	Yellow (Y)
405-531	Green-Yellow (GY)
532-646	Green (G)
647-752	Blue-Green (BG)
753-864	Blue (B)
865-1001	Purple-Blue (PB)
1002-1132	Purple (P)
1133-1269	Red-Purple (RP)

The largest error for the learned filters (16.0%), (sample 853, blue (B), Hue 10, Value 2.5, Chroma 4), is lower than for the subspace method (17.3%), (sample 1058, purple (P), Hue 5, Value 2.5, Chroma 6). The average reconstruction error for the subspace method is 1.9% and for the learned filters 2.3%.

Fig. 4 shows that the high error peaks are repeated after every 10-20 samples. We found that the corresponding spectra had low intensity information. The learned filters don't have enough information for these low intensity spectra. In the Munsell book, every color page has this kind of spectrum and because of the sampling order in the measurement the error peaks are repeated after every 10-20 samples. Fig. 5 a) and b) shows the spectra with the largest reconstruction errors. Fig. 5 c) shows the typical spectrum for the average reconstruction error using learned filters (sample 465, Green-Yellow (GY), Hue 5, Value 8, Chroma 8).

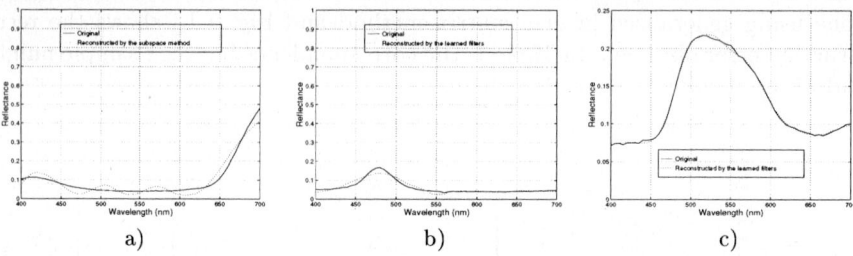

Fig. 5. Spectra with the largest errors, reconstructed by a) subspace method (plate 5P V2.5 C6), b) learned filters (plate 10B V2.5 C4). In c) is the typical spectrum with average reconstruction error using learned filters (plate 5GY V8 C8).

Finally we made a numerical comparison with the filters in Ref. [10]. To be comparable, we used the Munsell data at the wavelength area from 381 nm to 776 nm, at 5 nm intervals. We trained 6 filters and reconstructed the whole Munsell database with them. The Munsell database was also reconstructed by the filters produced in [10]. The average percentage error for our filters was 4.6% and for the filters in Ref. [10] the error was 6.1%.

6 Conclusions

In this study we presented a new unsupervised method to produce color filters for optical pattern recognition. The competitive neural network found the centers of color clusters in color space. The weight vectors from the neural network were used as color filters.

The experimental part of this paper showed that the learned filters are comparable to the basis vectors of the subspace method. Both of them span similar color spaces and the reconstruction errors with the Munsell data were small. It can be seen that the largest reconstruction errors in Fig. 4 are in the sample range 800-1269. The subspace method has largest errors in sample range 800-1050, containing the purple-blue (PB) and purple (P) colors in the Munsell color category. For the learned filters the largest errors are in the sample range 1050-1269, i.e. purple (P) and red-purple (RP) colors in the Munsell color category. Purple color spectra have usually a very flat region in the wavelength range 500-600nm and then a narrow peak or peaks in blue and red region. Both methods have problems to recover flat spectra in the range 500-600nm, as can be seen in Fig. 5. The reason is that both the basis vectors and learned filters vary in the wavelength range 500-600nm and therefore they cannot represent this flat region accurately.

For a smaller spectral region than the Munsell colors the number of filters can be decreased. In this case the number of learning cycles needed in the learning phase can also be decreased. In this study we produced filters for the Munsell spectra. The strategy of our method is general and can be used for special applications in which the filters have to be designed for a special class of spectra.

The learned filters with positive coefficients can be directly used in optical components, like in liquid crystal spatial light modulators. The optical implementation of these filters is under investigation.

Acknowledgements: M. Hauta-Kasari and W. Wang were financially supported by the Monbusho (Japanese Ministry of Education, Science, Sports and Culture). M. Hauta-Kasari was also supported by grants from Emil Aaltonen Foundation and Wihuri Foundation. He is on leave from the Department of Information Technology, Lappeenranta University of Technology, Finland. The spectral database is available at the www-server of Lappeenranta University of Technology, http://www.it.lut.fi/research/color/lutcs_database.html.

References

1. P.K. Kaiser and R.M. Boynton, *Human Color Vision, Second Edition*, Optical Society of America, Washington DC, 1996.
2. G. Wyszecki and W.S. Stiles, *Color science : concepts and methods, quantitative data and formulae*, Wiley, New York, 1982.
3. M.S. Drew and B.V. Funt, Natural Metamers, *CVGIP: Image Understanding* **56**, 1992, 139-151.
4. J.P.S. Parkkinen, J. Hallikainen, and T. Jaaskelainen, Characteristic spectra of Munsell colors, *J. Opt. Soc. Am. A.* **6**, 1989, 318-322.
5. T. Jaaskelainen, J. Parkkinen, and S. Toyooka, Vector-subspace model for color representation, *J. Opt. Soc. Am. A.* **7**, 1990, 725-730.
6. *Munsell Book of Color, Matte Finish Collection*, Munsell Color, Baltimore, USA, (1976).
7. S. Usui, S. Nakauchi, and M. Nakano, Reconstruction of Munsell color space by a five-layer neural network, *J. Opt. Soc. Am. A.* **9**, 1992, 516-520.
8. T. Jaaskelainen, S. Toyooka, S. Izawa, and H. Kadono, Color classification by vector subspace method and its optical implementation using liquid crystal spatial light modulator, *Optics Communications* **89**, 1992, 23-29.
9. N. Hayasaka, S. Toyooka, and T. Jaaskelainen, Iterative feedback method to make a spatial filter on a liquid crystal spatial light modulator for 2D spectroscopic pattern recognition, *Optics Communications* **119**, 1995, 643-651.
10. R. Lenz, M. Österberg, J. Hiltunen, T. Jaaskelainen, and J. Parkkinen, Unsupervised filtering of color spectra, *J. Opt. Soc. Am. A.* **13**, 1996, 1315-1324.
11. M. Österberg, *Quality Functions for Parallel Selective Principal Component analysis*, Ph.D. dissertation, Linköping University, Linköping, Sweden, 1994.
12. E. Oja, *Subspace Methods of Pattern Recognition*, Research Studies Press, Letchworth, England, 1983.
13. S. Grossberg, *Studies of the Mind and Brain*, Reidel Press, Drodrecht, Holland, 1982.
14. T. Kohonen, *The Self-Organizing Maps*, Springer-Verlag, Berlin Heidelberg, 1995.
15. S. Haykin, *Neural Networks*, Macmillan College Publishing Company, New York, 1994.

Foveated Vision for Scene Exploration

Naoki Oshiro[1], Atsushi Nishikawa[2], Noriaki Maru[3] and Fumio Miyazaki[2]

[1] University of the Ryukyus, 1 Senbaru, Nishihara-cho, Okinawa 903-01, JAPAN
[2] Osaka University, 1-3 Machikaneyama-cho, Toyonaka city, Osaka 560, JAPAN
[3] Wakayama University, 930 Sakaedani, Wakayama 640, JAPAN

Abstract. In this paper, foveated vision for scene exploration is implemented. The peripheral and central vision are the basic capabilities of foveated vision. The informations obtained from the peripheral vision are used to determine the next gaze point. Due to the low resolution of the periphery, however, the determination is not always appropriate. To solve this problem, we propose to evaluate the target object by the central vision after gazing. We implement foveated vision based on the Log Polar Mapping (LPM) and construct an evaluation scheme of the target object in the central vision using LPM rotational-invariance. The peripheral vision is realized by Zero Disparity Filter for LPM stereo images. Some experimental results are also shown to demonstrate the effectiveness of the proposed method.

1 Introduction

A human possesses foveated vision and the retina has a non-uniform structure. The retina's resolution is high in the central area and this resolution continuously drops as you go into the peripheral area. As detailed information can be obtained only in the central area (*central vision*), in order to search in the scene, gaze process becomes necessary. Meanwhile, the information from the peripheral area is used to determine the next gaze point (*peripheral vision*). That is, using the peripheral vision, we determine the gaze point and the target object will be analyzed based on the central vision. These are indispensable elements once foveated vision is used, and both coordinated movements are required.

In light of this, a method for scene exploration by foveated vision is discussed. In the previous works, we have proposed a method for using Log Polar Mapping (LPM), which is a good approximation of the non-uniform structure of the retina, to extract and track a gaze object by the vision system [4]. In this paper, we implement foveated vision based on the LPM and apply to scene exploration.

Our scene exploration algorithm by foveated vision consists of two parts: (i) feature extraction process (*peripheral vision*) and (ii) feature evaluation process (*central vision*). We first detect rapid changes in image intensity levels as features. Then, the features in the neighborhood of the gaze point are picked up from them; this is realized by Zero Disparity Filter for LPM stereo images [4]. These features are selected as the candidate of the next gaze point. Due to the low resolution of the information obtained from the peripheral vision, however, the next gaze point is not always appropriate. To solve this problem, we propose to evaluate the feature after gazing. We construct this evaluation scheme of the target object by making use of the LPM rotational-invariance. As a result, each point selected as the next gaze point through peripheral vision is evaluated as being suitable or not for fixation. Some experimental results are also shown to demonstrate the effectiveness of the proposed method.

2 Log Polar Mapping in Our Implementation

In primate visual systems, the retina has a space variant structure; the resolution is high at the central region and becomes decreasingly low toward the periphery. The mapping of the retina into the visual cortex can be approximately represented by "Log Polar Mapping" (henceforth LPM) [7]. Formulation of LPM as follows:

$$w = \ln(z), \quad z = x + iy, \quad w = u + iv \quad (1)$$

where z is a point on the original image plane Z, and w is a point on the LPM image plane W. Denoting the image resolution of Z-plane by (w, h), and that of W-plane by $(u_{\max}, v_{\max})$ respectively, the transformation from $z = (x, y)$ to $w = (u, v)$ is represented by

$$u = u_{\max} \times \left(\frac{\rho - \rho_{\min}}{\rho_{\max} - \rho_{\min}} \right) = u_{\max} \times \frac{\rho}{\rho_{\max}}, \quad v = v_{\max} \times \frac{\mathrm{atan2}(y, x)}{2\pi} \quad (2)$$

where $\rho_{\max} = \ln(\max(w, h)/2), \rho_{\min} = \ln(1) = 0, \rho = \ln((x^2 + y^2)^{\frac{1}{2}})$. In this paper, LPM is performed in software (look up table form) [4], [8]:

$$w = \mathrm{LUT}(z) \quad (3)$$

where LUT means the look up table for the original image $Z(x, y)$.

LPM has the following characteristics: (i) space-variant structure (ii) functional vision (iii) precessing data reduction. Mathematically, LPM is (iv) rotation-invariant and (v) scale-invariant, so it is useful for pattern recognition [2]. Notice that these two properties ((iv) and (v)) are satisfied only in case that the projection of the target object is at/near the image center. Thus, if we want to apply these properties to the target object, we must gaze it. In this paper, we propose a "scene exploration" method by making use of these characteristics. We find a target object by using the functional vision of LPM (*peripheral vision*) and gaze the object (*central vision*). Furthermore, we evaluate the uniqueness of the gaze object by utilizing the rotation-invariance of LPM.

Next, we present the determination method of the border between the fovea and the periphery. Eq. 3 represents many-to-one correspondence. Therefore, when doing the LPM transformation (Eq. 2) by Eq. 3, sparsely transformed region ($u < u_{\mathrm{fovea}}$) exists in the fovea. To overcome this problem, the following methods have been proposed: (a) Separately generate the fovea [9], (b) Fill up the value obtained by the inverse LPM, (c) Formulate LPM as $w = \ln(z + \alpha)$ [8] where α is the singularity avoidance parameter. We cope with this by only using the peripheral LPM image data. To do this, we determine the border between the fovea and the periphery as follows (see Fig. 1):

1. Initially, we decide the resolution of Z-plane, and that of W-plane.
2. The minimal number of partitions in the rotational direction of the Z plane corresponds to $v_{\max}$, we have Therefore, letting r_{fovea} be the radius such that this partition are individually done, we have $r_{\text{fovea}} \approx v_{\max} \times \sqrt{2}/8$. This means that the diagonal length of the square whose peripheral length is $v_{\max}$ is regarded as the diameter of the circle.
3. Denoting the border line between the fovea and the periphery on the LPM image by $u = u_{\text{fovea}}$, we obtain

$$u_{\text{fovea}} = \text{int}\left(\frac{1}{u_{\max}} \times \frac{\ln(r_{\text{fovea}})}{\ln(w/2)}\right) \qquad (4)$$

where "int" indicates the rounding operator.

We only use the LPM image region such that $u \geq u_{\text{fovea}}$.

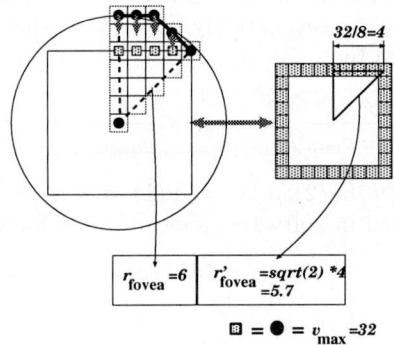

Fig. 1. Determination of the border between the fovea and the periphery

3 Peripheral Vision for Scene Exploration: Feature Extraction

In this section, we explain the function of *peripheral vision* for scene exploration. This corresponds to the feature extraction process for selecting the next gaze.

3.1 Periphery Filtering and Feature Extraction

The extracting method of scene features in our system based on the Zero Disparity Filter (ZDF) [3] which is a well-known method of using a binocular visual system to extract objects on the zero disparity area (*horopter*). The practical problem in applying the ZDF method is how to estimate binocular disparities.

To do this, a few methods have been proposed [1], [6], [10] . However, these are only applicable to uniform sampling images.

In our previous work, we have developed the LPM-ZDF method [4], which is the ZDF method applicable to LPM images. The interest points in the LPM-ZDF method are as follows: (i) stereo correspondence errors are reduced, (ii) the horopter of LPM spreads from the gaze point toward the periphery. (iii) wide and multiple disparity estimators can be worked in real-time. Details of the LPM-ZDF method can be found in [4].

We utilize the LPM-ZDF method to extract features from the periphery of LPM stereo images. It is summarized as follows: (i) Consider that stereo cameras gaze at a given point, (ii) Edges of the features located on the zero disparity area are picked up from LPM images by using the LPM-ZDF method, (iii) Continuous edge region are detected as the candidates of the next gaze point. They are selected in order of the amount of their area in terms of the pre-transformed original image.

3.2 Related Works

Generally speaking, gaze control algorithms are classified into two types: (i) context-dependent [5] and (ii) context-free. The latter uses general image features. For example, Yamamoto et al. [9] used an interest operator to detect geometrical image features such as corner or symmetry for selecting a sequence of gaze points. Our method also falls into the latter one. The major difference between Yamamoto et al.'s method and ours is that we only detect rapid changes in image intensity levels as features.

4 Central Vision for Scene Exploration: Feature Evaluation

In this section, we present the function of *central vision* for scene exploration. In our system, this corresponds to the feature evaluation process based on the rotational-invariance of LPM images.

The reasons of feature evaluation need as follows: we can select the next gaze point from the feature extraction results described in Section 3. However, there are the following problems: (a) the *ambiguity* of the extracted feature description due to the low resolution of the periphery (b) the *uncertainty* of the extraction results due to the same reason This means that we cannot always select the "appropriate" gaze point by the peripheral vision function. In foveated vision, the high resolution can be automatically realized by gazing process (*central vision*). Therefore, it is very reasonable to evaluate the feature object after gazing.

4.1 Feature Evaluation using Rotational-Invariance of LPM

We evaluate features by utilizing the rotational-invariance of LPM images. At first, we calculate the autocorrelation value for each image obtained by rotating the LPM image as follows:

1. Calculate the mean intensity μ:

$$\mu = \frac{1}{n} \sum_{u=u_{\text{fovea}}}^{u_{\max}-1} \sum_{v=0}^{v_{\max}-1} L_{u,v}, \quad n = (u_{\max} - u_{\text{fovea}}) \times v_{\max}. \quad (5)$$

2. Calculate the autocorrelation value c_i:

$$c_i = \sum_{u=u_{\text{fovea}}}^{u_{\max}-1} \sum_{v=0}^{v_{\max}-1} (L_{u,v} - \mu) \times (L_{u,v-i} - \mu) \quad (i = 0, 1, \ldots, c_{\max}-1), \quad (6)$$

where $c_{\max} = v_{\max}/2$, $L_{u,v}$ is the intensity value at $W(u,v)$ on the LPM image plane.

3. Normalize c_i by the maximal value $\max_i(c_i)$.

Then, we can judge the *uniqueness* of the feature by using the resulting values $\{c_i\}(i = 0, 1, \ldots, c_{\max} - 1)$. That is, the more biased the distribution of these autocorrelation values are, the more *unique* in the scene the feature on the LPM image is. Notice that the value c_0 is always 1 and the other values c_i are zero if the feature is the most *unique*. Thus, we use the sum of the autocorrelation values, denoted by S_c, as the criterion for *uniqueness*.

$$S_c = \frac{1}{c_{\max}} \sum_{i=0}^{c_{\max}-1} c_i \quad (7)$$

The minimal value of S_c is $1/c_{\max}$. The less *unique*, the greater S_c.

Figure 2 shows some examples. Each test images, we calculated their autocorrelation values (graphs in Fig. 2) and the criterion for *uniqueness* S_c. Original image size $(w, h) = (256, 256)$, and LPM image size $(u_{\max}, v_{\max}) = (34, 64)$ in these examples. (a) cross: the autocorrelation values take the shape with periodic iteration in the rotational direction. It can be regarded as the *non-unique* feature. (b) ellipse: almost all the autocorrelation values are near 1. It can be also regarded as the *non-unique* feature. (c) mnote: the autocorrelation values relatively take the shape of impulse. It is considered as the *unique* feature.

5 Experiment

5.1 System Configurations and Experimental Setup

The cameras tilt up and down together on a common platform, and pan independently from side to side, driven by three motors: one for the tilt platform and twin pan motors. Besides the cameras are mounted on a head which pans and slides independently, driven by two motors: one for the pan of the head and one slide motor. In the experiment, we use and control the three pan motors (eyes and neck) and tilt motors.

Stereo images sequences taken by the two CCD cameras are digitized into 512×480 pixel, 8 bit gray scale images and processed by the host computer (Sun

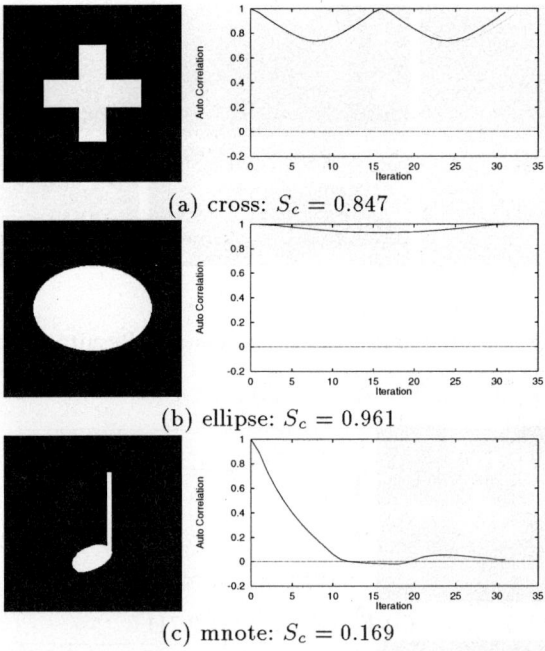

Fig. 2. Results of Feature Evaluation (Test Images)

SPARC station 10). The baseline length is 120 (mm) and focal length is 4.8 (mm) respectively. Details in relation to "Active stereo vision system" (motor control system, performance specifications etc.) can be found in [4].

In experimental setup, we assume the black background for simplicity. Initial gaze point was set right in front of the Vision Robot at a distance of 1000 mm, and three objects were put in the vicinity of the gaze point. Figure 3 shows the stereo images obtained by initial gazing process. Original image size $(w, h) = (256, 256)$, LPM image size $(u_{max}, v_{max}) = (34, 64)$ in this experiment.

5.2 Results

The proposed method has been applied to the actual scene. Figure 3 shows the feature extraction result; '+' indicates the feature points. We can see that all of the objects are appropriately picked up from the scene. As a result, three region (numbered as 1, 2, 3) were selected as the candidate of the next gaze point.

Figure 4 shows the inverse LPM images obtained after gazing at the selected feature region, in order of (a), (b), (c). The autocorrelation values (graphs in Fig. 4) and the criterion of *uniqueness* S_c are also shown. From these results, the feature region 1 (corresponding to Fig. 4(a)) is considered as the most *unique* feature.

Fig. 3. Stereo Images of after Initial Gaze Process and Results of Feature Extraction by Peripheral Vision

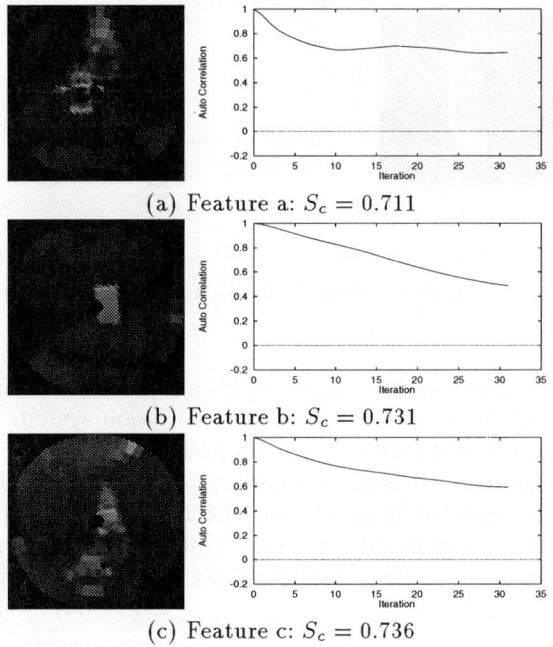

(a) Feature a: $S_c = 0.711$

(b) Feature b: $S_c = 0.731$

(c) Feature c: $S_c = 0.736$

Fig. 4. Results of Feature Evaluation by Central Vision

6 Conclusion

In this paper, we have proposed a method to realize "scene exploration" with an active stereo foveated vision system by making use of functional vision (*peripheral and central vision*). The foveated vision system was implemented in software by using LPM. Our exploration algorithm first detects the feature points based on the peripheral vision and evaluates them based on the central vision. Peripheral vision were implemented by the LPM-ZDF method that extracts features in the neighborhood of the gaze point. From this, it becomes possible to selectively extract from three dimensional space. By using the rotational-invariance of the LPM image, we calculated the autocorrelation value. As a result, we evaluated the feature by looking at the uniqueness in the rotation of the image taken from the central vision after gazing. In addition, we have shown the effectiveness of our method by the experiment.

Using the functional vision, the vision implementation is divided into two parts: where and what dichotomy. This process is important for solving the problems in the computer vision field such as reducing the amount of process data and process time. By making use of the space variant structure of foveated vision, we proposed a new evaluation method of the gaze object. This method can be used for object tracking problems.

Using stereo cameras, it is possible to selectively extract the feature object in three dimensional space, with respect to the distance. However, the selectiveness that is used is limited to only the process in the peripheral vision, and it is not used in the central vision. Since the relationship with the background can be considered to be a large influence, it is probably efficient to use the selectiveness in this process when evaluating the feature.

References

1. C. Capurro, F. Panerai and G. Sandini, "Dynamic Vergence", *IROS'96*, Vol. 3, pp. 1241–1248, 1996.
2. L. Massone, G. Sandini and V. Tagliasco, "'Form-Invariant' Topological mapping Strategy for 2D Shape Recognition", *Computer Vision, Graphics and Image Processing*, No. 30, pp. 169–188, 1985.
3. T. J. Olson and D. J. Coombs, "Real-Time Vergence Control for Binocular Robots ", *Int. Jounal of Computer Vision*, Vol. 7, No. 1, pp. 67–89, 1991.
4. N. Oshiro, N. Maru, A. Nishikawa and F. Miyazaki, "Binocular Tracking using Log Polar Mapping", *IROS'96*, Vol. 2, pp. 791–798, 1996.
5. R. D. Rimey and C. M.Brown, "Controlling Eye Movements with Hidden Markov Models", *Int. Journal of Computer Vision*, Vol. 7, No. 1, pp. 47–65, 1991.
6. S. Rougeaux, N. Kita, Y. Kuniyoshi, S. Sakane and F. Chavand, "Binocular Tracking Based on Virtual Horopters", *IROS'94*, Vol. 3, pp. 2052–2057, 1994.
7. E. L. Schwartz, "Computational Anatomy and Functional Architecture of Striate Cortex: a Spatial Mapping Approach to Perceptual Coding", *Vision Research*, Vol. 20, pp. 645–669, 1980.
8. R. Wallace, R. W. Ong, B. Bederson and E. Schwartz, "Space Variant Image Processing", *Int. Journal of Computer Vision*, Vol. 13, No. 1, pp. 71–90, 1994.
9. H. Yamamoto, Y. Yeshurun and M. D. Levine, "An Active Foveated Vision System: Attentional Mechanisms and Scan Path Covergence Measures", *Computer Vision and Image Understanding*, Vol. 63, No. 1, pp. 50–65, 1996.
10. Y. Yeshurun, E. L. Schwartz, "Cepstral Filtering on a Columnar Image Architecture: A Fast Algorithm for Binocular Stereo Segmentation", *IEEE Trans. on Pattern Analysis and Machine Intelligence*, Vol. 11, No. 7, pp. 759–767, 1989.

Evolutionary Methods Applied to Binocular Disparity Estimation

Carla L. Pagliari and Tim J. Dennis
Department of Electronic Systems Engineering
University of Essex
Wivenhoe Park, Colchester, CO4 3SQ, United Kingdom
clpagl@essex.ac.uk and tim@essex.ac.uk

Abstract. One problem on binocular disparity estimation is, given a point in one view of a scene, to find the homologous point in another view from an adjacent camera. In this work, we use the discrete cosine transform (DCT) with evolutionary methods as an approach to obtain the disparity map. The disparity estimation process is implemented by generating for each image block sets of DCT coefficients. The job of optimising the DCT coefficients, whose inverse transform gives the disparity map, is carried out by a optimisation technique inspired by natural evolution: the genetic algorithm. Matching is performed in the image domain using an intensity similarity measure.

1. Introduction

Most current applications of stereo image analysis lie in scene analysis for virtual reality and video-conferencing. Stereo algorithms aim to calculate the set of correct correspondent points between, at least, two images. The computation of the disparities is not a straightforward problem because alongside the small shift of image regions, many other variations may occur between the two views.

Genetic algorithms (GAs) were introduced by a research group leaded by John Holland [Holl75] at the University of Michigan. The mechanisms of natural evolution were duplicated in software for the solution of otherwise-intractable search and optimisation problems.

In this paper we discuss the use of GAs to guide a search in the DCT domain aiming to find the best set of DCT coefficients whose inverse discrete cosine transform (IDCT) will be the disparity values. GAs have been used elsewhere to obtain a disparity map [Vail94, Sait95], but using different approaches.

The paper starts with a brief review of the main stereo matching techniques followed by an introduction to GAs. In Section 4, related work is shown, while Section 5 describes the method itself and its use of the DCT as well as the adaptation of a GA to fit the desired goals. Experiments are described in section 6 and Section 7 gives the conclusions of the results obtained.

2. Stereo Matching Techniques

Classical methods of stereo matching lie within two major categories: area-based and feature-based [Barn82]. The area-based method uses a similarity criterion to obtain optimal statistical correlation between corresponding regions in a stereo image pair. The disparity maps produced by this method are dense. In contrast, feature-based stereo algorithms convert the intensity values to a set of features, thus obtaining depth at edges and corners which are generally sparse.

The phase- or frequency-based methods [Sang88, Jenk94] are an alternative approach for calculating binocular disparities. Image correspondence is carried out in the frequency domain by deriving Fourier-phase images from the raw intensity data or convolving the images with suitable band-pass responses like the Gabor and hypergeometric filters.

3. Genetic Algorithms

The genetic algorithm (GA) is a computer simulation based on the principle of Darwinian natural evolution whose goal is to identify the maximum/minimum of some objective function in a search space. A *population* of individual solutions, also known as *chromosomes*, of the space is constructed. A chromosome represents a proposed solution to the problem being solved and comprises a string of bits that may encode real numbers, integers or anything else that is relevant to the problem.

The simple GA uses three stochastic operators: *selection*, *crossover* and *mutation*, that are applied to the population, and successive *generations* are produced. The first operator ensures that the fitter individuals will have a higher probability to be selected and survive to breed the next generation. Crossover takes two chromosomes amongst those already selected to create a new one. This operator exchanges (or combines) regions of two selected chromosomes, therefore mixing the genetic material. Mutation may also be applied by randomly altering pieces of the chromosomes with the aim of maintaining diversity in the population.

In general, the search for an optimal solution begins with a randomly-generated population of chromosomes. Each generation will have a new set of chromosomes obtained from the application of the operators. To evaluate the quality of the individuals a *fitness*, or objective function, is defined according to the problem. At the same time, some lower fitness members are also selected in order to maintain diversity.

The algorithm terminates when a stop criterion is reached. More detail on GAs can be found in Goldberg [Gold89] and Davis [Davi91].

4. Related Work

Classical area-based disparity estimation methods maximise the correlation between windows of the left and right views. Kanade and Okutomi [Kana94] presented a method that selects adaptively the window size according to local variations of intensity and disparity. Saito and Mori [Sait95] proposed a method, based on the idea of [Kana94], to obtain the optimal window size by previously evaluating stereo pairs for different window sizes and generating disparity maps for all those window sizes. The values from this disparity maps space which optimised the evaluation were selected using GAs. The computation of the correlation measure and the subsequent maximisation of those values is an expensive computational process, since extensive search is required to build the configuration of the disparity maps space.

Both approaches tackled the problem of the usage of the appropriate window size at each position of the stereo pair in order to avoid the effects of geometric distortions and also to cover enough intensity variation [Barn82].

5. Describing the Algorithm

The idea is to make an initial estimate of the transformed components of the disparity map and then to use the GA to look for an optimal solution. As the DCT concentrates energy of a highly-correlated source [Jain89], not all the components have to be used in order to reproduce an accurate estimate of the disparity map. Another positive aspect of working in the DCT domain is that the coefficients are highly uncorrelated hence, one can be adjusted without disrupting the others.

A disparity map is a set of disparity vectors defined with respect to the spatial grid of points in one of the stereo images. Each disparity value is considered as a random variable. Considering that a disparity map is assumed to be highly correlated inside a delimited area (a window or a block), the DCT was used in order to estimate an initial disparity map.

Genetic algorithms (GAs) are a class of search technique that can handle large and possibly non-linear search spaces. Moreover, GAs are easy to parallelise and so far have achieved good results when applied to different classes of problems [Gold87].

Once an initial estimate of the disparity map is made, the job of the GA is to adjust the DCT coefficients whose IDCT will be the disparity map. For each block of the image, the technique generates a pseudorandomly initialised population of blocks of DCT coefficients. A desirable feature in this approach is to use large blocks in the DCT domain whenever possible, in order to embrace the different needs presented in Section 4.

5.1 Initial Block Generation

The estimated disparity map DCT is divided into i blocks $i = 0,1,...,Nb$, size $M \times M$ (Figure 1).

Then, an area-based block matching using an intensity similarity measure is applied to obtain roughly the disparities for each of the i blocks. These block disparities plus a Gaussian variance will be the DC components of the initial DCT coefficient disparity blocks.

The AC components are generated almost randomly. In fact, we assume that they have a Laplacian distribution over a block on the DCT domain [Smoo96]. Initial values of the estimated standard deviations of the AC components for the blocks are randomly generated and used in a transformation method to generate a Laplacian distribution for the AC coefficients [Press94].

5.2 The selection of the AC coefficients

The DCT energy compaction property offers the possibility of not using all the AC components within a block. We expect disparity always to be at least as or more highly correlated than the image itself, since a disparity (=depth) change will nearly always produce a luminance difference, while the converse is not necessarily true: consider a flat surface decorated with a detailed pattern. We use the idea of the quantisation matrices of the JPEG method [Wall91] to choose which AC coefficients to retain. This reduce selection speeds up the algorithm.

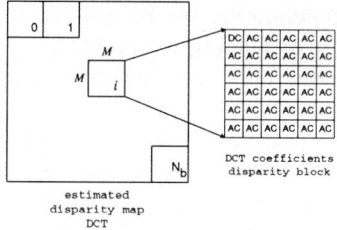

Figure 1. Estimated disparity map DCT and DCT coefficients disparity block.

5.3 Tuning the GA

The GA should be adapted to the specific characteristics of the problem. The following sub-sections show the encoding of the DCT coefficients as chromosomes, the definition of a set of genetic operators and the fitness function used to evaluate the performance of the individuals of the population.

Encoding

A problem-specific representation of the chromosome is used in which floating-point numbers represent the values of the DCT coefficients of each block (see Figure 2). Each transformed block is converted into a chromosome using a zig-zag scanning sequence [Wall91] in order to maintain approximate contiguity of the lower order DCT coefficients.

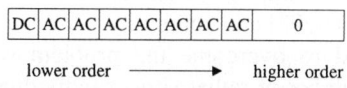

Figure 2. Chromosome representation.

Genetic Operators

Instead of generational replacement, we use the steady state technique in which only a few members of the population change. The selection is done using the stochastic universal sampling technique [Bake87]. The reproduction system guarantees that high-fitness individuals will survive for long periods, contributing their offspring to the population.

The crossover operator is useful because it explores promising areas in the search space, and is also responsible for fast convergence [Mich92]. We use a combination of arithmetical crossover and two-point crossover, in which we take two chromosomes and produce two children which are a linear combination of their parents.

The motivation behind arithmetical crossover is that if a chromosome presents some good values for its genes, it is possible for a linear combination of these values to do even better.

Random mutation is used in this GA. It substitutes one value for another chosen from a uniform random distribution of specified range, helping to maintain diversity in the population.

The Fitness Function

While programming the genetic algorithm itself is relatively straightforward, the most difficult and critical element is nearly always the design of the fitness function.

As the only window the population has on the outside 'world' it must reliably select 'good' individuals in the sense dictated by the problem.

The fitness function is based on a combination of two components: a correlation measure on disparity and disparity smoothness. The correlation measure is the zero mean sum of the squared differences (ZSSD). It shows good performance under uneven lightning conditions [Asch92]. The second component is the Euclidean distance between the disparity values at adjacent points in disparity space, thus invoking the continuity constraint [Faug93]. Therefore, the first component of the fitness function is given by:

$$\varepsilon(X, X+dx) = \sqrt{\frac{\sum_{\Psi \in W}((I_L(X+\Psi) - \bar{I}_L(X)) - (I_R(X+\Psi+dx) - \bar{I}_R(X+dx))^2}{(2N+1)(2M+1)}} \quad (1)$$

where, ε is the error (ZSSD), or fitness function representing the compatibility between corresponding points and continuity of the disparity maps, $X = (x, y)$ is the coordinate pair of each pel, $\bar{I}_L$ and $\bar{I}_R$ are the intensities of a pel on the left and right images respectively, $\Psi = (k, l)$, $\Psi \in W$, W is the window defined by $\{(k,l) | -N \leq k \leq N, M \leq l \leq M\}$, $\bar{I}_L$ and $\bar{I}_R$ are the mean values of intensity over a window W of the left and right images respectively, and dx is the horizontal disparity value.

The chromosomes with a small error are those with high levels of compatibility (photometric constraint) [Faug93] and continuity.

Fitness scaling is used to overcome the problem which often occurs when a population consists of a number of rather similar individuals with nearly equal fitness: effectively we stretch the fitness range at the higher levels, which helps prevent premature convergence. We use the sigma truncation formula [Gold89] to modify the fitness values.

5.4 The GA in Operation

The initial estimated transform coefficients of the disparity map of $N \times N$ images are the N_b blocks (Figure 1). A population of disparity blocks each with j DCT coefficients is generated. It is started by the block generation described in Section 5.1. Then, for each block i of the estimated disparity map DCT a population of j blocks is built. Next the i blocks are converted into chromosomes using a zig-zag sequence in order to ensure that the most significant (low order) AC coefficients will be together in order to confine actively initialised coefficients to DC and selected AC coefficients only. The remainder are set to zero. A maximum number of generations is chosen and the selection, crossover and mutation operators applied without disrupting the chromosome structure, since crossover occurs at coefficient boundaries only. Before each evaluation by the fitness function represented by (Eq. 1), the chromosome is reconverted to the block shape and the IDCT is applied. The fitness function is applied in the 'disparity domain', and iterations continue until the maximum number of generations is reached. This process is repeated for each block i of the estimated disparity map DCT (Figure 1). The points that are occluded (two-way matching is performed) or are considered to be unreliable (below a determined threshold level) will be marked as unmatched, and the others as matched. This information will be used further on by the interpolation process (Section 5.5).

5.5 Interpolation

The final stage is the use of a weighted least squares fitting with low order orthogonal polynomial functions as the interpolation process, [Wolb91]. The method takes the matched pels and uses them as seed values to extend the disparity estimate over the remaining unmatched pels by locally fitting polynomials of variable order.

6. Experiments

We show here the results of the method applied to a synthesised stereo image pair 'Wave' (Figure 3(a)), and to a real stereo image pair 'Pentagon' (Figure 4(a)); the latter obtained from: ftp://mindseye.berkeley.edu/pub/stereo.

This work assumes that any three-dimensional point is projected in the same image row, which can be seen as an simplification of the epipolar constraint [Faug93]. It is a reasonable assumption for near-parallel camera axes.

'Wave' (32×32) is a synthesised stereo pair generated by Gaussian noise and blurred. To create the disparity, a sine wave displacement was imposed on the right view. The algorithm finds the disparities (Figure 3(b)) using a 32×32 block size, which is the same size of the stereo image pair. It is also true that the disparity is smooth all over the image.

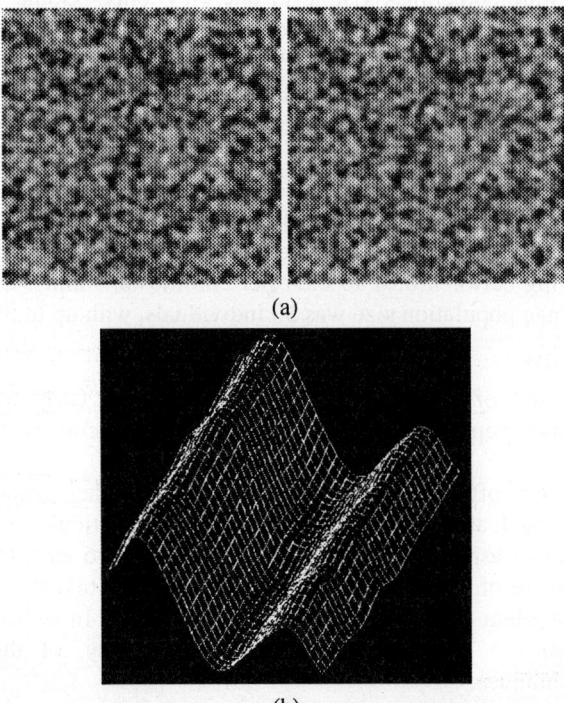

Figure 3. (a) 'Wave' stereo pair, (b) reconstructed disparity surface.

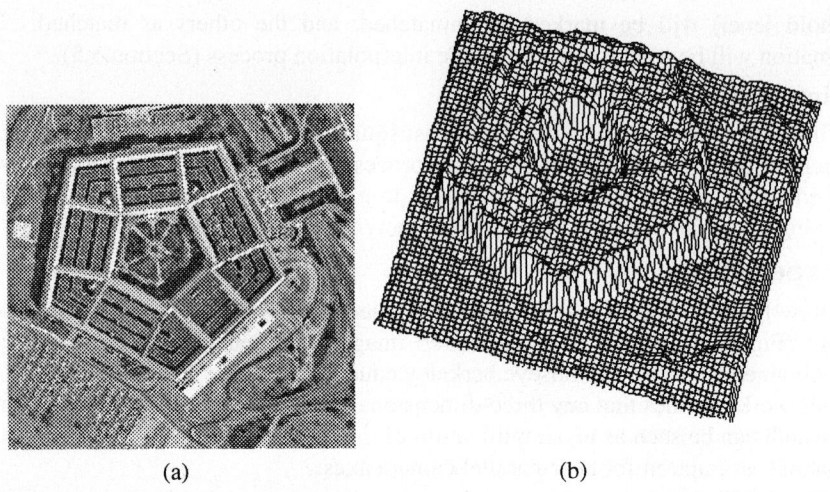

Figure 4. (a) Left image of the 'Pentagon' stereo pair, (b) reconstructed disparity surface.

The general shape of the 'Pentagon' stereo pair (512×512) is easily recovered including the bridge (Figure 4(b)). However, getting the finer detail is less straightforward. The depths between the concentric rings which make up the 'Pentagon' walls are sometimes visible in one view only. Some walls present high disparity gradient, and are present in just one image. As the pair has many fine details we use block size 8×8 to obtain the disparities.

The computational cost increases with the size of the DCT block as well as with the size of population and the number of generations chosen. However, a parallel implementation of the GA will reduce the computational cost of the method.

This method is using, initially, an algorithm to perform DCT and IDCT that accepts all block sizes even if is not as fast as the dedicated implementations.

The crossover probability in all simulations was set to 100%. A different mutation probability ranging between 0.01 to 0.04 per chromosome unit was assigned to each run and the average population size was 50 individuals, with up to 5000 generations.

7. Conclusions

The GA comprised of 'evolutionary' operators is able to infer, from a pseudo-randomly generated population, DCT components that generate the expected disparity map.

The method also offers a great deal of flexibility. Other genetic or evolutionary operators can be used, as well as other types of fitness functions.

Another positive aspect was the use of real numbers to encode the solution and, subsequently the use of real number operators which were revealed to be closer to the domain of the problem than bitstring operators [Gold89]. In addition, the use of real numbers simplifies a hybridisation, whenever necessary, of the GA with local optimisation techniques.

8. Acknowledgements

The first author acknowledges the support from IME and IPD, Ministério do Exército, Brazil, and from Conselho Nacional Científico e Tecnológico (CNPq), Brazil, under grant no. 201903/93-8.

9. References

[Asch92] Aschwanden and W. Guggenbühl, "Experimental results from a comparative study on correlation-type registration algorithms". In Förstner and Rudwiedel (eds.), Robust Computer Vision, 268-282, Wichmann, 1992.

[Barn82] S. T. Barnard and M. A. Fischler, "Computational Stereo", *ACM Computing Surveys*, 14(4), Dec, 1982, pp. 553-572.

[Davi91] L. Davis, *Handbook of Genetic Algorithms*, Van Nostrand Reinhold, 1991.

[Faug93] O. Faugeras, *Three-Dimensional Computer Vision: A Geometric Viewpoint*, MIT Press, Cambridge, Massachusetts, 1993.

[Gold87] D. E. Goldberg, "Genetic algorithms with sharing for multimodal function optimization", *Proc. of the Second Int. Conf. on Genetic Algorithms*, 1987, pp. 41-49.

[Gold89] D. E. Goldberg, *Genetic Algorithms in search, optimization and machine learning*, Addison-Wesley, 1989.

[Holl75] J. H. Holland, *Adaptation in Natural and Artificial Systems*, University of Michigan Press, Ann Arbor, 1975.

[Jain89] A. K. Jain, *Fundamentals of Digital Image Processing*, Prentice-Hall International, 1989.

[Jenk94] M. R. M. Jenkin and A. D. Jepson, "Recovering Local Surface Structure through Local Phase Difference Methods", *CVGIP*, vol. 59, 1994, pp. 72-93.

[Kana94] T. Kanade and M. Okutomi, "A stereo matching algorithm with an adaptive window: theory and experiment", *IEEE Trans. on Pattern Analysis and Machine Intelligence*, PAMI-16 (9), 1994, pp. 920-932.

[Mich92] Z. Michalewicz, *Genetic Algorithms + data Structures = Evolution Programs*, Springer Verlag, Berlin, 1992.

[Press94] W. H. Press et al., *Numerical Recipes in C*, Cambridge University Press, 1994.

[Sang88] T. D. Sanger, "Stereo disparity computation using Gabor filters", *Bio. Cybern.*, vol.59, 1988, pp. 405-418.

[Sait95] H. Saito and M. Mori, "Application of genetic algorithms to stereo matching of images", *Pattern Recognition Letters*, vol. 16, 1995, pp. 815-821.

[Smoo96] S. R. Smoot and L. A. Rowe, "Study of DCT coefficient distributions", Computer Science Division, University of California at Berkeley, Jan, 1996.

[Vail94] R. Vaillant and L. Gueguen, "Genetic Algorithms applied to Binocular Stereovision", *Lecture Notes in Computer Science,* Computer Vision -ECCV '94, vol. 801, pp. 193-198.

[Wall91] G. K. Wallace, "The JPEG Still-Picture Compress Standard", *Communications of the ACM*, Apr, 1991, vol. 34, no. 4, pp. 30.

[Wolb91] G. Wolberg, *Digital image warping*, IEEE Computer Society Press, Los Alamitos, CA, 1990.

Robust Epipolar Geometry Using Genetic Algorithm

Jinxiang Chai and SongDe Ma

National Laboratory of Pattern Recognition, Institute of Automation
Chinese Academy of Sciences Beijing, 100080, P.R.China
Email : chaij@prlsun3.ia.ac.cn, masd@prlsun2.ia.ac.cn

Abstract

Epipolar geometry is an important constraint to establish the correspondences in stereo vision. The 3 × 3 fundamental matrix describes the epipolar geometry between two uncalibrated images. In this paper, we formulate the epipolar geometry estimation as a global optimization probelm, and then we present a genetic algorithm for parameter searching. Experiments with simulated and real data show that our algorithm performs very well in terms of robustness to outliers, rate of convergence and quality of the final estimation.

Keywords: Fundamental matrix, Epipolar geometry, Global Optimization, Genetic algorithm.

1 Introduction

Matching different views of a single scene remains one of the most difficult problems in computer vision. Two images of the same scene, taken from two different viewpoints, are related by the so-called *epipolar geometry*. AS is known, epipolar geometry is the only available constraint information that could be gained from two uncalibtrated images. Recently, the applications of the epipolar geometry in computer vision cover $3D$ reconstruction(Faugeras, 1992; Shashua, 1993), camera self-calibration(Luong and Faugeras, 1992; Maybank and Faugeras, 1992), stereo analysis(Hartley and Gupta, 1993; Robert and Hebert, 1993), motion segmetation(Nishimura *et al*, 1990; Sinclair *et al*, 1994), image synthesis(Laveau and Faugeras, 1994), *etc*.

The 3 × 3 fundamental matrix **F**, which describes the epipolar geometry between two uncalibrated images, can be estimated by a certain number of corresponding points. A great deal of literature exists on this subject(Longuet-Higgins, 1981; Zhang *et al*, 1994; Hartley,1995; Xu and Zhang, 1996;). The commonly used estimation methods generally fall into four categories: linear method, nonlinear iterative method, normalized linear method and robust method, such as M-estimat or LMedS. In practice, the correponding points are inevitably corrupted by outliers and noise, thus robust epipolar geometry estimation must be achieved. By robustness, we mean that the performance of algorithm should not be affected significantly by small deviations from the assumed model and it should not deteriorate drastically due to outlier and noise. M-estimator or

LMedS is more robust than other three methods. The reason is that the influence of outliers has been reduced in M-estimator or they are simply discarded in LMedS.

Robust parameter estimation is usually formulated as a *non convex* optimization problem. Thus the goal of any robust parameter estimation methods is to find the global optimum from many local minima. However, the commonly used technique for optimization, is primarily based on gradient descent or random sample. Gradient descent can become stuck into local minima easily, which is far away from the true global optimum. Random sample, upon which almost all of highly robust estimator based, can avoid the local minima to some extent. However, it is potentially time-consuming when a large number of outliers are involved. Additionally, it is difficult to determine how many subsamples are needed to guarantee finding global optimum, if no *a priori* knowledge on outlier proportion is given.

The aim of this paper is how to use genetic algorithm to alleviate the above problems. Genetic algorithm (Holland, 1975; Goldberg, 1989; Michalewicz,1992) has the following features:

- It is a global optimization algorithm, which can effectively avoid falling in local minima.
- It is an evolution algorithm, which incorporates evolution idea, "fitter survival", into search process.

The rest of this paper is organized as follows. Section 2 devotes to the introduction of the epipolar geometry under full perspective. The application of GA in the epipolar geometry estimation is presented in Section 3. In section 4, we give our experiment's results. We conclude our paper in Section 5.

2 Epipolar Geometry under full perspective projection

In this section, We will describe the basic epipolar geometric constraint from two uncalibrated images in stereo vision. The full perspective projection model is used. Neither the intrinsic parameters of the images nor the extrinsic parameters are assumed to be known.

According to epipolar geometry constraint, the corresponding point $\mathbf{m}_i$ and $\mathbf{m}'_i$ must satisfy the following epipolar equation:

$$\tilde{\mathbf{m}}_i^T \mathbf{F} \tilde{\mathbf{m}}'_i = 0 \tag{1}$$

where $\tilde{\mathbf{m}}_i = [u_i, v_i, 1]^T$, $\tilde{\mathbf{m}}'_i = [u'_i, v'_i, 1]$. The above equation can be written as a linear and homogeneous equation:

$$\mathbf{u}_i^T \mathbf{f} = 0 \tag{2}$$

where

$$\mathbf{u}_i = [u_i u'_i, u_i v'_i, u_i, v_i u'_i, v_i v'_i, v_i, u'_i, v'_i, 1]^T$$
$$\mathbf{f} = [F_{11}, F_{12}, F_{13}, F_{21}, F_{22}, F_{23}, F_{31}, F_{32}, F_{33}]^T$$

Thus, given that 8 corresponding points are available, we will be able, in general, to determine a unique solution for matrix **F**, defined up to a scale factor,

3 Epipolar Geometry Estimation By Genetic Algorithm

Before using genetic algorithm to estimate fundamental matrix, we must creat an initial corresponding points set D between two images. The whole process can be summarized as follows:

- A corner detector (Harris, 1987) is first applied to each image to extract high curvature points.
- A classical correlation technique is used to establish matching candidate between the two images.
- Use genetic algorithm to search the optimum solution, *i.e.* the best *minimal subset*.
- According to Fundamental matrix computed by the best minimal subset, the *outliers* are detected and removed.
- The fundamental matrix is computed again using all *inliers*.

3.1 Proposed genetic algorithm-based approach to epipolar geometry eatimation

Now, we show how to apply genetic algorithm to estimate the fundamental matrix. The input to the algorithm is corresponding points set $D = \{\mathbf{y}_i | i = 1, \cdots, N\}$, where parameter vector $\mathbf{y}_i = (u_i, v_i, u'_i, v'_i)$ represents corresponding point. The output is the best minimal subset and an inlier subset which is consistent with this minimal subet.

3.2 Chromosome coding

Our chromosome coding is based on *minimal subset*, which is the smallest number of corresponding points necessary to define a unique parameter vector **f**. From Equation (2), we know that the size of minimal subset is eight. Obviously, there are C_N^8 minimal subsets in all. We assume that only parameter vector **f** computed by these minimal subsets are the candidate solutions to our estimation problem. Therefore, the search space is reduced greatly. The process of chromosome coding can be described as follows:

$$D \longrightarrow M_i = (\mathbf{y}_{i_1}, \cdots, \mathbf{y}_{i_8}) \longleftrightarrow A_i = (i_1, \cdots, i_8).$$

Where D is corresponding point set, M_i is minimal subset and A_i is encoded chromosome string.

3.3 Evaluation of chromosome

Another component necessary for a genetic algorithm is a **cost fuction**. This takes the chromosome definition and output a scalar which represents the fitness of that individual. The larger this scalar value the fitter the individual.

with a given chromosome $A_i = (i_1, \cdots, i_8)$, we can compute parameter vector **f** from the following 8 equations:

$$\mathbf{u}_k^T \mathbf{f} = 0 \qquad k = i_1, \cdots, i_8. \tag{3}$$

Then, for each corresponding point in set C, its *residual error* r_i can be computed as follows:

$$r_i^2 = d(\widetilde{\mathbf{m}}_i, \mathbf{F}\widetilde{\mathbf{m}}_i') + d(\widetilde{\mathbf{m}}_i', \mathbf{F}^T \widetilde{\mathbf{m}}_i) \tag{4}$$

where d is the point-to-line Euclidean distance expressed in pixels.
We further define the cost function $F(r_1, \cdots, r_N)$ as follows:

$$F(r_1, \cdots, r_N) = \sum_{i=1}^{N} s(r_i^2) \tag{5}$$

where s is step function; $s = 0$ if r_i is greater than or equal to the given threshold, and $s = 1$ otherwise.

We identify the best chromosome as that which maximizes the number of inliers. The large the number of inliers, the fitter the chromosome.

3.4 Genetic operator

- **Selection:** The *selection* operation selects two chromosomes for mating. In our works, th classical *roulette wheel* methods are implemented. During roulette wheel selection, two mates are selected for reproduction with probability values in direct proportion to their fitness(cost function) values. Therefore, fitter chromosomes will contributes, on a average, a greater number of offsprings in the succeeding generation whereas the least fit chromosomes will be eliminated in the selection process.
- **Crossover:** The *crossover* operation is a process by which new chromosomes are created from existing ones(parents) during reproduction. In our work, uniform crossover operator has been used with a probability of p_c. Specifically, each gene of the first offspring is randomly chosen between the parents, and the genes of the second offspring are complementary with respect to the random choice. An example, illustrating this operation, is shown in Fig.1. if crossover is not performed, the offspring are simple duplication of the parents.

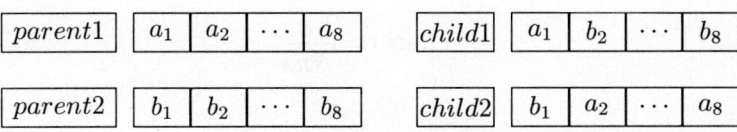

FIG.1: Uniform Crossover.

- **Mutation:** The role of the *mutation* operation is to introduce new genetic materials(genes) to the chromosomes with a probability of p_m, thus preventing the inadvertent loss of useful genetic material in earlier phases of evolution. an example, illustrating this operation, is shown in figure 2, where A new gene b_1 is introduced into the chromosome.

FIG.2: Mutation.

3.5 Outline of genetic algorithm

The main step of genetic algorithm can be described as follows:

Initialization Step: Select values for the following parameters: N_0(population size), p_c(crossover rate), p_m(mutation rate).
Step 1: Randomly Creat N_0 initial chromosomes, *i.e.* initial *population*, and evaluate their correponding cost function values according to Equation .
step 2: From the current *population*, two chromosomes are selected for reproduction using roulette wheel selection.
Step 3: Crossover and mutation operators are applied to the selected chromosomes, and new chromosomes are then generated.
Step 4: each new chromosome is evaluated by their cost function value.
Step 5: From all chromosomes in the current generation, select the N_0 best chromsome to constitue a new *population*.
Step 6: If the stopping criterion is not reached, Goto Step 2, otherwise return the best chromosome in the new population as the optimum solution.

4 Experimental Results

4.1 Comparing with random sample

Our genetic algorithm based estimation method is very similar to RANSAC. The difference between them lies in optimization technique. Unlike our algorithm, RANSAC use random sample to search optimal solution. Almost all of highly robust estimator based on random sample optimization technique. Thus, it is neccessary to compare genetic algorithm with random sample in computational efficency. The comparison results are shown in FIG.3, all of which are gained by hundreds of simulated data tests. Its evaluation criterion is defined as follows:

$$Ratio = \frac{N_{GA}}{N_{RA}}$$

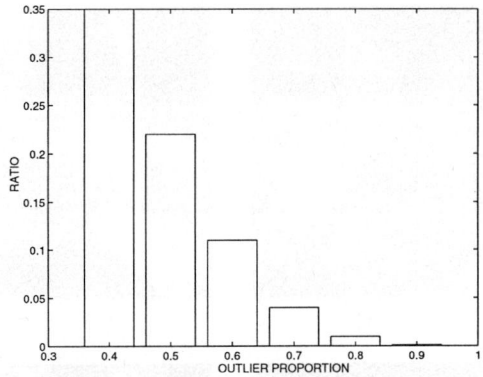

FIG.3. Ratio of the needed number of cost function evaluation between GA and RANSAC.

where N_{GA} and N_{RA} denote the number of cost fucntion evaluations respectively, when their cost function achieve the optimum value.

As is seen, GA is more efficient than random sample, especially when a large number of outliers are involved. The reason is that Genetic algorithm is an evolution algorithm, which incorporates the evolution idea, "fitter survival", into the search process. As mentioned early, a GA performs a multi-directional search by maintaining a population of potential solutions and encourages information formation and exchange between these direction. In a word, Genetic algorithm strikes a remarkable balance between exploration and exploitation of the search space. However, random sample is a typical example of a strategy which explores the search space ignoring the exploitations of the promising regions of the space.

4.2 Experiments with real images

For real image, we use one scene to test our method. Figure 4 shows synthetic scene **trancing**, the size of which is 512×512. The scene was generated by a ray tracing technique simulating two cameras placed by diagonally. Due to the numerous repetitive patterns in image, outlier proportion in the initial corresponding set is much more than a half. Thus, it is very difficult for LMedS to provide a useful results. However, our algorithm can cope with such case well. The recovered epipolar line is shown in Figure 4.3 and Fig 4.4, where we also label the eight corresponding points determined by best chromosome. All correponding points inconsistent with this best chromosome have been removed as *outliers*. The remaining correponding points, *i.e. inliers*, are used to recompute the fundamental matrix.

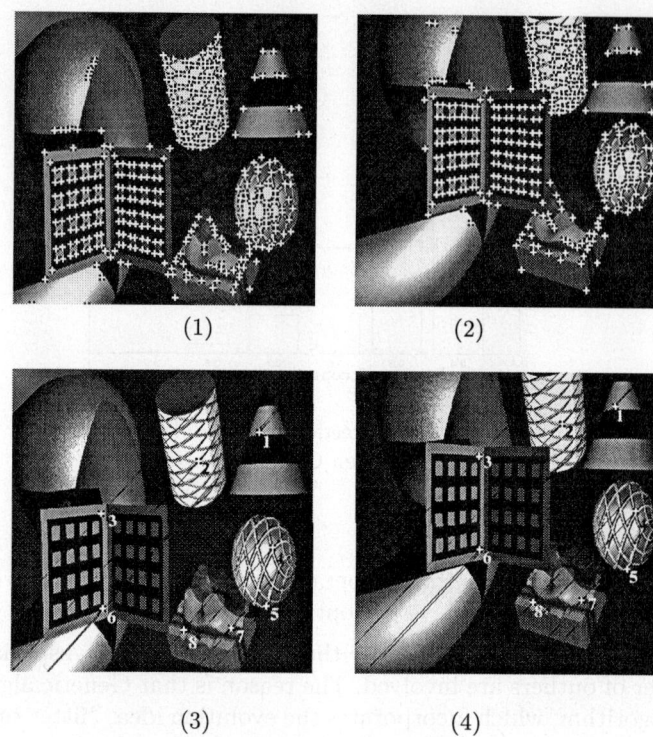

Fig.4 Epipolar geometry estimation: (1) corner points in left image. (2) corner points in right image. (3) recovered left epipolar line and the best minimal subset. (4) recovered right epipolar line and the best minimal subset.

5 Conclusion

In this paper, we propose a novel method for epipolar geometry eatimation from two uncalibrated images. Specifically, we have formulated it as a global optimization problem and presented a genetic algorithm-based estimation method. In summary, our new approach to epipolar geometry estimation has the following main features:

- It is very robust to outliers. In fact, As long as the number of inliers is higher than the minimum number to give a unique solution, our new approach can always gain the good results, which are not much affected by outlier proportions.
- Unlike general robust estimator, it is a global optimization algorithm, and thus can effectively escape from local optimum.
- In terms of computational efficiency, it is more efficient than random sample, the most commonly used method for highly robust estimator such as LMedS and RANSAC.

- It is a generic method for robust parameter estimation and can be used in various engineering problems.

References

1. Faugeras, O.D(1992). What can be seen in the three dimensions with an uncalibrated stereo rig ? In G.Sandini, editor, *Proceedings of the 2nd European Conference on Computer Vision* , *Santa Margherita Italy*, pages 563-578. Springer-Verlag.
2. Goldberg, D.E(1989). Genetic Algorithms in Search, Optimization and Machine Learning. Addison Wesley.
3. Harris, C(1987). A combined corner and edge detection. in *proceedings alvey conference*.
4. Hartely, R.I and R.Gupta(1993), Computing matched-epipolar projections. In *Proceedings of the Conference on Computer Vision and Pattern Recognition, New York*, pages 549-555.
5. Hartley, R.I(1995). In defence of the 8-point algorithm. *Proc. 5th Int Conf on Computere Vision, Boston MA*, pages 1064-1-70.
6. Holland, J(1975). Adaptation in natural and artificial systems. University of Michigan Press, Ann Arbor.
7. Laveau, S and O.D.Faugeras(1994). 3-D scence representation as a collection of images. In *Proc. International Conference on Pattern Recognition*, Jerusalem, Israel.
8. Longuet-Higgins, H.C(1981). A computer algorithm for reconstruction a scene from two projection. *Nature*,293:133-135.
9. Luong, Q.T, and O.D.Faugeras(1992). Self-calibrarion of a camera using multiple images. In *Proceedings of the 11th International Conference on Pattern Recognition, The Hag, Netherland*, pages 9-12.
10. Maybank, S.J and O.D.Faugeras(1992). A theory of self calibration of a moving camera. *International Jounal of Computer Vision* , 8(2):123-151.
11. Michalewicz, Z(1992). Genetic algorithms+data structures=evolution programs, Springer-Verlag.
12. Nishimura, E, Xu, G and S.Tsuji(1990). Motion segmentation and correspondence using epipolar constraint. In *Proceedings of the 1st Asian Conference on Computer Vision* , *Osaka, Japan*, pages 199-204.
13. Robert, L and M.Hebert(1993). Deriving orientation cues from stereo images . In *Proceedings of the Eurapean Conference on Computer Vision, Berlin, Germany*, pages 540-543.
14. Shashua, A(1993). Projective depth: a geometric invariant for 3d reconstruction from two perspective/orthographic views and for visual recognition. In*Proceedings of International Conference on Computer Vision* , *Berlin, Germany*, pages 583-590.
15. Sinclair, D, B.Boufama and R.Mohr(1994). Independent motion segmentationand collision predection for road vehicales. In *proceedings of the Conference on Computer Vision and Pattern Recognition* , *seattle, Washington*, USA, pages 958-961.
16. Xu, G and Z.Zhang(1996). Epipolar geometry in stereo, motion and object recognition. Kluwer Academic Publisher.
17. Zhang, Z, R.Deriche, O.D.Faugeras and Q.T.Luong(1994). A robust technique for matching two uncalibrated images through the recovery of the unknown epipolar geometry. *AI jounal*, **78**:87-119.

New Development of Stereo Vision: A Solution for Motion Stereo Correspondence

M. Xie

School of Mechanical & Production Engineering
Nanyang Technological University, Singapore 639798
Tel: (65) 799 5754 and Fax: (65) 791 1859
http://www.ntu.edu.sg/home/mmxie/research/index.htm

Abstract. This paper presents a new constraint, ie., *projective transformation constraint*, to solve the difficult problem of motion stereo correspondence in the context where the motion is dominated by rotation (for instance, 10 degrees of rotation with 1 cm of translation). Our observation here is that two consecutive images can be related to each other by a 2D projective transformation. Its coefficients are constants if the motion is a pure rotation (no use for depth recovery but helpful for 3D data fusion with rotating binocular vision and for active vision involving verging motion). The solution presented in this paper constitutes a closed form solution to motion stereo correspondence if the motion is a rotation or a quasi closed form solution if the motion is dominated by rotation (including translation needed for depth recovery). Experiments with real camera and real images prove the usefulness of the new constraint.

Keywords: Motion Stereo Correspondence, Projective Transformation

1 Introduction

Machine intelligence depends on machine's capability of doing perception of its internal/external environment. To use camera(s) to perceive a 3D scene is one of the main objectives of computer vision research. As for the task of 3D perception of an unknown 3D environment by using camera(s), it is achievable if reliable techniques exist for solving the problems of: (a) camera calibration, (b) stereo/motion correspondence and (c) structure/motion estimation. These three important topics of computer vision are inter-related. Now, significant progress has been made in the research areas of: (a) camera calibration and (b) structure/motion estimation. This has been reflected by the discovery of analytical solutions in these two areas ([1], [2]).

But, for the problem of stereo/motion correspondence, progress is still desirable in order to make computer vision to achieve a similar commercial success as for the case of computer graphics. So far, there is no discovery of any analytical solution for the problem of stereo/motion correspondence. To the best of our knowledge, the best known constraint in stereo/motion correspondence is the epipolar geometry. This constraint only allows the search for matches to be reduced from 2D space to 1D space. Therefore, epipolar geometry can not be considered as an analytical solution to stereo/motion correspondence.

In this paper, we introduce a new and useful constraint called "projective transformation constraint". This constraint can be concisely described by the following observation:

> **Observation**: If a camera is moved with a pure rotation from a position A to a position B, the image taken at A is related to the image taken at B by a 2D projective transformation of constant coefficients.

This powerful constraint is meaningful in practice for the following reasons:

1. Ease of Automatic Estimation of 2D Projective Transformation:
   A 2D projective transformation with constant coefficients can uniquely be determined with four pairs of matched points. It is a trivial task in computer vision to obtain, at run-time, four pairs of matched points if images contain information [5].
2. Need of Solving Motion Stereo Correspondence in Rotating Binocular Vision:
   The use of (pre-calibrated or no pre-calibrated) rotating binocular vision (ie., a pair of cameras mounted on a rotating head) to explore a 3D unknown scene requires a reliable technique to solve the motion stereo correspondence (either in 2D image space or 3D camera space). This is a prerequisite for performing the fusion of 3D data estimated from different positions of observation. In this case, the motion of the binocular vision can be a pure rotation because the depth recovery can be done with binocular vision itself.
3. Need of Solving Motion Stereo Correspondence in Active Vision Involving Verging Motion:
   Active vision is one of the promising research directions in computer vision [4]. In the case of active vision involving verging motion of the binocular cameras, one important task is how to recover the stereo correspondence of the two views [V1(t2), V2(t2)] at time t2 (after a verging motion) from the stereo correspondence of the two views [V1(t1), V2(t1)] at time t1 (before a verging motion). This task can be easily achieved if a simple and reliable method exists which allows to directly compute the correspondence between V1(t1) and V1(t2) as well as between V2(t1) and V2(t2). In this application, motion stereo correspondence needs to be solved regardless whether depth recovery can be done from motion stereo or not (depth can be recovered from the active vision's two views themselves).
4. Usefulness of Motion Stereo in Depth Recovery:
   To recover depth information, one can use either a pair of cameras (ie., binocular stereo) or a moving camera (motion stereo). In the latter case, if there is translational component in the motion, the depth map can be estimated from motion stereo correspondence (Of course, the precision depends on some factors including the magnitude of translation). For the special case of the motion dominated by rotation with weak translation (for instance,

10 degrees of rotation with 1 cm of translation), a 2D projective transformation is a good approximate solution to analytically solve motion stereo correspondence problem.

2 A Quasi Closed Form Solution For Motion Stereo Correspondence

In this section, we present the projective transformation constraint which constitutes an exact matching solution to motion stereo correspondence if the motion is a rotation or a quasi closed form solution if the motion is dominated by rotation with weak translation.

2.1 Statement of Motion Stereo Correspondence Problem

The correspondence problem in motion stereo can be shortly stated as how to identify, from two consecutive images, a pair of geometrical primitives (including image points) which are the projection of a same object primitive.

So far, there is no analytical solution having been found. All the existing algorithms for correspondence problem can only be qualified to be algorithmic solutions because they make use of constraints to perform the search for matches. The commonly used constraints are: (a) feature similarity (including image intensity), (b) epipolar geometry, and (c) figural continuity, etc.

Now, the question is whether we can find a closed form or quasi closed form solution to motion stereo correspondence problem ? In other words, we are interested in finding a matching function MF(.) which outputs the match (u_2, v_2) in one image, given an input (u_1, v_1) in its preceding image, that is:

$$(u_2, v_2) = MF(u_1, v_1). \tag{1}$$

We show in this paper that the answer to the above question is affirmative if the motion is a rotation or dominated by rotation with weak translation.

2.2 Projective Transformation Constraint

Here, we prove the following theorem which gives the matching function in Eq.1:

Theorem: In motion stereo, the match (u_2, v_2) in one image of a point (u_1, v_1) in its preceding image is determined by a 2D projective transformation as follows:

$$\begin{pmatrix} u_2 \\ v_2 \end{pmatrix} = MF(u_1, v_1) = \begin{pmatrix} \frac{m_{11} \bullet u_1 + m_{12} \bullet v_1 + m_{13}}{m_{31} \bullet u_1 + m_{32} \bullet v_1 + m_{33}} \\ \frac{m_{21} \bullet u_1 + m_{22} \bullet v_1 + m_{23}}{m_{31} \bullet u_1 + m_{32} \bullet v_1 + m_{33}} \end{pmatrix}. \tag{2}$$

The coefficients $\{m_{ij}, i, j = 1, 2, 3\}$ are constants if the motion is a pure rotation. If the motion is not a pure rotation, only (m_{13}, m_{23}, m_{33}) are not constants and depend uniquely on the depth (ie., Z coordinate of the corresponding 3D point).

Proof:

Suppose that the image formation by CCD camera is a perspective projection (pinhole model). Let's consider a camera moving from a position A to a position B with the motion transformation (R_c, T_c). We define:

- oxy: the coordinate system associated to the image plane.
- (x, y): the real coordinates of an image point.
- (u, v): the index coordinates (in terms of pixels) of an image point.
- $OXYZ$: the coordinate system associated to the camera with: (a) OX axis parallel to ox axis, (b) OY axis parallel to oy axis and (c) OZ axis being the optical axis of the camera.
- (X, Y, Z): the real coordinates of a 3D object point.
- (R_c, T_c): the motion of the moving camera with respect to a fixed world coordinate system.
- (R_o, T_o): the motion transformation of a static scene with respect to $OXYZ$. (R_o, T_o) is equal to the inverse of (R_c, T_c).

Consider a 3D object point P. If its coordinates are (X_1, Y_1, Z_1) with respect to $OXYZ$ when the camera is at A, then its image coordinates (x_1, y_1) will be:

$$x_1 = f \bullet X_1/Z_1 \quad \text{and} \quad y_1 = f \bullet Y_1/Z_1. \tag{3}$$

where "f" is the focal length of the camera.

When the camera moves to B, the coordinates of the point P will be (X_2, Y_2, Z_2), that is:

$$(X_2, Y_2, Z_2)^t = R_o \bullet (X_1, Y_1, Z_1)^t + T_o. \tag{4}$$

where "t" denotes the transpose of a vector or matrix, and the image coordinates (x_2, y_2) of P will be (when the camera is at B):

$$x_2 = f \bullet X_2/Z_2 \quad \text{and} \quad y_2 = f \bullet Y_2/Z_2. \tag{5}$$

If we denote $R_o = \{r_{ij}, i, j = 1, 2, 3\}$ and $T_o = (T_x, T_y, T_z)^t$, the combination of Eq.3, Eq.4 and Eq.5 yields:

$$\begin{cases} x_2 = \frac{f \bullet r_{11} \bullet x_1 + f \bullet r_{12} \bullet y_1 + f^2 \bullet r_{13} + f^2 \bullet T_x/Z}{r_{31} \bullet x_1 + r_{32} \bullet y_1 + f \bullet r_{33} + f \bullet T_z/Z} \\ y_2 = \frac{f \bullet r_{21} \bullet x_1 + f \bullet r_{22} \bullet y_1 + f^2 \bullet r_{23} + f^2 \bullet T_y/Z}{r_{31} \bullet x_1 + r_{32} \bullet y_1 + f \bullet r_{33} + f \bullet T_z/Z} \end{cases} \tag{6}$$

Let (u_1, v_1) and (u_2, v_2) be the index coordinates with respect to (x_1, y_1) and (x_2, y_2). Then, we have the following relations:

$$\begin{cases} x_1 = (u_1 - u_0) \bullet D_x \\ y_1 = (v_1 - v_0) \bullet D_y \end{cases} \text{and} \begin{cases} x_2 = (u_2 - u_0) \bullet D_x \\ y_2 = (v_2 - v_0) \bullet D_y \end{cases} \tag{7}$$

where (D_x, D_y) are the sizes of image pixel in x and y directions of the image plane, respectively, and (u_0, v_0) is the so-called optical center.

By applying Eq.7 to Eq.6, we obtain:

$$\begin{pmatrix} u_2 \\ v_2 \end{pmatrix} = \begin{pmatrix} \frac{m_{11} \bullet u_1 + m_{12} \bullet v_1 + m_{13}}{m_{31} \bullet u_1 + m_{32} \bullet v_1 + m_{33}} \\ \frac{m_{21} \bullet u_1 + m_{22} \bullet v_1 + m_{23}}{m_{31} \bullet u_1 + m_{32} \bullet v_1 + m_{33}} \end{pmatrix}. \tag{8}$$

with:

$$\begin{cases} m_{11} = (f \bullet r_{11} \bullet D_x + r_{31} \bullet u_0 \bullet D_x^2)/D_x \\ m_{12} = (f \bullet r_{12} \bullet D_y + r_{32} \bullet u_0 \bullet D_x \bullet D_y)/D_x \\ m_{13} = [f^2 \bullet r_{13} + f^2 \bullet T_x/Z - f \bullet r_{11} \bullet u_0 \bullet D_x - f \bullet r_{12} \bullet v_0 \bullet D_y + \\ \quad (u_0 \bullet D_x) \bullet (f \bullet r_{33} + f \bullet T_z/Z - r_{31} \bullet u_0 \bullet D_x - r_{32} \bullet v_0 \bullet D_y)]/D_x \\ m_{21} = (f \bullet r_{21} \bullet D_x + r_{31} \bullet v_0 \bullet D_x \bullet D_y)/D_y \\ m_{22} = (f \bullet r_{22} \bullet D_y + r_{32} \bullet v_0 \bullet D_y^2)/D_y \\ m_{23} = [f^2 \bullet r_{23} + f^2 \bullet T_y/Z - f \bullet r_{21} \bullet u_0 \bullet D_x - f \bullet r_{22} \bullet v_0 \bullet D_y + \\ \quad (v_0 \bullet D_y) \bullet (f \bullet r_{33} + f \bullet T_z/Z - r_{31} \bullet u_0 \bullet D_x - r_{32} \bullet v_0 \bullet D_y)]/D_y \\ m_{31} = r_{31} \bullet D_x \\ m_{32} = r_{32} \bullet D_y \\ m_{33} = f \bullet r_{33} + f \bullet T_z/Z - r_{31} \bullet u_0 \bullet D_x - r_{32} \bullet v_0 \bullet D_y \end{cases}$$

(9)

End of Proof

From Eq.9, we can make the following remarks:

- If the motion is a pure rotation, the parameters $\{m_{ij}, i, j = 1, 2, 3\}$ are all constants. The parameter m_{33} can be factorized out. Therefore, there are eight unknown in Eq.8. Four pairs of matched points allows to uniquely determine the parameters $\{m_{ij}, i, j = 1, 2, 3\}$ from a linear equation system.
- The translational components (T_x, T_y, T_z) are weighted by $1/Z$. In normal situation, the value of Z (depth) is much bigger than (T_x, T_y, T_z). If the motion is dominated by rotation with weak translation (for instance, 10 degrees of rotation with 1 cm of translation), the consideration of (m_{13}, m_{23}, m_{33}) being constants makes Eq.8 a good approximate solution to solve the motion correspondence problem.
- For some applications where the variation of depth is small (for instance, visual inspection in electronics and manufacturing industry), one can fix the value of Z with its average value in a particular scene. In this way, the parameters $\{m_{ij}, i, j = 1, 2, 3\}$ can be directly computed from Eq.9, if the stereo-vision system (its baseline can be big) is pre-calibrated during the setup stage.

3 Implementation and Experimental Results

We have implemented the algorithm on a SUN Sparc 20 workstation with a S2200 color frame grabber and a PULNIX color camera mounted on a pan-tilt device (Fig.1). The center of the physical lens has an offset of roughly 6 cm from the rotation axis. For the experiments reported here, the camera is rotated by 10 degrees each time. This produces roughly 1 cm of translation for the center of the physical lens (The camera's coordinate system is roughly centered at the center of the physical lens). Therefore, the motion of the camera is not a pure rotation.

In our experiments, the coefficients in Eq.2 are estimated by a least-squares method with the input of at least four pairs of matched points (These points are

Fig. 1. A Moving Camera Mounted On a Pan-Tilt Device: The center of the lens has an offset of roughly 6 cm from the rotation axis.

automatically matched by a program. There is no human intervention). Due to space limitation, we show the results with two consecutive images. Fig.2 shows the two consecutive images and the automatically established good matches with the algorithm presented in [7].

Fig.3 shows: a) the edge-map extracted from Image t1, b) the edge-map extracted from Image t2, c) the super-imposition of the computed matches with the edge points in Image t2 (the computed matches are obtained by taking the edge points of Image t1 as input and applying the estimated 2D projective transformation (ie., Eq.2). and d) the computed matches in Image t2 (the computation was done by taking the edge points of Image t1 as input).

4 Conclusions

The important conclusion of this paper is the following statement: "Two consecutive images are related to each other by a 2D projective transformation in a motion stereo configuration". This leads to the proposition of a quasi closed form solution to solve the motion stereo correspondence problem (pure rotation is difficult to realize in practice with real camera system). Experiments with real camera and real images show that the solution works. The significance of this work are three fold: (a) it is an efficient and simple method for solving motion correspondence in the context of rotating binocular vision where 3D data fusion depends on motion correspondence information, (b) it is a simple and reliable way of establishing motion stereo correspondence in the context of active vision where a pair of binocular cameras undergo verging motion, and (c) it is a good approximate solution to solve motion correspondence in the context of motion stereo where the motion do include rotation and weak translation. Therefore, it is obvious that motion stereo correspondence plays an important role in computer vision not only for helpful support to active and dynamic vision but also

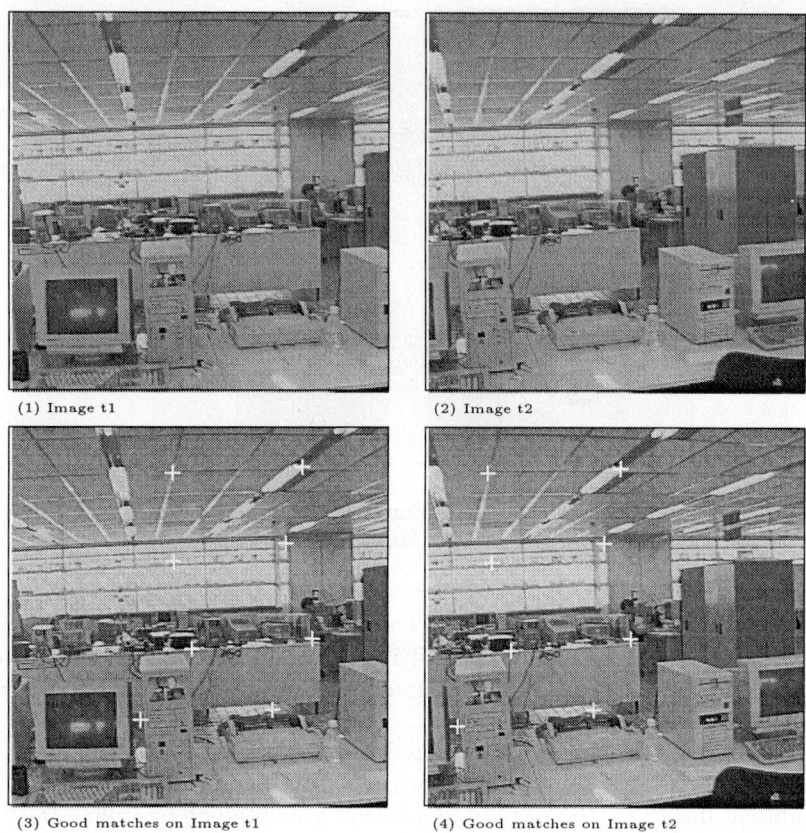

Fig. 2. Two consecutive images with automatically established good matches.

for depth recovery (if the motion is not a pure rotation) within certain limit of acceptable precision.

Acknowledgment This work is sponsored by the Ministry of Education (Singapore) under the grant 4-25003 (ARC 1/94).

References

[1] K. Kanatani, Group-Theoretical Methods in Image Understanding, *Springer-Verlag*, Berlin, p103-194 & 278-354, 1989.

[2] O. Faugeras, Three-Dimensional Computer Vision: A Geometric Viewpoint, *MIT Press*, Cambridge, p33-69 & p165-40, 1993.

[3] P. F. McLauchlan and D. W. Murray, Active Camera Calibration for a Head-Eye Platform Using the Variable State-Dimension Filter, *IEEE Trans. on PAMI*, Vol.18, No.1, p15-22, January, 1996.

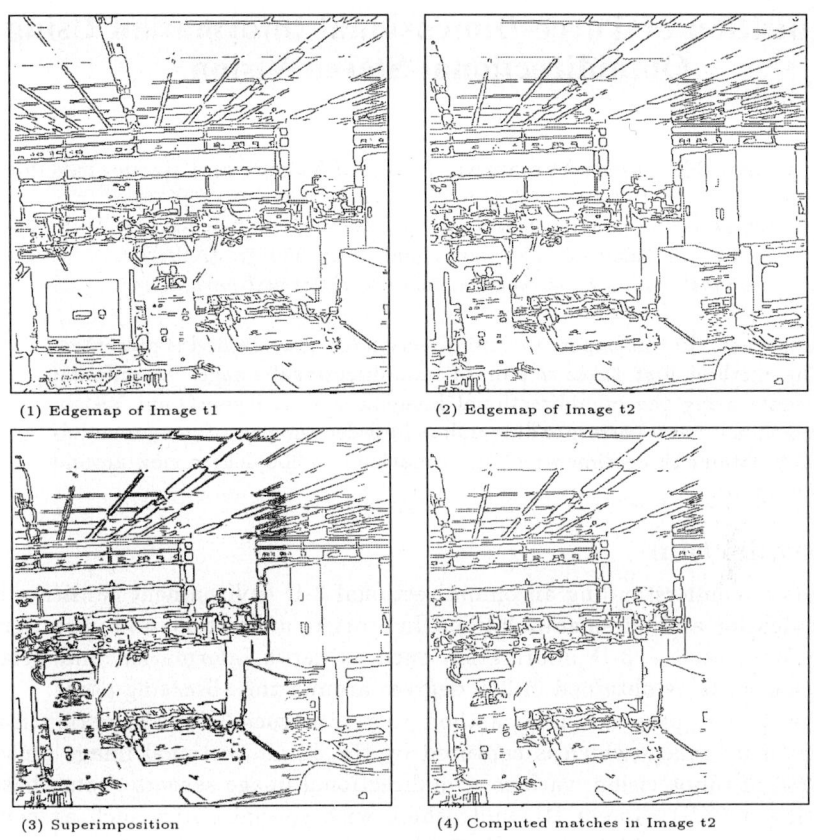

(1) Edgemap of Image t1
(2) Edgemap of Image t2
(3) Superimposition
(4) Computed matches in Image t2

Fig. 3. Results.

[4] T. Vieville, E. Clergue, R. Enciso, and H. Mathieu, Experimenting with 3D vision on a robotic head, *Robotics and Autonomous Systems 14*, p1-27, 1995.
[5] Z. Zhang, R. Deriche, O. Faugeras, and Q. T. Luong, A robust technique for matching two uncalibrated images through the recovery of the unknown epipolar geometry, *Artificial Intelligence*, Vol.78, p87-119, October 1995.
[6] M. Xie, A cooperative strategy for the matching of multi-level edge primitives, *Image and Vision Computing*, Vol.13, No.2, p89-99, March, 1995.
[7] M. Xie and L. Y. Liu, Color Stereo Vision: Use of Appearance Constraint and Epipolar Geometry For Feature Matching, *2nd Asian Conference on Computer Vision*, Vol.1, p282-286, Dec 6-8, 1995.
[8] J. You, E. Pissaloux, W. P. Zhu, and H. A. Cohen, Efficient Image Matching: A Hierarchical Chamfer Matching Scheme Via Distributed System, *Real Time Imaging*, Vol.1, No.4, p245-259, October, 1995.

Acquisition of Three-Dimensional Information Using Omnidirectional Stereo Vision

Atsushi Chaen, Kazumasa Yamazawa, Naokazu Yokoya and Haruo Takemura

Graduate School of Information Sciences, Nara Institute of Science and Technology
8916-5 Takayama-cho, Ikoma-shi, Nara 630-01, JAPAN
e-mail: {yamazawa, yokoya, takemura}@is.aist-nara.ac.jp

Abstract. In this paper, an omnidirectional stereo method is presented. The method first takes a pair of omnidirectional images at different heights using the omnidirectional imaging sensor, HyperOmni Vision, and then detects corresponding points between the pair of images and finally obtains three-dimensional information of a 360-degree view around the sensor.

1 Introduction

The needs for understanding an omnidirectional 3-D environment is raising for robot navigation and tele-manipulations. In order to understand omnidirectional 3-D environment, the 3-D information such as depth information and height information must be obtained in 360 degrees around the observing point.

In this paper, proposed is a method to reconstruct 3-D information from omnidirectional image which is captured by the omnidirectional image sensor. In the area of robot vision, various omnidirectional image sensors, some based on rotating a CCD camera [1], and others with special optics such as hemispherical or hyperboloidal mirrors[2, 3, 8], are being proposed. In our research, omnidirectional images are captured by HyperOmni Vision sensor[3], which uses a hyperboloidal mirror, because its eyesight and optical characteristics are suitable for our purpose.

In this research, stereo vision[4], the major passive method which does not affect the measured environment, is used. Stereo vision has been studied very well, because the stereo images are easily captured and do not require any special light source[5, 6]. In our method, two omnidirectional images at different heights are captured. Then, the omnidirectional 3-D information is computed by matching corresponding feature points in the omnidirectional stereo images. The method can simultaneously acquire the direction, depth and height of observed objects in a 3-D environment.

In the following sections, HyperOmni Vision, the algorithm to compute 3-D information, and the experiment using real images are described. We also show standard binocular stereoscopic images computed from the 3-D information as an example of applications of our method.

2 Omnidirectional Image Sensor : HyperOmni Vision

HyperOmni Vision is an omnidirectional visual sensor[3] which uses a hyperboloidal mirror and captures the image around the sensor at video-rate. The

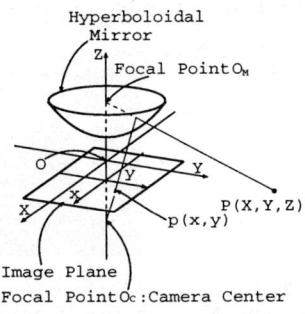

(a) Configuration of HyperOmni Vision   (b) Overview of HyperOmni Vision

Fig. 1. Omnidirectional image sensor HyperOmni Vision.

sensor obtains a circular omnidirectinal view of the surroundings including the ground(floor) and objects on it around the sensor itself. Moreover, the optics are based on perspective projection which is different from most other omnidirectional image sensors.

In Fig.1, the optics and appearance of the sensor are shown. The sensor are constructed of a revolving hyperboloidal mirror of two sheets and havs two focuses O_M and O_C. The camera lens center is located at the focus O_C and directed towards the focus O_M. In this situation, the following relation is established. All rays gathering at the inner focus O_M are reflected by the hyperboloidal mirror and gather at the outer focus O_C. Therefore, the scene is projected on the hyperboloidal curved surface and is captured as a mirrored image by the CCD camera.

The relationship between a point $P(X, Y, Z)$ in the 3-D environment and its projection $p(x, y)$ onto the omnidirectional image is described by the Eq.(1), using the hyperboloidal surface parameters (See Figs.2 and 3).

$$\left. \begin{array}{rl} Z &= \sqrt{X^2 + Y^2} \tan \alpha + c \\ \alpha &= \tan^{-1} \frac{(b^2 + c^2) \sin \gamma - 2bc}{(b^2 - c^2) \cos \gamma} \\ \gamma &= \tan^{-1} \frac{f}{\sqrt{x^2 + y^2}} \\ \tan \theta &= Y/X = y/x \end{array} \right\} \quad (1)$$

where, b, c : Parameters of hyperboloidal mirror,
$\quad \alpha$: Angle between the mirror focus and P,
$\quad \gamma$: Angle between camera lens center and p,
$\quad \theta$: Azimuth angle of P.

Fig. 4 shows an example of the omnidirectional image captured by the HyperOmni Vision sensor. As shown in Fig. 4, the 360-degree view surrounding the camera is captured. One of the characteristics of the image is that the direction (azimuth angle) to a target object in the 3-D environment around the Z axis, which passes through two focal points O_M and O_C, is identical to the direction (azimuth angle) towards the projected target object around the center of the

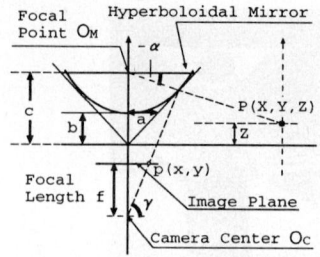 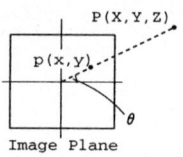 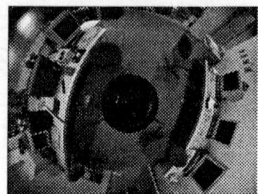

Fig. 2. Relation between P and its projection p. **Fig. 3.** Relation of azimuth angles of p and P. **Fig. 4.** Example of omnidirectional image.

image. Moreover, the image captured is the perspective projection of the scene from the focal point O_M onto the mirror surface. Therefore, we can generate an ordinary perspective camera image looking at the scene in an arbitrary direction from O_M by using geometric transformations.

3 Omnidirectional Stereo Vision

3.1 Omnidirectional stereo imaging

There are two possible placements of the HyperOmni Vision for stereo imaging. One is horizontal placement. The other is vertical placement. In the horizontal placement, the stereo view is rather restricted because each sensor occludes the other sensor's eyesight. Also the searching area and search path become substantially complicated one. On the other hand, in case of the vertical placement, the lower HyperOmni Vision sensor appears in the image captured by the upper sensor. Therefore, in this area, we can not perform stereo matching. However, excluding this area, omnidirectional stereo images for matching detection can be acquired all around the sensors. Vertically placing two sensors on Z axis in Fig. 1(a), the epipolar geometry constraint applies on radial lines around the center of the image. Thus, we can perform stereo matching on the lines. This situation corresponds to the traditional vertical binocular stereo vision. From these reasons, we have decided to place two HyperOmni Vision sensors vertically, at different heights, for capturing omnidirectional stereo images.

3.2 Matching in omnidirectional stereo images

The placement of the sensors for stereo image acquisition is illustrated in Fig. 5. Let the image captured by the lower sensor be I_l, and the image captured by the upper sensor be I_u. The position of the projected point p in an omnidirectional image can be denoted by a polar coordinate system as $p(r,\theta)$, where r is the distance between p and the image center, and θ is the direction to p from the image center (See Fig. 6).

In a pair of vertical omnidirectional stereo images, the projection of $P(X,Y,Z)$ in a 3-D environment onto omnidirectional images I_l and I_u have the same directional angle θ. In this situation, there is a relation as follows.

$$r_2 = r_1 - d, \tag{2}$$

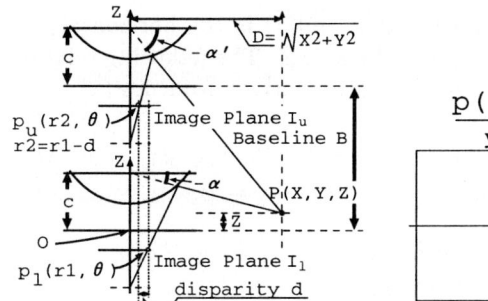 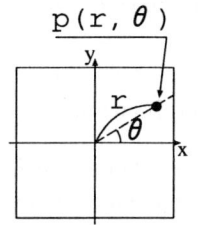 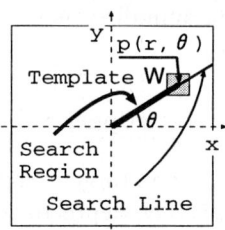

Fig. 5. Relation of omnidirectional stereo images.

Fig. 6. Polar coordinate system.

Fig. 7. Range of the corresponding point search.

where $p_l(r_1, \theta)$ and, $p_u(r_2, \theta)$ are the projections of P onto a stereo image pair, and d is the disparity of the projected points.

3.3 Corresponding point search by template matching

As is described in Section 3.2, with the reference image I_l, excluding the image center where the camera itself appears, the area of interest in I_l always appears in the upper omnidirectional image I_u. In this situation, there is an epipolar constraint on the radial line with the same azimuth angle in stereo pair. Therefore, the stereo matching seach can be limited to small areas in one dimension.

As is shown in Fig. 7, the search area in the omnidirectional image I_u is set on the radial line with the same azimuth angle with the pixel of interest (r, θ) in the omnidirectional image I_l. On the line, the search area is restricted to the inner region where d is in the range of $(0 < d \leq r)$. Therefore, the following evaluation function can be defined.

$$E(r, \theta; d) = \sum_{r, \theta \in W} \{I_u(r-d, \theta) - I_l(r, \theta)\}^2, \quad (3)$$

where $I_l(r, \theta)$ and $I_u(r, \theta)$ denote pixel intensity values of pixels (r, θ) in omnidirectional images.

Equation (3) is an evaluation function based on computing the sum of squared differences in the searching window region for pixel of interest (r, θ) in the omnidirectional image I_l. Searching window W actually used is a square region. $d(0 < d \leq r)$ is a disparity of corresponding points on the radial line with azimuth angle θ which meets the epipolar constraint. The problem here is to find d which minimizes Eq.(3) for each pixel (r, θ) in the image I_l.

3.4 Acquisition of 3-D information

The relationship between the disparity and the height and depth of the target object in the 3-D environment is described as follows.

First, let us define coordinate systems(See Fig. 5). The origin of the 3-D world is located at the origin of the lower HyperOmni Vision. The origin of the upper HyperOmni Vision is at $(0, 0, B)$ in the 3-D world coordinate system, considering the camera base line length B. Depth $D (= \sqrt{X^2 + Y^2})$ is the horizontal distance between the target point $P(X, Y, Z)$ and the axis Z. The following relation between omnidirectional stereo pair I_l and I_u can be derived from Eq.(1).

- Lower HyperOmni Vision, $Z = D \tan \alpha + c.$ (4)
- Upper HyperOmni Vision, $Z = D \tan \alpha' + c + B,$ (5)

where Z: The target heights in 3-D environment,
D: The depth to the target in 3-D environment,
B: Camera base line length,
c: Hyperboloidal mirror parameter,
α, α': view angle between the mirror focus and the target.

From the relationship above, the depth D can be computed using the camera baseline length B, the target view angle α, and upper mirror angle α' as follows:

$$D = \frac{B}{\tan \alpha - \tan \alpha'} \quad (6)$$

Angles α and α' are calculated from the disparity d and the r coordinates of pixels of interest $p_l(r_1 \theta)$ and $p_u(r_2, \theta)$ with the following equation.

$$\alpha = \tan^{-1} \frac{(b^2 + c^2) \sin(\tan^{-1} \frac{f}{r_1}) - 2bc}{(b^2 - c^2) \cos(\tan^{-1} \frac{f}{r_1})},$$

$$\alpha' = \tan^{-1} \frac{(b^2 + c^2) \sin(\tan^{-1} \frac{f}{r_1 - d}) - 2bc}{(b^2 - c^2) \cos(\tan^{-1} \frac{f}{r_1 - d})},$$

(7)

where b, c: parameters of hyperboloidal mirror,
f: focal length of camera lens.

From the equation above, we can see that the depth D can be calculated from r_1 and disparity d. Thus, the height Z is also calculated by using Eq.(4).

4 3-D Information Extraction: Experiment

4.1 Experiment

Figure 8 shows the example of a pair of omnidirectional stereo images used in the experiment. These are taken under the same lighting condition in a room. Figure 8(a) is the upper omnidirectional image and Figure 8(b) is the lower one. The images have 256 gray levels and consist of 640 × 486 pixels. The camera baseline length of HyperOmni Vision is set to the 242.0(mm). For corresponding point search, the lower image (Fig. 8(b)) is chosen for the reference image. The searching windows is a 5 × 5 rectangular region. The searching area is on the epipolar radial line, and the range of search is limited to the inside of the pixel of interest excluding the area where the camera itself appears.

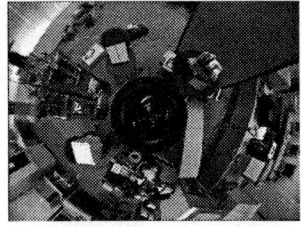 

(a) Upper image (b) Lower image

Fig. 8. A pair of omnidirectional stereo images.

Extraction of disparity Figure 9 shows an image of disparity computed from a pair of stereo images in Fig. 8 using the proposed method. In this figure, the intensity represents the disparity.

As we can observe in Fig. 9, the disparity is extracted in all range of 360–degree view, centering around the HyperOmni Vision sensor, by the proposed method. Quantitative evaluation of the correspondence accuracy is not yet done. But, It has enough accuracy to see the relative position of each object. However, we can observe that wrong correspondences are established at the edges which have the same directions as the epipolar line, and in an area where there are few changes of brightness and textures.

Results of 3-D information acquisition By using the relationship between disparity and 3-D information, which is described in Section 3.4, we show the results of depth and height calculation from estimated disparities. Table 1 shows the hyperboloidal mirror parameters and the focal length used for the computation.

Figure 10 shows the results of the depth calculation in omnidirection. Figure 11 shows the results of height calculation. The results are displayed on the reference omnidirectional image (lower omnidirectional image) as an intensity value goes up as the depth and the height value go up.

In order to observe the effectiveness of omnidirectional 3-D information for understanding the real 3-D environment, we show the depth and height information overlaid on a planar perspective projection image which is converted from a part of the omnidirectional image. This perspective projected image corresponds to the image captured by an ordinary camera placed at the hyperboloidal mirror's focal point. Figure 12 is the perspective projection image converted from the omnidirectional image. Figures 13 and 14 respectively show the depth and

 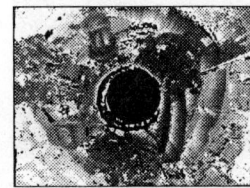 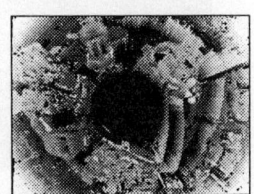

Fig. 9. Omnidirectional disparity image.

Fig. 10. Omnidirectional depth image.

Fig. 11. Omnidirectional height image.

Table 1. Parameters of Hyper-Omni Vision.

a	42.1 mm	$c = \sqrt{a^2 + b^2}$	60.0 mm
b	42.7 mm	focal length f	5.9~47.2 mm

Table 2. Parameters for computing binocular stereo images.

inter-ocular distance B_b	60.0 mm
focal length f_b	300.0 mm

height information in the transformed perspective image.

Reconstruction of binocular stereo images We can compute the distance L from the hyperboloidal mirror focus O_M to the target point P. Here stands Equation (8), with L, inter-ocular distance B_b, focal length f_b, and binocurla disparity d_b. When B_b and f_b are given, we can compute d_b for each corresponding pixels in binocular stereo images. In Fig. 15, shown are binocular stereo images which are generated from the image in Fig. 12 by using Eq. (8). Table 2 shows the parameters used for the computation.

$$d_b = B_b \frac{f_b}{L} \qquad (8)$$

By generating common stereo images depending on a user's view direction, a user can perceive the position and shape of the targets and understand the 3-D environment in some virtual reality applications

4.2 Discussion

It was confirmed that omnidirectional 3-D information is acquired by the experiment and the effectiveness of the proposed method was shown. The problem in this method is discussed below. The conventional stereo vision can obtain accurate correspondences only at clear edges or in areas with rich texture in images. Also, wrong correspondences may occur in the area where the similar features repeats. As we can see from our experimental results, in omnidirectional stereo vision, the same tendency can be observed. For solving this problem, corresponding point search should further be investigated.

The problems peculiar to the omnidirectional stereo are followings. The projection of the same targets in the 3-D environment have different sizes in the omnidirectional stereo image pair. Therefore, the 3-D area included in searching window of the reference image is different from that of the target image. We consider that the further investigation is required regarding the selection of the searching window's size.

Fig. 12. Computed perspective image.

Fig. 13. Depth image (perspective).

Fig. 14. Height image (perspective).

Fig. 15. A pair of computed binocular stereo images.

5 Conclusion

In this paper, an omnidirectional stereo method is presented, which can obtain three-dimensional information simultaneously in all directions. The method first takes a pair of omnidirectional images at different heights using the omnidirectional imaging sensor, HyperOmni Vision, and then determines correspondences between the pair of images and finally obtains three-dimensional information. Experimental results show the feasibility of the proposed method with real omnidirectional images. The computation of common horizontal binocular stereo images from a pair of omnidirectional images is also demonstrated for showing the possibility for virtual reality applications. This method can be useful for tele-operation or tele-control of the robots in remote sites.

The followings are the problems which still remain. First, dynamic sizing of the searching window must be exploited, in order to match the comparable areas in a searching window and a reference window. The selection of window shape which is suitable for the omnidirectional image is also required. Finally, planed are the development of a real time algorithm and system for generating stereo image pairs with horizontal parallax from omnidirectional stereo pair.

References

1. H.Ishiguro, M.Yamamoto and S.Tsuji: "Omni-directional Stereo for Making Global Map", *Proc. IEEE 3rd Int. Conf. on Computer Vision*, pp.540–547, 1990.
2. J.Hong, X.Tan, B.Pinette, R.Weiss and E.M.Riseman: "Image-based Homing", *Proc. IEEE Int. Conf. on Robotics and Automation*, pp.620–625, 1991.
3. K. Yamazawa, Y. Yagi and M. Yachida: "Obstacle detection with omnidirectional image sensor HyperOmni Vision", *Proc. Int. Conf. on Robotics and Automation*, Nagoya, IEEE, pp. 1062–1067 (1995).
4. B.K.P.Horn: *ROBOT VISION*, The MIT Press, 1986.
5. Y.Ohta and T.Kanade: "Stereo by Intra- and Inter-scanline Search Using Dynamic Programming", *IEEE Trans. Pattern Anal. Machine Intell.*, Vol.PAMI-7, No. 2, pp.139–154, 1985.
6. T.Kanade and M.Okutomi: "A Stereo Matching Algorithm with an Adaptive Window : Theory and Experiment", *IEEE Trans. Pattern Anal. Machine Intell.*, Vol. 16, No. 9, pp.920–932, 1994.
7. A.Chaen, K.Yamazawa, N.Yokoya, H.Takemura: "Omnidirectional Stereo Vision Using HyperOmni Vision", *Technical Report of IEICE*, IE96-122, pp.99–106, 1997 (in Japanese).
8. Y. Yagi, S. Kawato and S. Tsuji: "Collision avoidance using omnidirectional image sensor (COPIS)", *Proc. IEEE Int. Conf. on Robotics and Automation*, pp.910-915, April 1991.

Error Analysis in Stereo Vision*

R.S.Ramakrishna[†] B.Vaidyanathan [‡]

Abstract

This paper presents a new algorithm for error analysis in stereo vision based measurements. A framework for investigating the effects of various parameters has been proposed. It may also be applied for analysing coupled multivariate regression systems after suitable modifications.

1 Introduction

The maximum attainable accuracy with a prescribed set of parameters is of primary concern in stereo vision based measurements. The accuracy of the estimation depends on both the calibration- and the triangulation phases.

Several investigators have considered the case of plane parallel stereo in this context [2, 5, 6, 7]. In all this work, simulation results have been provided with a view to emphasize the importance of error analysis in stereo vision system design. A loose empirical bound for the uncertainty region is given in [5]. A stochastic technique is presented in [8]. A general approach for estimating the *pdf* of a function of several quantized variables, using a Taylor's series expansion of the function in terms of the distribution of the variables is reported in [9].A similar approach is persued in[10] wherein a first order expansion of the multivariate function is utilized. They requre that the calibration-triangulation function must be available in a closed form. A different measure of accuracy of calibration and triangulation, called the normalized stereo calibration error (NSCE), is proposed in [3]. This too has the drawback of [8]. A computational projective geometric approach has been proposed in [11], [12]. It also uses derived variances to improve the estimation of the parameters is described.

The need for a formal framework for investigating the quality of stereo vision based measurements is thus clearly felt. This paper proposes a *new* perturbational approach for estimating tight bounds on measurement error considering inaccuracies introduced during calibration and triangulation. The approach

---

*Supported in part by Board of Research in Nuclear Sciences, India, vide Grant No.25/1/88-G

[†]Information & Communication Deparatment, Kwangju Institute of Science & Technology, Kwangju, Korea 506-712.

[‡]Formerly Research Scholor, IIT., Bombay, India.

highlights the influence of camera geometry and the choice of calibration points on the final measurement. With this technique, the component of errors in the measurement due to each of the various inputs can be estimated under various conditions.

2 The Camera Model

The common camera model involving rigid body transformation followed by pinhole projection can be represented by a set of linear homogeneous equations. It has been shown in [1] that the addition of uncertainty in image center, scale factor, nonnormality of optical axis to the image plane etc., does not change the form of the model. This model can be described by

$$a_{11}X_W + a_{12}Y_W + a_{13}Z_W + a_{41}X^I X_W$$
$$+ a_{42}X^I Y_W + a_{43}X^I Z_W$$
$$+ a_{14} + a_{44}X^I = 0$$
$$a_{21}X_W + a_{22}Y_W + a_{23}Z_W + a_{41}Y^I X_W$$
$$+ a_{42}Y^I Y_W + a_{43}Y^I Z_W$$
$$+ a_{24} + a_{44}Y^I = 0$$

These equations map the 3-D world coordinates $\bar{X}_W = [X_W\ Y_W\ Z_W]^T$ to its image coordinates $\bar{X}^I = [X^I\ Y^I]^T$. Also, a_{ij} are the parameters of the camera system.

In the calibration phase, points with known 3-D coordinates are imaged and the system parameters are estimated from these. Those parameters for which calibration is effected are called the calibration parameters. If p points with known 3-D coordinates $\bar{X}_{Wi}$, $1 \leq i \leq p$ are imaged to their image coordinates $\bar{X}^I_{Li}$ by the left camera and $\bar{X}^I_{Ri}$ by the right camera, then for each of the p points, the model can be described compactly by [1],

$$P_L \bar{B}_L = \bar{q}_L \tag{1}$$
$$P_R \bar{B}_R = \bar{q}_R \tag{2}$$

$\bar{B}_L$ and $\bar{B}_R$ can be estimated in a least squares sense. This completes the calibration process.

If A is a point whose images in the left and right cameras are $\bar{X}_{Lai}$ and $\bar{X}_{Rai}$ respectively, then one can obtain $\bar{X}_{Wa}$, the back projection of A, during the triangulation phase. Indeed, it can be shown that

$$\begin{bmatrix} b_{L11}+X^I_{La} & b_{L12}+b_{L42}X^I_{La} & b_{L13}+b_{L43}X^I_{La} \\ b_{L21}+Y^I_{La} & b_{L22}+b_{L42}Y^I_{La} & b_{L23}+b_{L43}Y^I_{La} \\ b_{R11}+X^I_{Ra} & b_{R12}+b_{R42}X^I_{Ra} & b_{R13}+b_{R43}X^I_{Ra} \\ b_{R21}+Y^I_{Ra} & b_{R22}+b_{R42}Y^I_{Ra} & b_{R13}+b_{R43}Y^I_{Ra} \end{bmatrix} \begin{bmatrix} X_{Wa} \\ Y_{Wa} \\ Z_{Wa} \end{bmatrix} = \begin{bmatrix} -b_{L14}-b_{L44}X^I_{La} \\ -b_{L24}-b_{L44}Y^I_{La} \\ -b_{R14}-b_{R44}X^I_{Ra} \\ -b_{R24}-b_{R44}Y^I_{Ra} \end{bmatrix}$$

$$\cdots \tag{3}$$

Thus, one obtains

$$M\bar{X}_{Wa} = \bar{n} \quad (4)$$

where M and $\bar{n}$ are formed from $\bar{B}_L$, $\bar{B}_R$, $\bar{X}^I_{La}$ and $\bar{X}^I_{Ra}$.

The following errors need consideration:

1. Errors $\delta \bar{X}_{Wi}$, $1 \leq i \leq p$ in the world coordinates of the calibration points.

2. Errors $\delta \bar{X}_{ji}$, $1 \leq i \leq p$, $j \in \{L, R\}$ in the image coordinates of the calibration points.

3. Erors in the image coordinates of point A whose coordinates are to be triangulated for, $\delta \bar{X}^I_{ja}$, $j \in \{L, R\}$.

In the above context, the error estimation problem is one of obtaining $\delta \bar{X}_{Wa}$.

Incorporating these into the calibration equations, viz., eqns.(1), (2) the calibration–triangulation process can now be described by,

$$(P + \delta P)\bar{B}_L = \bar{q} + \delta \bar{q}$$
$$(M + \delta M)\bar{X}_{Wa} = \bar{n} + \delta \bar{n}$$

This is seen to be a perturbation problem of an overdetermined linear system. Several techniques have been reported for estimating the effect of inaccurate coefficients and RHS in a linear system [13, 14, 15, 16].

The primary difficulty in applying these techniques to the problem on hand is that the errors in the calibration parameters have to be obtained individually so as to apply them in the triangulation phase. Moreover, the structure of the calibration equations can be made use of. A closed form expression for the errors in each of the coordinates of $\bar{X}_W$ of a point would be highly desirable.

The next section presents a technique for estimating the sensitivity coefficients of the solution w.r.t each of the perturbed elements, using a Taylor's series expansion of eqns.(1), (2) and (4).

3 The Perturbation Algorithm

The problem of computing the error in the final measurement was shown to be one of perturbing the calibration–triangulation equations w.r.t. the calibration inputs $\bar{X}^I_{ij}$, $1 \leq i \leq p$, $j \in \{L, R\}$; $\bar{X}_{Wi}$, $1 \leq i \leq p$ and the image coordinates of the triangulation point. Assuming that the errors in the 3-D coordinates of the calibration points are small enough to be ignored, the triangulation equation, viz., eqn.(4) can be expanded in a Taylor's series to obtain

$$\Delta X_{Wa} = \frac{\partial X_{Wa}}{\partial b_{mn}} \Delta b_{mn} + \cdots +$$
$$\frac{\partial X_{Wa}}{\partial X_{La}^I} \Delta X_{La}^I + \frac{\partial X_{Wa}}{\partial X_{Ra}^I} \Delta X_{Ra}^I +$$
$$\frac{\partial X_{Wa}}{\partial Y_{La}^I} \Delta Y_{La}^I + \frac{\partial X_{Wa}}{\partial Y_{Ra}^I} \Delta Y_{Ra}^I \quad (5)$$

where the b_{mn} term occurs for all the calibration parameters. Similar expressions can be obtained for Y_{Wa} and Z_{Wa}. But, from the calibration equations viz., eqns. (1) and (2)), for each calibration parameter b_{mn}, $j \in \{L, R\}$, one has

$$\Delta b_{mn} = \frac{\partial b_{jmn}}{\partial X_{ji}^I} + \cdots, \; p \; terms$$
$$+ \frac{\partial b_{jmn}}{\partial Y_{ji}^I} + \cdots, \; p \; terms \quad (6)$$

The equations finally lead to :

$$P \begin{bmatrix} \frac{\partial b_{L11}}{\partial X_{L1}^I} \\ \vdots \end{bmatrix} = \begin{bmatrix} 0 \\ \vdots \\ -X_{Wi} - Y_{Wi} b_{L42} - Z_{Wi} b_{L43} - b_{L44} \\ \vdots \\ 0 \end{bmatrix} \leftarrow 2(i-1) - th \; row$$

for $1 \leq i \leq p$.

All the partial derivatives w.r.t. X_{Li}^I can be computed from the above equation. The process can be repeated to compute the partial derivatives of the calibration parameters w.r.t. X_{Ri}^I, Y_{Li}^I and Y_{Ri}^I, $1 \leq i \leq p$.

At this fully perturbed stage, the variation in the calibration parameters due to image point errors $\Delta \bar{B}_L$ can be found by,

$$\Delta \bar{B}_j = S_{Cj} \bar{\mathcal{X}}_j^I \quad (7)$$
$$\Delta \bar{\mathcal{X}}_j^I = [\; \cdots \; \delta X_{ji}^I \; \delta Y_{ji}^I \; \cdots \;]^T$$

for $j \in \{L, R\}$, $1 \leq i \leq p$. Here, S_{Cj} will be called the *sensitivity* matrix.

The next phase is to propagate the perturbations in the calibration parameters to triangulation. Similar computations lead to:

$$\Delta \bar{X}_{Wa} = S_{TCL} \Delta \bar{B}_L + S_{TCR} \Delta \bar{B}_R + S_{TI} \Delta \bar{\mathcal{X}}_a^I$$

where $\triangle \bar{B}_L$ and $\triangle \bar{B}_R$ are estimated from eqn. (7) and

$$\triangle \bar{\mathcal{X}}_a^I = \begin{bmatrix} \delta X_{La}^I & \delta Y_{La}^I & \delta X_{Ra}^I & \delta Y_{Ra}^I \end{bmatrix}^T$$

is the variation in the image coordinates of the point A. In combination with eqn.(7), the combined variation in the 3–D coordinate estimate due to variation in the calibration points and the image coordinates of the triangulated point becomes

$$\begin{aligned} \triangle \bar{X}_{Wa} &= S_{TCL} S_{CL} \triangle \bar{\mathcal{X}}_L^I + S_{TCR} S_{CR} \triangle \bar{\mathcal{X}}_R^I + S_{TI} \triangle \bar{\mathcal{X}}_a^I \\ &= S_L \triangle \bar{\mathcal{X}}_L^I + S_R \triangle \bar{\mathcal{X}}_R^I + S_{TI} \triangle \bar{\mathcal{X}}_a^I \end{aligned} \quad (8)$$

This completes the perturbation process as eqn.(8) represents a closed form expression for the error in the measured coordinate of A in terms of the errors in the image points both during calibration and triangulation. *It is to be noted that in this technique, the effect of perturbation of every element of the coefficient matrix on every element of the solution can be found and this is NOT the case with other techniques mentioned earlier.*

Also, the process of finding the sensitivity coefficients is not computation intensive as the pseudoinverses of P and M are to be found only once. These have to be computed in any case for calibration and triangulation.Also, the same concept can be used to propagate perturbations to any number of stages thereby extending the utility of the approach.

4 Discussion

The *LHS* of the error expression in eqn.(8) represents the effect on each of the elements of the solution vector. Now,

1. This can be viewed as defining a polyhedron of uncertainty around the measured point if the bounds on the independent variables are given.

2. Each of the input variables can be considered to be a random variable with known distribution and the error distribution can be determined.

In the case of (1) above, each sensitivity vector will represent a pair of parallel planes about the estimated point–one for the lower bound and one for the upper. The *aspect ratio* of the polyhedron formed by all the sensitivity vectors will determine if the error tends to occur along particular directions. Here, the difference between the actual region of uncertainty and the one given by the sensitivity equation can be seen. In the work reported in [8], the faces of the polyhedron representing a sensitivity vector are not parallel as in the case of the first order expansion. In the case of the plane parallel stereo, the region of uncertainty is seen to be $ABCD$ in Fig.1. On the other hand, the region $A'B'C'D'$ indicates the rejected terms(of the Taylor's series). The set of feasible solutions of the linear programming phase of [16] would fall within the region $ABCD$.

The authors have employed a dynamic programming approach for computing the farthest corners of this area of uncertainty, defined by the sensitivity coefficients, which might be of interest in many applications.

A similar computation can be performed in order to determine the direction of minimum error. The appropriate directions of errors in the inputs are also a byeproduct of these computations. The corresponding algorithm is shown below. The simplicity of the algorithm is noteworthy. This is obviously important in stereo vision applications.

```
Let:
    npars : Number of Parametrs
    Signs[npars] : Sign of Parameter for Maximum Error
    Sensitivity[npars][3] : Sensitivity Vectors

Algorithm :
    Xmax[3]=0;
    for i=1 to npars
    {
       Xplus=Xmax+Sensitivity[i]; (Vector Sum)
       Xminus=Xmax-Sensitivity[i];(Vector Sum)
       if(norm(Xplus)>=norm(Xminus))
               Signs[i]=+1;
          else
               Signs[i]=-1;
       Xmax=Xmax+Signs[i]*Sensitivity[i];(Vector Sum)
    }
```
Algorithm for Computing the Direction of Maximum Error.

If case(2) applies, then one might assume that the calibration points are specified by,
$$\bar{X}_L^I + \bar{x}_L^I \bar{X}_R^I + \bar{x}_R^I \bar{X}_a^I + \bar{x}_a^I$$

where $\bar{x}$ are normally distributed random errors with zero mean and covariance matrices $\Sigma_L, \Sigma_R, \Sigma_I$. If the errors in the image coordinates are independent, then the covariance matrix will be diagonal.

These can be inserted into the calibration sensitivity equation (eqn.(7)) to obtain
$$\triangle \bar{B}_j = S_{Cj} x_j^I, j \in \{L, R\}$$

Then, $\triangle \bar{B}_j, \triangle \bar{B}_j$ are normally distributed with covariances $S_{CL} \Sigma_L S_{CL}^T$ and $S_{CR} \Sigma_R S_{CR}^T$ and mean zero [17]. Again from eqn.(8) it follows that
$$\triangle \bar{X}_{Wa} = S_L \bar{x}_L^I + S_R \bar{x}_R^I + S_{TI} \bar{x}_a^I.$$

Assuming that $\bar{x}_L^I$, $\bar{x}_R^I$ and $\bar{x}_a^I$ are independent, the covariance matrix of the estimate is

$$\Sigma_{\hat{X}} = S_L \Sigma_L S_L^T + S_R \Sigma_R S_R^T + S_I \Sigma_I S_I^T$$

The three factors in the *RHS* of the above equation represent the effect of left camera calibration errors, right camera calibration errors and the error in the image coordinate of the triangulated point. Hence, this can be used to analyse the effects of the errors independently.

Also, the surfaces of constant probability are given by

$$\bar{X}_W^T \Sigma_{\hat{X}} \bar{X}_W \leq c^2$$

which is an ellipsoid with c^2 following a *Chi-square* distribution.

If e_1, e_2 and e_3 are the eigenvectors of $\Sigma_{\hat{X}}$ corresponding to eigenvalues $\lambda_1, \lambda_2, \lambda_3$ respectively, then the axes of the ellipsoid are given by $\frac{c}{\sqrt{\lambda_i}}$ in the directions $e_i, i = 1, 2, 3$.

5 Results

The perturbation algorithm was applied to a stereo vision system for analysing the effect of various parameters of the camera system on the measurement error. Initially, a number of camera geometries were simulated for a number of configurations of input points. In each case, the input image coordinates were perturbed with both uniformly distributed errors and also normal errors. It was found that the predicted direction of maximum error (DME) using the above dynamic programming approach was acceptable eventhough the error in it was much higher than the actual errors. In the case of normal errors, the prediction of the confidence interval was satisfactory. The dot product between the predicted DME and the actual DME is given in Table 1. The values are shown for each component of the error i.e., left camera calibration, right camera calibration etc. The number of estimates of ten points falling within 90% and 99% confidence interval is given in Table 2. For each point, 6000 samples were considered. These are representative of the actual experiments and it can be seen that the algorithm performs satisfactorily in both the cases.

6 Conclusion

A new approach for perturbing the stereo calibration–triangulation equations with error in input points has been presented. An expression for the errors in each of the triangulated coordinates, amenable to both deterministic and probablistic analyses has been derived. A dynamic programming approach for estimating the direction of maximum error has been given in the deterministic case.

References

[1] B.Vaidyanathan, R.S.Ramakrishna and N.R.Vishwambharan, *Stereo Vision Based Robotic Control*, Technical Report, IIT Bombay.

[2] E.McVay *et al* Some Accuracy and Resolution Aspects of Computer Vision Distance Measurements, *IEEE Trans. Pattern Analysis and Machine Intelligence,* PAMI 4(6), pp.646–649, 1982.

[3] J.Weng and M.Herniou, Camera Calibration with Distortion Models and Accuracy Evaluation,*IEEE Trans. Pattern Analysis and Machine Intelligence,* 14(10), pp.965–980, 1992.

[4] B.Kamgar–Parsi and B.Kamgar–Parsi, Evaluation of QUantization Error in Computer Vision, *Proceedings of IEEE–CVPR,* pp.52–60, 1988.

[5] R.Y.Tsai, A Versatile Camera Calibration Technique for High Accuracy 3–D Machine Vision Using Off–the–Shelf Cameras, *IEEE Jornal of Robotics and Automation,* Vol.RA–3(4), pp.323–344, 1987.

[6] A.Verri and V.Terra, Absolute Depth Estimate in Stereopsis, *Journal of Optical Society of America,* 3(3), 1986.

[7] J.J.Rodriguez and J.K.Aggarwal, Stochastic Analysis of Stereo Quantization Error, *IEEE Trans. on Pattern Analysis and Machine Intelligence,* PAMI 12(5), 1990.

[8] S.Blostein and T.Huang, Error Analysis in Stereo Determination of 3–D Point Positions, *IEEE Trans. on Pattern Analysis and Machine Intelligence,* PAMI 9(6), 1987.

[9] B.Kamgar–Parsi and B.Kamgar–Parsi, Quantization Error in Computer Vision, *Proceedings of Image Understanding Workshop,* 1988.

[10] P.W.Wong, On Quantization Error in Computer Vision, *IEEE Trans. on Pattern Analysis and Machine Intelligence,* PAMI 13(9), 1991.

[11] K.Kanatani, Unbiased Estimation and Statistical Analysis of 3–D Rigid Motion from Two Views, *IEEE Trans. on Pattern Analysis and Machine Intelligence,* PAMI 15(1), 1993.

[12] K.Kanatani, Statistical Analysis of Focal Length Calibration Using Vanishing Points, *IEEE Trans. on Robotics and Automation,* RA 8(6), 1992.

[13] G.H.Golub and C.H.Van Laon, An Analysis of the TOtal Least Squares Problem, *SIAM Journal of Numerical Analysis,* 17(6), 1980.

[14] G.W.Stewart, On the Perturbation of Pseudoinverses, Projections and Linear Least Squares Problems, *SIAM Review,* 19(4), 1977.

[15] G.W.Stewart, Matrix Perturbation Theory, Academic Press, Boston 1990.

[16] J.E.Cope and B.W.Rust, Bounds on Solutions of Linear Systems with Inaccurate Data, *SIAM Journal of Numerical Analysis*, 16(6), 1979.

[17] R.A.Johnson and D.W.Wichern, Applied Multivariate Statistical Analysis, Prentice Hall, Englewood Cliffs, New Jersey, 1982.

Errors In			
Calib(L)	Calib(R)	Image	Total
0.9972	1.0000	0.9976	0.9950
0.9988	1.0000	0.9977	0.9994
0.9999	0.9992	0.9973	0.9993
0.9972	0.9998	0.9963	0.9994
0.9996	0.9994	0.9957	0.9993
0.9992	0.9996	0.9963	0.9995
0.9991	0.9993	0.9965	0.9995
1.0000	1.0000	0.9964	0.9998
0.9995	0.9998	0.9958	0.9994
0.9955	0.9999	0.9965	0.9995

Table 1: DME Dot Products

Percentage of Points Under	
90% Ellipsoid	99% Ellipsoid
88.980	99.090
87.900	98.900
88.680	98.850
83.700	98.940
86.400	99.040
88.680	98.940
88.870	99.050
89.220	98.970
93.800	98.940
87.470	99.150

Table 2: Confidence Intervals

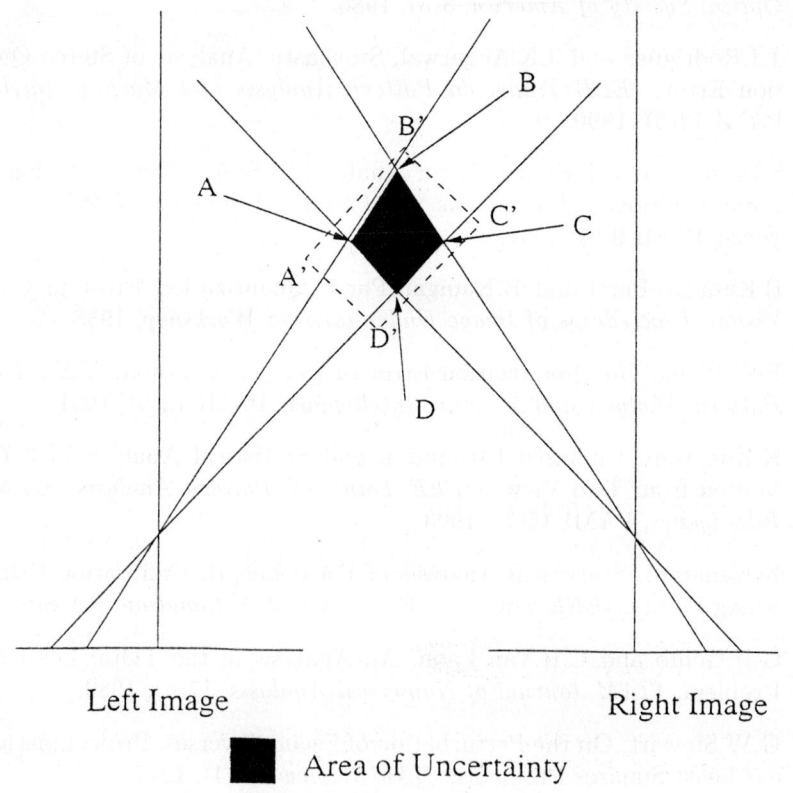

Figure 1: Geometrical Interpretation of Error Expression

Detecting Targets in SAR Images: A Machine Learning Approach

Qi Zhang   Zoran Duric   Ryszard S. Michalski*

George Mason University
4400 University Dr., Fairfax, VA 22030, USA

*Also with GMU Departments of Computer Science and Systems Engineering, and the Institute of Computer Sciences, Polish Academy of Science

Abstract

This paper describes a novel application of the MIST methodology to target detection in SAR images. Specifically, a polarimetric whitening filter and a constant false alarm rate detector are used to preprocess a SAR image; then the AQ15c learning program is applied to learn and detect targets. Encouraging and impressive experimental results are provided.

KEYWORD: Learning in vision, target detection in SAR images.

1 Introduction

Research on computer vision reveals that it is necessary for a flexible and robust vision system to incorporate learning capabilities. Now this line of research has become an active area (Michalski et al., 1992; Bhanu & Poggo, 1994). This paper is focused on using machine learning methods to detect targets in SAR images.

A synthetic aperture radar (SAR) is an all-weather imaging device, able to provide good images of what it has detected even in fog, clouds, or darkness in which optical sensors are useless. Detecting and recognizing targets in SAR images is of much significance in military and civil applications. Successful detection of targets in SAR images is difficult because there is a large amount of noise in image data. This paper presents a novel machine learning approach to this problem by using the MIST/AQ method (Michalski et al., 1996). In particular, a SAR image is first processed by a polarimetric whitening filter (PWF) to improve image quality (Novak et al., 1990), and then input to a constant false alarm rate (CFAR) detector for screening its natural clutter and detecting potential targets within it (Ravid & Levanon, 1992). The output from the CFAR detector is used by AQ15c (Michalski. et. al., 1986; Wnek et. al., 1996) for training and detection. The 1ft x 1 ft resolution SAR images we used were collected in Stockbridge New York, by the Lincoln Laboratory at MIT and provided to us by DARPA. The experimental results were very encouraging and impressive.

2 Related Work

Target detection in SAR images is difficult because of large amount of noise in image data. Kreithen et al. (1993) used the polarimetric whitening filter and a CFAR detector to preprocess SAR images. After grouping sets of clustered pixels (potential targets) which were seemingly from the same targets, attribute values were generated for each of them and a quadratic distance for each potential target was calculated and compared to a threshold determined previously by experiments. Obviously, a single threshold is not so understandable or flexible as symbolic knowledge descriptions. Besides, determining which pixels came from a target or not and then grouping them is often difficult or impossible due to noise. Burl et al. (1994) used the matched filter technique to detect potential volcanoes in Venus SAR images. Matched filters for each kind of volcano were constructed from training volcano examples and then were applied to scan an image pixel by pixel to locate potential volcanoes. The matched filter is possibly subject to rotation, size and other vision condition changes. Further, using matched filter to scan a whole image and computing the degrees of match with each constructed filter is time-consuming.

Application of machine learning techniques to target detection in images is relatively new. Rong and Bhanu (1996) adopted the reinforcement learning method. The training and testing were directly performed on raw FLIR (forward-looking infrared radar, not SAR) images without any transformation. They divided an image into many rectangulars which were the input unit to a learning system. Directly training and testing on raw data could consume more time and generate harder learning problems and dividing the image into small areas for training could lead to incompleteness of target information.

3 MIST Methodology

Among the most important research goals of applying machine learning methods to vision problems is to gain better understanding of matching appropriate learning methods to appropriate vision problems. MIST (Multilevel Image Sampling and Transformation) has been developed as a general methodology for applying machine learning methods to vision problems (Michalski et al., 1996). The purpose of MIST is to provide a researcher with an environment in which a variety of machine learning methods and approaches can be flexibly applied to a wide range of vision problems.

The MIST methodology works in two modes: *training mode* and *interpretation mode* (Figure 1). In the training mode, four steps are needed and, based on training performance, possibly repeated for better results. *Event generation* is to generate examples for learning or testing. *Description space generation* means that a trainer assigns concept names to areas in training images that contain concepts or objects to be learned. In the interpretation mode, three steps are executed. *Transformation application* is to apply the Image Knowledge Base (IKB) to examples generated from testing images to produce a transformed version which, for instance, could be a segmented image. The output is annotated symbolic images (ASI). In an ASI, areas or objects that correspond to the recognized concepts in the original image are marked by symbols (e.g., colors) denoting these concepts and linked to *concept annotations* (text containing additional information about that concept, such as degree of certainty of recognition, properties of the concepts, relation to other concepts, etc.). The output ASI in one level can be the end result (one-level training) or input to later levels (multilevel training) for better results (e.g., repeated training on the same natural scene image) or for other tasks (e.g., target recognition). The image knowledge base (IKB) contains prior or learned descriptions or visual concepts, image processing operators, attribute extraction operators and background knowledge relevant to image interpretation.

Among the advantages of this methodology are the ease of applying and testing diverse learning methods in a uniform manner, the potential for implementing advanced and complex learning processes, the natural way of interpreting images.

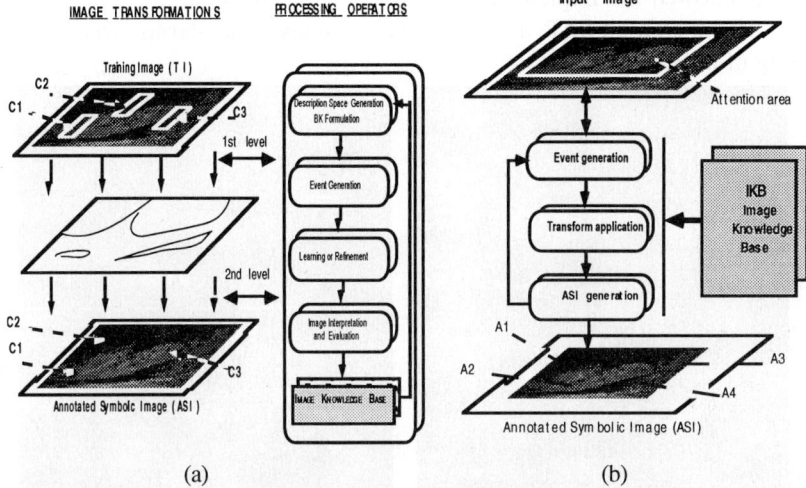

Fig. 1. The MIST methodology (a) training mode, (b) interpretation mode.

4 Learning to Detect Targets

Guided by the MIST methodology, we designed a MIST/AQ system which consists of three sequential phases: enhancing image quality by a PWF, screening by a CFAR detector, and target learning and detection by AQ15c (Figure 2). The first two stages provide the preprocessing needed for further target detection and recognition.

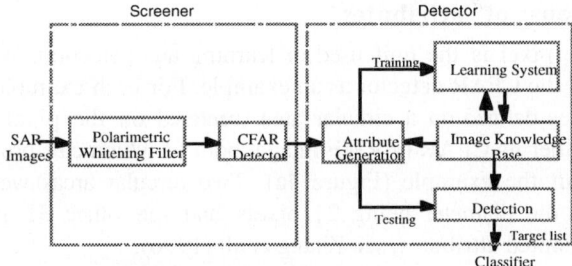

Fig. 2. The system architecture.

4.1 Polarimetric Whitening Filter

The technique called the polarimetric whitening filter (PWF) (Novak et al., 1990) improves SAR image quality in two aspects: minimization of the amount of speckle (noise) and sharpening the edges of image objects. For the details of this technique, see Novak et al.(1990) and for its application to this work, see Zhang et al. (1996).

4.2 CFAR Screening

Various constant false rate alarm (CFAR) algorithms (e.g., Ravad & Levanon, 1992; Wang et al., 1994) take a SAR image as input and perform a *screening* function, i.e., detecting potential targets in SAR images by examining intensities of radar returns and thereby providing a massive reduction of natural clutter (grass, trees etc.). Figure 3b is the CFAR image (i.e., processed by a CFAR detector) of Figure 3a. In our work, we followed the implementation presented in (Wang et al., 1994). Note that due to noise inherent in SAR images, there are many false alarms which passed the screening of the CFAR detector.

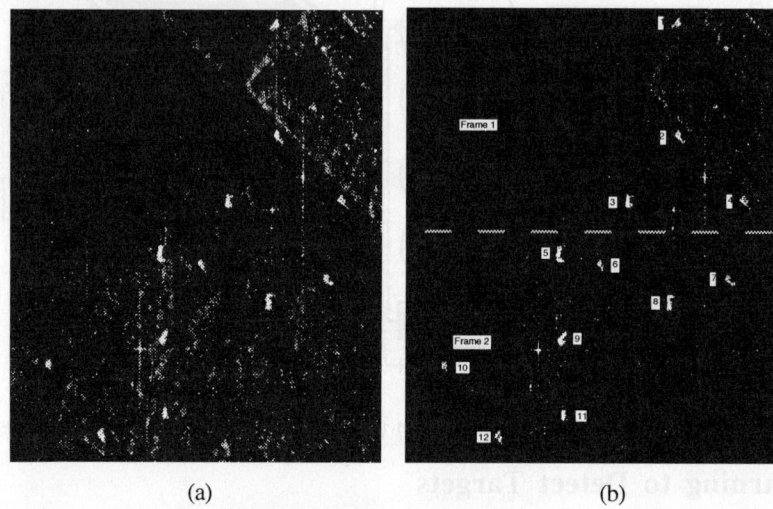

(a) (b)

Fig. 3. An exemplary (a) PWF image and (b) CFAR image containing targets: 2,11: M60-tank; 3,7,9: M48-tank; 4,10: M84-APC; 6: M113-APC; 12: M59-APC; 5,8: M55-howitzer; 1: unknown.

4.3 Learning and Detecting

4.3.1 Definitions of attributes

In our work, the pixel is the unit used in learning and detection. We consider each pixel that passed the CFAR detector as an example. For each example, produce a set of attribute values defined on a circular area (centered on the pixel or example of interest) from either its CFAR or PWF image, both of which can provide descriptive information about the example (Figure 4a). Two circular areas were used in this work, one with the diameter being 21 pixels and the other 31 pixels. For the definitions of adopted attributes, see Zhang et al. (1996).

4.3.2 Classification of training examples and discretization

Before learning, training examples must be classified as target and non-target examples. This is not easy, since it is impossible to accurately decide whether the pixels on the border of a target are target examples or non-target examples. Thus, there might be some misclassified examples for training. We adopted a simple rule: pixels connected (8-connectivity) to a target are target examples and otherwise non-target examples (Figure 4b).

Considering the characteristics of SAR image data, we adopted the Chi-merge discretization scheme (Kerber, 1992) and applied it to extracted raw data from CFAR or PWF images. The following are exemplary examples after discretization (attributes ending in "1" represent attributes defined on a circle with ta diameter of 21 pixels, those ending in "2" are for circles with a 31-pixel diameter):

Target examples
(1) [power1=49] [sd1=24] [fractal1=7] [area1=3] [wrfr1=8] [power2=25] [sd2=10] [fractal2=6] [area2=3] [wrfr2=4]
(2) [power1=47] [sd1=24] [fractal1=7] [area1=2] [wrfr1=9] [power2=24] [sd2=10] [fractal2=7] [area2=3] [wrfr2=5]

Non-target examples
(1) [power1=6] [sd1=0] [fractal1=0] [area1=0] [wrfr1=0] [power2=0] [sd2=0] [fractal2=0] [area2=0] [wrfr2=0]
(2) [power1=0] [sd1=0] [fractal1=0] [area1=0] [wrfr1=2] [power2=0] [sd2=0] [fractal2=0] [area2=0] [wrfr2=0]

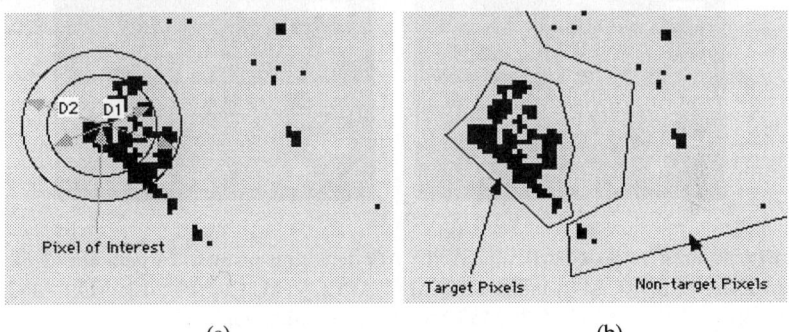

(a)   (b)

Fig. 4. Enlarged target 4 in Fig 3b. (a) Attributes of a pixel are defined on circular areas; (b) Classification of target and non-target examples.

4.3.3 Training

The SAR image acquired during a single imaging process by an airplane or satellite is called a *pass*, which can be divided into smaller portions called *frames*. Image objects are partitioned into two classes: targets and non-targets. The following are exemplary AQ rules acquired by taking pixels in frame 1 of Figure 4 as training examples.

Target examples
Rule1 [sd1=13..57] & [area1=0..10] & [wrfr1=0..9] & [power2=19..99] & [sd2=7..23] & [fractal2=3..15] & [wrfr2=0..8]   (t-weight:842, u-weight:107)
Rule2 [area1=10..12] & [area2=0..20]   (t-weight:742, u-weight:18)

Non-target examples
Rule1 [area1=0..9] & [area2=0..14] & [wrfr2=0..3] (t-weight:1561, u-weight:890)
Rule2 [power1=0]   (t-weight:632, u-weight:8)

The t-weight in the above rules represents the total number of examples covered by a rule and the u-weight the number of examples uniquely covered by that rule. The larger weights are, the more stronger the pattern in a learned rule represents.

5 Experimental Results and Discussion

We tested all our SAR images whose ground truth is known. The experimental results are summarized in Table 1.

experiment	# training examples	# testing examples	# learned AQ rules	accuracy (pixel-based)	accuracy (target-based)
1	3069	1706	12	98%	100%
2	9284	4915	19	98%	100%

Table 1. Summary of experimental results.

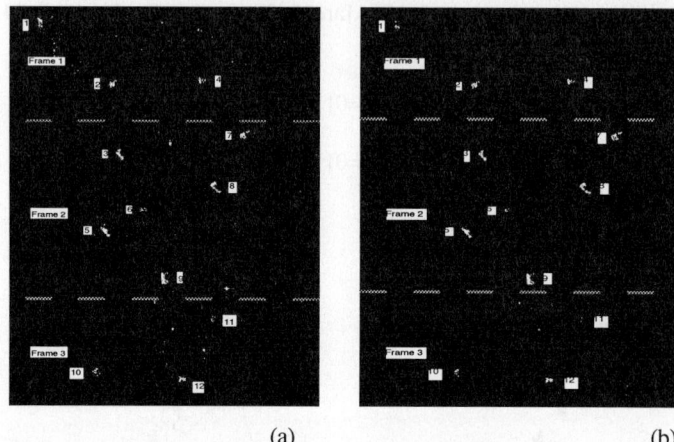

(a)   (b)

Fig. 5. Pass 7: (a) CFAR image, (b) AQ detection results. Targets are: 2,11: M60-tank; 3,7,9: M48-tank; 4,10: M84-APC; 6: M113-APC; 12: M59-APC; 5,8: M55-howitzer; 1: unknown.

In the first experiment, training was executed on Frame 1 in Figure 3b and tested on Frame 2 in Figure 5a. The goal of this experiment was to see the performance of the rules learned in one pass but applied to a *another* place of a *different* pass.

In the second experiment, in addition to Frame 1 in Figure 3b, the data in Frame 2 in Figure 6a was also used for training, in which there is a power-line tower. Frame 1 in Figure 6b was used for testing. As can be seen, AQ detection results were excellent (Figure 6b). Further, the rules learned in this experiment were retested in Figure 5b and the results were almost the same. Note that immense number of false alarms in Figure 6b indicate the difficulty in determining and grouping pixels which is the way adopted by Kreithen et al. (1993).

The first two experiments indicate that the MIST/AQ approach was able to capture the patterns among data and that learned knowledge was successfully applied to testing images, even to untrained targets. The results of the second experiment are interesting and important because it shows the necessity of learning. It can be easily seen from Figure 6b that it is virtually impossible to remove non-target examples by using vision techniques such as the size filter or majority voting.

An outstanding aspect in our experiments is that all targets remained after AQ detection while false alarms were maximally reduced, almost to zero. Another thing worth mentioning is that even though there was noise in our training data the results were still satisfactory. A possible way of avoiding data noise and acquiring fewer rules is selecting only some typical target pixels (e.g., pixels in the center of a target) for learning. However rules learned in this way might not capture the various data patterns near or on the border of a target. Because of this, the spatial distribution (important for target recognition) of pixels of detected targets would be damaged. Further, some targets would probably disappear. Experiments proved this analysis (not shown here). In our results, the number of learned rules was small and almost all targets were well preserved in their spatial distribution of pixels.

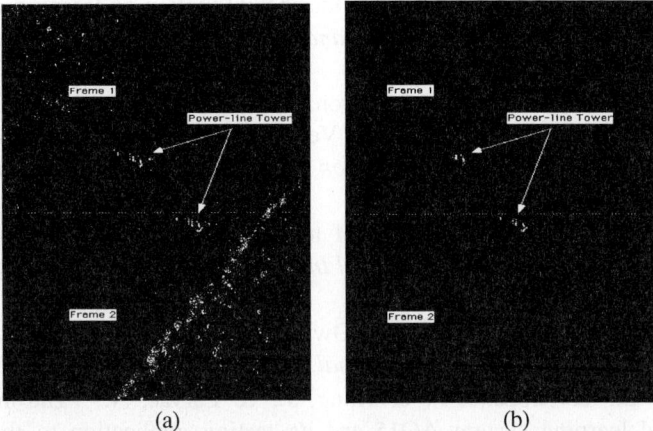

Fig. 6. Pass 5: (a) the CFAR image; (b) AQ detection results.

Our system for target detection is a good combination of vision and machine learning techniques. The noise reduction by the PWF and the screening from a CFAR detector maximally reduce and improve data needed as input to a learning method, in contrast to directly performing learning and testing on raw images (Rong & Bhanu, 1996).

6 Conclusion

This paper describes a novel application of the MIST methodology to target detection in SAR images. The presented MIST/AQ system for target detection in SAR images consists of: using the polarimetric whitening filter to enhance the quality of SAR images, applying a CFAR detector to screen natural clutter to maximally reduce the unnecessary information for training, and learning and detecting targets by AQ15c. The contributions of our approach can be summarized: the pixel-based operations in our approach avoid the problem of determining and grouping pixels or building templates; utilizing vision techniques maximally reduces and cleans data necessary to learning so that learning is more likely to succeed; false alarms are maximally reduced so as to provide a classifier with a list of potential targets of good quality. Experimental results presented are very promising and clearly show the effectiveness of the MIST/AQ method for solving this problem.

Acknowledgements

We would like to thank Dr. Maloof for his help in this work, Dr. Kaufman for his technical support in preparing image data, Jim Mitchell for his comments. This research was conducted in the Machine Learning and Inference Laboratory at George Mason University. The Laboratory's activities are supported in part by the Defense Advanced Research Projects Agency under grant F49620-95-1-0462, administered by the Air Force Office of Scientific Research, in part by the National Science

Foundation under grants DMI-9496192 and IRI-9020266, and in part by the Office of Naval Research under grant N00014-91-J-1351

References

Bhanu, B & Poggio, T., "Introduction to the special section on learning in computer vision", *IEEE Transactions on pattern analysis and machine intelligence*, vol. 16, no. 9, 1994.

Burl, M.C., & Fayyad, U.M., & Perona, P., & Smyth, P., & Burl, M.P., "Automating the hunt for volcanoes on Venus", *Proceedings 1994 IEEE Computer Society Conference on Computer Vision and Pattern Recognition*, Seattle, WA, pp. 302-309, 1994.

Kerber, R., "ChiMerge: discretization of numeric attributes", *Proceedings of the Tenth National Conference on Artificial Intelligence*, San Jose, CA, pp. 123-128, 1992.

Kreithen, D.E., & Halversen, S.D., & Owirka, G.J., "Discriminating targets from clutter", *The Lincoln Laboratory Journal*, vol. 6, no. 1, pp.25-52, 1993.

Michalski, R.S., & Mozetic, I., & Hong, J., & Larvac, N., "The multipurpose incremental learning system AQ15 and its testing application to three medical domains", *Proceedings of the 5th National Conference on Artificial Intelligence.*, 1986.

Michalski, R.S., & Rosenfeld, A. & Aloimonos, "Machine Vision and Learning: a report on the NSF/ARPA workshop on machine learning and vision", Harpers Ferry, WV., 1992.

Michalski, R.S., & Zhang, Q., & Maloof, M.A. & Bloedorn, E., "The MIST methodology and its application to natural scene interpretation", *Proceedings of Image Understanding Workshop*, Palms Springs, CA., 1996.

Novak, L.M., & Burl, M.C., Chaney, & R.D., & Owirka, G.J., "Optimal processing of polarimetric synthetic-aperture radar imagery", *The Lincoln Laboratory Journal,* vol. 3, no. 2, pp.273-290, 1990.

Ravid, & Levanon, N., "Maximum-likelihood CFAR for Weibull background", *IEE Proceedings-F*, vol. 139, no. 3. June, 1992, pp.256-264, 1992.

Rong, S. & Bhanu, B., "Reinforcement learning for integrating context with clutter models for target detection", *Proceedings of Image Understanding Workshop*, Palms Springs, CA, pp.1389-1394, 1996.

Wang, Y, & Chellappa, R., & Zheng, Q., "Detection of point targets in high resolution synthetic aperture radar images", *Proceedings of IEEE International Conference on Acoustics, Speech and Signal Processing,* vol. 5, pp9-12, 1994.

Wnek, J., & Kaufman, K., & Bloedorn, E., & Michalski, R.S., "Inductive learning system AQ15c: the method and user's guide", Reports of the Machine Learning and Inference Laboratory, MLI 95-4, George Mason University, Fairfax, VA., 1995.

Zhang, Q., & Duric, Z., Michalski, R.S., "Target detection in SAR images: the MIST/AQ method", MLI 96-28, Reports of Machine Learning and Inference Laboratory, George Mason University, Fairfax, VA., 1996.

Precise Matching by Robust Estimation of Deformation and Local Coherence

Zhong-Dan LAN [*], Roger MOHR and Long QUAN

Laboratoire Gravir, Projet Movi,
Inria, 655 avenue de l'Europe, 38330 Montbonnot France
zhong-dan.lan@imag.fr, zhong-dan.lan@mailexcite.com

Abstract. In this paper, we present a linear method that incorporates information from neighboring pixels for sub-pixel matching. Two algorithms are presented. Both rely on a rough initial estimate of the disparity. The first one is optimized for pairs of images requiring negligible window deformation. The second method is slower but more general and more precise. It is applicable for large window deformation and eliminates false initial matches using robust estimation of the local affine window transformation. The first algorithm attains a precision of 0.05 pixels for interest points and 0.06 for random points in the translational case. For general case, if the deformation is small, the second method gives an accuracy of 0.05 pixels; while for large deformation, it gives an accuracy of about 0.06 pixels for points of interest and 0.10 pixels for random points.

Key-words : Correlation, Precision, Robustness, Convolution, Matching.

1 Introduction

Visual correspondence is a major issue in vision [12]. It underlies both motion and stereo. Correlation techniques are widely used for it [1, 2, 17, 8, 3, 14, 11]. In Aschwanden [2], the problems of photometric variation and geometric distortion are treated. In Okutomi [14], the choice of window size is optimized to minimize the matching uncertainty. In Zabih [17], Bhat [3] and Lan [8, 10], *partial occlusion* (called *factionalism* by Zabih) is considered, using respectively non-parametric and robust statistical methods, while in Lotti [11], it is treated by constraining the windows by contours.

In this paper, we present a direct linear correlation method, which attains sub-pixel accuracy by incorporating information from neighboring pixels.

The paper is organized as follows: we first introduce our correlation model and explain how we get sub-pixel information from the correlation coefficients,

---

[*] Currently at *Max Planck Institute for Cognitive Neuroscience, Inselstrasse 22-26, 04103 Leipzig, Germany.*

taking into account the estimated uncertainty in the displacements. The 'fast' (rigid translation based) algorithm is given in this section. When there is a non negligible deformation, we estimate it robustly from a rough dense disparity map, and then estimate the local translation. This gives the 'robust' algorithm. experiments on some real images. For each algorithm, experiments on real images are shown. Both of the two algorithms can be followed by a post-processing, which incorporates the epipolar geometry and improves quantitatively the results. Finally, we give conclusions and some suggestions for future work.

2 Pure translation case: the Fast Algorithm

In this section, we consider the case where there is a pure translation between the two images to be matched, *i.e.* the window deformation is negligible. The algorithm is very fast and gives quite precise results. We first show that a signal translation can be approximated by a convolution, then we show that estimating the best convolution mask gives us the translation. To allow estimates to be fused, we also estimate their uncertainties. Finally some experimental results validate the approach.

2.1 Sub-pixel translation by convolution

We introduce our signal matching model in the one-dimensional case, which can be naturally extended to two dimensions.

Suppose we have two discrete signals f_1 and f_2, sampled from F_1 and F_2. F_1 and F_2 are defined on the real line $\mathcal{R}$, f_1 and f_2 are defined only at integer values.

The *matching* problem is the following:

Given f_1 and f_2, find $t \in \mathcal{R}$, such that: $F_1(x) \sim F_2(x+t) = F_2 * \delta_{-t}$.

Here $*$ means a convolution and $\sim$ means equality up to a signal transformation, for instance an offset (O), a scale effect (S), and a noise, *i.e.*: $F_1(x) = SF_2(x+t) + O + \epsilon$. In this paper, we concentrate on *precise matching*, for which an initial approximate match is given. Therefore, we suppose that $t \in [-1, 1]$. As we only know the values of f_1 and f_2, we are led to look for $t \in [-1, 1]$, such that $f_1 \sim f_2 * \delta_{-t}$.

If the function f_2 is regular, a piecewise linear interpolation is adequate: $f_2(i+\epsilon) \approx (1-\epsilon)f_2(i) + \epsilon f_2(i+1)$ for $0 \le \epsilon \le 1$. This is the convolution with $(1-\epsilon)\delta_0 + \epsilon\delta_{-1}$, an approximation to $\delta_{-\epsilon}$ for the signal f_2.

More generally, if we have b_i, λ_i, such that $\sum_i b_i = 1$, then we can have $\sum_i b_i f_2(x + \lambda_i) \approx f_2(x + \sum \lambda_i b_i)$, *i.e.* $\sum_i b_i \delta_{-\lambda_i}$ is an approximation of $\delta_{-\sum_i \lambda_i b_i}$ for a sufficient smooth function f_2.

f_1 and f_2 are only defined on integer values, so, instead of δ_{-t}, we propose to seek a linear combination of δ_{-1}, δ_0 and δ_1, $b_1\delta_{-1} + b_0\delta_0 + b_{-1}\delta_1$ ($b_{-1} + b_0 + b_1 = 1$), such that $f_1 \sim f_2 * (b_1\delta_{-1} + b_0\delta_0 + b_{-1}\delta_1)$. Once the b_i ($i = -1, 0, 1$) are found, we approximate $b_1\delta_{-1} + b_0\delta_0 + b_{-1}\delta_1$ by $\delta_{b_{-1}-b_1}$ and the displacement is estimated to be $b_1 - b_{-1}$.

In the signal matching problem, a rescaling effect and offset are usually also allowed. So we are led to estimate a_1, a_0, a_{-1} and O, such that:

$$f_1 = f_2 * (a_1\delta_{-1} + a_0\delta_0 + a_{-1}\delta_1) + O + \epsilon. \tag{1}$$

We solve this problem by least square minimization and we then define $b_i = a_i / \sum_{j=-1}^{j} a_j (i = -1, 0, 1)$, and again we estimate the displacement as $b_1 - b_{-1}$.

As the problem is symmetric, we could also have $f_2 = f_1 * (a'_1\delta_{-1} + a'_0\delta_0 + a'_{-1}\delta_1) + O' + \epsilon.$, we define $b'_i = a'_i / \sum_{i=-1}^{1} a'_i$ and the sub-pixel disparity is $b'_{-1} - b'_1$.

After having got the two estimates of the displacement, we can fuse them using their uncertainty to improve the result. Lack of space, the estimation of their uncertainty is omitted.

The approach extends naturally to two dimensions and we can thus get the sub-pixel displacement between two two-dimensional signals.

2.2 Link to other methods

We discuss the link between our model and other approaches. As this is just a theoretical analysis, we limit the attention to the one-dimensional case.

Consider the one-dimensional model (1): $f_1 = f_2 * (a_1\delta_{-1} + a_0\delta_0 + a_{-1}\delta_1) + O + \epsilon$. If we suppose that there is no offset ($O = 0$) and no rescaling ($a_1 + a_0 + a_{-1} = 1$), we get : $f_1 = f_2 * (a_1\delta_{-1} + (1 - a_1 - a_{-1})\delta_0 + a_{-1}\delta_1) + \epsilon$, i.e: $f_1 = a_1 f_2 * (\delta_{-1} - \delta_0) + a_{-1} f_2 * (\delta_1 - \delta_0) + f_2 + \epsilon$.

If we take $f_2(x+1) - f_2(x) = (f_2 * (\delta_{-1} - \delta_0))(x)$ and $f_2(x) - f_2(x-1) = (f_2 * (\delta_0 - \delta_1))(x)$ as two approximations of $F'_2(x)$, then we get : $f_1(x) \approx f_2(x) + (a_1 - a_{-1})F'_2(x)$.

This is Förstner's model [4]. So, our model can be seen as a generalization of Förstner's by taking into account the intensity offset and rescaling.

However, there are some more general models : that of Ackermann [1] or Gruen [6]. These two models consider the deformation (supposed affine) between two image windows.

More precisely, Ackermann modeled the corresponding intensities of the two images by an affine relation, $(f_1(x) = r + hf_2(ax + b))$, while Gruen used a simple offset $(f_1(x) = r + f_2(ax + b))$.

These two methods work iteratively; they do not give a precise result after the first iteration step. In fact, to implement their models, $f_2(ax+b)$ is approximated by $f_2(x) + (a - 1)xf'_2(x) + bf'_2(x)$, this approximation is good only when both b and $(a - 1)x$ are small. If the initial match is good enough, b is small, but $(a - 1)x$ might be large.

These analysis show that the window deformation can not be approximated very well from the image intensity relations alone. While window deformation occurs, we propose firstly estimating this window deformation by robust statistics and then estimating the displacement between the deformed windows. This method will be presented in section 3.

2.3 Experimental results

For experiments, we used a planar scene because a precise homography can be established between the two images in this case which can be used for evaluating. The homography is accurately estimated by matching targets [15]. The epipolar constraint and region of interest constraints are also used in the initialization stage to accelerate the process and reduce ambiguity [2, 8]. Validation is carried out on the points of interest [13] and some randomly selected general points.

For each experiment, the initialization is made by the *zncc* (*zero mean normalized cross correlation*) method [2]. The matching error for a pixel is defined to be the distance in the image between the pixel found and the pixel computed by the accurate homography. The *outlier percentage* is computed, where outlier means a match which is more than one pixel away from the accurate position. For good matches, we compute the mean of the errors. Window dimension varies from 7×7 to 15×15.

Consider the pair of images *Bolino* displayed in Figure 1 (a) and (b). The motion between the two images is essentially a translation with almost no rotation and zoom effects. We match the interest points [13] and some randomly chosen points also.

For the interest points in image *Bolino*, using our method with a precision of about 0.07 pixels is achieved. For general points, a precision of about 0.09 pixels is achieved. From such matchings, epipolar geometry can be accurately estimated. Re-projecting the matched points on the epipolar line improves then the accuracy, and a precision of about 0.05 pixels for interest points and 0.06 pixels for general points is obtained.

For a window of dimension 11×11, the average time per point is about 0.003 second on Sun Sparc 10.

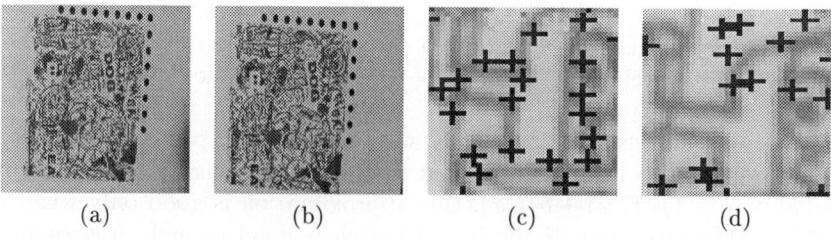

Fig. 1. The image pair *Bolino* and the points for test. (a) left image *Bolino1*; (b) right image *Bolino2*; (c) points of interest of *Bolino1* in (a); (d) randomly chosen points of *Bolino1* in (a).

3 General case: a robust algorithm

Unfortunately, the above algorithm is only applicable for pairs of images where the window deformation is negligible. To handle window deformations, we have developed a robust algorithm that is more general, and also more precise.

In this section, we develop a precise correlation algorithm that uses a rough dense correspondance map obtained by a global matching method as dynamic programming [5, 7]. We first try to recover the window deformation, and then estimate the local translation on the deformed window.

Let us assume the world is locally planar. As the patch seen through the correlation window is small, parallel projection is a good approximation and under these conditions, windows match through an affine transformation.

Given dense point correspondences (P_1, P_2) of two images I_1 and I_2, in a window around each point, we first seek A and t, such that $P_2 \approx AP_1 + t$, the affine transformation (A, t) captures approximately the window deformation; then we estimate the precise translation between $I_1(P_1)$ and $I_2(AP_1 + t)$, using the method developed above. Section 3.1 describes the robust estimation of the local affine window deformation, section 3.2, describes the local displacement estimation. This algorithm which considers also the window deformation, is called the *robust algorithm*.

3.1 Computing the affine transformation

Given an initial correspondance between two images of the same scene, under the assumption that the correspondance is locally affine [1, 6], we estimate the transformation and refine the correspondance.

Suppose the two two-dimensional point sets are $P_i = (x_i, y_i)^t$ and $P'_i = (x'_i, y'_i)^t$ $(i = 1 \ldots n)$. We look for $A = \begin{pmatrix} a_{11} & a_{12} \\ a_{21} & a_{22} \end{pmatrix}$ and $t = \begin{pmatrix} t_x \\ t_y \end{pmatrix}$, such that :

$$\begin{cases} a_{11}x_i + a_{12}y_i + t_x = x'_i \\ a_{21}x_i + a_{22}y_i + t_y = y'_i \end{cases}$$

for $i = 1 \ldots n$.

The method must be robust because of possible false matches. We use the Least Median Square [16] method (*LMedS*) to robustly find an initial affine transformation, detect outliers, and then perform least-square minimization on the inliers to get a precise result.

The *LMedS* method is implemented using Random Sampling [16]. One parameter to be set is the number of random samplings n. In our experiments, n is set to be 35.

3.2 How to compute the sub-pixel displacement

Once we have found the affine transformation between the two windows, we can reproject the points P to P', that is to say, replace P' by $AP + t$, called the *corrected match*. Through this correction step, the matches which are not coherent with their neighbors (in the local affine sense) are corrected, which makes the algorithm *robust* to false matches.

Now, suppose that we have two corresponding two-dimensional point sets $P_1(i,j)$ and $P_2(i,j)$, $P_2(i,j) = AP_1(i,j) + t$, where A is a 2×2 matrix and t is a two-dimensional vector.

Let I_1 and I_2 be the two images. $M_1(i,j) = I_1(P_1(i,j))$ and $M_2(i,j) = I_2(P_2(i,j))$, i.e., M_1 is the intensity window in the first image and the M_2 is the deformed window in the second one. The same technique as described in section 2.1 is applied to get the sub-pixel displacement between M_1 and M_2.

3.3 Experimental results

Lack of space, we only show results on the images 2. Matching is done between the pair *New1* and *New2*, the pair *New1* and *New3*, the pair *New1* and *New4*, on interest points and randomly chosen points.

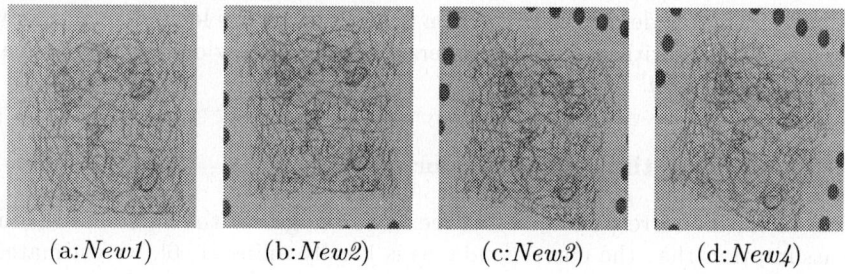

(a:*New1*) (b:*New2*) (c:*New3*) (d:*New4*)

Fig. 2. Images *New*.

Image pair	*New1* and *New2*	*New1* and *New3*	*New1* and *New4*
Geometric change	$S = 1 : 0.9, r \approx 0$	$S=1 : 0.9, r = 20$	$S1 : 0.9, r = 30$
Initialization	$F = 18\%, P = 0.53$	$F = 37\%, P = 0.56$	$F = 37\%, F = 0.57$
Our method	$IP: F = 0\%, P = 0.04$	$F = 0\%, P = 0.06$	$F < 1\%, P = 0.07$
	$RP: F = 0\%, P = 0.06$	$F < 1\%, P = 0.09$	$F < 1\%, P = 0.11$
Ackermann/Gruen's method	$IP: F \approx 1\%, P = 0.04$	$F > 20\%, P = 0.08$	$F > 50\%, P > 0.2$
	$RP: F > 15\%, P = 0.06$	$F > 50\%, P > 0.2$	$F > 50\%, P > 0.4$

Table 1. Results on images *New*. S: Scale, r: rotation (degrees). F: false matches, P: precision (pixels). IP:interest points, RP: randomly chosen points.

We present first tests performed on the two images *New1* and *New2* (Figure 2) from two views of a planar patch, where there is a small zoom effect (about

1 : 0.9). For these two images, there are 18% false matches in the initial disparity map and the average precision is about 0.53 pixels. For a large enough window, false matches are all eliminated and the precision is very good for both interest points and random points. For points of interest, a precision of 0.04 can be attained, and for random points, a precision of 0.06 can be attained. Ackermann/Gruen's methods work less well as displayed in the second column of table 1. Matching results on others pairs are also shown in the table 1.

Our method is robust to window deformation and false initial matches, the precision got is high also; the classical methods are not robust and much slower (30 times).

4 Conclusion

For image matching, sub-pixel accuracy is feasible at reasonable cost. If the corresponding window positions differ mainly by a translation, then sub-pixel matching is obtained using a first order development of the image around a given position. Using this method, the computation is linear and the precise correspondance is obtained within a 0.07 pixels accuracy for interest points and 0.09 pixels for randomly selected points.

If the stereo images do not correspond by translation, all methods based on window translation would fail. We propose firstly estimating the disparity map with pixel accuracy. Due to the severe imaging conditions, it is obvious that many of these matches will be false. We suggest therefore using *robust statistics* to compute the affine window deformation from these initial matches. Thanks to the *robust statistics*, false matches are corrected from the robustly estimated affine window transformation. From there the same kind of linear interpolation allows to estimate the sub-pixel disparity.

Many experiments were performed using different planar patches for which the ground truth could be estimated very accurately. Experiments on non planar patches were also performed and are evaluated using indirect measures [9]. They are not all shown owing to lack of space. In these experiments, we showed that even in the case of reasonable rotation and scaling between the two stereo images, accuracy of about a 1/10th of a pixel can be reached. In all the cases, the false matched percentage is negligible, even if there is a lot of false matches in the initialization.

One essential difference between our 'robust' algorithm and the classical ones is that ours *first* robustly estimates the *geometry transformation* between the two windows, and *then* estimates linearly the local displacement between the transformed windows from the *intensity transformation*. One the other hand classical ones [1, 6] consider the geometry transformation and the intensity transformation together. They are iterative and not robust to false initialization and large window deformations, since their recovery of the window transformation is not robust.

Many visual processes require precise matching: for instance this is the case for camera self calibration, or for image mosaicing and for photogrammetry. The evaluation of practical influence of such improved matching on these applications is undergoing.

References

1. F. Ackermann. Digital image correlation: performance and potential application in photogrammetry. *Photogrammetric Record*, 64(11):429–439, October 1984.
2. P. Aschwanden and W. Guggenbühl. Experimental results from a comparative study on correlation-type registration algorithms. In Förstner and Ruwiedel, editors, *Robust Computer Vision*, pages 268–282. Wichmann, 1992.
3. D. N. Bhat and S. K. Nayar. Ordinal measure for visual correspondence. In *Proceedings of the Conference on Computer Vision and Pattern Recognition, San Francisco, California, USA*, pages 351 – 357, San Francisco, California, June 1996.
4. W. Forstner. Mathematical models, accuracy aspects and quality control. *International Archives of Photogrammetry*, 24(3):176–189, June 1982.
5. D. Geiger, B. Ladendorf, and A. Yuille. Occlusions and binocular stereo. In G. Sandini, editor, *Proceedings of the 2nd European Conference on Computer Vision, Santa Margherita Ligure, Italy*, pages 425–433. Springer Verlag, 1992.
6. A.W. Gruen. Adaptative least squares correlation: a powerful image matching technique. *S. Afr. Journal of Photogrammetry, Remote Sensing and Cartography*, 14(3):175–187, 1985.
7. S.S. Intille and A.F. Bobick. Disparity-space images and large occlusion stereo. In *Proceedings of the 3rd European Conference on Computer Vision, Stockholm, Sweden*, pages 179–186. Springer-Verlag, 1994.
8. Z. D. Lan, R. Mohr, and P. Remagnino. Robust matching by partial correlation. In *Proceedings of the sixth British Machine Vision Conference, Birmingham, England*, pages 651–660, September 1995.
9. Z.D. Lan. *Méthodes robustes en vision : application aux appariements visuels*. Ph.d thesis, Institut National Polytechnique de Grenoble, 1997.
10. Z.D. Lan and R. Mohr. Robust location based partial correlation. Inria Rhone Alpes, January 1997. accepted by CAIP97.
11. J.L. Lotti and G. Giraudon. Adaptive window algorithm for aerial image stereo. In *Proceedings of the 12th International Conference on Pattern Recognition, Jerusalem, Israel*, pages 701–703, October 1994.
12. D. Marr. *Vision*. W.H. Freeman and Company, San Francisco, California, USA, 1982.
13. J.A. Noble. Finding corners. *Image and Vision Computing*, 6(2):121–128, 1988.
14. M. Okutomi and T. Kanade. A locally adaptive window for signal matching. *International Journal of Computer Vision*, 7(2):143–162, 1992.
15. P. Remagnino, P. Brand, and R. Mohr. Correlation techniques in adaptative template matching with uncalibrated cameras. In *Vision Geometry III, SPIE's international symposium on photonic sensors & control for commercial applications*, volume 2356, pages 252–253, October 1994.
16. P.J. Rousseeuw and A.M. Leroy. *Robust regression and outlier detection*, volume XIV of *Wiley*. J.Wiley and Sons, New York, 1987.
17. R. Zabih and J. Woodfill. Non-parametric local transforms for computing visual correspondance. In *Proceedings of the 3rd European Conference on Computer Vision, Stockholm, Sweden*, pages 151–158. Springer-Verlag, May 1994.

Active Viewpoint Control for Shape from Occluding Contours

Takashi Akutsu[1], Kenichi Arakawa[1] and Hiroshi Murase[2]

[1] NTT Human Interface Labs, 3-9-11 Midori-Cho, Musashino-shi Tokyo , Japan
[2] NTT Basic Research Labs, 3-1 Morinosato Wakamiya, Atsugi-shi Kanagawa, Japan

Abstract. The contour of an object contains a lot of useful information. For reconstructing the shape of the object's surface, a sequence of contours extracted from object images taken from controlled viewpoints is often used. However, there are many factors that cause errors in reconstruction. One factor - the intervals between the viewpoint positions - can produce serious errors. Taking images at finer intervals decreases these errors, but increases the time, as well as other costs, of taking and processing images. This paper introduces a method for controlling the intervals of viewpoint positions, which enables the errors caused by those intervals to remain within a previously decided fixed limit and can reduce the required number of viewpoint positions.

1 Introduction

One of the main subgoals of computer vision is to extract the three-dimensional shape of an object from images. For this purpose, the contour of the object is often used, because it contains important information in the images (see e.g. [8], [10],[11], or [15]), and many methods for detecting contours have been proposed (see e.g. [1]). Additional useful information can be obtained from the contours by actively controlling the positions of the viewpoint (see Fig. 1) as will be shown later, but the theory relating this information to different viewpoint positions is still not sufficiently well established.

Some studies have shown that an object's surface can be accurately reconstructed using the higher order terms of the contour changes (see [7] or [17]). Adding other information, such as the shading or texture of the object, can produce fairly good results for a concave surface or a high-quality result in real time (see [5],[6], or [19]). On the other hand, appropriate models for describing reconstructed areas has been studied in [18]. And various useful applications have been considered based on them (see e.g. [2],[3], or [9]). Those studies show that fairly good reconstruction techniques have already been developed, but some basic problems still remain. One of the most important problems in this area is

---

* We thank Dr. K. Kogure and Mr. E. Mitsuya of NTT Human Interface Labs and Dr. K. Ishii of NTT Basic Reserch Labs for their help and encouragement. We also thank the anonymous reviewers and other persons who gave good comments and advice.

how to reduce errors in the reconstructed results, together with how to extract contours appropriately e.t.c.

The reconstruction errors are caused by errors in the parameter values chosen for the camera lens, in the camera positions and poses, from quantization into pixels, from the intervals at which images are taken, and by other noise. Especially, the error from the intervals at which images are taken is a peculiar problem to the reconstruction using contours. The importance of this problem has gradually been recognized recently (see [4]). The error from the intervals can be reduced, of course, by taking images at finer intervals, but this increases processing time and cost. We focus our discussion on this problem, aiming to reduce the error along with processing time and cost.

For this aim, which global viewpoint path to choose is surely important. However, two different paths reconstruct different parts out of all the reconstructible areas, while which part is more important depends on the application, and the whole path for all areas can be very complex to implement (see [12] or [13]). In this paper, we propose a method to control the intervals among viewpoint positions on a given global path. Our approach is locating the viewpoint positions efficiently by using the upper limit of the error determined by the object's shape and viewpoint positions.

Sect. 2 explains the geometrical relationship between the surface of an object and the contour from a viewpoint position (see [6]) and estimates the reconstruction errors in preparation for the later discussion. In Sect. 3, our viewpoint control method is presented. The simulations are covered in Sect. 4.

To clarify the relation between object shape and our method's effectiveness, we carried out simulations not only for typical cylindrical objects, but also elliptic spheres, transforming from the right sphere to very long, narrow elliptic sphere. Here the global viewpoint path was a plain circle. The results show the effectiveness of our method and the shapes for which it is not so effective. For instance, our method is not effective for a right sphere, but it is very effective for more general shapes. Tests with real object images, and further study of the bounds of the applicable classes of object shapes remain as future work.

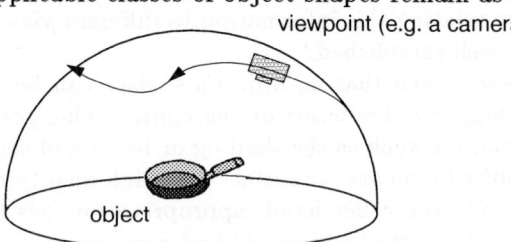

Fig. 1. Active viewpoint control.

2 Geometrical Preparation

Some notation and equations will be introduced here, for the later calculation of the object's (surface) shape from its image contours and the viewpoint positions. As shown in Fig. 2, let there be object D, surface S, and a point, r, on the surface.

The normal vector for S at r is n, the viewpoint is v, and the unit vector parallel to the ray passing from v to r is Q. The distance between v and r is λ. Thus,

$$r(s,t) = v(t) + \lambda(s,t)Q(s,t), \quad (1)$$

where t is time and s is the length on a visible rim curve (the inverse image of a contour on the object surface) such that for each t_o, $r(s, t_o)$ expresses each contour as a function of s.

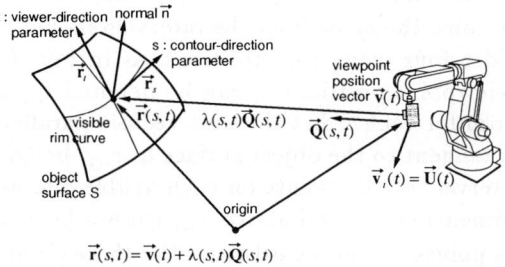

Fig. 2. Notation.

2.1 Taking Images with Fine Enough Intervals

Here, let's assume the surface is smooth, because a continuous surface can be approximated with an arbitrary small difference using a smooth surface (see [16]). Hence, we can take $r(s,t)$ with some smooth parametrization, e.g. the epipolar parametrization, in which the partial derivative of $r(s,t)$ on t, $\frac{\partial r}{\partial t} = r_t$, is parallel to Q, i.e., $r_t \times Q = 0$. We denote the partial derivative of Q, n, v, and λ using the abbreviated notation above. With this notation, for each s and t, the normal vector $n(s,t)$ is perpendicular to $Q(s,t)$ and is also perpendicular to $Q_s(s,t)$, which is linearly independent of $Q(s,t)$. Hence, $n = \frac{Q \times Q_s}{|Q \times Q_s|}$. In reconstructing the shape of surface S by determining points $r(s,t)$, the values of $\lambda(s,t)$ are necessary for calculation by (1). By taking the partial derivative on t and the inner product with n for each side of (1), we get (see [6]),

$$\lambda = -\frac{U \cdot n}{Q_t \cdot n}, \quad (2)$$

when $v_t(t) = U(t)$. By using (1) and (2), theoretically, all reconstructible areas in their entirety can be reconstructed using some viewpoint path (see [12]).

2.2 Error Estimation

In Sect. 2.1, the reconstruction method for all reconstructible areas on the object's surface has been provided. But, in that discussion, object images were handled assuming that images were taken finely enough, and the error related to the intervals was not considered. Here, the relationship between viewpoint positions and reconstruction errors should be shown.

Let's consider the 2-dimensional model in Fig. 3. There are four viewpoint positions $v_1 = v(t_1)$, $v_2 = v(t_2)$, $v_3 = v(t_3)$, $v_4 = v(t_4)$ from which object images are taken. Denoting the points on the visible rim from viewpoint positions v_i as r_i for each i, let the unit vectors parallel to ray l_i from v_i to r_i, which is tangent to the object surface at r_i, be Q_i for each i, and let $r'^{(1,2)}$, $r'^{(2,3)}$, and $r'^{(3,4)}$ be the respective points where l_1 and l_2, l_2 and l_3, and l_3 and l_4 cross.

The values v_1, v_2, v_3, and v_4 and Q_1, Q_2, Q_3, and Q_4 can be observed. In order to calculate $r(s, t_i)$ for each i, we need $\lambda(s, t_i)$; however, the values observed above do not provide them, but just determine $r'^{(1,2)}$, $r'^{(2,3)}$, and $r'^{(3,4)}$ instead. The error can be made among them, i.e. $r(s, t_i)$ and r'.

In three dimensions, the error from the intervals should be made in the same way. Again, consider four viewpoint positions, as in Fig. 4. The points on the visible rim from viewpoint positions v_i can be denoted $r_{i,j}$ as the sequence of j along the visible rim for each i. Let the unit vectors parallel to the ray $l_{i,j}$ from v_i to $r_{i,j}$, which is tangent to the object surface at $r_{i,j}$, be $Q_{i,j}$ for each i,j. Here, we assume the intervals of the points on each visible rim are fine enough. Just as with the two-dimensional model above, $r_{i,j}$ cannot be calculated. Instead, we calculate the cross points r' and regard the set of these points with lines between them as part of the exterior boundary of the object surface. Cross points r' are composed of two sets: one with the form $r'^{(i,i+1,i)}_{(j,k,j+1)}$ (case I), and the other with the form $r'^{(i+1,i,i+1)}_{(j,k,j+1)}$ (case II). The former is the cross point of $l_{i+1,k}$ and the plane extended between $l_{i,j}$ and $l_{i,j+1}$, while the latter is that of $l_{i,k}$ and the plane extended between $l_{i+1,j}$ and $l_{i+1,j+1}$.

In case I,

$$d_2 = d^{(2,3,4)}_{(I,j,k,j+1)} = r'^{(i+2,i+1,i+2)}_{(l,k,l+1)} - r'^{(i,i+1,i)}_{(j,k,j+1)}, \tag{3}$$

$$d_1 = d^{(1,2,3)}_{(I,j,k,j+1)} = h - r'^{(i,i+1,i)}_{(j,k,j+1)}, \tag{4}$$

where h is the closest point to $r'^{(i,i+1,i)}_{(j,k,j+1)}$ on the line $m^{i}_{(j,j+1)}$ which passes through $r'^{(i+1,i,i+1)}_{(n,j,n+1)}$ and $r'^{(i+1,i,i+1)}_{(n',j+1,n'+1)}$.

In case II, $d_1 = d^{(1,2,3)}_{(II,j,k,j+1)}$ and $d_2 = d^{(2,3,4)}_{(II,j,k,j+1)}$ can be calculated in a symmetrical way as d_2, d_1 in case I. From (3) and (4), the upper limit of the error is given by

$$E^{sup}_{r'^{(i,i+1,i)}_{(j,k,j+1)}} = \frac{|d_1 \times d_2|}{|d_2 - d_1|}. \tag{5}$$

3 Viewpoint Control Method

This section describes our interactive method for reconstructing an object's surface within the intended error limit. A viewpoint path can be chosen freely on a sphere of radius R from the origin surrounding the object surface. The methods in [13] would also give a global viewpoint path on such a sphere. Now we take some great circle on the sphere as a simple path.

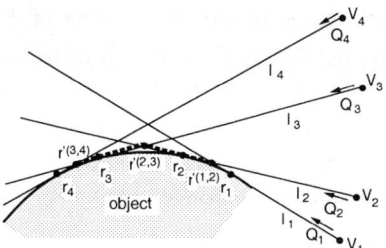

Fig. 3. Upper limit of the error (in 2dimensions).

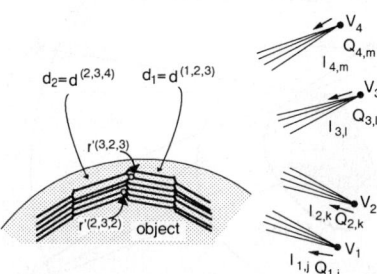

Fig. 4. Upper limit of the error.

Consider the situation where the object surface is within r ($< R$) of the origin, i.e. included in a sphere of radius r as in Fig. 1. Denote the admissible upper bound of the reconstruction error as E_{ad}. First, we fix standard distance D_{std} between adjoining viewpoint positions with the angle τ made between them at the origin as

$$D_{std} = 2R\sin\frac{\tau}{2} = \frac{\sqrt{2}E_{ad}\sqrt{\sqrt{1+\frac{4r^2}{E_{ad}^2}}-1}}{r}R. \qquad (6)$$

D_{std} is chosen such that the sphere of radius r (as an object) can be reconstructed within E^{sup} E_{ad} by the same interval of D_{std}. Using (6), the control procedure is as follows:

Step. 1. Control the viewpoint along a global path and take images consecutively at v_1, v_2, v_3, and v_4 at an interval in (6).

Step. 2. For the prepared four points, v_1, v_2, v_3, and v_4, calculate the upper limit of the errors at $r'^{(2,3,2)}_{(j,k,j+1)}$, $E^{sup}_{r'^{(2,3,2)}_{(j,k,j+1)}}$, and at $r'^{(3,2,3)}_{(j,k,j+1)}$, $E^{sup}_{r'^{(3,2,3)}_{(j,k,j+1)}}$, according to (5) and take their maximum $E^{sup}_{r'(2,3)} = \max\{E^{sup}_{r'^{(2,3,2)}_{(j,k,j+1)}}, E^{sup}_{r'^{(3,2,3)}_{(j,k,j+1)}}\}$. If the resulting value $E^{sup}_{r'(2,3)}$ is smaller than E_{ad}, then record $r'^{(2,3,2)}_{(j,k,j+1)}$ and $r'^{(3,2,3)}_{(j,k,j+1)}$ as valid reconstructed points and take $v_2^0 = v_2$. Step 2 is then finished. However, if the resulting value $E^{sup}_{r'(2,3)}$ is larger than E_{ad}, change the viewpoint to one between v_2 and v_3, i.e. v_2^1 as in Fig. 5. Choose v_2^1 as the point on the great circle between v_2 and v_3 such that the angle at the origin made between v_2 and v_3 and the angle between v_2 and v_2^1 are in the proportion of $E_{r'(2,3)}^{sup}$ to $\frac{|d_2|E_{ad}}{|d_2-d_1|}$. Then, calculate the error for v_1, v_2, v_2^1, and v_3, which is to be smaller than E_{ad} (the term $\frac{|d_2|}{|d_2-d_1|}$ is necessary for this). If the error for v_2^0, v_2^1, v_3, and v_4 is larger than E_{ad}, change the viewpoint between v_2^1 and v_3 again to another viewpoint position, v_2^2, chosen in just the same way as above. Repeat this selection until we get to some $n \geq 1$ such that the error for v_2^{n-1}, v_2^n, v_3, and v_4 is smaller than E_{ad}.

Step. 3. If the reconstructed points come to the end of the path, stop. Otherwise, take v_5 at the interval D_{std} and repeat the procedure from Step. 2 using v_2^n, v_3, v_4, and v_5 as the prepared four points.

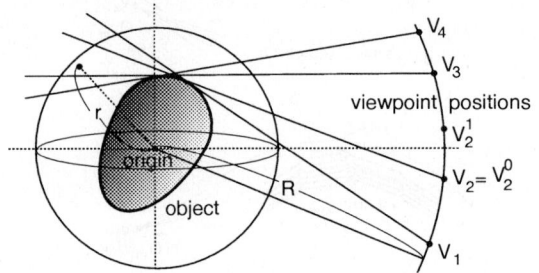

Fig. 5. Object and viewpoint control.

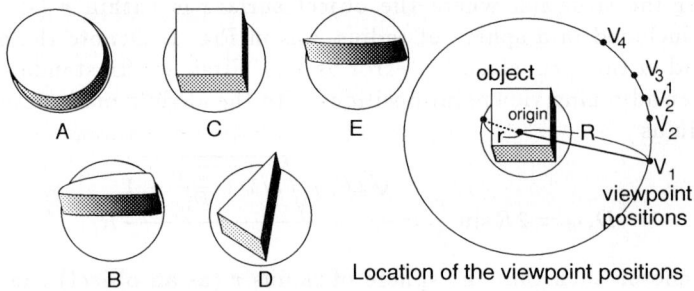

Fig. 6. Sample objects used for comparing our method with fixed case.

4 Experiment and Results

In first simulations, we chose objects A, B, C, D, and E (see Fig. 6), which have various shapes in the viewing direction from viewpoint positions on the global path. The global path was set to the great circle of radius R which was included in the plane whose normal was parallel to each object's axis. In the simulations, the values R, r, and E_{ad} were set to 300.0, 50.0, and 1.0 respectively, as a realistic location and situation.

The maximum reconstruction errors $E_{r'(2,3)}^{sup} = \max\{E_{r'(2,3,2) \atop (j,k,j+1)}^{sup}, E_{r'(3,2,3) \atop (j,k,j+1)}^{sup}\}$ used in our method were determined by the points on the visible rim at which the normal of its tangent plane was almost parallel to the directions along the viewpoint path, i.e. where $n \times v_t$ was very small using the notation in Sect. 2, because (5) shows the reconstruction errors are large where the angle of the rays from adjoining viewpoint positions is large. To examine the effect of our method, it is enough for the objects shapes to have variety in the direction for which $n \times v_t$ is very small. The setting above was selected as the most typical and simplest one to suit this purpose.

Fig. 7 shows the minimum number of images taken for each object when the viewpoint changed at a fixed interval within the admissible reconstruction error E_{ad}, and the result of our method for the setting above.

Fig. 8 compares the minimum number at a fixed interval with that of our method on elliptic spheres transforming from the right sphere to very long and narrow ones. In this test, R, r, and E_{ad} were also set to 300.0, 50.0, and 1.0 respectively. We used elliptic spheres with a long axis 49.0, and changed the other two short axes as in Fig. 8 keeping them equal.

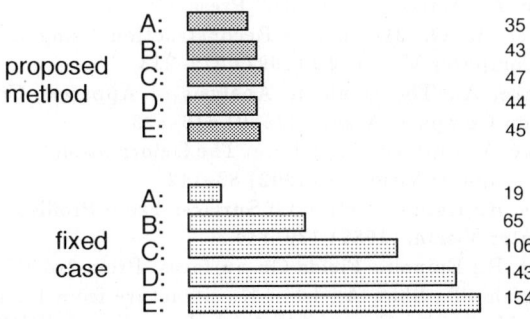

Fig. 7. Results of comparing our method with fixed case for the sample objects.

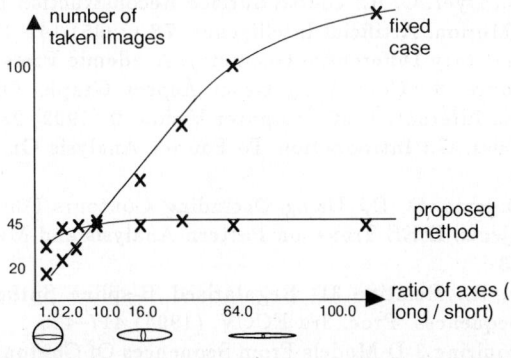

Fig. 8. Results of comparing our method with fixed case in elliptic spheres.

5 Conclusions and Discussions

The results in Fig. 7 and 8 show that our method greatly reduces the required number of viewpoint positions at which images must be taken for objects whose surface curvature is changing to some extent in the viewing direction from the global path, while limiting the upper bound of the reconstruction errors.

In this paper, we use the exterior boundary of the object surface as the result. With this approach, it is very easy to combine the results of the reconstruction of the same area reconstructed at different points on a global viewpoint path; we just take the intersection of the reconstructed object.

References

1. Ballard, D. H., Brown, C. M.: Computer Vision. Prentice-Hall. (1982)
2. Blake, A., Brady, J. M., Cipolla, R., Xie, Z., Zisserman, A.: Visual Navigation Around Curved Obstacles. Proc. IEEE Internat. Conf. on Robotics and Automation. **3** (1991) 2490–2495
3. Blake, A., Yuille, A.: Active Vision. MIT Press. (1992)
4. Boyer, E., Berger, M. O.: 3D Surface Reconstruction Using Occluding Contours. Internat. J. of Computer Vision. **22** (1997) 219–233
5. Cipolla, R., Blake, A.: The Dynamic Analysis of Apparent Contours. Proc. 3rd Internat. Conf. on Computer Vision. (1990) 616–623
6. Cipolla, R., Blake, A.: Surface Shape From The Deformation Of Apparent Contours. Internat. J. of Computer Vision. **9** (1992) 83–112
7. Giblin, P., Weiss, R.: Reconstruction Of Surfaces From Profiles. Proc. 1st Internat. Conf. on Computer Vision. (1987) 136–144
8. Giblin, P., Weiss, R.: Epipolar Fields On Surfaces. Proc. 3rd ECCV. (1994) 14–23
9. Kato, K., Nakanishi, T., Shio, A., Ishii, K.: Structure from Image Sequences Captured through a Monocular Extra-wide Angle Lens. Proc. CVPR. (1994) 919–924
10. Koenderink, J. J.: What Does The Occluding Contour Tell Us About Solid Shape? Perception. **13** (1984) 321–330
11. Koenderink, J. J.: Solid Shape. MIT Press. (1990)
12. Kutulakos, K. N., Dyer, C. R.: Global Surface Reconstruction By Purposive Control Of Observer Motion. Proc. CVPR. (1994) 331–338
13. Kutulakos, K. N., Dyer, C. R.: Global Surface Reconstruction By Purposive Control Of Observer Motion. Artificial Intelligence. **78** (1995) 147–177
14. O'Neill, B.: Elementary Differential Geometry. Academic Press. (1966)
15. Petitjean, S., Ponce, J.: Computing Exact Aspect Graphs Of Curved Objects: Algebraic Surfaces. Internat. J. of Computer Vision. **9** (1992) 231–255
16. Stein, E. M., Weiss, G.: Introduction To Fourier Analysis On Euclidean Spaces. Princeton. (1971)
17. Vaillant, R., Faugeras, O. D.: Using Occluding Contours For Recovering Shape Properties Of Objects. IEEE Trans. on Pattern Analysis and Machine Intelligence. **14** (1992) 157–173
18. Zhao, C., Mohr, R.: Relative 3D Regularized B-spline Surface Reconstruction Through Image Sequences. Proc. 3rd ECCV. (1994) 417–426
19. Zheng, J. Y.: Acquiring 3-D Models From Sequences Of Contours. IEEE Trans. on Pattern Analysis and Machine Intelligence. **16** (1994) 163–178

Point Selection: A New Comparison Scheme for Size Functions (With an Application to Monogram Recognition)

Massimo Ferri, Patrizio Frosini, Alberto Lovato and Chiara Zambelli

Dip. di Matematica, Università
Piazza Porta S.Donato, 5
I-40127 Bologna ITALY
E-mail: ferri, frosini@dm.unibo.it

Abstract. A new paradigm for the comparison of size functions is presented; it stresses the relevance of the "angular points" of the functions, and gives greater value to the stable ones. A simple example of classification of monograms (88 elements in the training set, 88 in the test set, for 22 classes, with a hit rate of 78%) is given, and a current enhancement of the experiment is described.

1 Introduction

Size function theory (see next Section) is slowly affirming its validity as a mathematical tool for representing and comparing shapes. Its advantages are the capability of formalizing qualitative concepts of shape, and the standard form assumed by size functions in spite of the different criteria adopted by the user. An intrinsic difficulty is the choice of a distance, or of a similarity measure between size functions; another one is the recognition of the relevant features of size functions of a training set.

The present work tries to solve both problems, by defining a "training matrix", a "test matrix", and a particular operation on them. Call *training* (respectively *test*) *size function* the size function of an element of the training (resp. test) set. Through the aforementioned matrices, only those parts — of a test size function — are taken into account, which match a stable feature of the training size functions.

A simple example, which applies this paradigm, completes the paper: classification of monograms.

2 Size functions

Size functions are modular shape descriptors. For extensive references on the subject, see [8, 14]. Here we just recall the main concepts.

Shape — in our opinion — is not simply a quality of a set of points in space or in a plane. It is rather a property of such a set, together with a real function defined on it; e.g, "bumpiness" is the behaviour of the function "distance from the center of mass" on the points of the considered object, "coarseness" is the behaviour of "curvature". We call a continuous real function $\varphi : \mathcal{M} \to \mathbf{R}$ defined on a subset $\mathcal{M}$ of a Euclidean space a *measuring function*.

The *size function* of the pair $(\mathcal{M}, \varphi)$ is a function $\ell_\mathcal{M} : \mathbf{R}^2 \to \mathbf{N} \cup \{\infty\}$. For each pair $(u, v) \in \mathbf{R}^2$, consider the set of points on which φ is worth $\leq u$. Two such points are then considered to be equivalent if they either coincide, or can be connected in $\mathcal{M}$ by a path, on whose points φ is worth $\leq v$. Then $\ell_\mathcal{M}(u, v)$ counts the equivalence classes so obtained.

For an example of the meaning of a size function, see Figure 1a, where a plane curve $\mathcal{M}$ is depicted; φ is taken to be the distance from point B. Fig. 1b represents the corresponding size function: The value $\ell_\mathcal{M}(u, v)$ informs us on the number of classes of points on the curve, which have distance $\leq u$ from B and can be joined together by walking on the curve without exceeding distance v. In other words, it is the number of those maximal arcs of $\mathcal{M}$ within distance v from B, which contain at least one point not farther than u from the same point.

The size function of a curve (or more generally, of an image) can be thought of as a sort of transform, which stresses only the aspect selected while choosing the measuring function. Modularity is assured by this freedom of choice. Invariance of the measuring function under a transformation implies invariance of the corresponding size function under the same transformation [12]. Of course, in practical cases only a discrete set of points is given (pixels of an image or, in the application we are going to report, sample points of a contour). For technical reference on the computation of size functions, see [6, 7, 13].

Another advantage of modularity is the possibility to use different "viewpoints" (i.e. measuring functions) at the same time; we make them cooperate by getting a fuzzy characteristic function from each, and then taking the average. The measuring functions can be different occurrences of the same function type — as will be the case for the example at the end of this paper — or be totally unrelated, so bringing in different aspects and classification criteria — as we are presently trying on the same problem.

3 Angular points

Whatever the measuring function, the outcoming size functions always have the same "format": it is determined by a finite set of points of $\mathbf{R}^2$ (called *angular points*). They are those intersection points of the discontinuity lines of the size function, at which the vertical value difference of the size function for slightly greater u is bigger than the one for slightly lesser u.

In the example of Fig. 1b, we have (a, d) and (b, c) as angular points. These convey all relevant information about the size function. Note that there are other interesting points. One is (a, ∞); this corresponds with the minimum value of φ; in general, such points correspond to as many connected components of $\mathcal{M}$. Another interesting (but not angular) point is (e, e), corresponding to the maximum value of φ. Anyway, we concentrate on angular points and their multiplicity, i.e. the difference of the vertical differences on the right and on the left of the point.

4 Point selection

A distance between size functions may be defined as a point–by–point difference (in absolute value) of the functions themselves. We used a slightly modified version of this with a fair success in other projects. A progress has been the use of a Hausdorff distance between the sets of angular points of two size functions (see [2]).

In the present work, we concentrate on the fact that angular points don't all have — so to say — the same relevance for the recognition process. We realized this first when dealing with occlusion. Size function are, by their very nature, global invariants. All the same, by comparing the size functions of differently occluded images of the same object, we noted the constant presence of a particular set of angular points. We observed the same phenomenon in those classification problems where the objects are stricltly referrable to a prototype (leaves of the same tree [14], on–line handwritten letters [4], signatures): Some contingent features, e.g. the "tail" of the signature of the same writer, may differ strongly from case to case, so giving angular points in very different positions. Other features don't vary that much, and the corresponding angular points are very stable. This may not be the case with other classification problems, where the objects can be referred to a common description, but not actually to a prototype (e.g., leukocytes [5], free–hand drawings [2], and also the sign language [11]).

Preprocessing of the object may work when the varying feature is cospicuous as a signature tail (even forcing a research group to erase tails before classifying [1]), but variations may be more deeply hidden, and difficult to get rid of. So we have preferred to act on the size functions, instead.

We divide the (u, v)–plane into 32×32 cells. Now, take a fixed measuring function, and the size functions of the members of the training set, belonging to a fixed class. We define the 32×32 *training matrix* of the class. Each angular point of each size function gives a contribution to the matrix, built by a Gaussian mask centered on the cell to which the angular point belongs; the mask has sum one, and a variance which increases with the distance from the diagonal $u = v$; the values are also multiplied by the multiplicity of the angular point. The final training matrix (r^i_j) is the average of the ones coming from the various objects of the training set, belonging to that class. Of course, cells below $u = v$ have null entry, and are inessential. Angular points in the cells just above this diagonal are not taken into account, since they are too much affected by discretization noise. The smoothing (by Gaussian mask) turns out to be necessary, since angular points do not occur, in the various size functions, at exactly the same position; this is true at least for the monograms of the experiment reported later on, and is to be expected for all images of natural origin.

For each single element of the test set, we build a completely analogous *test matrix* (s^i_j), with the only difference that angular points falling in the leftmost column or in the topmost row are not taken into account.

Given a test matrix, we now compute the likelihood of the corresponding object to belong to the given class. First, we sum

$$p = \sum_{i,j} min\{r_j^i, s_j^i\}.$$

The likelihood is then

$$\frac{p}{\sum_{i,j}(r_j^i + s_j^i) - p}.$$

An entry of the training matrix will be high, if several size functions in the training set have an angular point in the corresponding cell. If the size function of the test set has an angular point in that cell too, then this will give a good contribution to p (less so if it is in a nearby cell). So, the numerator of the fraction evaluates the correspondence of the angular points of the test size functions with the stable ones of the training set.

As for the meaning of the denominator, imagine that the training set is built by just one element, and that each angular point has multiplicity one, and gives a sharp contribution of 1 to its cell (with no Gaussian smoothing); then the denominator would just count the total number of cells occupied by the union of angular points of the test and training size functions. This has been introduced in order to avoid an unjustified stress on size functions with many angular points.

5 An application: Monogram recognition

The problem of monogram recognition is a nice one for testing our new paradigm, for the following reasons:

- Monograms are "natural" objects, so well suited for analysis by size functions.
- All the same, the elements of the training set can be seen as sort of true "prototypes".
- Variability is sufficient to make superposition methods ineffective.
- Context cannot help recognition.

We have gathered four elements for the training set and four for the test set, from each of 22 subjects. Acquisition was by a hand–held scanner and a PC, from black ink monograms. The only preprocessing is dilation. The program (In Visual Basic and C) extracts the outline (i.e. the outer contour) of the monogram under study; the outline will be the only input to the very classification process. This choice makes us loose much information, as can be seen in Figures 2a and 2b, where the left colums show the monograms of the training sets of "cri" and of "dan" respectively, and the right columns present the relative outlines. This choice was supported by previous experiences on tools [9] and on signatures [10]. A connected curve is a good test, because the analysis of angular points becomes simpler, and this makes it easier for us to understand what has happened. A further, deeper investigation presently under study involves the entire monogram. (More about this later in this section).

Subject	# 1	# 2	# 3	# 4	Hit/Tot
amm	1	1	2	1	3/4
clau	1	1	1	1	4/4
cri	1	1	1	1	4/4
cris	1	1	1	2	3/4
dan	1	1	1	1	4/4
dari	3	1	1	1	3/4
eric	1	2	7	12	1/4
fab	1	1	1	1	4/4
gui	1	5	3	4	1/4
luc	1	1	1	1	4/4
luis	1	1	1	1	4/4
lup	7	1	1	1	3/4
mai	1	8	2	1	2/4
mic	1	1	1	1	4/4
moca	1	1	1	1	4/4
moni	1	1	2	1	3/4
monz	1	2	1	1	3/4
pat	1	1	1	1	4/4
pie	1	2	1	1	3/4
ric	1	1	1	1	4/4
rim	1	1	1	1	4/4
stef	18	5	11	4	0/4

Table 1. Monogram recognition

Simplification was applied also in the choice of measuring functions. Unlike in other researches, where we strived to differentiate measuring functions, we have adopted five versions of the same one, i.e. distances from points [3]. The five points are the barycentre of the monogram and the middle points of the edges of the horizontal rectangle circumscribed to the monogram. Again, this choice was meant to simplify the analysis of the experiment.

Table 1 reports the results. The first four numerical columns show the placing of the four elements of the test set; e.g., the figure 3 reported in the first column of "dari" means that the first monogram of the test set of this subject has not been recognized, and that there are two other subjects to whom the program would rather ascribe the monogram as more likely. The last column resumes the results as a fraction of hits over total.

We are particularly surprised — and deceived — by the totally negative performance on "stef", whose monograms look easy to recognize to a human eye: see Figure 3a (training left, test right). Conversely, a fairly good 3/4 was reached both by the monograms of "lup" and "monz", which look rather similar to each other (Figures 3b and 3c respectively). The overall hit ratio is 78%.

Since this classification problem turns out to be of interest in itself, we are implementing — at the moment of revising the paper — a more complete analysis which takes the whole monogram into account. This is done by computing a sort of Radó transform of the monogram: For each of four directions we scan the monogram by parallel and adjacent straight lines. For each line we sum the number of black pixel present in a five pixel wide strip. After suitable normalization, this number is assigned to a vertex of a linear graph (otherwise said, to a pixel of an incident segment). Then the size function is computed; comparison is carried out as in the outline case. The two "orthogonal" classifications yield likelyhoods, the mean of which will be the final one.

We are confident that use of the whole monogram (not just the outline) and cooperation of different measuring functions, can give competitive results.

6 Conclusions

We have proposed a new paradigm for comparing size functions. It consists on stressing those angular points which are approximately in the same position in the size functions of most elements of the training set. This seems to be more efficient, and also faster, than other comparison methods.

An application to monogram recognition, although designed with big simplifications (use of the mere outline), gives acceptable results.

Future developments will surely be: integration of different measuring functions and use of the whole image in monogram classification with Radó-like measuring functions; possibly, application of the method to retinal images, to faces, and to other natural images for which a fixed (but not strictly geometric) model is given.

Acknowledgements

Research accomplished under support of CNR–GNSAGA, MURST, ASI, and within an agreement with Elsag Bailey.

We wish to thank S. Gallina, P. Donatini, and above all E. Porcellini for their great help.

References

1. Bajaj, R., and Chaudhuri, S., "Signature verification using multiple neural classifiers", *Pattern Recognition* 30 (1997), 1–7.
2. Collina, C., Ferri, M., Frosini, P., and Porcellini, E., "SketchUp: Towards qualitative shape data management", in *Proc. ACCV '98*, Hong Kong (1998).
3. Ferri, M., and Frosini, P., "Range size functions", in *Proc. of the SPIE on Vision Geometry III*, Boston, Mas., Vol. 2356 (1994), 243–251.
4. Ferri, M., Gallina, S., Porcellini, E., and Serena, M., "On-line character and writer recognition by size functions and fuzzy logic", in *Proc. ACCV '95*, Singapore, vol. 3 (1995), 622–626.

5. Ferri, M., Lombardini, S., and Pallotti C., "Leukocyte classification by size functions", in *Proc. 2nd IEEE WACV*, Sarasota (1994), 223–229.
6. Frosini, P., "Discrete computation of size functions", *J. Combin. Inf. Syst. Sci.* 17 (1992), 232–250.
7. Frosini, P., "Connections between size functions and critical points", *Math. Meth. Appl. Sci.* 19 (1996), 555–569.
8. Frosini, P., "Size theory as a topological tool for computer vision", in: *"From segmentation to interpretation and back: Mathematica methods in computer vision"*, Eds.: T. Moons, E. Pauwels, L. Van Gool (in print).
9. Mokhtarian, F., "Silhouette–based isolated object recognition through curvature scale space", *IEEE Trans. on PAMI* 17, 5 (1995), 539–544.
10. Nouboud, F., "Handwritten signature verification: A global approach", in "Fundamentals in handwriting recognition" Ed.: S. Impedovo, NATO ASI Series F, 124, Springer Verlag (1994), 455–459.
11. Uras, C., and Verri, A., "On the recognition of the alphabet of the sign language through size functions", in *Proc. 12th ICPR*, Jerusalem (1994).
12. Verri, A., and Uras, C., "Invariant size functions", in *Applications of invariance in Computer Vision, Lecture Notes on Computer Science*, J.L. Mundy, A. Zisserman, D. Forsyth eds., vol. 825, Springer–Verlag (1994), 215–234.
13. Verri, A., Uras, C., "Computing size functions from edge maps", *Internat. J. Comput. Vision* (1997).
14. Verri, A., Uras, C., Frosini, P., and Ferri M., "On the use of size functions for shape analysis", *Biol. Cybernetics* 70 (1993), 99–107.
15. Xu, C., and Velastin, S.A., "A comparison between the standard Hough transform and the Mahalanobis distance Hough transform", in: *Proc. ECCV '94*, Ed.: J.-O. Ecklundh, Stockholm, Vol. 1 (1994), 95–100.

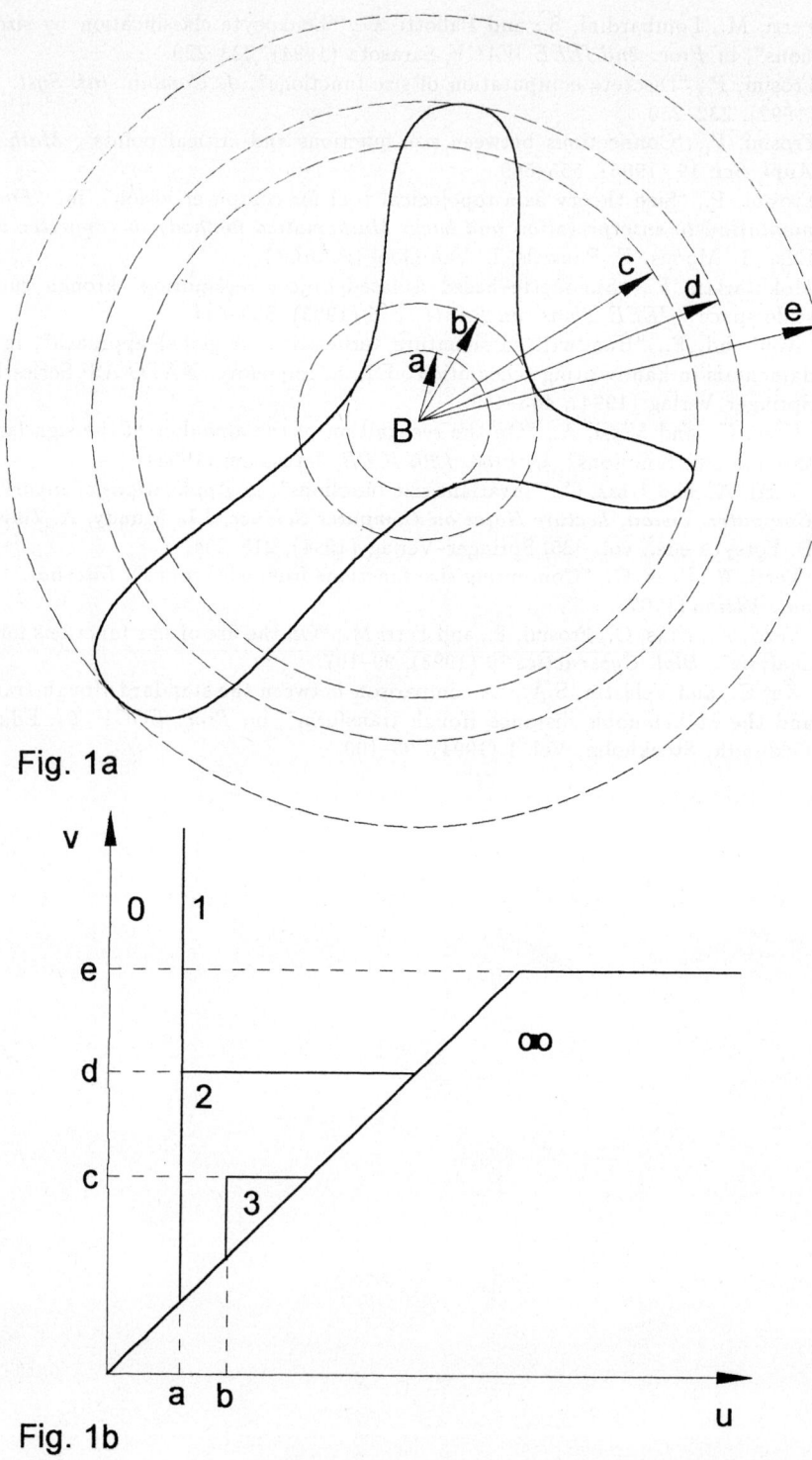

Fig. 1a

Fig. 1b

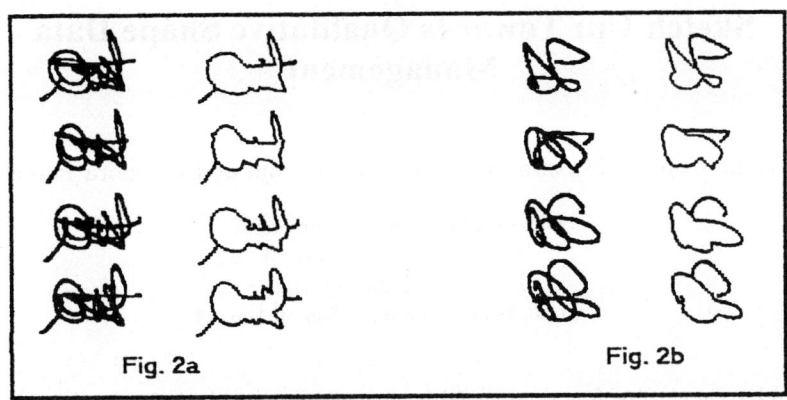

Fig. 2a

Fig. 2b

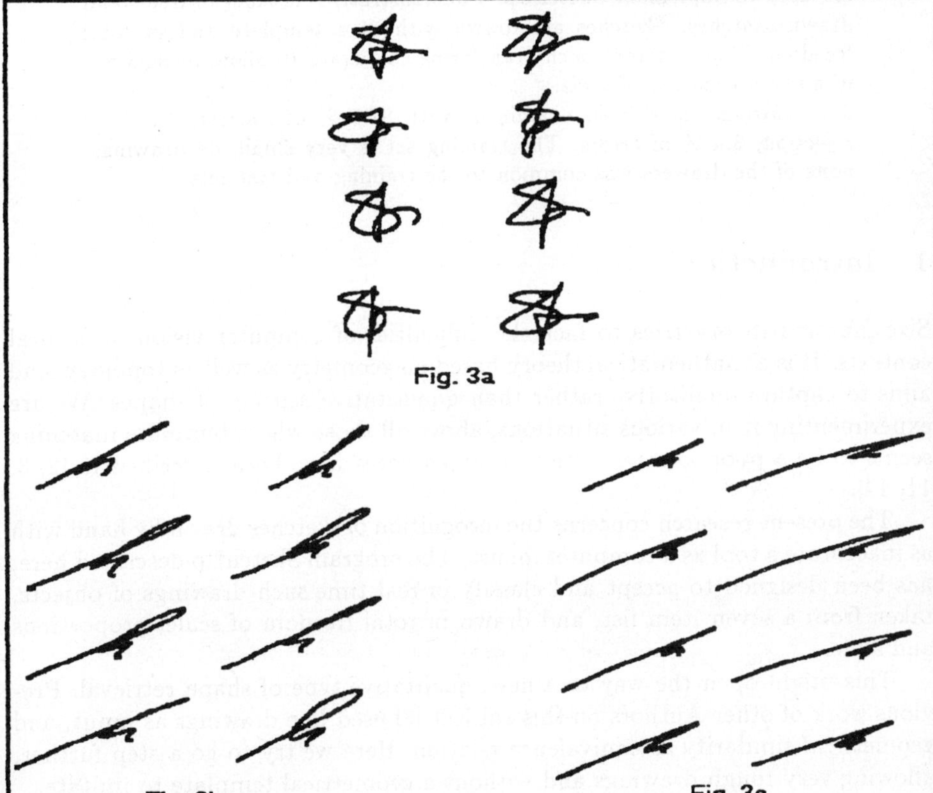

Fig. 3a

Fig. 3b

Fig. 3c

Sketch Up: Towards Qualitative Shape Data Management

Costantino Collina, Massimo Ferri, Patrizio Frosini and Eleonora Porcellini[1]

Dip. di Matematica, Università
Piazza Porta S.Donato, 5
I-40127 Bologna ITALY
E-mail: `ferri, frosini@dm.unibo.it`

Abstract. Size functions, a fairly new class of topological transforms, are used to implement SketchUp, a demonstrative classifier of free-hand drawn sketches. Sketches are drawn without a template and in total freedom. Four different such transforms cooperate to allow recognition in a seven item list of tools.
207 drawings have been examined, with 87.9 % of success, 8.7 % of rejection, 3.4 % of errors. The training set is very small: 66 drawings; none of the drawers was common to the training and test sets.

1 Introduction

Size function theory tries to face the difficulties of computer vision in natural contexts. It is a mathematical theory based on geometry as well as topology, and aims to capture qualitative rather than quantitative aspects of shapes. We are experimenting it in various situations, above all those where template matching seems to be a poor solution. Previous experiences have been described in [2, 3, 11, 14].

The present research concerns the recognition of sketches drawn by hand with as inaccurate a tool as a computer mouse. The program SketchUp described here, has been designed to accept and classify in real time such drawings of objects, taken from a seven item list, and drawn in total freedom of scale, proportions and style.

This might open the way to a new, qualitative type of shape retrieval. Previous work of other Authors on this subject [8] used line drawings as input, and geometrical similarity as equivalence relation. Here we try to go a step further, allowing very rough drawings and without a geometrical template to imitate.

2 Size functions

Size functions are modular shape descriptors. For extensive references on the subject, see [7, 14]. Here we just recall the main definitions.

A *size function* of a subset $\mathcal{M}$ of a Euclidean space is a function $\ell_\mathcal{M} : \mathbf{R}^2 \to \mathbf{N} \cup \{\infty\}$. Its definition depends entirely on the choice of a *measuring function*, i.e. a continuous function $\varphi : \mathcal{M} \to \mathbf{R}$. (For sake of clarity, think of φ as the ordinate function). For each pair $(u,v) \in \mathbf{R}^2$, consider the set of points on which φ is worth $\leq u$. Two such points are then considered to be equivalent if they either coincide, or can be connected in $\mathcal{M}$ by a path, on whose points φ is worth $\leq v$. Then $\ell_{\mathcal{M}}(u,v)$ counts the equivalence classes so obtained.

For an example of the meaning of a size function, see Figure 1, where a plane curve $\mathcal{M}$ is depicted. The value $\ell_{\mathcal{M}}(u,v)$ informs us on the number of classes of points on the curve, which have ordinate $\leq u$ and can be joined together by walking on the curve without exceeding ordinate v (three in the example). In other words, it is the number of those maximal arcs of $\mathcal{M}$ lying under level v, which contain at least one point under level u.

The size function of a curve (or more generally, of an image) can be thought of as a sort of transform, which stresses only the aspect selected while choosing the measuring function. Modularity is assured by this freedom of choice. Invariance of the measuring function under a transformation implies invariance of the corresponding size function under the same transformation [12]. Of course, in practical cases only a discrete set of points is given (pixels of an image or, in our case, sample points of a character). For technical reference on computation of size functions, see [5, 6, 13].

Another advantage of modularity is the possibility to use different "viewpoints" (i.e. measuring functions) at the same time; we make them cooperate by getting a fuzzy characteristic function from each, and then taking the average.

3 Distance and fuzzy characteristic functions

Whatever the measuring function, the outcoming size function always has the same "format": it is determined by a finite set of points in the plane (called *angular points*), i.e. particular intersection points of the discontinuity lines of the size function. A distance between size functions may be defined as a point-by-point difference (in absolute value) of the functions themselves. We used a slightly modified version of this with a fair success in other projects. Here we have preferred a Hausdorff distance of the sets of angular points.

The fuzzy characteristic function of a test drawing with respect to N given classes $\mathcal{B}_j (j = 1, \ldots, N)$ of objects is computed as follows. For each measuring function $\varphi^i (i = 1, \ldots, M)$ let ℓ^i be the corresponding size function of the drawing; let c_j^i be the least distance from ℓ^i to the size functions of elements of the training set, belonging to class $\mathcal{B}_j$, with respect to the same measuring function φ^i. Then let $p_j'^{\,i} = 1/(c_j^i)^2$. Then define the characteristic functions of the drawing, with respect to φ^i, as the normalizations

$$p_j^i = \frac{p_j'^{\,i}}{\sum_{k=1}^{N} p_k'^{\,i}}.$$

In order to get the final characteristic functions p_j, i.e. the likelihood of the drawing to refer to class $\mathcal{B}_j$, first compute $\bar{p}_j = \frac{1}{M} \sum_{i=1}^{M} p_j^i$, then

$$\tilde{p}_j = \bar{p}_j - min\{\bar{p}_k | k = 1, \ldots, N\}$$

and finally
$$p_j = \frac{\tilde{p}_j}{\sum_{k=1}^{N} \tilde{p}_k}.$$
Of course, the case of an object belonging to the training set is treated separately, in order to avoid division by zero.

We are currently considering different comparison methods, in order to keep into account the different sizes and shapes of the clusters formed by the size functions of the training set in their parameter space, and in order to stress the relevant features of the size functions themselves.

4 Classification

The classification is simply the assignment of the test drawing to the class which has got the highest likelihood p_j or its rejection. Rejection is of two sorts. Suppose that for a certain measuring function φ^m the least distance is from an object of the training set, belonging to class $\mathcal{B}_n$; suppose, on the other hand, that this distance exceeds half the diameter of the class itself. If this happens for at least two measuring functions, then the drawing is rejected as "unadmitted". A second type of rejection occurs if the two highest final characteristic functions differ too little; in this case the drawing is considered to be "ambiguous".

More sophisticated classification methods are presented in [1], and might be applied also to SketchUp, or rather to a true application.

5 The program SketchUp

For the program to be easily used on a portable computer, SketchUp was written in a Windows environment and tested with a PC based on a Pentium (133 MHz) microprocessor. The program consists of a graphical interface, within which the user can draw a sketch by use of a mouse. The objects to be sketched have been shown in an initial picture, but are not visible while the user draws. This is expressly done in order to give the user the highest freedom in drawing; he/she must not be biased during the experiment. The initial picture (true photographs) are there mostly for bypassing language problems.

The objects to be drawn are: nut, hammer, scissors, drill, screwdriver, wrench, compass. The user is allowed to draw them with any inclination and any size. The user has various possibilities to correct the drawing. When the sketch is complete, a virtual key takes to the contour extraction (see Fig. 2); this is performed in a variable time of the magnitude order of one second. Use of the outer contour (also called "outline" or "silhouette") is widespread for retaining essential features [9, 10]. The contour is then shown to the user (see Fig. 3), who has the possibility to resume drawing for further corrections. Hitting another virtual key starts the classification process; this is performed again in almost real time. It all ends with a new image (see Fig. 4), which displays, for each object class, a number representing the computed fuzzy characteristic function, together with a bar of length proportional to the number itself. The bottom part of the screen displays either the name of the object resulting from the classification or the indication of rejection.

A comment about the use of just the outline: This choice was made in order to make it clear that the method does not take into account additional (and precious) information contained in the interior. This is clearly shown to the user, before he/she is allowed to proceed to the classification. T-junctions, number of connected components and other features of the sketch are ignored by the program; all the same, the recognition rate is fairly high. Of course, we would (and maybe will) integrate the analysis of the interior if we had to build a true application; but for a demonstrative program as SketchUp, we have preferred to show the user how little of his/her drawing is sufficient for the size functions to recognize it.

6 Measuring functions

Four measuring functions have been used: distance from barycentre, distance from "north pole", distance from "south pole", curvature. All of them are defined on the contour. This is computed by a simple algorithm: Start from a frame pixel and scan the drawing until you find the first black pixel; then follow locally the drawing, always sticking at the boundary pixels of the drawing, until you step on the first found pixel again. Then close the loop and draw it. This way, only the outmost contour is extracted (and shown to the user).

The computation of the centre of mass B of the contour is straightforward, as also the computation of the distance from the single contour pixels to the barycentre B. This is the first measuring function.

The second and third measuring functions are similar to the first, in that they are distances from the single contour points to two fixed points C, D ("north" and "south" pole of the drawing) of the plane. These ones are related to point S introduced in [4] ($P_i (i = 1, \ldots, n)$ are the points of the contour) for $k = 2$.

$$C, D = B \pm \frac{10 \sum_{i=1}^{n} (P_i - B) \|P_i - B\|^2}{n \max_i \|P_i - B\|^2}.$$

As a fourth measuring function we have adopted a sort of curvature. This is computed, for each contour point, by taking the line connecting it to the one which is 27 points apart, and computing the angle between this line and the analogous one of the following point.

7 Experimental results

We have used a training set of 66 drawings and a test set of 207, performed by several people, none of whom was common to both sets. Table 1 reports the results. Globally, we have had 87.9 % hit, 8.7 % rejection and 3.4 % miss. Given

Object	#	Miss %	Rejection %	Hit %
Nut	15	0	6.7	93.3
Hammer	44	0	18.2	81.8
Scissors	44	9.1	6.8	84.1
Drill	21	14.3	14.3	71.4
Screwdriver	27	0	0	100
Wrench	40	2.5	2.5	95
Compass	16	0	6.2	93.8

Table 1. Sketch recognition

the small size of the training set, and the enormous variability of the drawn sketches, we are extremely satisfied by this success rate.

Unfortunately, we were not able to find a similar research for our results to compare with; the already quoted [8] uses more rigid drawings (and — on the other hand — a much wider database).

A reasonable objection to our research could be that it does not separate the contribution of the measuring functions, of the cooperation method, of the distance, and of the very tool of size functions. It does, anyway, show the modularity of size theory. The good feed-back we got by distributing SketchUp is persuading us to extend the experiment in two opposite directions: Integration with measuring functions working on the interior, in order to improve the performance; and selection of only one function, in order to allow a straight comparison with other (standard) methods to be applied to the same data sets.

8 Conclusions

SketchUp is a demonstrative program which classifies free-hand drawn sketches by size functions of their outer contours. The leading idea is that of providing a new, entirely qualitative tool for shape data management. A fairly high percentage of success is attained even with a very small training set. Better performances are expected with use, in future, of more sensitive distances.

Acknowledgements

Research accomplished under support of CNR-GNSAGA, MURST, ASI, and within an agreement with Elsag Bailey.

We wish to thank Andrea Padovano, who has started the programming of the user interface. We are very grateful to the many participants in the ACCV '95 who volounteered to draw most of the sketches we have used in this research.

References

1. Ferri, M., Frosini, P., Lovato, A., and Zambelli, C., "Point selection: A new comparison scheme for size functions (With an application to monogram recognition)", in *Proc. ACCV '98*, Hong Kong (1998).
2. Ferri, M., Lombardini, S., and Pallotti, C., "Leukocyte classification by size functions", in *Proc. 2nd IEEE WACV*, Sarasota (1994), 223-229.
3. Ferri, M., Gallina, S., Porcellini, E., and Serena, M., "On-line character and writer recognition by size functions and fuzzy logic", in *Proc. ACCV '95*, Singapore, vol. 3 (1995), 622-626.
4. Ferri, M., and Frosini, P., "Range size functions", in *Proc. of the SPIE on Vision Geometry III*, Boston, Mas., Vol. 2356 (1994), 243-251.
5. Frosini, P., "Discrete computation of size functions", *J. Combin. Inf. Syst. Sci.* 17 (1992), 232-250.
6. Frosini, P., "Connections between size functions and critical points", *Math. Meth. Appl. Sci.* 19 (1996), 555-569.
7. Frosini, P., "Size theory as a topological tool for computer vision", in: *"From segmentation to interpretation and back: Mathematica methods in computer vision"*, Eds.: T. Moons, E. Pauwels, L. Van Gool (in print).
8. Mehrotra, R., and Gary, J.E., "Similar-shape retrieval in shape data management", *Computer* 28 (1995), 57-62.
9. Mokhtarian, F., "Silhouette-based isolated object recognition through curvature scale space", *IEEE Trans. on PAMI* 17, 5 (1995), 539-544.
10. Nouboud, F., "Handwritten signature verification: A global approach", in "Fundamentals in handwriting recognition" Ed.: S. Impedovo, NATO ASI Series F, 124, Springer Verlag (1994), 455-459.
11. Uras, C., and Verri, A., "On the recognition of the alphabet of the sign language through size functions", in *Proc. 12th ICPR*, Jerusalem, IEEE Computer Soc. Press (1994), 334-338.
12. Verri, A., and Uras, C., "Invariant size functions", in *Applications of invariance in Computer Vision, Lecture Notes on Computer Science*, J.L. Mundy, A. Zisserman, D. Forsyth eds., vol. 825, Springer-Verlag (1994), 215-234.
13. Verri, A., Uras, C., "Computing size functions from edge maps", *Internat. J. Comput. Vision* (1997).
14. Verri, A., Uras, C., Frosini, P., and Ferri M., "On the use of size functions for shape analysis", *Biol. Cybernetics* 70 (1993), 99-107.

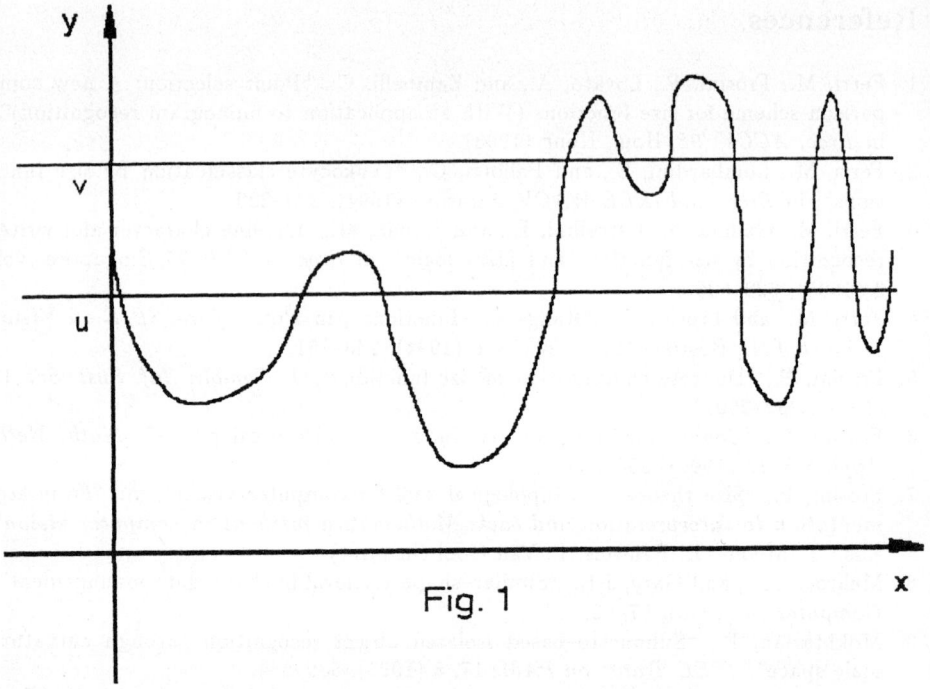

Fig. 1

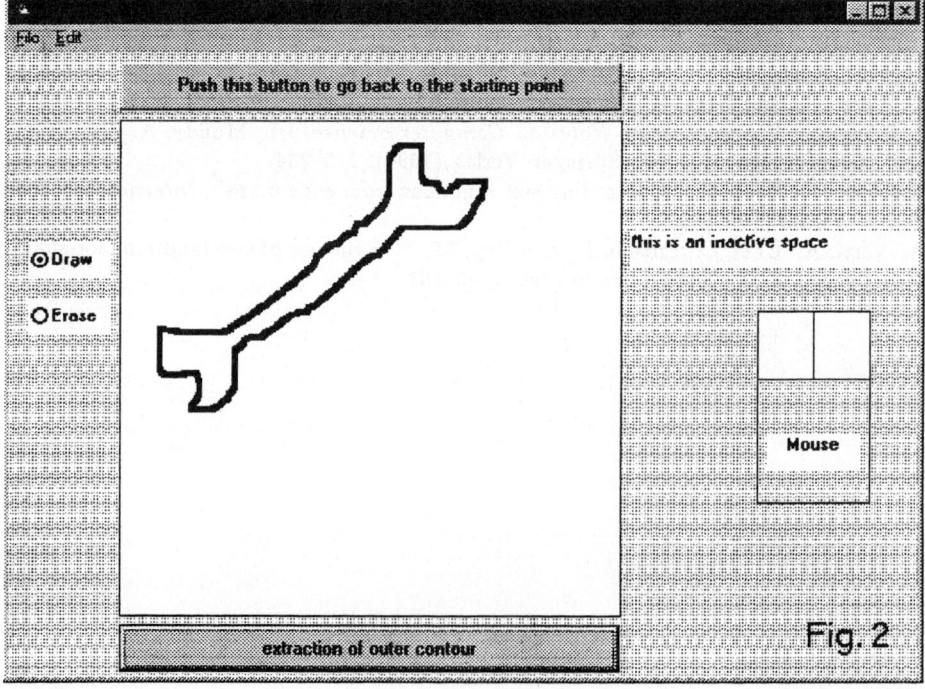

Fig. 2

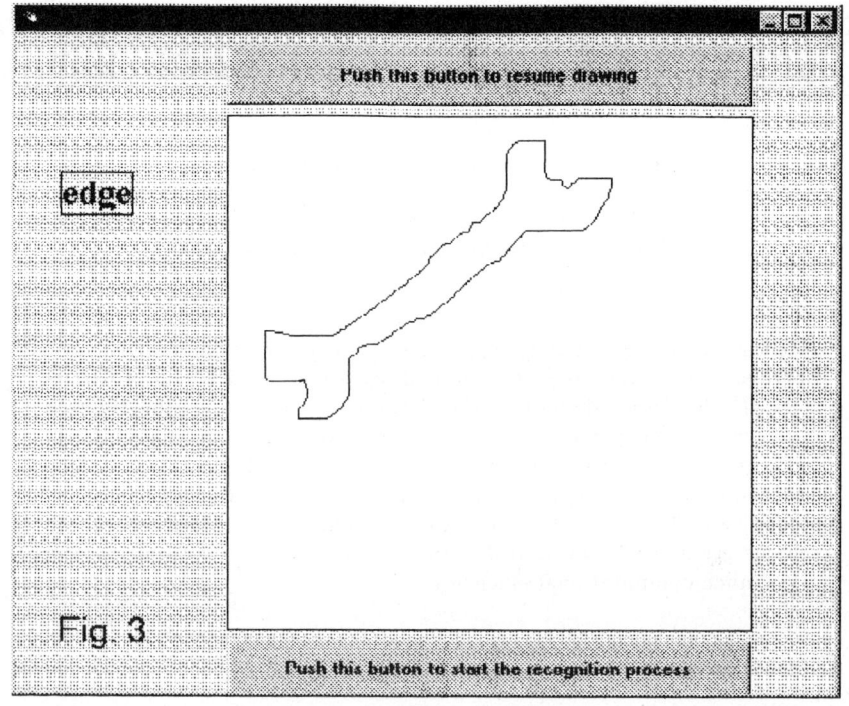

Fig. 3

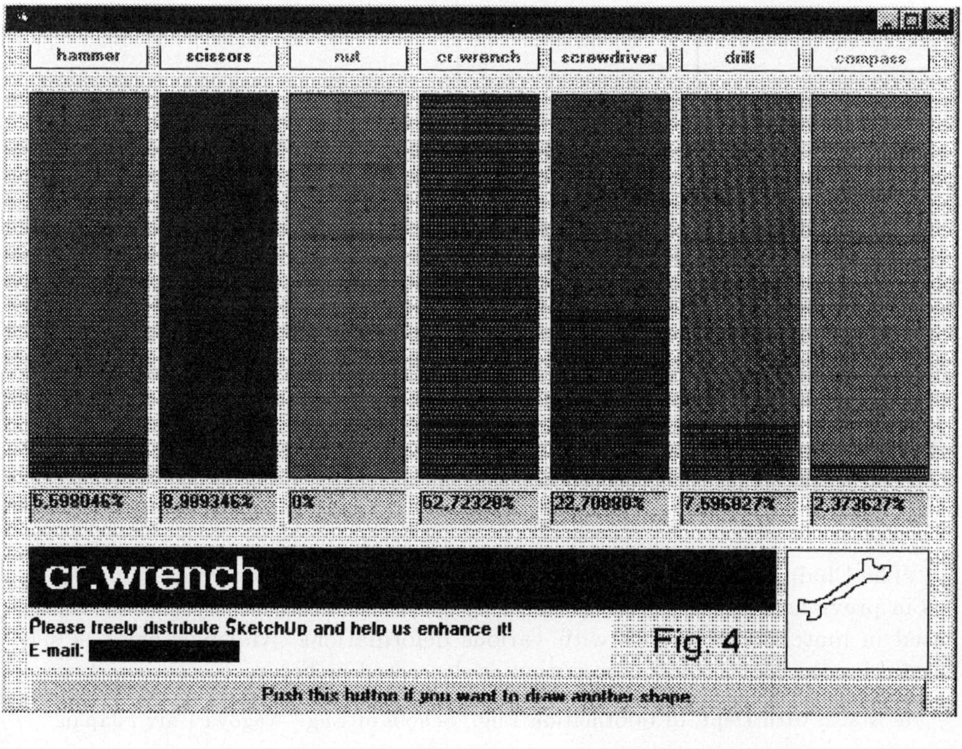

Fig. 4

Robust Matching and Hierarchical Recognition of 2-D Shapes Using "Chain of Circles"

Jae-Moon Chung and Noboru Ohnishi [*]

Bio-Mimetic Control Research Center, RIKEN
Shimoshidami, Moriyama-ku, Nagoya, 463 Japan

Abstract. Based on a resulting medial axis configuration of 2-D shapes, previously we proposed a new shape description called Chain of Circles (CoCs). This paper shows how the CoCs representation is robustly used in matching of shapes with various deformations. The shape dissimilarity between a pair of CoCs is first coarsely calculated via the matching process, then finely calculated between specific and salient parts of the shapes, and finally used for recognition. This hierarchical recognition scheme seems to be in accord with the human shape recognition and gives much computational efficiency.

1 Introduction

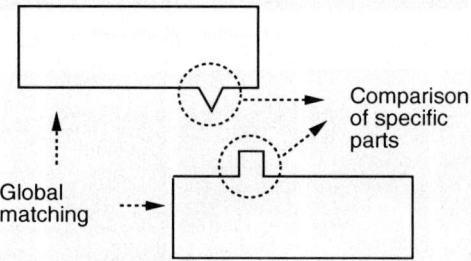

Fig. 1. Recognition by hierarchical matching: as the first step, two shapes are globally matched using their coarse CoCs descriptions for computational efficiency, and if the values of the dissimilarity vector between them are smaller than a threshold value, then the comparison of specific parts characterizing the shapes is made using fine CoCs descriptions for the parts for final recognition.

The Chain of Circles (CoCs) [1] is a new 2-D shape description proposed by us in previous. In this paper, we show how the CoCs is robustly and efficiently used in matching of shapes with various deformations. And how the CoCs is useful for the hierarchical recognition as described in Figure 1 is shown, too.

---

[*] He is also with Dept. of Information Eng., School of Eng., Nagoya Univ., Japan

2 Chain of Circles: define & control

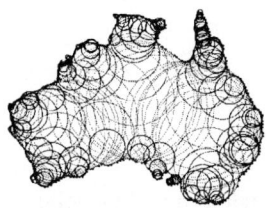

Fig. 2. A collection of the maximal disks (MDs) obtained for the inner part of a map of Australia. A maximal disk has its center at a point on the medial axis of the boundary curve.

The medial axis generated by applying Medial Axis Transform [2] can be considered as locus of center of the maximal disk (MD) as shown in Figure 2. Based on the MD, we proposed CoCs representation and its control as briefly described in Figures 3 and 4.

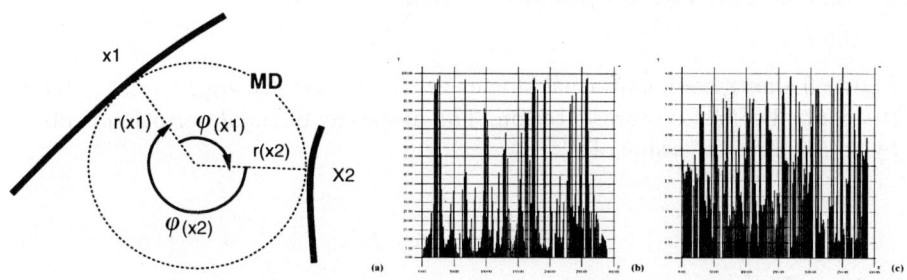

Fig. 3. (a) Definition of chain of circles [1], represented by (b) $r(x)$ and (c) $\phi(x)$ along contour arc length x of the shape shown in Figure 2.

2.1 Hierarchical Recognition

Figure 5 shows how the hierarchical recognition of shapes is performed by the control of CoCs representation. We can see that in spite of distinct differences between fine representations of salient parts of two shapes, the two coarse descriptions become almost similar to one another.

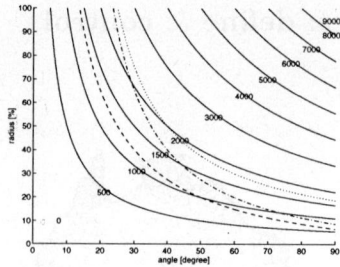

Fig. 4. Control of CoCs [1]. Each numeral on solid lines indicate Γ values when both r_c and ϑ_c are 0. The dotted line is for (Γ=1500, r_c=0, ϑ_c=−10), the dashed line (1500, 10, 0) and the dot-dash line (1500, 10, −10).

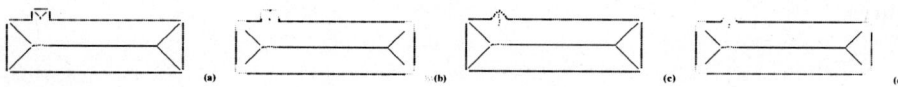

Fig. 5. Hierarchical recognition of shapes: (a)(c) fine and (b)(d) coarse representations of two shapes illustrated in Figure 1, each of which is for $\Gamma_{min} = 0$ and 1500.

3 Hierarchical Representation

Figure 6 shows how nonformal shapes are hierarchically approximated based on the control of CoCs representation. The hierarchy of the description is obtained by controlling the value of Γ_{min}.

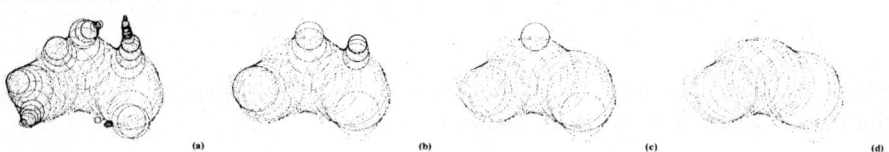

Fig. 6. The hierarchical approximation of the map of Australia, each of which is displayed as the collection of MDs used, and CoCs representations using $r(x)$ and $\phi(x)$. The values of Γ_{min} and the numbers of contour points are (a) 1000, 98 (b) 2000, 46 (c) 3000, 32 and (d) 4000, 23, respectively.

4 Matching and Recognition

Figures 7 and 8 and Table 1 show shapes used in the experiment, matching results and dissimilarity values obtained via the matching process, respectively.

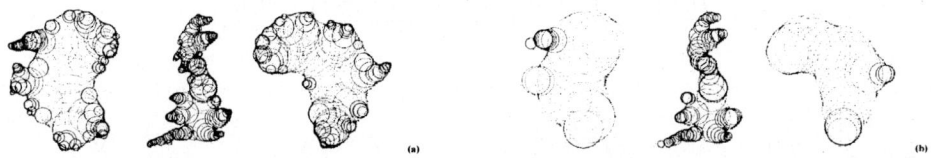

Fig. 7. Collections of MDs to the maps of rotated Australia, England and Africa, for (a) $\Gamma_{min} = 0$ and (b) $\Gamma_{min} = 2000$, respectively. Rotated Australia is used to investigate the effect of spatial quantization arising from the shift of sampling points on the contour curve.

Fig. 8. The results of matching of the map of Australia to (a) its rotated images, (b) the maps of Africa and (c) England. In each of the matches, squares show the contour points determined in matching among the points used in each CoCs representation ($\Gamma_{min} = 2000$). The lines connecting the squares show matched points between two contour lines.

4.1 Noisy Images

Figures 9 and 10 and Table 2 show shapes used in the experiment, matching results and dissimilarity values obtained via the matching process, respectively.

Table 1. The dissimilarity vectors between the map of Australia. The columns of Aus, $r(A)$, Afr and Eng are for self-matching of Australia, matching of Australia to rotated Australia, Africa and England, respectively.

	$\Gamma_{min}=2000$	$\Gamma_{min}=3000$
Aus	(0,0,0)	(0,0,0)
$r(A)$	6.8(0.33,0.74,0.57)	5.2(0.15,0.72,0.67)
Afr	16.1(0.71,0.67,0.12)	14.1(0.77,0.59,0.22)
Eng	40.0(0.95,0.22,0.21)	27.6(0.86,0.32,0.36)

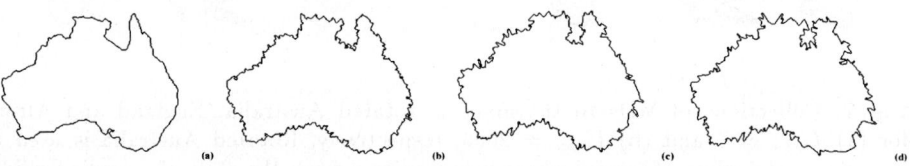

Fig. 9. Contour images of the map of Australia for (a) no noise, (b) noise = 2, (c) noise = 4 and (d) noise = 6.

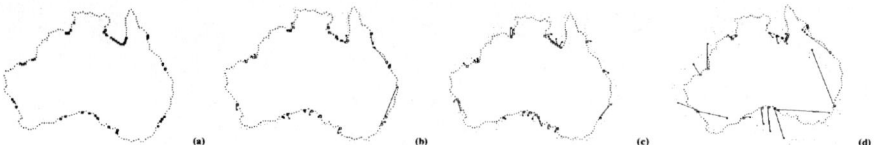

Fig. 10. The results of matching of the map of Australia to (a) the image of Australia with no noise (self-matching), (b) noise = 2, (c) noise = 4 and (d) noise = 6.

Table 2. The dissimilarity vectors between the map of Australia and the shapes shown in Figure 10 when the values of Γ_{min} are 2000 and 3000. The columns of Aus, $n2(A)$, $n4(A)$ and $n6(A)$ are for self-matching of Australia (no noise), matching of Australia to the image of Australia with noise = 2, noise = 4 and noise = 6, respectively.

	$\Gamma_{min}=2000$	$\Gamma_{min}=3000$
Aus	(0,0,0)	(0,0,0)
$n2(A)$	10.5(0.71,0.65,0.26)	9.9(0.77,0.56,0.28)
$n4(A)$	12.2(0.54,0.80,0.23)	10.7(0.62,0.76,0.14)
$n6(A)$	18.8(0.78,0.57,0.22)	13.4(0.46,0.57,0.67)

4.2 Skewed Images

Here, the skewed image means a contour image which is obtained by a camera from a contour image written on a plane in 3-D space. The degree of skew depends on the difference in parallel between the contour image plane and the image plane of the camera.

Figures 11 and 12 and Table 3 show shapes used in the experiment, matching results and dissimilarity values obtained via the matching process, respectively.

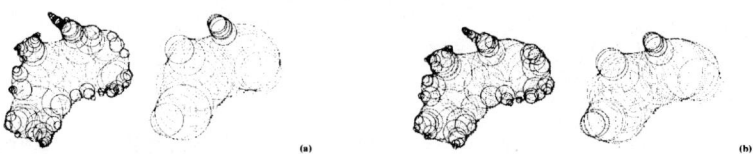

Fig. 11. Collections of MDs for skewed images of the map of Australia for $\Gamma_{min} = 0$ and 2000: (a) a skewed image and (b) a severely skewed image.

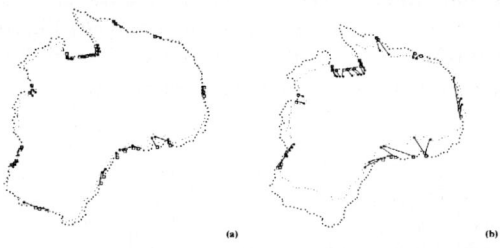

Fig. 12. The results of matching of the map of Australia to (a) the skewed image and (b) severely skewed image, respectively.

5 Hierarchical Recognition

Figures 13 and Table 4 show images used to show an example of the hierarchical recognition and the results. This example may be considered as an implementation of the concept of hierarchical matching and recognition addressed in Figure 1.

Table 3. The dissimilarity vectors between the map of Australia and the shapes shown in Figure 12 when the values of Γ_{min} are 2000 and 3000. The columns of Aus, $s(A)$ and $ss(A)$ are for self-matching of Australia, matching of Australia to the skewed image and severely skewed image, respectively.

	$\Gamma_{min}=2000$	$\Gamma_{min}=3000$
Aus	(0,0,0)	(0,0,0)
$s(A)$	9.8(0.70,0.67,0.24)	8.3(0.39,0.55,0.73)
$ss(A)$	14.3(0.50,0.62,0.59)	10.9(0.77,0.55,0.31)

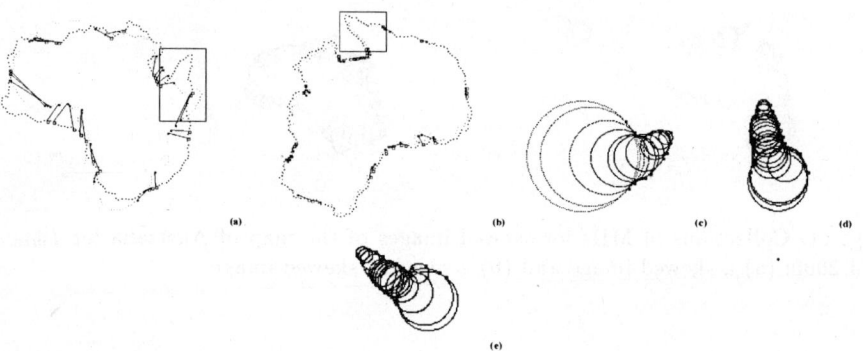

Fig. 13. Hierarchical matching. Rectangular blocks denote selected corresponding parts of which relations are obtained from (a) Figure 9(b) and (b) Figure 13(a), respectively. Whole CoCs representations ($\Gamma_{min} = 0$) for the selected parts for (c) Africa, (d) Australia and (e) the skewed image of Australia, respectively.

6 Discussion

Based on the experimental results, we discuss conclusions about the proposed method.

Figure 8 shows that proposed method is effective for the matching of different shapes as well as the same shapes. From Table 1 we can see that the magnitude of dissimilarity vector decreases along with the increase of similarity between shapes.

Figures 10 and 12 show how the salient parts of shapes are successfully matched despite various deformations in the shapes. As we expected Tables 2 and 3 display that the magnitude of dissimilarity vector increases according to the degree of shape deformation.

Figure 14 shows of relations of the magnitudes of dissimilarity vectors to several matched shape pairs. From this figure we can see that it is difficult to differentiate between Aus and Afr by using their highly approximated CoCs

Table 4. The dissimilarity vectors of matching results shown in Figure 13(a) and (b), and the selected corresponding parts. The second row is for the global matching results, and the third row is for the matching results of corresponding parts. The third row of Afr and $s(A)$ columns show the matching results between CoCs representations (c) and (d), and (d) and (e) of Figure 13, respectively.

	whole($\Gamma_{min}=2000$)	part($\Gamma_{min}=0$)
Afr	16.1(0.71,0.67,0.12)	13.3(0.74,0.66,0.08)
$s(A)$	9.8(0.70,0.67,0.24)	2.6(0.67,0.72,0.15)

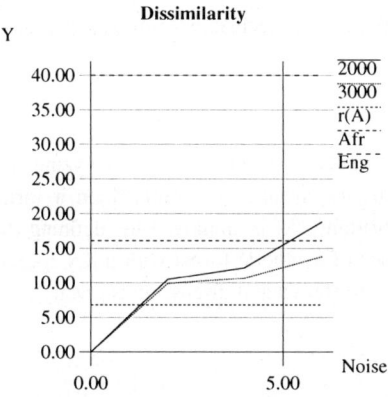

Fig. 14. The graph showing dissimilarity magnitudes of the map of Australia (Aus) to $r(A)$, Afr and Eng without noise, and to noisy images of Aus in cases of $\Gamma_{min} = 2000, 3000$.

representations. However it is possible to differentiate the two shapes by comparing the salient parts of the two shapes, as shown in Figure 13 and Table 4.

As a future work, we are going to investigate the effect of each element of the dissimilarity vector between shapes and its use for recognition.

References

1. Jae-Moon Chung and Noboru Ohnishi, "Chain of circles for matching and recognition of planar shapes," *Technical Report of IEICE*, vol. IE96-126, pp. 23–30, Feb. 1997, and *Proc. IJCAI-97*, pp.1482–1487.
2. H. Blum, "A transformation for extracting new descriptors of shape," In W. Wathen-Dunn, editor, *Models for perception of speech and visual form*, MIT Press, 1967.

Finding the Center of Rotational Symmetry from Noisy Forms

Hyoung Seop KIM, Nachi MOTOMURA, Seiji ISHIKAWA

Department of Mechanical and Control Engineering
Kyushu Institute of Technology
Sensuicho 1-1, Tobata, Kitakyushu 804, Japan

Email:kim@is.cntl.kyutech.ac.jp / ishikawa@is.cntl.kyutech.ac.jp

Abstract. This paper presents a technique for analyzing noisy symmetric forms. The approximate center of rotational symmetry is defined on a form which has noisy rotational symmetry. The genetic algorithm(GA) is employed for defining the center. Performance of the GA is examined by a number of synthetic forms with noisy rotational symmetry. Experimental results based on real images are shown and discussion is given.

1 Introduction

There are a number of objects which have symmetric forms in our life. Automatic symmetry analysis has therefore been one of the primary concerns in the pattern recognition community. Various symmetry analysis techniques have been proposed to date[1,2], but they do not have much practical significance with real objects which often have incomplete symmetry. There are many forms which have approximate (or incomplete) symmetry, *e.g.*, human faces, partially broken forms which originally had symmetry, symmetric forms with asymmetric textures. Even an image digitizer may change a symmetric form into asymmetric by quantization. All of these are understood as noisy symmetric forms or the forms that possess potential symmetry[5] in this paper.

Examples of noisy rotational symmetry are illustrated in Fig.1. Fig.1(a) is an image of a flower which originally had 12-fold rotational symmetry. On the other hand, a plate with the picture of tree branches is given in Fig.1(b). If one looks the plate as a gray value image, it does not have rotational symmetry, though it has 8-fold rotational symmetry with its contour. If the original rotational symmetry is recognized with the both cases, their intrinsic shape structure can be made clear, which may contribute to exact understanding of the form interested and also to the reconstruction of the original form when it has some broken parts. The application of the proposed technique may therefore include pattern recognition, robot vision, archaeology, *etc.* Visual psychologists may as well find interest in the technique for

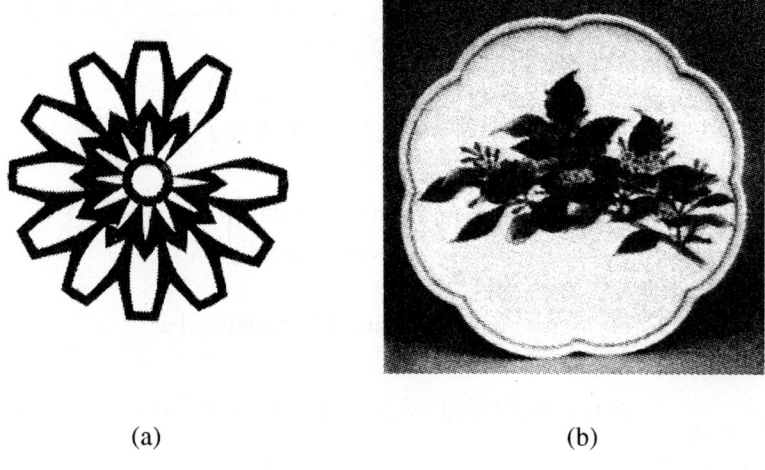

(a) (b)

Fig. 1. Examples of noisy rotational symmetry.

investigating human symmetry perception, for instance.

In spite of its importance, analysis of noisy symmetric forms has not yet paid much attention. Almost all techniques proposed to date fail in analyzing those noisy symmetric forms, since their theories are based on symmetry without noise, *i.e.*, complete symmetry. Some attempts have been reported [3,4] to analyze such noisy symmetric forms, but they are not very powerful so that they can be applied to practical forms in our life. The technique proposed in [5] can deal with those practical forms with noisy symmetry. They employ the mirror reflected image of the original and superpose it onto the original image to find the position which minimizes a certain evaluation function. They discuss only axial (or mirror) symmetry, though.

In this paper, a technique is proposed for identifying the center of rotational symmetry on the forms with noisy rotational symmetry. The problem of finding the center of rotational symmetry is described in terms of the optimization problem and the genetic algorithm is employed for the optimization. Experimental results using real images are shown and discussion is given.

2 Formulation as an Optimization Problem

A form with n-fold rotational symmetry is obviously congruent to itself after the rotation by $2\pi k/n$ ($k=1,2,\ldots,n-1$) around its center. Similar procedure is employed in the proposed technique.

Let us denote a form by $F(x, y)$, $(x, y) \in R$, and a form yielded by rotating F by θ ($0 < \theta < 2\pi$) around an arbitrary point (x_0, y_0) by $F'_\theta(X, Y)$, $(X, Y) \in R'$, *i. e.*,

$$F'_\theta(X, Y) = F(x, y) \qquad (1)$$

where

$$\begin{pmatrix} X \\ Y \end{pmatrix} = T \begin{pmatrix} x - x_0 \\ y - y_0 \end{pmatrix} + \begin{pmatrix} x_0 \\ y_0 \end{pmatrix},$$

$$T = \begin{pmatrix} \cos\theta & -\sin\theta \\ \sin\theta & \cos\theta \end{pmatrix}.$$

The difference form $\Delta(x, y)$ made by F and F'_θ is defined as

$$\Delta(x, y) = |F(x, y) - F'_\theta(x, y)|, \quad (x, y) \in R \cup R'. \tag{2}$$

This $\Delta(x, y)$ is to be minimized employing a certain evaluation function. In the performed experiment, two kinds of evaluation functions are employed.

The minimum difference of gray values is taken into consideration in the first place. The following evaluation function is to be minimized;

$$D = \sum_{(x,y) \in R \cap R'} |F(x, y) - F'_\theta(x, y)| \Big/ S \tag{3}$$

where

$$S = n\{(x, y) \mid (x, y) \in R \cap R'\}.$$

Here the number of the elements in the set A is denoted by $n(A)$.

The second index to be considered is the minimum variance of the difference form $\Delta(x, y)$ which scatters on an image plane. Sum of the variances to the x direction and to the y direction, denoted by D, is minimized. Let us define $\Delta_y(x)$, $\Delta_x(y)$, $S \equiv \Delta$ by

$$\Delta_y(x) = \sum_y \Delta(x, y),$$

$$\Delta_x(y) = \sum_x \Delta(x, y),$$

$$S \equiv \Delta = \sum_{x,y} \Delta(x, y).$$

Then

$$D = D_x + D_y \tag{4}$$

where

$$D_x = \frac{1}{S} \sum_{(x,y) \in R \cup R'} (x - \bar{x})^2 \Delta_y(x),$$

$$D_y = \frac{1}{S} \sum_{(x,y) \in R \cup R'} (y - \bar{y})^2 \Delta_x(y),$$

and

$$\bar{x} = \sum_x x\Delta_y(x)/S,$$
$$\bar{y} = \sum_y y\Delta_x(y)/S.$$

3 Optimization by the Genetic Algorithm

To realize the minimization of D in Eq.(3) and Eq.(4), the genetic algorithm(GA) is employed in both cases.

In the GA employed, the center of rotation (x_0, y_0) and the angle θ such that $\theta \doteq 2\pi k/n$ $(k=1,2,\ldots,n-1)$ are optimized. A chromosome defined is composed of three 10bit genes which are the x coordinate, the y coordinate, and the angle θ. The number of individuals are 30 whose fitness values are calculated by Eq.(3) for the case (a) and by Eq.(4) for the case (b). The probability of reproduction of individual I_i is defined by

$$P(I_i) = \frac{f(I_i)}{\sum_{j=1}^{30} f(I_j)}, \qquad (5)$$

where $f(I)$ is a fitness value of individual I. For crossover, one point crossover is employed and each bit of the gene mutates by the rate of 0.02. Finally, the employed GA stops its alternation of generations when the individual having the highest fitness values remains unchanged during 30 generations.

4 Experimental Results

The experiment is performed on a workstation. Analyzed image data are captured by a digital camera or an image scanner and transferred to a workstation(SS20) via a personal computer. The captured images are represented by 256×256 pixels and 256 gray levels. The programs employed for the analysis are coded by C.

In order to examine the performance of the GA employed, 30 synthetic images are prepared which contain 10 squares, 10 regular triangles, and 10 circles, each of the three basic forms have some noise ranging from 1% to 10%. This noise makes the forms rotationally asymmetric. The distance between the obtained center of rotational symmetry of a noisy form and the ideal center of rotational symmetry of the original noise-free form is normalized by the diameter of the circumscribed circle of the form concerned. This normalized value is employed as a relative error of the obtained center. With respect to the 30 synthetic images, about 0.8% of mean relative error was observed. In Fig.2, two of those synthetic images are shown with their difference images on the right-hand side. In both cases, the amount of added noise is

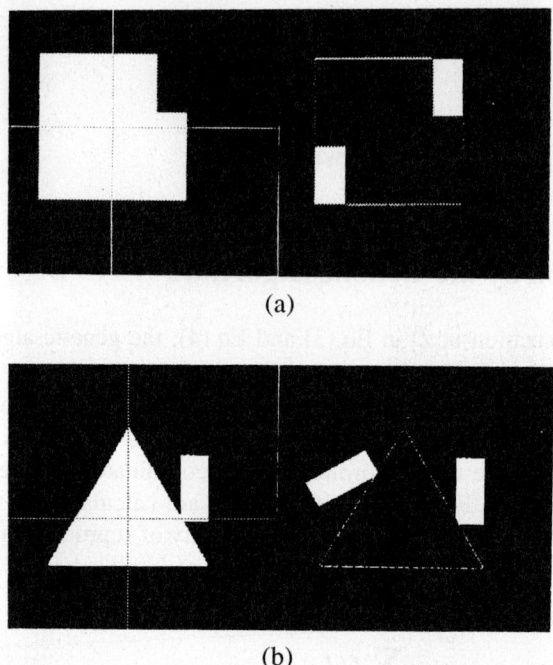

Fig. 2. Noisy regular polygons and the obtained difference images.

8% of the area of the noise-free form that is a square for Fig.2(a) and a regular triangle for Fig.2(b). The obtained center is indicated by the intersection point of the horizontal and the vertical lines.

Some experimental results are given in Fig.3. Fig.3(a) and (b) are synthesized images. We assumed 5-fold and 12-fold original rotational symmetry with respective forms and the exact centers were obtained. The rest four results are based on real images. Fig.3(c) is a partially broken ceramic bowl with no texture but peripheral shade on its inner surface, whereas the plate in Fig.3(d) has an asymmetric drawing on the surface which makes the plate itself asymmetric when it is viewed as a gray-scale image. The resultant images show nice performance of the proposed technique. Finally Fig.3(e) is a top view of a soft-drink can and it is partially crushed in Fig.3(f). In both cases, persuasive results were obtained.

In the performed experiments, no difference was observed with the location of extracted centers irrespective of the employment of Eq.(3) or Eq.(4), though better performance was expected for those forms with heavier noise.

5 Discussion

The experimental results were satisfactory not only with synthetic forms but

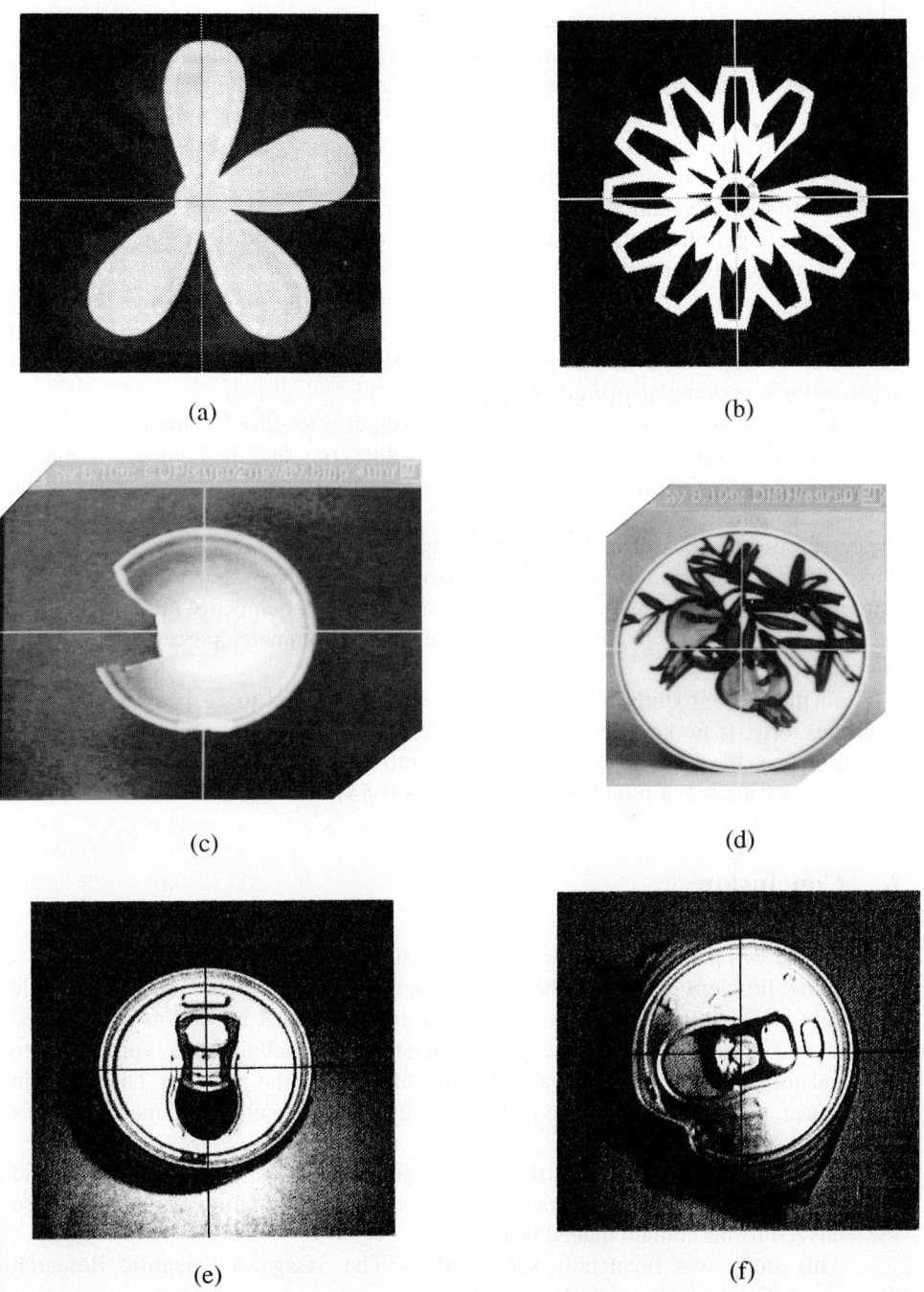

Fig. 3. Experimental Results.

also real forms. The extracted centers well coincide with what we expect visually.

As was mentioned in the former section, two evaluation functions, *i.e.*, Eq.(3) and Eq.(4), showed no difference in the results in the performed experiment, though we expected better performance of Eq.(4) in the case of larger amount of noise. It remains for further study to examine the respective effects of the two functions in the context of heavy noise added to the forms.

The present technique only finds the center of noisy rotational symmetry at the moment. To get the knowledge on possible n-fold rotational symmetry, actual rotation needs be performed around the obtained center of rotational symmetry, which may need certain exhaustive trials employing the angle θ that is finally obtained by the execution of the GA. As an alternative way, a smart technique might be found by employing the complex moments[6] which offer a novel technique for discovering n of complete n-fold symmetry[4].

The proposed technique may have contribution to intrinsic understanding of practical forms by computer, since, unlike other reported techniques, it allows various kinds of noise with respect to analyzed forms. It may offer a powerful shape analysis technique to pattern recognition and computer vision. One might as well expect application to archaeology where shape recovery is one of the main interests. The original form could be recovered from its defective shape employing the obtained center of symmetry. A future intelligent robot equipped with the proposed technique will be able to imitate the way of human symmetry perception which is very flexible.

In the present study, rotational symmetry is assumed to be possessed by the analyzed form. If no knowledge is available in advance on its shape, the moment of inertia matrix of the form needs be taken into consideration. Its eigenvalues are employed for aiming at possible types of symmetry[4].

6 Conclusion

A technique was proposed for extracting the center of rotational symmetry on a noisy form. The problem was described as an optimization problem and the genetic algorithm was employed. The technique was applied to real as well as synthesized images and satisfactory results were obtained. The technique is applicable to practical forms, since it allows various types of noise in the analysis. Therefore, in this respect, the present technique is discriminated from any other symmetry analysis techniques reported to date.

Some topics still need further investigation such as deciding n of n-fold symmetry and examining performance of the employed evaluation functions in case the analyzed forms contain indeed heavy noise.

This study was financially supported by The Sasagawa Scientific Research Grant from The Japan Science Society.

References

1. Eades, P.: "Symmetry finding algorithms", *Computational Morphology*, North-Holland, Amsterdam, 41-51, 1988.
2. Minovic, P., Ishikawa, S., Kato, K.: "Three-dimensional symmetry identification", *Memoirs of The Kyushu Inst. Tech.*, **21**, 1-16, 1992.
3. Marola, G. : "On the detection of the axes of symmetry of symmetric and almost symmetric planar images", *IEEE Trans. Patt. Anal. Mach. Intell.*, **PAMI-11**, 104-108, 1989.
4. Minovic, P., Ishikawa, S., Kato, K.: "Symmetry identification of a 3-D object represented by octree", *IEEE Trans. Patt. Anal. Mach. Intell.*, **PAMI-15**, 5, 507-514, 1993.
5. Ishikawa, S., *et al.*: "A method of analyzing a shape with potential symmetry and its application to detecting spinal deformity", *Computer Vision, Virtual Reality and Robotics in Medicine*, 465-470, Springer, 1995.
6. Abu-Mostafa, Y. S., Psaltis, D.: "Image normalization by complex moments", *IEEE Trans. Patt. Anal. Mach. Intell.*, **PAMI-7**, 1, 46-55, 1985.

Recognition in Wavelet-Compressed Imagery*

Wei Hu[1] and W. Brent Seales[2]

[1] Department of Math and Computer Science, Houghton College,
Houghton, New York 14744, USA
[2] Computer Science Department, University of Kentucky,
Lexington, Kentucky 40506, USA

Abstract. In this paper we analyze a method for classifying images which have been compressed using a biorthogonal wavelet transformation. The goal is to formulate a pattern recognition algorithm over a wavelet-compressed image set without requiring that the image set be decompressed. This paper extends previous work [8] which studies how to recognize objects in images compressed using orthogonal transforms, as in the JPEG/MPEG compression standards.

1 Introduction

Compressed images make storage and transmission more efficient, but complicate the task of automatic content analysis of imagery. Content analysis such as pattern recognition is usually performed on image pixels, either before compression or after decompression. Since pixel data is not directly available in compressed images, however, it is less efficient to recover pixels before processing than to attempt to analyze and process images in their compressed form.

In this paper we show that the task of pattern recognition can be formulated on wavelet-compressed data and that the result of classifying wavelet-compressed images is comparable to performing recognition in the pixel domain. When the compression transformation is orthogonal, pattern recognition is robust and efficient as shown in [8]. The biorthogonal wavelet transformation, however, is not orthogonal and therefore does not follow directly.

First we establish the theoretical basis for pattern recognition in wavelet-compressed imagery, and report the results of our experiments using a biorthogonal wavelet compression scheme. Our experiments show that it is possible to perform pattern classification in the compressed imagery. We show that the computational savings can be substantial and that the classification results are comparable to the same classification task performed directly on pixel images.

---

* We gratefully acknowledge support from the National Science Foundation (grant number IRI-9308415)

2 Wavelet Transforms

If a function $\psi(x) \in L^2(R)$ satisfies

$$C_\psi = \int_{-\infty}^{\infty} \frac{|\hat{\psi}(s)|^2}{|s|} ds < \infty$$

where $\hat{\psi}$ is the Fourier transform of ψ, then ψ is called a wavelet. A function $\psi(x) \in L^2(R)$ is an orthogonal wavelet if the set $\{\psi_{j,k}(x)\}$ of the functions defined by

$$\psi_{j,k}(x) = 2^{j/2}\psi(2^j x - k)$$

where $-\infty < j,k < \infty$ are integers, forms an orthogonal basis of $L^2(R)$ (the multiplication by $2^{j/2}$ is needed to make the basis orthonormal). The integer j determines the dilation, while k specifies the translation. Significant research has been done to construct $\psi(x)$, the so-called "mother wavelet", for which such a basis exists [2].

Given an orthogonal wavelet function $\psi(x)$, any function $f(x) \in L^2(R)$ can be written as

$$f(x) = \sum_{j=-\infty}^{\infty} \sum_{k=-\infty}^{\infty} c_{j,k} \psi_{j,k}(x)$$

where the transform coefficients are given by the inner product

$$c_{j,k} = <f, \psi_{j,k}> = 2^{j/2} \int_{-\infty}^{\infty} f(x)\psi(2^j x - k) dx$$

In real applications, it is too expensive to compute all the wavelet coefficients $c_{j,k}$ one by one using the inner product. There is a multi-resolution form [7] which relates the $c_{j,k}$ values for different scale values of j. As a result, we can use the information of $c_{j,k}$ to get the next level $c_{j-1,k}$.

In the discrete case, given a digital signal $f(i)$, low-pass filter $h_0(i)$ and high-pass filter $h_1(i)$, we can perform two-channel sub-band coding, which yields the two half-length sub-band signals

$$g_0(k) = \sum_i f(i) h_0((-i + 2k)) \tag{1}$$

and

$$g_1(k) = \sum_i f(i) h_1((-i + 2k)) \tag{2}$$

Reconstruction can be accomplished by

$$f(i) = 2 \sum_k [g_0(k) h_0((-i+2k)) + g_1(k) h_1((-i+2k))] \tag{3}$$

Note that equations (1) and (2) use the same filters for decomposition and reconstruction, which is one of the basic properties of the orthogonal wavelet transform. When the wavelet is biorthogonal, however, two different pairs of filters are required: one for decomposition and the other for reconstruction.

A basic question is how to get the low-pass filter h_0 and the high-pass filter h_1. Daubechies [2] shows that if the low-pass filter h_0 satisfies

$$\sum_k h_0(k) = 1 \quad \text{and} \quad \sum_k h_0(k)h_0(k+2l) = \delta(l),$$

then there exists a high-pass filter h_1 such that

$$h_1(k) = (-1)^k h_0(-k+1)$$

and we can have the following "scaling" function $\phi(x)$ and "mother wavelet" function $\psi(x)$:

$$\phi(x) = \sqrt{2} \sum_k h_0(k)\phi(2x-k), \quad \psi(x) = \sqrt{2} \sum_k h_1(k)\phi(2x-k).$$

A major disadvantage of the orthogonal wavelet is its asymmetry. To apply the filters to an image, usually the image must be padded, which can cause artifacts at the wavelet sub-band borders. Therefore symmetric filters are more desirable, but from the definition, a symmetric wavelet can not be both compactly supported and orthogonal (except for the Harr wavelet). However, biorthogonal wavelets can be constructed [1] to be symmetric and compactly supported. With biorthogonal wavelets, two different pairs of filters, (h_0, h_1) and $(\tilde{h}_0, \tilde{h}_1)$, must be used in decomposition and reconstruction. If h_0 and $\tilde{h}_0$ are two finite filters satisfying

$$\sum_i h_0(i)\tilde{h}_0(i+2k) = \delta(k) \quad \text{and} \quad \sum_i h_0(i) = \sum_i \tilde{h}_0(i) = 1$$

then there exist two pairs of functions $(\phi(x), \psi(x))$ and $(\tilde{\phi}(x), \tilde{\psi}(x))$. They are related by

$$\psi(x) = \sqrt{2} \sum_k h_0(k+1)\phi(2x-k) \quad \text{and} \quad \tilde{\psi}(x) = \sqrt{2} \sum_k \tilde{h}_0(k+1)\tilde{\phi}(2x-k)$$

They are biorthogonal, which means that

$$< \psi_{j,k}, \tilde{\psi}_{l,m} > = \delta(j,l)\delta(k,m)$$

The above equality shows that $\psi_{j,k}$ and $\tilde{\psi}_{l,m}$ are orthogonal at different levels and also at different positions for the same level. Also $\{\psi_{j,k}(x)\}$ constitute a frame in $L^2(R)$, their dual frame is given by $\{\tilde{\psi}_{j,k}(x)\}$. That is, for any $f(x) \in L^2(R)$, we have

$$C_1 \|f\|^2_{L^2(R)} \le \sum_{j,k} | < f, \psi_{j,k} > |^2 \le C_2 \|f\|^2_{L^2(R)} \qquad (4)$$

and

$$C_3 \|f\|^2_{L^2(R)} \le \sum_{j,k} | < f, \tilde{\psi}_{j,k} > |^2 \le C_4 \|f\|^2_{L^2(R)} \qquad (5)$$

We call the constants C_1, C_2, C_3 and C_4 frame constants for the forward and backward biorthogonal wavelet transform.

3 Recognition Based on Principal Components

There are many algorithms for pattern recognition [4]. We focus on Principal Component Analysis (PCA), based on the Hotelling Transform (refer to [8] for a more detailed explanation of this method). Given a set of image templates $\{f_1, f_2, \ldots, f_n\}$ and another image f, all with $N \times N$ pixels, PCA provides a strategy for efficiently classifying f as one particular template from the set. The principal components of an image set are constructed from the covariance matrix C of mean-adjusted image templates, written as vectors:

$$C = UU^t = [f_1 - m, f_2 - m, \ldots, f_n - m] \begin{bmatrix} (f_1 - m)^t \\ (f_2 - m)^t \\ \vdots \\ (f_n - m)^t \end{bmatrix} \tag{6}$$

The size of matrix C is $N \times N$. The N eigenvalues and eigenvectors of C form an orthogonal basis of R^N. The orthogonal matrix of ordered eigenvectors

$$\Lambda = \begin{bmatrix} \Lambda_1^t \\ \Lambda_2^t \\ \vdots \\ \Lambda_N^t \end{bmatrix}$$

is the Hotelling transform. Λ can be used as a map from the image space to itself by

$$p = \Lambda(f - m)$$

Since Λ is invertible, we can rewrite the above expression as

$$f = \Lambda^t p + m \tag{7}$$

We can also write each template as

$$f_i = \Lambda^t p_i + m$$

where $p_i = \Lambda(f_i - m)$. Because Λ is orthogonal, we see that

$$\|f - f_i\| = \|p - p_i\| \tag{8}$$

Equation 8 reveals that if we want to compare two images f_i and f, it suffices to compare p_i and p.

In order to have equality in (7), we need to use all N eigenvectors. But if we choose to have only k with $k \ll N$ to form a Hotelling matrix $\tilde{\Lambda}_k$, which is of size $k \times N$, then

$$\frac{1}{N}\|f - (\tilde{\Lambda}_k^t p + m)\|^2 = \sum_{i=k+1}^{N} \lambda_i$$

Therefore we have

$$f \approx \tilde{\Lambda}_k^t p + m \qquad f_i \approx \tilde{\Lambda}_k^t p_i + m \tag{9}$$

and
$$\|f - f_i\| \approx \|p - p_i\| \qquad (10)$$

Note that in this case, p and p_i are of size k.

In reality we always try to achieve the balance between smaller k (fast in time) and bigger k (smaller error in classification). This method provides us the control of these parameters.

4 Classification of Transformed Data

The biorthogonal wavelet transform is not orthogonal but satisfies Eqs. (4) and (5). One of the benefits of an orthogonal transform is that it preserves the eigenvalues of the correlation matrix from the PCA computation (Eqn. 6) and gives a direct relationship between the eigenvectors of an image set and its corresponding, transformed image set [8]. The question considered here is how the eigenvalues and eigenvectors of an image set will change after applying the biorthogonal wavelet transform, since our classification method depends directly on the eigenvalues and eigenvectors of the transformed images.

We approach this problem using basic results in matrix theory [3, 5]. If U in equation (6) has singular values

$$\sigma_1(U) \geq \sigma_2(U) \geq \ldots \sigma_n(U) \geq 0$$

then since $C = UU^t$, the covariance matrix in (6), has eigenvalues

$$\lambda_1(C) = \sigma_1(U)^2 \geq \lambda_2(C) = \sigma_2(U)^2 \geq \ldots \lambda_n(C) = \sigma_n(U)^2 \geq 0$$

Now we denote W as the forward biorthogonal wavelet transform, and we want to estimate the eigenvalues of $WCW^t = WU(WU)^t$ in terms of the eigenvalues of C. Our main result shows how the eigenvalues of matrix C change after applying a forward biorthogonal wavelet transform. The following estimate indicates that the change of eigenvalues of the covariance matrix C after applying the forward biorthogonal wavelet transform is up to a constant, and this constant is determined by the frame constants of W. For any $k \leq n$,

$$\|W^{-1}\|^{-2} \sum_{i=1}^{k} \lambda_i(C) \leq \sum_{i=1}^{k} \lambda_i(WCW^t) \leq \|W\|^2 \sum_{i=1}^{k} \lambda_i(C) \qquad (11)$$

See [6] for a derivation of this estimate (11).

We can consider how the directions of the eigenvectors get changed. Using (4), we can have

$$C_1|X - Y| \leq |W(X - Y)| \leq C_2|X - Y|$$

which can be rewritten as using the inner product notation

$$C_1^2 <X-Y, X-Y> \;\leq\; <W(X-Y), W(X-Y)> \;\leq\; C_2^2 <X-Y, X-Y>$$

If we simplify the above inequality, we can see that

$$< X, Y > - (C_2^2 - C_1^2)(|X|^2 + |Y|^2) \le < WX, WY >$$
$$\le < X, Y > + (C_2^2 - C_1^2)(|X|^2 + |Y|^2) \tag{12}$$

Estimate (12) implies that if W is orthogonal, then the angle between X and Y remains the same after applying W. And estimate (12) also shows that the error introduced by using a biorthogonal transform compared to the orthogonal transform is $(C_2^2 - C_1^2)(|X|^2 + |Y|^2)$. This error can be small if the frame constants C_1 and C_2 are close to each other. For example, in one experiment, we calculated $C_1 = 0.068089$ and $C_2 = 0.092627$, for 180 gray-scale images, each image at resolution $= 256 \times 256$.

Suppose $CX_i = \lambda_i(C)X_i$ and $WCW^t Y_i = \lambda_i(WCW^t)Y_i$. We need to study the difference between X_i and $W^t Y_i$. From equation (33) in [8], we know $D^t Y_i = X_i$, where D stands for DCT. We can prove the following estimate under the assumption of $\|C^{-1}\| < \lambda_1(WCW^t)^{-1}$ [6]:

$$\sum_{i=1}^{k} |(X_i - W^t Y_i)| \le \frac{1}{1 - \|C^{-1}\lambda_1(WCW^t)\|} \{I + II\} \tag{13}$$

where

$$I = \max_{1 \le i \le k} |C^{-1} X_i| \max\{|\|W^{-1}\|^2 - 1|, \|W\|^2 - 1|\} \sum_{i=1}^{k} \lambda_i(C)$$

$$II = \|C^{-1}\|\|W^{-1}\|\|\|W\|^2 - 1| \sum_{i=1}^{k} \lambda_i(WCW^t)|Y_i|.$$

Estimate (13) shows that when the frame constants are close to one, which means W is close to being orthogonal, the difference between X_i and $W^t Y_i$ is small. We can say that under the assumption of (13) the biorthogonal wavelet transform is close to an orthogonal transform like the DCT in terms of the changes of the eigenvalues and eigenvectors between the original and transformed images. The key facts that enable us to get the estimates in (11), (12) and (13) are biorthogonality of W and the special structure of the covariance matrix, which is UU^t.

5 Experimental Results

We have performed pattern classification tests using the wavelet transform implemented in the EPIC system [9].

We have run a number of experiments to classify images of faces from a training set of 180 images (20 distinct poses of 9 faces). The curves in Figure 1 show average mis-classification rate over different combinations of poses. The misclassification rate eventually becomes high as binsize (quantization levels)

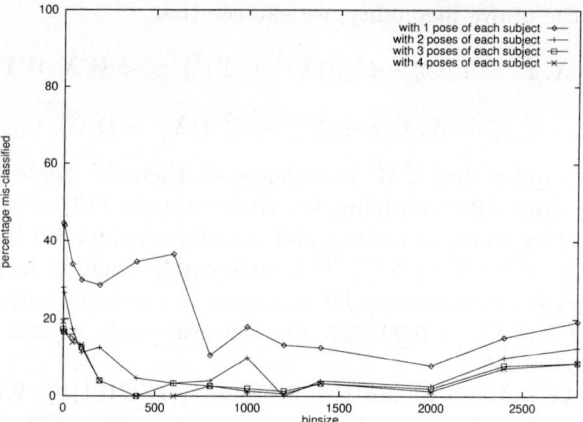

Fig. 1. Mis-classification rate as a function of binsize.

increase. For a substantial range, however, ... ss rate low. This indicates that it is possible to accurately classify images over large compression ratios.

We have obtained other data as well which can be found in [6], such as a comparison to pattern-classification in DCT-compressed imagery, as well as a measure of how eigenvalues and eigenvectors change as a function of quantization.

We analyze how the eigenvalues and eigenvectors are changed at each step of the wavelet transform. We know that scaling preserves the eigenvalues and eigenvectors, so only three steps affect the eigenvalues and eigenvectors. These three steps are applying the forward biorthogonal wavelet transform, quantization and applying the backward biorthogonal wavelet transform. The affects of applying the biorthogonal wavelet transforms are mainly determined by the frame constants C_1, C_2, C_3 and C_4 in (4) and (5). In our experiments, we compute these frame constants and note that they are relatively small values, which means that the transformation behaves very close to an orthogonal one. In one experiment, for example, they were $C_1 = 0.068089, C_2 = 0.092627, C_3 = 10.601637$ and $C_4 = 14.515660$. Note that $C_1 = C_4^{-1}$ and $C_2 = C_3^{-1}$.

6 Conclusions

In this paper we have shown that images can be classified in their wavelet-compressed form, where the wavelet transform is the biorthogonal wavelet. Classification rates are stable over a large range of quantization levels. We formulate the classification problem as a principal component representation in the wavelet-compressed domain. We have shown that although the biorthogonal wavelet transform is not strictly orthogonal, the principal components of the wavelet-compressed are perturbed only slightly from the principal components of the image-domain data.

References

1. A. Cohen, I. Daubechies, and J.-C. Feauveau. Biorthogonal bases of compactly supported wavelets. *Comm. Pure Appl. Math.*, 45:485–560, 1992.
2. I. Daubechies. Orthogonal bases of compactly supported wavelets. *Comm. Pure Appl. Math.*, 41:909–996, 1988.
3. G. Golub and C. Van Loan. *Matrix Computation*. The John Hopkins University Press, 1984.
4. R.C. Gonzales and R.E. Woods. *Digital Image Processing*. Addison-Wesley, 1993.
5. R.A. Horn and C.R. Johnson. *Topics in Matrix Analysis*. Cambridge University Press, 1991.
6. W. Hu and W. B. Seales. Biorthogonal wavelets and object recognition in the compressed domain. Technical report, Computer Science Dept., University of Kentucky, Lexington, Kentucky, 1997.
7. S. Mallat. A theory for multiresolution signal decomposition: The wavelet representation. *IEEE Trans. Pattern Analysis and Machine Intelligence*, 11:674–693, 1989.
8. W.B. Seales, M.D. Cutts, C.J. Yuan, and W. Hu. Object recognition in compressed imagery. *Image and Vision Computing*, to appear, 1998.
9. E.P. Simoncelli and E. H. Adelson. Efficient pyramid image coder (epic), 1996. Distribution available from *ftp://ftp.cis.uppen.edu/pub/eero/epic.tar.Z*.

Fast Image Template and Dictionary Matching Algorithms

Sung-Hyuk Cha

Information Technology R&D Center, SamsungSDS

Abstract. Given a large text image and a small template image, the Template Matching Problem is that of finding every location within the text which looks like the pattern. This problem, which has recieved attention for low-level image processing, has been formalized by defining a distance metric between arrays of pixels and finding all subarrays of the large image which are within some threshold distance of the template. These so-called metric methods tends to be too slow for many applications, since evaluating the distance function can take too much time.

We present a method for quickly eliminating most positions of the text from consideration as possible matches. The remaining candidate positions are them evaluated one by one against the template for a match. We are still guaranteed to find all matching positions, and our method gives significant speed-ups.

Finally, we consider the problem of matching a dictionary of templates against a text. We present methods which are much faster than matching the templates individually against the input image.

Key words: *Template Matching, Metrics Similarity Methods, Filtration Methods, Dictionary Matching.*

1 Introduction

Recent techniques have facilitated the solution of previously infeasible queries such as, "Find all images which look like this picture," or "Find all images which contain a tumor like this" [2]. Although there is no universal definition for the term *"looks like"*, the idea that one image looks like another is usually modeled in terms of a *similarity measures*. In this setting, an image looks like another if the distance between them is less than some threshold value. Many similarity measures have been proposed in the literature [1, 2, 7]. For most of these definitions, no sufficiently fast algorithm for finding good matches is known [7].

Jain, Murthy, and Chen [7] classified similarity measures between images into three groups: *metric based, set-theoretic based,* and *decision-theoretic based*. The algorithm introduced in this paper uses a metric based method. We assume that digital images are represented by rectangular arrays of picture elements called *pixels*. A pixel consists of a grey level represented numerically. Let I be an $n \times n$ text and T an $m \times m$ template. Let $S_{i,j}$ be the $m \times m$ submatrix of I with upper-left-hand corner at position (i, j). Jain, Murthy, and Chen defined the *M-difference* between $S_{i,j}$ and T as $M(T, S_{i,j}) = \sum_{k=0, l=0}^{m,m} |I(i+k, j+l) - T(k, l)|$, that is, as the sum of the absolute values of differences in corresponding grey levels. The smaller the *M-difference*, the more similar $S_{i,j}$ is to T. We define the *Image Template Matching Problem (ITM)* as

Input: An image I, a template T, and a threshold value d.
Output: All (i,j)'s such that $M(S_{i,j}, T) \le d$.

The straightforward method to solve the image template matching problem is to compute the *M-differences* at all $S_{i,j}$'s. The time complexity of this brute forced method is $O(m^2 n^2)$ which is too slow for users to wait for the outputs. Barnea and Silverman proposed the *sequential decision technique* [1]. This saves some computational time, decreasing the running time typically by rejecting all $S_{i,j}$'s whose *M-differences* exceed the pre-determined threshold before all pixels are examined.

We take a different approach. Suppose we have a function M' such that $M'(S_{i,j}, T) \le M(S_{i,j}, T)$ and M' can be computed for all (i,j) very quickly. Then we can use M' to *filter* the positions of the text, eliminating from consideration all (i,j) such that $M'(S_{i,j}, T) > d$. We call the surviving position *candidates*. We can then evaluate M itself at each of the candidates. We will describe several different filtering functions M' and show that we can achieve significant speed-up over the sequential decision technique.

Another standard similarity measure technique to solve the template matching problem is called *cross-correlation* [5]. The basic correlator [8] and the statistical correlator [6] methods are criticized because of their computational expense [1]. The standard *cross-correlation* uses the product instead of the difference between $S_{i,j}$ and T. It finds all $S_{i,j}$'s such that $\prod_{k=0,l=0}^{m,m} S_{i,j}(k,l) T(k,l) \ge d$. This multiplication is computed fast in the *Fourier* domain [4]. The time complexity is $O(n^2 (\log n)^2)$. The *cross-correlation* comes from the Euclidean distance $d(I,T)$ squared, given by $d(I,T) = \sum_{k=0,l=0}^{m,m} (I(i+k,j+l) - T(k,l))^2, i,j = 0, \cdots, n-m$. It expands to $d(I,T) = \sum_{k=0,l=0}^{m,m} I(i+k,j+l)^2 - 2I(i+k,j+l)T(k,l) + T(k,l)^2$. The third term, $T(k,l)^2$, is constant for all positions of I's, the first term, $I(i+k,j+l)^2$, is assumed to be approximately constant, and only the multiplication term is left. Barnea and Silverman reported that the metric based differencing method has the advantage of simplicity of feature definition and speed of computation over the conventional correlation technique [1].

Matching many Templates In addition to the single template matching problem, certain applications, especially in template-based computer vision, raise the *multiple template matching problem*, also known as *image dictionary matching problem* or *IDM*. A set of templates is called an *Image Dictionary*, $D = \{T_1, T_2, \cdots, T_t\}$ where t is the number of templates. One would like to find all occurrences of a set of templates within a large image. It is defined as follows:

Input: An image I, a dictionary $D = \{T_1, T_2, \cdots, T_t\}$, and a threshold value d.
Output: All (p, i, j)'s such that $M(S_{i,j}, T_p) \le d$.

A naïve way to search an image dictionary is to run each template on each image independently, although this will typically be about t time slower than a single run. We introduce modified *trie* data structures, called the *H-trie* and the H^+-*trie*, as well as the *H-table* to accelerate the time in the candidate selection stage for the image dictionary matching problem.

Previous works on the *IDM* problem include the *cross-correlation* method and a multilevel sequential decision technique. The *cross-correlation* method reduces the cost appreciably since the transformation into *Fourier Domain* only needs to be calculated once. Ramapriyan developed a multilevel sequential decision technique to solve the *IDM* [3]. Instead of matching each template of the image dictionary, the templates are partitioned and a representative template is defined for each of the partitions. However, this method has the difficulty of deciding representative templates and minimizing error probability.

2 Image Template Matching Algorithm

There are two stages: candidate selection and verification stages.

2.1 A Candidate Selection Stage

A candidate set, C, is a subset of S which contains all matches. Candidates are selected using *a candidate selection function*. The input for a candidate selection function is S, and the output is C. There are three essential properties about good candidates. First, C must contain all matches to the template. Second, C must be computed in $o(m^2n^2)$. Last, the expected $|C|$ must be small. The first property ensures the correctness of the algorithm, and the second and third properties guarantee the speed up. We present some of candidate selection functions such as *h-distance*, *diffsum*, and *combined function*, and discuss which function satisfies above all three properties.

An *h-distance* function : Figure 1 represents *intensity histograms* of two images. The total number of pixels in both intensity histograms is the same

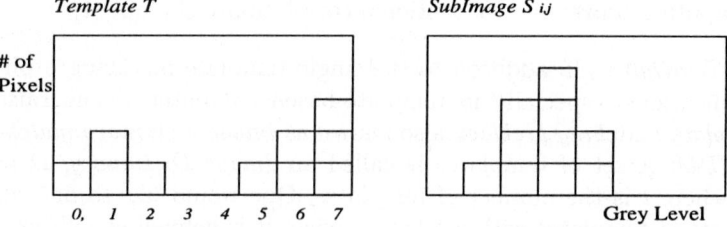

Fig. 1. Intensity Histograms

$(= m^2)$. Operations are *moveleft*(x) and *moveright*(x), and each costs 1. The shape of the histogram of the sub-image of I can be transformed to that of the template by using some operations. The total number of pixel movements necessary to transform shapes from one histogram of a sub-image to that of the template is the edit distance. The minimum edit distance is called h-*distance*, $\sum_{k=0}^{r-1} |\sum_{l=0}^{k}(L_l(S_{i,j}) - L_l(T))|$ where $L_l(x)$ is the number of pixels on grey level l of x. For example of Figure 1, the h-*distance* is 21.

Theorem 1. *The* h-*distance is the minimum edit distance between intensity histograms.*

Proof. In order to transform a histogram a to b, one has to move $|L_a(k) - L_b(k)|$ to the left or right if $L_a(k)$ is smaller or larger respectively at the level k. This must hold for all levels, $\sum_{k=0}^{r-1} |L_a(k) - L_b(k)|$, which is h-*distance*.

Computing h-*distance* takes $O(r)$ and the candidate selection stage using the h-*distance* function can be accomplished either in $O((m+r)n^2)$ time with $O(r)$ space required or in $O(rn^2)$ with $O(rn)$ space required where r is the number of grey level. In the former process, a mask moves along the image by subtracting the leftmost column of the previous masked area and adding the right most column of the next masked area on the image. The later one computes $h(p_{x,y})$ by $h(p_{x-1,y}) + h(p_{x,y-1}) - h(p_{x-1,y-1}) + g(p_{x-1,y-1}) - g(p_{x-1+m,y-1}) - g(p_{x-1,y-1+m}) + g(p_{x+m-1,y+m-1})$, where $h(p)$ and $g(p)$ are the intensity histogram and the grey level value at pixel p. Both processes run faster than $O(m^2n^2)$ as long as $r < m^2$.

Theorem 2. h-distance$(S_{i,j}, T) \leq$ M-Difference$(S_{i,j}, T)$ *by Theorem 1.*

The set of matches, M, must be a subset of C. The theorem 2 guarantees that positions whose M-*difference* $\leq d$ are always in C.

Alternative functions : Another candidate selection function is the absolute value of difference between sum of pixels in $S_{i,j}$ and sum of pixels in T, called $diffsum(S_{i,j}, T) = |\sum_{k=0,l=0}^{m,m} S_{i,j}(k,l) - \sum_{k=0,l=0}^{m,m} T(k,l)|$.

Theorem 3. $diffsum(S_{i,j}, T) \leq$ M-difference$(S_{i,j}, T)$.

Proof. $|\sum_{k=0,l=0}^{m,m} S_{i,j}(k,l) - \sum_{k=0,l=0}^{m,m} T(k,l)| \leq \sum_{k=0,l=0}^{m,m} |S_{i,j}(k,l) - T(k,l)|$, by triangle inequality.

The candidate selection stage using $diffsum(S_{i,j}, T)$ is accomplished linearly. Both h-*distance* and *diffsum* have drawbacks; the former one has higher time complexity and the later one has a larger number of candidates. The theorem 4 ensures why the later one has more candidates.

Theorem 4. $diffsum(S_{i,j}, T) \leq h$-*distance*$(S_{i,j}, T)$.

Proof. $|\sum_{k=0}^{r-1}(k \times L_k(S_{i,j})) - \sum_{k=0}^{r-1}(j \times L_k(T))| \leq \sum_{k=0}^{r-1} |\sum_{l=0}^{k}(L_l(S_{i,j}) - L_l(T))|$, by triangle inequality.

By combining two functions, we build a better function. The $diffsum(S_{i,j}, T)$ is used to select *pre-candidates*, $C'_{i,j}$'s, and the $h\text{-}distance(C'_{i,j}, T)$ is used to select the final candidates, $C_{i,j}$'s. Pre-candidates are computed in $O(mn^2)$, not $O(n^2)$ because we build the histogram for each sub-image. The final candidates are computed in $O(c'r)$ where $c' = |C'|$. The time complexity is $O(mn^2 + c'r)$. The number of candidates is the same as that of using $h\text{-}distances$ and candidates are computed faster.

2.2 Candidate Verification Stage

The sequential decision technique is used to verify candidates C. Those $C_{i,j}$'s whose *M-differences* exceed the threshold, d, are rejected before all pixels are examined. The time complexity for the candidate verification stage is $O(cm^2)$ where c is the total number of candidates. The total running time for the algorithm is either $O((m+r)n^2 + cm^2)$, $O(rn^2 + cm^2)$, $O(n^2 + cm^2)$, or $O(mn^2 + c'r + cm^2)$ depending on candidate selection functions. In the best case, when $c = 0$, the running time complexities are simply the same as the time complexity for the candidate selection stage. The worst case is when all S_i's are matches. In this worst case, the running time is $O(m^2n^2)$ for all pre-processes. No theoretical analysis for the expected running time has been studied yet.

2.3 Experiment

Two *MRI* head images (256 × 254) are tested on the *SPARCstation10* machine to measure the running time of the image template matching algorithm. The template (48 × 47) comes from one image by cropping the interesting part, a tumor. The system locates the similar parts within the other image. The Figure 2 shows running times of various methods in different resolutions, r.

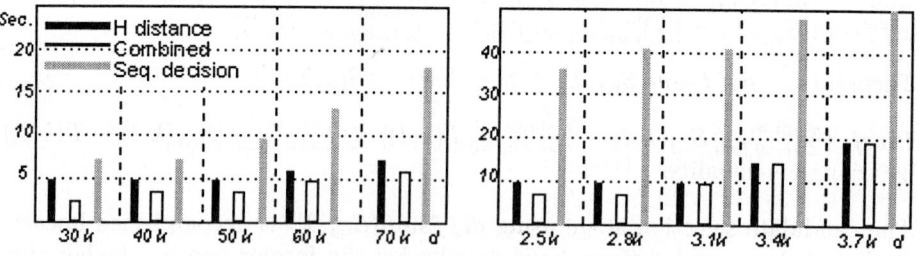

Fig. 2. Running times of Various ITM Methods

Observation 5. The smaller threshold value, d, the faster algorithms run.

The smaller threshold value means the less number of candidates.

Observation 6. There is the maximum grey level point, r which gives the minimum running time of the algorithm using the h-*distance* function.

If r is too small, the number of candidates is large causing no speed up. If r is too large, on the other hand, the candidate selection stage takes long time.

3 Image Dictionary Matching Algorithms

The new algorithms to solve the *IDM* problem consist of the same stages as the *ITM* algorithm. The difference is that there is more than one template in the *IDM* problem. We present three data structures to find candidates for each template quickly.

H-trie : An *H-trie* is a tree whose paths represent intensity histograms of a template. The height is the number of grey levels. Each internal node contains the number of pixels in the grey level. The root contains the pointers to the number of pixels in the grey level 0 for all templates and the leaves are the names of templates. The *h-trie* for the example is shown in Figure 3. After building an

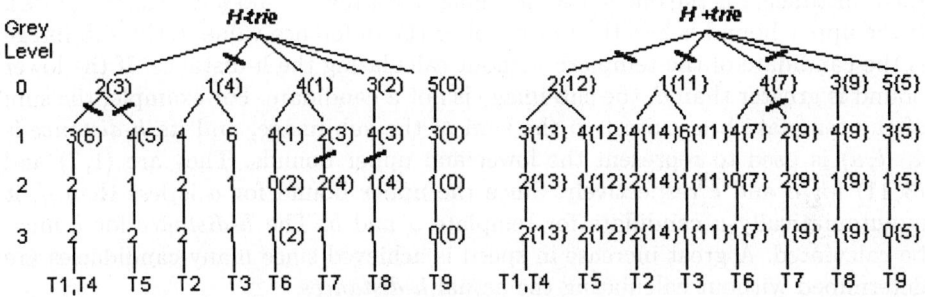

Fig. 3. h-trie and h+trie

h-trie, do a depth first search looking for nodes whose h-$dist(i)$ exceeds d. If we find such a node, prune the branch and move to its parent. If a leaf is reached, then the sub-image is a candidate for that template. The method using the *H-trie* is superior to the method that computes *H-distances* individually because there are some branches or edges shared by several templates.

H^+-trie : To further increase speed in computing candidates, we modify the *H-trie* structure, called an H^+-*trie*. It saves the minimum sum of child templates in every node as in Figure 3. If the minimum sum is larger than $\sum_{i=0, j=0}^{n,n}(S_l(i,j))+d$, then that edge is pruned. The number of operations is reduced because edges are pruned not only when the *h-distance* exceeds d, but also when *diffsum* does.

H-table : We precalculate *diffsum* and *h-distances* among templates in the h-table as shown in Figure 4. Sum is $\sum_{i=0,j=0}^{m-1,m-1} T(i,j)$. Suppose $d = 7$, then it is

Histograms

H(T$_i$)'s a = 1 3 1 1 4 0 0 0 0 0
b = 1 3 1 1 4 0 0 0 0 0
c = 1 3 1 1 4 0 0 0 0 0
d = 1 3 1 1 4 0 0 0 0 0
e = 1 3 1 1 4 0 0 0 0 0

H(S$_i$)'s = 1 3 1 1 4 0 0 0 0 0

r = 10, size of template = 10

H-table

Sum	a	b	c	d	e
a	24	0	3	10	19	46
b	27	3	0	7	16	43
c	34	10	7	0	9	36
d	43	19	16	9	0	27
e	70	49	43	36	27	0

Fig. 4. histograms for a dictionary and a sub-image and the h-table

not necessary to look up those templates whose sums are greater than the sum of the sub-image $+d$ or less than the sum of the sub-image $-d$. In the example, the sum of the sub-image is 29. Therefore, only *h-distances* for a, b, and c may need to be calculated. We use the upper and lower bounds for *h-distances* to further eliminate candidates: $|h(y,t) - h(x,y)|$ and $h(y,t) + h(x,y)$, where $h(x,t)$ denotes the h-distance, t is the one of the sub-images of a text, x and y are the templates. If the upper bound is less than or equal to the difference value d, the sub-image is the candidate of the template without calculating the h-distance. If the lower bound is greater than d, the sub-image is not a candidate. For example, the sum of the template b is closest to the sum of the sub-image, and its *h-distance* is 4. $\langle l, u \rangle$ is used to represent the lower and upper bounds. They are $\langle 1, 7 \rangle$ and $\langle 3, 11 \rangle$ for *a* and *c* respectively. Since the upper bound for *a* is less than *d*, it is automatically a candidate for template *a* and *b*. The *h-distance* for *c* must be calculated. A great increase in speed is achieved since many candidacies are determined without calculating the actual *h-distances*.

Lemma 7. $h(y,t) + h(x,y)$ *is the upper bound of* $h(x,t)$, *by triangle inequality.*

Lemma 8. $|h(y,t) - h(x,y)|$ *is the lower bound of* $h(x,t)$.

Proof. In case of $h(y,t) \geq h(x,y)$, then $h(y,t) > h(x,y) + h(x,t)$. y can be edited to t by editing it to x then to t. It takes $h(x,y) + h(x,t)$. It contradicts the theorem 1. In case of $h(y,t) < h(x,y)$, then $h(x,y) > h(x,t) + h(y,t)$. x can be edited to y by editing it to t then to y. It takes $h(y,t) + h(x,t)$. It also contradicts the theorem 1. Therefore, $|h(y,t) - h(x,y)|$ is the lower bound of $h(x,t)$.

Lemma 9. *if* $|Sum(x) - Sum(t)| > d$, *then* $h(x,t) > d$.

Proof. Let s be a histogram whose all levels are 0's. $|h(s,x) - h(s,t)| \leq h(x,t)$. $Sum(x) = h(s,x)$. $|Sum(x) - Sum(t)| \leq h(x,t)$.

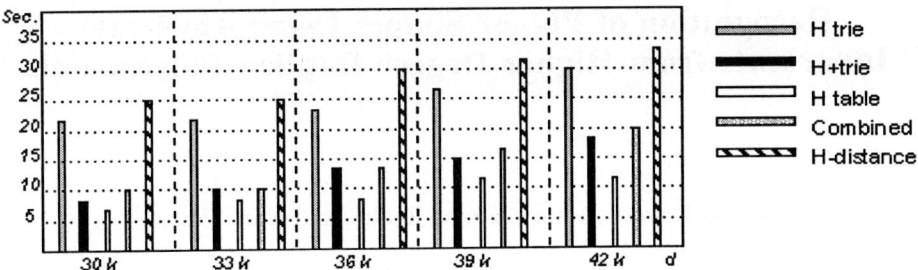

Fig. 5. Running times of Various IDM Methods

3.1 Experiment

We tested 10 templates against an *MRI* head image. Templates (50 × 50) are tumor images cropped from different images. Figure 5 shows the running time comparisons. The fourth and last methods are using the combined functions and the *h-distance* function respectively for each template. The naïve method takes more than 3 minutes.

Observation 10. The running time for *IDM* problem using the *h-distance* or combined functions is not $k\times$ the running time for *ITM* problem using the *h-distance* or combined functions.

References

1. Barnea and Silverman. A class of algorithms for fast digital image registration. IEEE Trans. Comput., 21:179–186, 1972.
2. Tzi-cker Chiueh. Content-based image indexing. In Proceedings of the Twentieth International Conference on Very Large Databases, pages 582–593, Santiago, Chile, 1994.
3. Ramapriyan H. K. a multilevel approach to sequential detection of pictorial features. IEEE Trans. Comput., 25:66–78, 1976.
4. Wayne Niblack. Digital Image Processing. Prentice hall International (UK), Birkeroed, Denmark, 1986.
5. Wayne Niblack. Comments on cross correlation vs. differencing methods. Personal Comunication, 1996.
6. W. K. Pratt. Correlator techniques of image registration. IEEE Trans. Aerosp. Electron. Syst., 10:353–358, 1974.
7. Peter L-J Chen Ramesh Jain, S.N. Jayaram Murthy. Similarity measures for image databases. SPIE, 2420:58–65, 1995.
8. R.Y. Wong. Sequential pattern recognition as applied to scene matching. PhD. thesis, Univ. of Southern California, L.A. Dec., 1976.

Recognition of Planar Shapes Using Algebraic Invariants from Higher Degree Implicit Polynomials

Satish Kaveti[1], Eam Khwang Teoh[2], and Han Wang[2]

[1] German Singapore Institute
Nanyang Polytechnic
10 Science Center Road
Singapore 609079

[2] School of Electrical and Electronic Engg
S1 Block
Nanyang Technological University
Nanyang Avenue
Singapore 639798

Abstract. Higher degree implicit polynomials and moments have been used earlier for characterizing shapes. However, computation of invariants of implicit polynomials of degree greater than four has been known to be a complex problem. In this paper, an computationally efficient method for obtaining affine invariants from coefficients of higher degree implicit polynomials has been proposed. The algorithm is based on tensors and unlike the previous works, it is not based on partial derivative forms or symbolic computation. The algebraic invariants from the higher degree implicit polynomials can be directly used for obtaining moment invariants.

1 Introduction

Invariants are view-independent features which can be used for recognition of objects. The use of invariants allows the use of indexing based approach, which is computationally less expensive than the graph matching based approaches, e.g. interpretation tree search. Algebraic invariants from the coefficients of higher degree implicit polynomials can be used effectively to describe free form surfaces, as coefficients provide finite and complete set of descriptors of the global shape of the object.

Obtaining algebraic invariants from higher degree implicit polynomials is a classical problem and has been solved earlier using symbolic methods. Using matrices, Taubin [1] proposed a efficient and general approach for obtaining algebraic covariants and invariants from higher degree implicit polynomials. The approach also provides useful methods for obtaining covariants and joint covariants/invariants. However, due to matrix representation, the number of affine invariants which can be determined are rather limited. The theory of Lie derivative provides a general framework for determination of invariants, wherein the invariants are formulated as solution to a system of partial differential equations [2]. These system of PDFs may not be very convenient to solve.

The tensor based approach for obtaining invariants from a single implicit polynomial is more general than the approach by Taubin [1,3]. The approach can also be used for determining joint invariants of implicit manifolds.

The main emphasis of this paper is on methods for determining algebraic affine invariants and its applications for characterizing curved bounded planar shapes. In our approach, the affine invariants are obtained as determinants of tensors and the concept is applicable to polynomials of arbitrary degree and dimension.

The paper is organized as follows. In Section 2, notation used for represention of implicit polynomials is given. Section 3 describes the methods to obtain tensor notation for coefficients of higher degree implicit polynomials. Determinants (affine invariants) of coefficient tensors are discussed in Section 4.

2 Basic Definitions

In this section, the basic symbols used for representation of higher degree implicit polynomials will be introduced. The analysis has been done within a general framework and can be used for implicit polynomial of any degree and dimension.

Consider a n-dimensional space, wherein the co-ordinates are expressed as $\mathbf{x} = [x_1, \ldots, x_n]$. For the linear parameterization, the implicit polynomial can be expressed as a linear combination of monomials in the elements of $\mathbf{x}$. Mathematically,

$$f(\mathbf{x}) = \phi(\mathbf{u}, \mathbf{x}) = u_1 X_1(\mathbf{x}) + \ldots + u_r X_r(\mathbf{x}) = \mathbf{u}^t \mathbf{X}(\mathbf{x}), \tag{1}$$

where $\mathbf{X}(\mathbf{x}) = [X_1(\mathbf{x}), \ldots, X_r(\mathbf{x})]^t$ and $\mathbf{u} = [u_1, \ldots, u_r]^t$. The $X_i(\mathbf{x})$'s are monomials and can be expressed as

$$X_i(\mathbf{x}) = \mathbf{x}^{\boldsymbol{\alpha}_i} = x_1^{\alpha_{i_1}} \ldots x_n^{\alpha_{i_n}}, \tag{2}$$

where $\boldsymbol{\alpha}_i = [\alpha_{i_1}, \ldots, \alpha_{i_n}]$, is called as the *multi-index* of $X_i(\mathbf{x})$. In a general context, the multiindex will be denoted by $\boldsymbol{\alpha}$. The *size of the multi-index* $\boldsymbol{\alpha}_i$, denoted by $|\boldsymbol{\alpha}_i|$, is defined as,

$$|\boldsymbol{\alpha}_i| = \sum_{k=1}^n \alpha_{i_k}, \text{ and } \alpha_{i_k} \geq 0 \text{ for all } i \text{ and } k \tag{3}$$

and the *degree d* is given by,

$$d = \max_i |\boldsymbol{\alpha}_i|, i = 1, \ldots, r. \tag{4}$$

Instead of u_i, the coefficients of the polynomial can be denoted as $p_{\boldsymbol{\alpha}_i}$, or simply $p_{\boldsymbol{\alpha}}$. For ordering the monomials of the polynomial, lexicographical ordering has been used.

A *form*, also known as *homogeneous polynomial*, is a polynomial for which all the terms have the same size. A homogeneous polynomial of degree d, $\psi_d(\mathbf{x})$, can be mathematically written as :

$$\psi_d(\mathbf{x}) = \sum_{|\alpha|=d} \binom{d}{\alpha} p_\alpha \mathbf{x}^\alpha \tag{5}$$

where $p_\alpha \in \mathbb{R}$. This representation is consistent with the classical invariant theory and allows moment invariants to be determined directly from the expressions of algebraic invariants [4].

The *leading form*, $l_d(\mathbf{x})$, of the implicit polynomial of degree d is the sum of all the terms $u_i X_i(\mathbf{x}), i = 1, \ldots, h_{d,n}$, where X_i's are monomials with multiindices of size d. The coefficients, u_i's, of the leading form are invariant to translations and are the best choice for obtaining the rotational component. The use of $l_d(\mathbf{x})$ for determination of translation, allow separation of the problems of determination of orientation and location of the object.

In this paper, affine invariants which are homogeneous polynomial $I(p_{\alpha_1}, p_{\alpha_2}, \ldots, p_{\alpha_h})$ of the coefficients are explored.

3 Tensor representation of implicit polynomials

3.1 Tensors for higher degree implicit polynomials

Consider again a n-dimensional space wherein the co-ordinates are denoted by $\mathbf{x} = [x_1, \ldots, x_n]$. To obtain the tensor representation, a sequence of monomials using the kronecker product of $\mathbf{x}$ is generated. For a polynomial of degree d,

$$\mathbf{x}_{[1,d]} \stackrel{\triangle}{=} \underbrace{\mathbf{x} \otimes \ldots \otimes \mathbf{x}}_{d \text{ times}} = [x_1^d \; x_1^{d-1} x_2 \; \ldots \; x_n^d]^t, \tag{6}$$

and $\mathbf{x}_{[1,d]}$ is a column vector containing $l \stackrel{\triangle}{=} n^d$ terms. The vector, $\mathbf{x}_{[1,d]}$, will be referred to as the *prolonged monomial vector*. The expression of $\mathbf{x}_{[1,d]}$ depends on whether ordinary or homogeneous co-ordinates are used. For ordinary co-ordinates, $\mathbf{x}_{[1,d]}$ contains monomials of size d, and, for homogeneous co-ordinates all the monomials from size 0 to d are included in $\mathbf{x}_{[1,d]}$. Let the multiindex of the j-th term in $\mathbf{x}_{[1,d]}$ be γ_j for $j = 1, \ldots, n^d$. From the multinomial theorem, there will be multiple occurrences of some monomials in $\mathbf{x}_{[1,d]}$ and the number of occurrences of monomials of multiindex $\gamma = [\gamma_1, \ldots, \gamma_n]$ is,

$$k_\gamma \stackrel{\triangle}{=} \binom{|\gamma|}{\gamma_1, \ldots, \gamma_n}, \tag{7}$$

which is same as the coefficient of $\mathbf{x}^\gamma$ in the expansion of $(x_1 + \ldots + x_n)^d$. The repeated occurrences of some of the monomials in $\mathbf{x}_{[1,d]}$, implies that the γ_j's are not distinct. However, there is a property which is unique for each element in

$\mathbf{x}_{[1,d]}$ which is based on expressing the index of each element in $\mathbf{x}_{[1,d]}$ as a vector of length d. Each element of $\mathbf{x}_{[1,d]}$ is of the form $x_{j_1} x_{j_2} \ldots x_{j_d}$ where $j_k \in [1,n]$ for $k = 1, \ldots, d$, and x_{j_m} is obtained from the m-th $\mathbf{x}$ in the kronecker product of Eq. (6). The index, $\boldsymbol{\eta}_j \triangleq [j_1, \ldots, j_d]$, which is a vector of size d, is unique for each element of $\mathbf{x}_{[1,d]}$. The relationship between j and $\boldsymbol{\eta}_j$ is given by,

$$j = j_1 + (j_2 - 1)n + (j_3 - 1)n^2 + \ldots + (j_d - 1)n^{d-1} \qquad (8)$$

which is exactly the same as arithmetic in radix n.

Finally, using the monomials in $\mathbf{x}_{[1,d]}$ the expression for the implicit polynomial is obtained. Collecting the coefficients corresponding to the monomials in $\mathbf{x}_{[1,d]}$ and dividing them by k_γ to account for the multiple occurrences of associated monomials, coefficient vector $\mathbf{q}_{[1,d]}$ is defined as,

$$\mathbf{q}_{[1,d]} \triangleq [q_1 \ldots q_l]^t \triangleq [p_{\gamma_1}, \ldots, p_{\gamma_l}]^t, \qquad (9)$$

where $l = n^d$, and the implicit polynomial can be written as,

$$f(\mathbf{x}) = \mathbf{q}_{[1,d]}^t \mathbf{x}_{[1,d]}, \qquad (10)$$

where $\mathbf{x}$ is expressed in homogeneous co-ordinates. For ordinary coordinates, the above equation gives a homogeneous polynomial. Each element of $\mathbf{q}_{[1,d]}$, $q_j = p_{\gamma_j}$, has a unique d-D index (which is same as index $\boldsymbol{\eta}_j$ of monomial, $\mathbf{x}^{\gamma_j}$). This gives an equivalent representation of $\mathbf{q}_{[1,d]}$ as a symmetric tensor of rank d with element indices given by $\boldsymbol{\eta}_j$, $j = 1, \ldots, l$. Throughout our discussion, $\mathbf{q}_{[1,d]}$ will refer to vector representation of the coefficient tensor, $\mathbf{K}_{[1,d]}$, of rank d and dimension n. The elements of the tensor will be denoted by $K_{[1,d]i_1,\ldots,i_d}$, or simply $K_{i_1,\ldots,i_d}$ when the tensor under discussion is clear from the context.

3.2 Alternative tensor representation

Every homogeneous polynomial of degree d can be represented as a tensor of rank d. A important property of implicit polynomial is that it can be represented in alternative tensor forms and each tensor form provides additional set of invariants. In this section, these alternative tensor forms of higher degree implicit polynomials will be discussed.

Let us consider again an homogeneous polynomial of degree $d = d_1 \cdot d_2$ in a n-dimensional space. The monomials of the homogeneous polynomial can be obtained by

$$\mathbf{x}_{[d_1,d_2]} \triangleq \underbrace{\mathbf{x}_{[d_1]} \otimes \ldots \otimes \mathbf{x}_{[d_1]}}_{d_2 \text{ times}} \qquad (11)$$

where $\mathbf{x}_{[d_1]}$ is a vector of monomials of degree d_1 lexicographically ordered and defined as,

$$\mathbf{x}_{[d_1]} = \left\{ \sqrt{\binom{d}{\alpha}} \mathbf{x}^\alpha : |\alpha| = d_1 \right\} \qquad (12)$$

and $\mathbf{x}_{[d_1,d_2]}$ is a column vector containing $(h_{n,d_1})^{d_2}$ terms. This form of representation results in a tensor of rank d_2. By substituting the monomials $\mathbf{x}^{\alpha}$ in $\mathbf{x}_{[d_1,d_2]}$ by p_{α}, $\mathbf{q}_{[d_1,d_2]}$ is obtained, and the polynomial can be expressed as,

$$\mathbf{q}^t_{[d_1,d_2]} \mathbf{x}_{[d_1,d_2]}.$$

The expressions can be quite lengthy for higher degree implicit polynomials. To make formulation neater, the homogeneous polynomial of degree $d = d_1 d_2$ will be written as,

$$\mathbf{K}_{[d_1,d_2]} \odot^{d_2} \mathbf{x}_{[d_1]} \tag{13}$$

where $\mathbf{K}_{[d_1,d_2]}$ is a tensor of rank d_2.

4 Affine invariants

4.1 Determinants of tensors

Determinants have been known to be invariants since the development of classical invariant theory during the nineteenth century. Cayley had used a theoretical construct called hyperdeterminants for generating the invariants. In this section, a concise approach for obtaining these determinants will be discussed.

Determinants are used in our approach for obtaining affine invariants from implicit polynomials. The approach is suitable for finding invariants for polynomials of even degree only. For odd degree polynomials, the invariants are found by squaring the polynomial to obtain an even degree polynomial.

Computation of determinant The determinant of a matrix is usually recursively defined in terms of cofactors. The determinant can be more concisely defined in terms of permutation groups. The determinant of a $n \times n$ matrix, $\mathbf{A}$, is defined as [5],

$$\Delta(\mathbf{A}) \triangleq \sum_{\sigma \in \mathcal{N}} \mathrm{sgn}(\sigma) a_{1\sigma(1)} a_{2\sigma(2)} \cdots a_{n\sigma(n)}, \tag{14}$$

where a_{ij} is the (i,j)-th element of $\mathbf{A}$, and $\mathcal{N}$ is the set of all permutations of size n. Furthermore,

$$\mathrm{sgn}(\sigma) = \begin{cases} 1 & \text{if } \sigma \text{ is even} \\ -1 & \text{if } \sigma \text{ is odd} \end{cases} \tag{15}$$

For a general case of a tensor, $\mathbf{K}_{[1,d]}$, of rank d and dimension n, the determinant is given by,

$$\Delta(\mathbf{K}_{[1,d]}) \triangleq \sum_{\sigma_1 \in \mathcal{N}, \cdots, \sigma_{d-1} \in \mathcal{N}} \left(\prod_{k=1}^{d-1} \mathrm{sgn}(\sigma_k) \right) a_{1\sigma_1(1)\ldots\sigma_{d-1}(1)}$$
$$a_{2\sigma_1(2)\ldots\sigma_{d-1}(2)} \cdots a_{n\sigma_1(n)\ldots\sigma_{d-1}(n)},$$

where $a_{j_1,\ldots,j_d}$ is the $(j_1,\ldots,j_d)$-th element of $\mathbf{K}_{[1,d]}$. The algorithm requires determination of sign of the permutation.

The affine invariant of tensor forms, $\mathbf{K}_{[d_1,d_2]}$, for which $d_2 = 2k+1$, i.e., d_2 is an odd number, cannot be obtained directly. It is important to note that all the invariants so obtained need not be necessary independent. However, the main advantage of the above procedure is that it gives more number of invariants compared to earlier approaches and at a lesser computational cost. In some cases, it is also possible to determine the independence of the invariants.

5 Results

As indicated in the previous works, it is very difficult to obtain good fits for higher degree implicit polynomials. The terrain on which the minimization has to be done is non-convex, and the presence of noise makes the matters worse. The presence of noise results in a large number of 'shallow' local minima, and complex methods are required to dislodge the intermediate solutions from the incorrect local minima. An interested reader is advised to refer to [6, 3] for relevant details on implicit polynomial fitting. The details of our fitting algorithm can be found in [7].

In our experiments, fourth degree implicit polynomials are used for fitting closed planar curves. In this paper, we shall not investigate the Euclidean invariants but only affine invariants from fourth degree implicit polynomial fits. The Euclidean invariants have been investigated in [8, 7].

The test data was obtained from the edge pixels of the bitmap images of curved objects drawn by hand using a PC based drawing package, CorelDraw. The bitmap images of our test images are shown in Fig. 1.

Table 1: Affine invariant for heart image

Fig no.	Fig. description	$\Delta(\mathbf{K}_{2,2})$
1(e)	Heart at reference position	0.2536
1(f)	Elongated heart	0.2679
1(g)	Skewed heart	0.2532
1(h)	Wide heart	0.2418
1(k)	Real image - heart	0.2475

The affine invariant for the heart and butterfly image with different affine distortion are shown in Tables 1 and 2 respectively. From the tables it is clear that the invariant is quite reliable for recognition. In cases where the image library is huge, our method can be used for initial pruning of the search space.

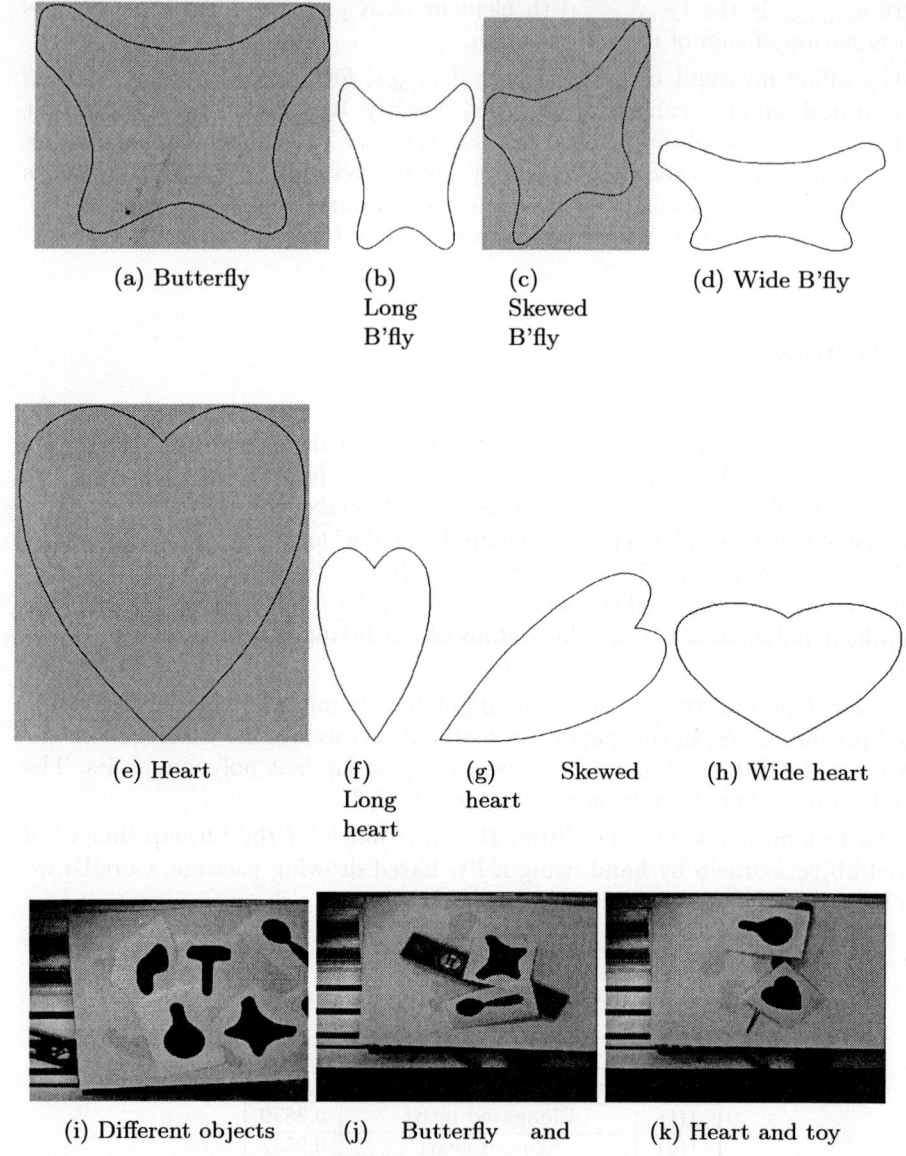

Fig. 1: Test images used in the experiments

Table 2: Affine invariant for butterfly image

Fig no.	Fig. description	$\Delta(\mathbf{K}_{2,2})$
1(a)	Butterfly at ref. position	0.3297
1(b)	Elongated B'fly	0.3324
1(c)	Skewed B'fly	0.3286
1(d)	Wide B'fly	0.3262
1(i)	Real image - B'fly	0.3564
1(j)	Real image - B'fly	0.3511

6 Conclusion

Using the tensor approach, it is possible to obtain shape invariants and pose estimates from higher degree implicit polynomials. The proposed approach does not require any form of symbolic computations for obtaining complete set of algebraic invariants. The tensor based representation presented in the paper is more general than matrix based approach suggested by Taubin.

References

1. Gabriel Taubin, "Recognition and positioning of rigid objects using algebraic and moment invariants", Tech. Rep. LEMS-80, Brown University, December 1990.
2. L Van Gool, T Moons, E Pauwels, and A Oosterlinck, "Vision and lie's approach to invariance", *Image and Vision Computing*, vol. 13, no. 4, pp. 259–277, May 1995.
3. Gabriel Taubin, "Estimation of planar curves, surfaces, and nonplanar space curves defined by implicit equations with applications to edge and range image segmentation", *IEEE Trans. on Pattern Analysis and Machine Intelligence*, vol. 13, no. 11, pp. 1115–1138, 1991.
4. Ming Kuei Hu, "Visual pattern recognition by moment invariants", *IRE Transactions on Information Theory*, vol. 8, no. 2, pp. 179–187, February 1962.
5. William C. Brown, *Matrices and Vector Spaces*, Marcel Dekker, Inc., New York, 1991.
6. Paul D. Sampson, "Fitting conic sections to very scattered data : An iterative refinement of the Bookstein algorithm", *Computer Graphics and Image Processing*, vol. 18, pp. 97–108, 1982.
7. Satish Kaveti, "Free form object recognition and pose estimation using higher degree implicit polynomials", Tech. Rep., Nanyang Technological University, March 1995.
8. Satish Kaveti, Eam Khwang Teoh, and Han Wang, "Efficient algorithms for obtaining algebraic invariants from higher degree implicit polynomials for recognition of curved objects", *Pattern Recognition Journal*, 1997.

Object Recognition and Orientation via Zernike Moments

Samer M. Abdallah[1], Eduardo M. Nebot[2], and David C. Rye[2]

[1] Department of Engineering, Faculty of Engineering and Information Technology,
The Australian National University, Canberra ACT 0200, Australia
[2] Department of Mechanical and Mechatronic Engineering,
The University of Sydney, NSW 2006, Australia

Abstract. The *magnitude* of the Zernike moments has been successfully used in previous works [1, 9, 13, 19] as rotation invariant feature for object recognition. The *phase angle* of the Zernike moments however, was ignored in the past. In this paper, this *phase angle* is used to determine the orientation of the identified object. Furthermore, it is demonstrated that saving of 65% on computation time of the Zernike moments can be achieved by using the *explicit forms* of their radial polynomials. The effectiveness of the approach is illustrated by tests on real images of industrial objects.

1 Introduction

Many objects can be adequately represented by a two-dimensional (2D) view and a height (thickness) along their third dimension. Examples of such objects include industrial tools and parts like gears, washers, retaining clips, etc. Characters of the alphabet are another example of 2D shapes. These objects can be treated as planar shapes, hence reducing the recognition process to a 2D problem. This paper deals with the recognition and pose determination of such planar objects. We present a vision system capable of identifying 2D un-occluded objects in a scene under similarity transformations. The system also solves for the pose (translation, scale, and rotation) of the identified objects.

The image of an unknown object is first normalised with respect to scale and translation using its low order regular moments [1, 9]. The magnitudes of a set of orthogonal complex moments, Zernike moments [1, 9, 13, 19], are then employed as rotation invariant features. These features serve as input to an artificial neural network classifier which identify the unknown object. Once recognised, the object's pose can be determined; its size and position are found from its zeroth and first order regular moments, while its orientation angle is computed from the *phase angle* of its Zernike moments; previous works [1, 9, 13, 19] have successfully utilised the *magnitude* of the Zernike moments as rotation invariant features for planar shape recognition. The *phase angle* of the Zernike moments however, has been ignored in the past.

The efficiency of computing the Zernike moments is also addressed in this paper. The *explicit forms* of the radial polynomials of the Zernike moments are used to save more than 65% on computation time.

2 Zernike Moments

Moment invariant techniques were first introduced by Hu [6], and more recently used by Maitra [15]. Hu's moments are defined over the monomial basis set $\{x^p y^q\}$ which is not orthogonal, hence the reconstruction of the image is computationally expensive due to the information redundancy included in these moments. Teague [19] suggested the use of orthogonal moments to overcome the problems associated with the regular moments. Zernike moments [1, 5, 10, 12, 14, 19] belong to the set of such orthogonal moments. Other moment-based techniques used in un-occluded object recognition include Legendre [19], pseudo-Zernike [11, 20], rotational [4, 18] and complex [2] moments. A survey of moment-based techniques for object recognition is found in [17]. Comparative studies [16, 20] established the superiority of Zernike and pseudo-Zernike moments in terms of noise sensitivity, information redundancy, and image representation ability.

The Zernike moments are defined over a set of complex polynomials which forms a complete orthogonal set over the unit disk $x^2 + y^2 \leq 1$. This set of Zernike polynomials, ZP, was first introduced by Zernike [22] and can be denoted by

$$ZP = \{V_{nm}(x,y) \mid x^2 + y^2 \leq 1\}. \tag{1}$$

The form of the Zernike polynomial basis of order n and repetition m is

$$V_{nm}(x,y) = R_{nm}(x,y) \exp(jm\theta), \tag{2}$$

with $n \in N^+, m \in N \mid (n - |m|)$ even and $|m| \leq n$, $\theta = \arctan(y/x)$, and the radial polynomial, $R_{nm}(x,y)$ is defined as

$$R_{nm}(x,y) = \sum_{s=0}^{(n-|m|)/2} (-1)^s \frac{(n-s)!}{s!\left(\frac{n+|m|}{2}-s\right)!\left(\frac{n-|m|}{2}-s\right)!} \left(x^2 + y^2\right)^{(n-2s)/2}. \tag{3}$$

Note that $R_{nm}(r)$, the polar coordinate representation ($x = r\cos\theta, y = r\sin\theta$) of $R_{nm}(x,y)$, is a polynomial of degree n in r containing terms in $r^n, r^{n-2}, \ldots, r^{|m|}$. The *explicit forms* of the radial polynomials up to the twelfth order are given in tables 1a and 1b for even and odd n's respectively. The computation time of the Zernike moments can be reduced dramatically by using the *explicit forms* of $R_{nm}(r)$ instead of equation 3 as will be shown later.

The Zernike moments are the projections of the image function $f(x,y)$ onto the orthogonal basis functions $V_{nm}(x,y)$. The Zernike moment of order n with repetition m is simply a complex number given by

$$A_{nm} = \frac{n+1}{\pi} \sum_{x=0}^{N-1} \sum_{y=0}^{M-1} f(x,y) \left[V_{nm}(x,y)\right]^*, \tag{4}$$

with $x^2 + y^2 \leq 1$, and $*$ denoting the complex conjugate. Replacing equation 2 into equation 4, A_{nm} can be expressed as [1, 9],

$$A_{nm} = \frac{2n+2}{\pi} \left[\sum_{x=0}^{N-1} \sum_{y=0}^{M-1} f(x,y) R_{nm}(x,y) \cos m\theta - j \sum_{x=0}^{N-1} \sum_{y=0}^{M-1} f(x,y) R_{nm}(x,y) \sin m\theta \right]. \tag{5}$$

2.1 Rotational Properties of Zernike Moments

An important property of Zernike moments is that their values for an image and for its rotated version have a simple relationship. If $f(r,\theta)$ is the polar coordinate representation of the image $f(x,y)$, and $f^r(r,\theta^r)$ is the image rotated through an angle α, then, $f^r(r,\theta^r) = f(r,\theta-\alpha)$. Let A_{nm} and A^r_{nm} be the Zernike moments of the images $f(r,\theta)$ and $f^r(r,\theta^r)$ respectively. It can be shown [1,9] that

$$A^r_{nm} = A_{nm} \exp(-jm\alpha). \qquad (6)$$

It is clear from equation 6 that Zernike moments acquire only a phase shift on image rotation, and hence their magnitudes remain constant. The *magnitudes* of the Zernike moments $|A_{nm}|$ are therefore rotation-invariant features. These features were first used by Khotanzad and Hong [5,7–10] for image representation. They however ignored the *phase angle* of the Zernike moments. This *phase angle* can be used to solve for the orientation of the object in the scene.

Once an object is recognised using the invariant Zernike features, the object's orientation can be determined by solving equation 6 for α. Using Euler's formula, one can write equation 6 as

$$|A^r_{nm}| \exp(j\theta^r) = |A_{nm}| \exp(j\theta) \exp(-jm\alpha), \qquad (7)$$

and noting that $|A^r_{nm}| = |A_{nm}|$, equation 7 can be simplified to

$$\exp(jm\alpha) = \exp(j(\theta - \theta^r)). \qquad (8)$$

Solving equation 8 for α we obtain

$$\alpha = \frac{(\theta - \theta^r)}{m} + \frac{k(360°)}{m} \pmod{360°}, \qquad (9)$$

Table 1. *Explicit forms* of the radial polynomials $R_{nm}(r)$ for $n \leq 12$.

m\n	1	3	5	7	9	11
1	r	$3r^3-2r$	$10r^5-12r^3+3r$	$35r^7-60r^5+30r^3-4r$	$126r^9-280r^7+210r^5-60r^3+5r$	$462r^{11}-1260r^9+1260r^7-560r^5+105r^3-6r$
3		r^3	$5r^5-4r^3$	$21r^7-30r^5+10r^3$	$84r^9-168r^7+105r^5-20r^3$	$330r^{11}-840r^9+756r^7-280r^5+35r^3$
5			r^5	$7r^7-6r^5$	$36r^9-56r^7+21r^5$	$165r^{11}-360r^9+252r^7-56r^5$
7				r^7	$9r^9-8r^7$	$55r^{11}-90r^9+36r^7$
9					r^9	$11r^{11}-10r^9$
11						r^{11}

(a) Even n.

m\n	0	2	4	6	8	10	12
0	1	$2r^2-1$	$6r^4-6r^2+1$	$20r^6-30r^4+12r^2-1$	$70r^8-140r^6+90r^4-20r^2+1$	$252r^{10}-630r^8+560r^6-210r^4+30r^2-1$	$924r^{12}-2772r^{10}+3150r^8-1680r^6+420r^4-42r^2+1$
2		r^2	$4r^4-3r^2$	$15r^6-20r^4+6r^2$	$56r^8-105r^6+60r^4-10r^2$	$210r^{10}-504r^8+420r^6-140r^4+15r^2$	$792r^{12}-2310r^{10}+2520r^8-1260r^6+280r^4-21r^2$
4			r^4	$6r^6-5r^4$	$28r^8-42r^6+15r^4$	$120r^{10}-252r^8+168r^6-35r^4$	$495r^{12}-1320r^{10}+1260r^8-504r^6+70r^4$
6				r^6	$8r^8-7r^6$	$45r^{10}-72r^8+28r^6$	$220r^{12}-495r^{10}+360r^8-84r^6$
8					r^8	$10r^{10}-9r^8$	$66r^{12}-110r^{10}+45r^8$
10						r^{10}	$12r^{12}-11r^{10}$
12							r^{12}

(b) Odd n.

where $k \in \{0, 1, 2, \ldots, m-1\}$. Note that α is a function of the phase angles θ and θ^r, and it measures the orientation of an object $(f^r(x,y))$ with respect to that of the matching object $(f(x,y))$ from the model base.

The rotational properties of Zernike moments are illustrated by comparing the Zernike features of the six differently oriented images of figure 1. The results of the comparison are displayed in table 2. The invariance of the Zernike features—magnitude of the Zernike moments, is clear from the entries of the columns labelled $|A_{nm}|$ and $|A_{nm}^r|$. Also shown in table 2 is a comparison of the calculated rotation angles α with the actual rotation angles (0° (original image), 30°, 60°, 90°, 180°, 270°).

With $\alpha \in [0, 360)$, equation 9 has m solutions corresponding to $k = 0, 1, 2, \ldots, m-1$, with only one of these solutions corresponding to the correct object orientation. This ambiguity can be solved by choosing $m = 1$, i.e. working with Zernike moments of odd n's and with $n \geq 3$. This reduces the number of solutions of equation 9 to a single $\alpha \in [0, 360)$ corresponding to $k = 0$ as shown by the highlighted rows in table 2.

It should be clear that Zernike moments with $m = 0$ cannot be used with equation 9. Also due to translation normalisation [1, 9], A_{11} is zero for all images, and hence cannot be used to calculate α.

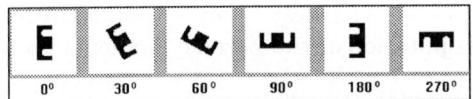

Fig. 1. Six rotations of object number 1 of figure 2a.

Table 2. Magnitudes of the Zernike moments of various orders for different image orientations of the object of figure 1. Also shown, is a comparison of the actual rotation angles and the calculated rotation angle α.

		0°	30°			60°			90°			180°			270°		
n	m	$|A_{nm}|$	$|A'_{nm}|$	α	k	$|A'_{nm}|$	α	k	$|A'_{nm}|$	α	k	$|A'_{nm}|$	α	k	$|A'_{nm}|$	α	k
2	2	27.49	27.20	31.78	0	28.34	61.41	0	28.27	90.01	0	26.63	180.01	1	25.84	270.00	1
3	1	1.67	1.78	42.53	0	1.36	62.15	0	1.50	94.34	0	1.45	182.53	0	1.62	268.26	0
3	3	0.37	0.54	29.89	1	0.43	67.57	1	0.33	87.50	1	0.32	178.55	2	0.36	270.98	0
4	2	93.88	93.27	31.62	1	96.85	61.41	1	97.45	90.01	1	93.44	180.01	1	89.87	270.00	0
4	4	4.87	4.86	32.87	1	5.55	61.59	1	5.54	90.01	1	4.78	180.01	2	4.15	269.99	3
5	1	5.73	6.39	44.46	0	4.76	67.89	0	5.11	95.39	0	5.15	182.90	0	5.81	267.73	0
5	3	4.96	5.38	29.76	0	5.29	61.51	0	4.41	89.64	1	4.17	179.56	1	4.70	269.94	2
5	5	0.37	0.36	33.72	0	0.37	57.87	0	0.40	90.17	1	0.34	180.10	2	0.31	269.93	3
6	2	136.40	136.74	31.34	0	141.31	61.48	0	144.67	90.03	0	143.51	180.03	1	135.21	270.00	1
6	4	23.99	23.90	32.54	0	27.15	61.46	1	27.33	90.01	1	24.04	180.00	2	20.83	269.99	3
6	6	0.38	0.41	24.19	0	0.16	50.31	0	0.03	95.39	1	0.14	180.50	3	0.46	269.88	4

3 Zernike Moments Computation

Let A_{nm}^* be the complex conjugate of A_{nm}. Because $R_{n,-m}(x,y) = R_{n,m}(x,y)$, hence $A_{n,-m} = A_{nm}^*$ and $|A_{nm}| = |A_{n,-m}|$. One can therefore work with the set

$$\{|A_{nm}|, \text{ with } n \geq 0, \ 0 \leq m \leq n, \text{ and } (n-m) \text{ even}\}$$

as far as the defined Zernike features are concerned. The number of these features, N_{zf}, corresponding to order 0 through n can be computed from

$$N_{zf} = \begin{cases} 1/4(n+2)^2 & n \text{ even} \\ 1/4(n+2)^2 - 1 & n \text{ odd} \end{cases} \quad (10)$$

To compute the Zernike moments, the image $f(x,y)$ is first normalised with respect to translation and scale using the regular geometrical moments of the image [1, 12]. Then, the centre of the image is taken as the origin, and pixel coordinates are mapped to the unit disk $x^2 + y^2 \leq 1$. The Zernike moments are then evaluated by applying equation 5 to those pixels falling inside the unit circle. The radial polynomials are calculated from their *explicit forms* given in table 1 rather than from equation 3, reducing the computation time by more than 65% (using a 66 MHz 486 DX2 PC). This drastic reduction is achieved by simply noting that the *explicit form* of the radial polynomial $R_{nm}(r)$ is a polynomial of degree n in r containing terms in $r^n, r^{n-2}, \ldots, r^{|m|}$ (see table 1); the terms $r^2, r^3, \ldots, r^n$ are calculated only once and stored in a look-up-table to be used in the computation of the Zernike moments up to the n-th order.

4 The Model Base

The objects used in this work consists of the 15 industrial parts shown in figure 2a. They are picked from a mechanical construction kit that consists of standard components such as gears, wheels, links, strap coupling, retaining clips, washers, etc. For each object, we consider 24 different 64 × 64 binary images corresponding to combinations of six rotations (0°, 30°, 60°, 90°, 180°, 270°), two scales (1 and 0.5), and two translations ((0,0) and (5,5)). This brings the total number of utilised data set to 360 images. Figure 2b shows the 24 different images of object number 1. The translation- and scale-normalised images are displayed in figure 2c.

5 Neural Network Classifier

Classification is the final step in the vision process. Neural networks (NNet) have recently emerged as an important tool for solving industrial vision problems [21]. A multi-layer perceptron with a back propagation learning algorithm is used as a shape classifier in this work. The back propagation algorithm implemented uses individual adaptive learning rates for the weights making the convergence of the algorithm significantly faster [3].

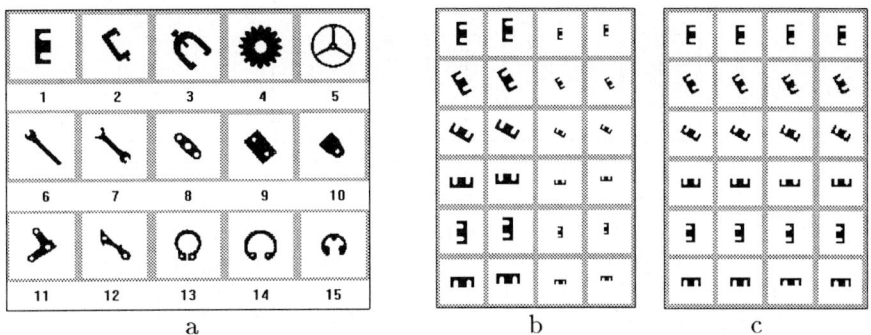

Fig. 2. (a) The 15 industrial parts used in the experiment. (b) The 24 rotated, scaled and translated images of object number 1 from (a). (c) Regular moments are used to normalise the images from (b) with respect to scale and translation.

The magnitudes of the Zernike moments, ranging from the second order up to the eleventh order, form 40-dimensional feature vectors which serve as inputs to the classifier. Fifteen outputs, corresponding to the 15 classes (objects), emerge from the neural network. The classifier is initially trained such that the output neuron that fires the most (output $\approx$ 1) determines the class to which an object belongs. All the other output neurons should fire near the zero level for sharp discrimination.

For each object, a set of 24 different images, as described in section 4, are used. A randomly selected subset of 12 images, is employed for training, whilst the full set of 24 images is used for testing the neural network. Figures 3a and 3b show the training and testing results respectively. On the x-axis, the first 12 (or 24 for testing) inputs are the 40-dimensional feature vectors representing the 12 (or 24 for testing) sample images of object 1 used for training (or testing). The second 12 (or 24 for testing) inputs correspond to the feature vectors of object 2, and so on.

The curves of figure 3, labelled as 1, 2, 3, ... , 15, represent the respective firing levels of the fifteen output neurons. Output neuron 1 corresponds to object 1, output neuron 2 corresponds to object 2, and so on. To further clarify the results, figure 3c shows a more detailed graph of the training outcomes for the first three objects; this is a magnified view of the curves labelled 1, 2 and 3 shown in figures 3a. It is clear from figure 3b that full classification of the 360 images is achieved.

6 Conclusions

This paper described a planar object recognition system. Invariant object recognition under similarity transformations was achieved by using the rotation invariant magnitudes of the Zernike moments, and by utilising the regular moments to normalise the images for translation- and scale-invariance.

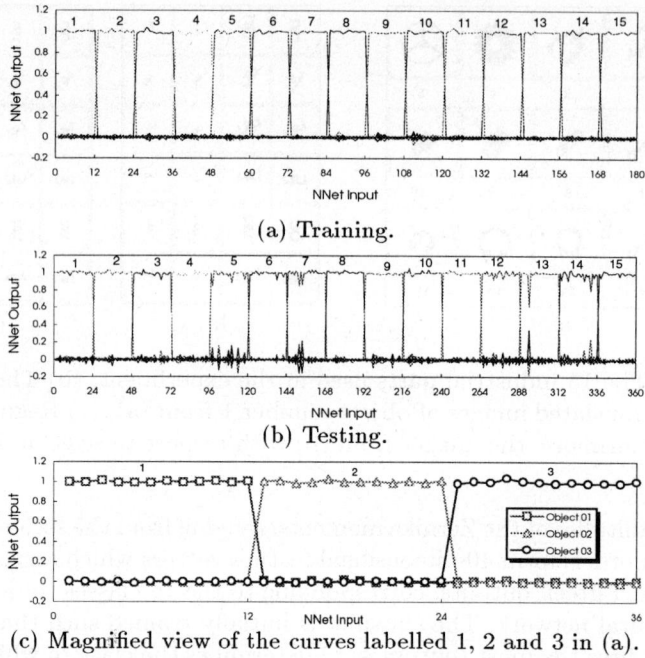

Fig. 3. Neural network results.

The model base used in the experiment consisted of real images of 15 industrial parts, with 24 differently rotated, scaled and translated images of each part for a total of 360 images. Using 40 Zernike features and a neural network classifier, full classification of the 360 images was achieved.

It was also demonstrated that the *phase angle* of the Zernike moments can be used to determine the orientation of identified objects. Coupled with the object translation and scale computed from the regular moments, the full similarity transformation between the model and the image is then known, and hence the pose of the object is determined. The similarity transformation could also be used for verification by backpojection of the model onto the image to reject false positives.

Furthermore, savings of more than 65% on computation time was achieved by using the *explicit forms* of the radial polynomials of the Zernike moments.

References

1. Abdallah S.M., Nebot E.M., and Rye D.C. 2-D object recognition via Zernike moments. In *Proc. International Conference on Mechatronics and Machine Vision in Practice*, pages 100–105. IEEE Computer Society Press, 1994.
2. Abu-Mostafa Y.S. and Psaltis D. Image nornalization by complex moments. *IEEE Transactions on PAMI*, 7(1):46–55, 1985.

3. Agamennoni O., Chessari C., Romagnoli J.A., Barton G.W., and Bourke K. A neural network based prediction scheme for an industrial propathene reactor. In *World Congress on Neural Networks*, volume I, pages 195–200, 1994.
4. Hossak WJ Boyce JF. Moment invariant for pattern recognition. *Pattern Recognition Letters*, 1:451–456, 1983.
5. Hong Y.H. *Invariant Image Recognition by Zernike Moments*. PhD thesis, Southern Methodist University, Dallas, Texas, USA, December 1988.
6. Hu M.K. Visual pattern recognition by moments invariant. *IRE Transactions on Information Theory*, IT-8:179–187, 1962.
7. Khotanzad A. and Hong Y.H. Rotation invariant pattern recognition using Zernike moments. In *Proc. International Conference on Pattern Recognition*, pages 326–328, 1988.
8. Khotanzad A. and Hong Y.H. Zernike moments based rotation invariant features for pattern recognition. In *Intelligent Robots and Computer Vision*, volume 1002, pages 212–219. SPIE, 1988.
9. Khotanzad A. and Hong Y.H. Invariant image recognition by Zernike moments. *IEEE Transactions on PAMI*, 12(5):489–497, 1990.
10. Khotanzad A. and Hong Y.H. Rotation invariant image recognition using features selected via a systematic method. *Pattern Recognition*, 23(10):1089–1101, 1990.
11. Khotanzad A. and Liou J.J.-H. A neural network system for 3-d object recognition and pose estimation from a single arbitrary 2-d view. In *Applications of Neural Networks III*, volume 1002, pages 107–118. SPIE, 1992.
12. Khotanzad A. and Lu J.-H. Classification of invariant image representations using a neural network. *IEEE Transactions on Acoustics, Speech and Signal Processing*, 38(6):1028–1038, 1990.
13. Lisboa P.J.G. and Perantonis S.J. Invariant digit recognition by Zernike moments and third-order neural networks. In *Proc. International Conference on Artificial Neural Networks*, pages 82–85, 1991.
14. Lisboa P.J.G. and Perantonis S.J. Invariant pattern recognition using third-order networks and Zernike moments. In *Proc. International Joint Conference on Neural Networks*, pages 1421–1425, 1991.
15. Maitra S. Moments invariant. *Proceedings of the IEEE*, 67(4):697–699, 1979.
16. Nebot E.M., Abdallah S.M., and Rye D.C. Two-dimensional object recognition and classification. In *Proc. Digital Image Computing: Techniques and Applications*, pages 842–849. Australian Pattern Recognition Association, 1993.
17. Prokop R.J. and Reeves A.P. A survey of moment-based techniques for unoccluded object representation and recognition. *Computer Vision, Graphics and Image Processing*, 54(5):438–460, 1992.
18. Smith F.W. and Wright M.H. Automatic ship photo interpretation by the method of moments. *IEEE Transactions on Computers*, C-20(9):1089–1095, 1971.
19. Teague M.R. Image analysis via the general theory of moments. *Journal of the Optical Society of America*, 70(8):920–930, 1980.
20. Teh C.H. and Chin R.T. On image analysis by the methods of moments. *IEEE Transactions on PAMI*, 10(4):496–513, 1988.
21. White K. Is there a future for vision with neural nets? *SME Vision*, 6(3):1–4, 1989.
22. Zernike F. Beugungstheorie des schneidenverfahrens und seiner verbesserten form, der phasenkontrastmethode. *Physica*, 1(8):689–704, 1934.

A Study of Zernike Moment Computing

Simon X. Liao[1] and Miroslaw Pawlak[2]

[1] The University of Winnipeg, Winnipeg, Manitoba, Canada, R3B 2E9
[2] The University of Manitoba, Winnipeg, Manitoba, Canada, R3T 5V6

Abstract

In this paper, we address a detailed analysis of the accuracy of Zernike moment computing in terms of its discretization error. It is found that there is an inherent limitation in the precision of computing the Zernike moment due to the geometric nature of a circular domain. This is explained by relating the accuracy issue to a celebrated problem in analytic number theory of evaluating the lattice points within a circle. We illustrate our theory by various numerical studies including the image reconstruction.

1 Introduction

Moments can provide characteristics of an object that uniquely represent its shape and have been extensively employed as the invariant global features of an image in pattern recognition and image classification since 1960s[3].

Generally, the moment features of an image are invariant under translation, scale change, and rotation only when they are computed from the original analog images. For the digitized images, however, the invariant properties are satisfied only approximately. Nevertheless, the error analysis of inaccuracies in moment computing has been rarely addressed.

Traditionally, geometric moments $\{M_{pq}\}$ have been utilized. However, the set of functions $\{x^p y^q\}$ is not orthogonal and consequently the recovery of an image from its $\{M_{pq}\}$ is computationally expensive. Teague[9] suggested the use of orthogonal moments to overcome the shortcomings associated with the geometric moments. He proposed to use the orthogonal moments defined in terms of the Legendre and Zernike polynomials. The advantage of the Legendre and Zernike moments has been demonstrated in terms of their reconstruction and classification power [4], [6], [7], [10]. Despite their good reconstruction properties, however, the Legendre moments are not invariant to linear operations. The Zernike orthogonal moments have been suggested that have better reconstruction properties along with the build in invariance to linear transformations, which have been demonstrated theoretically and supported by a number of experimental studies [4], [10].

In this paper, the discretization error analysis for the Zernike moments is carried out. Our main result reveals an inherent limitation in the accuracy of the Zernike moment computing related to the fact that the Zernike moments are confined to a unit circle rather than a square. This is explained by using the celebrated problem in the analytic number theory referred to as the lattice points of a circle, due originally to Gauss. The numerical accuracy of the Zernike moments computing is also examined. Some new techniques for increasing the numerical accuracy of the Zernike moments are proposed. Based on the progress made in Zernike moments computing, the inverse moment problem of reconstruction of an image from a finite set of Zernike moments is examined.

2 Zernike Moments

The Zernike moments have been frequently utilized for a number of image processing and computer vision tasks[4], [6], [10]. The (p,q) order Zernike function is defined over the unit disk as

$$V_{pq}(x,y) = R_{pq}(\rho)\exp(jq\theta), \quad x^2 + y^2 \leq 1, \tag{1}$$

where $\rho = \sqrt{x^2 + y^2}$ is the length of the vector from the origin to the pixel (x,y), and $\theta = \arctan(y/x)$ is the angle between the vector and the x axis. In (1), the radial polynomial $R_{pq}(\rho)$ is given by

$$R_{pq}(\rho) = \sum_{l=0}^{(p-|q|)/2} (-1)^l \frac{(p-l)!}{l!(\frac{p+|q|}{2}-l)!(\frac{p-|q|}{2}-l)!} \rho^{p-2l}. \tag{2}$$

The completeness and orthogonality of $\{V_{pq}(x,y)\}$ allow us to represent any square integrable image function $f(x,y)$ defined on the unit disk as

$$f(x,y) = \sum_{p=0}^{\infty} \sum_{q=-p}^{p} \tau_p A_{pq} V_{pq}(x,y), \tag{3}$$

where $p-|q|$ =even, $\tau_p = \frac{p+1}{\pi}$ is the normalizing constant, and A_{pq} is the Zernike moment of order p with repetition q, i.e.,

$$A_{pq} = \iint_D f(x,y) V_{pq}^*(x,y) dx dy. \tag{4}$$

The polar coordinates version of A_{pq} takes the following form

$$A_{pq} = \int_0^{2\pi} \int_0^1 f(\rho\cos\theta, \rho\sin\theta) R_{pq}(\rho) \exp(-jq\theta) \rho d\rho d\theta. \tag{5}$$

The fundamental feature of the Zernike moments is their rotational invariance. If $f(x,y)$ is rotated by an angle α, then we can obtain that the Zernike moment A'_{pq} of the rotated image is given by

$$A'_{pq} = A_{pq} e^{-jq\alpha}. \tag{6}$$

Thus, the magnitude of the Zernike moment can be used as a rotationally invariant image feature.

These favorable properties of the Zernike moments are valid as long as one uses a true analog image function. In practice, however, one has to deal with a discrete data yielding deterioration of the performance of the Zernike moment descriptors. In fact, a major problem with the use of the Zernike moments is that the digital version of A_{pq} employs the summation with square pixels over a circular region. This is in contrast to the situation for geometric moment descriptors employing rectangular integration regions. We can define the following discrete version of A_{pq} over the pixel set $\{(x_i, y_j), 1 \leq i \leq M, 1 \leq j \leq N\}$

$$\widehat{A}_{pq} = \sum_{x_i^2+y_j^2 \leq 1} \sum h_{pq}(x_i, y_j) f(x_i, y_j), \qquad (7)$$

where

$$h_{pq}(x_i, y_j) = \int_{x_i - \frac{\Delta x}{2}}^{x_i + \frac{\Delta x}{2}} \int_{y_j - \frac{\Delta y}{2}}^{y_j + \frac{\Delta y}{2}} V_{pq}^*(x, y) dx dy \qquad (8)$$

represents the integration of $V_{pq}^*(x, y)$ over the (i, j) pixel. Since the image function is defined over the unit disk, the summation in (7) takes into account only those pixels whose centers fall completely inside the circle. It is important to observe that for the Zernike moments, we have

$$\widehat{A}_{pq} \neq A_{pq} \qquad (9)$$

even if the image function is constant on D.

Hence, there is an inherent error in computing $\widehat{A}_{pq}$ related to the circular nature of the support of $\{V_{pq}(x, y)\}$. We refer to such an error as the geometric error. For a general image function, the error between $\widehat{A}_{pq}$ and A_{pq} can be decomposed as follows

$$E_{pq} = \widehat{A}_{pq} - A_{pq} = E_{pq}^g + E_{pq}^n, \qquad (10)$$

where E_{pq}^g is the geometric error, and E_{pq}^n is the numerical error related to the need of numerical integration in (8).

3 Error Analysis

The geometric error in the Zernike moment computing, E_{pq}^g, can be written as

$$E_{pq}^g = \mu \Big(\sum_{x_i^2+y_j^2 \leq 1} \sum h_{pq}(x_i, y_j) - \pi \Big), \qquad (11)$$

where

$$\mu = \pi^{-1} \iint_D f(x, y) dx dy.$$

The geometric error, E_{pq}^g, is not equal to zero due to the fact that if the center of a pixel falls inside the border of the unit disk $\{(x, y) : x^2 + y^2 \leq 1\}$, this pixel is used in the computation of the Zernike moments; otherwise, the pixel

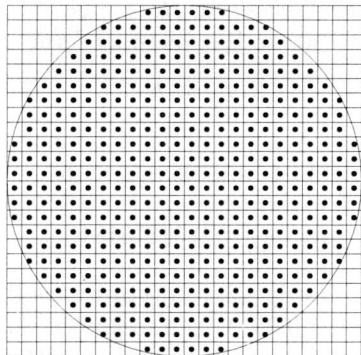

Fig. 1. Lattice-point approximation of a circular region.

is discarded. Figure 1 shows the union of the pixels whose centers fall inside the unit circle. Note that some pixels are not entirely inside the circle; on the other hand, some parts of the circle are not covered by the pixels.

In our case, the unit disk is located in the $[-1, 1] \times [-1, 1]$ square which is composed of n^2 pixels, i.e., $\Delta x = \Delta y = 2/n$. The evaluation of the size of the error in (11) is not a trivial problem. In order to get further insight into the behavior of the geometric error, let us consider the $(0,0)$ order Zernike moment. Recalling that $V_{00}^* = 1$ and that the area of a single pixel is $4/n^2$, we have

$$E_{00}^g = \mu\left(\frac{4}{n^2}K(n) - \pi\right), \tag{12}$$

where

$$K(n) = \sum\sum_{x_i^2 + y_j^2 \leq 1} 1 \tag{13}$$

denotes the number of the points $\{(x_i, y_j) : 1 \leq i \leq n, \ 1 \leq j \leq n\}$ inside the unit circle. The geometric error, $R(n)$, can be determined by

$$R(n) = \frac{4}{n^2}K(n) - \pi. \tag{14}$$

It is crucial to know the size of $R(n)$, i.e., how fast $R(n)$ tends to zero as $n \to \infty$.

It turns out that quantity $K(n)$ has been extensively examined in the analytic number theory with the relation to the so-called lattice points of a circle problem due originally to Gauss.

Let $K(x)$ be the number of integer points (i, j) inside or on the circle $i^2 + j^2 = x$. For large x, the value of $K(x)$ is approximately πx, the area of the circle. Let

$$R(x) = K(x) - \pi x \tag{15}$$

be the difference between $K(x)$ and πx. Gauss' problem on the number of integer points inside a circle is to determine the correct order of the magnitude of $R(x)$

as $x \to \infty$. It is known after Gauss that $R(x) = O(x^\vartheta) = O(x^{\frac{1}{2}})$. This result was improved by Sierpinski in 1906 to the following form

$$R(x) = O(x^{\frac{1}{3}}). \tag{16}$$

It was further improved by Walfisz(1927), Titchmarsh(1935), Hua(1942), and most recently Iwaniec and Mozzochi(1988)[5] with

$$\vartheta = \frac{7}{22} = 0.318181818....$$

In the other direction, Hardy's[2] conjecture says that the value of ϑ can be arbitrary close to $\vartheta = 1/4 = 0.25$. It is also known that $\vartheta = 1/4$ is impossible. This still remains an open problem in analytic number theory.

Comparing our case defined in (14) to the lattice points problem, we observe that x in (15) is equivalent to $(\frac{n}{2})^2$. Therefore

$$K(n) = \pi(\frac{n}{2})^2 + O(n^{2\vartheta}). \tag{17}$$

Comparing (17) with (14), we obtain the following evaluation for $R(n)$

$$R(n) = O(n^{-2(1-\vartheta)}). \tag{18}$$

With the latest result due to Iwaniec and Mozzochi[5], (18) takes the following form

$$R(n) = O(n^{-\frac{15}{11}}). \tag{19}$$

The result in (19) applied to (12) yields

$$E_{00}^g = O(n^{-\frac{15}{11}}). \tag{20}$$

For the (p,q) order Zernike moment, the geometric error is at least of order as in (20). In fact, the formula in (11) can be viewed as a weighted version of the lattice points in a circle problem. Therefore E_{pq}^g should tend to zero at least as fast as E_{00}^g in (20).

	n=24	n=36	n=48	n=64	n=128
n^{-1}	0.0416667	0.0277778	0.0208333	0.0156250	0.0078125
$n^{-\frac{15}{11}}$	0.0131188	0.0075468	0.0050979	0.0034437	0.0013382
$n^{-\frac{3}{2}}$	0.0085051	0.0046296	0.0030070	0.0019531	0.0006905

Table 1. Range of geometric errors for several commonly used image sizes.

Table 1 displays the values of geometric error for the cases of $n^{-1}, n^{-\frac{11}{15}}$, and $n^{-\frac{3}{2}}$ for several common image sizes.

The numerical error, E_{pq}^n, is caused by the fact that the factor $h_{pq}(x_i, y_j)$ appearing in the definition of the Zernike moment $\widehat{A}_{pq}$ requires the use of some cubature formulas. In fact, for moderate and large values of (p,q), the formula

$$h_{pq}(x_i, y_j) = \int_{x_i - \frac{\Delta x}{2}}^{x_i + \frac{\Delta x}{2}} \int_{y_j - \frac{\Delta y}{2}}^{y_j + \frac{\Delta y}{2}} V_{pq}^*(x,y) dx dy \tag{21}$$

needs to be evaluated by integration rules. The simplest strategy, one dimensional rule, is to approximate $h_{pq}(x_i, y_j)$ as

$$h_{pq}(x_i, y_j) \simeq \Delta x \Delta y V_{pq}^*(x_i, y_j). \tag{22}$$

Traditionally, to achieve sufficient accuracy, one has to increase the number of nodes in each pixel to reduce the approximation error in computing $h_{pq}(x_i, y_j)$ by using some well known cubature formulas[1].

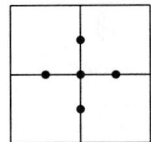

Fig. 2. 5-dimensional cubature formula.

One simple formula which can be adopted to increase the approximation accuracy is shown in Figure 2. We can determine the weights of the 5-dimensional cubature formula[6] as

$$C_5(f) = \frac{4}{3}\{-f(0,0) + f(0,0.5) + f(0.5,0) + f(0,-0.5) + f(-0.5,0)\}. \tag{23}$$

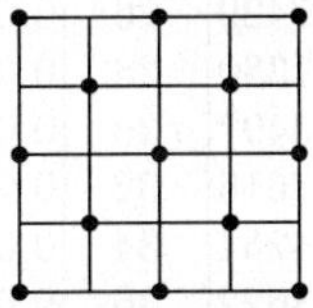

Fig. 3. 13-dimensional cubature formula.

The number of nodes in each pixel can be increased further to achieve even higher accuracy. An example is to use the 13-dimensional cubature formula, which is illustrated in Figure 3. The weights of the 13-dimensional cubature formula can be determined as

$$\begin{aligned}C_{13}(f) = \frac{1}{45}\{&120f(0,0) + 8[f(1,1) + f(1,-1) + f(-1,-1) + f(-1,1)]\\&+16[f(0.5,0.5) + f(0.5,-0.5) + f(-0.5,-0.5) + f(-0.5,0.5)]\\&-9[f(0,1) + f(1,0) + f(0,-1) + f(-1,0)]\},\end{aligned} \tag{24}$$

which is called as the 13-dimensional cubature formula.

4 Image Reconstruction and Experimental Results

In order to assess the performance of a set of image descriptors, one can reconstruct an image with the following reconstruction algorithm

$$\widehat{f}_T(x,y) = \sum_{p=0}^{T} \sum_{q=-p}^{p} \tau_p \widehat{A}_{pq} V_{pq}(x,y), \tag{25}$$

where $p - |q| =$ even, and $\widehat{A}_{pq}$ is defined in (7).

An image of cross is employed as the testing image. It is of the size of 36×36 pixels and the range of graylevels for each pixel is 64, where all of the pixels representing cross have the gray level 24 and the background has the value 48. Figure 4 illustrates this image projected onto the unit disk.

Fig. 4. The original image used in image reconstruction.

The normalized mean square error

$$\overline{e^2} = \frac{\iint_D |f(x,y) - \widehat{f}(x,y)|^2 dxdy}{\iint_D [f(x,y)]^2 dxdy} \qquad (26)$$

is adopted here as a measure of the accuracy of the reconstructed images.

Order	1-D	5-D	13-D	Order	1-D	5-D	13-D
2	0.072390	0.072331	0.072339	22	0.010101	0.011519	0.009751
4	0.069251	0.069534	0.068524	24	0.009699	0.011046	0.008771
6	0.058068	0.057908	0.057429	26	0.009525	0.009721	0.007481
8	0.032009	0.031680	0.031380	28	0.010752	0.010374	0.007456
10	0.023649	0.024230	0.023497	30	0.008122	0.007861	0.005360
12	0.017415	0.018582	0.017616	32	0.013553	0.008110	0.007015
14	0.013938	0.015461	0.014254	34	0.019827	0.009616	0.009271
16	0.013504	0.014889	0.013622	36	0.020814	0.009661	0.008801
18	0.012743	0.014119	0.012953	38	0.019986	0.010293	0.007855
20	0.011768	0.012804	0.011301	40	0.034095	0.012175	0.012275

Table 2. The error $\overline{e^2}(\gamma)$ for the testing image utilizing three different integration rules.

Figure 5 illustrates the reconstructed images of Figure 4 from the 1-dimensional formula and the proposed 5-dimensional and 13-dimensional cubature formulas. The first row shows the patterns reconstructed from the traditionally calculated Zernike moments, and the next two rows show the images reconstructed from the Zernike moments computed from the new 5-dimensional and 13-dimensional formulas. From left to right, the three rows display the patterns reconstructed from moments of order 12, 16, 20, 24, 28, 32, 36, and 40, respectively. Table 2 shows the error $\overline{e^2}$ for increasing values of the moment order.

Table 2 and Figure 5 indicate that the proposed 5-dimensional and 13-dimensional cubature formulas perform better than the simplest 1-dimensional formula of the traditional Zernike moment method, visually and numerically.

5 Conclusion

In this paper, we have studied the accuracy problem for Zernike moments determined from discrete data observed on the $n \times n$ pixels image domain. It has

Fig. 5. Reconstructed patterns of the original image.

been observed that two kinds of errors determine the accuracy of computing of Zernike moments, i.e., the numerical error related to the usual need of calculating accurately two dimensional integrals and the geometric error being a distinctive feature of Zernike moments. We have shown that the geometric error plays the critical role for the accuracy of the Zernike moment computing. We also performed the image reconstruction from Zernike moments. As expected, compared with the traditional method, the proposed 5-dimensional and 13-dimensional formulas provide better qualities, visually and numerically.

References

1. H. Engels, *Numerical Quadrature and Cubature*. London: Academic Press Inc., 1980.
2. G.H. Hardy and E. Landau, *"The lattice points of a circle"*, Proc. Royal Soc. (A), 105 (1924), pp. 245-258.
3. M.K. Hu, *"Visual problem recognition by moment invariant"*, IRE Trans. Inform. Theory, vol. IT-8, pp. 179-187, Feb. 1962.
4. A. Khotanzad and Y.H. Hong, *"Invariant image recognition by Zernike moments"*, IEEE Trans. Pattern Anal. Mach. Intelligence PAMI-12, 1990, pp. 489-498.
5. H. Iwaniec and C.J. Mozzochi, *"On the divisor and circle problems"*, Journal of Number Theory, 29 (1988), pp. 60-93.
6. S. X. Liao, *Image Analysis by Moments*, Ph. D. dissertation, The University of Manitoba, 1993.
7. S. X. Liao and M. Pawlak, *"On image analysis by moments"*, IEEE Trans.Pattern Anal. Machine Intell., vol. PAMI-18, pp. 254-266, 1996.
8. M. Pawlak, *"On the reconstruction aspects of moment descriptors"*, IEEE Trans. Information Theory, vol. 38, No. 6, pp. 1698-1708, November, 1992.
9. M.R. Teague, *"Image analysis via the general theory of moments"*, J. Optical Soc. Am., vol. 70, pp. 920-930, August 1980.
10. C.H. Teh and R.T. Chin, *"On image analysis by the methods of moments"*, IEEE Trans. Pattern Anal. Machine Intell., vol. PAMI-10, pp. 496-512, July 1988.
11. F. Zernike, *"Beugungstheorie des Schneidenverfahrens und seiner verbesserten Form, der Phasenkontrastmethode"*, Physica, vol. 1, p. 689-701, 1934.

Query Expansion by Raw Image Features and Text Annotations in Image Retrieval

Kok F. Lai[1], Hong Zhou[2] and Syin Chan[2]

[1] Information Technology Institute, Singapore 117685
[2] Nanyang Technological University, Singapore 639798

Abstract. In this paper, we investigate the effect of using several color and texture features in query expansion. We assume that the initial queries are text-based and are matched against the text annotations. The system performs a two-pass retrieval by expanding the initial query using text and image features from the initial retrieved set, assuming that it contains more relevant images than non-relevant ones. We performed extensive experiments using 1019 color pictures with captions, relevance judgments and queries supplied by a national archives agency.

1 Introduction

With the collections of digital images growing at a rapid rate, providing easy access to image database has become a significant service. Unlike text retrieval [1] which has already been used widely and successfully, image retrieval is a relatively new research area.

A considerable number of image retrieval systems [2, 3, 4] rely solely on text annotations found in titles, authors, captions and descriptive labels. In these systems, retrieval is based on computing the overlap of the query terms with the text annotations of the images. Nevertheless, it has been reported [5] that users querying an image collection tend to be much more specific in their requests and information needs, than when querying a text database. Moreover, most text annotations tend to be short. These imply that a simple term-by-term match between query terms and annotations terms may not be effective in some applications.

In contrast, image retrieval systems that are based purely on image features are still in the preliminary stages of development. As image processing and computer vision techniques that provide domain independent recognition of image content are still in their infancy, these systems rely mainly on low-level image features. Being poor descriptors of perceptual information and semantic similarity, these low-level features have exhibited limited effectiveness in general retrieval tasks (such as "find images containing computers and printers"). In fact, researchers from the computer vision community have shown [6] that in the absence of *a priori* knowledge or an imposed model, most low-level vision tasks are ill-posed problems.

It is generally believed that the combination of text annotations and image features holds the best promise in image retrieval. Several attempts in this direction include using image texture to generate text annotations [7], and combining text captions with image understanding modules to identify human faces [8].

This paper presents our experiments on combining text annotations and several raw image features in image retrieval. We adopt a two-pass retrieval approach. In the first pass, users specify their information needs using text-based queries and the system retrieves an initial list of candidate images. In the second pass, the system uses the text annotations, color and texture features of these candidate images to expand the initial query and further refine the retrieval. We believe that this approach allows the users to interact with the retrieval system in a natural and intuitive manner, while retaining the ability to exploit image features.

The paper is organized as follows: Section 2 describes the image features used in this study. Section 3 describes the experimental set-up and the image collection. Sections 4 and 5 present the results, discussions, conclusion and future works.

2 Image Features for Query Expansion

The primary objective of the experiments is to investigate the effectiveness of using raw image features in query expansion as compared to text annotations. We have chosen four different image features in color, grayscale and texture for our experiments. These are described in the subsequent paragraphs.

2.1 Color Histograms

Color histograms are obtained by summing the number of pixels having similar values in the RGB (Red, Green, Blue) components. Let $\hat{h}_r(i)$ be the number of pixels with R value equal to i. Similarly, let $\hat{h}_g(i)$ and $\hat{h}_b(i)$ be the corresponding numbers for the G and B components. Also, $i = 0, 1, 2, \ldots, N_h - 1$ where N_h is the total quantization steps and is equal to 256 in an 8-bit/component display system.

We can reduce the length of the histogram features by using only the most significant bits of the RGB component values. For example, to obtain $N_h = 2^p$, we use only the p most significant bits. This allows us to study the effect of using 4, 8, 16, 32 and 256 quantization levels which correspond to $p = 2, 3, 4, 5$ and 8 respectively. The overall histogram feature vector can be represented by **h**.

2.2 DC Coefficients from Discrete Cosine Transform

Image retrieval based on JPEG or MPEG compressed data has been studied by many researchers [9, 10], where discrete cosine transform (DCT) coefficients of 8×8 blocks are commonly used as the image features. In our experiments, we use the zero-frequency (DC) components of DCT (of the luminance component) as the query expansion feature.

Let N_c be the total number of DCT blocks in an image. Because of the large resolution (768 × 512) of our images, there are too many 8 × 8 blocks, *i.e.*, $N_c = 96 \times 64 = 6144$ blocks. Grouping 16 such blocks (4 in horizontal direction and 4 in vertical direction) and averaging their DC coefficients, we obtain a feature vector **c** of length $N_c = 24 \times 16 = 384$. We also use $N_c = 12 \times 8 = 96$ and $6 \times 4 = 24$ in our study.

2.3 Multiresolution Simultaneous Autoregressive Models

Multiresolution Simultaneous Autoregressive (MRSAR) [11] is a texture feature used by Photobook [12] to characterize spatial interactions among neighboring pixels. It computes texture features for a second-order simultaneous autoregressive model (on grayscale component) over three scales (resolution levels 2, 3, 4) on overlapping 25 × 25 subwindows.

Each SAR model can be expressed by five model parameters, thus the fetures from the multiresolution (three levels) SAR model are appended to form the feature vector **s** of length 15.

2.4 Local Binary Pattern

Local Binary Pattern (LBP) is a simple texture feature originally proposed by Wang and He [13]. Each of the eight connected neighbors in a 3 × 3 window are compared with the center pixel. The comparison result for each center pixel is an 8-bit binary number. Based on these 8-bit binary numbers, a histogram of 256 bins is constructed and used as the LBP feature vector, **b**.

3 Experimental Set-Up

3.1 Image Collection

The image collection consists of 1019 pictures supplied by a national archives agency. The image resolution is either 512 × 768 or 768 × 512. Each image is assigned one or more categories by archivists during the acquisition process. A total of 51 categories exist in the image collection, and these category headings formed the basis of our relevance judgment. Each image is also accompanied by a short text caption of approximately 70 to 300 bytes which is supplied by the archivists. After discarding stopwords, there are 1161 unique index terms in the collection. 51 sets of keywords which constitute the initial queries to the system are also supplied by the archivists.

3.2 Object Representation and Query Expansion

Each image is represented as a composite vector as follows:

$$\mathbf{d} = \begin{bmatrix} \mathbf{t}/|\mathbf{t}| \\ \mathbf{x}/|\mathbf{x}| \end{bmatrix} \qquad (1)$$

$\mathbf{t}$ is the text feature vector: $\mathbf{t} = [f(1), f(2), \ldots, f(N_t)]^T$, where $f(j)$ is a function of term frequency for term j. N_t is the total number of unique text terms in the collection, and is equal to 1161 in the image collection used.

In the same token, $\mathbf{x}$ is the image feature vector. $\mathbf{x}$ will be substituted by $\mathbf{h}$, $\mathbf{c}$, $\mathbf{s}$ or $\mathbf{b}$ depending on the feature used.

The initial query $\mathbf{q}_0$ is similarly represented as a vector where

$$\mathbf{q}_0 = \begin{bmatrix} \mathbf{t}_0/|\mathbf{t}_0| \\ 0 \end{bmatrix} \qquad (2)$$

since all initial queries are text-based.

The expanded query is given by

$$\mathbf{q} = \mathbf{q}_0 + \alpha \frac{\mathbf{f}^m}{|\mathbf{f}^m|} \qquad (3)$$

where $\mathbf{f}^m$ is the feedback vector using the the top m images retrieved from the initial query, and α is a weighting factor.

3.3 Evaluation

Similarity between an image $\mathbf{d}$ and a query $\mathbf{q}$ is given by $s = \mathbf{d} \cdot \mathbf{q}$. For each query, images are ranked in descending order of similarity.

We used the standard evaluation mechanism in the series of TREC evaluations for document retrieval [14]. This includes *recall*, a measure of the ability of a system to present all relevant items, and *precision*, a measure of the ability of the system to present only relevant items.

Finally, the term *average precision* refers to the 11-point average value of precision for recall levels $0.0, 0.1, 0.2, 0.3, \ldots, 1.0$.

4 Results and Discussions

4.1 Baseline Experiments

We first performed a baseline experiment using only the initial query and no expansion. The 11-point average precision over 51 initial queries is 0.310400.

We then compute the mean of the various feature vectors using only the images that are relevant to each of the 51 queries. Thus we obtain five different query vectors, each being the mean obtained from different features ($\mathbf{t}$, $\mathbf{h}$, $\mathbf{c}$, $\mathbf{s}$, $\mathbf{b}$). The average precisions obtained are shown in Table 1.

Table 1. Average precisions using mean feature vectors of relevant images

Feature type	Text	Histogram	DC	MRSAR	LBP
Avg. precision	0.851940	0.203158	0.224745	0.171518	0.192284

The table clearly shows that text captions are very good in identifying the relevant images. It also suggests that all the image features used, including the very expensive MRSAR, are poor discriminators in general image retrieval tasks.

4.2 Experiments on Query Expansion

In these experiments, we perform a two-pass retrieval by first retrieving an initial list of candidate images using text queries. We then expand the query using the top m documents and weights α as shown in equation (3).

Effects of Varying Number of Feedback Images. Figure 1 shows the results where α is fixed at 1.0, and m is varied.

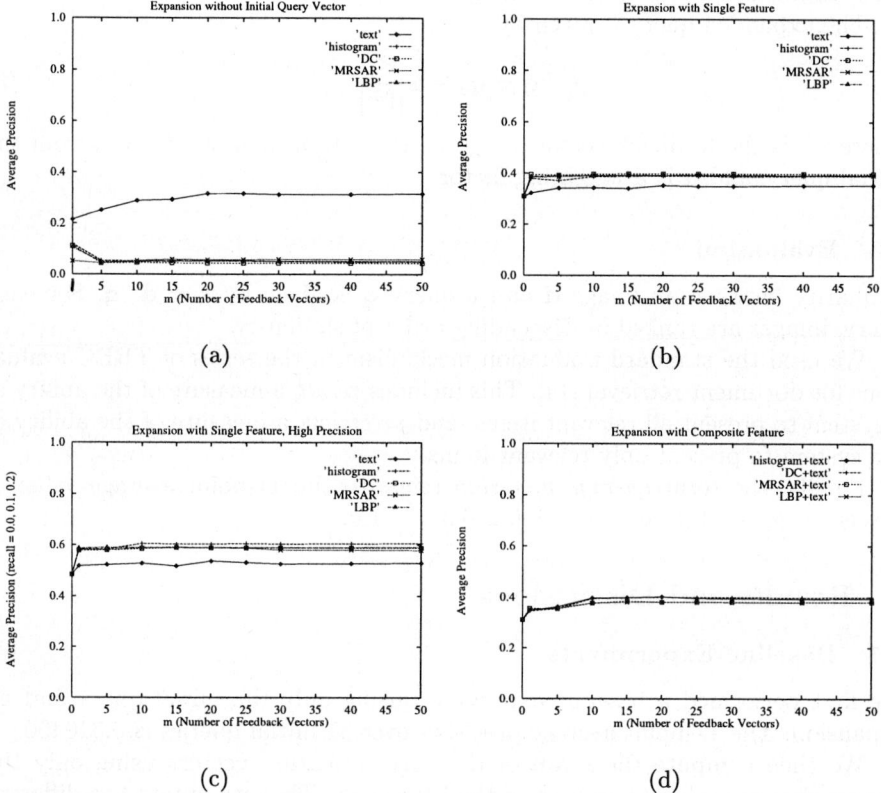

Fig. 1. Query expansion using text and image features by varying the number of images used in feedback

Figure 1(a) shows the results obtained when the expansion are performed without using the initial query vector, *i.e.*, $\mathbf{q}_0$ is omitted in equation (3). This is similar in concept to the experiment that produced Table 1, except that we now assume the top-m retrieved images are relevant.

For query expansion using the text feature, the results are generally in agreement with other similar experiments in text retrieval, *i.e.*, results are improved with more images used for feedback, although there is a diminishing rate of return. Using 50 images to expand the queries, the result is improved by about

48% compared to using one image. The graph also shows that the four image features perform poorly when used by themselves in query expansion, which is in agreement with the results in Table 1.

Figure 1(b) shows an interesting result when text or image features are combined with the initial query during expansion. All image features performed substantially better when compared with text annotation. This shows that while the raw image features performed poorly when used by themselves, they improve averaged precision more than text annotation when used in query expansion.

Figure 1(c) shows that the above findings hold even if the average precisions are computed at the high precision region only (recall = 0.0, 0.1, 0.2).

Figure 1(d) shows the results when both text and image features are used in query expansion. Comparing with Figure 1(b), it shows that using additional text captions in query expansion did not improve the results.

Effects of Varying Weight of Expansion Vector. We also investigate the effects when m is fixed at 10 and α is varied.

Figure 2(a) shows that when the values of α are very small ($\alpha < 0.5$), there is very little performance difference for all the features. When the expansion vector is as important as the initial query ($\alpha = 1.0$), image features perform better than text when used in feedback. With larger values of α, however, the performance of image features deteriorate rapidly. This is equivalent to using an image feature as the primary retrieval vector. With larger α values, the two texture features, MRSAR and LBP, appear to yield better performance than the color and grayscale features. Figure 2(b) confirms that the performance is consistent at high precision levels.

Figures 2(c) and (d) show the performance of the composite feedback features (text and image) under different weighting conditions. There appears to be a weak maximum at low values of α ($\alpha < 1.0$), but all composite features produce similar performance.

Effects of Varying Acquisition Parameters of Image Features. We have performed similar experiments by varying various parameters in the acquisition of image features. These include varying the quantization levels of the color histograms $N_h = 4, 16, 32, 256$, and the number of DCT blocks $N_c = 96, 384$. We found that the results are similar to those reported in Figure 1 and Figure 2 which were obtained with $N_h = 8$ and $N_c = 24$.

5 Conclusion and Future Works

We have presented a series of experiments on query expansion using text, color, grayscale and texture features. The following are some important findings:

- By themselves, raw image features cannot infer semantic similarity. This is evident from the poor results when initial text-based queries are discarded.

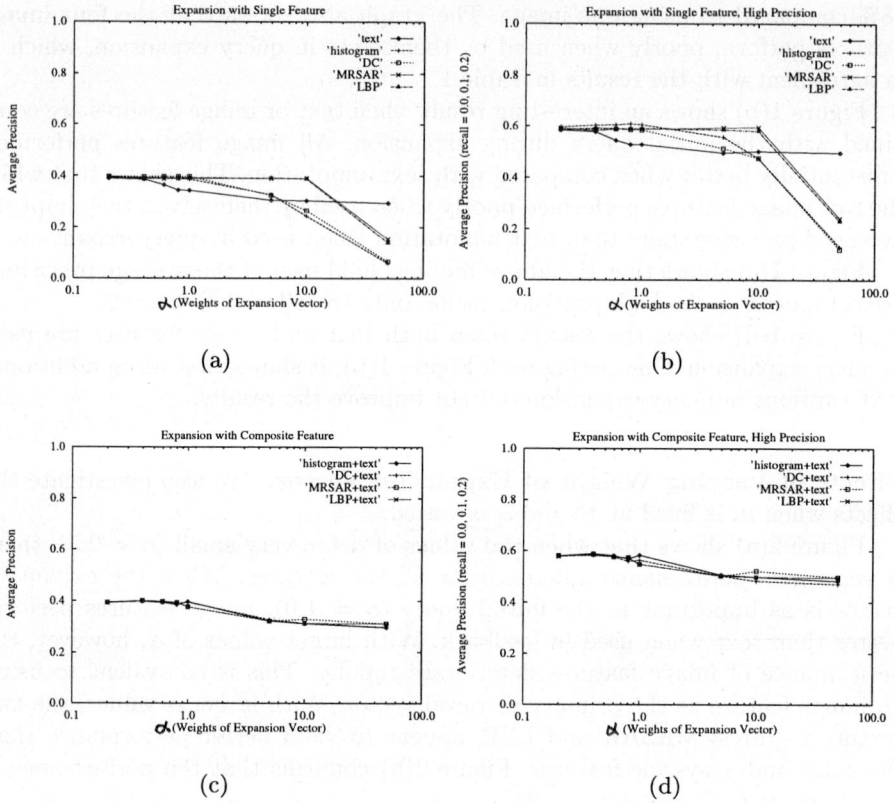

Fig. 2. Query expansion using text and image features by varying weights of expansion vector

- Using image features for query expansion improves the results more significantly than using text annotations. The average improvements are about 25% as compared to about 10% for text.
- The above observations hold at high precision levels, as well as at all recall levels.
- The more expensive image feature (MRSAR) did not produce substantial improvements in performance over other simpler image features.
- The performance are not sensitive to the acquisition parameters of the image features.
- Combining image features and text in query expansion (together with the initial query vector) yield better results than using text only. However, there is no significant difference compared to using image features only in query expansion.

These findings have been obtained using an image collections of 1019 pictures. Certainly we do not claim that they will generalize to all types of image collection. Nevertheless, we believe the above findings are significant enough to warrant further investigation. Our future works will consist of the following:

- Investigate the reasons behind the behavior of image features in query expansion;
- Allow other researchers to verify the findings and perform related experiments by making this image collection widely available.

References

1. G. Salton, *Automatic Text Processing*, Addison-Wesley, 1989.
2. A.F. Smeaton and I. Quigley, "Experiments on Using Semantic Distances Between Words in Image Caption Retrieval," *Proc. 19th ACM SIGIR Conf. on R&D in Information Retrieval*, Zurich, pp. 174-180, 1996.
3. H. Zhou, S. Chan and Kok F. Lai, "Multilingual Information Retrieval System," *Multimedia Storage and Archiving Systems*, SPIE Proc Vol. 2916, Boston, 1996, pp. 33-45.
4. H. Chen, B. Schatz, T. Ng, J. Martinez, A. Kirchhoff and C. Lim, "A Parallel Computing Approach to Creating Engineering Concept Spaces for Semantic Retrieval : The Illinois Digital Library Initiative Project," *IEEE Trans. Pattern Analysis and Machine Intelligence*, 18(8), 1996, pp. 771-782.
5. P.G.B. Enser, "Query Analysis in a Visual Information Retrieval Context," *Journal of Document and Text Management*, 1(1), 1993.
6. T. Poggio and V. Torre, "Ill-posed Problems and Regularization Analysis in Early Vision, " *Proceedings of AARPA Image Understanding Workshop*, 1984, pp. 257-263.
7. R. W. Picard and T. P. Minka, "Vision Texture for Annotation", *Multimedia Systems*, Vol 3, 1995, pp. 3-14.
8. Rohini K. Srihari, "Automatic Indexing and Content-Based Retrieval of Captioned Images," *IEEE Computer*, September 1995, pp.49-56.
9. M. Shneier and M. Abdel-Mottaleb, "Exploiting the JPEG Compression Scheme for Image Retrieval," *IEEE Trans. Pattern Analysis and Machine Intelligence*, 18(8), 1996, pp. 849-853.
10. V. Kobla, D. Doermann and K. Lin, "Archiving, Indexing, and Retrieval of Video in the Compressed Domain," *Multimedia Storage and Archiving Systems*, SPIE Proc Vol. 2916, Boston, 1996, pp. 78-89.
11. J. Mao and A.K. Jain, "Texture Classification and Segmentation Using Multiresolution Simultaneous Autoregressive Models," *Pattern Recognition*, vol.25, No.2, 1992, pp.173-188.
12. A. Pentland, R. W. Picard and S. Sclaroff, "Photobook: Tools for Content-Based Manipulation of Image Databases," *International Journal of Computer Vision*, 18(3), 1996, pp233-254.
13. L. Wang and D. He, "Texture Classification using Texture Spectrum," *Pattern Recognition*, Vol.23, No.8, 1990, pp. 905-910.
14. D.K. Harman, "The First Text REtrieval Conference (TREC-1)," *Information Processing and Management*, 29(4), 1993, pp. 411-414.

Montage : An Image Database for the Fashion, Textile, and Clothing Industry in Hong Kong

Tak Kan Lau and Irwin King

Department of Computer Science and Engineering
The Chinese University of Hong Kong
Shatin, New Territories, Hong Kong
{tklau,king}@cse.cuhk.edu.hk

Abstract. The fashion, textile, and clothing industry is a main constituent in Hong Kong. In this industry, handling a large amount of images is an important task in various phases, for example, the designing, sourcing, and merchandising phase. We develop an image database system called, Montage for managing and retrieving these visual information efficiently and effectively. Montage is an image database supporting content-based retrieval by *color histogram*, *sketch*, *texture*, and *shape*. One important feature of Montage is the *Open Architecture* design which makes the system *extensible*, *customizible*, and *flexible*. There are two aspects of this open architecture design: (1) Open DataBase Connectivity (ODBC) and (2) plug-in framework which we will discuss in more details. Moreover, we describe an experimental Java system enabling internet access to Montage. In the paper, we also present an experiment to evaluate the performance of several query methods.

1 Introduction

The fashion, textile, and clothing industry is a main constituent in Hong Kong's manufacturing sector which is one of the largest employers in Hong Kong. In this industry, handling a large number of images is an important issue in various phases such as the designing, sourcing, and merchandising phase. For example, fashion designers often refer to previous designs from a large collection of designer sketches when creating new designs; consumers and retailers have to purchase apparel merchandise from catalogs.

In the past, we use traditional databases to handle images. These databases use keywords for image retrieval which poses difficulties for the end users without special training. First, different users may use different words to describe an image. Second, even when a standardized vocabulary is used, it is still hard to depict the image clearly and precisely.

In order to manage such a large amount of images efficiently and easily, image databases [4, 1, 5] are emerged. These databases support content-based retrieval which lets us retrieve images by image content. By using visual information, we can describe images more accurately and easily.

We develop an image database system named Montage for the fashion, textile, and clothing industry in Hong Kong on the Windows NT and Windows 95 platforms. The goals of the system are:

- *Open Architecture Design*: Open architecture allows users to customize or extend the system to their own needs. *Open DataBase Connectivity* (ODBC) and *plug-in framework* are the two main aspects in the design.
- *Internet Solution*: An experimental Java system enables internet access to Montage. The system becomes platform independent by using Java applet programming. Moreover, users from all over the world may retrieve images in the server database.
- *Content-Based Retrieval*: Montage supports content-based image retrieval by using color histogram, sketch, texture, and shape. Users can retrieve images easily and efficiently in the system by using image content.

In the rest of this paper, we discuss the system architecture and the internet solution of Montage in Section 2 and Section 3 respectively. Section 4 describes the details of the query methods. Section 5 shows how to use classification trees to pre-organize the images. We present an performance experiment on the query methods in Section 6. In section 7, we discuss some possible applications of Montage for the industry. A conclusion is drawn in Section 8.

2 System Architecture

Montage uses an open architecture system design. It contains three main layers: *Core Layer*, *Extension Layer*, and *Application Layer*. Each layer consists of different modules as shown in Fig. 1(a). First, the core layer contains all the necessary components of the system for basic image processing and retrieval. Second, the extension layer contains modules for system customization. Third, application programs can be built on top of the system in the application layer.

2.1 Catalog Module

Catalog module is responsible for database manipulation such as creating new catalog, inserting images into a catalog, and retrieving images from a catalog. It uses a master file to store all the records of an image catalog. Moreover, the master file also keeps the header information of the file of the thumbnail images extracted from the database images for later display.

2.2 Image Processing Module

Image processing module is responsible for the image processing operations for the system. It provides basic image processing functions such as image transformation, filtering, resizing, color adjustment, 2D object drawing, and the object-oriented operations. With these functions, users can edit images stored in the database and create new images.

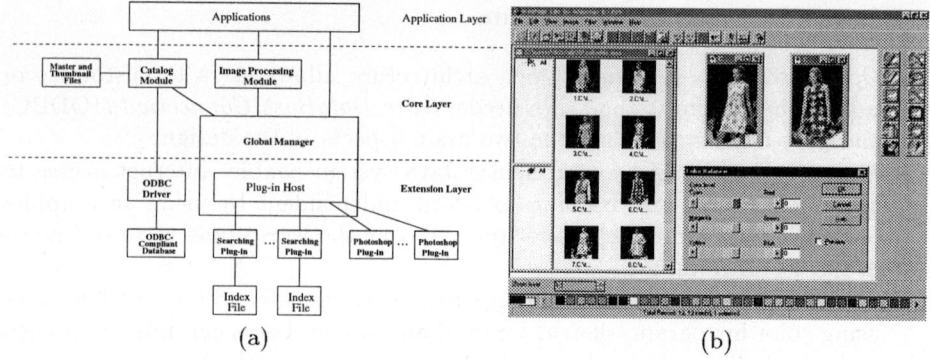

Fig. 1. (a) The system architecture. (b) Image editing and image browsing.

2.3 Global Manager

Global manager communicates with the main modules of the system. It is in charge of global resource management. It also takes part in the plug-in framework of the system. The main tasks of global manager are: (1) keeping the global status of the system such as current foreground color, (2) managing the global utilities such as MeasureUnit, (3) supplying global user interface such as tool boxes and color palettes, and (4) providing a plug-in host environment.

2.4 ODBC driver and the ODBC-compliant Database

Open DataBase Connectivity (ODBC) is a uniform interface standard used to access ODBC-compliant databases. By using a specific ODBC driver, ODBC-compliant databases can be accessed by the same set of ODBC calls.

In Montage, we use an ODBC-compliant commercial database for alphanumeric data manipulation. We use ODBC to access the database which is used for non-image related processing and image query by keywords. ODBC is useful for the users who already have their own databases to manage image data. By using ODBC, those users can retain all the data in our system and are able to retrieve images by features. Moreover, our system gains a set of text-related operations. In short, Montage complements other ODBC-compliant databases.

2.5 Plug-in Host and Plug-in Modules

We use a plug-in framework to make Montage extensible, maintainable, and flexible. The plug-in framework has two main components: *Plug-in Host* and *Plug-in Modules*. We can enhance or customize the system by means of plug-in modules. Plug-in host is used to provide the necessary system interfaces for the plug-in modules to interact with Montage.

Montage supports two main kinds of plug-ins :

- *Photoshop Plug-ins* : Many Photoshop plug-ins can be found on the market and they provide various image processing functions. We can enhance the image processing module by adding these plug-ins to Montage. As a result, we can reduce the functionality of the image processing module and leave most of the operations to plug-ins. Hence, the system becomes more flexible.
- *Searching Plug-ins* : A searching plug-in corresponds to an image searching method (see Section 4 for details). By adding searching plug-ins, new searching methods can be used by the system. Basically, each searching plug-in is in charge of the following tasks:
  - *Extracting feature*: When a new image is inserted into the database, the catalog module calls all the loaded searching plug-ins to extract specific features from the image for later retrieval.
  - *Indexing*: In order to reduce access time and narrow down the search space in image searching, different indexing methods such as R-tree [3] are used in different searching plug-ins.
  - *Searching images*: The searching plug-in calculates search keys for the queries of the searching method. It compare the features of the images in the database with the search keys to produce the results of the queries.
  - *Providing user interface*: Each searching plug-in provides the user interface to specify queries for the searching method (see Fig. 2).

2.6 An Application Program

There is an application program built in the application layer. Basically, it tightly integrates two sub-programs: *Image Catalog* and *Image Editor*. Image catalog provides a user-friendly interface to access the image database whereas image editor allows users to modify images in the database and create new images. Using this application program, users can browse and edit images at the same time (see Fig. 1(b)). This feature is very essential. For example, fashion designers can refer to the previous designs while they are creating new designs.

3 Internet Solution

The Internet is growing rapidly in recent years and has become one of the best ways for people to communicate around the world. Hence, it is essential to make Montage internet accessible.

Apart from the main system described in Section 2, we are prototyping an experimental Java system to enable internet access to Montage. By using Java applets, the system becomes platform independent. In the experimental system, the searching engine and the image database are located at the server side. Users from the client sides may use Java-compliant WWW browsers to execute the Java applets and specify queries for remote image retrievals. The queries are then sent to the server for processing and the results are returned and displayed on the browsers.

4 Query by Image Contents

4.1 Query-by-Color-Histogram

Query-by-color-histogram uses the global color distribution of an image for image retrieval (see Fig. 2(a)). We make use of RGB and Hue values of the pixels to calculate a 46-bucket color histogram for each image in the database. Then, the histogram will be indexed by R-tree. After pre-processing all the images, we can specify a query image and perform nearest-neighbor search to find images with overall color distribution similar to the query image.

4.2 Query-by-Sketch

The query-by-sketch method makes use of the regionalized color information of the images for retrieval. For each image in the database, we first partition the image into 400 non-overlapping (20 × 20) same-size partitions. Then, we quantize the RGB color space evenly from 16.7 million to 4096 colors. We find the modes R_m, G_m, B_m, and a Hue value calculated from the three modes for each partition. An 1,600 feature vector is then formed by collecting the modes and the Hue values from all the partitions of the image.

A sketch pad is used to specify sketch queries as shown in Fig. 2(b). Based on the color assigned to each cell, we calculate a key feature vector for an query according to the method mentioned in last paragraph. We then compare the key feature vector with the feature vectors of the images in the database to find out the result of the query.

4.3 Query-by-Texture

The query-by-texture method uses statistical methods to compute textural features for texture matching (see Fig. 2(c)). For each image in the database, we first quantized the intensity space to 16 gray levels. Then, we compute six features: *smoothness SM, angular second moment ASM, contrast CON, correlation COR, inverse element difference moment $IEDM$*, and *entropy EN* form the image [6, 2]. Based on the six values, we calculate a representative feature value F for the image by the formula $F = \lambda_1 \cdot SM + \lambda_2 \cdot ASM + \lambda_3 \cdot CON + \lambda_4 \cdot COR + \lambda_5 \cdot IEDM + \lambda_6 \cdot EN$ where $\lambda_1, ..., \lambda_6$ are the pre-defined weighting factors. Given a query texture, its extracted feature value will compare to the feature values of the images in the database. Images having smaller differences in feature values are selected as the result of the query.

4.4 Query-by-Shape

The query-by-shape method uses the shape information in an image for retrieval (see Fig. 2(d)). For each image, the method uses the user-specified boundary points of the shapes in the image to calculate a feature vector by using Fast Fourier Transform. A user may then draw a polygon as an query to find the images with shapes similar to the query polygon.

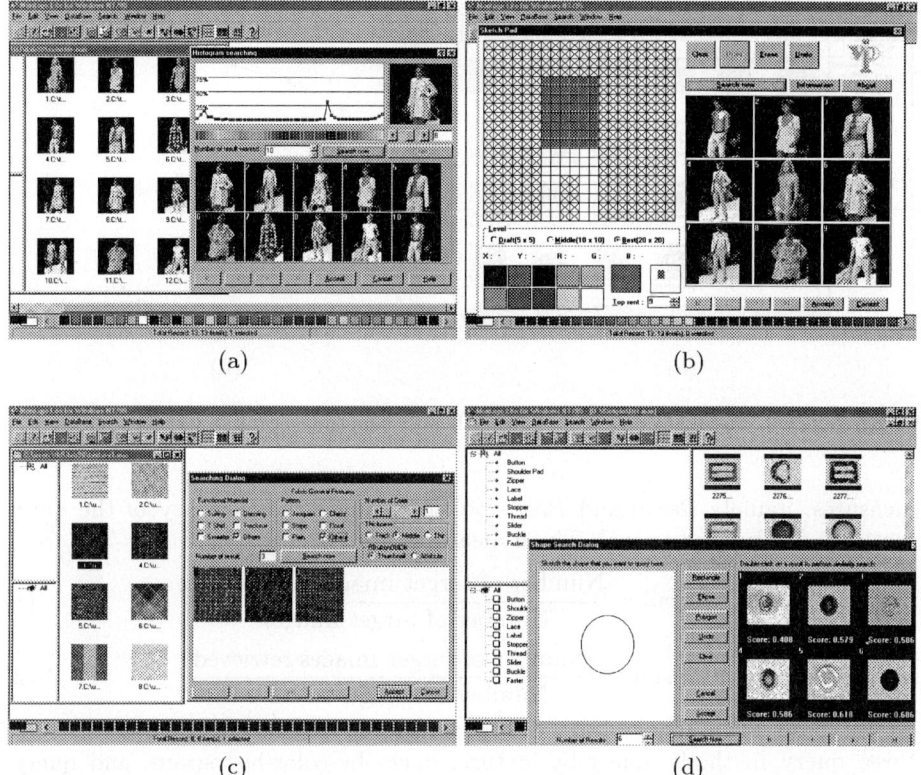

Fig. 2. (a) Query by color histogram. (b) Query by sketch. (c) Query by texture. (d) Query by shape.

5 Classification Tree

One way to allow users to pre-organize the image information is the use of classification trees (see Fig. 3(a)). Each image catalog keeps a classification tree for image classification. A classification tree is a hierarchy of image classes. Every image in the catalog belongs to one or more user-defined classes. With the classification tree, images retrieval become more efficient. If we want to find some images belonging to certain classes, we can first specify the classes in the classification tree in order to reduce the search space. Searching can then be performed on a smaller database to obtain an accurate result faster.

6 Performance Experiment

We present an experiment on Montage to evaluate the performance of several query methods on a fabric database. It was conducted on a 32M-RAM Pentium-100 PC using the Windows 95 operating system. We used two performance

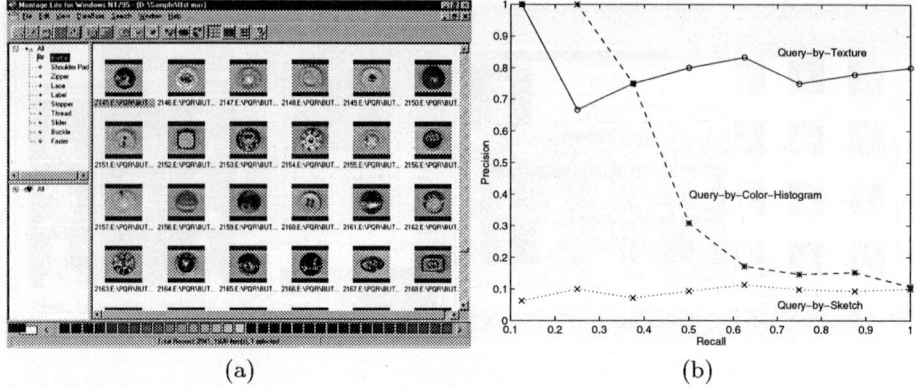

Fig. 3. (a) A classification tree. (b) Recall and Precision results of the experiment.

measures, namely *Recall* and *Precision*, to evaluate the accuracy of the query methods for image retrieval. These measures are given by :

$$\text{Recall} = \frac{\text{Number of target images retrieved}}{\text{Number of target images}} \quad (1)$$

$$\text{Precision} = \frac{\text{Number of target images retrieved}}{\text{Number of images retrieved}} \quad (2)$$

We used a database of 100 fabric images to evaluate the performance of three query methods: query-by-texture, query-by-color-histogram, and query-by-sketch for texture searching. We did not test the query-by-shape method because it was obviously not suitable for this kind of database. All the images were in GIF format with 128 × 128 pixels. We tested the methods to retrieve 8 pre-selected target fabric images. In the experiment, we selected one of the target images as the query for the first two query methods. Based on this target image, we specified a sketch query for query-by-sketch. It took 130 seconds to pre-processing the 100 images. Their Recall and Precision performance was shown in Fig. 3(b).

For an ideal case, the precision would be 1 for every recall. For recall values over 0.5, both query-by-color-histogram and query-by-sketch have precision values less than 0.4. Therefore, they are not good for texture searching. For query-by-texture, we can see that the precision is up to 0.8 even for high recall values. Hence, it was a good method for locating similar fabric images.

We did similar experiments on different kinds of databases and found that different query methods are good in different cases. Therefore, Montage included the four query methods for different applications and users.

7 Applications

Montage is a content-based image retrieval database and it is potentially useful for the fashion, textile, and clothing industry. Two possible applications are:

- *Office Automation*: Companies may manipulate and manage their visual information in electronic format rather than in hard copy format. In the fashion industry, for example, fashion designers may store their previous designs in Montage so that they may refer to these designs quickly and easily. Moreover, they may create new designs at the same time.
- *Global Sourcing*: Global sourcing is an important issue in the textile and apparel industry. Manufacturers often need to find different suppliers from a large printed catalog for different resources used in the manufacturing phase. Using Montage to organize the suppliers' information, manufacturers can make use of the graphical interface to specify queries for effective retrievals.

8 Conclusion

We have developed an image database system called, Montage for the fashion, textile, and clothing industry in Hong Kong. It supports content-based retrieval by color histogram, sketch, texture, and shape. The system provides an open architecture for the third-party developers by means of ODBC and plug-in framework. We have also implemented an experimental Java system for internet access to Montage. With these features, Montage is potentially useful for the industry.

9 Acknowledgment

This work is supported in part by Hong Kong's Industry Grant #AF/17/95 (2427 01300) and a RGC Grant #CUHK 485/95E(CU95513). The authors would like to thank Prof. Ada Fu, Prof. Laiwan Chan, Prof. Lei Xu, and all the members of the VIP Lab (http://www.cse.cuhk.edu.hk/~viplab) for their contributions to the system described here.

References

1. J.R. Bach, C. Fuller, A. Gupta, A. Hampapur, B. Horowitz, R. Humphrey, R. Jain, and S. Chiao-Fe. The Virage Image Search Engine: an open framework for image management. In *Proceedings of the SPIE - The International Society for Optical Engineering*, volume 2670, pages 76–87, 1996.
2. R.C. Gonzalez. *Digital Image Processing*. Addison Wesley, 1992.
3. A. Guttman. R-tree: A Dynamic Index Structure for Spatial Searching. *ACM SIGMOD*, 14(2):47–57, 1984.
4. W. Niblack, R. Barber, W. Equitz, M. Flickner, E. Glasman, D. Petkovic, P. Yanker, C. Faloutsos, and G. Taubin. The QBIC project: querying images by content using color, texture, and shape. In *Proceedings of the SPIE - The international society for optical engineering*, volume 1908, pages 173–87, 1993.
5. J.R. Smith and S. F. Chang. VisualSEEk: a fully automated content-based image query system. In *ACM Multimedia '96*, November 1996.
6. F. Tomita and S. Tsuji. *Computer analysis of visual textures*. Kluwer Academic Publishers, Boston, 1990.

Auto Cameraman Via Collaborative Sensing Agents

Qian Huang, Yuntao Cui, Supun Samarasekera, and Michael Greiffenhagen

Siemens Corporate Research, 755 College Road East, Princeton, NJ 08540, USA

Abstract. In this paper, we propose a new multiple sensing agent based scheme for an *automated cameraman*. It is capable of 1) constantly monitoring the visual events in a global surrounding, 2) dynamically, based on the detected visual events, determining the monitoring strategy. These heterogeneous agents are coupled in a unique way to work not only *asynchronously* but also *collaboratively* via a *facilitator*. Such collaborative behavior leads to more effective solutions to some of the very difficult problems such as occlusion.

1 Introduction

With the fast growing sensing and video technologies, it is increasingly becoming more and more economical to use video-based systems in domains such as surveillance, monitoring, security, telecommunications, and remote services. One important feature of an automate surveillance system is to track the moving objects.

In the literature, most tracking systems rely on a single visual sensor, either a stationary camera [5, 10, 12] or a camera which actively pursuits the target [6, 7, 11] to obtain the motion information. Because the amount of information provided by a single visual sensor is usually local and limited, these approaches often have difficulties.

There are also multiple sensor based tracking systems in the literature. These approaches introduce redundancy among sensors to increase the reliability [1, 2, 8, 9]. Individual sensors in these systems work individually in a local way (in the sense that they do not communicate with each other). They cooperate in a very loose fashion: a central unit collects the votes from different sensors and makes final decisions using some kind of loss minimization scheme.

In this paper, we propose a scheme which uses multiple sensors. The overall system is a decentralized system like [9]. While the tracking mechanism consists of multiple sensing agents that can run in parallel in an asynchronous mode, even more importantly, they are uniquely coupled during tracking.

2 Overall Setup

In this section, we give an overview of our decentralized yet collaborative scheme based on multiple heterogeneous sensing agents. Two different visual sensors are used: *a peripheral sensor* with fish-eye like lens that takes care of the global monitoring/tracking tasks and *a foveal sensor* that is capable of pan, tilt, and zoom that takes care of the focused monitoring/tracking tasks. Fig. 1 shows our current lab setup.

A hemispherical mirror is installed on the ceiling facing down the area to be monitored, underneath which, a camera with normal lens is placed, looking into the mirror. The two coupled together serves as a peripheral sensor (which is equivalent to a fish-eye camera but much more cost effective) to capture the overall view of the hemisphere of the space to be monitored. A foveal camera is installed at any convenient locations within the monitored area.

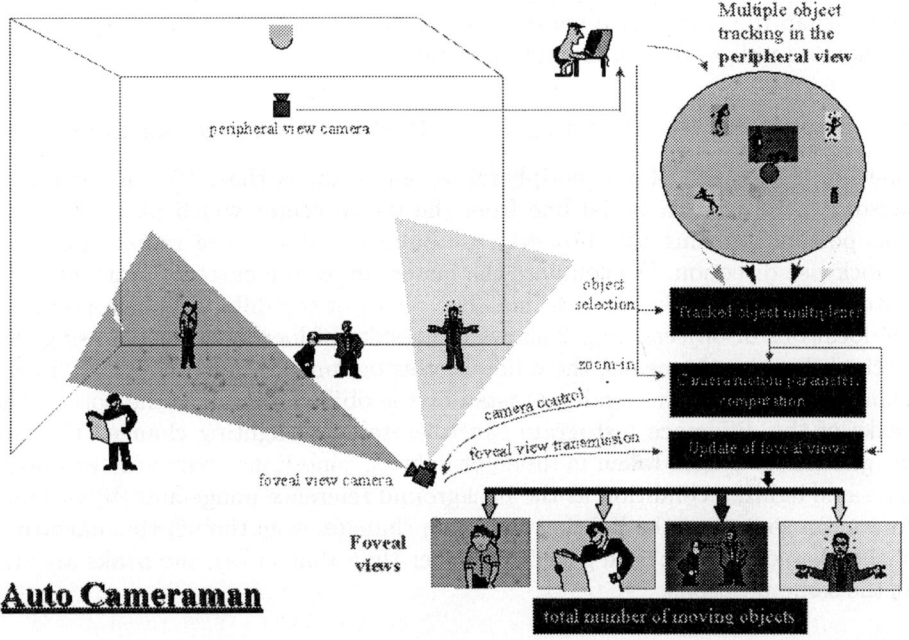

Fig. 1. The overall set up of a peripheral sensor and a foveal sensor.

2.1 Peripheral Agent, Foveal Agent and Facilitator

The major role of a peripheral agent associated with the peripheral sensor is to simultaneously keep track of all moving targets within its view. Fig. 4 gives an example of a peripheral view from a real scene. A real time peripheral tracking mechanism is presented in Section 3.

In general, the foveal agent directs the foveal sensor to follow the moving person based on the tracking result of the peripheral sensor. However, in some special cases, it can track by its own as illustrated in Section 4.

Even though peripheral and foveal agents act independently, they are coordinated by the facilitator whose task is to reach a decision about the actions of its governed sensing agents based on the information currently gathered from them. The details of how the facilitator works are presented in Section 5.

3 Multiple Object Tracking By Peripheral Agent

Since the peripheral camera is stationary, locating people in the scene can be done by a simple background subtraction (e.g. [12]). Most existing approaches compare the current frame with a reference background image pixel by pixel. Instead, our method uses radial profiles. Our experiments show that it is more robust than the pixel-based background subtraction method in dealing with the shadows as well as the small lighting change.

3.1 Localizing People Using Color Profiles Along Radial Lines

One unique feature of the peripheral sensor setup is that, for each standing person, there exists a radial line from the image center which passes through that person. Utilizing this property, we build a histogram of radial profiles in a clockwise direction. We compare the histogram of the current frame with the histogram of the background frame. The peaks of the difference histogram are the locations of objects. Fig. 2 shows the results of localizing people using this method. Besides the peaks, the difference histogram also has some small fluctuation largely due to the shadows casted by the objects as well as the noises. The peaks of the difference histogram can tolerate small lighting change. The two images in Fig. 2 were taken in different lighting conditions, where (a) was using the same lighting condition as the background reference image and (b) was not. As we can see, when the lighting condition changes, even though the magnitude of the histogram in (b) are generally higher than that in (a), the peaks are still at the same positions.

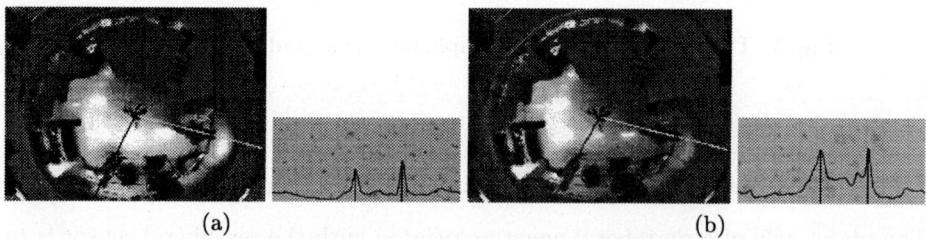

(a)   (b)

Fig. 2. The results of the localization are shown using radial lines (left) corresponding to the peaks in the difference histogram (right). (a) Same lighting condition. (b) Different lighting condition.

3.2 Multiple People Tracking in a Peripheral View

Let p_i^t be the angular position of the ith person at time t and ω_i^t be the angular velocity of the ith person at time t. We employ a simple linear Kalman Filter to track the object [4]. The state vector contains the angular position and the angular velocity (p_i^t, ω_i^t).

We find the measurement p_i^{t+1} using the following procedures. Let $Q = \{q_1, q_2, \cdots, q_n\}$ be the angular positions of people found by our localization scheme as stated in the previous subsection at time $t + 1$. We denote $\hat{p}_i^{t+1} = p_i^t + \omega_i^t$. Then, we have $p_i^{t+1} = q_j$ such that for $\forall k \neq j$, we have $\|\hat{p}_i^{t+1} - q_j\| < \|\hat{p}_i^{t+1} - q_k\|$. This means that p_i^{t+1} is the nearest neighbor of the prediction $\hat{p}_i^{t+1}$ in Q.

4 Tracking By Foveal Agents

The foveal sensor which we use is a Sony EVL-D30 camera. This camera is driven by the final decision of the facilitator to follow the target person. The images acquired by this camera during the pursuit are used directly by the foveal agent to track the object. The tracking is accomplished by tracking the face of the target.

4.1 Locating Human Faces

In our implementation, we use the normalized color model to locate faces as [7, 13], where a learned Gaussian distribution characterized by its mean (μ) and its covariance matrix (Σ) is used to find facial pixels.

While this model characterizes the skin color well, there still exists variation in the mean μ between different races. To overcome this problem, both [7] and [13] let the mean μ (obtained from samples of all races) adapt to the current person. This means that each time a new person appears, the system has to manually locate the face of that person, compute the statistics of that person, and then update the mean μ. In this paper, we present a n ew adaptive approach based on a mean shift algorithm which can avoid the manual initialization.

4.2 Mean Adaptation

Let p be a Gaussian distribution of the normalized red r and green g of the skin color of a person and let μ_g be the mean obtained from the training samples. If we can estimate the gradient of p, we can use the steepest ascent algorithm to move μ_g in the direction of the gradient to the mean of that person.

A simple nonparametric estimation method of the density gradient was presented in [3]. Let $p(\mathbf{x})$ be the probability density function of n-dimensional feature vectors and $S_\mathbf{x}$ be a sphere of radius r centered at $\mathbf{x}$. It has been shown in [3] that the mean of the samples in the sphere $S_\mathbf{x}$ is

$$E(\mathbf{x}|\mathbf{x} \in S_\mathbf{x}) - \mathbf{x} = \frac{r^2}{n+2} \frac{\nabla p(\mathbf{x})}{p(\mathbf{x})}.$$

From the above equation, we know that the vector difference between the local mean and the center of the sphere is proportional to the gradient of the probability density at **x**. Thus, we can shift the center of the sphere along the opposite direction of the difference vector until it converges. Once we update the mean, we can use it to locate the facial pixels.

In order to locate multiple faces in the image, we repeat the mean shift processes. Each time a face is located, we mark off the pixels on this face, so that they would not interfere with the formation of the new mean. Fig. 3 shows the results of locating three faces that have significant variations in their skin colors.

Fig. 3. The results of locating faces are shown using the rectangulars.

5 Collaboration Between Peripheral and Foveal Agents

Both peripheral and foveal agents interact with the facilitator, which coordinates and controls the exchange of messages. The facilitator is designed to handle the following situations:

1) *Ambiguity in the peripheral view*, when more than one persons are standing closely along the same radial line so that they appear overlapped in a peripheral view. In this case, we pass the control to the foveal agent.

2) *Occlusion in the foveal view*, which occurs if both persons and the foveal camera are all aligned in the peripheral view. Then we use the the peripheral agent only.

3) *None of the above*, both results are used.

6 Experimental Results

We conducted our experiments in an indoor office. Fig. 4 shows the view of the office from the peripheral camera. In our experiments, people were allowed to move freely in the office. The experiments were designed to demonstrate the effectiveness of the collaborative behavior between peripheral and foveal agents in tracking under four different scenarios.

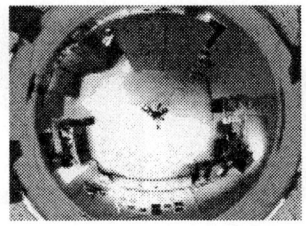

Fig. 4. The view of the office from the peripheral camera.

6.1 Scenario 1: Single Object

This is the simplest case where the room has only the target object. We show the tracking results in Fig. 5. For each time interval, we show the frames from both the peripheral camera and the foveal camera.

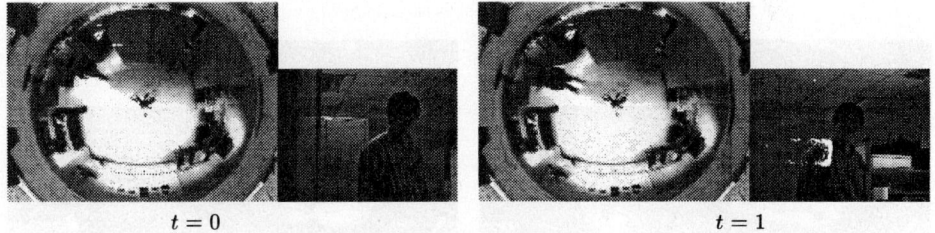

$t = 0$ $t = 1$

Fig. 5. The tracking results when the room has only the target object.

6.2 Scenario 2: Occlusion in the Foveal Camera

This time, the new person walked in front of the target person from the point of view of the foveal camera. The peripheral agent detected the situation and informed the facilitator to disregard the result of the foveal agent. Once the two persons were separated, the tracking returned to its normal routine. Fig. 6 shows the tracking results.

6.3 Scenario 4: Ambiguity in the Peripheral Camera

In this case, two persons were moving in the same radial line in the peripheral view as illustrated in 7 ($t = 3$ and $t = 4$). The peripheral agent detected the ambiguity situation defined in Section ??. The peripheral agent set the confidence of its own decision to zero and passed the control to foveal agent. The results are shown in 7.

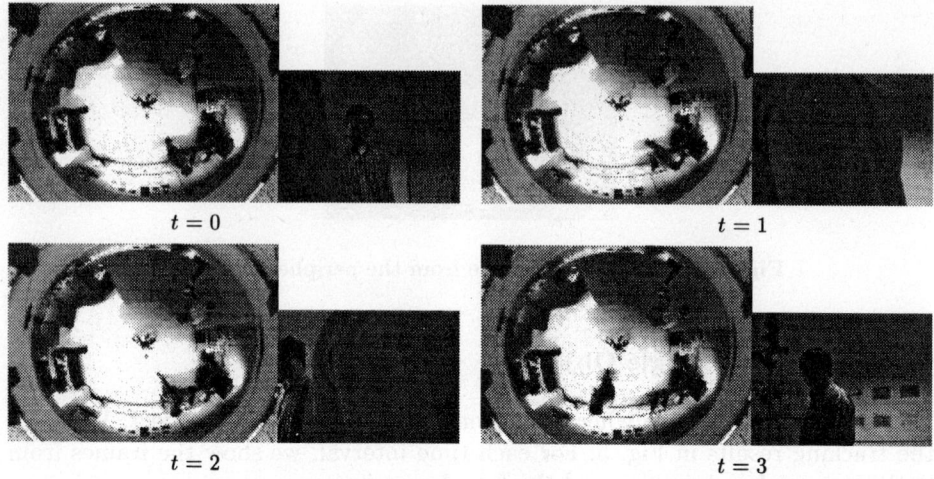

Fig. 6. The tracking results when the target is occluded by the others.

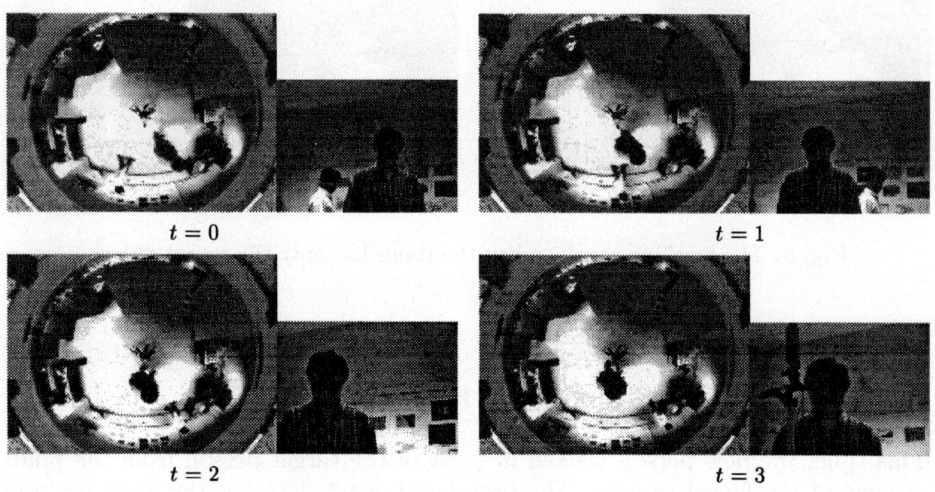

Fig. 7. The tracking results of the ambiguity scenario.

7 Conclusions

In this paper, we describe a novel setup which uses a peripheral sensor and a foveal sensor to perform the indoor monitoring task. The scheme is designed based on the needs from many real world applications in surveillance, telecommunications, distance learning, home care, and transportation.

References

1. P. Allen and R. Bajcsy. Two sensors are better than one: example of vision and touch. In *Proc. of 3rd International Symposium on Robotics Research*, pages 48–55, Gouvieux, France, 1986.
2. A.M. Flynn. Combining ultra-sonic and infra-red sensors for mobile robot navigation. *International Journal of Robotics Research*, 7:5–14, 1988.
3. K. Fukunaga and L.D. Hostetler. The estimation of the gradient of a density function with applications in pattern recognition. *IEEE Transactions on Information Theory*, 21:32–40, 1975.
4. A. Gelb. *Applied Optimal Estimation*. MIT Press, Cambridge, MA, 1974.
5. H. Gu, Y. Shirai, and M. Asada. Mdl-based segmentation and motion modeling in a long image sequence of scene with multiple independently moving objects. *IEEE Transaction on Pattern Analysis and Machine Intelligence*, 18:58–64, 1996.
6. D. Murray, K. Bradshaw, P. McLauchlan, and I. Reid. Driving saccade to pursuit using image motion. *International Journal of Computer Vision*, 16:205–228, 1995.
7. N. Oliver, S. Pentland, F. Berard, and J. Coutax. Lafter: Lips and face tracker. Technical Report 396, MIT Media Lab Perceptual Computing Group, 1996.
8. T.J. Olson and F.Z. Brill. Moving object detection and event recognition algorithms for smart cameras. In *Proc. of DARPA Image Understanding Workshop*, volume 1, pages 159–175, New Orleans, LA, 1997.
9. B.S. Rao and H.F. Durrant-Whyte. A decentralized bayesian algorithm for indentification of tracked targets. *IEEE Transaction on Systems, Man, and Cybernetics*, 23(6):1683–1698, 1993.
10. J. Segen and S. Pingali. A camera based system for tracking people in real time. In *Proc. of International Conference on Pattern Recognition*, volume C, pages 63–67, Vienna, Austria, 1996.
11. T. Uhlin, P. Nordlund, A. Maki, and J.O. Eklundh. Towards an active visual observer. In *Proc. of International Journal Conference on Computer Vision*, pages 679–686, Boston, MA, 1995.
12. C. Wren, A. Azarbayejani, T. Darrel, and A. Pentland. Pfinder: real-time tracking of the human body. Technical Report 353, MIT Media Lab Perceptual Computing Group, 1995.
13. J. Yang and A. Waibel. A real-time face tracker. In *Proceedings of the 3rd IEEE Workshop on Appkications of Computer Vision*, pages 142–147, Sarasota, FL, 1996.

Dynamic Adaptive Data Structures for Semantic Analysis and Synthesis of Video Information

Doc.Nat.Sci., Prof. V.V. Alexandrov, E.V. Laikov, Dr. B.E. Frenkel
St.-Petersburg, Russia, 14-linia 39
St.-Petersburg's Institute for Informatic and Automation

Abstract

The wide range of image processing methods such as wavelets and fractal decomposition or texture analysis usually miss the vital components - the semantic structure.

In our approach to the structuring of video-data, which was realised in SAI (Semantic Analysis of Images) software package, we apply the adaptive dynamic data structure for object fitting hierarchical analysis of video-data. This package presents the new software tools to reveal the interrelated network of context independent semantics from the initial data structure.

The scientific basis of our approach is the localisation of semantically important areas through the new rating principle and following iterative synthesis of hierarchical trees to create the coherent structure of selected fragments. The investigations lead us to the idea that the distribution of the fragments (F) between the levels (l) of this tree correlates with the growth law $F_l \approx l^{-0.618}$, an analogy to the empirical Zipf laws.

Adaptive dynamic data structure, as the result of this process, contains the image fragments which are essentially important for the following processing.

Most evident application of semantic decomposition is the preliminary structurization of video-data for subsequent application of object identification routines and target-oriented discriminating compression of images.

The objectives and main components of the Semantic Image Analysis software.

The semantic decomposition is considered to be one of the most promising branches of image structuring. The Analysis software package has been designed for the object fitting hierarchical analysis of video-data [1]. This package presents the new software tools to reveal the interrelated network of context independent semantics from the initial data structure.

The scientific basis of our approach is the localisation of semantically important areas through the new rating principle and following iterative synthesis of hierarchical trees to create the coherent structure of selected fragments. Adaptive dynamic data structure as the result of this process will contain the image fragments which are important for the person or automatic programs conducting the image processing.

The internal data structures of SAI package is oriented to the practical realisation of the object-fitting method for video-data approximation. The reason for the object-fitting organisation of semantically important areas is a new rating principle for data structuring.

Our implied rating principle is oriented on hyperbolic distribution correlated with the growth low of the developing systems - the universal principle applicable both to the various nature phenomena and the human artwork. The most known manifestation of this principle is the Zipf law for social and economical structures [2].

The application of the most fragment localisation methods requires the initial division of the image into the set of fragments. The main problem at this non-trivial stage of the image analysis is the absence of general methods for the reasonable fragments selection. We use the semantic approach to overcome this difficulty [3,4,5].

The development of SAI package oriented to reveal the independent semantics contained in a source data. The program task to build the contextual-independent hierarchical network of flat and spatial structures distinguishes this package from other classes of software. As the future development of this direction we plan to extend the analysis package with the image synthesis tools. The sum of this methods would allow to apply the new approaches of the image perception and the development of semantically oriented compression system.

The practical application of the SAI programs is defined by its orientation to a problem of effective and economical data storage. The plan to solve this problem by the selection and storing only the semantically important data.

As the result of the image analysis investigations, we developed the iterative method for objects approximation via the multilevel image fragmentation [1,5]. The hierarchy of the source image fragments forms the original adaptive dynamic data structure which could be used to preserve the semantic relations of the independent image parts. The elements of this data structure should be meaningful and recognisable for the human operator or automatic programs working with the image.

For the solution of this problem the following algorithm was used. The source image is presented as the RGB colour-divided matrix. The pixels of the same colour form the fragments of the first level image subdivision.

At the following iterations the program combines the fragments whose colour components is of little divergence, i.e. the difference between them fits into the selected range. The alternative variants suppose the independent structure building for each colour components or grey-scale converted image. The special recursive procedure controls the connections between the fragments and uniform processing of the whole picture.

This process results in the inter-related semantic network of the image fragments that is the pyramidal adaptive structure for dynamic image representation.

The source image I^* is transformed into the specialised computer representation I as the set of $a \times b$ pixels.

For the effective processing it is necessary to change the image structure once more. The image I is presented as the hierarchy of fragments. The m-th level of this hierarchy $L_m = \{F_p | p \in D_m\}$ contains the fragments, defined by the set of fragment indexes D_m, where $I = \bigcup_{p \in D_m} F_p$. The links between the fragments connect them into the indexed tree with specific organisation.

One element of the m-th level will be defined as F_p^m. The fragments of the first level are the separate pixels of image $I : L_1 = \{F_p^1 | 1 \leq p \leq a*b\}$.

The process of fragment construction validates the following equations for the fragments number in different levels:
$$\# L_i > \# L_{i+1}; \ \# L_{255} = 1$$
$$\forall p \in D_m \ \exists q \in D_{m+1} : F_p^m \subset F_q^{m+1}.$$

The object localisation in terms of this structure is realised by the location of the network fragment with suitable node attributes (colour, luminosity, space or more complex fragment characteristics). The essential problem arises when no coherent network fragment corresponds to the needed fragment. This predicament is the main obstacle to the complete automation of the image analysis in this approach.

To overcome this difficulties the SAI package allows the user the compulsory manual restructurisation of the semantic structure in the necessary cases. This possibility could be considered as the way to insert the pragmatic orientation to image analysis.

The main objective of our software is the assistance in the search of practical solutions for general image analysis and task oriented restructurisation of video-data. This possibility is reached by means of the location of the semantic information and following processing of the meaningful image fragments.

The basic elements of SAI package are:
•The method to construct the fragments with the uniform colour and luminosity characteristics.
•The secondary fragment arrangement, allowing to design the hierarchical inter-fragment relations.
•Links rearrangement mechanics, allowing to insert the pragmatic elements.

The perspective applications of this technology are the development of the following procedures:
•Semantic relation analysis with contextual image processing as the task oriented video-compression.
•Tree structure synthesis as the replication of real objects structures.
•The morphological image classification.

The analysis of different computer formats lead us to conclude that most effective formats preserve the natural image structure.

The pixel oriented methods are artificial. They destroy most part of the essential image characteristics - the general composition, 2D and 3D object shapes, image dynamic and contextual information.

The pixel image representation requires the line ordered description of pixel characteristics. The image compression, restoration, analysis, and object location utilities require the different data organisation. This group of image formats includes the vector image representation, fractal compression, wavelets decomposition and the number of special image formats, storing only the important information, the most known instance of this group is the optic character recognition programs, transforming the source image into the lines of text.

According to the large transmissions of information required in the satellite image processing, thematic search in video data bases etc. there is a necessity in the effective automatic methods for image analysis in order select the objects and sets of image fragment which carry the semantic load.

The selection of semantically important components.

The SAI package employs the method for the location of real objects through their iterative approximation by the elements of adaptive dynamic data structure.

The source image is presented as RGB matrix. It is possible to handle the different colour planes in complex or separately. The network of fragments forms through the unification of the fragments with the similar attributes. This procedure defines the hierarchical links between the fragments, which could be noted as the planar or, generally, the spatial relations, the base of the adaptive dynamic structure for image representation. The fragments of the first level are pixels of the source image. We combine two fragments when the difference between their attributes fits in some range. This procedure repeats iteratively with the continually enlarging range. The iterative application of that procedure makes this process different from the direct reducing of the image palette.

The image fragment we define as the node of the tree structure associated with the set of attributes. The fragment attributes are the average colour and brightness values and the set of pixels inside the fragment area, ensuring the graphic form of the given fragment.

That allows to organise an object-oriented identification of semantically important areas of the image. The object location in terms of our new structure is defined as the set of the fragments constructing the needed object.

The criteria for the object selection.

The automatic selection of the meaningful objects is really the serious task. However, it is possible to define some formal criteria allowing to select the recognisable artwork and natural photographs from the artificial senseless images. There is no absolute known principle but we can use the empirical heuristics allowing to solve this task with the good probability. The toolkits using this criteria should be more adequate to the processing images.

Our version of this heuristic is the Growth law. The growth law suppose the hyperbolic distribution of the fragments between the levels of the adaptive dynamic structure used in the SAI package. The empirical data state the this is the good criteria to divide the natural and artificial images. The only known exception from this rule is the escape-time fractal images [2]. The cause of this fact is the iterative process of fractal creation.

The following graphs display the direct application of the growth law to the sample images.

Fig. 1. Sample image "axe" in different semantic levels.

The similar graphs show good tests not only for photographic but also for the artificial images depicting the real objects

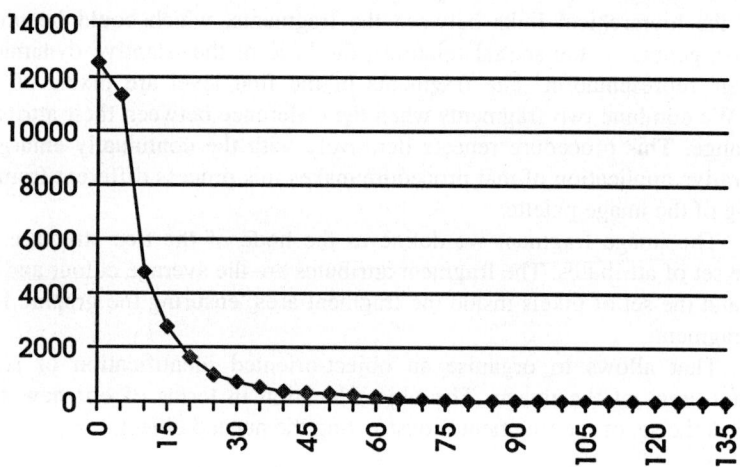

Fig. 2. a) Corresponding graph demonstrates the number of the fragments in first 135 levels.

Object location utilities.

The multilevel fragmentation is the effective data representation for various object location algorithms. The links between fragments shows the neighbouring semantically important areas. The modification of the fragment tree allows to modify the resulting tree and reach more exact object approximation.

Each level of the tree of fragments can be considered as an alternative image interpretation.

Adaptive dynamic tree assume the object search as the combination of its attributes fitting in the corresponding ranges with the following detailed analysis of all applicable areas. This makes possible to use the automatic learning algorithm when the set of needed objects is given and it is necessary to find the corresponding attribute ranges. The learning process could be organised in the following way. The program user selects the appropriate level of fragment hierarchy and points the set of the suitable areas. This areas could be defined through the combinations of the corresponding fragments. The program computes the characteristics of the located fragments and relations between them and constructs the formal criterion for the search of the similar objects.

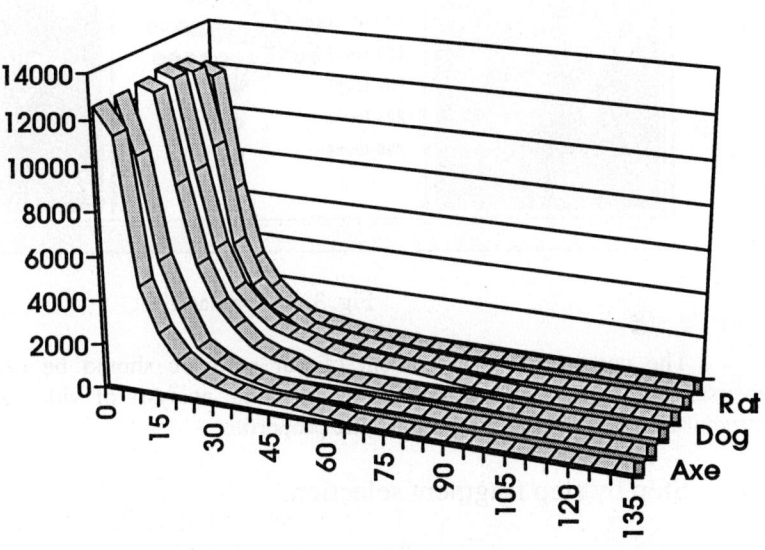

Fig. 2. b) The similar results for fragments distribution for other test images.

Practical realisation.

The SAI package is realised as the MS-Windows/95 application. It gives the possibility to process the true color images, to create and demonstrate the different levels of the adaptive fragment structure, to modify this structure in the interactive mode, to search and recognise objects through the sets of their attributes. The current version 1.1 is the result of the first stage of our project aimed to develop the software able to locate the characteristic areas on images. This version is oriented to the arbitrary colour images with large areas of similar colour.

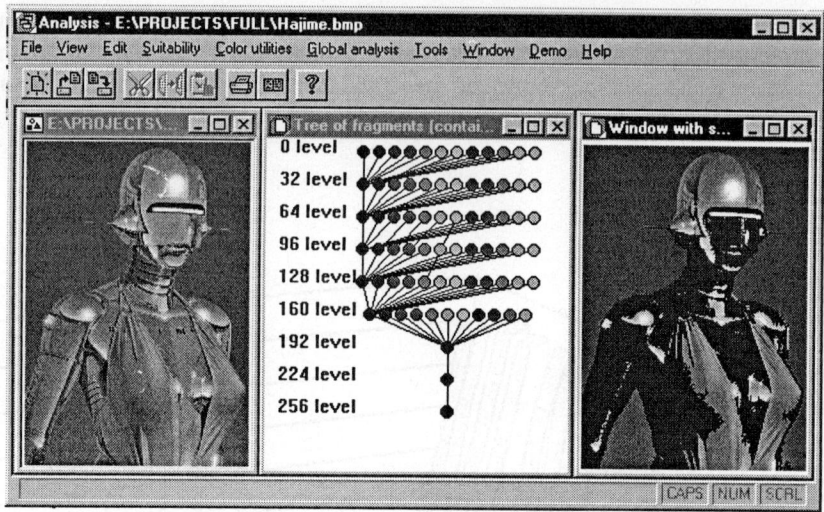

Fig. 3. SAI package.

The perspective development of our software should be oriented to the universal learning system utilised for the dynamic analysis of video-data and their restructurisation according to the semantic properties.

Step by step fragment selection.

To demonstrate the possibility of fragment location we use the following example. The source image is the fragment of the earth surface fragment (Fig. 4. a). The task is to divide the ground from the sea.

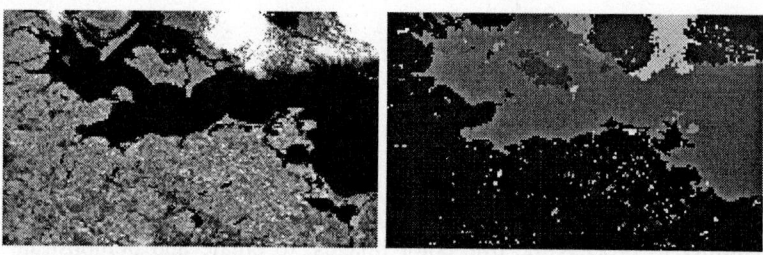

Fig. 4. a) Fig. 4. b)

First of all we should create the tree of fragments.

Next we select the appropriate level (64-th) of our tree structure where the needed areas approximated by the fragments of similar scale. According to our task, the small scale fragments (consisting from one or two pixels) are not essential. To see the different fragments distinctively we use the contrast mapping of this level (Fig. 4. b).

From this mapping of semantically important areas we can point out any of the "ground" fragments (it is shown in blue - the 1-st node in the 64-th level of our adaptive dynamic data structure).

To select all land surface we need to edit the fragment tree and unify all ground fragments.

Possible application of semantic decomposition.

The semantic decomposition implied in SAI package gives the effective way to find two important image characteristics - the structure of the image and the set of image fragments corresponding to the real world objects.

This allows the effective organisation of the storing, compression and fast associative search of images presented through the adaptive dynamic data structures. This property makes useful the application of the developed method to various fields of image analysis.

Publications

1. Alexandrov V.V., Frenkel B.E., Kharinov M.V. *Object-fitting Approach to Image Representation, Computer Applications in Industry / Proc. Fourth IASTED International Conference*, Cairo, Egypt (ARE), 1995.
2. Benoit B. Mandelbrot, Fractals: form, chance and dimension, W.H. Freeman and Company, San Francisco,1977.
3. Alexandrov V.V., Gorsky N.D. *Computer and Human Vision Systems, Machine GRAPHICS and VISION, Proceedings of 2nd GKPO Conference*, Naleczow,May - 18-23, -1992
4. Alexandrov V.V., Gorsky N.D. *From Human To Computers: Cognition throw Visual Perception*, World Scientific, Singapore, 1991. 203 p.
5. Alexandrov V.V., Gorsky N.D. *Can a Computer Vision System Work Like a Human One // International Journal of Imaging Systems and Technologies*, 1991, v. 3.
6. Demin Wang, Veronique Haese-Coat, Joseph Ronsin *Shape Decomposition and Representation Using a Recursive Morphological Operation* / Pattern Recognition, November 1995, v. 28, num. 11, pp. 1783-1792.

Recognition of Simple Curved Surfaces from 3D Surface Data

Dr Alan M. McIvor and Peter T. Waltenberg

Industrial Research Limited
P. O. Box 2225, Auckland
New Zealand

Abstract. This paper is concerned with the robust identification of patches from simple geometric surfaces such as planes, spheres, and cylinders. This is done from 3D data acquired using a range scanner, along with principal quadric estimates at each point. Properties of the Gaussian and mean curvature (e.g., their sign) at points on surfaces of the target class are used to reject all points that could not be on such a surface. Each remaining point is then mapped to the parameters of the unique surface in its class that contains it, if it were on such a surface. Region growing and clustering techniques are used to identify those points actually on a surface of the target class and to extract the surface parameters.

1 Introduction

This paper is concerned with the segmentation of smooth, connected surface patches identified in range data into surfaces of simple shape classes (planes, spheres, cylinders), and the estimation of the parameters of the identified surface instances. It assumes that the principal quadric has been estimated at each surface point [1, 2], and this information is used to map each point to the parameters of the surface it lies on, if it is indeed of a given class. Clusters in this parameter space are identified, thereby segmenting the surfaces. The use of principal quadric data as part of the shape recovery process differentiates this method from others such as deformable surfaces [3], scan line grouping [4], tripod operators [5], and least-squares fitting [6, 7].

2 Planar Surface Patches

At all points on a planar surface, both principal curvatures (and hence the Gaussian and mean curvatures) are zero. Although the principal frame is not well-defined, the principal quadric is. Points at which both principal curvatures are zero are called *planar points*. A smooth surface patch containing only planar points is a planar surface. Thus planar surface patches can be extracted by detecting points at which both principal curvatures are zero and grouping them into connected regions.

The operation of such a planar surface detection scheme is illustrated in Figure 1, which shows the planar patches (corresponding to the cube's faces) extracted from 3D data off a cube.

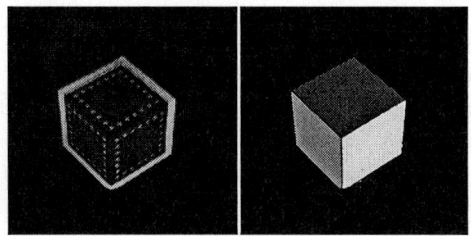

Fig. 1. The image on the right shows the connected planar regions identified in 3D data taken from the scene on the left.

The performance of the planar surface extraction scheme was evaluated using a Monte Carlo simulation [8]. Results are presented in Figure 2 as a function of the standard deviation of the noise added to the stripe data. For all values of standard deviation, the largest extracted planar region was used for calculating the statistics. The first graph shows the number of data points in the region. The other graph plots the RMS standard deviation of the surface normal perturbations [9, pg 281], where the RMS standard deviation is defined as the square root of the average of the perturbations' two eigenvalues. The bias in the estimator was measured by the angle between the true surface normal and the estimated one. In the worst case it was 12°, but in general it was insignificant.

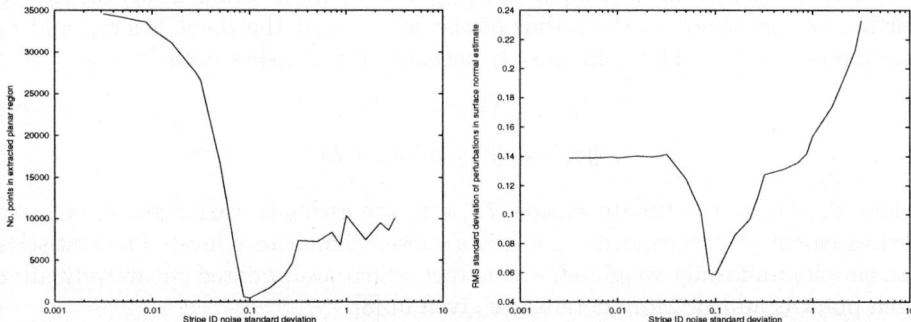

Fig. 2. Performance of planar region extraction method, as measured by the number of data points in the extracted planar region (left), and the RMS standard deviation of the normal perturbations.

The results in Figure 2 suggest very stable performance at low levels of stripe noise, with a rapid fall off in performance as the stripe id noise standard deviation increases from 0.01 to 0.1. There is some improvement beyond this.

3 Spherical Surface Patches

Points on a spherical surface patch are *umbilic*. Patches consisting of umbilic points, all having the same principal curvatures, are spherical surface patches. As such, it can be seen that planar surface patches are special cases of spherical surface patches. Spherical surface patches can be segmented by first identifying points which satisfy $|\kappa_1 - \kappa_2| < \epsilon$. These are then segmented into regions of spatially constant principal curvature, which correspond to spherical surface patches. This can be done with region growing or Bayesian classification techniques as described below.

3.1 Region Growing

The "Basic Connected Graph Traversal Algorithm" [10, pg 102] was used as a basis for a region growing method. Points on the boundary of the already extracted spherical region are added to it if they are "similar" enough to the underlying sphere, based on the current estimate of the sphere's parameters (radius and centre). Similarity is measured as follows, where the ith data point is being considered and there are n points in the already extracted spherical region. The parameters of the sphere on which the data point lies, given that it is an umbilic point, are estimated by

$$r_i = \frac{1}{|H_i|} \quad (1)$$

$$\mathbf{c}_i = \mathbf{x}_i + \frac{1}{H_i}\mathbf{n}_i \quad (2)$$

where H_i is the mean curvature at the point, $\mathbf{x}_i$ its location in space, $\mathbf{n}_i$ the surface normal there, r_i the radius of the sphere that the point lies on, and $\mathbf{c}_i$ the sphere's centre. The point is only accepted if it satisfies both

$$(r_i - \bar{r}_n)^2 \leq \alpha^2(\check{r}_n + \acute{r}) \quad (3)$$

$$\|\mathbf{c}_i - \bar{\mathbf{c}}_n\|^2 \leq \alpha^2(\check{\mathbf{c}}_n + \acute{\mathbf{c}}) \quad (4)$$

where $\bar{r}_n$, etc, are estimate means, $\check{r}_n$, etc, are estimate variances, $\acute{r}$, etc, are measurement error variances, and α is a chosen significance level. The statistics are simple, uniformly weighted, estimators which are updated recursively after each point is added. More details are given in [8].

A Monte Carlo simulation was used to measure the performance as a function of system noise. The number of points in the largest extracted region, and its estimated radius are shown in Figure 3. At small noise standard deviation levels, approximately a quarter of the underlying spherical region is extracted in the largest spherical patch (there were 45225 points total). Accurate radius estimates are also obtained; the systematic bias is due to the smoothing inherent in the principal quadric estimation. In the central part of the noise standard deviation range, around the value 0.1, the size of the extracted (connected) regions gets smaller but good estimates of the radius (and centre) are still generated. Above noise levels of 1.0, the performance is poor by all measures.

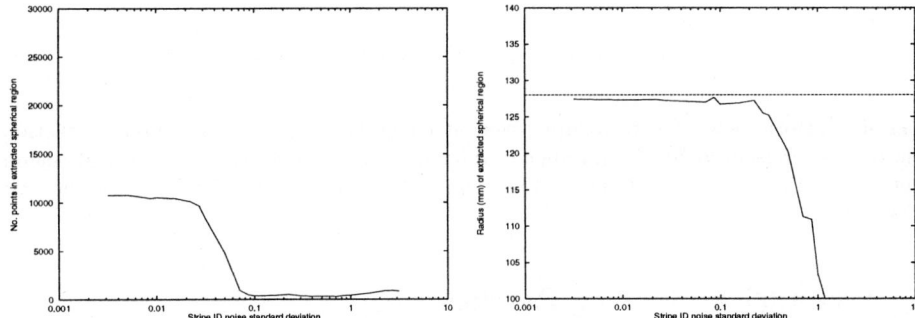

Fig. 3. Performance of the region growing based spherical region extraction method, as measured by the number of data points in the extracted spherical region (left), and the radius of the extracted spherical region (right). The true radius was 128mm.

3.2 Unsupervised Learning

The Basic Isodata Procedure [11, pg 201] was used to cluster the data points into regions of constant principal curvature.

The clustering algorithm requires the number of clusters, c, to be specified as input. This, however, is not known. It is determined by repeating the clustering for $c = 1, 2, 3, \ldots$, and looking at how the sum-of-squared-error criterion $J_e(c)$, used by the clustering algorithm to determine the optimal partitioning, varies with c. Termination tests as defined in [11, pg 242] are not appropriate here because the data often consists of one or more clusters against a background scattering of "noise" points. This leads to small decreases in $J_e(c)$ as the "noise" points are partitioned into small clusters of their own. Hence a suitable criterion is the smallest value of c which satisfies

$$\frac{J_e(c+1)}{J_e(c)} > 1 - \epsilon \quad (5)$$

where a good value for ϵ was determined to be 0.01.

A Monte Carlo simulation was used to measure the performance as a function of system noise. The number of points in the largest extracted region, and the estimated radius are shown in Figure 4. These results show that the Bayesian classification method is significantly more robust at high noise levels than the region growing based method.

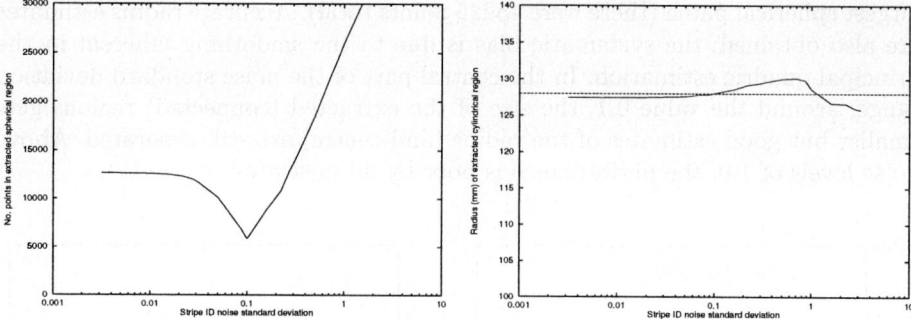

Fig. 4. Performance of the unsupervised learning based spherical region extraction method, as measured by the number of data points in the extracted spherical region (left), and the radius of the extracted spherical region (right). The true radius was 128mm.

4 Cylindrical Surface Patches

Given a point and the principal quadric there, the parameters of the cylinder it must lie on can be uniquely determined as follows (for the ith data point):

1. Radius estimate r_i

$$r_i = \frac{1}{2|H_i|} \qquad (6)$$

2. Axis direction estimate $\mathbf{d}_i$

$$\mathbf{d}_i = \pm \mathbf{r}_i \qquad (7)$$

where $\mathbf{r}_i$ is the principal direction in which the principal curvature is 0.

3. Unique point on axis $\mathbf{p}_i$

$$\mathbf{x}_a = \mathbf{x}_i + \frac{1}{2H_i}\mathbf{n}_i \qquad (8)$$

$$\mathbf{p}_i = \mathbf{x}_0 + (\mathbf{I} - \bar{\mathbf{d}}_n \bar{\mathbf{d}}_n^T)\mathbf{x}_a \qquad (9)$$

where $\mathbf{n}_i$ is the surface normal, $\bar{\mathbf{d}}_n$ is the current estimate of the axis direction, and $\mathbf{x}_0$ is an arbitrary fixed point.

All points on a cylindrical surface are parabolic points (i.e., the Gaussian curvature is zero) with non-zero mean curvature. That is, one but not both principal curvatures is zero. This is used to segment out potentially cylindrical regions for further segmentation into cylindrical regions. As with spherical surface segmentation, region growing and unsupervised learning approaches were implemented and their performance measured using Monte Carlo simulation [8]. Results for the region growing method are shown in Figure 5. The underlying cylindrical region contained 123392 points, so the extracted regions contain approximately

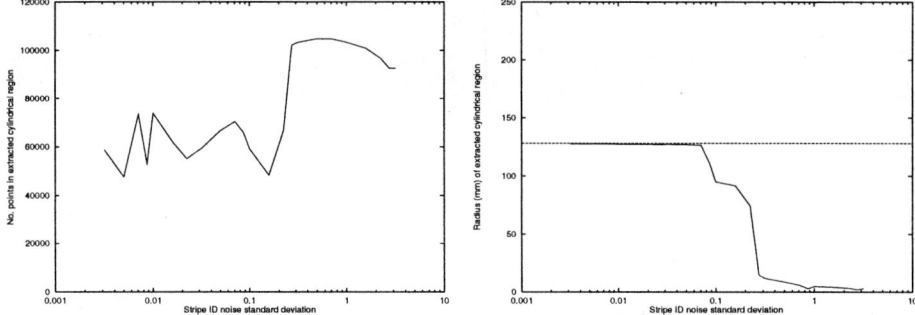

Fig. 5. Performance of the region growing based cylindrical region extraction method, as measured by the number of data points in the extracted cylindrical region (left), and the radius of the extracted cylindrical region (right). The true radius was 128mm.

half. Results for the unsupervised learning approach are shown in Figure 6. For low stripe id noise standard deviation, almost all points in the cylinder are extracted and a good estimate of the cylinder's radius is produced. However, there is a sharp drop-off in performance between standard deviation levels of 0.01 and 0.1, after which only cylinder fragments are extracted.

The performance on real data is illustrated in Figure 7, which shows the largest cylindrical region extracted. The radius of the extracted cylinder was 39.50mm compared to the actual value of 37.76mm estimated by other means.

5 Conclusion

This paper has considered the recovery of shapes from a collection of standard surface classes. The starting point is a set of data points that correspond to a smooth surface patch, i.e., a preliminary segmentation along occlusion boundaries and surface folds has been performed. The methods developed for shape recovery use the information in the principal quadric at surface points, as well as the point location information. This has resulted in robust methods that have two phases. Firstly, points that could not possibly be on such a surface because of their curvature properties are eliminated. Then, the remaining points are examined in more detail to identify the parameters of the surface that they are on. Region growing and clustering (i.e., unsupervised learning) techniques are used to segment the points into connected sets corresponding to surfaces of the various types. The methods were tested on real and synthetic data. The latter was used to measure the performance as a function of system noise.

Three classes of surface type are considered: planar, spherical, and cylindrical surfaces. Planar surfaces are characterised by containing planar umbilic points. Such points are detected by thresholding the principal curvatures. Then the resulting point set is segmented into connected components using a labelling algorithm.

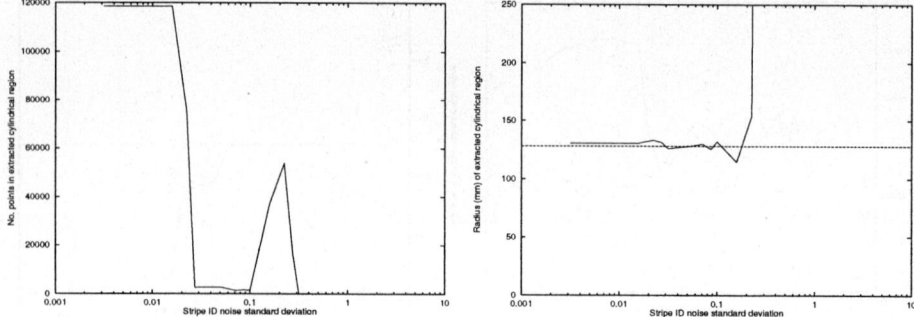

Fig. 6. Performance of the unsupervised learning based cylindrical region extraction method, as measured by the number of data points in the extracted cylindrical region (left), and the radius of the extracted cylindrical region (right). The true radius was 128mm.

Spherical surfaces contain only umbilic points. Such points are detected by thresholding the difference between the principal curvatures. The points then need to be segmented into connected regions containing points all with the same principal curvature. Region growing and clustering algorithms are considered for this task. The clustering method was found to be the more robust of the two.

Cylindrical surfaces contain only parabolic points. They are detected by thresholding the smaller magnitude principal curvature. The other principal curvature gives the cylinder radius, and its principal direction is parallel to the cylinder axis. They need to be grouped on the basis of their radius and cylinder axis (direction and position) to segment out connected cylindrical regions. Both region growing and clustering methods were considered for doing the segmentation. The region growing method was found to be more reliable than the clustering method.

All of the methods have a number of parameters that control their operation and whose value must be selected. The initial point rejection phase requires principal curvature thresholds. The region growing method requires initial estimates of the surface parameters, variance values for them, and a significance value for generating the validation gate. The clustering method requires a significance level for deciding how many clusters are present. Appropriate values for some of these are data dependent. Of those that are, the initial surface parameter estimates are easily generated from the starting point or a small patch around it. The others are the thresholds for the principal curvatures, which clearly are dependent on the noise in the stripe id measurements.

Acknowledgement

This work was funded by the New Zealand Foundation for Research, Science, and Technology under Contract CO8410 Objective 2 "3D Vision".

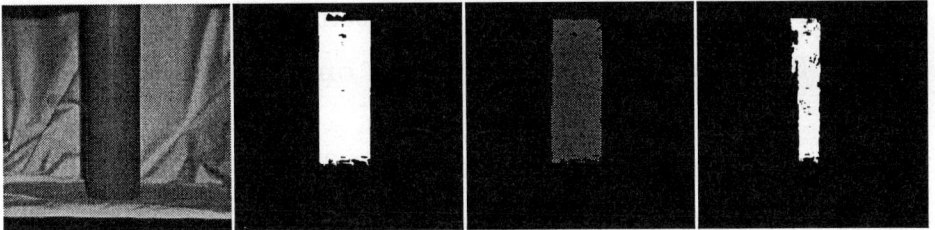

Fig. 7. Example of cylindrical region extraction. From left to right: An image of a scene containing a piece of pipe; A mask indicating the largest connected region of parabolic points; The largest cylindrical region extracted by region growing; The largest cylindrical region extracted by Bayesian classification.

References

1. Alan M. McIvor and Robert J. Valkenburg. A comparison of local surface geometry estimation methods. *Machine Vision and Applications*, 10(1):17–26, 1997.
2. Frank P. Ferrie, Jean Lagarde, and Peter Whaite. Darboux frames, snakes, and super-quadrics: Geometry from the bottom up. *IEEE Transactions on Pattern Analysis and Machine Intelligence*, 15(8):771–784, August 1993.
3. Demetri Terzopoulos, J. Platt, A. Barr, and K. Fleischer. Elastically deformable models. *Computer Graphics*, 21(4):205–214, 1987.
4. Xiaoyi Jiang and Horst Bunke. Fast segmentation of range images into planar regions by scan line grouping. *Machine Vision and Applications*, 7:115–122, 1994.
5. Frank Pipitone. Rapid recognition of elementary surface shapes in cluttered range images using tripod operators. In *MVA'94 IAPR Workshop on Machine Vision Applications*, pages 21–25, Kawasaki, Dec 13–15 1994.
6. Narayan Sriranga Raja and Anil K. Jain. Obtaining generic parts from range images using a multiview representation. *CVGIP: Image Understanding*, 60(1):44–64, July 1994.
7. B. Parvin and G. Medioni. B-rep object description from mulitple range views. *International Journal of Computer Vision*, 20(1/2):81–112, October 1996.
8. Dr Alan M. McIvor and Peter T. Waltenberg. Recognition of simple curved surfaces from 3D surface data. IRL Report 690, Industrial Research Limited, March 1997. URL ftp://ftp.vision.irl.cri.nz/pub/doc/1997/irl-report-690.ps.
9. Kenichi Kanatani. *Geometric Computation for Machine Vision*. Number 37 in The Oxford Engineering Science Series. Claredon Press, Oxford, 1993.
10. T. Pavlidis. *Algorithms for graphics and image processing*. Computer Science Press, Springer-Verlag, Berlin – Heidelberg, 1982.
11. R. O. Duda and P. E. Hart. *Pattern Classification and Scene Analysis*. Wiley, New York, 1973.

A Recursive Fitting-and-Splitting Algorithm for 3-D Object Modeling Based on Superquadrics

Hongbin Zha   Tsuyoshi Hoshide   Tsutomu Hasegawa

Dept. Intelligent Systems, Kyushu University
6-10-1 Hakozaki, Higashi-ku, Fukuoka, 812, JAPAN

Abstract. In the paper, we propose a systematic approach to object modeling by combining superquadric-fitting and segmentation into an interactive algorithm. It is assumed that the input data are a discrete description of the whole close-surface (CS) of the object. Using the data as input, the method is a top-down, recursive procedure as follows: At first, it finds an initial approximation of the object by fitting a single superquadric to the whole CS data. The residual errors are examined to pick up data points locating in concave regions and far away from the fitted superquadric. A dividing plane is then extracted from the selected points to partition the original data set into two disjoint subsets, which are approximated, respectively, further by the same fitting-and-splitting process. This process is repeated until the whole data are decomposed into a number of primitive superquadrics each with a satisfactory accuracy.

1 Introduction

Recently, a lot of methods have been proposed for solving the problem of automatic object modeling from multi-view range images ([4], [14]). However, they are usually still a set of unorganized 3-D data vast in number and represented relying on certain selected viewpoints. To build a model-base easy to use for the model retrieval purposes, we have to transform the data further into some compact, structural description based on global, parametric surface representation schemes.

There existed a number of schemes that can be utilized for such descriptions. Among them, superquadrics are most preferable because of the favorable trade-off between its flexibility in shape approximation and relative simplicity in required computations ([1], [9]). However, superquadrics have yet some serious drawbacks in representing objects having large shape irregularity and complex part structures. To overcome the difficulty, two strategies have been proposed and widely used for different vision tasks. The first is to incorporate global or local deformation to the descriptions based on a single superquadric ([13]). However, the approach has also weaknesses including its bad convergence when a large number of deformations are embodied simultaneously in a single superquadric.

The other approach is to derive a structural description by first segmenting an object into several primitives and then modeling each primitive with single superquadrics ([3], [5], [6], [12]). Until now, the segmentation is customarily carried out by some clustering algorithm mainly based on detected 3-D 1st and 2nd order edges ([3], [7]). Obviously, objects with smooth shape changes can not be handled adequately here due to the inherent limitations in these segmentation methods.

In this paper, we propose a new method of describing an object by integrate together the superquadric fitting and segmentation interactively. In our method, we assume that 3-D data on the whole object surface (close-surface: CS) have been obtained. We first fit a single superquadric to the whole CS data set, and then, split the object into two parts by employing a dividing plane that is defined in terms of error distribution in the preceding superquadric fitting. To better the partitioning, the resultant data subsets undergo a refinement to get a locally optimal clustering. After that, a child superquadric is fitted to each part, and the fitting-and-splitting procedure is repeated until the whole object is described by primitive superquadrics all with a reasonable accuracy.

As compared with the previous methods, our algorithm can be applied in modeling of axisymmetric objects, irrespective of a linear or curved axis. Some results of experiments using real 3-D image data are presented in the paper.

2 Superquadric-Based Modeling

In the section, we give a brief overview on superquadrics and some basic principles in its fitting to 3-D image data.

2.1 Superquadrics and Deformation

A superquadric is an equation which is defined as an extension of basic quadric surfaces. A short review on the definition and classification of original superquadrics can be found in [1]. In our method, we use superquadric ellipsoids, which are an extension of elliptic surfaces, defined by the implicit function as follows

$$\left[\left\{\left(\frac{x}{t_x \cdot a_1}\right)^{\frac{2}{\epsilon_2}} + \left(\frac{y}{t_y \cdot a_2}\right)^{\frac{2}{\epsilon_2}}\right\}^{\frac{\epsilon_2}{\epsilon_1}} + \left(\frac{z}{a_3}\right)^{\frac{2}{\epsilon_1}}\right]^{\epsilon_1} = 1, \qquad (1)$$

where

$$t_x(z) = 1 - k_x \frac{z}{a_3} \; ; \; t_y(z) = 1 - k_y \frac{z}{a_3}. \qquad (2)$$

a_1, a_2 and a_3 are called scale parameters which affect the sizes of the superquadric along the x, y and z coordinates, respectively, in the object-centered coordinate system. ϵ_1 and ϵ_2 are called shape parameters, which determine the squareness of the cross sections in the latitudinal (xz plane) and longitudinal (xy plane) directions. We show some examples of superquadrics with changing parameters ϵ_1 and ϵ_2 in Fig.1(a),(b) and (c).

We introduce global tapering deformation to enhance modeling capabilities of a single superquadrics. t_x and t_y are tapering function, and k_x and k_y are called tapering parameters. Three typical objects by using tapering deformation are illustrated in Fig.1(d),(e) and (f).

2.2 Superquadric Inside-Outside Function

From eq.(1), we define the inside-outside function as

$$F(x,y,z) = \left[\left\{\left(\frac{x}{t_x \cdot a_1}\right)^{\frac{2}{\epsilon_2}} + \left(\frac{y}{t_y \cdot a_2}\right)^{\frac{2}{\epsilon_2}}\right\}^{\frac{\epsilon_2}{\epsilon_1}} + \left(\frac{z}{a_3}\right)^{\frac{2}{\epsilon_1}}\right]^{\epsilon_1}, \qquad (3)$$

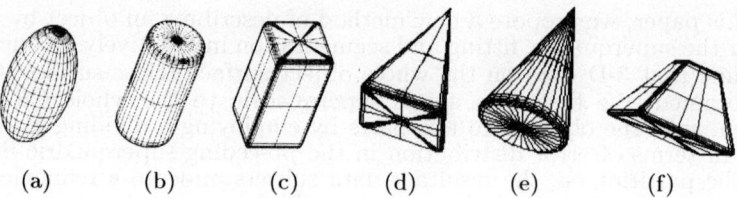

Fig. 1. Examples of superquadrics

which judges where a given point $r_i = (x_i, y_i, z_i)^T$ lies relative to the superquadric surface. If $F(x_i, y_i, z_i) = 1$, r_i lies exactly on the surface of superquadrics. If $F(x_i, y_i, z_i) > 1$, r_i lies outside of the volume, and if $F(x_i, y_i, z_i) < 1$, it is inside.

2.3 Recovery of Superquadrics

Our goal here is to obtain a superquadric function for a given set of 3-D image data. In our cases, we assume that we have acquired, as the data set, location measurements of the 3-D points on the whole object surfaces. We refer to them as the close-surface (CS) data, and they can be derived by registering and integrating multi-view range images ([4]). With CS data available, our task is to estimate a group of superquadric parameters in eq.(1) in an optimization manner.

To define the criterion function for the optimization, we use a minimum volume inside-outside error of fitting ([2]) as follows. Suppose we have N 3D surface points

$$r_i = (x_i, y_i, z_i)^T \quad (i = 1, 2, \cdots, N). \tag{4}$$

Using the inside-outside function, the error between the all data and the model is

$$E_{eof} = \frac{1}{N} \sum_{i=1}^{N} \{\sqrt{a_1 a_2 a_3}(1 - F(x_i, y_i, z_i))\}^2. \tag{5}$$

If $E_{eof} = 0$, all data lie on the surface of the superquadric, namely, we obtain a suitable model for the image. Minimizing eq.(5) by using a least square method, we can obtain a group of the superquadric parameters.

In the definition, $\sqrt{a_1 a_2 a_3}$ is introduced to prevent the parameters from being overestimated. By multiplying it here, we are forced to recover the minimum volume of the inside superquadric, and this will results in a viewpoint-invariant representation.

Also, because eq.(1) is defined in an object-centered coordinate system, the superquadric must undergo another transformation to align with the CS data, which are usually represented with respect to some viewpoint-centered coordinate system. Six parameters are necessary here, with displacements u_x, u_y and u_z for translation and Eular angles α, β, ψ for rotation. In total, we need to estimate 13 parameters in the parameter vector

$$q = (a_1, a_2, a_3, \epsilon_1, \epsilon_2, k_x, k_y, u_x, u_y, u_z, \alpha, \beta, \psi)^T \tag{6}$$

from the CS data. The Levenverg-Marquardt method ([10]) is employed here to implement the nonlinear least squares minimization.

3 Fitting-and-Splitting Segmentation Algorithm

As stated earlier, it is difficult to model a complex object accurately by a single superquadric. In such cases, the most straightforward measure is to have a two-step procedure by first segmenting the object into some component parts and then fitting superquadrics to each part. This section provides detailed explanation on an algorithm. We developed on the basis of the fitting-and-splitting principles.

3.1 Fitting-and-Splitting Algorithm

The whole procedure is outlined as follows: 1) Fit a single superquadric to the whole CS data; 2) Split the data into two disjoint subsets by a dividing plane. To improve the results, data in the two subsets are further selected in a small number of refinement fitting steps. 3) Fit a superquadric to each part. This fitting-and-splitting procedure is repeated until the whole object is described with a satisfactory accuracy.

3.2 Superquadric Splitting and Dividing Planes

In general, superquadrics themselves are not good at modeling objects with concave structure. This property is utilized here to define dividing planes by searching out such concave portions in the CS data with respect to the superquadric fitted at the last step.

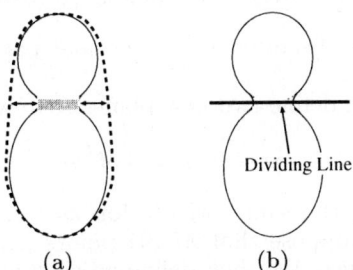

Fig. 2. Segmentation by using a dividing line in a 2-D case: (a) 2-D contours of an object; (b) its partitioning with a dividing line.

Fig.2 illustrates an example showing the use of dividing lines in a 2-D case. In Fig.2(a), the sensed object contours are shown by continuous solid curves. Given the contours, we first fit them with a single superquadric that is shown by dotted curves. As is evident in the figure, concave portions featured by large fitting error can be detected in the crooked regions indicated here by short arrows. We then draw a line in the regions, as shown in Fig.2(b), to divide the whole contours into two disjoint parts, which can be approximated individually by two small superquadrics. We call the superquadric fitted to the original data the parent superquadric, and the resulting two small ones the child superquadrics.

The concept of 2-D dividing lines can be extended to 3-D space as dividing planes. A dividing plane is represented here by a twofold set $\mathcal{P} = \langle \boldsymbol{l}_p, \boldsymbol{n}_p \rangle$, where $\boldsymbol{l}_p = (x_p, y_p, z_p)^T$ is the location vector, and $\boldsymbol{n}_p = (o_x, o_y, o_z)^T$ is the normal vector of the plane. Thus, the plane to be determined is written by

$$(\boldsymbol{u} - \boldsymbol{l}_p)^T \boldsymbol{n}_p = 0, \tag{7}$$

where $u = (x, y, z)^T$ is any point in 3-D space.

Determination of l_p: At first, we compute the error distribution between the CS data and the approximate superquadric. For any point $r = (x, y, z)^T$ in the data set, the error is estimated here by another distance measure as

$$d_p(x, y, z) = \frac{1 - F(x, y, z)}{\|\nabla F(x, y, z)\|}, \tag{8}$$

where ∇ is the Laplacian operator and $F(x, y, z)$ is the inside-outside function defined by eq.(5). Since deep concave points usually locate inside of the parent superquadric, the above distance provides a direct measurement on the concavity, and we can choose the point with the largest d_p as l_p.

In practical implementation, to ensure reliability of the plane positioning, we select the largest N_p points according to the values of d_p. Then, the selected points are clustered to remove outliers and l_p is determined as the center of the largest cluster.

Determination of n_p: Then we search in an optimization manner the most suitable orientation of the plane passing through the chosen location vector. Let the normal be represented as

$$n_p = (\sin\phi\cos\theta,\ \sin\phi\sin\theta,\ \cos\phi)^T. \tag{9}$$

The plane orientation is discreted in the normal parameter space (ϕ, θ) with $0 \leq \phi \leq \pi/2$ and $0 \leq \theta \leq 2\pi$.

For a plane $\langle l, n \rangle$, we define two new planes represented, respectively, by

$$(u - l)^T n = \pm \delta l^T n. \tag{10}$$

They are planes having the same orientation as $\langle l, n \rangle$ but locating above or below at a distance δ. Suppose that M 3D points (x_j, y_j, z_j), $j = 1, 2, \cdots, M$, fall between the two planes. We then define an evaluation function as

$$O_r = \frac{1}{M} \sum_{j=1}^{M} d_p(x_j, y_j, z_j), \tag{11}$$

where $d_p(x, y, z)$ is given in eq.(8). The normal maximizing the above function is chosen as the desired orientation n_p of the dividing plane.

3.3 Refinement of Segmentation

We can split an object into two parts by the use of the obtained dividing plane. As shown by an example in Fig.3, however, this segmentation is not always appropriate. When we segment the object shown in Fig.3(a) with a dividing plane derived by our algorithm, the CS data are not divided into subsets, shaded with different gray levels in Fig.3(b), as proper as we expect. Therefore, a refinement process on the approximate segmentation is necessary.

The segmentation refinement is accomplished here by a recursive and locally optimal algorithm. Suppose we have divided the CS data into two subsets $D_1^{(1)}$

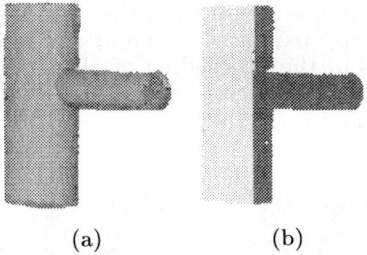

Fig. 3. An example of false segmentation: (a) the original CS data; (b) approximate segmentation.

and $D_2^{(1)}$ by the above fitting-and-splitting process. We first fit superquadrics to each part respectively, and denote the child superquadrics as $\mathcal{S}_1^{(1)}$ and $\mathcal{S}_2^{(1)}$. After that, we re-classify data points in each subset according to their proximity with $\mathcal{S}_1^{(1)}$ and $\mathcal{S}_2^{(1)}$. If a data point lies exactly in the inside of one child superquadric, it is assigned to the superquadric; otherwise, its distances to the child superquadrics are computed and it is assigned to the closest one.

After the re-classification, superquadrics are fitted once again to the two new sets of data points. The re-classifying process is repeated to result in $\mathcal{S}_k^{(t)}$ ($k = 1, 2; t = 1, 2, \cdots$) for new $D_k^{(t)}$ ($k = 1, 2; t = 1, 2, \cdots$) until the error values for the component superquadrics are all small enough.

4 Experimental Results

In this section, we present results of two experiments using real range data. All range images were obtained by a range finder system manufactured in our laboratory ([8]).

We first take images of an object from several specified viewpoints to realize a complete view coverage and then integrate the images into a close-surface point-based description ([14]).

The first experiment is carried out mainly to show the capacity of our method in extracting dividing planes and splitting parent CS data sets. Fig. 4(a) show the close-surface data of an object with two cylinders intersecting with each other, the same one as in Fig.3(a). The parameters required in extracting the dividing plane are set as

$$N_p = 15; \quad \delta = 10.0. \tag{12}$$

As mentioned earlier, after one time of fitting-and-splitting process, we get two child superquadrics, as shown in Fig.4(b), which result in a false segmentation and description. In this figure, the partitioned data subsets are shaded by different gray levels with the child superquadrics plotted at the corresponding locations. The algorithm then go into the refinement phase to produce improved segmentations illustrated in Fig.4(c), and (d) which is the final partition with a satisfactory accuracy. Table 1 shows the fitting errors of the child superquadrics, changes in the numbers of data points in each CS data subsets, and CPU times (UltraSparc·140MHz) cost in the refinement steps.

The second experiment aims to show the flexibility of our method in modeling objects with smooth shape changes. The object to be modeled is a banana-like bent object, as shown in Fig.5(a).

After two times of fitting-and-splitting, the object is segmented into three parts which are all well approximated by child and grandchild superquadrics. The segmentation results in the proceeding of algorithm are given in Fig.5(b), (c), and (d). The errors for each single superquadrics and the consumed CPU times for each step are listed in Table 2.

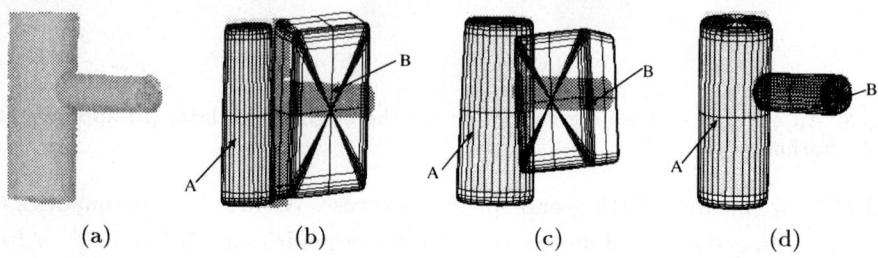

(a)　　　　　　(b)　　　　　　(c)　　　　　　(d)

Fig. 4. Experiment for an object consisting of intersecting cylinders: (a) original CS data; (b) initial rough splitting; (c) splitting after two times of refinement; (d) final splitting with fives times of refinements.

Table 1. The first experiment: errors and data numbers in the refinement process.

	Initial Splitting		3rd Splitting		5th Splitting	
Part Label	A	B	A	B	A	B
Number of Data	3536	2547	4162	1921	4754	1329
Error of Fitting	0.018	0.483	0.011	0.456	0.009	0.003
CPU Time(sec.)	28.01		58.80		102.67	

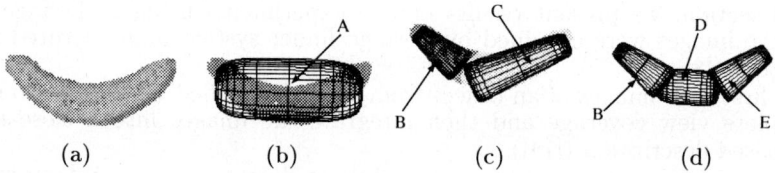

(a)　　　　　　(b)　　　　　　(c)　　　　　　(d)

Fig. 5. Experiment for a banana-like object (a) original CS data; (b) initial superquadric fitting; (c) first time of segmentation; (d) second time of segmentation.

Table 2. The second experiment: errors of fitting and used CPU time.

	1st Fitting	2nd Fitting	3rd Fitting
Part Label	A	(B, C)	(B, D, E)
Errors of Fitting	0.297	(0.0144 , 0.056)	(0.0144 , 0.00712 , 0.00957)
CPU Time (sec.)	12.61	50.58	82.02

5   Conclusions

In this paper, we proposed a method of creating superquadric-based descriptions of 3-D objects from a set of 3-D closed-surface data. It is a recursive fitting-and-splitting algorithm and has a great description capability for modeling axisymmetric objects with smooth shape changes.

In the method, we do not use any high-order geometric features such as surface normals, curvatures, and so on, and hence the method is robust to noises in the raw range images. As is evident in the experiment results, our method is able to produce an object description with a satisfactory accuracy in a reasonable time.

As the future works, we have to incorporate more other splitting strategies in our method to deal with objects with more complex structures. Multi-resolution descriptions based on superquadric fitting are another problem we have to address for building a dynamic model-base used in real-time targeting or recognition.

References

[1] A.H. Barr, "Superquadrics and Angle-Preserving Transformations", IEEE Computer Graphics Application, 1(1), pp.11-23, 1981
[2] A.D. Gross, T.E. Boult, "Error of Fit Measures for Recovering Parametric Solids", Proc. 2nd Int. Conf. on Computer Vision, pp.690-694, 1988
[3] A. Gupta, R. Bajcsy, "Volumetric Segmentation of Range Images of 3D Objects Using Superquadric Model" CVGIP:Image Understanding, 58(3), pp.302-326, 1993.
[4] A. Hilton, On Reliable Surface Reconstruction from Multiple Range Images, Technical Report VSSP-TR-5/95, Dept. EEE, Univ. of Surrey, 1995.
[5] T. Horikoshi, S. Suzuki, "3D Parts Decomposition from Sparse Range Data using Information Criterion", Proc. IEEE Computer Vision and Pattern Recognition, pp.168-173, 1993.
[6] S. Kumar, D. Goldgof, "Model Based Part Segmentation of Range Data – Hyperquadrics and Dividing Planes", Proc. Workshop on Physics-Based Modeling in Computer Vision, pp.17-23, 1995.
[7] C. Liao, G. Medioni, "Surface Approximation and Segmentation of Objects with Unknown Topology", 1996 Proc. Image Understanding Workshop, pp.1033-1040, 1996
[8] N. Okada, T. Nagata, "A Range Finder System with a Controllable Laser Slit Marker", Proc. 1st Asian Conf. Computer Vision, pp.145-148, 1993.
[9] A. Pentland, "Perceptual Organization and the Representation of Natural Form", Artificial Intelligence, 28, pp.293-331, 1986
[10] W.H. Press, S.A. Teukolsky, etc., Numerical Recipes in C, Cambridge University Press, 1988
[11] F. Solina, R. Bajcsy, "Recovery of Parametric Models from Range Images: The Case for Superquadrics with Global Deformations", IEEE Trans. Pattern Analysis and Machine Intelligence, 12(2), pp.131-147, 1990.
[12] F. Solina, A. Leonardis, A. Macerl, "A Direct Part-Level Segmentation of Range Images Using Volumetric Models", Proc. of IEEE Int. Conf. on Robotics and Automation, pp.2254-2259, 1994.
[13] D. Terzopoulos, D. Metaxas, "Dynamic 3D Models with Local and Global Deformations: Deformable Superquadrics", IEEE Trans. Pattern Analysis and Machine Intelligence, 13(7), pp.703-714, 1991.
[14] H. Zha, K. Morooka, T. Hasegawa, T. Nagata, "Active Modeling of 3D Objects: Planning on the Next BEst Pose (NBP) for Acquiring Range Images", Proc. Int. Conf. Recent Advances in 3-D Digital Imaging and Modeling, pp.68-75, 1997

Learning and Recognizing 3D Objects by Using Partial Planar Curve Matching Method

Jin Jia[1] and Keiichi Abe[2]

[1] Graduate School of Science and Engineering, Shizuoka University, Japan.
[2] Dept. of Computer Science, Shizuoka University, Japan

Abstract. A scheme for learning and recognizing 3D objects from their 2D views is presented. The scheme proceeds in two stages. In the first stage, we try to learn a model automatically from 2D training images of different objects which have similar shape of components and similar adjacency relation between the components. In the second stage, the generated model is used to recognize the learned object. We use partial planar curve matching method to implement our scheme. We tested the approach on recognizing objects from images of complex real scenes and got satisfactory results.

1 Introduction

Learning and recognizing 3D objects is a complex subject in computer vision. A large variety of methods have been proposed for the task of visual object learning and recognizing. Among these methods, three main classes can be distinguished, as summarized by Ullman[1]: *Invariant properties and feature spaces*; *Parts and structural descriptions*; *Alignment approach.* Our approach lies in parts and structural description approach which relies on the decomposition of objects into constituent parts and the capturing of the invariant properties which are relations among parts. There are several reasons for us to go along with this approach. Firstly, since we try to generate a model automatically from training examples of several kinds of objects, to find invariant properties is difficult. Secondly, since we deal with non-rigid objects, we can not use alignment approach either.

We try to develop a system which can learn a model of similar objects from 2D training images automatically, and then use the model to recognize the learned object or some objects similar to the learned ones from a complex scene. In our system, we just handle those objects such that they can be decomposed into parts and relations among the parts are easily extracted and easily described. Chairs have been chosen here as objects for processing.

The outline of the system is shown in Fig. 1. As shown in Fig. 1, inputs of the system in the learning stage are gray level images of chairs. These images are then fed to a preprocessing procedure which consists of segmenting images and putting the results of segmentation into order. Then all 2D images of a 3D object are divided into several clusters. Thus only several 2D clusters are used to describe the object instead of using all images. In the next step, the clusters

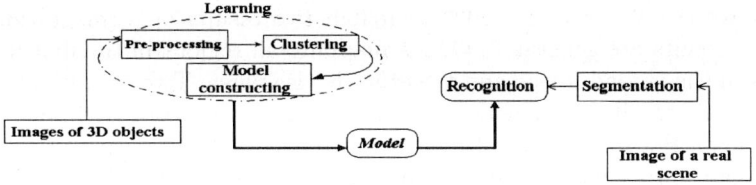

Fig. 1. The outline of the system.

of different objects are matched each other and a common model (prototype) representing different objects is constructed. This model is used to recognize the objects from an image of a real complex scene.

The rest of this paper is arranged as follows. In section 2, we will describe preprocessing procedure. In section 3 we describe how to divide input images into clusters. We will introduce model construction in section 4 and recognition approach in section 5. Finally, we show some results of experiments.

2 Preprocessing

Preprocessing includes segmentation[2] and putting results of segmentation into order. In Fig. 2(a) and 2(b), two segmentation results of an object observed from two directions are shown. From these figures, we can explain why we do not use the results of segmentation directly.

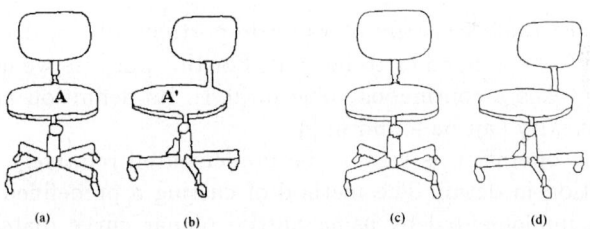

Fig. 2. (a) and (b) show the segment results; (c) and (d) show the segmentation results after "correction."

Because of the lighting condition and colors of the object itself, we get different segmentation result. The "difference" means that after matching two results of segmentation, a pair of matched regions may have very different appearance features and the change of appearance features is not due to change of observing directions. For example, the region A in Fig. 2(a) matches region A' in Fig. 2(b). But they have very different shapes, because regions A' includes parts of the pillar and the leg. For constructing a model from these images, we have to "correct" the results of segmentation.

First we use Wang's method[3] to match two results of segmentation. Two matching results are generated. One is mapping all regions in the first image to the ones in the second image and the other is vise versa. Two operations may be applied to the results of segmentation based on the results of matching: *Splitting* and *Merging*. Between a pair of matched regions, splitting means to divide the larger region into several parts and among generated parts, there is one which is similar to the smaller region. Merging means to merge some parts connecting with the smaller region so that the merged region is similar to the larger one. We use several rules listed below for deciding whether merging or splitting should be applied. For two candidate images, S_1 denotes regions in one image and S_2 denotes regions in the other image. Here we assume that the images depict the object observed in two near directions.

Rule 1 The splitting operation is only applied to "stable" matched regions. "Stable" matched regions means those regions that matched each other both in first to second mapping and in second to first mapping.

Rule 2 If features of two matched regions are obviously different, one of them needs to be split. Here we pay attention to those features such as convexity, area, length of major axis and minor axis.

Rule 3 If two regions in one image match the same region in the other image, then these two regions will be merged if the convexity of merged region is not less than the convexity of any region before merged.

Rule 4 If there is a region in one image which does not match any region in the other image, it will be merged with one of regions connecting with it, if the convexity of the merged region is not less than the convexity of any regions before merged. Otherwise the region remains unchanged.

We can notice that *convexity* plays an important role on deciding whether two regions are to be merged or to be split. For this purpose we use a convexity measure which takes a continuous value in $[0, 1]$. Its definition and the way of calculating convexity can be found in [4].

The merging operation is easy to be implemented relatively, so we discuss splitting operation in detail. The method of cutting a predefined shape from a given region is implemented by using partial planar curve matching (PPCM) method[5]. This method is a way of detecting common parts in two 2D curves. The reason why we can use it in our approach is that the shape of two regions observed from different directions does not change much if the observing direction does not change much. The main idea of PPCM method is to use dynamic programming technique to detect common parts between two planar curves from their total-curvature graphs. See [5] for detail.

The cutting of a region is implemented by using PPCM method and feature points extraction method. After finding matched segments (consecutive point sets) in two regions, a transformation of the form shown in Eq.(1), can be decided uniquely by each matched segment.

$$\begin{bmatrix} x' \\ y' \end{bmatrix} = s \cdot \begin{bmatrix} cos\theta & sin\theta \\ -sin\theta & cos\theta \end{bmatrix} \begin{bmatrix} x \\ y \end{bmatrix} \qquad (1)$$

where s is the scaling factor and θ is the rotation angle. Because there may be several matched segments between two regions, so the transformation is not unique. The best transformation for cutting is chosen as follows. Each transformation is applied to all points on one region curve and calculate the difference between two region curves R_1 and R_2 by using Eq.(2).

$$d(R_1, R_2) = df(R_1^t, R_2)/s \qquad (2)$$

where $d(R_1, R_2)$ denotes the distance between R_1 and R_2, s is the scaling factor in Eq.(1). R_1^t denotes R_1 after being transformed and $df(X, Y)$ is defined as

$$df(X, Y) = \sum_{x \in X} (\min_{y \in Y} dist(x, y)) \qquad (3)$$

$dist(x, y)$ denotes the distance between points x and y. The transformation which gives the least $d(R_1, R_2)$ is used for cutting. Notice that $d(R_1, R_2) \neq d(R_2, R_1)$. If $d(R_1, R_2) > d(R_2, R_1)$, then R_1 will be split and otherwise R_2 will be split.

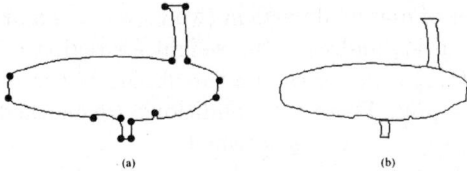

Fig. 3. (a) shows result of feature points detecting by using N-code method, (b) shows result of the regions being cut.

Then we extract feature points from the region that is to be split, suppose R_2 here. The result of feature points extraction using N-code method[6] is shown in Fig. 3(a). The contour of the region is divided at each feature point into several curve segments. Then we calculate the distance between each curve segment C_i and R_1 by calculating $d(C_i, R_1)$. The segment whose distance to R_1 is obviously large will be split from R_2. The result of splitting is shown in Fig. 3(b).

The corrected segmentation results are shown in Fig. 2(c) and 2(d). We use corrected segmentation results for further processing.

3 Clustering

In our system, a 3D object is described by its 2D views. We noticed that the shape of parts of an object and the relation between these parts do not change much if the object is observed from two near directions, so we can divide all 2D images into several groups. In each group, we generate a cluster and save it in memory for recognizing. The clustering is implemented by considering both shape change of parts and adjacency relations change between parts.

For each part of the object, we can define measurement of shape change between two views. Let P_1 and P_2 be one part observed from two near directions and define their difference as

$$diff(P_1, P_2) = df(P_1^t, P_2)/s + df(P_2, P_1^t) \qquad (4)$$

$df(X, Y)$ is defined in Eq.(3).

The measurement of change of adjacency relations between two views V_1 and V_2, denoted by $rd(V_1, V_2)$, is defined similarly to the relational distance[7] between these two views.

After defining the measurement of changes of shape and adjacency relation, we can make clusters from 2D images of an object. Here we suppose 2D views of an object is arranged in order of viewing direction. Firstly, if $rd(V_m, V_n)$ is less than a predefined threshold and $rd(V_m, V_{n+1})$ is larger than the threshold, then the views from V_m to V_n will be grouped into a cluster. Secondly, if the next equation is satisfied, the views from V_m to V_n will be grouped into a cluster.

$$\sum_i^N \Upsilon(d(R_{m,i}, R_{n,i})) \times w_i < \alpha \quad \text{and} \quad \sum_i^N \Upsilon(d(R_{m,i}, R_{n+1,i})) \times w_i > \alpha \qquad (5)$$

where N is the number of matched parts in two views, α is a predefined threshold, $R_{m,i}$ denotes region i in V_m and w_i is the weight for region i. $\Upsilon(d(D_{m,i}, D_{n,i}))$ is equal to 1 if $d(D_{m,i}, D_{n,i})$ is larger than a threshold. At the moment, the weights for all regions are set equal. These two conditions are checked simultaneously. A cluster is made when any of them is satisfied.

An example of obtained groups are shown in Fig. 4.

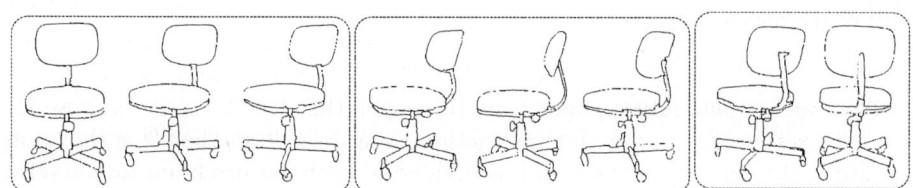

Fig. 4. Result of cluster of several views of a chair

After clustering, let C be a cluster generated from view V_m to view V_n and let P_i be a component of it. We use PPCM method again to generate the shape of P_i in C. First the component in one view is transformed according to (1), here we suppose P_i in V_m is transformed to V_n. Let Cur_m and Cur_n denote the total curvature of P_i in V_m after being transformed and V_n, respectively. The total-curvature of the component in the cluster is calculated simply by Eq.(6).

$$Cur_C = \frac{Cur_m + Cur_n}{2} \qquad (6)$$

Then according to Cur_C, the shape of component P_i in the cluster can be generated. See Fig. 5 for an illustration.

Fig. 5. Generate a cluster component from two sample parts

4 Model generation

Till now, one 3D object is described by several 2D clusters. These clusters can be regarded as a model of the object. We want to extend the model by combining clusters generated from similar objects so that the new model can be used not only to recognize one kind of object, but also to recognize various similar objects. This part of work can be regarded as an extension of R. Basri's work[8], but we deal with more complex cases than he did.

The model generation is also implemented by using PPCM method. After 2D views of each object having been grouped into several clusters, each cluster is described by a hierarchical graph structure. If there is an edge between nodes n_i and n_j, it means two regions stored in the nodes are connecting. A common model is generated by matching among these hierarchical graphs of each kind of objects.

Suppose C_r and C_s denote graphs of two clusters of different objects. Firstly, we start to find the most similar part in C_s for a component P_{ri} stored in C_r by using Eq.(2). The condition of judging whether two nodes are matched well is done not only by examining Eq.(2), but also examining the features of two regions stored in the nodes. We pay special attention to those features such as convexity, length of major axis and minor axis. Then we construct adjacency relations. For a pair of matched regions, a new region is generated by using Eq.(6). The adjacency relations between two pairs of matched nodes are stored into model graph by using *or* relation, if there are unmatched nodes between them. See Fig. 6 for an illustration.

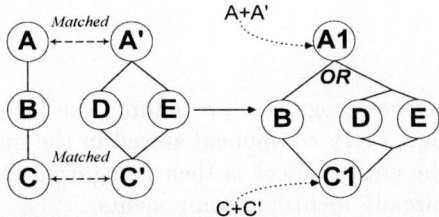

Fig. 6. Generate a model based on matching two different kinds of chairs

After a model being generated based on two objects, it can be used to match with other kinds of object. Fig. 7 shows a few views of three similar objects, which are used to generate a model. The result of model generation is illustrated in Fig. 8. Only those components and adjacency relations which exist in most of objects are stored in the model.

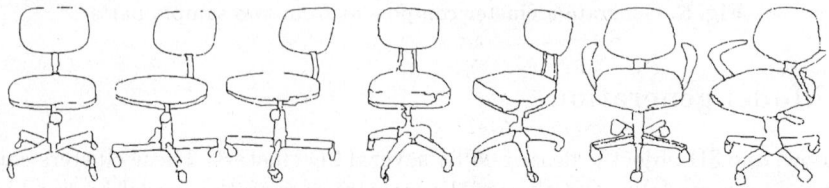

Fig. 7. Objects used to generate the model shown in Fig. 8.

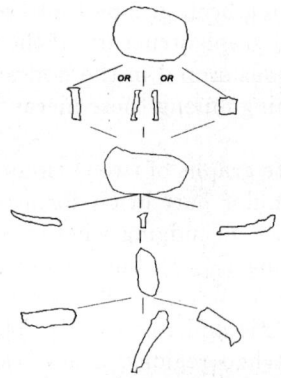

Fig. 8. The model generate from the objects shown in Fig. 7.

5 Recognition

In this section we describe recognition procedure. The recognition works in two stages. In the first stage, every component stored in the model is identified and in the second stage the entire object is then recognized in terms of distinctive relations among the already identified components.

The input is a gray level image and it is segmented first. Then a component can be identified from the result of segmentation of input image, denoted R, by using PPCM method. Among all regions in R, we choose several potential candidates P_j in R which minimize $d(C_i, P_j)$, where $d(X, Y)$ is defined in Eq.(2) and C_i is the component in the model that we want to detect. After all components stored in the model have been detected, those regions that the relations between them are similar to the relations between components in the model is regarded as a result of recognition. How similar are the relations of detected regions to the relations of components in the model can be measured by using relational distance between two views as we mentioned in section 3.

Fig. 10(a) shows the segmentation result of an input image. Fig. 10(b) shows three regions in the image which are most similar to a region stored in the top node of the model. Fig. 10(c) shows result of detecting another component in the middle node of the model. Fig. 10(d) shows the final result.

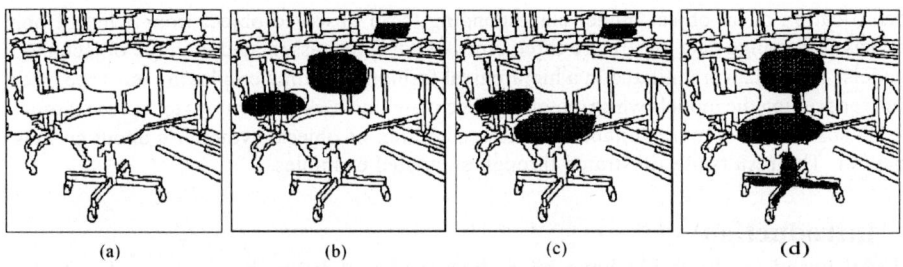

Fig. 9. (a) shows the segment result of an input image; (b) shows result of finding one component in the model; (c) shows result of find another component in the model. (d) shows the final result: a chair is detected.

References

1. Ullman, S.: *High-level vision*, The MIT Press, 1995
2. Sugiyama, T., Abe, K.: Image segmentation using both edge and region information. Proc. of IAPR Workshop on Machine Vision Application, MVA'94, Kawasaki, Japan, (1994) 127–130.
3. Wang, C., Abe, K.: Region correspondence by inexact attributed planar graph matching. Proc. 5th ICCV, (1995) 440–447.
4. Held, A., Abe, K.: On approximate convexity. Pattern recognition letters. Vol.**15** No.**6** (1994) 611–618
5. Pikaz, A., Dinstein, I.: Matching of partial occluded planar curves. Pattern Recognition. Vol.**28** No.**2** (1995) 199-209
6. Abe, K., Morii, R., Nishida, K., Kadonage, T., Comparison of methods for detecting corner points from digital curves — a preliminary report. Proc. 2nd Int'l Conf. on Document Analysis and Recognition. (1993) 854–857
7. Haralick, R., M., Shapiro, L., G.: Computer and robot vision. Chap. 18, Addison-Wesley Publishing, (1993)
8. Basri, R.: Recognition by prototype. Proc. Conf. on Computer Vision and Pattern Recognition. (1993) 161–167

Contour Matching Technique for 3D Object Recognition Using Kalman Filter

M. Hanmandlu
Dept. of Electrical Engg.
I.I.T. Delhi,
New Delhi - 110 016. INDIA
e-mail : mhmandlu@ee.iitd.ernet.in

V. Shantaram
Dept. of Computer Sc. & Egg.
M.S.Ramaiah Institute of Tech.
Bangalore - 560 054. INDIA

Abstract. This paper presents a contour matching technique using the Kalman filter for the identification of an object model corresponding to an observed object from a list of object models from range data. There are three types of edge data associated with the object and the models. These data are utilized in a hierarchical fashion each time employing one type of edge data for pruning the models while matching. The matching uses quaternions to represent rotation in 3D space and is more suitable for the recognition of objects symmetric about an axis of rotation. The results are illustrated through simulated examples.

1 Introduction

Model based methods [1] have been discussed widely in the context of 3D object recognition [3]. In the model based recognition problem, there is a given image of an object O' which is assumed to be a member of a collection of possible objects O_1. ...O_n known a priori. Our problem is to identify the object O' with one of the elements in the list of object models and also to determine its position and orientation. This problem is fundamental to computer vision and has been discussed repeatedly in the literature. Instead of using surface data [2,5,7,8], we can use the edge data in the representation of the object and the models. The approach described here differs from Bolles et al. [2] who have used cylindrical features, Fisher [6] who has used curvature class and Grimson and Lozano-Perez [7] who have used straight line features in the matching process.

In the present work we are concerned with the edge data derived from range images after segmentation [8] and our objective is to use this data for shortlisting a model belonging to the object and then compute its position and orientation. We do not make use of the surface data unlike others as reported in the literature [2,6] where the edge data is also used in addition to surface data for the purpose of representation of 3D objects. Lin and Jungthirapanich [9] have used invariants of 3D contours derived from elliptic Fourier descriptors as features for 3D object recognition. These are implicit functions of axis lengths (i.e., minor and major) as well as the angles defining the relative orientation.

The proposed approach to the object identification problem is based on the notion of rigidly embedded curve C defined by the geometry of the object O and fixed at the origin so that when the object undergoes any rotation and translation (i.e. Euclidean motion) the curve C also moves with it. The object models are stored under some generic categories. In order to find the closest match of the given object in the existing models we first shortlist the models by matching the outer boundary of the object in the scene with outer boundaries (step edges) of stored models and then the inner (semi step and ramp) edges of the observed object are matched with the inner edges of the shortlisted models to calculate the necessary rotation and translation information. Both the outer and inner edges are matched using a fast contour matching technique employing the Kalman filter. Since contour matching involves point sequences, we assume that the scene object and its corresponding model are of the same size.

Organization of this paper is as follows: Contour matching technique using Kalman filter is presented in section 2, while the recognition scheme is given in section 3. The results of implementation and conclusions are relegated to sections 4 and 5 respectively.

2 Contour Matching Technique

Here the points on the boundary C' of an observed image is matched with the points on the boundary C of a model object O assuming both C' and C are parameterized by an arclength s. Since C' and C may not start at the same point, they may be separated by an arc length s_0, which we refer to as an offset. The matching calls for the determination of the offset s_0 and Euclidean motion E for which curves $EC'(s)$ and $C(s+s_0)$ are closest to each other. To be more specific we represent each of the curves C' and C by a sequence of n evenly spaced points. Let these sequences be u_j, v_j, j= 1, ..., n corresponding to the observed contour C' and model contour C respectively. To simplify matching we initially assume that the whole of the boundary O' is visible and no offset calculation is required. Matching then amounts to finding the Euclidean motion E between the sequences by minimizing F.

The measure is therefore

$$F = \min_{E} \sum_{j=1}^{n} |Eu_j - v_j|^2 \qquad (2.1)$$

For proper matching we also translate O' so that the centroid of the contour lies at the origin, such that

$$\sum_{j=1}^{n} u_j = 0 \qquad (2.2)$$

Next, we express **Eu** as **Ru** + **a** where **R** is a rotation matrix and **a** is translation vector. In view of this, (2.1) becomes :

$$F = \min_{R,a} \sum_{j=1}^{n} |Ru_j + a - v_j|^2 \qquad (2.3)$$

2.1 Computation of Translation

To obtain the translation parameter **a** consider the above measure and expand to obtain

$$F = \min_{R,a} \left[\sum_{j=1}^{n} |v_j|^2 + n|a|^2 - 2\sum_{j=1}^{n} a \cdot v_j \right.$$

$$\left. + \sum_{j=1}^{n} |u_j|^2 + 2\sum_{j=1}^{n} a \cdot Ru_j - 2\sum_{j=1}^{n} Ru_j \cdot v_j \right] \qquad (2.4)$$

But,

$$\sum a.Ru_j = a.R\left(\sum u_j\right) = 0 \qquad (2.5)$$

Here **a** and **R** appear independently in **F** so that we can minimize their contributions separately. To minimize over **a**, we simply use

$$a = \frac{1}{n}\sum_{j=1}^{n} v_j \qquad (2.6)$$

as this will eliminate the terms involving **a**, i.e.,

$$n|a|^2 \text{ and } -2\sum_{j=1}^{n} a.v_j$$

2.2 Computation of Rotation

As **a** has been found from the model data, this can be subtracted from v_j to force the centroid of the scene data to be located at the origin. Accordingly, the measure F now becomes:

$$F = \min_{R} \sum_{j=1}^{n} |Ru_j - v_j'|^2 \qquad (2.7)$$

where
$$v_j' = v_j - a \qquad (2.8)$$

Having removed **a** from the minimization process, the measure has to be minimized with respect to rotation **R** which relates v'_j to u_j by the following equation when **R** is known exactly:

$$v_j' = Ru_j \qquad (2.9)$$

Instead of following the approach in [10] for finding **R** which minimizes **F**, we make use of quaternions for representing a rotation in 3D space. As we know that any three dimensional vector **v** is identified with quaternion (o,v) and the product **Rv** can be identified as the product of two quaternions :

$$Rv = q \otimes v \otimes \bar{q} \qquad (2.10)$$

This result is used to minimize **F** as shown below.

2.3 Minimization of F

Consider the measure **F** and treat u_j and v_j' as quaternions in (2.7)

$$F = \min_{R} \sum_{j=1}^{n} |Ru_j - v_j'|^2 \qquad (2.11)$$

Using the relation (2.10) in (2.11), we obtain

$$F = \min_{q} \sum_{j=1}^{n} |q \otimes u_j \otimes \bar{q} - v_j'|^2 \qquad (2.12)$$

Since $|q|^2 = 1$, we can multiply both sides with $|q|^2$ resulting in;

$$F = \min_{q} \sum_{j=1}^{n} |q \otimes u_j - v_j' \otimes q|^2 \qquad (2.13)$$

Using quaternion multiplication [4], (2.13) can be written as:

$$F = \min_{q} \sum_{j=1}^{n} |u_j^- q - v_j'^{+} q|^2 \qquad (2.14)$$

$$= \min_{q} \sum_{j=1}^{n} |Aq|^2 \qquad (2.15)$$

Where

$$A = u_j^- - v_j'^{+} = \alpha^- - \beta^+$$

$$\alpha^- = \begin{bmatrix} \alpha_0 & -\alpha^t \\ \alpha & \alpha_0 I - \bar{\alpha} \end{bmatrix}, \beta^+ = \begin{bmatrix} \beta_0 & -\beta^t \\ \beta & \beta_0 I + \bar{\beta} \end{bmatrix}$$

$$\bar{\alpha} = \begin{bmatrix} 0 & -\alpha_3 & \alpha_2 \\ \alpha_3 & 0 & -\alpha_1 \\ -\alpha_2 & \alpha_1 & 0 \end{bmatrix}$$

We now invoke the Kalman filtering with the performance index

$$I = \sum \|Z - HX\|^2 \qquad (2.16)$$

For the minimization of **F**, Comparing (2.15) and (2.16), we find the following correspondences:

$$Z = 0, \quad H = A, \quad X = q \qquad (2.17)$$

The outcome of the Kalman filtering is the converged value for the state **X**. From the elements of q, **R** is computed as follows [11]:

$$R = \begin{bmatrix} q_0^2 + q_1^2 - q_2^2 - q_3^2 & 2(q_1 q_2 - q_0 q_3) & 2(q_1 q_3 + q_0 q_2) \\ 2(q_1 q_2 + q_0 q_3) & q_0^2 - q_1^2 + q_2^2 - q_3^2 & 2(q_2 q_3 - q_0 q_1) \\ 2(q_1 q_3 - q_0 q_2) & 2(q_2 q_3 + q_0 q_1) & q_0^2 - q_1^2 - q_2^2 + q_3^2 \end{bmatrix} \qquad (2.18)$$

If O' is partially occluded, then we have to match the sequence u_j, $j = 1, ..., n$ to each of the contiguous sub-sequences v'_{j+d}, $j = 1,..., n$ of the circular sequence v'_j, $j = 1, ..., m$, where d is the offset between the starting points of two sequences.

Here we can appropriately assume $m > n$, otherwise the partial periphery of O' can not match O. For each d, (2.11) can be written as

$$F_d = \min_R \sum_{j=1}^{n} |Ru_j - v'_{j+d}|^2 \qquad (2.19)$$

for $d = 0,1,...,m-1$

and in view of (2.14), (2.19) can be expressed as:

$$F = \min_q \sum_{j=1}^{n} |(u_j^- - v'^{/+}_{j+d})q|^2 \qquad (2.20)$$

with $A = u_j^- - v'^{/+}_{j+d}$. Thus the elements in $v_j^{/+}$ have to be shifted by **d**. After obtaining the solution, we find **R** from (2.18).

3 The Proposed Recognition Scheme

In our model based recognition scheme, models of objects are stored in data base through their outer boundaries and the inner edges. Recognition of an object takes place by matching the outer boundary and the inner edges of an observed object with corresponding outer boundary and inner edges of models. Matching with both the outer boundary and inner edges is necessary because two different 3D objects may have the same outer boundary but different inner edges.

We have implemented the segmentation procedure of [8] in two parts, the first part generates the outer boundary with consists of step edges and the second part generates the semi step and roof edges. In this procedure, the first equidistance contours (EDCs) are found by slicing the range data at equal increments of depth. Next, three types of critical points are found. The first type yields the step edges in which the points separate the object from the background. The second type yields the semi step edges where two surfaces appear to meet but they do not. The third type yields the ramp edges at which two surfaces meet. These are found using the gradient operator [8]. This procedure can be implemented in parallel to extract outer boundary and inner edges simultaneously.

To reduce the computational effort, first only the outer boundary of the input segmented scene is matched with the outer boundaries of all the models. Using boundary matching discussed in the previous section, we find the minimum value of error measure for best possible match between each model and the scene. The error measure is taken to be the trace of the state error covariance matrix obtained from the Kalman filter. The model whose outer boundary yields the minimum error is selected, and the inner edges of this model are matched with the inner edges of input segmented image to complete the recognition process.

4 Results of Implementation

We have considered range images of three different objects for our study. These range images are passed on to the segmentation routine. The segmentation is done in two stages with the first stage giving the step edges or the outer boundary and the second stage giving the semi-step and roof-edges. The original range images and the boundary edges of the images and segmented images are shown in Figs 1-3.

The scene data has been generated by applying a rotation to the corresponding model data, i.e., outer boundary and inner edges after translating the associated centroid to the origin. We have generated 3 input scenes for our study. Thus, input scene 1 is obtained from model 2 after it is rotated by an angle of 30^0 about the Z axis. Input scene 2 is obtained from model 1 after it is rotated by an angle of 180^0 about the Z axis. Likewise the third input scene is generated. Thus there is no translation between the scene and the model data. The actual rotation matrices to generate scene1, scene2 and scene3 are :

$$\begin{bmatrix} 0.866 & -0.5 & 0 \\ 0.5 & 0.866 & 0 \\ 0 & 0 & 1 \end{bmatrix}, \begin{bmatrix} 1 & 0 & 0 \\ 0 & -1 & 0 \\ 0 & 0 & -1 \end{bmatrix}, \begin{bmatrix} -0.966 & 0.259 & 0 \\ 0.259 & -0.966 & 0 \\ 0 & 0 & 1 \end{bmatrix}$$

It may be mentioned that in the proposed technique, the translation component may be computed first before proceeding with the computation of rotation component. In this way the two computations are delinked. The centroids of the first, the second and the third images are (84, 26, 176), (61, -64,-191), and (-74, -45, 199). These centroids are then shifted to the origin resulting in the model data. The scene data are created by multiplying model data with the above rotation matrices.

Since $Z = 0$ in(2.17), we generate the initial values for the state vector X by a random number generator. The initial state error covariance matrix is chosen as Diag{10}. Results of matching the outer boundaries of three input scenes with various models are given in Table 1.

Table 1 : Results of contour matching

Model No	Scene 1 Quaternions a_o, a_1, a_2, a_3 (Error measure)	Scene 2 Quaternions a_o, a_1, a_2, a_3 (Error measure)	Scene 3 Quaternions a_o, a_1, a_2, a_3 (Error measure)
1	0.6451535 0.1293007 0.4498127 0.0994000 (7.2142×10^5)	0.0000 1.0000 0.0000 0.0000 (0.0000×10)	0.2923059 -0.2684609 -0.2066246 -0.7231419 ($6.73.9 \times 10^5$)
2	0.9329098 -0.0000367 0.0002674 0.2501780 (8.345×10^1)	0.0077303 0.6522407 0.0718296 0.4707478 (7.2277×10^5)	0.2175508 0.1591400 -0.3609342 -0.7169622 (5.5084×10^5)
3	0.7753804 0.3937158 0.1153466 0.0764765 (5.502708×10^5)	0.2079023 -0.5784629 0.3961512 -0.2090081 (7.49116×10^5)	0.1293551 0.0000393 -0.0000919 -0.9829775 (5.7299×10^1)

From the table we find that the model 2 is shortlisted to scene 1, model 1 is shortlisted to scene 2 and model 3 to scene 3. The respective computed rotation matrices are:

$$\begin{bmatrix} 0.8072 & -0.4673 & -0.0008 \\ 0.46773 & 0.8072 & -0.0003 \\ 0.0008 & -0.0002 & 0.9327 \end{bmatrix}, \begin{bmatrix} 1 & 0 & 0 \\ 0 & -1 & 0 \\ 0 & 0 & -1 \end{bmatrix}, \begin{bmatrix} -0.9497 & 0.2538 & -0.0006 \\ -0.2538 & -0.9497 & 0.0004 \\ -0.0005 & 0.0005 & 0.9830 \end{bmatrix}$$

5 Conclusions

In this paper, the boundary matching technique in [10] has been reformulated to encompass the recoenition scheme for 3-D objects. However, the basic approach is quite different. In our scheme, outer boundary of the scene is matched with the outer boundaries of all the listed models. When the number of models increases the proposed method of point to point matching becomes computationally prohibitive. We have to necessarily go in for signatures which are invariant to translation and rotation. The computed values of rotation are quite comparable with the actual rotation values.

We have used edges in our model based recognition scheme. The edge data reduces the total amount of range data needed for the purpose of object recognition. A recognition system based on edge description, however, is not robust and can not distinguish between very similar objects. Thus to make a robust system, both edge and surface information should be used. The edge information can be effectively used for shortlisting of models for the purpose of matching and surface information can be used for final recognition.

The technique would work only with the points that do not move as the viewpoint changes. Thus, the boundary matching is intended for the recognition of surfaces generated about an axis of rotation. For the recognition of general surfaces we need the wireframe models of the objects.

References

1. Besl, P.J. and Jain, J.C., Three Dimensional Object Recognition, ACM Computing Surveys Vol.17, 1975, pp.75-145.
2. Bolles, R.C., Horaud, P.D. and Hannah, M.J., 3DPO : A Three Dimensional Part Orientation System, Proceedings of 8th IJCAI, 1983, pp.116-120.
3. Chin, R.T. and Dyer, C.R., Model Based Recognition in Robot Vision, ACM Computing Surveys, Vol.18, 1986, pp.67-108.
4. Chou, J.C.K. and Kamel, M., Quaternions Approach to Solve the Kinetic equation of rotation, Aa Ax = Ax Ab of Robotic manipulator, Proceedings of IEEE Int. Conf. Robotics and Automation, 1988, pp.656-662.
5. Faugeras, O.D. and Hebert, M., The Representation, Recognition and Locating of 3-D Objects, Int. J. Robotics Research, Vol.3, 1986, pp. 27-52.
6. Fisher, R.B., **From Surfaces to Objects : Computer Vision and Three Dimensional Analysis**, John Wiley & Sons Ltd. 1989.
7. Grimson, W.E.L, and Lozano-Perez, T., Model Based Recognition and Localization from Sparse Range Data or Tactile Data, Int. J. Robotics Research, Vol.3, 1984, pp.3-35.
8. Hanmandlu, M., Rangaiah, C., and Biswas, K.K., Quadrics Based Matching

Technique for 3D Object Recognition, Image and Vision Computing, Vol.10, 1992, pp.578-588.
9. Lin, C.S. and Jungthirapanich, C., Invariants of three dimensional contours, Pattern Recognition, Vol.23, 1990, pp.833-842.
10. Schwartz, J. and Sharir, M., Identification of Partially Observed Objects in Two and Three Dimensions by Matching Noisy Characteristic Curves, The Int. J. Robotics Research, Vol. 6, 1986, pp. 29-43.
11. Zhang, Z. and Faugeras, O.D., Determining Motion from 3D Line Segments : A Comparative Study, Image and Vision Computing, Vol. 9, 1991, pp. 10-19.

 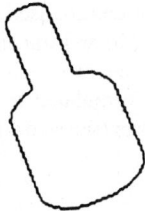 

Fig.1   original range image; (b) boundary edges; (c) segmented image

 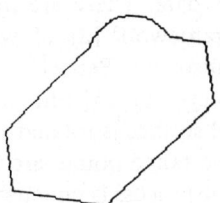 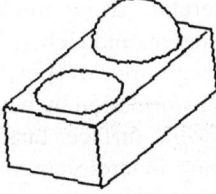

Fig.2   original range image; (b) boundary edges; (c) segmented image

 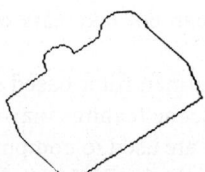 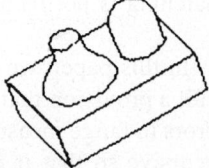

Fig.3   original range image; (b) boundary edges; (c) segmented image

Kalman Filter Based Matching Technique for 3D Object Recognition

M. Hanmandlu
Dept. of Electrical Engg.
I.I.T. Delhi,
New Delhi - 110 016. INDIA
e-mail : mhmandlu@ee.iitd.ernet.in

V. Shantaram
Dept. of Computer Sc. & Engg.
M.S.Ramaiah Institute of Tech.
Bangalore - 560 054. INDIA

Abstract. A recursive matching technique which uses the Kalman filtering for the recognition of 3D objects is presented. We make use of model based methodology in which both the models and scenes are assumed to be described by quadratic equations. The parameter matrices involved in the matrix form of quadratic equations are used in the matching process. The model features derived from these matrices are rotated and translated so as to match with those of the scene. The features consist of Euler parameters which represent the orientation of a surface and translation vector. The matching is formulated so as to apply the Kalman filter equations and the trace of the error covariance matrix (the error measure) guides the search process in pairing a model with a scene.

1 Introduction

Recognition of 3D objects from range images is an active field of research [14,1] since range data is insensitive to occlusion and type of illumination [5]. Model based methodology is widely employed in the recognition of objects as it reduces the requirement for multiple views [2]. In this paper, we present a model based 3D object recognition method using range data. There are many techniques available in the literature using model based framework [2] of which mention may be made of Faugeras and Hebert [5] and Grimson and Perez [7]. Both these methods deal with the recognition of polyhedral objects. In [5] the matching consists of finding a transformation by which a model surface is rotated and translated so as to match with a scene surface. In [7] the sparse range points are matched with the object models using an interpretation tree. Search process is cut down by applying certain constraints. Fisher [6] uses curvature class and structural associations in the matching. Suk and Bhandarker [14] have discussed recognition and localization of polyhedral and curved objects using Hough transform and Hough clustering. Hanmandlu et al. [9] have presented a quadrics based matching technique for the recognition of curved objects following the framework of Faugeras and Hebert [4]. The matching is similar to that in [4] except that the features are different and derived from quadrics. However, the matching is not recursive and it can not take care of errors in the features.

In this paper, we propose a Kalman filter based solution to the matching problem with a provision that model and scene features may contain error as a consequence of errors in range measurements that are used to compute features. The matching is made recursive so that it is helpful to obtain final rotation and translation in a recursive manner.

The organization of the paper is as follows : Section 2 deals with the segmentation and representation of scene surfaces and section 3 is devoted to the matching process. The search strategy is given in section 4. Section 5 presents a case study. Finally, section 6 gives conclusions.

2 Segmentation and Representation of 3D Surfaces

2.1 Segmentation

Different segmentation techniques have been discussed in [1]. However, we have followed a simple approach as described in [9]. In this, the range image is sliced to produce equidistant contours, using which three types of critical points are determined. The first type yields the step edges which separate the object from the background. The second type yields the roof edges which separate one surface from another. The third type yields the semi-step edges at which two surfaces appear to meet but they do not.

2.2 Representation of a surface

A curved surface of quadric type is represented by the following quadratic equation

$$ax^2 + by^2 + cz^2 + dxy + exz + fyz + gx + hy + iz + j = 0 \tag{1}$$

In matrix form, the above equation becomes

$$[x\ y\ z]\begin{bmatrix} a & d/2 & e/2 \\ d/2 & b & f/2 \\ e/2 & f/2 & c \end{bmatrix}\begin{bmatrix} x \\ y \\ z \end{bmatrix} + \begin{bmatrix} g \\ h \\ i \end{bmatrix}^t \begin{bmatrix} x \\ y \\ z \end{bmatrix} + j = 0 \tag{2}$$

or, $X^t Q X + C X + D = 0$ \hfill (3)

where $X = [x\ y\ z]^t$

Defining $\bar{a} = [a\ b\ c\ d\ e\ f\ g\ h\ i\ j]^t$
$\bar{X} = [x^2\ y^2\ z^2\ xy\ xz\ yz\ x\ y\ z\ 1]^t$

We rewrite the eqn. (1) as

$$\bar{X}^t \bar{a} = 0 \tag{4}$$

If $\bar{X}$ is constructed from the coordinates $[x_i, y_i, z_i]$ of i th point, then eqn. (4) becomes:

$$\bar{X}_i^t \bar{a} = 0 \tag{5}$$

If the fitted function $\bar{X}_i^t$ does not satisfy eqn. (5) because of errors, then eqn. (5) is modified as

$$e_i = \bar{X}_i^t \bar{a} \tag{6}$$

In fitting a quadric to the range data of a surface, we form an objective function which has as its argument the error function over n data points. Defining $\rho(e_i)$ as the error metric, the objective function is given by

$$\zeta(\bar{a}) = \sum_{i=1}^{n} \rho(e_i) \tag{7}$$

where $\rho(e_i) = e_i^2$

The gradient vector $\Delta \zeta(\bar{a})$ is

$$\Delta \zeta(\bar{a}) = \sum_{i=1}^{n} \partial \frac{\rho(e_i)}{\partial a_k} = \sum_{i=1}^{n} \partial \frac{\rho(e_i)}{\partial e_i} \cdot \frac{\partial e_i}{\partial a_k} \quad , k=1...10 \tag{8}$$

We define the derivative of the error metric as

$$\psi(e_i) = \frac{\partial \rho(e_i)}{\partial e_i} \tag{9}$$

Using this definition, eqn. (8) is rewritten as :

$$\Delta \zeta(\bar{a}) = \sum_{i=1}^{n} \frac{\partial e_i}{\partial a_k} \psi(e_i), \quad k=1\ldots 10$$

$$= \left[\sum_{i=1}^{n} \frac{\partial e_i}{\partial a_1} \psi(e_i), \ldots, \sum_{i=1}^{n} \frac{\partial e_i}{\partial a_{10}} \psi(e_i) \right] \tag{10}$$

For n data points $\rho(e)$ is computed as

$$e^t e = \bar{a}^t X^t X \bar{a} = \bar{a}^t R \bar{a} \tag{11}$$

where $X = [\bar{x}_1, \ldots, \bar{x}_n]^t$; $R = X^t X$

The solution vector $\bar{a}$ is approached by proceeding down a direction that is conjugate to the previous direction using the algorithm of Fletcher and Reeves [12], which requires a starting point, function values at a point and gradient value. In order to test the efficacy of this method, a surface is generated with a given coefficient vector and from the generated points on the surface, the coefficient vector is recomputed.

If an object being considered has N surfaces, then we would have N coefficient vectors describing the entire object. The classification of surfaces is also possible with the calculation of invariant indices from the coefficients [8,13]. This classification provides qualitative constraints about the type of surfaces so that the matching can be attempted on the same type of surfaces. From the coefficients, we form the parameter set [Q, C, D] to extract the feature vector required in the matching.

As described in [9], the eigenvalues of the matrix Q give the length of principal axes and the corresponding eigenvectors give the orientation of the surfaces. We require a transformation to convert Q into a diagonal matrix. This transformation corresponds to a rotation matrix that rotates the coordinate axes of the original surface such that the rotated axes are orthogonal to the old ones. Thus this rotation matrix represents the orientation of the surface.

3 Matching Process

In the model based object recognition, an object model is defined with respect to an object oriented coordinate system and an observed scene is defined with respect to a viewer oriented coordinate system. A model contains all the information about its surfaces whereas a scene contains the information about a few surfaces that are visible. Each surface is represented by the parameter set [Q, C, D]. Our objective is to find a transformation J that rotates and translates a model surface so that it is compatible with a scene surface with respect to a measure of consistency. Let $[Q_m, C_m, D_m]$ be the parameters of a model surface and $[Q_s, C_s, D_s]$ be the corresponding parameters of a scene surface. Let J consist of both rotation R and translation T. In [4] it has been shown that the following relations are satisfied by the model and scene parameters after the application of R and T on model parameters :

$$Q_s = R Q_m R^t \tag{12}$$

$$C_s = C_m + 2 R^t T \tag{13}$$

$$D_s = T^t Q_m T + D_m \tag{14}$$

In eqn. (12), we replace Q_m by R_m, and Q_s by R_s, to obtain :
$$R_s = R R_m R^t \tag{15}$$
where R_m and R_s stand for orientations of model and scene surfaces respectively.
We need two equations to find two unknowns R and T. In [9] eqns. (15) and (14) have been used whereas we use here eqns. (15) and (13) as this set of equations would make the estimation of R and T to be recursive. Equation (14) can perhaps be used for validation of results.

If there are N_s scene surfaces and N_m model surfaces represented by the feature vectors U_{si} and U_{mj} respectively, then the measure of consistency is devised as :
$$G(M) = \operatorname*{Min}_{J} \sum (U_{si} - JU_{mj})^2 \tag{16}$$
subject to the condition $S_{si} = S_{mj}$
where M is the matching of pairs (i,j) and S_{si}, S_{mj} denote the types of scene and model surfaces respectively. In view of the selected eqns. (15) and (13), the above matching criterion is rewritten as :
$$G(M) = \left[\sum_i (R_{si} - RR_{mj} R^t)^2 + w \sum_i (C_{si} - C_{mj} - 2R\,^tT)^2 \right] \tag{17}$$
for $S_{si} = S_{mj}$
where w is a weighting factor. A model surface is assumed to be rotated first followed by translation, before comparing with a scene surface. A j th model surface is represented by $U_{mj} = \{R_{mj}, C_{mj}\}$ and the i th scene surface is represented by the feature vector $U_{si} = \{R_{si}, C_{si}\}$. S_{si} and S_{mj} are the qualitative constraints required to limit the search between the surfaces of the same type. While assuming the form of eqn. (17), we have neglected the effect of translation T on the first sum $F_e = \sum_i (R_{si} - RR_{mj} R^t)^2$. Since we are making rotation first, the effect of T on F_e does not arise. Now, we compute R by minimizing the sum F_e. But for estimating T, we need the minimization of the second sum $F_t = \sum (C_{si} - C_{mj} - 2R\,^tT)^2$. However, in this case the errors in the estimation of R are reflected in the estimation of T. We can choose an appropriate value for w to minimize the effect of errors in T due to R.

3.1 Estimation of Rotation

As noted above, rotation matrix R is estimated by minimizing F_e. Since R contains nine nonlinear terms, the minimization becomes difficult. If a rotation of an angle θ is assumed about an arbitrary axis to yield R then this R can be represented by a unit quaternion containing only four parameters known as Euler parameters [3]. Conversion of R into unit quaternion is given in [11].

3.1.1 Minimization of F_e
Consider the sum
$$F_e = \sum_i (R_{si} - RR_{mj} R^t)^2 \tag{18}$$
Now, multiplying F_e by $\|R\|^2$, we obtain
$$F_e \|R\|^2 = F_e = \sum_i (R_{si} R - RR_{mj})^2 \tag{19}$$

Replacing R_{si}, R_{mj} and R by equivalent Euler parameter l_{si}, l_{mj} and l respectively in eqn. (19) we obtain

$$F_e = \sum_i (l_{si} \otimes l - l \otimes l_{mj})^2 \qquad (20)$$

Using the result of quaternion multiplication from [3], eqn. (20) is rewritten as [10]:

$$\begin{aligned} F_e &= \sum_i (l_{si}^+ l - l_{mj}^- l)^2 \\ &= \sum_i (Al)^2 \end{aligned} \qquad (21)$$

where $A = l_{si}^+ - l_{mj}^-$

We now minimize this F_e using the Kalman filter which employs the following performance index :

$$I = \sum_i \|Z_i - H_i X\|^2 \qquad (22)$$

Comparing eqn. (21) with eqn. (22), we obtain the following correspondences :

$$Z_i = 0; \quad X = l \text{ and } H_i = A \qquad (23)$$

The Kalman filter estimates the state X given the initial values of state vector, the measurement matrix H_i and state error covariance matrix P; the trace of this matrix gives the error involved in the matching of rotational components (R_{si}, R_{mj}). It may be noted that eqn. (22) has to be minimized for each i and H_i is fixed during the estimation of X.

3.2 Estimation of Translation

Having obtained R, we now consider the minimization of the second sum F_t given by

$$F_t = \sum_i (C_{si} - C_{mj} - 2R\,{}^tT)^2 \qquad (24)$$

for the estimation of T. Comparing eqn. (24) with eqn. (22), we establish the following correspondences.

$$Z_i = C_{si} - C_{mj}, \quad H_i = 2R\,{}^t, \quad X = T \qquad (25)$$

As mentioned above, the Kalman filter has to be run for each i to estimate the state X and trace of the state error covariance matrix gives the error involved in the matching of translational components (C_{si}, C_{mj}).

Use of correspondence relations in eqns. (23) and (25) makes the estimation of R and T recursive while pairing a scene surface with a model surface. This facility is the outcome of considering the eqn. (13) in place of eqn. (14) to estimate T.

4 Search Strategy

The matching problem leads to a combinatorial explosion if the constraints are not imposed. The first constraint is that we allow matching to take place between the surfaces of the same type. This requires the classification of surfaces. For this, we can use invariant indices computed from the coefficients of the quadratic equation [13]. Next, we can use the connectivity matrix as the second constraint while deciding about the likely model surfaces to be matched to a particular scene surface. Since the rotation matrix is represented by equivalent Euler parameters, the feature vectors now consist of $U_{si} = (l_{si}, C_{si})$ and $U_{mj} = (l_{mj}, C_{mj})$. We have used an exhaustive search along with the connectivity constraints. At the first level, we make arbitrary pairings of model and scene surfaces. At the second level, pairings are evaluated on the basis of error measures obtained from the traces of state error covariance matrices during estimation. However, these pairings are made subject to the satisfaction of connectivity constraints. This procedure is continued until all the scene surfaces find matches from the set of model surfaces. When a particular pairing gives a larger error, it is eliminated and not pursued further.

5 Results of a Case Study

We have considered the range data of a composite object consisting of a small cylinder on the top of a larger cylinder. We have used edge based segmentation for demarcating visible surfaces. Symmetry property is not made use of here, as a result, each cylinder is supposed to have front and back surfaces. The segmented image is shown in Fig. 1. To carry out the matching process, a CAD model of the object is built up as shown in Fig. 2, whose surfaces are modeled as quadrics. The connectivity matrices are given in Table 1 for both the model and the scene. All the surfaces are labeled. The range data for each surface is taken for determining the coefficients of quadratic equation. We then form the parameter sets (Q_{si}, C_{si}, D_{si}) and (Q_{mj}, C_{mj}, D_{mj}) corresponding to each scene surface and model surface respectively. From Q_{si} and Q_{mj}, we compute eigenvector matrices R_{si} and R_{mj} which correspond to orientation of scene and model surfaces respectively. Next, we replace these by the equivalent Euler parameters l_{si} and l_{mj}. Now the feature vectors consist of (l_{si}, C_{si}) and (l_{mj}, C_{mj}).

Table 1: Connectivity matrices and results of matching

Model connectivity matrix							Scene connectivity matrix				Solution Matching	Kalman error measure
1	1	0	0	1	1	0	1	1	0	0	(1, 3, 7, 6)	16.3289
1	1	0	0	1	1	0	1	1	1	0	(2, 3, 7, 6)	19.6199
0	0	1	1	0	1	1	0	1	1	1	(3, 1, 7, 6)	22.2236
0	0	1	1	0	1	1	0	0	1	1	(4, 3, 7, 1)	19.2777
1	1	0	0	1	0	0					(5, 3, 7, 6)	19.6362
1	1	1	1	0	1	0					(6, 3, 7, 5)	31.9626
0	0	1	1	0	0	1					(7, 3, 2, 1)	19.1587

In the matching process, we treat planes as a special case of quadrics. Note here that we have not used any qualitative constraints such as type of surfaces during matching for the reasons cited in [9]. Next, the matching is done by following an exhaustive search using the connectivity of the surfaces for guiding the search. In the matching process the first scene surface is assigned to the first model surface and then we proceed to find matches for the other scene surfaces. This requires matching for the next three levels as we have a total of 4 scene surfaces. At the end of this, error is computed. In order to avoid the possible error at the first level because of arbitrary assignment, we need to repeat the entire matching by pairing the first scene surface with the 2nd, 3rd and so on upto 7th model surface. The errors generated for each such matching are shown in Table 1. From this Table the selected final solution consists of visible surfaces 1, 3, 7, 6 of the model, viz., bottom front cylinder surface, top front cylinder surface, top plane and middle plane respectively as identified in the model surfaces. This solution has been obtained for w of 0.5.

6 Conclusions

A Kalman filter based matching technique is presented for the recognition of 3D curved objects from the range images. Each surface is assumed to be a quadric where a plane is treated as a special case. From the coefficients of a quadratic equation that fits a quadric, feature vectors are generated. They consist of Euler parameters and translation vectors. The feature vectors of model after rotation and translation would be compatible with those of scene. Rotation and translation components are estimated recursively using the Kalman filter and the resulting errors are utilized in selecting the pairings of scenes and models. The use of connectivity matrices has drastically reduced the search process. However, qualitative constraints have not been used due to lack of an efficient method for classification of quadrics.

References

1. Besl, P.J. and Jain, R.C., Three - Dimensional Object Recognition, Computing Surveys, Vol. 17, No. 1, 1985, pp. 75-145.
2. Chin, R.T. and Dyer, C.R., Model-Based Recognition in Robot Vision, ACM Computing Surveys, Vol. 18, No. 1, 1986, pp. 67-108.
3. Chou, J.C.K. and Kamel, M., Quaternions approach to the Kinematic Equation of Rotation, $A_a A_x = A_x A_b$ of a Robotic Manipulator, Proc. IEEE Int. Conf. on Robotics and Automation, Philadelphia, 1988, pp. 656-662.
4. Faugeras, O.D. and Hebert, M., A 3D Recognition and Positioning Algorithm using Geometrical Matching between Primitive Surfaces, Proc. 8th IJCAI, Karlschule, Germany, 1983, pp. 116-120.
5. Faugeras, O.D. and Hebert, M., The Representation, Recognition and Locating 3D Objects, Int. J. Robotics Research, Vol. 13, 1986, pp. 155-167.
6. Fisher, R.B., **From Surfaces to Objects : Computer Vision and Three Dimensional Analysis**, Wiley, New York, 1989.
7. Grimson, W.E.L. and Lozano-Perez, T., Model-Based Recognition and Localization from Sparse Range data or Tactile data, Int. J. Robotics Research, Vol. 3, 1984, pp. 3-35.
8. Hall, E.L., Tio, J.B.K., McPherson, C.A., Draper, C.S. and Sadjadi, F.A., Measuring Curved Surfaces for Robot Vision, Computer, Vol. 15, No. 12, 1982, pp. 42-54.
9. Hanmandlu, M., Rangaiah, C. and Biswas, K.K., Quadrics based matching technique for 3D Object Recognition, Image and Vision Computing, Vol. 10, No. 9, 1992, pp. 577-588.
10. Hanmandlu, M. and Shantaram, V., Contour Based Matching Technique for 3D Object Recognition using the Kalman filter, Proc. of 3rd Asian Conference on Computer Vision, Hong Kong, Jan. 8-11, 1998.

11. Paul, R.P., **Robot Manipulators : Mathematics, Programming and Control**, MIT Press, Cambridge, MA, 1981.
12. Press, W., Flannery, B., Teukolsky, S. and Vetterling, W., **Numerical Recipes in C : The Art of Scientific Computing**, Cambridge, 1986.
13. Sadjadi, F.A. and Hall, E.L.,Three Dimensional Moment Invariants, IEEE Trans. on Pattern Analysis and Machine Intelligence., Vol. 2, 1980, pp. 127-136.
14. Suk, M. and Bhandarkar, S.M., **Three - Dimensional Object Recognition from Range Images**, Springer-Verlag, Tokyo, 1992.

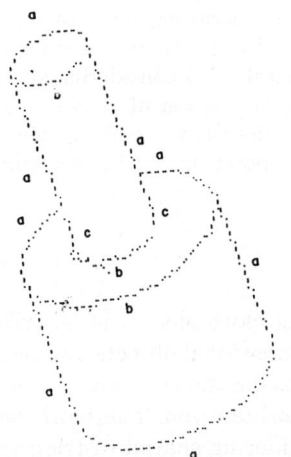

Fig.1

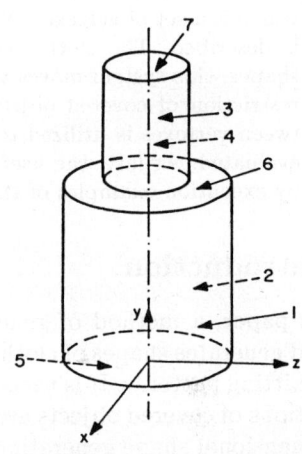

Fig.2

A Generating Method for 3-dimensional Knitting Cloth Shapes

Tatsushi Funahashi[1]   Tsuyoshi Miyazaki[2]
Masashi Yamada[1]   Hirohisa Seki[1]   Hidenori Itoh[1]

[1] Department of Artificial Intelligence and Computer Science
Nagoya Institute of Technology
Gokiso-cho, Showa-ku, Nagoya, Japan, 466
[2] NEC Corporation
tfuna@juno.ics.nitech.ac.jp

Abstract. In this paper, a method of generating the 3-dimensional knitting clothes shapes is described. First, the method of representation of the cloth is described. In this method, a cloth is represented by a mesh, which is a set of vertexes. Next, a method of generating the cloth shapes is described. The cloth covers 3-dimensional objects. To generate the shapes, this system moves vertexes of the mesh with considering spatial restriction of covered objects. In addition, the action of repulsion between vertexes is utilized to prevent shrink the shapes, and its effect is evaluated. At last, the usefulness of the proposed method is confirmed by execution examples of the system.

1 Introduction

In this paper, a method of generating knitting cloth shapes is described. This method generates shapes of cloth covering 3-dimensional objects and pastes them with knitting patterns. It is necessary for generating the cloth shapes that spatial restrictions of covered objects are taken into consideration. Therefore, we present a 3-dimensional shape generating method considering spatial restrictions of covered objects.

A cloth is represented by a mesh of $m \times n$ vertexes. In process of generating the cloth shapes, vertexes move under spatial restrictions of covered objects and length of linkage between vertexes.

Knitting patterns are generated by Itoh's method [1][2]. 3-dimensional knitting pattern is obtained by pasting the generated shapes with the generated knitting pattern.

2 A Generating Method for the 3-dimensional Cloth Shapes

2.1 Representation of the Cloth

A cloth is represented by a mesh of $m \times n$ vertexes. The vertexes are arranged in 3-dimensional space. A vertex $V_{i,j}$ is linked with $V_{i-1,j}$, $V_{i+1,j}$, $V_{i,j-1}$, and $V_{i,j+1}$ (See Figure 1).

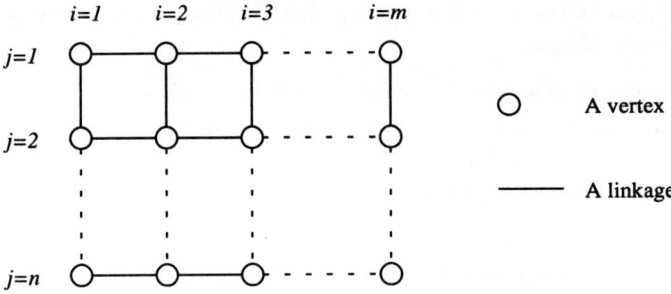

Fig. 1. Data structure of mesh representing the cloth

Each linkage is given a length.

2.2 Movement of Vertexes

In this section, a movement of vertexes is defined. This movement is applied in order to generate the cloth shape covering 3-dimensional objects.

The moving vector $v_{i,j-move}$ of a vertex $V_{i,j}$ is defined by equation (1).

$$v_{i,j-move} = \alpha \left\{ \sum_{(k,l)\in \mathcal{R}} \frac{v_{i,j,k,l}}{|v_{i,j,k,l}|} \left(\frac{1}{1+e^{-(|v_{i,j,k,l}|-l_{i,j,k,l})}} |v_{i,j,k,l}| - \beta l_{i,j,k,l} \right) \right.$$
$$\left. + v_{external} \right\}, \qquad (1)$$

where

- $\mathcal{R}$ is the set regrouping all couples (k, l) such as $V_{k,l}$ linked to $V_{i,j}$,
- $v_{i,j,k,l} = \overrightarrow{V_{i,j}V_{k,l}}$,
- $l_{i,j,k,l}$ is a given length of linkage between $V_{i,j}$ and $V_{k,l}$,
- α is a moving coefficient,
- β is a repulsion coefficient,
- $v_{external}$ is input of an external force.

$v_{external}$ represents an external force. For example, gravitation, force of the wind and so on. The moving coefficient α affects amount of movement at a time. If α is too large, interferences between vertexes are happened and the cloth shape does not become stable. If α is too small, it takes long time to become stable. The repulsion coefficient β affects repulsion between vertexes. If β is zero, repulsion is not affected.

2.3 An Algorithm of Generating Cloth Shapes Covering 3-dimensional Objects

Here, an algorithm of generating cloth shapes is presented. This algorithm takes account of spatial restrictions of covered objects while it iterates movement of vertexes, and then generates a cloth shape.

This algorithm is as follows.

1) Set initial state.
   - Arrange mesh in the 3-dimensional space.
   - Select fixed vertexes among the mesh.
   - Arrange covered objects in the 3-dimensional space.
2) Change of environment.
   - Move the fixed vertexes, at will.
   - Move or deform the covered objects.
3) Move vertexes except fixed vertexes by equation (1). If a vertex will be in the covered object, move the vertex to the nearest outer point of the object.
4) Iterate step 3) until every vertex becomes stable, (that is until every vertex becomes stable).
5) Return to 2) or end.

3 A Transformation Technique from Stitch Symbols to Knitting Patterns

We have already proposed a method of generating knitting patterns from stitch symbols [1][2]. The stitch symbols (see Figure 2(a)) are JIS(Japanese Industrial Standards), which shows a basic stitch pattern.

This method, first, generates an initial string diagram (see Figure2(c)) from a pattern knitting diagram, where the pattern knitting diagram is a combination of stitch symbols (see Figure2(b)). In this process, the string diagram database as Figure 3 is used. Stitch symbols are replaced with corresponding string diagrams by referring the string diagram database.

This method, next, generates a knitting pattern from the initial string diagram. In this process, cross points (where two strings cross) in the initial string diagram are moved by tension of string. When cross points are moved, topology of string diagram does not change.

An generating example is shown Figure 4. Figure 4(b) is a knitting pattern generated from the pattern knitting diagram of Figure 4(a).

4 A Method for Pasting the Cloth Shapes with Knitting Pattern

3-dimensional knitting cloth shapes are generated by pasting the 3-dimensional cloth shapes with knitting patterns. This process is shown in the following steps 1), 2), 3) and 4). (See Figure 5).

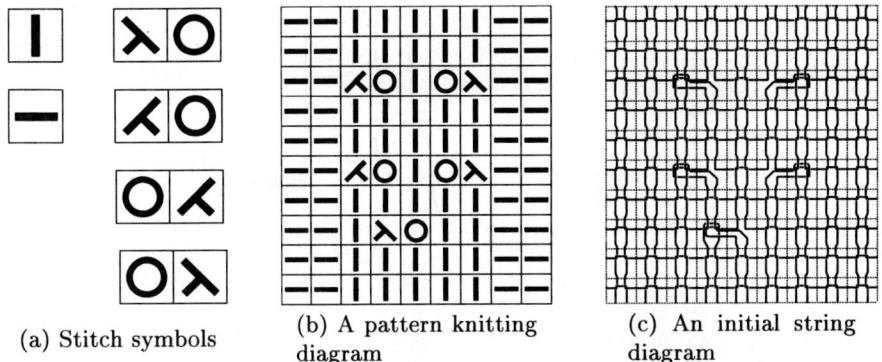

Fig. 2. Stitch symbols, A pattern knitting diagram, An initial string diagram

(a) Stitch symbols (b) A pattern knitting diagram (c) An initial string diagram

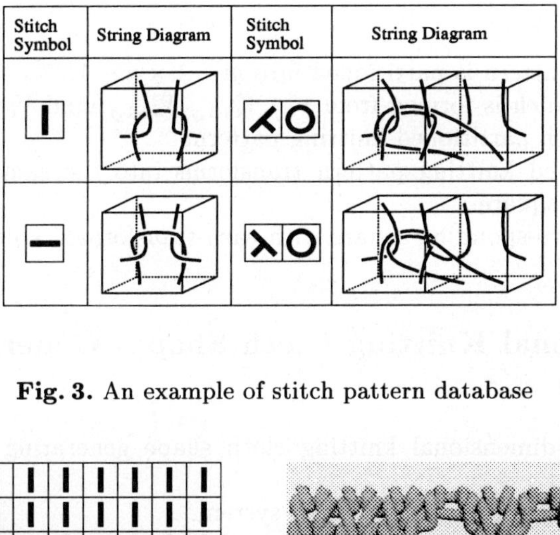

Fig. 3. An example of stitch pattern database

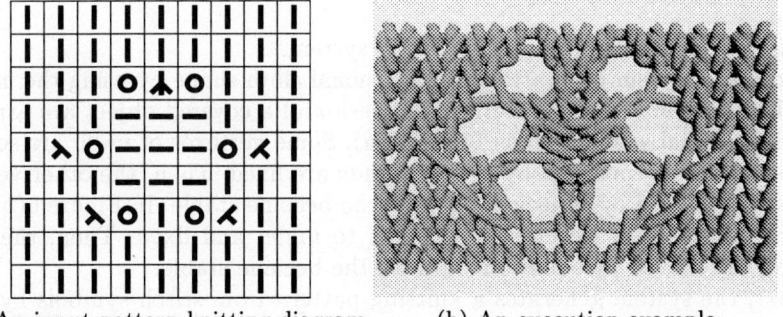

(a) An input pattern-knitting diagram (b) An execution example

Fig. 4. A generating example of knitting pattern

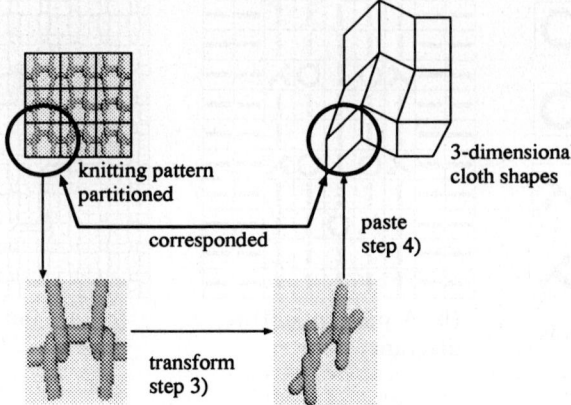

Fig. 5. Process of generating a 3-dimensional knitting cloth shape

1) The knitting pattern is partitioned into $m - 1 \times n - 1$.
2) The square, which is formed from $V_{i,j}, V_{i+1,j}, V_{i+1,j}$, and $V_{i+1,j+1}$, is corresponded to each partitioned knitting pattern.
3) Each partitioned knitting pattern transforms into the same shape of the corresponding square.
4) Pasting the corresponding square with each transformed knitting pattern.

5 3-dimensional Knitting Cloth Shapes Generating System

In this section, 3-dimensional knitting cloth shape generating system is described.

Figure 6 illustrates the scheme of the system.

First, the system generates a 3-dimensional cloth shape by using the method described section 2. In this method, A mesh and a covered object are arranged in 3-dimensional space such as Figure 6(a). Some vertexes of mesh are selected and moved to any position by user and they are fixed. Then, the other vertexes of mesh are moved by equation (1) until the become stable. In Figure 6(b), four corner vertexes were selected and moved to under and fixed. Then, the other vertexes are moved by equation (1) until the became stable.

Next, the system generates a knitting pattern from stitch symbols by using the method described section 3. Figure 6(d) shows a generated knitting pattern from stitch symbols of Figure 6(c).

Finally, the system pasts the 3-dimensional cloth shapes which the generated knitting patterns by using the method described section 4 and generates a 3-dimensional knitting cloth shape such as Figure 6(e).

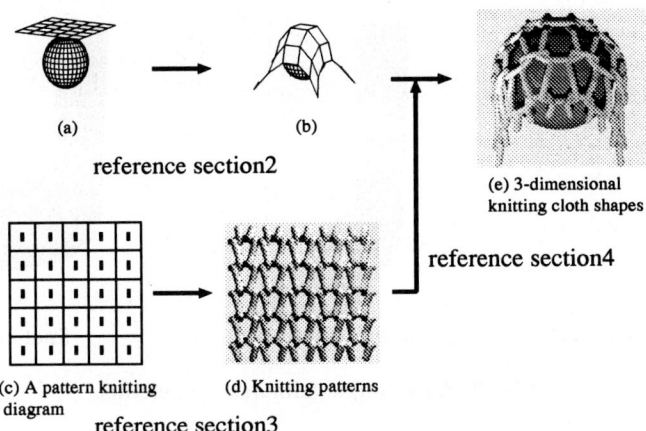

Fig. 6. A completed the 3-dimensional knitting cloth shape generating system

6 Examples

In Figure 7, the elbow is simulated. The mesh has 50 × 20 vertexes. A vertex $V_{i,j}$ is linked with $V_{i-1,j}$, $V_{i+1,j}$, $V_{i,j-1}$ and $V_{i,j+1}$. In addition, $V_{i,1}$ is linked with $V_{i,20}$. The mesh form is a cylinder in initial state. The covered objects are two cylinders, they means the upper arm and the forearm. At change of environment a cylinder, that means the forearm, is raised little by little. In this example, parameters are $\alpha = 0.1$, $\beta = 0.5$, $l_{i,j,k,l} = 5$ and $v_{external}$ is zero vector.

In Figure 8, the sleeve is simulated. The mesh has 30 × 20 vertexes. A vertex $V_{i,j}$ is linked with $V_{i-1,j}$, $V_{i+1,j}$, $V_{i,j-1}$ and $V_{i,j+1}$. In addition, $V_{i,1}$ is linked with $V_{i,20}$. The mesh form is a cylinder in initial state. The covered object is a cylinder that means the forearm. $v_{external}$ is down force that means gravitation. Given length of linkage at a cuff is shorter than other place. In this example, parameters are $\alpha = 0.1$, $\beta = 0.5$, $l_{i,j,k,l} = 2$ at a cuff, $l_{i,j,k,l} = 5$ at other place and $v_{external} = (0, 0, 0.2)$.

7 Effect of Repulsion Coefficient β

Figure 9 shows two generated knitting cloth shapes. The left side and right side patterns are generated under $\beta = 0.0$ and $\beta = 0.5$ respectively. When $\beta = 0.5$, the knitting pattern has loose. However when $\beta = 0.0$, the knitting pattern has no loose.

Table 1 shows total length of string constituting knit and its ratio to the total length of initial state. It's obvious that when $\beta = 0.0$ the total string length of knitting pattern become a much shorter.

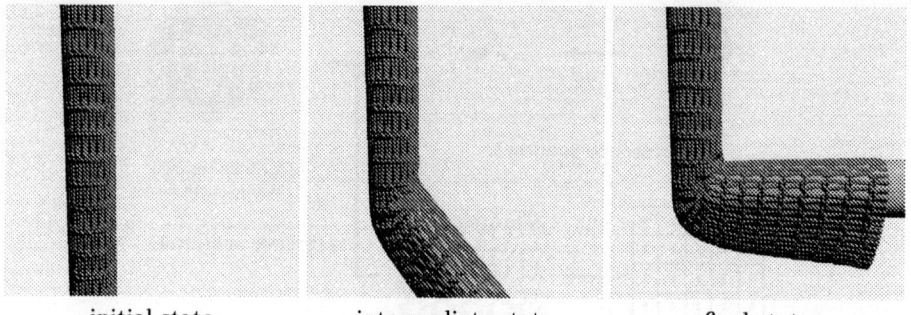

initial state   intermediate state   final state

Fig. 7. An example (simulate the elbow).

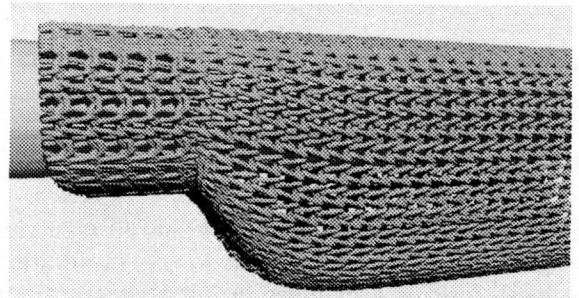

Fig. 8. An example (simulate the sleeve).

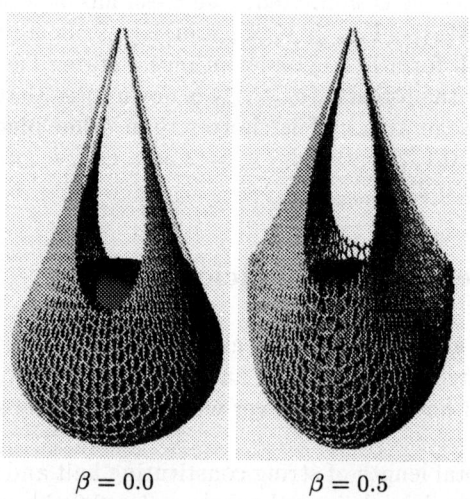

$\beta = 0.0$ $\beta = 0.5$

Fig. 9. Comparison between $\beta = 0.0$ and $\beta = 0.5$

Table 1. The total string length and ratio to the length of initial state

	initial state	$\beta = 0.0$	$\beta = 0.5$
total length	15954	14337	16557
ratio	100%	89.9%	103.8%

8 Conclusion

In this paper, we propose a method of generating knitting cloth shapes. In this method, the cloth shape is represented by mesh and vertexes in mesh moved under spatial restrictions of covered objects. In addition, repulsion of between vertexes be utilized to prevent shrinking the shape. The generated shapes are pasted with knitting pattern generated from stitch symbols.

In a future work, we will examine a method of generating loose of the inside bend of his elbow.

References

1. Itoh, Y., Yamada, M., Miyazaki, T., Seki, H. and Itoh, H: Processing for Knitting Patterns Using a Representation Method for 3D String Diagrams, Trans. of IPSJ, Vol 37, No. 2,pp.249-258(1996).
2. Itoh, Y., Yamada, M., Miyazaki, T., Kunitachi, T., Fukumura, Y., Seki, H. and Itoh, H.: A Transformation Technique from Symbolic Media to 3-dimensional Patterns for Knitting, *Proceedings of the Multimedia Japan 96*, pp. 338-345(1996).
3. Yamada, M., Itoh, Y., Seki, H. and Itoh, H.: An Implementation of a Knit-Pattern Generating System for Supporting Knit Designing, Trans. of IPSJ, Vol 36, No. 11,pp.2728-2735(1995).
4. Yamada, M., Budiart, R., Itoh, H. and Seki, H.: A String Diagram Transformation Process using a Generic Algorithm - A Cat's Cradle Diagram Generating Method -, Proc. of PRICAI'94, Vol 1, pp.429-434(1994).
5. Yamada, M., Budiart, R., Itoh, H. and Seki, H.: A Cat's Cradle String Diagram Display Method Based on a Genetic Algorithm, FORMA Vol.9, No.1, pp.11-28(1994).
6. Yamada, M., Budiart, R., Itoh, H. and Seki, H.: An Implementation of a Knit Designing System Based on a Genetic Algorithm, Proc. of ICARCV'94, Vol.1, pp.1277-1281(1994).
7. R. Budiarto, M. Yamada, H. Seki, K. Itoh, T. Miyazaki and H. Itoh.: A System for Simulating and Display Magic Game in Cat's Cradle and a Characterization Method of Its String State, FORMA Vol.12 No.1 pp.75-89(1997).
8. Letter Symbols for Knitting Stitch, JIS L 0201(1978).

A Fast Mesh Deformation Method to Build Spherical Representation Models of 3D Objects

Antonio Adán*, Carlos Cerrada**, Vicente Feliu***
* E.U. Informática UCLM. SPAIN (aadan@inf-cr.uclm.es)
** E.T.S.I. Industriales UNED. SPAIN (ccerrada@ieec.uned.es)
*** E.T.S.I. Industriales UCLM. SPAIN (vfeliu@ind-cr.uclm.es)

ABSTRACT

In this paper a new method to build more efficiently spherical representation models of 3D objects is presented. Our approach is applied to a specific representation called LSR/GSR which is a hybrid spherical model where local and global object features are stored. The main contribution of this work is the rearrangement of the iterative mesh deformation process required in any conventional spherical method, in such a way that the high computational cost usually involved in this stage is significantly reduced. A new intrinsic data structure defined over the original mesh is introduced to improve the deformation process: the *Modeling Wave Structure* (MWS). Thanks to this new concept the deformation process can be seen as a very fast wave propagation phenomenon where displacement of each MWS node only affects to a reduced and fixed set of *Neighbor* nodes. The method has been applied to synthetic and real 3D closed objects of different shapes. Experimental results for both synthetic range data and data coming from a gray range finder sensor are shown in the paper.

1. Introduction. Problem Description

One of the fundamental problems in 3D computer vision is how to represent the object models. Among other modeling techniques, spherical representations exhibits certain advantages when curved objects as well as free-form objects are considered, while regular objects can also be modeled without loss of generality. In the spherical representation models an intrinsic coordinate system is defined and object surface features can be expressed in terms of that system. Then a mapping function is required to map features from the object closed surface to the standard unit sphere [8].

Spherical representation models have experimented a significant evolution since the first applications in computer vision field [4]. Present time there are in course 3D computer vision research activities in topics such as object recognition, pose determination, 3D shape similarity [11], 3D shape synthesis [12], where spherical models play an important role.

The EGI (*Extended Gaussian Image*) can be considered one of the first developed representations. The EGI is only applicable to convex objects. CEGI (*Complex Extended Gaussian Image*) was introduced later on. CEGI can deal with translation estimation and with non-convex objects [9], [10]. Subsequently, in [1],

[5], object recognition and image segmentation tasks are accomplished through the EGI representation family.

Development of SAI (*Spherical Attribute Image*) [2], [3] come to solve the problem existing in EGI family of having the same representation for several different objects. The connectivity property preserves the relative spatial relation among surface patches.

Recently, we have defined a hybrid representation of the latest called LSR/GSR (*Local Spherical Representation/Global Spherical Representation*) [6],[7]. Our LSR/GSR model uses a mesh deformation algorithm which is an iterative process repeated until a convergence criteria is satisfied, the same as SAI and all other mentioned methods do. Since a search process involving a great number of points must be accomplished for each iteration, global computational cost is too expensive.

In the process of constructing a LSR/GSR model there is a subprocess wiht a singular objetive:to deform the tessellated sphere T_i in order to fit the surface of a normalized and closed object S, where S is defined by a set of range data points We call T_r to the final deformed mesh.

From the computational point of view, this is the more expensive process, due to two main factors. First, it is an iterative process which usually takes between 10 to 15 iterations until the convergence criteria are satisfied. On the other hand, for each iteration, a search algorithm over S must be carried out, being ORD(S) a high number.

This paper is focused in the rearrangement of the iterative mesh deformation process as a way to reduce the high computational cost this stage involves. Our approach take advantage of what we call *Modeling Wave Structure (MWS)*, a new intrinsic data structure defined over the original standard spherical mesh, which is used to propagate the node movements from the initial to the final mesh. This structure exhibits helpful properties to solve the object recognition and 3D pose determination problems. The iterative process can be almost suppressed thanks to the decoupling of the mesh deformation process in two independent problems: the approximation problem and the regularity problem.

2. Formal Definition of the Original Mesh

In this section we introduce the first concept of the spherical representation concerning to the semi-regular mesh building process through a tessellated unit sphere. In our case, mesh geometry is a fundamental factor for the forward discussion. From now on, we call T_i to this initial mesh.

A natural discrete representation of a surface is a set of points or nodes, or tessellation, such that each node is connected to each of its closest neighbor. We use a set of nodes over a unit sphere such that each node has exactly three neighbors (see figure 2). This process can be seen in [2].

Below, we formally define the *Neighbor Relation* and some properties of the initial mesh T_i.

Definition 1. Let $T_i = \{N_1, N_2, \ldots N_n\}$ be the initial semi-regular mesh with n nodes and let D be the length of the edges of T_i. We define the *Neighbor Relation* V, $V: T_i \to T_i * T_i * T_i$ as $V(N_j) = \{N_{j1}, N_{j2}, N_{j3}\}$ where $d(N_j, N_{ji}) \equiv D$, $\forall$ i=1..3, j=1..n., and d is the euclidean distance in R^3. Nodes N_{j1}, N_{j2} and N_{j3} are called *Neighbors* of N_j.

Due to the T_i geometric structure, next properties are verified: 1) Each T_i node has exactly three Neighbors and 2) All node *Neighbors* are located in a spherical equilateral triangle where the node is located as center of the three vertices.

3. Mesh Deformation Process

The mesh deformation process can be decomposed in two conceptual stages. First, the original range data must be normalized is such a way that the object can be enclosed in a unit sphere. Also an approximation of the initial tessellated sphere to the normalized object range data must be carried out at this stage. In a second stage mesh is deformed until regularity constrains are verified. These two stages are presented in this section.

3.1. Normalization and Aproximation

Figure 1 schematically shows this stage. It is performed in two phases. We can describe *Normalization* (f_n) and *Approximation* (A) processes as follows:

In the Normalization a translation and a scale transformation are carried out with the original range data defined in a world coordinate system. Original points are translated to a coordinate system located at the center of gravity of the input data set. The scale factor is such that transformed data points are inside a unit sphere.

Let $T_i\{N_{i1}, N_{i2}, \ldots N_{in}\}$ be the initial mesh, $T_f\{N_{f1}, N_{f2}, \ldots N_{fn}\}$ be the final mesh after the *Approximation* process is performed, and let $S\{P_1, P_2, \ldots P_k\}$ be the normalized surface points. The method applied for approximation of T_i to S is carried out in a single step. Each node N_{ij} moves towards its associated point P_j in a fixed direction $\bar{n}_{ij}$ (unitary vector which is perpendicular to the plane defined by the three Neighbors). In practical P_j is searched into a solid angle around $\bar{n}_{ij}$.

Note that the use of this search algorithm through a reduced solid angle represents a meaningful cost reduction with respect to the conventional methods [1],[3], in which the search is accomplished over all object points.

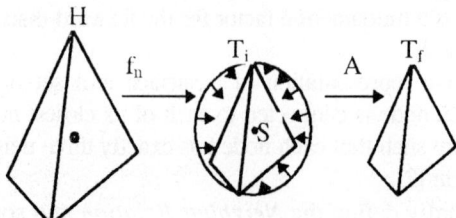

Fig. 1. Schematic representation of the *Normalization* (f_n) and *Approximation* (A) processes.

3.2. Mesh Regularization

Once the previous deformation stage has finished, T_f must be manipulated in order to make the regularity constraints to be satisfied. We will call T_r to the mesh at the final process. Convergence criteria will be stated from two points of view. First, T_r must be as close as possible to S. Second, T_r must be as regular as possible. To help in the task of criteria verification the new intrinsic data structure MWS defined over the original mesh T_i is introduced.

Modeling Wave Structure Formal Definition

Considering T_i as an initial mesh as described in section 2, the following new definitions can be introduced previous to derive the MWS concept.

Definition 2. Node N is a *Initial Focus* if $N \in T_i$.
Note that from this definition every node can be *Initial Focus* and $ORD(T_i)$ *Initial Focuses* can be defined.

Definition 3. Let N be an *Initial Focus*. A *Wave Front Set* $F=\{F^1, F^2,... F^q\}$ is an ordered set of T_i partitions or *Wave Fronts* F^i i = 1..q, verifying the following equations: 1) $F^1 = \{N\}$ 2) $F^{i+1} = V(F^i) - \bigcup_{m=1}^{i-1} F^m$ $\forall$ i:1..q-1

F^{i-1} and F^{i+1} are the *Neighbor Wave Fronts* to F^i.

Considering the last two definitions and due to the T_i geometric structure, the following properties can be introduced:

1. Every node does not have any *Neighbor* in its own *Wave Front*.
2. All the *Wave Fronts* are disjunct sets.
3. Every *Wave Front* F^i has two *Neighbor Wave Fronts* F^{i-1}, F^{i+1} $\forall$ i=2...q-1, while F^1 and F^q has a single *Neighbor Wave Front*.

Definition 4. Let us assume that $F=\{F^1, F^2,... F^q\}$ is the T_i *Wave Front Set*. We define the *Next Neighbor Front* relation as $V_F: F^i \to F^{i+1}$ $\forall$i:1..q-1, where:
$$N' \in V_F(N) \Leftrightarrow N' \in V(N) \text{ and } N \in F^i, N' \in F^{i+1},$$

Definition 5. Let T_i be a semi-regular mesh. We define the *Modeling Wave Structure of Initial Focus* N_0 as the $\{N_0, F, V_F\}$ tern.

Figure 2 shows a view of the tessellation and the nodes of a unit sphere grouped by *Wave Fronts*.

Node Movement Propagation Through the MWS

The mesh deformation process can be seen as an iterative propagation procedure where every node of the mesh moves to the new position following an ordered sequence, called propagation direction. The propagation direction is imposed

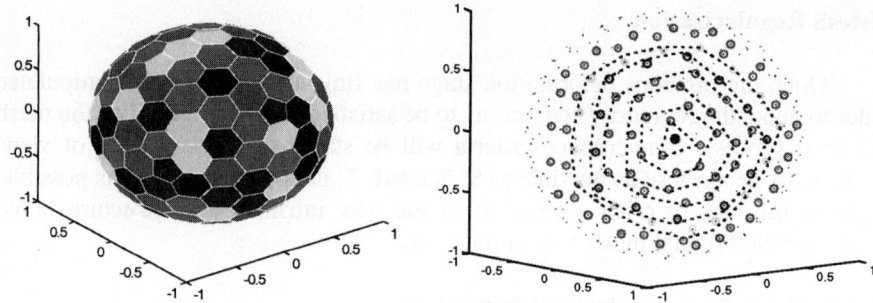

Fig. 2: Tessellated unit sphere with 320 nodes (left) and *Wave Front Set* (rigth).

by the *Wave Front Set* F of the MWS. The whole procedure is formally presented in next paragraphs.

Propagation Algorithm

Let T_f be the mesh after the approximation phase is carried out. Let us introduce the parameter t to establish the status of the mesh at intermediate instants of the iterative deformation process. Then T_f can be represented now as $T_f(0) = \{N_1(0), N_2(0), ... N_n(0)\}$ where t=0 is used to mean the starting point of the deformation process. Let $T_f(t)$ be the deformed mesh at the instant t and let $\bar{d}(t)$ be the corresponding mean of mesh inter-node distances.

Definition 6. We define the *Propagation Step* at instant t as the set of incremental movements experimented by all the nodes of the MWS from instant t to t+1 (see figure 3). The algorithm to evaluate new positions can be stated as follows ($\forall j=1,...m, \forall i=2...q-1$):

I) If $d(N_j^{i+1}(t), N_k^i(t+1)) > \bar{d}(t)$

$$\overrightarrow{ON}_j^{i+1}(t+1) = \overrightarrow{ON}_j^{i+1}(t) + \beta_1(t) \cdot \left(d(N_j^{i+1}(t), N_k^i(t+1)) - \bar{d}(t) \right) \cdot \bar{u}_{jk}$$

where $N_j^{i+1} \in V_F(N_k^i)$, u_{jk} is the unit vector in the direction from N_j^{i+1} to N_k^i, and β_1 is an increasing empirical function defined in the [0,1] interval.

II) If $d(N_j^{i+1}(t), N_k^i(t+1)) < \bar{d}(t)$

$$\overrightarrow{ON}_j^{i+1}(t+1) = \overrightarrow{ON}_j^{i+1}(t) + \beta_2(t) \cdot \left(\bar{d}(t) - d(N_j^{i+1}(t), N_k^i(t+1)) \right) \cdot \bar{u}_{jr}$$

where $N_j^{i+1} \in V_F(N_k^i)$, $N_r^{i+2} \in V_F(N_j^{i+1})$, u_{jr} is the unit vector in the direction from N_j^{i+1} to N_r^{i+2}, and β_2 is another increasing empirical function defined in the [0,1] interval.

The iterative process should finish when regularity constraints are verified. Following criteria can be adopted to finish with the iterative process: 1) $\bar{d}(t)$ convergence, 2) Minimum value of std($\bar{d}(t)$) reached. 3) Minimum Interquartile

Range (IQR) over $T_f(t)$ edges reached, 4) Acceptable fitting margins of $T_f(t)$ to S. Properties of the node movement propagation method can be consulted in [16].

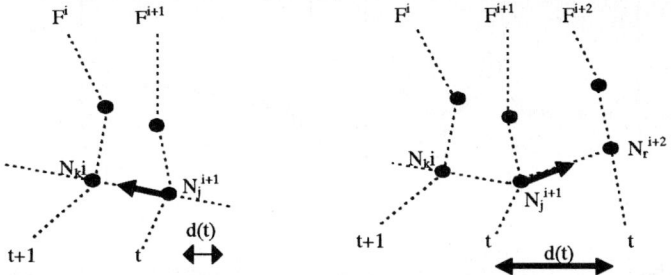

Fig. 3: Behavior of the algorithm for node movement propagation

4. Experimental Setup and Results

The proposed method has been tested both with synthetic and real range data corresponding to closed 3D objects. Range data from real objects have been acquired using a gray range sensor [13],[14] available in our laboratory.

Our experimental setup deals with objects whose standard dimensions are around 5 to 10 cm. Objects are located over a controlled turntable in front of the static GRF. Mean range data error from real objects measurements is around 2mm and polyhedral and free-form shapes, both convex and concave, have been considered. Figure 4 shows different objects at three different *Propagation Steps*.

The new mesh deformation method has been applied to the mentioned set of objects and it has been compared with a conventional iterative mesh deformation process like the used at [6]. The following results have been observed for the new approach:

1. Computation time has been reduced around 80% with respect to our conventional process and similar reductions are observed with respect to other published model generation methods [11].

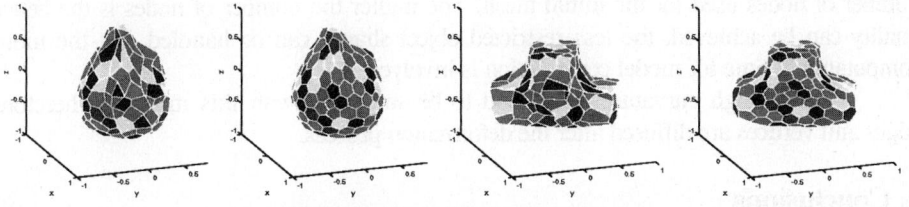

Fig. 4: $T_f(0)$ and T_r for two objects.

2. A reduction of the standard deviations of the $\bar{d}(t)$ values are experimented in most of the cases. It means an improvement of the mesh regularity because a higher percentage of the mesh edges have lengths near the mean value (Figure 5). As it can be observed IQR values are significantly reduced for consecutive Propagation Steps, which means that a high percentage of edges are near the $\bar{d}(t)$.

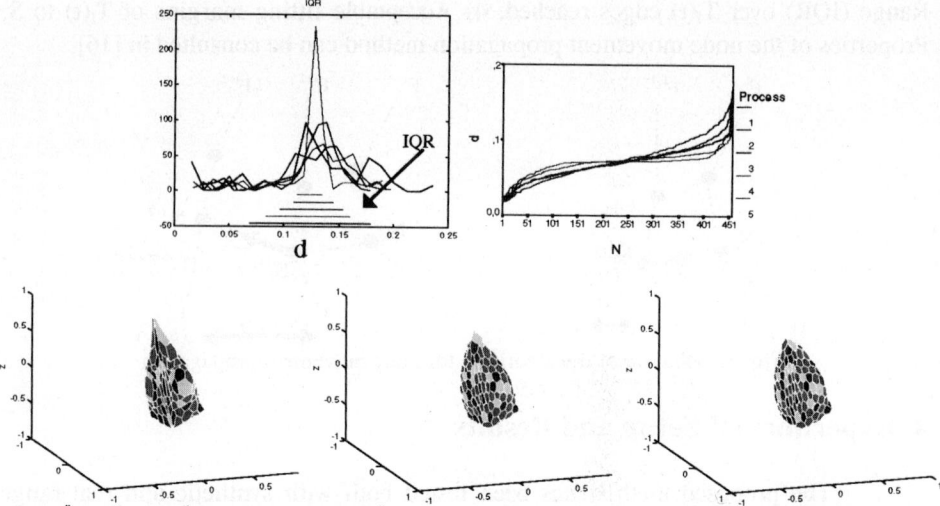

Fig.5: Object example at different instants: a) Histogram of mesh edge lengths (upper left) and IQR ($[P_{85}-P_{15}]$). b) Edges ordered by length (upper right). c) Deformed mesh.

3. It has been verified that GSR/LSR models obtained for different MWS are quite similar, what means that the modeling method is independent of the chosen wave structure.

4. This new structure denoted as MWS has been used in 3D model generation but can be naturally extended to applications such as 3D object recognition and 3D pose determination with a little effort [16].

5. The proposed method presents some limitations and inconveniences. It only copes with closed objects and objects without holes (genus zero) and object surfaces should not have abrupt changes. But this inconvenience also exists for previous methods.

On the other hand, quality of the approximation of T_i to S depends on the number of nodes used for the initial mesh. The higher the number of nodes is the better quality can be achieved, the less restricted object shapes can be handled, but the more computational time for model construction is involved.

Finally, high curvature areas tend to be smoothed with this method. Therefore edges and vertices are diffused after the deformation process.

5. Conclusions

A new method to fit a quasi-regular mesh to an object surface has been established. The method is based on the spherical representation models of 3D objects and a comparison with the previous existing methods has been carried out. The main advantage of the proposed method is the rearrangement of the iterative mesh deformation process required in this kind of representation, obtaining significant computation time reductions, as well as other features that have been exposed in the paper.

The iterative fitting process is almost suppressed because approximation and regularity stages are solved separately as two sequential and independent processes. In fact, we have defined a new intrinsic data structure called *Modeling Wave Structure* (MWS) which is responsible of the computational cost reduction. Deformation stage can be seen as a wave propagation phenomenon over the nodes of the MWS constructed from the original object range data. MWS gives a path followed in the computation of the displacements corresponding to each mesh node at every propagation step.

The used experimental platform and the obtained results have also been presented in the paper. Both synthetic and real range data have been considered. The method shows good behavior for polyhedral objects as well as for free form ones. Actually this method is been applied to solve the 3D object recognition and 3D pose determination problems, obtaining significant computation time reductions.

REFERENCES

[1] H. Delingette, M. Hebert, K. Ikeuchi. "A Spherical Representation for the Recognition of Curved Objets" *International Conference Computer Vision, (iccv)* Berlin 1993

[2] M. Hebert, K. Ikeuchi, H. Delingette." A Spherical Representation for Recognition of Free-Form Surfaces". *IEE Transactions on pattern analysis and machine intelligence"*, vol 17, no 7. July 1995

[3] K. Higuchi, H. Delingette, M. Hebert, K. Ikeuchi "Merging Multiple Views Using a Spherical Representation".*Proceedings IEEE Workshop on CAD-based Vision"*. Feb 1994.

[4] K. Ikeuchi, N. Hebert "Spherical Representation: From Egi to Sai" *Technical Report. CMU.* October 1995.

[5] H.Delingete, M.Hebert, K.Ikeuchi. "Shape Representation and Image Segmentation Using Deformable Surfaces", *Image And Vision Computing*, 10(3) Abril 1992

[6] A.Adán. M.Bachiller, C.Cerrada, V.Feliu. "An Approach to Object Tracking Based on a Simplified 3D Geometric Model". *Proceedings IEEE/SCM Computational Engineering in Systems Applications.* France. July 1996.

[7] A.Adán, C.Cerrada, V.Feliu, M.Bachiller. "3d Object Recognition Approach Based On Shape Discrimination Under A Spherical Model Representation". *Seventh National Symposium on Pattern Recognition and Image Analysis.* Barcelona. Spain. April 1997.

[8] B.K.Horn. "Robot Vision". *MIT. Press.* 1986

[9] S. Bing Kang, K. Ikeuchi. "Determining 3-D Object Pose Using the Complex Extended Gaussian Image". *Proceedings Computer Vision And Pattern Recognition (Cvpr)*, 1991

[10] S.Bing Kang, K.Ikeuchi. "The Complex EGI: A New Rpresentation for 3-D Pose Determination". *IEEE Trans. On Pattern Analysis And Machine Intell.*, Vol 15, No 7 , July 1993

[11] H. Shum, M. Hebert, K. Ikeuchi. "On 3D Shape Similarity". Tech. Rep. CMU. Nov.1995

[12] H. Shum, M. Hebert, K. Ikeuchi. "On 3D Shape Synthesis". Tech. Rep. CMU. Nov. 1995

[13] K.Sato, S.Inokuchi ."Range Imaging System Utilizing Nematic Liquid Crystal Mask" .*Proceedings IEEE First International Conference On Computer Vision*, London, June 1987

[14] K. Sato, S.Inokuchi . "Three-Dimensional Surface Measurement by Space Encoding Range Imaging". *Journal Robotic Systems*, Vol 2(1), 1985

[15] H.Delinguete. "Simplex Meshes: a General Representations for 3D Shape Reconstruction" *INRIA* report 2214, 1994.

[16] A. Adán. "Modelado 3D de Formas Libres Orientado a Reconocimiento y Posicionamiento de Objetos Tridimensionales". PhD. Thesis. Nov. 1997

Semi-Automatic 3D Object Digitizing System Using Range Images

C. Schütz, T. Jost and H. Hügli

Institute of Microtechnology, University of Neuchatel,
rue Breguet 2, CH-2000 Neuchatel, Switzerland

Abstract. Manual object digitizing is a tedious task and can be replaced by 3D scanners which provide an accurate and fast way to digitize solid objects. Since only one view of an object can be captured at once, several views have to be combined in order to obtain a description of the complete surface. In this paper a digitizing system is proposed which captures and triangulates views of a real world 3D object and semi-automatically registers and integrates them into a virtual model. This process is divided into three steps. First, an object is placed at different poses and its surfaces are sensed by a range scanner. Then, the different surfaces are aligned automatically starting from a pose estimate entered interactively. Finally, the overlapping triangle meshes of the registered surfaces are fused in order to obtain one unique mesh for the entire object.

1 Introduction

The increasing use of virtual object representations for various applications creates a need for fast and simple object digitizing systems. Therefore, 3D surface digitizers get used more and more since the model construction with a standard modeler is a quite tedious task especially for objects of arbitrary shape.

Range scanners give direct access to the 3D geometric information of object surfaces. They allow an accurate digitizing of an object surface at low cost and high speed. However, since most objects self occlude, one acquisition captures only a subpart of the entire object surface. Therefore, there exists a need to combine several range scanner views into one unique object representation.

The view combination is straightforward if the object is moved in a well known coordinate system like a rotation table: the relative transformation of two acquisitions is known. However, this implies a sophisticated mechanical system used to orientate the object or the scanner and to measure its pose. To avoid such complex pose systems, we propose to work with views from unknown object pose. The idea is to combine views based on the sole features of the geometric measurements.

We present a digitizing system which captures and triangulates views of a real world 3D object and finally registers and integrates them into a virtual model. The following steps have to be performed to combine the acquired surfaces.

An object to be digitized is placed in different poses on the acquisition field. The surface points measured by a range scanner are triangulated in 2.5D. The different measurements have to be transformed into a common frame. An interactive interface allows the operator to roughly align the acquired surfaces in 3D space. The precise surface registration is calculated with an automatic registration algorithm which matches the surfaces precisely by minimizing the distance between the common surface parts. In a last step called mesh fusion, the aligned triangle meshes are fused into an unique mesh for the entire object. Here, an erosion process eliminates the redundant surface parts and the remaining triangle meshes are joined with triangles by a gap filling algorithm.

A novel aspect of the presented system is the fact that the view registration and the integration modules are linked together and working completely in 3D space. Table 1 compares our work to systems proposed by other authors.

surface based registration	3D mesh fusion	2.5D mesh fusion
[TUR], [BES]	[RUT], [PIT], [HIL]	[TUR], [SOU]
presented system		

Table 1. Comparison of digitizing systems

The next section presents the global architecture of the 3D digitizing system. Its two main modules are detailed in the following sections and its effectiveness is shown in a further section devoted to experimental results on real objects.

2 System Architecture

The digitizing system consists of two blocks: view digitizing and view integration. The view digitizing block generates a virtual view of the observed object surface. The view integration block iteratively integrates each new virtual view in the virtual model under construction. This allows an incremental construction of the virtual model.

The view digitizing block measures the points of the visible object surface, filters the range data and triangulates the surface points. The resulting output is a triangle mesh representing the virtual view of the real object. Section 3 explains the detailed implementation of the modules used for the view digitizing.

The view integration block combines the virtual views and builds one virtual object becoming an entire model of the real object. The view integration block is composed of the view registration block and the mesh fusion module. The view registration block aligns the different views using interactive pose estimation and automatic registration. The mesh fusion module combines the individual triangle meshes into a new global mesh covering the union surface of the single meshes. Section 4 presents the different methods used for these modules. Fig. 1 gives an overview of the modules and the data flow during the digitizing process.

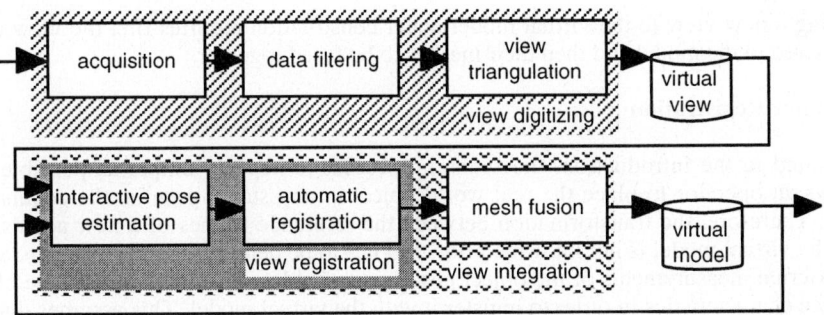

Fig. 1. Digitizing system architecture

3 View Digitizing

Range scanners give access to accurate and fast scanning of an object shape without any need of contact or visual marks. In our laboratory, the geometry of the object surface is acquired by a range scanner working on the principle of space coding with projected stripe pattern and triangulation. The coordinates measurements are arranged in a two dimensional array corresponding to the CCD camera image of the range scanner and can be visualized as a range image where the pixel intensity represents the camera to object distance as shown for a duck toy in Fig. 2. The virtual object can be easily textured since both images are represented in the same coordinate frame and the intensity and range information do correspond for every pixel.

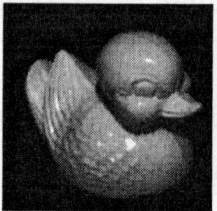 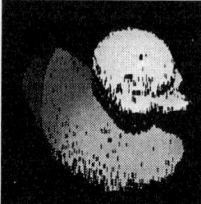

Fig. 2. Intensity and range images of a duck toy

The black pixels in the range image represent non-valid data (holes). They are found where the measurement confidence is low and are often caused by shadow regions or failed stripe coding. The filter module uses morphological operations such as erosion and closing to remove speckles and to fill small holes of non-valid data. Furthermore, the measurement noise present in the range data is filtered with a Gaussian filter.

Since the measurements are ordered in a regular grid namely the range image the triangulation of the surface becomes straightforward as proposed by [RUT]. Adjacent valid pixels are connected by triangles which results in a triangle mesh for the visible object surface. Checking the validity of the range points is not sufficient to avoid bad triangles. Other authors [TUR] [RUT] showed that additional checks are necessary to avoid the connection of range points separated by a discontinuity step in the range image. We have to ensure that occluded parts are not covered by triangles. Only triangles with small edges and an angle below 80° between the triangle normal vector and the sensor view axis are valid.

4 View Integration

Adding a new view to the virtual model under construction requires first the view to be registered to the model and then their meshes to be fused together.

4.1 View Registration

As stated in the introduction, the scanning process is kept as simple as possible and allows an operator to place the real world object in any stable pose on the acquisition field. Therefore, the transformation between the reference frames of a new acquisition and the virtual model is not known a priori and has to be determined in a first step. Since no external measurements of the object pose are available, we have to rely on the object surface characteristics in order to register it with the virtual model. This assumes that the virtual model and the new object view have at least some surface area in common which allows to establish correspondences between them.

View registration is performed in a two step process: interactive rough pose estimation and automatic matching.

Interactive Rough Pose Estimation.

Since human perception easily identifies corresponding surface parts for any object type and shape, we use an interactive graphic interface that permits an operator to enter a pose estimate for the two objects to be aligned. Both the virtual model and the new view are rendered in 3D and can be manipulated in all six degree of freedoms using a space mouse as input device. However, even a sophisticated object rendering and pose manipulation hardware is not sufficient to align the objects precisely. In fact, there is no measure apart from the visual feedback indicating the quality of the surface matching. Therefore, the interactive interface only provides a rough pose estimate to be used as a starting pose for the automatic precise registration. Fig. 3 shows an example of two roughly aligned surfaces used as starting configuration for the automatic registration.

Fig. 3. Roughly aligned views of a cat toy

Automatic Registration.
As stated before the precise alignment of two 3D surfaces is a tedious task if it has to be done manually. Besl proposed an automatic surface registration algorithm called ICP which avoids this problem [BES]. This algorithm registers two surfaces starting from an initial pose estimate. The algorithm proceeds iteratively. First, it pairs every point of one surface called P with the closest point of an other surface called X. These pairs of closest points are used to calculate the rigid transformation (**R**, **t**) which minimizes the mean square coupling distance or error. The surface P is then translated and rotated by the resulting transformation and the algorithm starts again with the closest point coupling.

This algorithm has been shown to converge fast but not necessarily towards the optimal solution. A good starting configuration for the two surfaces P and X is preliminary to a successful convergence. However, the range of successful starting configurations is quite large (see [HUG] and Fig. 3) which does not impose difficult constraints to the operator when entering a pose estimate for P and X.

In the original ICP algorithm the surface P is a subpart of X which is not the case in our application where both surfaces contain data not present in the other. The ICP algorithm needs therefore to be modified as proposed by Turk [TUR]. Closest points which are too far apart are not considered to be corresponding points and marked as invalid so they have no influence during the error minimization. The modified ICP algorithm is defined as follows:

- input: Two 3D surfaces P and X containing respectively N_p and N_x vertices.
- output: Transformation (**R**, **t**) which registers P and X
- iteration:
1. Build the set of closest point pairs (**p**, **x**):

$$\forall \mathbf{p} \in P \text{ find } \mathbf{x} \in X \text{ with } d_k = \min\left(\|\mathbf{p} - \mathbf{x}_j\|^2\right), j \in [1, \ldots N_x] \quad (1)$$

2. Weight every closest point pair (**p**, **x**) by applying the following distance threshold:

$$w_k = \begin{cases} 1 & d_k < (c \cdot s \cdot r)^2 \\ 0 & \text{else} \end{cases} \quad \text{with } d_k = \|\mathbf{p}_k - \mathbf{x}_k\|^2 \text{ and } k \in [1, \ldots, N_p] \quad (2)$$

3. Find the rigid transformation $(\mathbf{R}, \mathbf{t})$ that minimizes the mean square error

$$e(\mathbf{R},\mathbf{t}) = \frac{1}{W}\sum_{N_p} w_k \|\mathbf{R}\mathbf{p}_k + \mathbf{t} - \mathbf{x}_k\|^2 \quad \forall \text{pairs of } (\mathbf{p}_k, \mathbf{x}_k) \text{ and } W = \sum_{N_p} w_k \quad (3)$$

4. Apply the transformation $(\mathbf{R}, \mathbf{t})$ to P

The decision threshold for a valid coupling distance is set to the product $(c \cdot s \cdot r)^2$ where s equals the range scanner sampling distance and r equals the range image subsampling factor. The constant c allows to control the convergence of the automatic matching. It is set to a relatively large value at the beginning when the two surfaces are far apart. When the surfaces are superimposed the value of c can be lowered so only similar points are coupled which results in higher matching precision.

In order to verify the registration quality and to stop the iteration, a pertinent matching quality measure is needed. The minimization error $e(\mathbf{R}, \mathbf{t})$ corresponding to the mean μ of the square distances is a measure generally used to qualify the matching. Another statistical measure which has been used successfully to qualify matched surfaces in object recognition [SCH] is the deviation σ of the square distances indicating the matching regularity. The matching is stopped if the sum of μ and σ does not change any more.

$$\mu = \frac{1}{N_p}\sum_{N_p} d_k \qquad \sigma = \sqrt{\frac{1}{N_p - 1}\sum_{N_p}(d_k - \mu)^2} \quad (4)$$

In order to detect cases where only very few points are coupled, matchings with a high number of coupled points on the surface P are selected, as proposed by Krebs [KRE] and expressed by a high value of the coupling measure ε.

$$\varepsilon = \frac{W}{N_p} 100 \quad \text{with } W = \sum_{N_p} w_k \quad (5)$$

As mentioned before, the two surfaces should have enough common data points for successful matching. 30 to 50 % of common surface has been observed to be a good amount.

4.2 Mesh Fusion

There exist several methods to integrate registered surfaces acquired from different views [TUR] [RUT] [PIT] [SOU] [HIL]. They differ mainly in how they treat the redundant overlapping zone of the two registered surfaces and can be separated into two groups: partial erosion and complete retriangulation of the surface points.

Methods using a partial erosion approach [PIT] [TUR] erode the overlapping surfaces until the overlap disappears. The two triangle meshes are then linked at their frontiers in order to have one unique mesh for the union of the two surfaces. Other authors [HIL] [SOU] [RUT] discard the mesh information from the triangulated views if calculated at all and retriangulate the overlapping zone or even the complete point set.

Since the object views can be easily triangulated using the range image structure, we opt for the partial erosion approach which keeps intact as much as possible of the triangle mesh structure. We propose a new mesh fusion algorithm that benefits from the closest point relationships established during the geometric matching. There is no need to run an extra routine to erode overlapping surfaces and to detect the surface frontiers as done in other work [PIT] [TUR].

The following features characterize the proposed mesh fusion algorithm. We refer to the same surfaces P and X as introduced for the geometric matching, where the vertices on P are coupled with points on the triangles of X.

1) **overlap detection:** The registration algorithm calculates for the vertices on surface P the closest points on the triangles of the surface X. Closest points with an Euclidean distance below a defined threshold are coupled. During the geometric matching iterations, the coupled points on P converge to the overlapping area of the two surfaces P and X.
2) **overlap remove:** The redundant part of the surface P is deleted by removing the triangles with one or more coupled vertex. The remaining meshes are separated by a gap defined by a frontier on P and X.
3) **frontier detection:** Triangles where the geometric matching coupled only one vertex are connected to the frontier on P. Actually, the two non-coupled vertices build an edge of the frontier on P. The frontier on X is detected by a closest point search.
4) **gap filling:** The gap enclosed by the two frontiers is filled iteratively with triangles. Vertices on the two frontiers are used as candidates to build a filling triangle. The triangulation does not need projection into tangential planes which allows a correct triangulation of sharp edges. Triangles with a maximal opening angle are constructed in order to have an optimal approximation.

The implementation details are discussed below and illustrated by examples shown in the figure Fig. 4.

Overlap detection

The closest point routine of the automatic registration module marks the vertices on surface P which overlap surface X. This results in a set C_V of coupled vertices with $C_V = \{ \mathbf{p}_k \in P | w_k = 1 \}$ and therefore $C_V \subseteq P$.

Overlap remove

The remove process eliminates all the vertices member of C_V from the surface P. The clipped surface $P_C = \{ \mathbf{p}_k \in P | \mathbf{p}_k \notin C_V \}$ is separated from the surface X by a gap since C_V is slightly larger than the actual overlap area due to a coupling threshold which is not zero.

Frontier detection

A frontier is defined by the frontier list F which contains the ordered vertices of the surface P_C which limit the gap created by the above overlap remove process. The list F is build as follows: During the automatic matching the list

$$T_F = \left\{ \mathbf{t}_l = \{ \mathbf{v}_{l,0}, \mathbf{v}_{l,1}, \mathbf{v}_{l,2} \} \middle| \mathbf{v}_{l,i} \in P \text{ and } \sum_{i=0}^{2} w_{l,i} = 1 \right\}$$

containing the triangles with only one coupled vertex is established. For every triangle in T_F, the vertices which are not coupled are inserted to

$$F = \{ \mathbf{v}_{l,i} \in P | \mathbf{v}_{l,i} \in \mathbf{t}_l \text{ with } \mathbf{t}_l \in T_F \text{ and } w_{l,i} = 0 \}.$$

Such a frontier list is established for every frontier on P_C.

Gap filling

The gap between the two surfaces X and P_C is filled with triangles in order to join the two meshes. The different frontiers on P_C delimiting these gaps are processed sequentially. The filling process is initialized for a frontier on P_C with the search of the first vertex $\mathbf{x}_N$ on the frontier of X. To do so, the first two vertices $\mathbf{f}_0$ and $\mathbf{f}_1$ of the frontier list F are selected and the nearest point $\mathbf{x}_N$ to $\mathbf{f}_0$ and $\mathbf{f}_1$ on the frontier of X is calculated. Then, the first bridge triangle joining the two frontiers is constructed with the vertices x_N, f_0 and f_1 as shown in Fig. 4. The frontier list F is updated by setting its first vertex f_0 equal to x_N.

The following algorithm fills the gap iteratively starting with the above initialization. Two candidate vertices are selected to build the next bridge triangle. One is f_2, the third vertex in the frontier list F and the other one is x_C, the next vertex on the frontier of X. These two candidates form together with the vertices f_0 and f_1 of F the next potential bride triangles as shown in Fig. 4. The candidate which encloses the maximal angle is selected in order to obtain a regular triangulation. The frontier list F is updated with the new vertices as follows: f_0 is set equal to x_C if x_C is chosen or f_1 is removed from F if the candidate f_2 is selected. The candidate selection starts again with the modified frontier list and the above procedure is applied until F contains only two vertices.

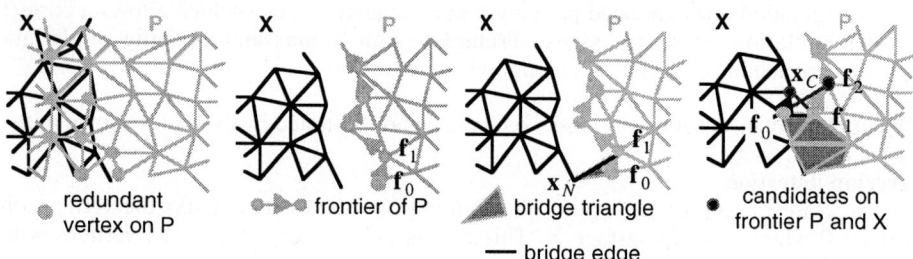

Fig. 4. Mesh fusion showed for an example

5 Results

The described 3D object digitizing system has been implemented and used to create virtual models from several real world objects. Results are presented here for two toy objects. Six views have been used to create the cat toy model whereas ten views have been merged to obtain the rabbit model. The cat object consists of 9700 points and about 19000 triangles whereas the rabbit object contains 15000 points and 30000 triangles. The successful reconstructed objects are shown in Fig. 5.

Fig. 5. Digitized cat and rabbit toy

The range finder assigns an object color to every vertex. If the triangle mesh is fine enough the object texture is maintained, as shown for a rabbit toy.

As discussed above, the object views to be assembled need common surface parts with enough geometric structure in order to allow the automatic matching to converge to a stable solution. For example for a box, a view should contain at least three faces in order to find a stable alignment of two views. The proposed digitizing system is especially suited for objects of complex free-form shape.

6 Conclusions

The presented digitizing system permits to construct models from free-form 3D objects. It integrates object views acquired by a range scanner. In order to provide simplicity and full flexibility during the acquisition, the system does not need any information about the object poses. The various views are integrated by view registration which combines an interactive rough view registration step followed by an automatic precise matching. A new mesh fusion algorithm combines the meshes into a global one.

The particular interest of the system consists in the adequate combination and linking of modules of the digitizing process. For example, the overlap information obtained in the matching module is directly used to remove redundant mesh area.

7 Acknowledgments

This research has been funded by the Swiss national science foundation under project number 2100-43530.

References

[BES] P.J. Besl and N.D. McKay, "A Method for Registration of 3-D Shapes," Proceedings of IEEE Transactions on Pattern Analysis and Machine Intelligence (PAMI), vol. 14, no. 2, pp. 239-256, 1992.

[HIL] A. Hilton, A.J. Stoddart, J. Illingworth and T. Windeatt, "Marching triangles: range image fusion for complex object modelling," IEEE International Conference on Image Processing, Lausanne, vol. 2, pp. 381-384, 1996.

[HUG] H. Hugli, Ch. Schutz, "How well performs free-form 3d object recognition from range images?," Intelligent Robots and Computer Vision XV, Algorithms, Techniques, Active Vision and Materials Handling, SPIE, Boston, Vol. 2904, 1996, pp. 66-74.

[KRE] B. Krebs, P. Sieverding and B. Korn, "Correct 3D Matching via a Fuzzy ICP Algorithm for Abitrary Shaped Objects," DAGM96 - Mustererkennung, Heidelberg, 1996.

[PIT] R. Pito, "Mesh Integration Based on Co-Measurements," IEEE International Conference on Image Processing, Lausanne, vol. 2, 1996, pp. 397-400.

[RUT] M. Rutishauser, M. Stricker, M. Trobina, "Merging Range Images of Arbitrarily Shaped Objects," Proceedings of the IEEE Conference on Computer Vision and Pattern Recognition (CVRP), Seattle, 1994, pp. 573-580.

[SCH] Ch. Schutz, H. Hugli, "Towards the recognition of 3D free-form objects," Intelligent Robots and Computer Vision XIV, Algorithms, Techniques, Active Vision and Materials Handling, SPIE, Philadelphia, Vol. 2588, pp. 476-484, 1995.

[SOU] M. Soucy, D. Laurendeau, "A General Surface Approach to the Integration of a Set of Range Views," IEEE Transactions on Pattern Analysis and Machine Intelligence, vol. 17, no. 4, 1995, pp. 344-358.

[TUR] G. Turk, M. Levoy, "Zippered Polygon Meshes from Range Images," Proceedings ACM Siggraph, Orlando, 1994, pp. 311-318.

Image Guided Surgical Systems

Eric Grimson

Artificial Intelligence Laboratory
Massachusetts Institute of Technology
Cambridge, MA
U.S.A.

ABSTRACT

Imagine giving a surgeon an "X-Ray vision" ability, that is, allowing him to look at a patient and see through skin, bone, fat to visualize internal structures, like tumor or blood vessels, exactly as they occur within the patient. Imagine allowing him to insert instruments through narrow openings in the body, yet still be able to visualize the full anatomical context around the instrument. Such capabilities would enable a surgeon to better plan procedures, to navigate through delicate procedures and to evaluate the effectiveness of surgical procedures, e.g "have I removed all of the tumor". This talk will describe such an image guided surgical system, currently in regular use at Brigham and Women's Hospital in Boston for neurosurgical cases. The system combines a suite of computer vision techniques, and allows a surgeon to visualize internal structures in registration with the patient, to interactively probe points within the patient and see the full anatomtical context of his position, and to evaluate progress of the surgical procedure. Related techniques allow a clinician to noninvasively map functional areas of the cortex, such as motor cortex, in order to provide the surgeon with registered information about critical structures during surgery.

Technical Evaluation of Biometric Systems

Brigitte Wirtz

Siemens AG, Corporate Technology, Human-Machine Interaction
Otto-Hahn-Ring 6, 81739 München, Germany
E-mail: brigitte.wirtz@mchp.siemens.de

1 Introduction

In the following paper basic considerations are made on the evaluation of biometric systems. Generally, three different evaluation aspects can be distinguished: legal, application oriented and system oriented aspects. In most cases, these can not be treated independently, since they may include some common factors. For instance, the time necessary for verification describes a system characteristic as well as being relevant for the acceptance of the approach within a given application.

The focus of this investigation is the judgment of the efficiency of biometric systems. The first two aspects mentioned above will only be considered if they are important for the technical system evaluation. A set of factors used as criteria for the technical efficiency will be described in the next sections. The choice and valuation of these factors can only be done application dependently. The list of criteria does not claim to be complete, it should rather be interpreted as a basis for a reasonable definition of biometric test scenarios.

The central parameters of a quantitative evaluation are error rates and system characteristics. Other criteria like discriminative power are discussed, though without showing how to measure them. Such measurements partially depend on the respective biometrics [Og95], thus no exact common definition is known. Therefore, we propose a pragmatic approach for biometric evaluations, consisting of an exact description of the test design and the application scenario combined with well-defined characteristic measurements based upon uniform statistically significant data sets.

In the following, the emphasis lies particularly in the description of criteria suited for the description of biometric security, their means of calculation and their influence on the error rates in the sense of their verification efficiency. Even if those criteria are not explicitly weighted in a formal decision approach, they must at least be stated in conjunction with the obtained error rates.

2 Known Approaches

Some work on the evaluation of biometric systems has been done within the European ESPRIT project *BioTest*, at the Biometric Identification Research Center at the San Jose State University [Wa97a], [Wa97b], and by the biometrics working group of the German *TeleTrust* organization.

A few papers deal with aspects of biometric evaluations. [Wa97a] is the most complete work on features and the design of biometric system testing. An overview of application oriented evaluation methods is given in [Wa97b] and [Ri95]. [GMM96] describe a way to model biometric error curves for the calculation of error rates based on small test data sets. [CoFi96] develop a biometric evaluation model with different criteria and factors for characterizing the effectiveness of biometric systems, but without defining a general test scenario. [Og95] defines five characteristic measures for the evaluation of speaker verification systems (given by the complete error curves, the storage demand defined by the average value for different sizes of the enrollee databases, the speaker discrimination, the quality and the length of the speech data in the test), which are visualized in a single diagram with common scale. [Ho97] deals with the evaluation of large scale AFIS systems (Automated Fingerprint Identification Systems).

3 Evaluation Constraints

3.1 Biometric Information Contents

The general characteristics of a specific biometrics may provide possible limitations for its universal applicability, and therefore give hints for the general obtainable security. As stated above, the applicability and security of a biometric system depend heavily on the application scenario including its man-machine interface (see below).

Discrimination

For some biometrics general limits for the ability to separate different persons can be given because of the uniqueness of the respective biometric features, whereas for others a certain interchangeability of their features is known. Examples for the first case are especially the iris and fingerprint verification, where identity is, even for twins, practically impossible, because of the genetically determined development of these features. In [Wi97a] the probability of the accidental identity of two iris scans is estimated to be 10^{-52}. In Europe the equivalence of 12 pairs of minutiae is generally considered to be sufficient to identify a person. On the other hand a certain confusability of speech data might be present particularly for speakers of the same sex, potentially resulting in higher error rates for speaker verification systems.

The discriminative power of a specific biometric feature gives basic clues for the achievable verification efficiency. However, discriminative power is not the only requirement for a secure system, as the error rates also heavily depend on the application scenario and its man-machine interface. For example, the degradation of error rates for optical fingerprint systems in the presence of scanner dirt or variable finger positioning is well-known, despite the acknowledged discriminative power of the underlying features.

A problem is the quantitative evaluation of the discrimination ability by a measure that is independent of the specific verification approach. In [Og95] this is done for speaker verification, although the used metrics is intrinsically just another verification measure.

Natural Variability of Original Data

Generally speaking, biometric techniques are divided into behavioral and physiological approaches. Behavioral systems such as dynamic signature verification, keystroke dynamics or speaker verification must deal with inherent natural variations, which lead to variable input data, independently of the underlying data collection mechanism. On the other hand, physiological characteristics of a human being normally change only due to external influences or variations. In both cases, this results in variable data sets. This natural variability may be modified by application constraints and the test population (see below). The variability introduced to original data within a test should reflect the expected natural variability for the respective biometrics.

Exposure to Falsification

Different characteristics can be more or less easy to forge. If there is a systematic way to forge a biometric pattern, it must be included as an imitation attempt into the test database.

User Limitations

If users exist who might not be able to provide a specific biometric feature consistently, they might be excluded from a specific application (disabled or emotionally stressed persons for example). This limits the general applicability of the related biometrics.

3.2 System Parameters

The efficiency and complexity of a specific approach can be described by different parameters, their values being application dependent:

- Number of data sets necessary for enrollment of a person
- Storage demands for the reference data (compressed/uncompressed)
- Complexity of the enrollment process
- Complexity of one verification cycle
- Complexity of reference data adaptation (if enabled)
- Necessity of adaptation of the system in case of a change of the installation location (e.g. learning of new environment conditions)

3.3 Data Sets

Regarding the test data sets, the most important characteristics are uniformity, statistical significance, quality and structure of originals and forgeries.

3.3.1 Uniformity

A comparison of techniques described in literature is often impossible because of the disjunctive data sets used in the tests. Test databases are either not publicly available for most biometrics or can not be used. For example, NIST databases being used for the evaluation of fingerprint classification systems (AFIS) are only of limited value for the evaluation of fingerprint verification systems. One reason is the image quality - NIST data consist of scanned images -, which significantly differs from the one obtainable by on-line fingerprint scanners. Furthermore, there are only two quite different images available per person, not sufficient for a reliable *FRR* (false rejection rate) estimation.

Given a uniform test database, it can still be difficult to test different systems which need different data sets (e.g. face recognition with moving images vs. single images). An example is the FERET face database for face recognition systems (containing pictures taken by a photograph, frontal images and well-defined variations of normal position, temporal variability), which can only be used for systems that enroll with still images [PRD96].

In general, the comparison of different approaches of the same biometric class should only be performed on a uniform test database. For comparing different biometric techniques, only qualitative estimations of the degree of difficulty of the test databases can be made, since a quantitative comparison is hardly possible.

3.3.2 Statistical Significance

Size of the Data Set

An important parameter for designing a test database is the minimum number of data sets which are needed for achieving a statistically relevant estimate of the error rates. A good estimate is given by the formula [Sa70]

$$n_h = \left(\frac{z_\tau}{\delta}\right)^2 h(1-h),$$

where h is the estimated error probability, δ the maximum deviation of the estimated from the real error probability, and z_τ the respective value of the standard normal distribution of significance level τ. The estimated error probability deviates by less than δ with a security of $(1-\tau)$, if the data set holds at least n_h samples. For instance at least 4500 samples are needed, for $\tau = 0.05$ (security of 95%), $h = 3\%$ and $\delta = 0.5\%$.

Composition and Quality of the Data Set

- **Test Population**

The size of the data set is a necessary, but insufficient criterion for statistical significance. The data consisting of N samples for each of M persons have other statistic properties than NxM samples of one person. The achieved error rates can only be considered as reliable if the data set is an accidental collection, matching the real application-typical statistical composition. There is no general population for biometric systems [Wa97a]. In particular, test databases collected in a laboratory setting do not correspond to the data in a specific application scenario. If only a laboratory setup is possible, then the estimated error rates will be rather optimistic. Therefore, representative test data can only be acquired in a field test. However, normally only a few samples per person can be acquired in such tests, as opposed to the laboratory setting.

Since the error rate will depend on the test population, the validity of this group of persons should always be specified in addition to the error rate for valuable testing.

- **Application Scenario**

The application scenario, including usability and guidance for the user, directly influences the quality of test data and the achievable error rates. In general, supervised scenarios enable better error rates and exception handling than unsupervised ones. Public environments lead to other rates than ones restricted to a limited user circle, and limited application scenarios with comfortable user interfaces lead to smaller error rates than complex user scenarios. Inhouse tests of face recognition systems have shown that error rates for a constrained application scenario can be significantly lower than for a scenario closer to reality. Thus the application scenario should be specified in conjunction with the error rates.

- **Quality of Original Data**

Variations in the quality of original data can be caused by the natural variability of data (see above), by the application scenario or by the sensor or sensor-human interface. When investigating the *EER* (equal error rate, see below), all relevant variability should be covered, or the related value range should at least be described. For practical use, all natural variations of the original data must not lead to confusion with forgeries and thus should be included in the test data set. A basic prerequisite is the variation of the collection time for originals over a significant time range.

- **Forgeries**

The structure of forgeries decisively determines the evaluated effectiveness of a biometric system. For biometric forgeries, two basic kinds can be distinguished:

- **Accidental Forgeries/Replacements**, if other originals are used as forgeries. In general, such a test only represents the classification capability of the respective system. In the case of existing deliberate forgeries, this can lead to a strong overestimation of the system efficiency.

- **Skilled Forgeries** are deliberately generated with the intention to match the respective originals as closely as possible. Error rates calculated on deliberate forgeries will normally be significantly higher than those calculated on accidental forgeries (e.g. as shown for dynamic signature verification [Wi97b]). However, skilled forgeries will not be possible for all biometrics, and might demand high expense (e.g. an artificial head) or non-practicable scenarios (cut fingers). Especially for the behavioral biometrics falsifications are possible, and thus should be included in the testing procedure.

3.4 Verification Performance

3.4.1 Error Rates

Biometric systems basically have to classify a given pattern data set into two classes. The *false rejection rate* (*FRR*) gives the percentage of falsely rejected originals, whereas the *false acceptance rate* (*FAR*) gives the percentage of falsely accepted forgeries. A measure for the overall verification performance is the so-called *equal error rate (EER)*, e.g. the error rate where the *FRR* and the *FAR* are equal.

Ideally, the *EER* of a system should be zero, but in general this is not the case, due to the variability encountered in biometric data. Thus the error rate to be minimized depends on the application. For electronic banking, the *FRR* should be minimized at a tolerable *FAR*, since legitimate users will not accept being rejected on a regular basis. On the other hand, proving identity for the entrance into security relevant areas of a company demands minimizing the *FAR*, since here the entrance of unauthorized people has to be avoided by all means, whereas a repeated trial of legitimate users can be tolerated.

3.4.2 Error Curves

Figure 1 shows the typical shape of biometric error curves. It is assumed that the distance between the reference and the test pattern represents a measure for their similarity. The smaller the threshold, the fewer forgeries can break into the system, but the more originals will be rejected. At a high threshold, no legitimated originals will be rejected, but the more forgeries will be accepted at the same time. The *EER* is defined by the intersection point of the two curves. The hatched area below the *FRR* and *FAR* curves visualize the areas of accepted forgeries and rejected originals, respectively.

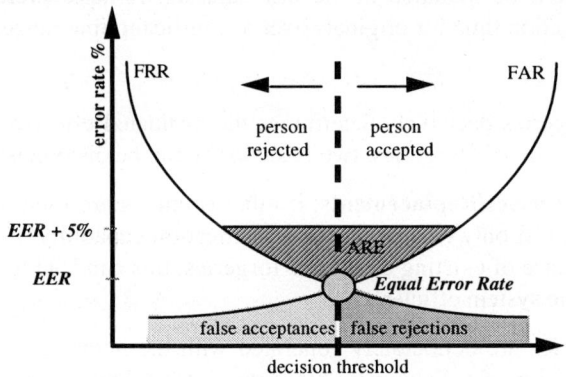

Figure 1: Typical error curves for biometric techniques

The determination of the *EER* is a theoretical evaluation of the efficiency of a system with given classified originals and forgeries. For real system use, the threshold has to be estimated from reference data according to the desired error rates and adapted if necessary. It must then be verified whether real error rates differ from the estimated ones.

3.4.3 Separability

The search for the ideal threshold can be difficult, if classifications for originals and forgeries lie close together. This problem gets worse within more complex application scenarios and can lead to significant differences between *FAR* respective *FRR* and the theoretical *EER* when only deviating a little from the ideal threshold. Systems having the same *EER* can significantly differ around the ideal point of lowest error rate. If the *FRR* and *FAR* curves show a flat shape around the *EER* area (as in Fig. 1), lower error rates will be achieved in practical use compared to the case where *FRR* and *FAR* increase immediately outside the ideal threshold domain. Characterizing a system therefore demands the examination of *FAR* and *FRR* around the ideal threshold. Besides the determination of the *EER* as the general system characteristic, both the *FAR* and *FRR* curves should be simultaneously visualized.

As an alternative, besides the *EER* a second measure *ARE* is proposed to describe the capability of a system to separate between originals and forgeries. *ARE* is defined as the area that is enclosed by the two error curves and the horizontal line y = *EER* + 5% (s. Fig. 1). If originals and forgeries are close together around the optimal threshold, this area is small. The larger this area will be, the simpler the optimal threshold for the system can be set. Given an identical *EER*, a system with a such defined larger separability is preferred.

In [Wi97a] a so-called *Decidability-Index* was also defined with the objective to achieve an *EER*-independent measure for the comparison of biometric techniques:

$$d' = \frac{\mu_2 - \mu_1}{\sqrt{\frac{\sigma_1^2 + \sigma_2^2}{2}}}$$

Herein μ_1, μ_2 and σ_1, σ_2 are mean values respective standard deviations of the original and forger distributions. Basically d' gives a measure for the general separability of these distributions as achieved by the underlying algorithm. The more those distributions differ, the greater is the separability and thus d'. However, the mean value and variance are only a good description for data with approximately normal distribution. Especially originals and forgeries normally do not carry identical distributions, thus the simplification assuming equal standard deviations as in [Wi97a] is not allowed in many cases. The advantage of the previously defined parameter *ARE* is the independence of the statistical distribution of the data.

3.4.4 Practical Generation of Error Curves and Characteristics

The *EER* curve visualized in Fig. 1 is defined for one original person. Averaging the curves for all persons yields the *EER* of the system, but requires normalized classification values. If verification measures are person dependent, the *EER* curves must be averaged after normalization.

Due to the discrete calculation of the error curves, the *EER* represents an estimate of the real *EER* that is derived as the linear interpolation of the respective neighboring *FAR* and *FRR* points. Actually this estimate could take on any value between the calculated *FAR* and *FRR*. The averaged system *EER* is thus again an estimate of the real system *EER*.

In [Og95] the so-called *ROC* (Receiver Operator Characteristic) curves are described as an alternative to *EER* curves to visualize the system characteristics. At any point, such curves show the percentage of false rejections as a function of the percentage of false acceptances. However the percentage numbers of rejection and acceptance can only be determined practically for discrete threshold values or alternatively via a parametric modeling of the *ROC*. Thus the threshold independence of *ROC* curves exists only formally.

4 Summary and Outlook

An overview of factors describing the efficiency of biometric systems has been given. Additionally, some criteria influencing the evaluation are described. In the future, a quantitative measurement of those factors combined with their weighting according to their relevance for a specific application could lead to a judgment of different systems by a formal decision method. Furthermore, evaluation techniques known from the image processing field should be considered [JaDo97].

Acknowledgments: I wish to express my thanks to my colleagues, Ernst Haselsteiner, Gerd Hribernig, Wolfgang Marius and Claudia Windisch (Siemens Austria), for fruitful discussions on the topic of biometric evaluation.

5 References

[CoFi96]   T.M. Corcoran, J.S. Fischer, *A Dynamic Model for the Evaluation and Selection of Biometric Technologies*, CardTech/SecurTech'96, 1996, pp. 431-449

[GMM96]   M. Golfarelli, D. Maio, D. Maltoni, *On the Error-Reject Tradeoff in Biometric Verification Systems*, Rapporto tecnico CSITE-022-96, University of Bologna, Oct. 1996, pp. 1-26

[Ho97]   R. Hopkins, *Benchmarking Very Large Scale Biometric Identity Systems,* CardTech/SecurTech'97, 1997, vol. 2, pp. 313-331

[JaDo97] A.K. Jain, C. Dorai, *Practising vision: Integration, evaluation and application*, Pattern Recognition, 1997, vol. 30, no. 2, pp. 183-196

[Og95]   John Oglesby, *What´s in a number? Moving beyond the equal error rate*, Speech Communication, 1995, vol. 17, pp. 193-208

[PRD96]   P.J. Phillips, P.J. Rauss, S.Z. Der, *FERET Recognition Algorithm Development and Test Results*, ARL-TR-995, October 1996

[Ri95]   D.R. Richards, *Rules of thumb for biometric systems*, Security Management, 1995, vol. 39, no. 10, pp. 67-71

[Sa70]   L. Sachs, Statistische Methoden, Springer Verlag, 1970

[Wa97a]   J.L. Wayman, *The Science of Biometric Technologies: Classifiying, Testing, Evaluating and Selecting*, CardTech/SecurTech, 1997, vol. 1, pp. 385-396

[Wa97b]   J.L. Wayman, *A Scientific Approach to Evaluating Biometric Systems using a Mathematical Methodology*, CardTech/SecurTech, 1997, vol. 1, pp. 477-492

[Wi97a]   G. Williams, *Comparing the Efficacy of Various Biometric Devices*, BiometriCon'97, 1997, pp. Q1-Q14

[Wi97b]   B. Wirtz, *Ein System zur dynamischen Unterschriftsverifikation*, 6. Handschriften-Symposium, BKA Wiesbaden, 1997, pp. 1-12

Face Recognition from Sequences Using Models of Identity

Stephen J. McKenna and Shaogang Gong

Dept. of Computer Science, Queen Mary and Westfield College, London.
E-mail: stephen@dcs.qmw.ac.uk

Abstract. A method for modelling and recognising facial identity is described within the context of an integrated system for face recognition in dynamic scenes. Recognition is based sequences rather than isolated images. Mixture models provide estimates of class-conditional probabilities and these are used to accumulate recognition confidence over time. Results are presented using data from the integrated system.

1 Introduction

The face recognition tasks considered in this paper are characterised by the availability of entire image sequences containing many face images of relatively small groups of individuals. Such data arise from the type of integrated approach to face recognition in dynamic scenes illustrated in Fig. 1. In such a system, the visual tasks of detection, tracking, alignment, normalisation, modelling and recognition of identity are performed by a set of co-operating modules with closed-loop feedbacks. Image sequences are processed in real-time to acquire normalised, aligned face image sequences using a combination of motion, colour and face appearance models [9, 10, 13]. This paper focuses upon the tasks of modelling and recognising identity within such a framework.

A common approach to performing recognition of identity in dynamic scenes is to attempt to isolate one (or a few) accurately aligned and normalised face images and to base recognition only upon these images. In contrast, the approach adopted here is to perform recognition based upon entire tracked face sequences. This allows temporal information to be exploited and avoids the rather arbitrary discarding of potentially informative face images. The tasks considered here require recognition to be performed using image sequences acquired and normalised automatically in poorly constrained dynamic scenes. These sequences are characterised by low resolution, large changes in scale and variable illumination conditions. Recognition based upon isolated images of this kind is highly inconsistent and unreliable. However, the poor quality of the image data can be compensated by accumulating recognition scores over time.

The use of entire sequences means that many images of a person can be acquired in a few seconds. Given sufficient data, it becomes possible to begin to model class-conditional structure, i.e. to estimate probability densities for each person. Such models of identity can allow more accurate recognition than the use of a few example face images.

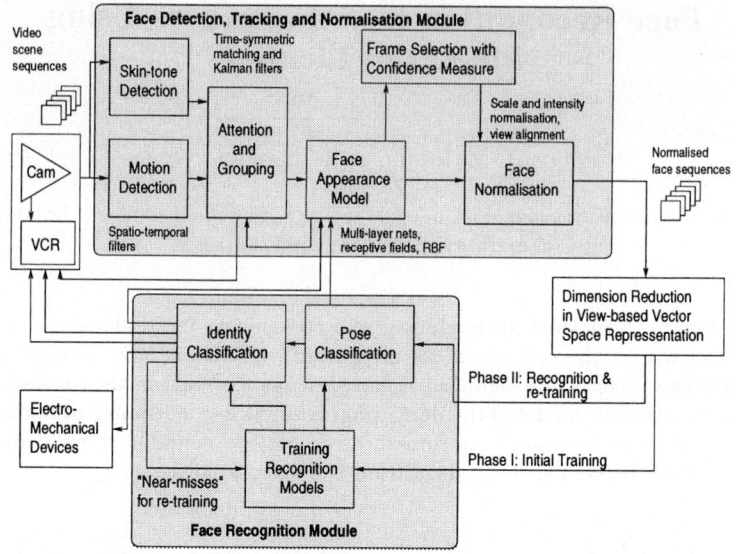

Fig. 1. An integrated framework for face recognition in dynamic scenes.

In sections 2 and 3, recognition tasks are defined and methods for modelling class-conditional densities using mixture models are presented for use in these tasks. Results and conclusions are given in sections 4 and 5.

2 Statistical models of identity

Given a database consisting of a set, $\mathcal{S}$, of N known people, different face recognition tasks can be envisaged. Four tasks are defined here as follows:

1. *Face classification*: The task is to identify the subject under the assumption that the subject is a member of $\mathcal{S}$.
2. *Known/unknown*: The task is to decide if the subject is a member of $\mathcal{S}$.
3. *Identity verification*: The subject's identity is supplied by some other means and must be confirmed. This is equivalent to task 2 with $N = 1$.
4. *Full recognition*: The task is to determine whether or not the subject is a member of $\mathcal{S}$, and if so to determine the subject's identity.

Fig. 2 illustrates the four recognition tasks defined above in a hypothetical face space $\mathcal{F}$, where $\mathcal{F}$ is assumed to contain all possible face images and to exclude all other images. Plotted in $\mathcal{F}$ are example faces for three different people[1]. Suitable decision boundaries for performing the recognition tasks are

---

[1] Several simplifications have been made for illustrative purposes: (1) the identities are likely to overlap significantly, (2) the space has high dimensionality and visualisation in two dimensions would reveal little or no structure and (3) several connected regions may be required per person in reality.

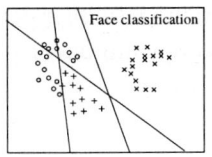 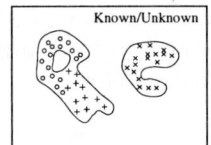 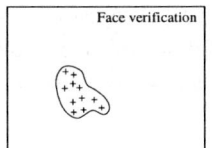 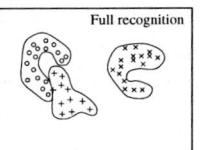

Fig. 2. Schematics illustrating the four face recognition tasks. Plotted in a hypothetical face space, $\mathcal{F}$, are example faces from three different people. Suitable decision boundaries are shown for each of the four face recognition tasks.

shown. The separability of face identities in $\mathcal{F}$ will depend upon the technique used to model $\mathcal{F}$. However, it is likely that each identity will form strongly non-convex regions in this subspace [4]. In the *face classification* task, all N classes can be modelled. In contrast, the other three tasks all suffer from the need to consider the class of unknown faces.

The approach argued for here provides a recognition framework that can be applied to any of the four tasks defined above. The main idea is to model a class-conditional density for each person in a representation space of relatively low dimensionality. Given such class-conditional densities, all four recognition tasks can be performed in a well-founded, statistical way (see [11] for further discussion). The method chosen to estimate these densities must be sufficiently general in order to model the highly non-convex distributions generated by different images of a face. It should also allow for a range of model complexity in order to model people for whom a relatively small amount of data are available. As more data are collected through recognition the model should be able to adapt to capture the underlying distribution more accurately.

The method selected here for density estimation was Gaussian mixture models. Modelling face classes as mixtures has several attractive characteristics: (1) density estimation is performed in a semi-parametric way so that the size of the model (number of mixture components) scales with the complexity of the data rather than with the size of the data set; (2) the method is sufficiently general to model complex, non-linear distributions given enough data; (3) the models can be constrained to obtain well-conditioned estimation given limited data; (4) when classification is performed, other models emerge as special cases of using Gaussian mixtures, e.g. nearest neighbour and nearest mean classification.

3 Modelling identity using Gaussian mixtures

Let each person, k, constitute a class C_k. A person's identity is modelled by estimating the class-conditional density, $p(\mathbf{x}|C_k)$, from n examples of that person's face, $\mathbf{x}_i$, $i = 1\ldots n$. This density takes the form of a mixture of M components:

$$p(\mathbf{x}|C_k) = \sum_{j=1}^{M} p(\mathbf{x}|j)P(j) \qquad (1)$$

The mixing parameter $P(j)$ represents the prior probability that the d-dimensional data, $\mathbf{x}$, was generated by mixture component j. Each mixture component is a Gaussian with mean $\boldsymbol{\mu}_j$ and covariance matrix $\boldsymbol{\Sigma}_j$:

$$p(\mathbf{x}|j) = \frac{1}{(2\pi)^{\frac{d}{2}}|\boldsymbol{\Sigma}_j|^{\frac{1}{2}}} e^{-\frac{1}{2}(\mathbf{x}-\boldsymbol{\mu}_j)^T \boldsymbol{\Sigma}_j^{-1}(\mathbf{x}-\boldsymbol{\mu}_j)} \qquad (2)$$

Expectation-Maximisation (EM) provides an effective maximum-likelihood algorithm for learning a Gaussian mixture model (see [11, 14] for further details).

Appearance-based face representations usually have high dimensionality and in practice fitting a mixture of Gaussians is often highly under-constrained due to limited data and the "curse of dimensionality". There are, however, at least three complementary approaches to making the modelling tractable.

Firstly, the number of parameters in the model can be reduced by constraining the form and the number of Gaussian mixture components. In the most general case, each Gaussian, j, has a covariance matrix, $\boldsymbol{\Sigma}_j$, which is completely determined by the data, i.e. $d(d+1)/2+d$ parameters must be estimated. If $\boldsymbol{\Sigma}_j$ is constrained to be a diagonal matrix then there are only $2d$ parameters to be determined. If $\boldsymbol{\Sigma}_j = \sigma_j \mathbf{I}$ for some $\boldsymbol{\Sigma}_j$ then the Gaussian is radially symmetric and there are only $d+1$ parameters to be determined. Finally, if $\boldsymbol{\Sigma} = \mathbf{I}$ then only the mean must be estimated (d parameters).

Secondly, the data set can be artificially enlarged by synthesising new *virtual* images for each person by using models of possible variations of a face image. In its simplest form, this approach can consist of applying a set of simple transformations to the images e.g. small translations, scalings, rotations and mirroring about the vertical axis. Noise can also be artificially added to the images. More complex models of deformation can also be employed for synthesis of virtual views, e.g. [2, 3].

Thirdly, the dimensionality of the face representation vectors can be reduced. A significant reduction is achieved by representing faces as vectors in the subspace of faces $\mathcal{F}$ rather than as image vectors in the space of all possible images. However, $\mathcal{F}$ is difficult to model.

In theory, if local correspondences can be established between all face images, face space can be approximated using linear vector spaces [3, 5]. In practice, however, establishing even a small set of feature correspondences between faces is highly problematic, especially at low resolution. In addition, the use of entire image sequences to perform real-time recognition places constraints on the computation which can reasonably be performed for each image. Attempts to use optical flow to establish correspondences becomes prohibitively expensive.

A representative data set containing a large number of different identities is needed in order to build a *generic* model of the face space. In practice, a *specific* approximation, $\mathcal{F}_\mathcal{S}$, is usually obtained from images in the set $\mathcal{S}$ of N known people. When N is small, $\mathcal{F}_\mathcal{S}$ is a poor approximation to $\mathcal{F}$. If a specific model is used, it must be updated each time the set $\mathcal{S}$ changes. Furthermore, any identity-specific models which make use of $\mathcal{F}_\mathcal{S}$ must also be updated. In contrast, a generic model need never be updated. An important point here is that *face classification* is easier to perform in $\mathcal{F}_S$ while *identity verification*, *known/unknown* and *full recognition* are best performed in a generic space, $\mathcal{F}$.

In this paper, the images provided by the tracking subsystem are approximately aligned frontal or near-frontal views. Linear models of face space were used for the purpose of dimensionality reduction in order to obtain tractable, although approximate, models. Principal Components Analysis (PCA) has been used to obtain face space models for face classification [5, 12, 16]. The models are computed without the use of any identity class information. PCA is therefore suitable for data sets with only a few example images per person and (or) large numbers of people. Linear discriminant analysis (LDA) has also been used [1, 6, 15]. It is able to preserve class linear separability when applied to data sets with many images per person and relatively few people. It is therefore suitable for computing specific face space models for face classification using many training images of a few people.

In experiments described in the next section, a large data set containing many different people with only a few images per person was used to compute a generic face space using PCA. The reduced dimensionality pattern vector representation, $\Omega(\mathbf{x})$, of a face image, $\mathbf{x}$, is obtained by projection onto the dominant eigenvectors of the covariance matrix formed from the training images [8]. This pattern vector can be 'normalised' by dividing its elements by the corresponding eigenvalues to obtain $\Omega_{\mathbf{norm}}(\mathbf{x})$. This gives the data equal variance along each principal component axis. Class-conditional densities can be modelled in a principal subspace by estimating either $P[\Omega(\mathbf{x})|C_k]$ or $P[\Omega_{\mathbf{norm}}(\mathbf{x})|C_k]$.

4 Experiments

In this section, experiments using Gaussian mixture models of identity are described. Face image data were acquired and normalised in a fully automated way by a face tracking system. The neural network model used to perform tracking was trained using 9000 example face images rotated by $\pm 10°$ and scaled to 90% and 110% [10, 9]. The normalised faces from the tracker therefore varied by at least these amounts in scale and rotation. Since the aim of these experiments was to compare methods for modelling identity rather than to optimise recognition accuracy, no attempt was made to reduce these variations. Eight subjects were tracked through relatively unconstrained indoor scenes as they walked towards a fixed camera. Overhead lighting resulted in variations in facial illumination. Fig. 3 shows three examples from the tracking process. The resolution of the area of the face tracked ranged from approximately 10×10 pixels to 80×80 pixels when the subject approached the camera. Two normalised face sequences were obtained for each subject. The first sequence of each subject was used for training and the second for testing. In total, there were 326 training and 296 test images. Fig. 4 shows some examples.

Face space was modelled by performing PCA. A specific model, $\mathcal{F}_S$, was computed from the training set. A generic model, $\mathcal{F}$, was computed using 644 other images originally used to train a face detection neural network in the tracking

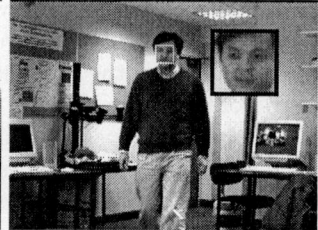

Fig. 3. Example frames from tracked sequences. The face bounding box is overlaid on each image. The extracted face images are shown inlaid.

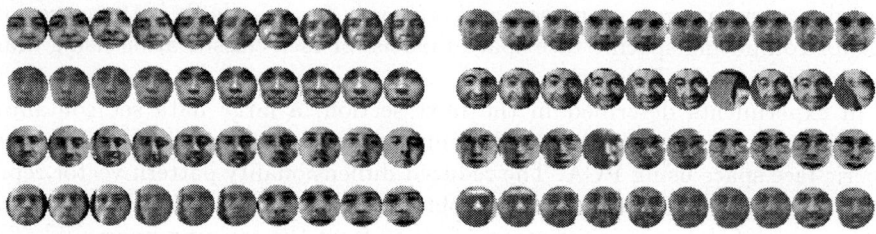

Fig. 4. A subset of the test data.

system. These images were highly suitable since they had similar variations in scale and rotation to the tracked data to be recognised. Training images were projected onto the dominant eigenvectors and each person's identity was modelled by estimating either $P[\Omega(\mathbf{x})|C_k]$ or $P[\Omega_{\mathbf{norm}}(\mathbf{x})|C_k]$ with Gaussian mixtures. The 8 mixture models' parameters were stored along with the eigenvectors and eigenvalues and subsequently used to perform classification of the test sequences.

Initially, both a specific and a generic eigen-space were computed using the first 40 eigenvectors. Table 4 shows a comparison of face classification using the specific and generic models. Identities were modelled by fitting a single radial Gaussian to each person's data to estimate $P[\Omega_{\mathbf{norm}}(\mathbf{x})|C_k]$. The percentage of images correctly classified per person along with the percentage of total images classified correctly are given. Sequence classification results are also given based upon (1) a majority vote (Maj.), i.e. the sequence is classified as the person with the most images and (2) the highest summed log probability (Pr.). The result illustrates the fact that the use of a generic face space which could be used to facilitate identity verification, known/unknown or full recognition, in turn makes face classification more difficult. However, if probabilities are accumulated over each sequence, performance becomes comparable. A reduction in the dimensionality of the generic face space from 40 to 20 did not result in any significant loss of accuracy. Face classification results using the 20-dimensional generic space are given in Table 2. Gaussian mixtures of various complexity were compared.

The first two methods in Table 2 used unnormalised pattern vectors. The methods $T - P$ and $T - P_{norm}$ used single radial Gaussians of equal variance resulting in a nearest-mean classifier which was equivalent to the eigenfaces method

Name	Face space	Person (% images correct)								Total %	Seq.	
		0	1	2	3	4	5	6	7		Maj.	Pr.
Radial	Specific	75	64	74	85	56	78	29	11	55.1	7	6
Radial	Generic	57	67	66	20	13	72	25	29	43.6	4	6

Table 1. Test set results with generic and specific face space models using 40 principal components. Identities were modelled by fitting a single radial Gaussian to each person.

Name	M	Σ type	$\frac{y_i}{\lambda_i}$	Tot. %	Seq.	
					Maj.	Pr.
T-P	1	σ_f	N	25.0	2	2
1-NN	n	$\sigma \to 0$	N	32.1	1	1
T-P$_{norm}$	1	σ_f	Y	46.3	4	4
Radial	1	σ_j	Y	44.3	4	4
Diag	1	Σ_d	Y	42.9	4	3
2-Rad	2	σ_j	Y	52.0	5	7
3-Rad	3	σ_j	Y	42.2	5	5
2-Diag	2	Σ_d	Y	41.9	4	5

Table 2. Test results with a 20-dimensional generic face eigenspace and identity mixture models. Column 2 indicates the number of Gaussians, M. Column 3 indicates the type of Gaussian where Σ_d denotes a diagonal covariance, σ_j an independent variance and σ_f a variance equal to that of all other components. A 'Y' in column 4 indicates that normalised pattern vectors were used.

of Turk and Pentland [16]. The use of normalised pattern vectors resulted in a significant improvement. The nearest neighbour classifier (1-NN) is equivalent to a mixture of Gaussians with one component per image and variances tending to zero. The remaining methods shown are mixture models of radial or diagonal covariance Gaussians with between 1 and 3 components. A mixture of 2 radial Gaussians provided the best performance of all the methods.

5 Conclusions

The face recognition tasks considered were characterised by poor resolution and variable lighting and arose within the context of an integrated system for recognition in dynamic scenes. The data sets consisted of image *sequences* of small groups of people. Four recognition tasks were defined: face classification, face verification, known/unknown and full recognition. All but face classification require consideration of the class of unknown people. As a consequence, identities should be modelled in a generic rather than a specific face space.

Gaussian mixtures provided an effective way to model identities as classconditional probability densities in face space. This approach to recognition results in a system which can learn and update identity models independently of one another. Model complexity adapts to the structure of the data and constrained models are obtainable when data is lacking. The eigenface method of [16] can be viewed as a special case and was outperformed by simple mixture models. It was shown that modelling identities using such models is beneficial given an appropriate level of mixture complexity. They provide estimates of class-conditional probability which, when accumulated over time, provide improved recognition performance. The results reported here are preliminary and evaluation on larger data sets is needed. Mixtures of principal components analysers or factor analysers should provide even more effective ways to constrain the mixture components [7].

References

1. P. N. Belhumeur, J. P. Hespanha, and D. J. Kriegman. Eigenfaces vs. fisherfaces: recognition using class specific linear projection. *IEEE PAMI*, 19(7):711–720, 1997.
2. D. Beymer and T. Poggio. Face recognition from one example view. Technical Report 1536, MIT AI Memo, September 1995.
3. D. Beymer and T. Poggio. Image representations for visual learning. *Science*, 272:1905–1909, 28 June 1996.
4. M. Bichsel and A. P. Pentland. Human face recognition and the face image set's topology. *CVGIP: Image Understanding*, 59(2):254–261, March 1994.
5. N. Costen, I. Craw, G. Robertson, and S. Akamatsu. Automatic face recognition: what representation ? In *ECCV*, pages 504–513, Cambridge, England, April 1996.
6. K. Etemad and R. Chellappa. Discriminant analysis for recognition of human face images. In *AVBPA*, pages 127–142, 1997.
7. G. E. Hinton, P. Dayan, and M. Revow. Modeling the manifolds of images of handwritten digits. *IEEE Trans. Neural Networks*, 8(1):65–74, 1997.
8. M. Kirby and L. Sirovich. Application of the karhunen-loeve procedure for the characterization of human faces. *IEEE PAMI*, 12(1), 1990.
9. S. J. McKenna and S. Gong. Tracking faces. In *Proc. 2nd Int. Conf. on Automatic Face and Gesture Recognition*, Killington, Vermont, US, October 1996.
10. S. J. McKenna, S. Gong, and J. J. Collins. Face tracking and pose representation. In *BMVC*, Edinburgh, Scotland, September 1996.
11. S. J. McKenna, S. Gong, and Y. Raja. Face recognition in dynamic scenes. In *BMVC*, 1997.
12. A. Pentland, B. Moghaddam, and T. Starner. View-based and modular eigenspaces for face recognition. In *CVPR*, 1994.
13. Y. Raja, S. J. McKenna, and S. Gong. Segmentation and tracking using colour mixture models. In *ACCV*, 1998.
14. R. A. Redner and H. F. Walker. Mixture Densities, Maximum Likelihood and the EM Algorithm. *SIAM Review*, 26(2):195–239, 1984.
15. D. L. Swets and J. Weng. Discriminant analysis and eigenspace partition tree for face and object recognition from views. In *Proc. 2nd Int. Conf. on Automatic Face and Gesture Recognition*, pages 192–197, 1996.
16. M. Turk and A. Pentland. Eigenfaces for recognition. *J. of Cognitive Neuroscience*, 3(1), 1991.

Enhancing Human Face Detection Using Motion and Active Contours

Kin Choong Yow and Roberto Cipolla

Department of Engineering,
University of Cambridge
Cambridge CB2 1PZ, England

Abstract. Recent advances in human face detection algorithms have seen varying degrees of success using numerous approaches. We identify that a feature-based approach is able to detect faces efficiently over large viewpoint and illumination variations. In this paper, we will enhance the approach by proposing the use of active contour models to detect the face boundary, and subsequently use it to verify face candidates. We present a method to initialize the active contour model, and show how the resulting information can be used to verify true face candidates. Further verification of the face hypothesis is achieved by checking for consistency with motion. We present results of testing the system under a wide range of imaging conditions, demonstrating its capability and robustness.

1 Introduction

Human face detection is an active research area with important applications in automatic face recognition, visual surveillance, and man-machine interface. Successful detection on fronto-parallel faces have been reported by Sung and Poggio [11] and Rowley et al. [8] using a neural network based approach on large datasets. A colour based approach have also being shown (Chen et al. [3] and Graf et al. [6]) to be very effective against an unknown face pose.

A closely related problem with human face detection is that of face localization. The output of the above-mentioned algorithms are only rectangular boxes, which is not sufficient when the locations of features and boundaries need to be known. To achieve face localization, the popular approach is to use active contours. Cootes and Taylor [5] have reported very good localization results using active shape models, and other researchers have also demonstrated success using snakes (Sobottka and Pitas [10], Wu et al. [12]). Unfortunately, their successes are currently limited only to a narrow range of fronto-parallel viewpoints.

In this paper, we will describe the selection and use of active contour models to detect face boundaries over a large range of viewpoints. The objective is twofold: we want to provide a more informative face detection output indicating the location of the facial features and the face boundary; and also we want to use the detected face boundary to verify the face hypotheses and reject false candidates. In addition, we will describe how we can exploit motion information to reduce processing time, and to increase the confidence of detecting partially occluded faces from the motion information.

2 Feature-Based Face Detection

Yow and Cipolla [13] have demonstrated that a feature-based approach can work well under variations in scale, orientation and viewpoint. The advantages of a feature-based approach are that it is robust in a wide range of viewpoints, even at profile views. Being a bottom-up approach it also maintains a small search space by reducing false or impossible groupings at early stages, resulting in efficient computation throughout the algorithm.

We will extend the work in [13] and make use of the output of the feature-based system directly. The features detected by the system will be used in our algorithm to initialize the active contour.

3 Extracting the Face Boundary

The face boundary is one of the hardest feature in a human face image to be extracted. This is because the face boundary is an occluding boundary obtained from projecting the 3D shape of the human head onto a 2D image. As such, active contours are probably the best choice for extracting the face boundary.

The main difficulties with active contours are that a good initial estimate must be available, and that the active contour will always converge onto a solution, whether it is the desired solution or a false optima. The following discussion focusses on the initialization of the snake. In our implementation we choose a cubic B-spline snake because of the speed and ease in specifying and moving the control points rather than the actual snake pixels (Cipolla and Blake [4]).

Many researchers have modelled the head or the face as an ellipse (Graf et al. [6], Sobottka and Pitas [10]), and the successes in their approaches have suggested that we can model our snake(s) as partial ellipse(s). We observe in Fig. 1 that we can indeed initialize our snake as two quarter ellipses sharing the same center, but with two different minor axis parameters (Fig. 1(c)).

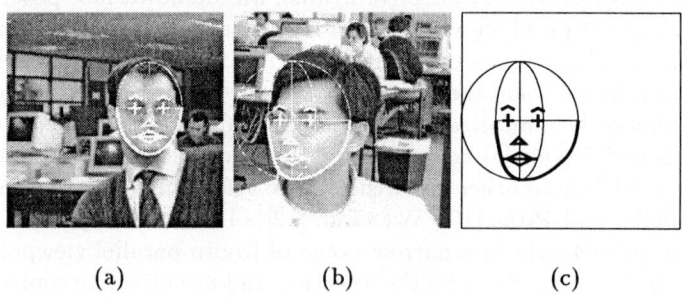

Fig. 1. Estimating the face boundary with partial ellipses. (a) Fronto-parallel view. (b) Non fronto-parallel view. (c) Partial ellipses used.

We choose the center of the partial ellipses to be the intersection point between the line joining the two eyes and the perpendicular from the line to the mouth (Fig. 2). The major axis r_3 is determined first by searching for a short segment of edges below the mouth which have orientations perpendicular to the major axis. Subsequently, we fix the major axis and initialize quarter ellipses with different values of the minor axis radii. If the edge orientation of the points along the ellipse correspond to the orientation of the tangent to the ellipse at those points, we choose those radii as the snake's minor axis parameters r_1, r_2.

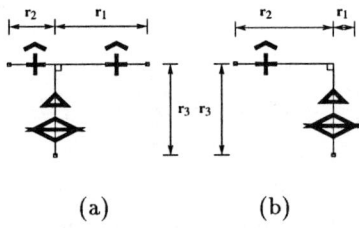

Fig. 2. (a) Ellipse parameters for a face. (b) Ellipse parameters in a profile view.

In the case where only one eye is present (e.g. profile view), the ellipse center is taken to be the intersection of the line joining the nose and mouth, and the perpendicular from the line to the eye (Fig. 2(b)). An illustration of the search for the ellipse parameters of the initial position of the snake is shown in Fig. 3.

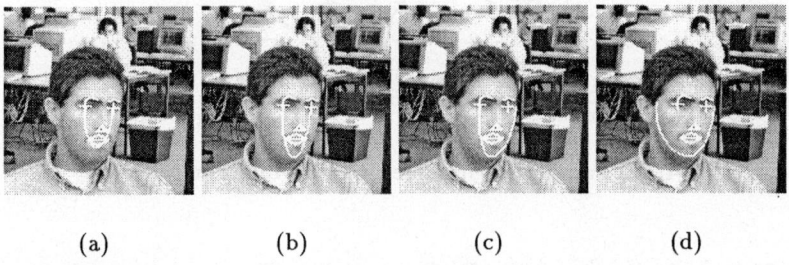

Fig. 3. (a) Ellipse drawn using the position of eyes and mouth as parameters. (b) Ellipse after searching along major axis for parameter r_3. (c) Ellipse after searching for minor axis parameter r_1. (d) Ellipse after searching for minor axis parameter r_2.

4 Face verification

For false candidates, there is no face boundary for the snake to lock on. A snake initialized on false faces when run to convergence will have a wrong shape or an impossible face configuration. As such, we can verify our face candidates by examining the shape, size or position of the snake after it has run to convergence.

Classical approaches of examining and recognizing curves have made use of geometric invariants (Zisserman et al. [14], Rothwell et al. [7]). However, our face boundary curve does not have enough structure to produce an invariant signature for recognition. Also, as the face boundary is an occluding boundary, we need to seek a non-linear mapping of the face to a canonical form.

We define a transformation $T(x,y) : (x,y) \rightarrow (u,v)$ based on thin-plate splines (Bookstein [2]) of the form:

$$T(x,y) = a_1 + a_2 x + a_3 y + \sum_{i=1}^{N} w_i U(|Z_i - (x,y)|) \;. \qquad (1)$$

where $U(r) = r^2 \log r^2$ is the thin-plate spline function, N is the number of landmark points, $Z_i = (x_i, y_i)$ are the N landmark points ($i = 1..N$), and the weights w_i sum to one. The computation of the coefficients in (1) involves the solution of two square linear systems of size $N+3$ (with the same matrix in each case). An algebraic treatment of this mapping is given in Bookstein [1].

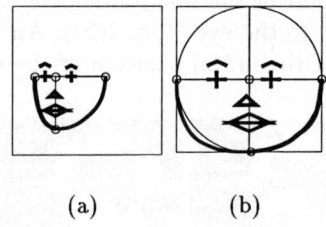

(a) (b)

Fig. 4. (a) Face in image frame. (b) Face in canonical frame.

We choose 4 landmark points for the mapping. The 4 points we chose are the midpoint of the snake, the 2 endpoints, and the intersection of the line joining the 2 endpoints with the perpendicular from the midpoint of the snake. These 4 points are mapped to a unit circle in the canonical frame (see Fig. 4).

The face verification process is then straightforward. By examining a set of training images and plotting the feature points and face boundary in the canonical frame, we can form clusters or regions in the canonical frame which the facial features or the face boundary must lie. When we verify a new face candidate, we simply map this face into the canonical frame and examine if the facial features and the face boundary fall in the desired regions.

Fig. 5 show some results of applying our face boundary detector to various images of faces in complex background. The time taken for the cubic B-spline snake with 30 control points to converge is approximately 80 ms on a SUN-Sparc20 workstation.

Fig. 5. Face boundary detection on different subjects.

5 Exploiting Motion

The results in the last section have shown a reliable detection of the face boundary and hence the face under different viewpoints. However, the processing time is slow and false detects are still likely. Several researchers have made use of image motion for face detection (Schiele and Waibel [9], Graf *et al.* [6]) and have demonstrated good results. We shall exploit the constraint of motion to increase processing speed and to further reduce false detects.

The simplest and the most straightforward way to detect motion is by image differencing. The main problem about image differencing is that for a uniformly coloured object, image differencing will only give response at the edge of the object, but not in its interior. However, since we are looking for high contrast features in the face, these features stands out very well under image differencing. We observe a large reduction of the number of feature candidates after image differencing, making the feature grouping process faster and more reliable.

5.1 Temporal filtering

We cannot assume, even at a dense sampling rate, that the face(s) in the image will not be occluded or disappear from the image. Therefore, instead of tracking a face in the image, we can expect better robustness if we run a detection algorithm on each frame and make use of correspondences from the previous frame to increase the confidence of accepting/rejecting face candidates.

To establish the correspondence between a face candidate in the current frame with the previous frame, we map both face candidates to the canonical frame, and measure the disparity between the feature positions and the face boundary location.

The quality of a match $q_{k,k-1}$ between a face in the kth canonical frame and a face in the $(k-1)$th frame is given by :

$$q_{k,k-1} = \frac{1}{1 + \sum w_i (\mathbf{x}_i(k) - \mathbf{x}_i(k-1))} \quad . \tag{2}$$

where $\mathbf{x}_i(k)$ are the position vectors of facial features and face boundary points i in the kth canonical frame. w_i are weights such that $\sum w_i = 1.0$. Matches with $q_{k,k-1}$ less than a certain threshold τ are deemed invalid and the face candidate in the kth frame will be discarded.

5.2 Feature Estimation

Sometimes not all the facial features are detected in the image. In such cases, we will examine all possible partial face groups in the current frame for a possible match. If the features are present in the previous frame but not in the current frame, any missing facial features can be determined from the canonical frame of the face by performing the inverse mapping. The centroid of the missing feature region in the canonical frame is mapped back into the image, giving us the feature points as desired.

6 Results

The proposed system is implemented on a SUNSparc20 workstation with an S2200 framegrabber board. The image size used is 288x192 pixels and they are captured by a Pulnix monochrome CCD camera. The time taken for the system to process an image in a sequence is about 2 seconds. The best frame rate we obtained for grabbing images to disk is about 3 frames per second.

Figure 6 shows some results of the detection of faces under conditions of strong directional illumination and poor camera contrast. The robustness of the gaussian derivative filters ensure that the facial features remain detected. The system is also shown to be able to detect faces of subjects wearing glasses.

Fig. 6. Results of face detection on images with multiple faces, strong directional illumination, and poor camera contrast.

Figure 7 shows some interesting results. The first image shows the detection of a hand drawn face. The result of this test raise some interesting questions: Do we want a system that can be fooled by a hand drawn face or a photograph? Or do we want a system that will detect only real faces? We believe that depending on the application, it is always better to err on the safe side.

The second image shows a failure to locate the face boundary accurately. This is due to non-uniform illumination which causes a strong edge on the face which stops the snake search too early. The third image shows a false detect in the image where the false features has remained in the correct face configuration for more than a single frame. This problem can be overcome by performing temporal

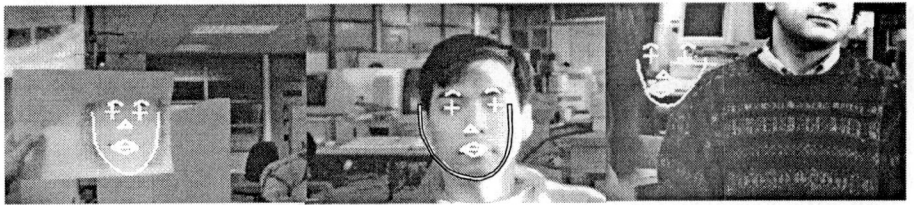

Fig. 7. Some interesting cases.

filtering over a larger number of image frames. But this will need faster processing power or equipment such as convolution hardware.

Table. 1 shows a summary of results obtained for 4 image sequences. These sequences represent tests under strong directional illumination, poor image contrast, multiple faces and expression, and hand drawn faces respectively. The image sequences can be viewed at

http://svr-www.cam.ac.uk/~kcy/Research.html

Table 1. Summary of results.

Sequence No.	Number of faces	True positives	False positives
Sequence 1	44	43 (97.7%)	1 (2.3%)
Sequence 2	32	28 (87.5%)	1 (3.1%)
Sequence 3	53	39 (73.6%)	2 (3.8%)
Sequence 4	20	14 (70.0%)	0 (0.0%)
Total	149	124 (83.2%)	4 (2.7%)

7 Discussion and Future work

The algorithm is still unable to detect faces that are too small (smaller than 60x60 pixels). This is a restriction imposed by the feature-based approach. To overcome this, we will have to look at exploiting other reliable features (such as skin colour or body silhouette) to improve the performance of the algorithm.

Our future work will be directed towards increasing the speed of the algorithm so that a real-time or a near real-time system can be implemented. One possible way of doing this is to build a multi-modal system in which the system can switch itself into the static or dynamic modes depending on the speed required to produce an output.

8 Conclusion

In this paper, we have addressed the issue of not just detecting a human face in an image, but also locating its internal features and its external boundary. The choice of a suitable active contour model and its initialization have led to positive results which can be applied directly to verify face hypotheses and reduce false detects. We have also exploited motion information for further verification, to reduce processing time, and to estimate missing features in the image.

References

1. F. L. Bookstein. Principal warps: Thin plate splines and the decomposition of deformations. *IEEE Trans. Patt. Analy. and Machine Intell.*, 11:567–585, 1989.
2. F. L. Bookstein. *Morphometric Tools for Landmark Data: Geometry and Biology*. Cambridge University Press, New York, 1991.
3. Q. Chen, H. Wu, and M. Yachida. Face detection by fuzzy pattern matching. In *Proc. 5th Int. Conf. on Comp. Vision*, pages 591–596, MIT, Massachusetts, 1995.
4. R. Cipolla and A. Blake. The dynamic analysis of apparent contours. In *Proc. 3rd Int. Conf. on Comp. Vision*, pages 616–623, Osaka, 1990.
5. T. F. Cootes and C. J. Taylor. Locating faces using statistical feature detectors. In *Proc. 2nd Int. Conf. on Auto. Face and Gesture Recog.*, pages 204–209, Vermont, 1996. IEEE Comp. Soc. Press.
6. H. P. Graf, E. Cosatto, D. Gibbon, M. Kocheisen, and E. Patajan. Multi-modal system for locating heads and faces. In *Proc. 2nd Int. Conf. on Auto. Face and Gesture Recog.*, pages 88–93, Vermont, 1996. IEEE Comp. Soc. Press.
7. C. A. Rothwell, A. Zisserman, D. A. Forsyth, and J. L. Mundy. Planar object recognition using projective shape representation. *Int. Journal of Comp. Vision*, 16:57–99, 1995.
8. H. A. Rowley, S. Baluja, and T. Kanade. Human face detection in visual scenes. Technical Report CMU-CS-95-158, CMU, July 1995.
9. B. Schiele and A. Waibel. Gaze tracking based on face-color. In *Proc. Int. Workshop on Auto. Face and Gesture Recog.*, pages 344–349, Zurich, 1995.
10. K. Sobottka and I. Pitas. Segmentation and tracking of faces in color images. In *Proc. 2nd Int. Conf. on Auto. Face and Gesture Recog.*, pages 236–241, Vermont, 1996. IEEE Comp. Soc. Press.
11. K. K. Sung and T. Poggio. Learning human face detection in cluttered scenes. In G. Goos, J. Hartmonis, and J. van Leeuwen, editors, *Computer Analysis of Images and Patterns*, pages 432–439. Springer-Verlag, New York, 1995.
12. H. Wu, T. Yokoyama, D. Pramadihanto, and M. Yachida. Face and facial feature extraction from color image. In *Proc. 2nd Int. Conf. on Auto. Face and Gesture Recog.*, pages 345–350, Vermont, 1996. IEEE Comp. Soc. Press.
13. K. C. Yow and R. Cipolla. Feature-based human face detection. *Image and Vision Computing*, 15(9):713–735, 1997.
14. A. Zisserman, D. A. Forsyth, J. L. Mundy, and C. A. Rothwell. Recognizing general curved objects efficiently. In J. L. Mundy and A. Zisserman, editors, *Geometric Invariance in Computer Vision*, pages 228–251. MIT Press, 1992.

Learning Identity and Behaviour with Neural Networks

A. Jonathan Howell and Hilary Buxton

School of Cognitive and Computing Sciences,
University of Sussex, Falmer, Brighton BN1 9QH, UK

Abstract. Radial Basis Function (RBF) networks are compared with other neural network techniques on a face recognition task for applications involving identification of individuals using low resolution video information. The RBF networks have been shown to exhibit useful shift, scale and pose (y-axis rotation) invariance after training, when the input representation is made to mimic the receptive field functions found in early stages of the human vision system. Extensions of the techniques to the case of image sequence analysis are described and a Time Delay (TD) RBF network is used for recognising simple movement-based gestures. Finally, we discuss how these techniques can be used in real-life applications that require recognition of faces and gestures using low resolution video images.

1 Introduction

We are tackling the unconstrained face recognition problem and the main issue discussed in this paper is how a face can be effectively recognised once it has been localised in an image or image sequence. This problem of automatic face recognition has stimulated lively debate and research in computer vision for many years, but it is only recently that techniques have become sufficiently robust to allow useful application systems to be developed.

The real-life problems we want to solve are related to identifying individuals and their intentions in a domestic setting. The development of intelligent environments has been highlighted recently by the 'Smart Rooms' projects [1] at the MIT Media Lab, which enable novel forms of interactive control for computer systems. Our particular focus is the role of adaptive learning techniques in recognising the individuals and simple movement-based gestures like head rotation. The unconstrained appearance of faces of individuals in the videoed scenes makes this a particularly difficult problem.

We know that recognising a face poses several severe tests for any visual system, such as the high degree of similarity between different faces, the great extent to which lighting conditions and expressions can alter the face, and the large number of different views from which a face can be seen. Indeed, variations in facial appearance due to lighting, pose and expression can be greater than those due to identity [2]. However, there must be some sufficiently invariant set of features that allow us to recognise familiar faces. Our automated face recognition systems must be robust with respect to variability and generalise over a wide range of conditions to capture essential similarities over view. In our work, we use adaptive learning techniques to find sets of invariant features for face classification and preprocessing schemes that can overcome the problem of lighting variation and multiple scales.

Our adaptive learning component is based on radial basis function (RBF) networks, which have been identified as valuable model by a wide range of researchers [3-5, 7, 6]. Their main advantages are computational simplicity and robust generalisation, supported by well-developed mathematical theory. RBF networks are suitable for practical vision applications [5] as they are good at handling sparse, high-dimensional data and because they use approximation to handle noisy, real-life data. Primate vision systems seem to use view-based representations for recognition [8]. This is captured by RBF techniques as a hidden unit is devoted to each view of the face to be classified. The output layer is trained to combine the views so that a single unit corresponds to an individual person.

In this paper, we first describe the RBF network model and then compare recent studies of neural network classification on the standard "Olivetti" database of faces. The RBF networks provide fast effective performance on this task. We extend the techniques to tackle image sequence data and use a Time-Delay RBF (TDRBF) network to recognise simple gestures. Finally, we discuss progress towards our target application of unconstrained face and gesture recognition from low resolution video data.

2 The RBF Network Model

We employed a Gaussian RBF network [3, 9], which has a supervised layer from the hidden to the output units, and an unsupervised layer, from the input to the hidden units. Individual radial Gaussian functions for each hidden unit simulate the effect of overlapping and locally tuned receptive fields. This is applied to the vector norm distance, $|\mathbf{i} - \mathbf{c}|$ between the input vector $\mathbf{i}$ and hidden unit centre $\mathbf{c}$. The output for the hidden layer is normalised [9].

Each training example was assigned a corresponding hidden unit, the image vector is used as its centre, as seen in recent work by [10]. This approach should not lead to overfitting because each image in the dataset contains unique 3-D information. Each hidden unit is given an associated σ width or scale value which defines the nature and scope of the unit's receptive field response; we use the mean Euclidean distance to all other units [11]. This gives an activation that is related to the relative proximity of the test data to the training data with a direct measure of confidence in the output of the network for a particular pattern. Patterns more than slightly different to those trained will produce very low (or no) output. The weights between the hidden and output layers are calculated using the matrix pseudo-inverse method [12] which allows an almost instantaneous 'training' of the network, regardless of size.

A major advantage of the RBF over other network models, such as the multilayer perceptron (MLP), is that a direct level of confidence is reflected in the level of each output unit. A discard measure was used on the tests to exclude low-confidence output; the proportion discarded and the subsequent generalisation rate are shown in the results. Low-confidence output was identified when the ratio between the highest and second highest outputs was below a certain value, i.e. the network exhibited "confusion" in the classification.

3 Comparing Face Recognition Techniques

The RBF network has been shown to provide robust classification even where data is noisy or partially missing [6]. Can this ability be used with complex 3-D objects such as faces, where the data varies in lighting, expression and pose? We wanted to make sure that the RBF network could distinguish a useful number of classes to ensure that it would be a practical technique for future applications. A suitable source of data to test this is the Olivetti Research Laboratory (ORL) database of faces (http://www.cam-orl.co.uk/facedatabase.html), which contains 400 images of 40 people with limited variation of lighting, facial expressions (such as open or closed eyes and smiling or not smiling) and facial details (such as glasses or no glasses). All the images were taken against a plain background, with tilt and rotation up to 20°, and scale variation up to 10%.

As this data has been used for a variety of face recognition techniques, it is possible to compare our results with other common approaches. Table 1 summarises the results from these papers, plus our own tests. Performance for systems with differing numbers of training images are given, together with times for the train and test (classification) stages (where available).

Table 1. Test generalisation (% correct) and processing times for various face recognition techniques used by various researchers using ORL Face Database.

Group	Technique	Images per Person					Processing Time	
		1	2	3	4	5	Training	Classification
Samaria	HMM	?	?	?	?	87	?	?
& Harter	pseudo 2-D HMM	?	?	?	?	95	?	4min
Lawrence	Eigenfaces	61	79	82	85	89	?	?
et al.	PCA + CN	66	83	87	88	92	?	?
	SOM + CN	70	83	88	93	96	4hr	<0.5sec
Lin et al.	PDBNN	?	?	?	?	96	20min	<0.1sec
Lucas	n-tuple	54	68	75	78	81	0.9sec	0.025sec
	cont n-tuple	73	84	90	93	95	0.9sec	0.33sec
	1-NN	?	?	?	?	97	0sec	1sec
Howell	RBF before discard	49	65	72	80	86	8sec	0.01sec
& Buxton	after discard	84	90	91	95	95	8sec	0.01sec

The original tests used **Hidden Markov Models** (HMMs) to encode feature information [13], using several subjective parameter selections. Later work using pseudo 2-D HMMs was able to improve this to 95%, but the complexity of this approach seems to count this out as a useful real-time technique.

Two groups [13, 14] tested the ORL database with the 'eigenface' [15, 16] approach. Both report performance of around 90%, though the latter found that they could only get this by using separate training vectors for each image. This is in contrast to [16], who averaged the eigenfaces for all images of each person in their tests. When tested with ORL data, this latter approach gave 74% for 5 training images per person [14].

A **self-organising map** (SOM) has been used, reducing the dimensions of the input representation, with a five-layer **convolutional network** (CN) to give translation and deformation invariance [14]. This was faster than the previous HMM approach, but still required several hours training time. The SOM was compared with PCA, showing the best results from all combinations, rather than average results (which are given for the other groups).

A **probabilistic, decision-based neural network** (PDBNN) [17], a modular network structure with non-linar basis functions (each sub-network similar to a HyperBF (HBF) network [4]) was able to train and classify much faster than the CN approach of [14], while reaching a similar level of performance.

The **continuous n-tuple classifier** [18] has been able to train and classify quickly and provide a high level of performance. The figures shown are for tests with 200 3-tuples (600 values) per image. Using 500 4-tuples (2000 values) per image improved recognition to 86% and 97% for the n-tuple and continuous n-tuple classifier respectively. Very high performance can even be obtained using a simple **1-nearest-neighbour** (1-NN) classifier [18]. The success of simple matching indicates how constrained the database is in terms of lighting and pose, as such techniques will not be invariant to such factors.

To use the ORL data with the **RBF network**, we applied Gabor filter preprocessing [19]. A simple discard measure, based on the relative magnitudes of the output units, was used to remove low confidence classifications. The RBF network approach was fast in training and the fastest in classification of all the published techniques. Our experiments were conducted on a moderately fast Sun SPARC 20 workstation. Test generalisation before discard was fairly poor in comparison to the other approaches, though the results were well above random (2.5%). For 5 training examples per person, discarding 39% of results allowed performance to be improved from 84% to 95%, which was comparable with the best of the other techniques. The RBF performance is sufficient for our target application where training data is relatively sparse and test data is abundant.

A particularly important point is that most other face recognition techniques took much longer than the RBF network in the training and testing of data. It is apparent that the RBF network provides a solution which can process test images in inter-frame periods on low-cost processors. The rest of this paper will be concerned with how to apply this network to our target application of identity recognition in unconstrained domestic environments.

4 Learning Invariance

Having established the RBF network as a suitable technique for our unconstrained face recognition task, we investigated its invariance characteristics. Our invariance experiments used the 'Sussex Database' [20], designed to test recognition abilities for faces in widely varying poses. The database has ten images of ten individuals showing the head and shoulders and taken in 10° steps from face-on to profile of the left side, 90° in all. The results showed that not only can the RBF network learn identity in spite of pose variations, but it can continue to be invariant to pose in the presence of other variations. In addition, it can learn an invariance to scale variation more easily than shift as shown by both the better classification performance and the lower discard rates, especially with the Gabor preprocessing [19].

5 The RBF Network with Image Sequences

We have been able to show that the RBF network can learn to be invariant to certain types of variations that can be expected in real-life face images. We need to be able to use these abilities in a less constrained environment [22]. In the context of videos of people moving around a room, which produces large numbers of images of each person in the environment, even high discard ratios of 80-90% are acceptable if the remaining output is of sufficiently high quality. We were able to collect suitable image sequences as a result of collaboration with Stephen McKenna and Shaogang Gong at Queen Mary and Westfield College, London, who are researching real-time face detection and tracking [21].

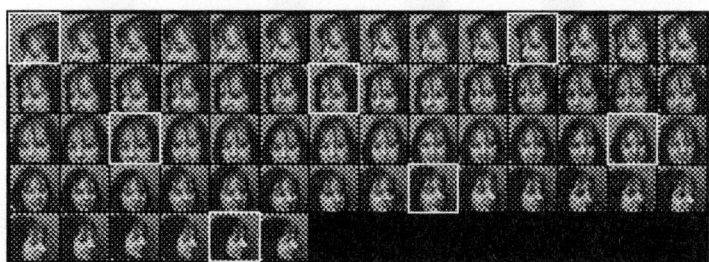

Fig. 1. A complete image sequence, after segmentation but before pre-processing (boxes indicate frames used for training).

We are currently working on a 'time window' integration level with the raw output and using image sequences from real life, where individuals will be present for significant periods. Techniques which take advantage of temporal coherence can be used to improve performance where periods of low confidence output can be discarded to create a coherent stream.

6 The RBF Network and Temporal Behaviour

To extend the investigation into temporal learning, we then asked whether the RBF could learn actions through time? To answer this, we treated our original 100-image database as 10 image sequences of a person rotating their head from side to side. We adapted our network model to use time-delays in order to process temporal context. This type of Time-Delay RBF (TDRBF) network had been used previously for speech recognition [23] and combines data from a fixed time 'window' into a single vector as input. Berthold, however took a constructive approach, combining the idea of a sliding input window from the standard TDNN network with a training procedure for adding and adjusting RBF units when required. We used a simpler technique, successful in previous work with RBF networks, which uses a fixed number of units, one for each example.

The network was trained with sequences of images from five people and tested with sequences of images from the other five people, so that generalisation will reflect learning of the temporal task rather than identity. Three types of rotation sequence were created, shown in Figs. 2(a–c). 1) **LR sequences** simulate a left to right head rotation, and are taken from a 'window' from all ten frames of five people. 2) **RL sequences** are identical to LR, except with the reverse rotation. 3) **Static sequences** simulate a fixed head position through time.

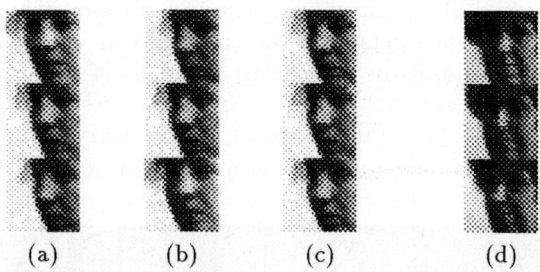

Fig. 2. Example image sequences of three side to side head rotations used for training the RBF network (time window of 3 frames): (a) LR training (b) RL training (c) Static training (d) LR test (different person).

Table 2 shows that the TDRBF network can learn the different types of movement and generalise with sequences from individuals not encountered during training. The mid-length time window gave the best generalisation performance and lowest discard proportions. The temporal task we used is very simplistic, as each image sequence only contains one standard-length movement. We have tested our trained network on a less constrained sequence, similiar to Fig. 1, with some success.

Table 2. Effect of time window size on generalisation for a TDRBF network in a 3 class problem, distinguishing LR, RL and static sequences.

Time Window	Initial %	% Discarded	% after Discard
10	87	20	100
8	89	13	100
6	83	21	100
4	77	37	100
2	63	56	93

The issue of the *time base* of actions, i.e. how fast or slow actions occur, would have to be taken into account in any real-life image sequences, as any movement would occur at a variety of speeds quite naturally. Although [23] used the integration layer to cope with shifts in time, the scale (time base) of events was not discussed. This type of variation can be handled by a recurrent network, or by training data which explicitly demonstrated the different classes of motion-based behaviour at a variety of speeds.

7 Conclusions

In Sections 1 and 2, we described the background to our investigations of the role of adaptive learning in face recognition. In Section 3, we showed that Radial Basis Function (RBF) networks performed well on the ORL comparison face recognition task, delivering both fast training and exceptionally rapid classification times. We were able to obtain 95% correct classification after training on half the views if a simple confidence metric is included. This is comparable with other state of the art techniques. In Section 4, we briefly discussed the invariance capabilities of the RBF networks with variations in pose, translation (shift) and scale. In Section 5, we briefly outlined the extensions required to tackle real-time video from a general purpose motion tracker.

In Section 6, we considered extensions to simple motion-based behaviour and our preliminary results seem very promising. The main points here are 1) the simple, deterministic 'training' of the TDRBF networks means that they are highly suited to on-line learning, 2) the shift invariance and ability to recognise features in time means they are capable of recognising simple behaviours, and 3) high levels of performance on the generalisation to new datasets that behave in similar ways means they are very useful for such practical dynamic vision tasks. The limitations of this technique are 1) the problem of the time-base which was not fully overcome even with the addition of an integration layer, and 2) the problem of defining the simple behaviours. The TDRBF networks are capable of distinguishing a 'quick turn' from a 'slow turn' as well as distinguishing whether the turn was to the right or the left, but it seems that more qualitative definitions of behaviour would best be tackled using more general recurrent networks. Nonetheless, the TDRBF networks are able to perform extremely well where there is a straightforward quantitative relationship between the data and the simple behavioural pattern to be learnt.

Future work will concentrate on the integration of the face tracking mechanism with the adaptive learning techniques discussed in this paper to progress to a full prototype system and extend to more general gestural and behavioural analysis [24, 25]. The key capability here is face and gesture recognition from low-resolution video sequences where there is a small set of known users. This seems to be a realistic goal if we use RBF techniques combined with good image representations as discussed in the main sections of this and the accompanying papers we have cited. It also clearly has great commercial potential for use in many interactive systems as well as in security and monitoring applications.

References

1. A. Pentland. Smart rooms. *Scientific American*, 274(4):68–76, 1996.
2. Moses, Y., Adini, Y., Ullman, S. Face recognition: the problem of compensating for illumination changes. *ECCV'94*, pp. 286–296, 1994.
3. J. Moody and C. Darken. Learning with localized receptive fields. *1988 Conn. Models Summer School*, pp. 133–143, 1988.
4. T. Poggio and F. Girosi. Regularization algorithms for learning that are equivalent to multilayer networks. *Science*, 247:978–982, 1990.
5. F. Girosi. Some extensions of radial basis functions and their applications in artificial intelligence. *Comp. & Mathematics Appl.*, 24(12):61–80, 1992.
6. S. Ahmad and V. Tresp. Some solutions to the missing feature problem in vision. *NIPS*, vol. 5, pp. 393–400, 1993.
7. S. Edelman, D. Reisfeld, and Y. Yeshurun. Learning to recognize faces from examples. *ECCV'92*, pp. 787–791, 1992.
8. D. I. Perrett and M. W. Oram. Neurophysiology of shape processing. *Image & Vision Computing*, 11:317–333, 1993.
9. J. Moody and C. Darken. Fast learning in networks of locally-tuned processing units. *Neural Computation*, 1:281–294, 1989.
10. D. J. Beymer and T. Poggio. Image representations for visual learning. *Science*, 272:1905–1909, 1996.
11. M. T. Musavi, W. Ahmad, K. H. Chan, K. B. Faris, and D. M. Hummels. On the training of radial basis function classifiers. *Neural Networks*, 5:595–603, 1992.
12. T. Poggio and F. Girosi. Networks for approximation and learning. In *Proceedings of IEEE*, volume 78, pp. 1481–1497, 1990.
13. F. S. Samaria and A. C. Harter. Parameterisation of a stochastic model for human face identification. *2nd IEEE W'shop Appl. Computer Vision*, 1994.
14. S. Lawrence, C. L. Giles, A. C. Tsoi, and A. D. Back. Face recognition: A convolutional neural network approach. *IEEE Trans. Neural Networks*, 8:98–113, 1997.
15. M. Kirby and L. Sirovich. Application of the Karhunen-Loève procedure for the characterization of human faces. *IEEE Trans. PAMI*, 12:103–108, 1990.
16. M. Turk and A. Pentland. Eigenfaces for recognition. *Journal of Cognitive Neuroscience*, 3:71–86, 1991.
17. S.-H. Lin, S.-Y. Kung, and L.-J. Lin. Face recognition/detection by probabilistic decision-based neural network. *IEEE Trans. Neural Networks*, 8:114–132, 1997.
18. S. M. Lucas. Face recognition with the continuous n-tuple classifier. *BMVC*, pp. 222–231, Colchester, 1997.
19. A. J. Howell and H. Buxton. Receptive field functions for face recognition. *Proc. 2nd Int. W'shop PAMONOP*, pp. 83–92, Faro, Portugal, 1995.
20. A. J. Howell and H. Buxton. Face recognition using radial basis function neural networks. *BMVC*, pp. 455–464, Edinburgh, 1996.
21. S. McKenna, S. Gong, and J. J. Collins. Face tracking and pose representation. *BMVC*, pp. 755–764, Edinburgh, 1996.
22. A. J. Howell and H. Buxton. Towards unconstrained face recognition from image sequences. *ICAFGR*, pp. 224–229, Killington, VT, 1996.
23. M. R. Berthold. A time delay radial basis function network for phoneme recognition. *IEEE ICNN*, vol. 7, pp. 4470–4473, Orlando, FL, 1994.
24. A. Bobick. Computers seeing action. *BMVC*, pp. 13–22, Edinburgh, 1996.
25. S. Nagaya, S. Seki, and R. Oka. A theoretical consideration of pattern space trajectory for gesture spotting recognition. *ICAFGR*, pp. 72–77, Killington, 1996.

Open Sesame! Speech, Password or Key to Secure Your Door?

Stéphane H. Maes , Homayoon S. M. Beigi
Human Language Technologies Group,
Speech Decoding Design Department,
IBM T.J. Watson Research Center
P.O. Box 218, Route 134,
Yorktown Heights, NY 10598, USA
e-mail: smaes@watson.ibm.com

Abstract

This paper reviews the state of the art in speaker recognition. It clarifies the different technical solutions that have been explored with some success as well as the challenges and limitations of current systems. It also describes the different functions and modalities involved in speaker recognition, where the terminology is still amazingly confused: specialist often uses the same terms for different conecpts. We review the classical techniques used in speaker recognition. Finally, we introduce the revolutionary concepts of speech biometrics. By discussing the impact of these new concepts, the maturity of speaker recognition is re-focussed.

1 Introduction

Access control to locations, services and resources is one of the oldest human goals. Except for some rare societies, private, tribal, corporate, official or national properties have always existed and required means to prevent violation, damages or losses. Jean-Jacques Rousseau even attributes the origins of all wars and injustices to the inception of private property. As for any human endeavor, more and more complex techniques have been developed or imagined. They can be classified into: hiding places: treasures; strongholds: fortress, armed protection, electronic protection, remote servers; key-based access: door, vaults, encryption; knowledge-based access: login, PIN, password; biometrics.

We could present numbers of legends and stories where biometrics play or could have played major roles. Obviously "Alibaba and the 40 thieves" is the most appropriate in this context: a secret speech command is used to open a door. ¿From the beginning all the ingredients for using speech as biometrics are present.....
Obviously Sesame was voice controlled, but in order to help Alibaba, the author could not equip the door with speaker verification or identification capabilities: no biometrics on Sesame! Sure, the author of the 1001 Nights would have needed a different scheme to let Alibaba enter! But what a wonderful security system it would have been! What an idea! Unfortunately, in 1992, Robert Redford showed us its modern weakness in "Sneakers"! So, is it a good or a bad concept? Will you ever equip your house with Sesame?

Hopefully, by the end of this paper, we will convince the reader of the importance and perspectives opened by "speech biometrics". By the same token we plan to radically change the meaning of these words.

2 Functions of speaker recognition

Speaker recognition is a generic term that encompasses all the activities involving matching a speech waveform to the identity of the speaker. Numbers of contradictory sub-divisions have been proposed in the literature. From paper to paper and specialist to specialist, it is very common to find the same terms used to denote diametrically opposite activities. The fact that even specialists or vendors systematically interchange the role of these categories almost from sentence to sentence does not help a broader audience to embrace the field of speaker recognition. Amazingly, this plethora of contradictions does not affect other biometrics. We conjecture that it should be attributed to a large extent to the additional confusion that exists between speech recognition and speaker recognition.

We distinguish three different functions, with some minor subdivision:

- speaker identification
- speaker verification
- speaker classification
- speaker enrollment

Our speaker recognition approach does not distinguish from a technology point of view between these different functions: our speaker recognition engines implement them all at once. Since June 1996, when the SRAPI - Speech Recognition API - committee created the SVAPI - Speaker Verification and Identification API - sub-committee, we made sure that these definitions and functions are all included in SVAPI [6]. This is especially important as the concept of speaker classification, which until recently was mostly ignored by the community. Without our effort, SVAPI would probably be limited to text-dependent/text-prompted speaker verification. There is no doubt that confusions exist even within the ranks of the proponents of speaker recognition!

2.1 Speaker identification

Speaker identification consists of identifying a speaker based on his or her voice. The speakers are already enrolled in the system. No identity claim is provided. We speak of closed-set speaker identification if we restrict the set of speakers to be identified to the enrolled speakers. If unknown speakers, not yet enrolled, must be rejectected by the system, we speak of open-set speaker identification. In terms of biometrics, speaker identification is a "many-to-many" recognition task. The decision alternatives are equal to the size of the enrolled speakers (+ 1 in open-set case). Therefore, the accuracy of speaker identification degrades as the size of the speaker population increases. As speaker identification accuracy does not yet compare to the performances of other biometrics, we can understand why, after for more than 30 years of research, speaker identification has not yet reached maturity! Especially as speaker identification engines are not yet able to cope with uncooperative speakers! Uncooperative users are those users who disguise their voice intentionally in order to avoid being identified e.g. candidates for social benefits who cheat the system by submitting with multiple applications. Besides classical speaker identification, some extensions exists with added functionality of providing N-best lists or confidence scores. In the former case, a speaker identification system returns a sorted list of N identities who match the best the current speaker. The latter case rather implies that the identifier will produce a confidence level for each enrolled speakers that he or she matches the current speaker. Within these frameworks, speaker identification is much closer to maturity.

The reader will note in the next session that when it comes to confidence levels and rejection as out-of-set, speaker identification and speaker verification share a common behavior. Indeed, although such strategy is computationally expensive, identification can be implemented by repeated verifications with each speaker in the enrolled population used for subsequent identity claims.

Open set speaker identification requires rejection features that can usually be directly used for verification purposes.

2.2 Speaker verification

Speaker verification consists of verifying the identity claim of a speaker based on his or her voice. The identity claim designates a speaker enrolled in the system. Otherwise, rejection is trivial. Concepts of open or closed sets are not relevant to speaker verification.

In terms of biometrics, speaker verification is a "one-to-many" recognition task. There are only two choices: accept or reject. Contrary to speaker identification, the accuracy of speaker verification is not directly dependent on the population size. However, as it is typical in biometrics, the estimate of this accuracy depends on the representation of the population samples used to evaluate the accuracy. In contrast to other biometrics, these estimators also strongly depend on the channel effects and noise corruption of the signal. In general, speaker recognition performances vary dramatically from matched conditions (same type of microphone, channel characteristics and background noise) to mismatched conditions. The difficulty to correctly accommodate the effects of these mismatches dramatically damage the performances of speaker verification engines. To a large extent, this explains the relatively limited deployment of speaker verification systems. The other major cause being due to the goat phenomena: the majority of errors committed by verification engines are concentrated over a small fraction of the population (a few percent at maximum). Unfortunately, from the point of view of an application developer, it is not acceptable to exclude from a service a portion of the population on the basis of unexplained and uncontrolled behavior of their voice! Hence, some systems were never completely deployed!

There are two types of errors: *false acceptance* (an imposter has been incorrectly granted access) and *false rejection* (the authorized user has been denied access). Depending on the application, false acceptance will often be critical. However, in the literature, the focus is more often on the *equal error rate*: the total error rate committed when the false acceptance rate is equal to the false rejection rate.

Besides classical speaker verification, we must also mention extensions where instead of hard accept or reject decisions, confidence levels are returned.

2.3 Speaker classification

Speaker classification consists of performing speaker recognition over an unknown number of unknown speakers. Usually, it means to be able to detect speaker changes, also called speaker separation, and index the resulting segments according to the identity.

This function is specifically speech related. Only portions of the concept are met in other biometrics. However, the capabilities that it offers to distinguish between different undeclared successive users of a system may also be implemented with other biometrics.

Errors are measured in terms of segmentation mistakes (segmentation points versus speaker changes, end-times in the middle of words instead of in silences), and grouping mistakes (segments of one speakers attributed to another speaker). Different sub-functions can be distinguished: speaker separation (speaker changes and regrouping segments of a same speaker) [14, 13, 5]; segment clustering [5]; speaker clustering (grouping speaker based on their similarities) [4, 5, 16, 18, 39]. Speaker clustering in unsupervised mode involves a bottom-up clustering of the model of different speakers. On the other hand, supervised speaker clustering usually leads to classes of speakers based on their gender, age, regional accent etc.

2.4 Speaker enrollment

In order to recognize the user based on his or her voice, first we need to acquire samples of the user's voice and create a model. Such models are usually called speaker models. Often, the models used for speaker identification differ from those used for speaker verification. By analogy to fingerprints, voice-prints refer to the minimum set of characteristics of a speaker required to create the speaker models used for identification and verification. We use the same models for identification as well as verification.

Of course in text-independent mode and with the appropriate technology, any voice sample can be used to create a voice-print. However, enrollment usually involves a strict procedure that the enrolling speakers must follow step by step: e.g. repeating words digits or sentences.

As for speech recognition, the principle is that there is no better enrollment data than more data! The more data that is available for a speaker the more accurate the voice-prints will be. Especially if this data can be collected over multiple mismatched conditions representative of the actual mismatches experienced during recognition.

3 Modalities of speaker recognition

There are multiple modalities in speaker recognition, i.e. different types of constraints imposed on the utterances used for enrolment or recognition. We distinguish between:

- text-dependent speaker recognition
- text-prompted speaker recognition
- text selected by user speaker recognition
- text-independent speaker recognition

3.1 Content-constrained speaker recognition

The first three categories may be defined as text-constrained speaker recognition.

Text-dependent speaker recognition The content of the testing utterance matches the content of the enrolment utterance. In other words, the engine knows explicitly what the user is saying. The text can be different from user to user and it is possible that multiple texts are associated with each speaker.

Text-prompted speaker recognition As in the previous case, the speaker recognition engine knows what the user is saying or supposed to say. The system asks the user to repeat a text, usually obtained by combining elementary keys or units; typically sequences of digits. During enrolment, the user is also prompted to repeat combinations of tokens.

Text selected by user speaker recognition The users select a password or some key sentences to repeat at recognition. During enrollment, the user is asked to repeat their selected utterance. At recognition he or she repeats the same sentence. Furthermore, this sentence can contain the actual identity claim of the user. For example, it can be achieved by implementing speaker verification with open set speaker identification.

Usage In practice, text constrained speaker recognition is only appropriate for verification tasks where a separate process devoted to text-constrained speaker recognition is acceptable. Speaker identification and classification can rarely accommodate such constraints: their applications usually require free speech capabilities.
Fraud remains easy when it comes to text-dependent speaker verification and even to a lesser extent text-selected by user speaker verification: play-backs or synthesis. Text-prompted speaker verification seems to solve these issues unfortunately, it is a major burden for the user. It also significantly slows down transactions.

3.2 Text-independent speaker recognition

Definition This category designates speaker recognition on free speech utterances. The content of the speech utterance is completely unknown to the engine and does not have to be related to the enrollment utterances. In true text- independent speaker recognition, the recognition can be done in a language different from the language used during enrollment.

Usage Contrary to text-constrained speaker recognition, text-independent speaker recognition presents the unique advantage of being applicable to any speech utterance. Speaker recognition can therefore be performed in the background of a regular conversation, request or transaction. Enrollment can be simplified: any speech from a speaker can be used to enroll this speaker.
By allowing the recognition to happen in the background of a transaction, with an IVR or an operator, the scenario of the transaction should forbid playbacks or syntheses. Indeed with the current technology it is not possible to generate the dialogs in real time. Also, synthesis effects can be reliably detected from the signal (e.g. pitch discontinuities).
Text-independent modality is mandatory for identification or classification.

3.3 Availability

Most of the current technology providers in speaker verification have opted for text-constrained speaker verification rather than text-independent speaker verification which is technically still more challenging. However, the non-obstrusive character of text-independent speaker verification presents multiple advantages including the capability to continuously process speech until a decision can be made.
Multiple research organizations have engaged in efforts to develop and improve text-independent speaker recognition. We are pursuing such research efforts and our speaker recognition engines are text-independent with a strong emphasis on its integration with our speech recognition engines.

4 Technology

The limited space available for this review paper forces us to limit the technical review of speaker recognition. We suggest that the reader consults the following papers for a more detailed discussion: [3, 7, 28, 11, 8, 9].
A speaker recognition system typically encompasses two elements: an acoustic feature extractor and a feature classifier.

4.1 Acoustic features

Acoustic features characterize the vocal tract characteristics of a speaker of one hand and the source properties (supra-segmental features, pitch etc) on the other hand. In contrast to speech recognition that in the vast majority of the systems uses Mel Frequency Cepstral Coefficients (MFCC), a whole variety of exotic or proprietary acoustic features are used for speaker recognition. The features usually considered to be the most appropriate are LPC cepstral coefficients. MFCCs are considered less efficient; however, systems developed by speech recognition providers often use them.

4.2 Classifiers

Content-contained systems Most of the techniques used can be classified into DTW (Dynamic Time Warping) which are template-based [11] and HMM (Hidden Markov Model) systems which are statistical models [27, 37, 36, 42]. HMMs significantly outperform DTW.

Text-independent system Most of the techniques used can be classified into feature classifiers and speaker-dependent speech decoding systems. The first category can be further subdivided into long-term statistics classifiers [10, 23, 22, 15] and frame-by-frame classifiers. The latter category encompasses VQ [40, 24, 25], ergodic HMM [29, 41] and speech recognition based systems [38, 12].

4.3 Rejection

Speaker verification and open set speaker identification require efficient rejection capabilities. Rejection is usually implemented by normalizing distances or likelihood measures obtained by competing models: cohorts or background population models.

Cohorts The cohort of a speaker is a set of competing speakers: the score of the candidate speaker is compared to the scores of other speakers in the cohort [35]. The candidate speaker is accepted only if his or her score significantly dominates over the scores of the speakers in the cohort.

In case of speaker-dependent cohorts, a cohort is composed of speakers in the population which are the closest to the candidate speaker. Once large enough cohorts are considered (typically more than 20 or 30 members), speaker-independent or randomly selected cohorts perform at least as well as speaker-dependent cohorts [30].

Recently a new concept of cohort has been proposed: unconstrained cohorts [1], obtained by selecting the cohort as a set of speakers close to the testing data rather than close to the training data (or model) associated to the candidate speaker. Unconstrained cohorts are reported to improve the robustness to mismatches, at least in text-dependent cases.

Background models Instead of competing with each of the models in a cohort, the background model is a unique speaker-independent model. Such models have become very common recently as multiple reports suggest improved performances with background models over cohorts [32].

4.4 Mismatch compensations

Mismatches currently present one of the biggest challenges in speaker recognition. Changes of microphones and handsets, telephone network non-stationarities, different compression/coding algorithms and different background noise significantly impact matches of voice-prints.

Different methods are used to handle mismatches:

- specific acoustic features (e.g. modulation model) [2, 21]
- feature compensation (e.g. cepstral mean subtraction) [31]
- speaker models under different mismatches (e.g. data from carbon, electret, cellular and GSM data to build one speaker model or close-talk and far field data) [17]
- model adaptation (e.g. PMC) [33]
- mismatch detection and threshold adaptation (e.g. hnorm, updated threshold, competing model selection) [32, 26]

5 Speech biometrics

5.1 Is speaker recognition mature?

Is speaker recognition mature? Can we reliably use it in products, field applications or actual real-life security systems? Can it be fooled? Inevitably, these questions arise whenever speaker recognition is considered. There is no simple answer. Yes, speaker recognition is reaching maturity. Where are we standing today?

Multiple unknown factors still exist: how unique are voice-prints? How unique are they when mismatched effects play a role? How can we handle uncooperative speakers? Can we handle uncooperative speakers? There is no doubt that text-dependent and text-prompted verification can be deployed in some specific environments or for some specific applications. Multiple commercial systems are already available. However, in most cases, these systems can only be used as secondary systems. Text-independent systems are technically more challenging. Performances are lower. But they open new perspectives and application possibilities, especially since text-dependent systems are much too prone to fraud by play-backs or synthesizers. We are developing a hierarchical speaker recognition system able to handle, in real-time, very large populations with no loss of accuracy. We have concentrated our efforts on developing engines, applications and toolkits based on text-independent technology fully integrated with speech recognition engines and applications. While it is obvious that voice-prints have a unique advantage over other biometrics as the primary vector of communication, especially for remote authentication (as microphone and handset are cheap and widely deployed). Furthermore, the explosion of speech recognition is opening the way for voice interaction with computers and machines and therefore dramatically increasing opportunities and needs for accurate speaker recognition capabilities. Unobstrusive speaker recognition becomes even more important.

5.2 A solution

Classical verification relies on three items: what you own, what you are and what you know. Key or card-based systems characterize what you own. PIN and password based systems rely on what you know. Biometrics and in particular speaker recognition rely on what you are. We develop a new patented approach called speech biometrics. Text-independent speaker recognition is used to acoustically identify or verify answers to questions prompted by the system. These questions can be randomly selected or follow a pre-defined sequence. Verification and identification rely on acoustic recognition and on the content of the answers to the questions. A typical scenario is the following. A real-time, low bit rate (lower than 4kB/s) client-server demonstration will be presented during the talk. When prompted, the user says his name. The name is recognized and it defines a subset of speakers with that name. Text-independent speaker identification extracts the identity claims used for verification. The system prompts randomly with questions (e.g. what is your mother maiden name, what is your favorite color?). Answers are decoded and matched with the answers provided during enrollment and text-independent speaker verification is performed to authenticate the acoustics of the answers. If all the results are positive, the user is authenticated, otherwise a separate procedure is started which can involve additional questions. Such approach leads to arbitrarily low error rates. Recently, following our approach, other authors have studied particular cases of speech biometrics: utterance verification for access control, password compliance [34, 20, 19].

Combination of speech recognition, dialog management and natural language understanding capabilities enable the system to perform customer identification and verification through a natural transaction. Actually, full automation of order taking or transactions becomes possible with high security/accuracy. Furthermore, because it becomes a natural part of the transaction, the procedure is unobstrusive and does not slow down the transaction.

6 Conclusions

Speaker recognition is an relatively old field. It has not yet reached maturity: too many questions are still opened, too many imperfections. However, speaker recognition can not be dissociated from speech recognition. With the growing demand and effort involved in speech recognition, speaker recognition is much demanded. No doubt, significant contributions are still needed in acoustic speaker recognition. However, integrated combinations of speech recognition capabilities with speaker recognition, dialog management and natural language understanding enable us to bridge the gap and deploy actual solutions with today's technology. The perspectives opened by speech biometrics are numerous and the experience gained from real-life deployment brings classical speaker recognition closer to maturity. The growth in interest in biometrics in general also impacts the future of speaker recognition and speech biometrics: nothing is more convenient for everyday transactions, especially remote transactions, than speech biometrics!

7 Acknowledgments

The authors want to acknowledge the key contributions Jeffrey Sorensen to IBM speaker recognition researchs and engines. Also many thanks to Radek Hampl and Dan Coffman for their useful input. Eventually, the authors are thankful to Ponani S. Gopalakrishnan for his continuous support.

8. References

[1] A. M. Ariyaeeinia and P. Sivakumaran. Analysis and comparison of score normalisation methods for text-dependent speaker verification. In *Proc. Eurospeech*, volume 3, pages 1379–1382, 1997.

[2] K. Assaleh, R. J. Mammone, and J. L. Flanagan. Speech recognition using the modulation model. In *IEEE Proc. ICASSP*, volume 2, pages 664–667, 1993.

[3] B. S. Atal. Automatic recognition of speakers from their voices. *Proc. IEEE*, 64:pp. 460–475, 1976.

[4] H. S. M. Beigi, S. Maes, and J. Sorensen. A distance measure between collections of distributions and its application to speaker recognition. In *Proc. ICASSP*, 1998.

[5] H. S. M. Beigi and S. H. Maes. Speaker, channel and environment change detection. In *Proceedings of WAC'98*, Anchorage, Alaska, May 1998. ISSCI.

[6] SRAPI Committee. URL=http://www.srapi.com/svapi.

[7] G. R. Doddington. Speaker recognition - identifying people by their voices. *Proc. IEEE*, 76(11):pp. 1651–1664, 1985.

[8] S. Furui. Automatic speech and speaker recognition, advanced topics. In C.-H. Lee, F. K. Soong, and K. K. Paliwal, editors, *An overview of speaker recognition technology*. Kluwer Academic Publishers, Norwell, MA, 1996.

[9] S. Furui. Recent advances in speaker recognition. In J. Bigun, G. Chollet, and G. Borgefors, editors, *Proc. Audio- and Video-based biometric person authentication*, pages 237–252. Springer-Verlag, 1997.

[10] S. Furui, F. Itakura, and S. Saito. Talker recognition by longtime averaged speech spectrum. *Trans. IECE*, 55A, 1(10):pp. 549–556, 1992.

[11] S. Furui and M. Sondhi, editors. *Advances in speech signal processing*. Marcel Dekker, New York, NY, 1991.

[12] J. L. Gauvain, L. F. Lamel, and B. Prouts. Experiments with speaker verification over the telephone. In *Proc. Eurospeech*, pages 651–654, 1995.

[13] P. Gopalakrishnan, R.Gopinath, S. Maes, M.Padmanabahn, and L. Polymenakos. Acoustic models used in the IBM system for the ARPAHUB 4 task. In *Proceedings of the Speech Recognition Workshop*, Arden House, Harriman, NY, February 1996. ARPA.

[14] P.S. Gopalakrishnan, R. Gopinath, S. Maes, M. Padmanabahn, L. Polymenakos, H. Printz, and M. Franz. Transcription of radio broadcast news with the IBM large vocabulary speech recognition system. In *Proceedings of the Speech Recognition Workshop*, Arden House, Harriman, NY, February 1996. ARPA.

[15] C. Griffin, T. Matsui, and S. Furui. Distances measures for text-independent speaker recognition based on mar model. In *IEEE Proc. ICASSP*, number 23.6, pages I–309–312, 1994.

[16] L. Heck and A. Sankar. Acoustic clustering and adaptation for improved speaker recognition. In *Proceedings of Speech Recognition Workshop*, Chantilly, VA, February 1997. ARPA.

[17] L. P. Heck and M. Weintraub. Handset-dependent background models for robust text-independent speaker recognition. In *Proc. ICASSP*, 1997.

[18] H. Jin, F. Kubala, and R. Scwartz. Automatic speaker clustering. In *Proceedings of Speech Recognition Workshop*, Chantilly, VA, February 1997. ARPA.

[19] O. Kimball, M. Schmidt, H. Gish, and J. Waterman. Speaker verification with limited enrollment data. In *Proc. Eurospeech*, volume 2, pages 967–970, 1997.

[20] Q. Li, B.-H. Juang, Q. Zhou, and C.-H. Lee. Verbal information verification. In *Proc. Eurospeech*, volume 2, pages 839–842, 1997.

[21] R. J. Mammone, X. Zhang, and R. P. Ramachandran. Robust speaker recognition, a feature-based approach. *IEEE Signal Processing Magazine*, 13(5):pp. 58–71, 1996.

[22] J.D. Markel and S. B. Davi. Text-independent speaker recognition from a large linguistically unconstrained time-spaced data base. *IEEE Trans. ASSP.*, ASSP-27(1):pp. 74–82, 1979.

[23] J.D. Markel, B. T. Oshika, and A. H. Gray. Long-term feature averaging for speaker recognition. *IEEE Trans. ASSP.*, ASSP-25(4):pp. 330–337, 1977.

[24] T. Matsui and S. Furui. Text-independent speaker recognition using vocal tract and pitch information. In *Proc. ICSLP*, number 5.3, pages 137–140, 1990.

[25] T. Matsui and S. Furui. A text-independent speaker recognition method robust against utterances variations. In *Proc. ICASSP*, number S6.3, pages 377–380, 1991.

[26] T. Matsui and S. Furui. Robust methods of updating model and a priori threshold in speaker verification. In *Proc. ICASSP*, number 13.1, 1996.

[27] J. M. Naik, L. P. Netsch, and G. R. Doddington. Speaker verification over long distance telephone lines. In *Proc. IEEE ICASSP*, volume S10b.3, pages 524–527, May 1989.

[28] D. O'Shaughnessy. Speaker recognition. *IEEE ASSP Magazine*, 3(4):pp. 4–17, October 1986.

[29] A. B. Poritz. Linear predictive hidden markov models and the speech signal. In *Proc. ICASSP*, number S11.5, pages 1291–1294, 1982.

[30] D. Reynolds. Speaker identification and verification using gaussian mixtures speaker models. In *ESCA Workshop on Automatic speaker recognition, identification and verification*, pages 27–30, 1994.

[31] D. A. Reynolds. The effect of handset variability on speaker recognition performance: experiments on the switchboard corpus. In *Proc. ICASSP*, pages 113–116, 1996.

[32] D. A. Reynolds. Comparison of background normalization methods for text-independent speaker verification. In *Proc. Eurospeech*, volume 2, pages 963–966, 1997.

[33] R. C. Rose, E. M. Hofstetter, and D. A. Reynolds. Integrated models of signal and background with applications to speaker identification in noise. *IEEE Trans. Sp. Audio Proc.*, 2(2):pp. 245–257, 1994.

[34] A. E. Rosenberg and S. Parthasarathy. Speaker identification with user-selected password phrases. In *Proc. Eurospeech*, volume 3, pages 1371–1374, 1997.

[35] A.E. Rosenberg, J. Delong, C. H. Lee, B. H. Juang, and F. K. Soong. The use of cohort normalized scores for speaker recognition. In *Proc. ICSLP*, pages 599–602, 1992.

[36] A.E. Rosenberg, C. H. Lee, and S. Gokeen. Connected word talker recognition using whole word hidden Markov models. In *Proc. IEEE ICASSP*, pages 381–384, 1991.

[37] A.E. Rosenberg, C. H. Lee, and F. K. Soong. Sub-word unit talker verification using hidden Markov models. In *Proc. IEEE ICASSP*, pages 269–272, 1990.

[38] M. Savic and S. K. Gupta. Variable parameter speaker verification system based on hidden Markov modelling. In *Proc. IEEE ICASSP*, pages 281–284, 1990.

[39] M. A. Sigler, U. Jain, B. Raj, and M. Stern. Automatic segmentation, classification and clustering of broadcast news audio. In *Proceedings of Speech Recognition Workshop*, Chantilly, VA, February 1997. ARPA.

[40] F.K. Soong, A. E. Rosenberg, L. R. Rabiner, and B. H. Huang. A vector quantization approach to speaker recognition. In *IEEE Proc. ICASSP*, pages 387–390, 1985.

[41] N. Z. Tishby. On the application of mixture AR hidden Markov models to text independent speaker recognition. *IEEE Trans. ASSP*, 39:pp. 563–569, 1991.

[42] Y.-C. Zheng and B.-Z. Yuang. Text-dependent speaker identification using circular hidden markov models. In *IEEE Proc. ICASSP*, pages 580–582, 1988.

A Unified Framework for Image-Derived Invariants

Yuan-Fang Wang and Ronald-Bryan O. Alferez
Department of Computer Science
University of California
Santa Barbara, CA 93106

Abstract: We propose a general framework for computing invariant features from images. The proposed approach is based on a simple concept of basis expansion. It is widely applicable to many popular basis representations, such as wavelets, short-time Fourier transform, and splines.

1 Introduction

Image features and shape descriptors that capture the essential traits of an object and are insensitive to environmental changes are ideal for recognition. The search for invariants (e.g., algebraic and projective invariants) is a classical problem in mathematics dating back to the 18th century. The need for invariant image descriptors has long been recognized in computer vision [3, 6]. Invariant features form a compact, intrinsic description of an object, and can be used to design recognition algorithms that are potentially more efficient than, say, the aspect-based approaches. Hence, it was even argued that object recognition is the search for invariants [6]. Invariant features can be designed based on many different methods and made invariant to rigid motion, general affine transform, scene illumination, occlusion, and projection. See [3, 6] for a comprehensive survey of the subject of invariants.

The proposed framework exploits both global and local information about shape and color, and is neither strictly global nor local. It has the advantage of tolerating a certain degree of occlusion (unlike global analysis) and does not require estimating high-order derivatives in computing invariants (unlike local analysis), whence is more robust. Furthermore, it enables a quasi-localized, hierarchical shape analysis which is unique among known invariant techniques. Unlike some current research on image invariants which concentrates on either geometry or illumination invariants, the proposed framework is very general and produces invariants which are insensitive to rigid motion, general affine transform, changes of parameterization and scene illumination, noise, occlusion, and perspective transform and view point change. Finally, we introduce the rational basis functions [4] to facilitate the analysis of invariants under perspective transform. Though rational basis functions, such as NURBS, are widely known in the computer graphics community, to the best of our knowledge they have not been widely used in computer vision.

2 A Framework of Image-Derived Invariants

We will illustrate the mathematical frameworks using a specific scenario of invariants for planar curves. The particular basis functions we use will be the wavelet bases and spline functions. Though the same framework can be easily extended to other bases such as the short-time Fourier analysis.

A word on the notational convention: matrices and vectors will be represented by bold-face characters, such as $\mathbf{M}$ and $\mathbf{V}$, while scalar quantities by plain-face characters such as S. 2D quantities will be in small letters while 3D quantities in capital letters. Hence, a 3D coordinate will be denoted as $[X, Y, Z]^T$ while a 2D coordinate as $[x, y]^T$.

We have considered variation in an object's image induced by rigid-body motion, general affine transform, changes in parameterization and illumination, and perspective transform. Due to the page limit, we will review briefly only the formulations for affine, luminance, and perspective invariants and present some preliminary results. Interested readers are referred to [5] for other invariants expressions and more results.

Rigid-Body Motion and Affine Transform Consider a 2D curve, $\mathbf{c}(t) = [x(t), y(t)]^T$, where t denotes the *affine* arc length (invariant under affine transform [2]), and its expansion onto the wavelet basis $\psi_{a,b} = \frac{1}{\sqrt{a}} g(\frac{t-b}{a})$ (where $g(t)$ is the mother wavelet [1]) as

$$\mathbf{u}_{a,b} = \int \mathbf{c}\psi_{a,b} dt, \quad \text{or} \quad \mathbf{c}(t) = \int_a \int_b \mathbf{u}_{a,b} \psi_{a,b} da db .$$

If the curve is allowed a general affine transform with the transformed curve denoted by:

$$\mathbf{c}'(t) = \mathbf{m}\mathbf{c} + \mathbf{t} = \mathbf{m} \begin{bmatrix} x(t) \\ y(t) \end{bmatrix} + \mathbf{t} ,$$

where $\mathbf{m}$ is any non-singular 2×2 matrix and $\mathbf{t}$ represents the translational motion, then we have

$$\begin{aligned}
\mathbf{u}'_{a,b} &= \int \mathbf{c}' \psi_{a,b} dt \\
&= \int (\mathbf{m}\mathbf{c} + \mathbf{t})\psi_{a,b} dt \quad = \int \mathbf{m}\mathbf{c}\psi_{a,b} dt + \int \mathbf{t}\psi_{a,b} dt \quad (1)\\
&= \mathbf{m}\int \mathbf{c}\psi_{a,b} dt + \mathbf{t}\int \psi_{a,b} dt = \mathbf{m}\mathbf{u}_{a,b} , \quad \text{or}
\end{aligned}$$

$$\mathbf{c}'(t) = \int_a \int_b \mathbf{u}'_{a,b} \psi_{a,b} da db .$$

Note that we use wavelet property $\int \psi_{a,b} dt = 0$ to simplify the second term in Eq. 1. Hence, the transformed curve can be generated using the transformed coefficients and the same wavelet bases, instead of transforming the curve point-by-point. This is an observation made in the computer graphics community on curves generated by the spline functions and associated control vertices [4]. In that sense, $\mathbf{u}_{a,b}$'s are equivalent to the control vertices in a spline curve.

If $\mathbf{m}$ represents a rotation (or the affine transform is a rigid-body motion of a translation plus a rotation), it is easily seen that invariant features can be derived using the ratio expression

$$\frac{|\mathbf{u}'_{a,b}|}{|\mathbf{u}'_{c,d}|} = \frac{|\mathbf{m}\mathbf{u}_{a,b}|}{|\mathbf{m}\mathbf{u}_{c,d}|} = \frac{|\mathbf{u}_{a,b}|}{|\mathbf{u}_{c,d}|} .$$

If the second term in Eq. 1 is not zero, but is a constant (e.g., for spline functions, the area under a spline basis integrates to a constant 1 for a uniformly spaced

knot vector [4]), then invariant expressions can still be derived, albeit in a slightly more complicated form:
$$\frac{|\mathbf{u}'_{a,b} - \mathbf{u}'_{c,d}|}{|\mathbf{u}'_{e,f} - \mathbf{u}'_{g,h}|} = \frac{|(\mathbf{mu}_{a,b} + \mathbf{v}) - (\mathbf{mu}_{c,d} + \mathbf{v})|}{|(\mathbf{mu}_{e,f} + \mathbf{v}) - (\mathbf{mu}_{g,h} + \mathbf{v})|} = \frac{|\mathbf{m}(\mathbf{u}_{a,b} - \mathbf{u}_{c,d})|}{|\mathbf{m}(\mathbf{u}_{e,f} - \mathbf{u}_{g,h})|} = \frac{|(\mathbf{u}_{a,b} - \mathbf{u}_{c,d})|}{|(\mathbf{u}_{e,f} - \mathbf{u}_{g,h})|},$$
where $\mathbf{v}$ denotes the constant second term in Eq. 1.

For invariants under general affine transform, many forms using ratios, cross ratios, and ratios of ratios have already been derived [3, 6]. For example, it is known that the cross ratio of four co-linear points are invariant under an affine transform, and the area of the triangle formed by any three $\mathbf{u}_{a,b}$ changes linearly in an affine transform (an invariant of weight 1 [3, 6]). Hence, an absolute invariance can be generated by using the ratio of two triangles: [1]

$$\frac{\begin{vmatrix} \mathbf{u}'_{a,b} & \mathbf{u}'_{c,d} & \mathbf{u}'_{e,f} \\ 1 & 1 & 1 \end{vmatrix}}{\begin{vmatrix} \mathbf{u}'_{g,h} & \mathbf{u}'_{i,j} & \mathbf{u}'_{k,l} \\ 1 & 1 & 1 \end{vmatrix}} = \frac{\begin{vmatrix} \mathbf{u}_{a,b} & \mathbf{u}_{c,d} & \mathbf{u}_{e,f} \\ 1 & 1 & 1 \end{vmatrix}}{\begin{vmatrix} \mathbf{u}_{g,h} & \mathbf{u}_{i,j} & \mathbf{u}_{k,l} \\ 1 & 1 & 1 \end{vmatrix}}. \quad (2)$$

Variation in Lighting Condition Another possible variation in the appearance is due to lighting: that objects can be illuminated by light sources of different numbers and types. To simplify the notation, in the following derivation we will consider three spectral bands of red, green, and blue. Generalizing to an n-band illumination model is straightforward. Let $\mathbf{L}(t)$ denote the perceived image color distribution along a curve, we have:

$$\mathbf{L}(t) = \begin{bmatrix} r(t) \\ g(t) \\ b(t) \end{bmatrix} = \int \begin{bmatrix} f^r(\lambda) \\ f^g(\lambda) \\ f^b(\lambda) \end{bmatrix} s(\lambda, t) d\lambda ,$$

where λ denotes the wavelength, and $f^r(\lambda)$ the sensitivity of the red sensor (similar interpretations for the green and blue channels). Using a Lambertian model, $s(\lambda, t)$ is

$$s(\lambda, t) = (\sum_{i=1}^{n} l_i(\lambda) \mathbf{N} \cdot \mathbf{N}_i) \rho(\lambda, t) + a(\lambda) , \quad (3)$$

where n is the number of light sources used, $l_i(\lambda)$ the source luminance spectral distribution, $\mathbf{N}$ the surface normal, $\mathbf{N}_i$ the incident direction for source i, $\rho(\lambda, t)$ the surface reflectivity, and $a(\lambda)$ the ambient light luminance.

When the lighting condition changes, because lights are moved, turned on, or turned off, or the ambient light intensity changes, we have
$$s'(\lambda, t) = (\sum_{j}^{m} l'_j(\lambda) \mathbf{N} \cdot \mathbf{N}'_j) \rho(\lambda, t) + a'(\lambda) ,$$
$$= \frac{\sum_{j}^{m} l'_j(\lambda) \mathbf{N} \cdot \mathbf{N}'_j}{\sum_{i}^{n} l_i(\lambda) \mathbf{N} \cdot \mathbf{N}_i} \left[(\sum_{i}^{n} l_i(\lambda) \mathbf{N} \cdot \mathbf{N}_i) \rho(\lambda, t) + a(\lambda) \right] + a'(\lambda) - \frac{\sum_{j}^{m} l'_j(\lambda) \mathbf{N} \cdot \mathbf{N}'_j}{\sum_{i}^{n} l_i(\lambda) \mathbf{N} \cdot \mathbf{N}_i} a(\lambda)$$
$$= c_1(\lambda) s(\lambda, t) + c_2(\lambda) , \text{ where}$$

---

[1] Note that there are many valid expressions for affine invariants. Some may require a smaller number of coefficients than that in Eq. 2. For example, when wavelet bases are used where $\int \psi_{a,b} dt = 0$, Eq. 2 can be simplified as
$$\frac{\begin{vmatrix} \mathbf{u}'_{a,b} & \mathbf{u}'_{c,d} \\ \mathbf{u}'_{e,f} & \mathbf{u}'_{g,h} \end{vmatrix}}{} = \frac{\begin{vmatrix} \mathbf{u}_{a,b} & \mathbf{u}_{c,d} \\ \mathbf{u}_{e,f} & \mathbf{u}_{g,h} \end{vmatrix}}{},$$
where only four coefficients are needed.

$$c_1(\lambda) = \frac{\sum_j^m l'_j(\lambda)\mathbf{N}\cdot\mathbf{N}'_j}{\sum_i^n l_i(\lambda)\mathbf{N}\cdot\mathbf{N}_i} \qquad c_2(\lambda) = a'(\lambda) - \frac{\sum_j^m l'_j(\lambda)\mathbf{N}\cdot\mathbf{N}'_j}{\sum_i^n l_i(\lambda)\mathbf{N}\cdot\mathbf{N}_i}a(\lambda)$$

capture the changes in the gain and offset in the two different lighting conditions. Following a path similar to that adopted by several researchers, we assume that surface reflectance functions are modeled as a linear combination of a small number of basis functions $s_k(\lambda)$, whence,

$$s(\lambda, t) = \sum_k \alpha_k(t) s_k(\lambda) ,$$

where $s_k(\lambda)$ denotes the k-th basis function for representing the surface reflectance properties, and $\alpha_k(t)$ is the space varying expansion coefficients. Then using an analysis which is similar to that employed in the affine case, we have

$$\begin{aligned}
\mathbf{u}_{a,b} &= \int_t \mathbf{L}\psi_{a,b}\,dt \\
&= \int_t\int_\lambda \begin{bmatrix} f^r(\lambda) \\ f^g(\lambda) \\ f^b(\lambda) \end{bmatrix} s(\lambda,t)\,d\lambda\,\psi_{a,b}\,dt \qquad\qquad = \int_t\int_\lambda \begin{bmatrix} f^r(\lambda) \\ f^g(\lambda) \\ f^b(\lambda) \end{bmatrix} (\sum_k \alpha_k(t)s_k(\lambda))\,d\lambda\,\psi_{a,b}\,dt \\
&= \sum_k (\int_\lambda \begin{bmatrix} f^r(\lambda)s_k(\lambda) \\ f^g(\lambda)s_k(\lambda) \\ f^b(\lambda)s_k(\lambda) \end{bmatrix} d\lambda)(\int_t \alpha_k(t)\psi_{a,b}\,dt) = \sum_k \begin{bmatrix} L^r_k \\ L^g_k \\ L^b_k \end{bmatrix} v^k_{a,b} \\
&= \begin{bmatrix} L^r_1 & L^r_2 & \cdots & L^r_k \\ L^g_1 & L^g_2 & \cdots & L^g_k \\ L^b_1 & L^b_2 & \cdots & L^b_k \end{bmatrix} \begin{bmatrix} v^1_{a,b} \\ v^2_{a,b} \\ \vdots \\ v^k_{a,b} \end{bmatrix} = \mathbf{L}_{rgb}\mathbf{v}_{a,b}, \quad \text{where} \\
&\begin{bmatrix} L^r_k \\ L^g_k \\ L^b_k \end{bmatrix} = \int_\lambda \begin{bmatrix} f^r(\lambda)s_k(\lambda) \\ f^g(\lambda)s_k(\lambda) \\ f^b(\lambda)s_k(\lambda) \end{bmatrix} d\lambda \quad \text{Similarly,} \\
&v^k_{a,b} = \int_t \alpha_k(t)\psi_{a,b}\,dt .
\end{aligned}$$

$$\begin{aligned}
\mathbf{u}'_{a,b} &= \int_t \mathbf{L}'\psi_{a,b}\,dt \\
&= \int_t\int_\lambda \begin{bmatrix} f^r(\lambda) \\ f^g(\lambda) \\ f^b(\lambda) \end{bmatrix} [c_1(\lambda)s(\lambda,t) + c_2(\lambda)]\,d\lambda\,\psi_{a,b}\,dt = \int_t\int_\lambda \begin{bmatrix} f^r(\lambda) \\ f^g(\lambda) \\ f^b(\lambda) \end{bmatrix} [c_1(\lambda)(\sum_k \alpha_k(t)s_k(\lambda)) + c_2(\lambda)]\,d\lambda\,\psi_{a,b}\,dt \\
&= \sum_k \{(\int_\lambda \begin{bmatrix} c_1(\lambda)f^r(\lambda)s_k(\lambda) \\ c_1(\lambda)f^g(\lambda)s_k(\lambda) \\ c_1(\lambda)f^b(\lambda)s_k(\lambda) \end{bmatrix} d\lambda)(\int_t \alpha_k(t)\psi_{a,b}\,dt) + \int_t\int_\lambda \begin{bmatrix} f^r(\lambda) \\ f^g(\lambda) \\ f^b(\lambda) \end{bmatrix} c_2(\lambda)\,d\lambda\,\psi_{a,b}\,dt\} \\
&= \sum_k \begin{bmatrix} L^{r'}_k \\ L^{g'}_k \\ L^{b'}_k \end{bmatrix} v^k_{a,b} = \begin{bmatrix} L^{r'}_1 & L^{r'}_2 & \cdots & L^{r'}_k \\ L^{g'}_1 & L^{g'}_2 & \cdots & L^{g'}_k \\ L^{b'}_1 & L^{b'}_2 & \cdots & L^{b'}_k \end{bmatrix} \begin{bmatrix} v^1_{a,b} \\ v^2_{a,b} \\ \vdots \\ v^k_{a,b} \end{bmatrix} \\
&= \mathbf{L}'_{rgb}\mathbf{v}_{a,b} .
\end{aligned}$$

Then it is easily shown that the following expression is invariant:

$$\begin{aligned}
&\frac{\left|[\mathbf{u}'_{a_1,b_1}\mathbf{u}'_{a_2,b_2}\cdots\mathbf{u}'_{a_k,b_k}]^T[\mathbf{u}'_{a_1,b_1}\mathbf{u}'_{a_2,b_2}\cdots\mathbf{u}'_{a_k,b_k}]\right|}{\left|[\mathbf{u}'_{c_1,d_1}\mathbf{u}'_{c_2,d_2}\cdots\mathbf{u}'_{c_k,d_k}]^T[\mathbf{u}'_{c_1,d_1}\mathbf{u}'_{c_2,d_2}\cdots\mathbf{u}'_{c_k,d_k}]\right|} \\
&= \frac{\left|[\mathbf{v}_{a_1,b_1}\mathbf{v}_{a_2,b_2}\cdots\mathbf{v}_{a_k,b_k}]^T\mathbf{L}'^T_{rgb}\mathbf{L}'_{rgb}[\mathbf{v}_{a_1,b_1}\mathbf{v}_{a_2,b_2}\cdots\mathbf{v}_{a_k,b_k}]\right|}{\left|[\mathbf{v}_{c_1,d_1}\mathbf{v}_{c_2,d_2}\cdots\mathbf{v}_{c_k,d_k}]^T\mathbf{L}'^T_{rgb}\mathbf{L}'_{rgb}[\mathbf{v}_{c_1,d_1}\mathbf{v}_{c_2,d_2}\cdots\mathbf{v}_{c_k,d_k}]\right|} \\
&= \frac{\left|[\mathbf{v}_{a_1,b_1}\mathbf{v}_{a_2,b_2}\cdots\mathbf{v}_{a_k,b_k}]^T\mathbf{L}^T_{rgb}\mathbf{L}_{rgb}[\mathbf{v}_{a_1,b_1}\mathbf{v}_{a_2,b_2}\cdots\mathbf{v}_{a_k,b_k}]\right|}{\left|[\mathbf{v}_{c_1,d_1}\mathbf{v}_{c_2,d_2}\cdots\mathbf{v}_{c_k,d_k}]^T\mathbf{L}^T_{rgb}\mathbf{L}_{rgb}[\mathbf{v}_{c_1,d_1}\mathbf{v}_{c_2,d_2}\cdots\mathbf{v}_{c_k,d_k}]\right|} \\
&= \frac{\left|[\mathbf{u}_{a_1,b_1}\mathbf{u}_{a_2,b_2}\cdots\mathbf{u}_{a_k,b_k}]^T[\mathbf{u}_{a_1,b_1}\mathbf{u}_{a_2,b_2}\cdots\mathbf{u}_{a_k,b_k}]\right|}{\left|[\mathbf{u}_{c_1,d_1}\mathbf{u}_{c_2,d_2}\cdots\mathbf{u}_{c_k,d_k}]^T[\mathbf{u}_{c_1,d_1}\mathbf{u}_{c_2,d_2}\cdots\mathbf{u}_{c_k,d_k}]\right|} .
\end{aligned} \qquad (4)$$

By using a ratio expression, we obtain a much simpler and computationally efficient form of invariant which does not require computing the color correlation matrix and the SVD of such a matrix as in some previously reported technique.

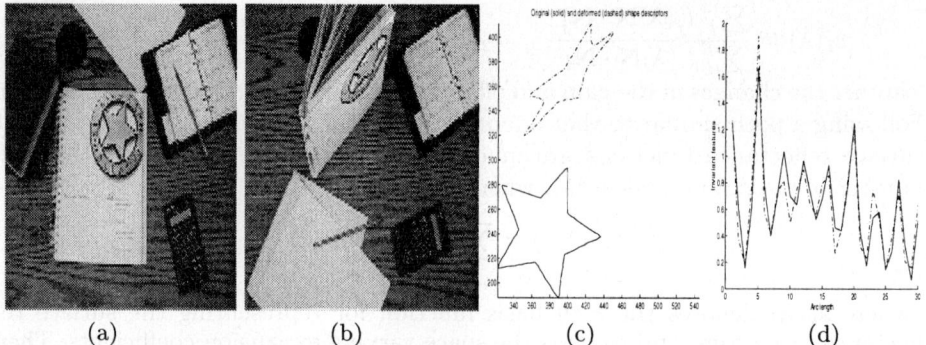

Fig. 1. (a) Original image, (b) comparison image, (c) original and deformed patterns after undergoing a rigid-body motion, and (d) invariant signatures of the original (solid) and transformed (dashed) patterns plotted along the contours.

Perspective Transform Perspective transform is a non-linear process involving a division in computing 2D coordinates. The projection process can be linearized by using the *rational* form of a basis function, such as NURBS [4]. Briefly, we represent a general 3D curve by decomposing it onto pre-selected bases. We will call the projection coefficients control vertices, following the convention used in computer graphics. The projection of a 3D curve is then represented by the projected control vertices and the *rational* bases. The problem of finding projective invariants is then a curve fitting problem: If a 2D curve results from the projection of a 3D curve, then it should be possible to interpolate the observed 2D curve using the projected control vertices and the rational spline bases and obtain a good fit. If that is not the case, then the curve probably does not come from the projection of the particular 3D curve. Hence, the error in curve fitting is a measure of invariance (In the ideal case, the error should be zero).

3 Experimental Results

Here we report some preliminary experimental results. Interested readers are referred to [5] for more detailed descriptions of the algorithms and results.

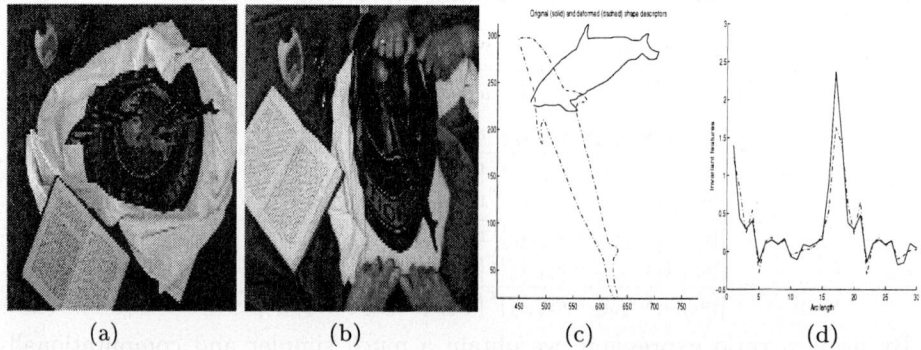

Fig. 2. (a) Original image, (b) comparison image, (c) original and deformed patterns after undergoing an affine transform, and (d) invariant signatures of the original (solid) and transformed (dashed) patterns plotted along the contours.

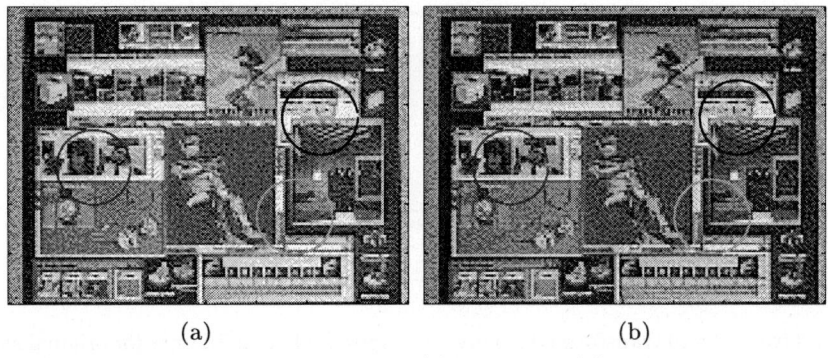

Fig. 3. The same mouse pad under (a) white and (b) blue illumination.

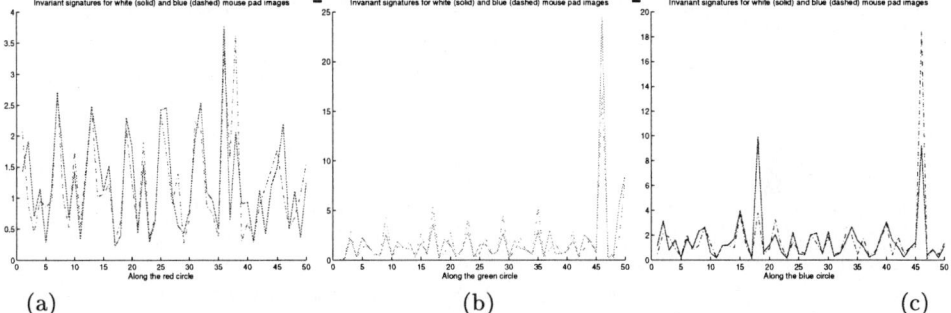

Fig. 4. R-g-b invariant signatures computed for the mouse pad under white (solid) and blue (dashed) illumination. Invariant signatures were computed on the red (a), green (b), and blue (c) circles in Fig. 3.

Rigid-body Motion Fig. 1 shows (a) a star pattern on a book cover and (b) the same pattern after undergoing a rigid-body motion in 3D (a rotation and translation of the book cover). The extracted patterns are shown in Fig. 1(c) as solid (original pattern) and dashed lines (transformed pattern). We use the second-order b-spline function of a uniform knot vector [4] in the basis expansion step. Fig. 1(d) shows the invariant signatures (based on Eq. 2) of the original (solid) and transformed (dashed) curves along the contours. (The starting point and traversal direction were manually picked.) As can be seen from the figure, the invariant signatures are quite consistent.

General Affine Transform Figs. 2(a) and (b) show a shirt with a dolphin imprint and a stretched and deformed version of the same imprint. Fig. 2(c) shows the extracted patterns. The invariant signatures are plotted in Fig. 2(d), and, again, they are quite consistent.

Change of Illumination To illustrate invariance under illumination changes, we placed different color filters in front of the light source. An example is shown below: Fig. 3 shows the same mouse pad under white and blue illumination. We randomly placed three circular curves–the red (left), green (bottom right), and blue (top right) curves in Fig. 3, and computed the invariant signatures along these three curves for both the images under white and blue illumination. Fig. 4(a), (b), and (c) show the invariant profiles computed from the white (solid)

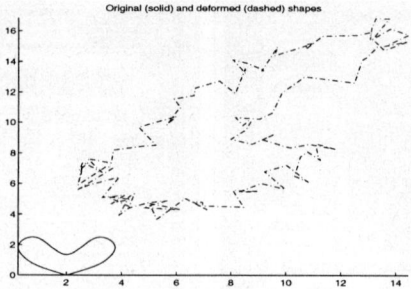

Fig. 5. Original and transformed shapes with noise added. Solid lines for original shapes and dashed lines for transformed and noise-corrupted shapes.

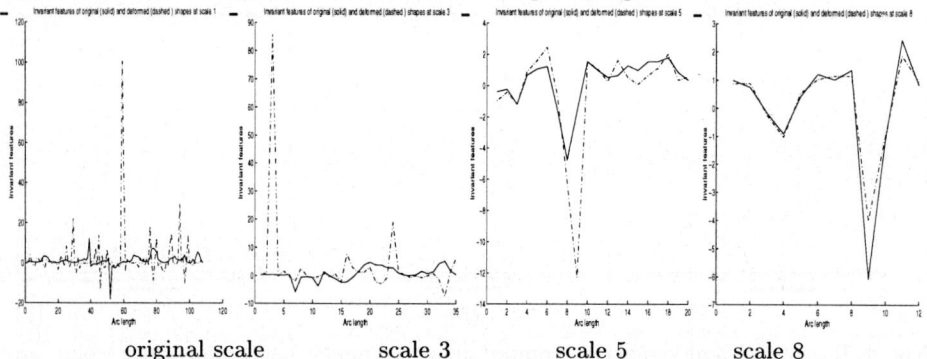

      original scale   scale 3   scale 5   scale 8

Fig. 6. Invariant shape descriptors at many scales for the original and deformed shapes with noise added. Solid lines for original shapes and shape descriptors and dashed lines for transformed and noise-corrupted shapes and shape descriptors.

---

and blue (dashed) illumination. As can be seen from the figure, they are quite consistent.

Hierarchical Invariant Analysis The additional degree of freedom in designing the basis function enables a hierarchical shape analysis. To illustrate, Fig. 5 shows the original (solid lines) and deformed (dashed lines) shapes with a significant amount of noise added. Our approach, which analyzes the shape at many scales locally, will discover the similarity which may manifest itself at different levels of detail. This is the case in Fig. 6, where shape similarity may not be apparent in the original shapes and shape descriptors, but eventually manifest itself from scale 5 onward. Traditional analysis relying on a single scale or requiring high-order derivatives of the contour function will have difficulty handling this and similar cases.

Perspective Invariants Fig. 7(a) and (b) show the canonical (head-on) view and another perspective of a package box, respectively. We extracted the arc-shape pattern from both images for verifying the perspective invariance. We display the predicted shapes superimposed on Fig. 7(b) after certain iterations of our algorithm. The predicted shape matched the real one well which verifies the perspective invariance.

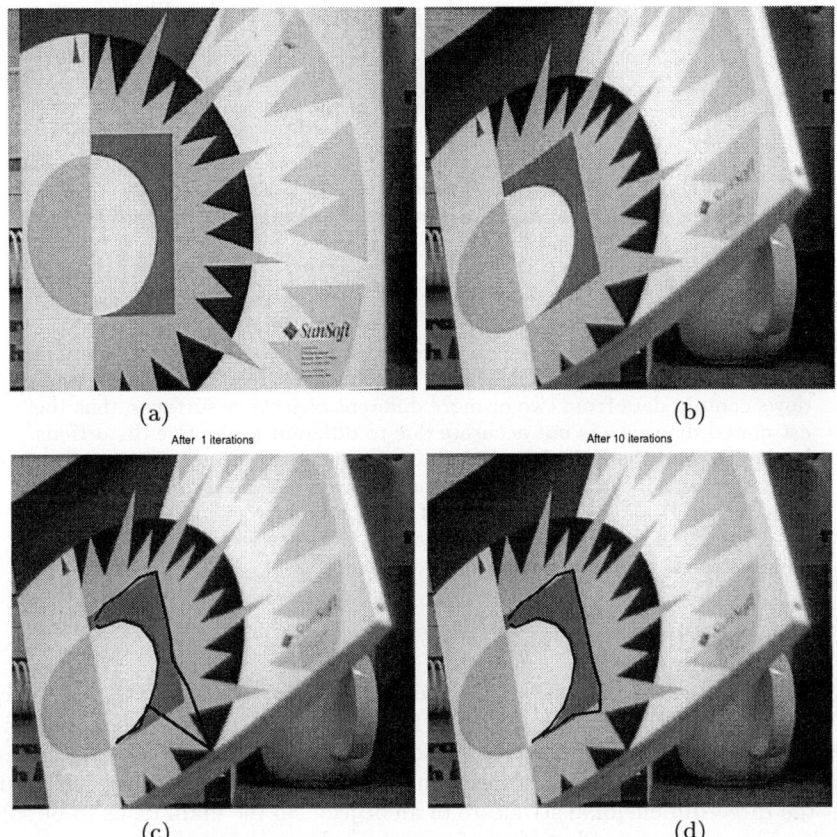

Fig. 7. (a) The canonical view, (b) another perspective, and (c) and (d) the estimated shapes after the 1st and 10th iterations.

4 Conclusion

We present a new framework for computing image invariants. The framework utilizes many desirable properties of wavelet and basis expansion techniques and is neither strictly global nor purely local. Furthermore, it is quite simple and straightforward to implement. Preliminary results on both real and synthetic images are very promising. These results demonstrate the tolerance to noise, change of luminance, and perspective distortion, and the ability for multi-scale analysis.

References

1. I. Daubechies. Orthonormal Bases of Compactly Supported Wavelets. *Commun. Pure Appl. Math.*, 41:909–960, 1988.
2. H. W. Guggenheimer. *Differential Geometry.* McGraw-Hill, New York, 1963.
3. T. H. Reiss. *Recognizing Planar Objects Using Invariant Image Features.* Springer-Verlag, Berlin, 1993.
4. D. F. Rogers and J. A. Adams. *Mathematical Elements for Computer Graphics, 2nd Ed.* McGraw-Hill, New York, NY, 1990.
5. Y. F. Wang. A New Framework for Image Invariants using Basis Expansion. Technical Report TRCS97-02, Dept. of Computer Science, UCSB, Feb. 1997.
6. I. Weiss. Geometric Invariants and Object Recognition. *Int. J. Comput. Vision*, 10(3):207–231, 1993.

Stereo Correspondences in Scale Space*

Christian Menard

Pattern Recognition and Image Processing Group,
Vienna University of Technology, Treitlstraße 3/183/2,
A-1040 Vienna, Austria, e-mail: men@prip.tuwien.ac.at

Abstract. A central problem in stereo matching using correlation techniques lies in selecting the size of the search window. Small windows contain only a small number of data points, and thus are very sensitive to noise and therefore result in false matches. Whereas large search windows contain data from two or more different objects or surfaces, thus the estimated disparity is not accurate due to different projective distortions in the left and the right image.
The new method introduces a continuous scale parameter for the matching process. It allows the adaption of the scale for every individual region and overcomes the drawbacks of fixed window sizes which is impressively demonstrated by the experimental results.

1 Introduction

An important fact in perceiving the *three-dimensional information* of a scene is that of scale. If the scale is too low for a certain problem it can be refined by humans by foveating interesting structures, or if necessary, moving closer to the interesting object. Through movement additional information can be acquired about the three-dimensional structure of an object. So the main tasks to be solved by vision algorithms are what kind of operators should be performed on the data. In order to reconstruct the three-dimensional information of an object, computer vision uses two different views of the same scene. This technique is known under the term *stereo-vision* or *binocular vision*. It is important for a vision system that interesting structures for certain scales are found by the stereo-vision process. Starting from the image acquisition, noise can be introduced for certain reasons into the images, which complicates the modeling of the used cameras and, of course, the search for the corresponding points in a stereo pair. But the success of the approach mostly depends on its ability to solve the problem of stereo matching, a topic upon which this work concentrates.

The paper is organized as follows: In the next section basic notations regarding the area-based stereo matching are given. Section 3 discusses the problem of the window size. A new adaptive matching method is proposed using a correlation scale-space. Experimental results are described in section 4.

2 Standard Area-Based Stereo Matching

The epipolar geometry is the starting point for this method. For a given pair of stereo images, the corresponding points are supposed to lie on the epipolar

---

* This work was supported by the Austrian Science Foundation (FWF) under the grant Nr. P09954-SPR.

lines [8]. Since a parallel camera alignment is used in this paper, the epipolar lines are the scan lines in both images. In Fig. 1 a synthetic stereo pair is depicted

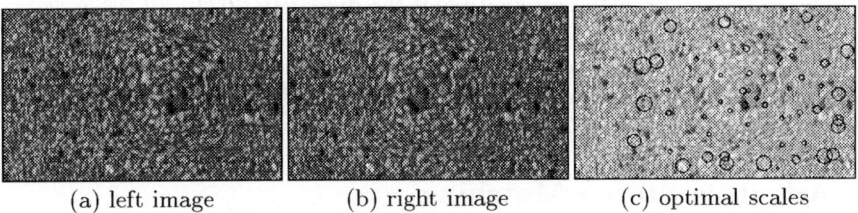

(a) left image (b) right image (c) optimal scales

Fig. 1. Synthetic stereo pair: *Pyramid* on a flat ground with natural texture added on the surface. (c) optimal scales for points of interest.

which consists of a pyramid on a plane ground with natural texture added on the surface. If for a given point (x_L, y_L) in the left image, a corresponding point (x_R, y_R) in the right image can be found, the three dimensional position of the object point can be computed with the additional information about the camera parameters. As a similarity measure the correlation of gray-level intensities is used. 1D-correlation $C(x_L, x_R, w)$ between two intervals of size w can be written as convolution with a rectangular function $\delta_{1/w}$ [7, 5].

$$C(x_L, x_R, w) = \frac{[I_L(x_L)I_R(x_R)] * \delta_{1/w} - \mu_L(x_L, w)\mu_R(x_R, w)}{\sqrt{[I_L^2(x_L) * \delta_{1/w} - \mu_L^2(x_L, w)][I_R^2(x_R) * \delta_{1/w} - \mu_R^2(x_R, w)]}}, \quad (1)$$

with $\mu(x, w) = I(x) * \delta_{1/w}(x)$, where $\delta_{1/w}(x) = \begin{cases} \frac{1}{w} & |x| \leq \frac{w}{2} \\ 0 & elsewhere \end{cases}$. (2)

The 2D-correlation $C(x_L, y_L, x_R, y_R, w)$ for two-dimensional regions can easily be extended from equation (1). For each point in the left stereo image the disparity information $D(x_L, y_L)$ is computed using the correlation function C as

$$D(x_L, y_L) = \begin{cases} x_L - x_R & for\ x_R = argmax\{C(x_L, y_L, x_R, y_R, w)\} > T \\ -1 & otherwise \end{cases}, \quad (3)$$

where T defines the minimal threshold for accepting a corresponding point. The fact that the maximum of C is below this threshold T may be caused by occlusion, highlights or depth discontinuities. The maximum of the correlation function is accepted if it is above the threshold T. In this case the position of the distinct maximum defines the corresponding point in I_R.

A central problem in finding correspondences lies in selecting the size of the search window w. Small windows contain only a small number of data points, and thus are very sensitive to noise and therefore result in false matches. Whereas large search windows contain data from two or more different objects or surfaces, thus the estimated disparity is not accurate due to different projective distortions in the left and the right image. The strategy is to find a corresponding point with the smallest window size w. There exist various works dealing with this problem; Levine et al. use an adaptive correlation window, the size of which varies inversely with the variance of the region which is currently considered [3]. In another adaptive approach Kanade and Okutomi proposed that the size and shape of the matching window is chosen adaptively on the basis of local evaluation of the variation in both the intensity and the disparity [2]. In all these works the search window is rectangularly shaped and the size is changed in a discrete way.

In the next section a new method is proposed by changing the size of the window in a continuous way, thus making it possible to determine an optimal size for a given region.

3 Adaptive Stereo Matching

The problem of changing the size of the search window during the matching process depends on the objects which are considered. If a window contains data from two or more objects or surfaces the correlation for that region does not show a clear maximum, unless the window is decreased and contains only data points from one single object. Another problem occurs when a search window contains occluded regions. In this case the computed disparity value is not correct. In order to find an optimal size of the search window the function C has to be modified in such way that the scale can be changed in a continuous way.

3.1 Gaussian Weighted Correlation

The products of intensity values in equation (1) are constantly weighted with the normalizing function $\delta_{1/w}$ (2). The formulation as convolution allows us to substitute $\delta_{1/w}$ by the Gaussian weights. In the following, the scale-space notation[2] is used, where the scale is defined by t. For a 1D-image $I : \mathbb{R} \mapsto [0,1]$ the Gaussian weighted local mean is defined by $\mu : \mathbb{R} \times \mathbb{R} \mapsto [0,1]$

$$\mu(x;t) = \int_{-\infty}^{\infty} I(x-\xi)g(x+\xi,t)\,d\xi = I(x) * g(x;t)\,, \text{ with } g(x;t) = \frac{1}{t\sqrt{2\pi}} e^{\frac{-x^2}{2t^2}}. \tag{4}$$

The Gaussian weighted standard deviations are defined by

$$\sigma_k^2(x;t) = I_k^2(x) * g(\bullet;t) - \mu_k^2(x;t) \qquad k = L, R\,, \tag{5}$$

and the covariance can be written as

$$\sigma_{LR}^2(x_L, x_R; t) = [I_L(x_L)I_R(x_R)] * g(\bullet;t) - \mu_L(x_L;t)\mu_R(x_R;t) \qquad k = L, R\,. \tag{6}$$

The correlation can be written as convolution with the Gaussian kernel in the one-dimensional case (4) as follows:

$$C_\Gamma(\bullet;t) = \frac{[I_L(x_L)I_R(x_R)] * g(\bullet;t) - \mu_L(x_L;t)\mu_R(x_R;t)}{\sqrt{[I_L^2(x_L) * g(\bullet;t) - \mu_L^2(x_L;t)][I_R^2(x_R) * g(\bullet;t) - \mu_R^2(x_R;t)]}}. \tag{7}$$

---

[2] The notation $C_\Gamma(\bullet;t)$ stands for $C_\Gamma(x_L, x_R; t)$ and $C_\Gamma(\bullet,\bullet;t)$ for $C_\Gamma(x_L, y_L, x_R, y_R; t)$.

For the two-dimensional case the two-dimensional Gaussian kernel is used and the 2D-weighted correlation function $C_\Gamma(\bullet,\bullet\,;t)$ can be extended from equation (7). Instead of using all the data points in the correlation window equivalently weighted, the window is weighted with the Gaussian kernel. The size of the search window can be controlled by the scale parameter t. The influence of pixels far from the center of the window diminish at a rate controlled by t. Furthermore the shape of the search window has changed from rectangular to circular. The function C_Γ defines a *Correlation Scale-Space (CSS)* for one point $I_L(x,y)$ in the left stereo image. In the *CSS* the similarity value is available at different scales driven by the parameter t of the scale-space kernel. The main advantages of the *CSS* compared to standard methods, such as the hierarchical approach, is that the scale can be changed in a continuous way. Furthermore in this representation all levels of scale are immediately accessible.

3.2 Optimal Scale Selection

In general situations it is not possible to know in advance at what scales interesting structures can be expected to appear. Size variations of image structures in a stereo pair can occur for several reasons:

- objects in the scene have different physical size;
- surface textures contain structures at different scales; and
- scale variations appear due to perspective distortions.

There are many ways to select the best scale for a given problem. A very interesting work in this field was presented by Lindeberg in which he describes the "scale-space primal sketch" [4]. An operator gives maximal output if its size is best tuned to the object. Other approaches study the variation of the information content over scale [1]. For the correspondence establishment it is possible so far to change the scale parameter t in a continuous way using the correlation function C_Γ. But the problem to be solved is to find the "best scale(s)" t_{opt} for certain regions in a stereo pair. Basically, the scale at which a maximum over scales is attained will be assumed to give information about the window size for that region. The maximum over scale for a region defines the optimal scale. In the next step the *CSS* for different placements of a point is analyzed: on a *plane* parallel to the image plane; on a *roofed surface*; near a *depth edge*; and in an *occluded area*. For these situations the correlation values at the corresponding

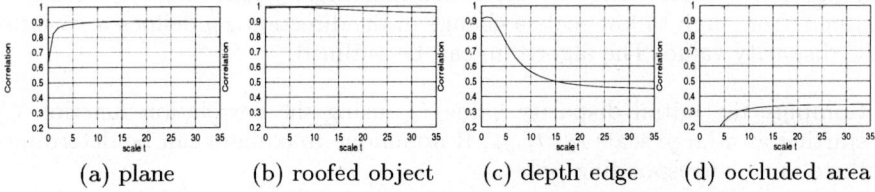

(a) plane   (b) roofed object   (c) depth edge   (d) occluded area

Fig. 2. Optimal scale values for different situations: (a) $C_\Gamma=0.91$ at $t=15.0$, (b) $C_\Gamma=0.99$ at $t=3.0$, (c) $C_\Gamma=0.92$ at $t=1.5$ and (d) $C_\Gamma=0.35$ at $t=35$.

position in the right stereo image are tested by tracing along the local maxima in the direction of the scanline from high to low scale. In Fig. 2 the correlation function at the corresponding point along scale t is visualized for these different situations.

Plane: For the plane lying parallel to the image plane there are only small variations along the correlation function. The correlation value decreases if no or not enough gray-level information is available in the correlation window. The scale-space maximum for this example lies at $t=15$, with $C_\Gamma(\bullet,\bullet\,;t) \in [0.6..0.9]$, which means that for this scale the correlation value can be maximized.

Roofed surface: In the case of a roofed surface, by tracing from large to smaller scale the value of the correlation increases successively until it decreases because of too little information in the correlation window. For this example a small scale obtains the highest correlation value at scale $t=3$, with $C_\Gamma(\bullet,\bullet\,;t) \in [0.95..0.99]$.

Depth Edge: Near a depth edge it is similar. At larger scales the correlation value is low because the window contains data from two objects which have different disparity values. By decreasing the scale parameter t the correlation value is maximized at a very low scale, since the window contains only data from one single object. The obtained scale for this situation lies at $t = 1.5$, with $C_\Gamma(\bullet,\bullet\,;t) \in [0.45..0.93]$.

Occluded area: In the last test the correlation value over scale for an occluded region is determined. It can be seen in Fig. 2 (d) that the maximum of the correlation function along scale is at $t=35$ with $C_\Gamma(\bullet,\bullet\,;t) \in [-0.1..0.35]$. This represents the highest scale and the correlation value decreases successively to lower scale. But the correlation value is very low anyway, thus the position of this maximum represents an average depth information of the surrounding regions, which contain the occluded area.

In order to establish correspondences, the function C_Γ is tested on the pyramid (Fig. 1). In the left image some points of interest are chosen for which the corresponding points are determined in the right stereo image. Every scale-space maximum is graphically illustrated by a circle centered at the point in the left stereo image for which the correspondence is established. The size of the circle corresponds to the scale of maximum correlation. The circles are superimposed on a bright copy of the left image. The result is visualized in Fig. 1 (c). The optimal scale of points on the surface of the pyramid is small, since the planes form an oblique angle with the image plane, thus not all points in the search window have the same disparity value, whereas for regions on the flat ground the selected scale value is large. One way to determine the optimal scale t_{opt} for finding correspondences is to determine the scale-space maximum by tracing from high to low scale along the correlation maxima. By following the correlation maximum from high to low scale a change in the direction x_R defines a variation in the disparity value. The algorithm can be outlined:

1. Compute the initial disparity value D_0 using the correlation function C_Γ starting at a large scale $t_0=t_{max}$. If no unique maximum can be determined there is no corresponding point.
2. Modify the scale with $t_{n+1} = t_n - \Delta t_n$.
3. Compute the new disparity value D_{n+1} by using the previously estimated disparity D_n.
4. Estimate new Δt_{n+1} using D_n, t_n.
5. Iterate steps 2.,3. and 4. until a global maximum along the scale is found or a maximum number of iterations is reached.

4 Experimental Results on Synthetic and Real Images

The adaptive matching algorithm using function C_Γ is applied on a set of synthetic and real images and is compared to the standard stereo approach.

As synthetic stereo pairs a *pyramid*, a *box* and a *sphere* on a flat ground are used in each case with texture added onto the surface. The disparity values for these synthetic sets of stereo pairs are known to compare the absolute accuracy of the matching method. For each test the accuracy is compared to the results

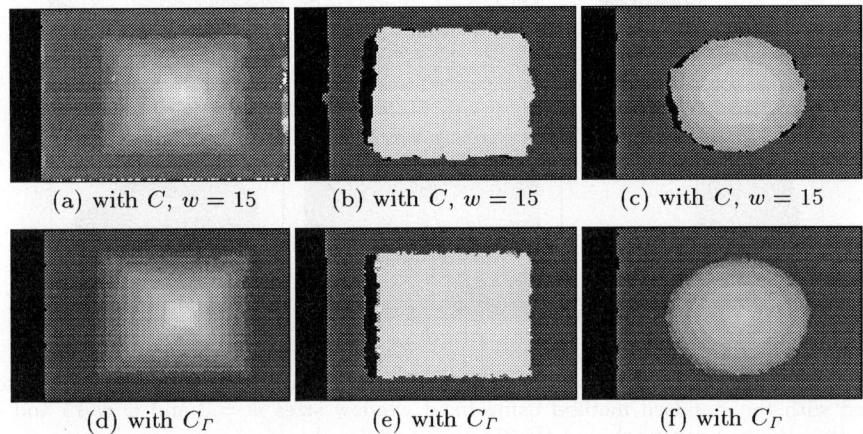

Fig. 3. Disparity maps for the test objects: The first row depicts the results computed with the standard stereo method using a fixed window size $w = 15$ and the second row using the adaptive approach.

computed with the standard matching method using the MSE^3. These stereo pairs are tested first with the standard stereo method using fixed window sizes $w = 3$, $w = 7$ and $w = 15$ and then with the adaptive matching method using the function C_Γ (Fig. 3). For flat surfaces a large search window obtains good results, whereas on depth edges the disparity values are blurred. The disparity

MSE	$MSE(C(w=3) - ideal)$	$MSE(C(w=7) - ideal)$	$MSE(C(w=15) - ideal)$	$MSE(C_\Gamma - ideal)$
Pyramid	280	203	266	50
Box	275	220	200	45
Sphere	320	280	303	65

Table 1. Difference between true disparity and the computed disparity.

values along depth edges are more accurate using small search windows, but the disparity over the complete image is very noisy. The adaptive approach obtains good results both on depth discontinuities and on flat surfaces. Table 1 illustrates the $MSE(C(w \in \{3, 7, 15\}) - ideal)$ and the $MSE(C_\Gamma - ideal)$. The adaptive matching method has the smallest MSE.

---

[3] The notation $MSE(< method > (< w >) - ideal)$ is used for the comparison. The MSE is only computed for the region which is visible in both stereo images.

This approach reduces two types of errors,
- large random errors all over the image caused by a small search window and
- systematic errors along depth discontinuities which occur when using large search windows.

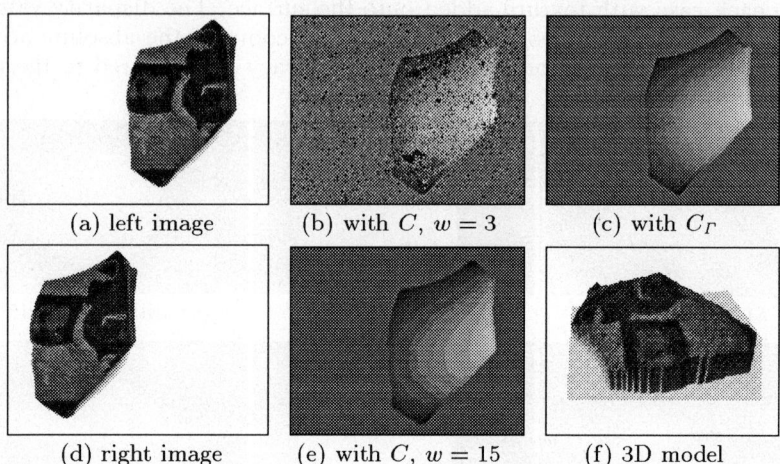

Fig. 4. Archaeological fragment 1: (a) and (d) stereo pair, (b)(e) disparity maps computed with the standard method using fixed window sizes $w = 3$ and $w = 15$ and (c) with the adaptive approach and (f) 3D model of the object.

Archaeological fragments are used as real objects [6]. The experimental setup consists of two 8Bit-CCD cameras with a resolution of 768 × 568 pixels and a 133-Pentium running OS Linux. About twenty different fragments were tested, each of them with different orientation parameters and under varying lighting conditions. The stereo pairs of two different fragments are depicted in Figs. 4 and Figs. 5 (a) and (d).

The curved surface of both fragments is recovered. The objects have a smooth surface without any false matches and the edge of the fragments is also recovered. For comparison the disparity values computed with the standard method using fixed window sizes are shown in Figs. 4 (b),(e) and Figs. 5 (b),(e). A small window size produces several mismatches on the surface of both fragments whereas large search windows smooth the disparity values. Together with both gray-level and disparity information three-dimensional models of the objects with the texture added on the surface can be created (Figs. 4 (f) and 5 (f)).

5 Summary

In this paper a new method detects the optimal scale to determine the corresponding region for each location in a given stereo pair. A correlation scale-space defines the scale in a continuous way. For each region in the stereo pair, depending on the gray-level and disparity information, the size of the search window can be changed adaptively in a continuous way by changing the scale parameter t of the correlation scale-space. Furthermore the shape of the search window has

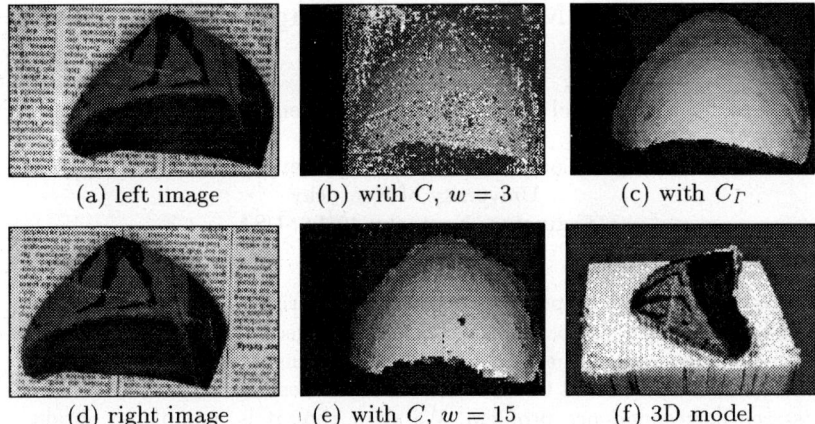

Fig. 5. Archaeological fragment 2: (a) and (d) stereo pair (b)(e) disparity maps computed with the standard method using fixed window sizes $w = 3$ and $w = 15$ and (c) with the adaptive approach and (f) 3D model of the object.

changed from rectangular to circular. Experience demonstrates that tilted regions with varying disparity values need a small scale, whereas flat regions with low gray-level variations favor a larger scale which produces a distinct maximum. The global scale-space maximum for a certain region, which maximizes the correlation value is defined as the optimal size of the search window. The adaptive matching strategy combines the benefits of small and large scales which is demonstrated by experiments with synthetic data and archaeological fragments.

References

1. M. Jaegersand. Saliency maps and attention selection in scale and spatial coordinates: An information theoretic approach. In *Proceedings Fifth Intern. Conf. on Computer Vision*, pages 195–202, Cambridge, MA, June 20-23 1995. MIT, IEEE. Catalogue no 95CB35744.
2. T. Kanade and M. Okutomi. A stereo matching algorithm with an adaptive window: Theory and experiment. *PAMI*, 16(9):920–932, September 1994.
3. M.D. Levine, D.A. O'Handley, and G.M. Yagi. Computer determination of depth maps. *Comput. Graphics Image Processing*, 2:131–150, September 1973.
4. L. Lindeberg. *Scale-Space Theory in Computer Vision*. Kluwer Academic Publishers, 1994.
5. C. Menard. *Robust Stereo and Adaptive Matching in Correlation Scale-Space*. PhD thesis, TU Wien, Institut für Automation, PRIP, Wien, 1996.
6. C. Menard and R. Sablatnig. Computer based Acquisition of Archaeological Finds: The First Step towards Automatic Classification. In Hans Kamermans and Kelly Fennema, editors, *Interfacing the Past. Computer Applications and Quantitative Methods in Archaeology. CAA95*, number 28, pages 413–424, Leiden, March 1996. Analecta Praehistorica Leidensia.
7. Azriel Rosenfeld and Avinash C. Kak. *Digital Picture Processing Volume 2*. Academic Press, Inc., 1982.
8. J. Weng. Camera calibration with distortion models and accuracy evaluation. *IEEE Trans. Patt. Anal. Machine Intell.*, 14:965–980, 1992.

Fast Stereo Matching in Compressed Video

Michael S. Brown and W. Brent Seales

Computer Science Department
University of Kentucky
Lexington, Kentucky 40506, USA

Abstract. In this paper we present new algorithms that exploit compressed image data to achieve coarse stereo reconstructions in a real-time environment. Others have shown large gains from processing image streams in compressed form, and we extend those results to address the stereo correspondence problem. We show that it is possible to obtain stereo matches between two frames of a compressed video stream in approximately the same time it takes to fully decompress just one image.

Keywords: stereo, compression, depth reconstruction, real-time, compressed-domain

1 Introduction

Compression is central to the problem of transferring video streams in real-time from a sensor across a network for processing. Compression reduces bandwidth and conserves storage space. Systems to analyze image data which are built on top of such real-time video environments must address the problem of manipulating compressed video. The manipulation that usually takes place is complete decompression of the video stream to obtain images in the pixel domain which are then processed with pixel-based methods.

In this paper we present an algorithm designed to obtain depth information from real-time stereo video streams. The novel aspect of the work is the ability to extract feature matches and obtain depth measurements without completely decompressing the data. The depth estimates we obtain are coarser than what can be recovered with pixel-domain images, but we show that the gain in computational savings can be enormous. In particular, we show that it is possible to obtain a set of matches from two frames of a compressed stereo video stream in approximately the same time it takes to fully decompress just *one* complete image. The savings result from avoiding decompression, operating on smaller amounts of data, and using algorithms for matching which take advantage of the information that is already present at certain stages of the compression pipeline.

The compressed domain is the form of the video data after compression. The pixel domain is the representation of the video as sets of pixel images. Conversion between these two domains is accomplished using the steps of the compression algorithm, transforming the data through multiple, intermediate phases on the way to the totally compressed output stream. These intermediate forms can be exploited, and often the data can be viewed in that form at a substantially smaller computational cost than is incurred for complete decompression. We use this approach to find matches in compressed video streams without totally decompressing the video.

There has been recent interest in processing multimedia data in its compressed form. A survey [8] covers some of the these approaches. The work of the Photobook project by Pentland et al. [4] uses compression schemes that preserve semantic content and are thus searchable. The new algorithm we present in this paper is most similar to multi-resolution stereo [2, 10] in the sense that our algorithm allows us to constrain stereo matches and perform real-time matching on 8×8 pixel boundaries, giving a coarser spatial resolution. The algorithm operates with standard compression schemes, providing substantial speedups without the need for dedicated hardware.

2 Image Compression

The ability to extract anything directly from compressed data depends on the particulars of the transformations that the data has undergone during compression. Most compression algorithms are a sequence of steps, each of which contributes to the decorrelation of the data. The JPEG standard, which we use in our prototype system, consists of the following key steps [11]: (1) Pixel range shifting; (2) Discrete Cosine transform (DCT) on 8×8 pixel blocks; (3) DCT quantization; (4) DCT "zig-zag" ordering; (5) Run-length coding; and (6) Entropy coding. Applying all the steps gives a totally-compressed JPEG image. Motion JPEG is essentially a continuous set of JPEG frames. Our algorithms operate on run-length-encoded (RLE) vectors, which is the data after step (5). Very little can be done with the data after step (6) since it is no longer byte-aligned. More information can be found in [1, 11, 3].

3 Finding Correspondence in Compressed Imagery

We construct a matching criterion in compressed video by extracting coarse edge information and combining that result with an area-based matching approach. In particular, we classify 8×8 JPEG blocks (in the compressed domain) according to edge orientation and intensity transition by examining frequency information available from the compressed stream. Searching for a region of blocks with similar edge information (regions of similar texture), we formulate a measure of similarity between areas in the stereo images. Using the epipolar geometry of the stereo rig, we reduce the search space by constraining the search to the epipolar line. Because of the coarseness of the edge information, the resolution of the disparity maps will lie on 8×8 boundaries, reducing the spatial resolution of the final 3D reconstruction.

Our approach follows these steps:

1. Cameras are assumed to be rectified (if this is not possible, warping transformations can be performed in the compressed data as in [9]).

2. Blocks with only low frequency components are not considered (this reduces the number of blocks in the matching stage).
3. Coarse features are identified: edge orientation and frequency activity based on the DCT coefficients define features
4. Matching based on feature extraction: uses a block-similarity responses within a small neighborhood

3.1 RLE Data Representation of DCT information

We only decompress frames to the run-length-encoded (RLE) stage of the pipeline. The RLE encoding describes the sparseness of the 8×8 DCT block and provides an efficient mechanism for summing AC values: one need not visit each position in the DCT block. From the RLE structure it is trivial to build an algorithm that traverses the blocks and their individual AC values. While traversing each 8×8 DCT block we also build an index pointing to the start of each block in the RLE buffer. This is necessary for classifying the block's texture.

3.2 Filtering by measuring AC activity

The first step of the compressed-domain stereo matching process is to determine which blocks have enough texture to be useful for matching. "Textureless" blocks, such as background, are very difficult to match because they provide no cues for similarity between the stereo pair. It is necessary to filter these blocks and determine which blocks are good candidates for matching.

The summation of the AC activity has been used [7] to filter out low-texture blocks:

$$\mathbf{f} = \sum_{u=0}^{7} \sum_{v=0}^{7} F(u,v)^2 \qquad (1)$$

where u and v are not both zero (the DC component is excluded), and $F(u,v)$ are the frequency values at in the DCT block at positions u and v. This simple equation sums the high-frequency information and acts as a high-pass filter for the 8×8 DCT blocks[7].

3.3 Texture Classification by Determining Edge Orientation

Blocks determined to have enough texture are used as candidates for matching. It is necessary to extract information from the blocks that can be used in a similarity constraint. Edge orientation can be determined using a method published by Shen and Sethi [7]. This method examines AC values to determine the coarse edge orientation and intensity transition in the blocks that pass the DCT filter. In order to achieve coarse edge classification it is only necessary to examine AC coefficients F01 and F10, where FXY refers to the DCT coefficient at position XY.

It has been shown that coefficient F01 depends upon intensity differences in the vertical direction between the upper and lower parts of the 8 × 8 image block. Similarly, F10 is determined by the intensity differences in the horizontal direction between the left and right parts of the 8 × 8 image block. By examining the sign and magnitude of these coefficients, we can determine if the edge is vertically dominant (|F01| > |F10|) and if the intensity transition of the block is dark to light (F01 < 0) or light to dark (F10 > 0). This method not only allows us to determine useful information about a block's texture, but because it relies solely on the frequency values of F01 and F10, its results remain stable at low JPEG qualities. F01 and F10 are not as severely quantized as the other frequency values when the quality factor is reduced.

For blocks that do not pass the filter and are determined to lack sufficient texture activity, we classify the block as solid and determine its intensity as either dark or light depending on its DC value.

3.4 Matching with Classified Edges

We now have a new representation of the compressed image. Textureless blocks have been filtered and are labeled only as dark or light depending on their DC value; edge orientation and intensity transition of each textured 8 × 8 block has been determined. With this new representation of the image, we address the correspondence problem at a lower resolution in the image pyramid, matching along 8 × 8 boundaries.

Using the epipolar geometry of stereo cameras, the correspondence problem is constrained to blocks on the epipolar line. We have rectified our cameras such that the epipolar lines are aligned with the scan lines of the images. This narrows the matching problem to finding corresponding blocks in two images that lie on the same row.

We now define a similarity response using an area-based technique. Given a block X in left image and a block Y in the right image, we look at the similarity of the edge orientation and intensity transition of this block and its eight neighboring blocks (N_8). Incrementing the response for blocks with alike edge orientation and alike intensity transition, we obtain a response to discern the correctness of the hypothesized matches.

The similarity response is the basis for addressing the matching problem. Given a block X in the left image, we examine the responses for blocks in the same row in the right image. When we find a high response that satisfies uniqueness and disparity constraints, we mark the block pair as matching.

After finding a valid match, we choose the center point of the two blocks as our stereo match. This gives a rough estimation of where the match actually lies. By choosing the center pixel, we are within a four-pixel neighborhood of the true match. In the case where the edge is offset from the center of a block, Shen and Sethi [7] show that it is possible to determine the offset by examining the signs and magnitudes of more AC values, thus enabling a better localization of the match. We can improve our results using this method to a localization of at least 2 pixels.

At this stage in the stereo pipeline, we operate using a coarse edge map. Other area-based algorithms may also be applied here to improve results. In a time-constrained (i.e., *Quality of Service*) environment, it may be reasonable to use several different area-based matching techniques which could be invoked depending on the time constraints at hand [6]. Selective deompression, for example, can be employed to improve localization.

4 Results

The results reported here were computed on a stereo stream of 15 640 × 480 images of a Coconut Head. Before edge classification, images are filtered to remove low texture activity blocks. This is done by measuring the AC activity for each block with the criterion in Eq. 1. The number of calculations needed to perform the block filtering is greatly reduced by using the RLE vector, which describes the spareness of the DCT coefficients. For the Coconut head sequence, at quality factor 75, we achieved an eight-times speed-up by filtering blocks with low AC activity. At quality factor 10 and below we get speedups in excess of twenty times [1].

Edge classification is performed using only 2 of the AC coefficients: F01 and F10. The amount of time for classification depends on the number of blocks that pass the filter. It takes takes only 3 comparison operations of the values of F10 and F01 to determine the edge orientation and intensity transition. The DC coefficient is examined for blocks that do not pass the filter. If the DC value (F00) is negative the block is considered dark. A positive value indicates a light block. F01 and F10 are not dramatically affected by the quantization process, and thus this coarse edge classification scheme remains reliable for lower quality factors. Edge orientation is classified as: *upward-sloping, downward-sloping, vertical*, and *horizontal*. Intensity transitions are labeled either *white to black* or *black to white*.

4.1 Matching

Matching is determined by the similarity of two blocks and their neighbors as described in Section 3.4. Incorrect matches which violate the disparity constraint are eliminated. The cameras in the experiments performed in this paper had a baseline of 11.3 cm. A pixel increase in disparity produced approximately a 8 mm shift in the world coordinate system of the 3D point. Therefore mismatches of one 8 × 8 block results in a 6.4 cm shift of the reconstructed 3D point. Mismatches, which are present in the figure 1, appear as outliers away from the primary shape of the object.

Figure 1 shows a comparison of a stereo stream matched and reconstructed in the compressed-domain versus the results from a traditional stereo method [5]. One can see that our compress-domain algorithm gives a coarser result, but the general shape of the object is preserved. Fifteen stereo pairs have been matched and reconstructed. The motion of the object was known and each reconstructed frame is rotated appropriately to render all frames in a common coordinate system.

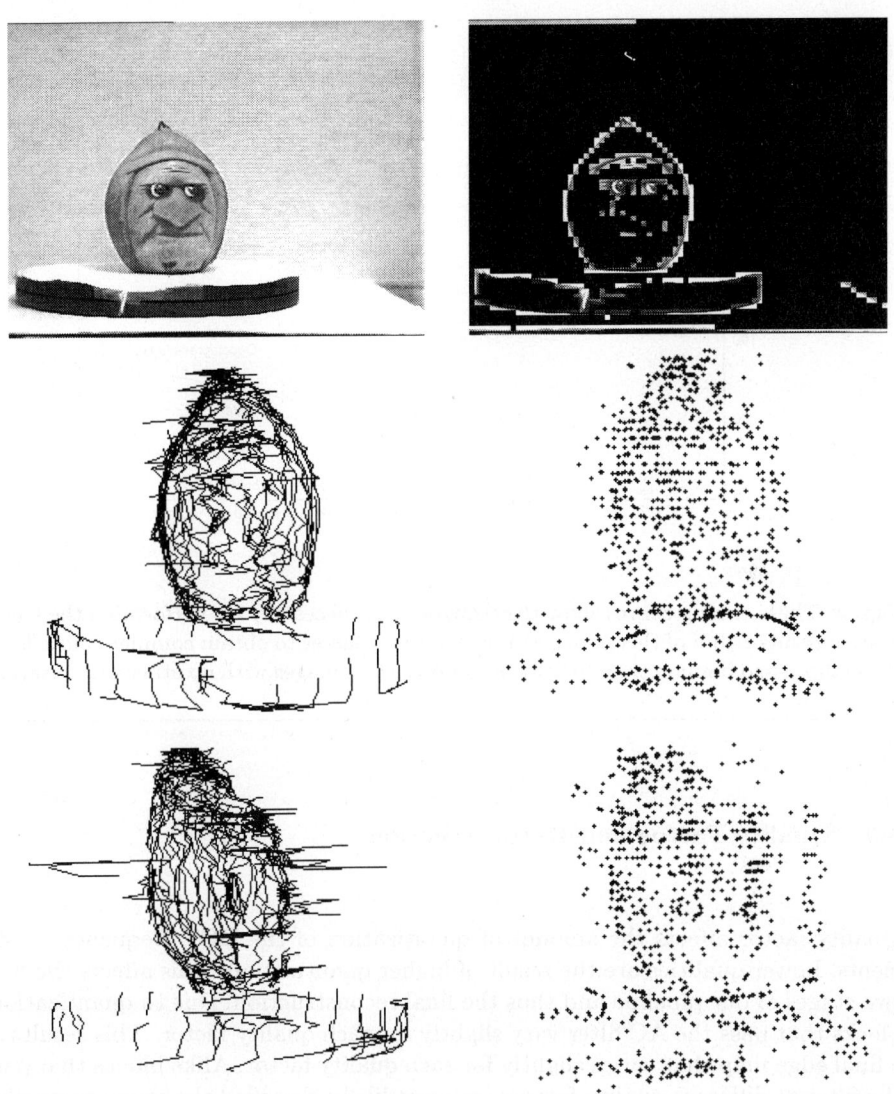

Fig. 1. The top images show a typical frame (left) and the result of filtering low-activity blocks (right). The middle two data sets show a reconstruction over all 15 frames from the Coconut sequence. The bottom two sets show a complete 15-frame reconstruction, rotated 90 degrees about the vertical axis. The reconstructions compare the result of traditional curve-based stereo, on the left, to our compressed-domain stereo, on the right.

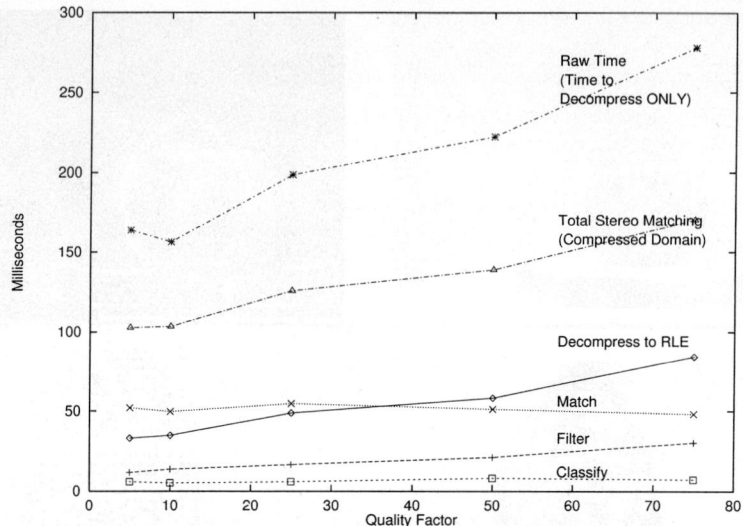

Fig. 2. These timing curves show the relative times necessary for performing the stereo matching algorithm of this paper. It is consistently faster to obtain complete matches in the compressed domain than to fully decompress the images with no other computation.

4.2 Quality Factor and Reconstruction

Quality factor affects the amount of quantization of the DCT frequency coefficients. Lower qualities are the result of higher quantization. This affects the first two stages of our pipeline and thus the final reconstruction. Due to quantization, blocks that pass the AC filter vary slightly for each quality factor. This results in a final edge map that varies slightly for each quality factor. Alike blocks that pass the filter at different quality factors are most likely classified the same, since edge classification relies only on the coefficients F01 and F10 which are not severely quantized. The overall change, however, is minor and the final reconstruction is quite stable at quality at factors 75, 50, 25, 10, and 5 [1].

Figure 2 shows the timings over all quality values for the sequence. The complete matching process is consistently less than the top curve which represents nothing but total decompression time. Timing was performed on an SGI Indigo2, with a MIPS R4400 processor, with 128 MB Memory. Times ares given in milliseconds. We modified a public domain JPEG library[3] to do the partial decompression. Full decompression is performed with the standard JPEG library distribution.

5 Conclusions

It is possible to obtain correct stereo matches from MJPEG and MPEG compresed video without fully decompressing the image stream. Our matches are oriented on the JPEG 8×8 block boundries, providing lower spatial resolution in a multi-resolution stereo pyramid, although in many contexts it may be preferable because of the efficiency gains. Our results show that a set of matches for two frames in a stereo sequence can be obtained from compressed video streams at approximately the same cost as fully decompressing *just one image*. We can improve spatial resolution by further examination of the frequency coefficients present in the compressed data and can add image warping in the compressed pipeline to eliminate parallel camera constraints. It is our goal to incorporate this technique for use with real-time vision systems under low bandwidth constraints.

6 Acknowledgements

We gratefully acknowledge the support of the National Science Foundation for this work under grants CDA-9320179, CDA-9502645 and IRI-9308415.

References

1. M.S. Brown and W. B. Seales. Fast stereo matching in compressed video. Technical Report 272-97, Computer Science Dept., University of Kentucky, Lexington, Kentucky, 1997.
2. Umesh R. Dhond and J.K. Aggarwal. Structure from Stereo-A Review. *IEEE Transactions on Pattern Analysis and Machine Intelligence*, 19(6):1489–1510, Nov. 1989.
3. T. Lane, P. Gladsonte, L. Ortiz, and J. Boucher. The Independent JPEG group's JPEG software, Febuary 1996. Distribution available at archie site: *ftp.uu.net*.
4. A. Pentland, R.W. Picard, and S. Sclaroff. Photobook: Content-based manipulation of image databases. *Int. Journal of Computer Vision*, 18(3), 1996.
5. L. Robert and O. D. Faugeras. Curve-based stereo: figural continuity and curvature. *Proc. 3rd Int. Conf. Computer Vision*, pages 57–61, 1990.
6. W.B. Seales, J. Lumpp, and M. Brown. Distributed wireless stereo reconstruction. In *IEEE Aerospace Applications Conference*, 1997.
7. B. Shen and I. Sethi. Direct feature extraction from compressed images. In *Proc. Int. Soc. for Optical Engr. vol. 2670, Storage and Retrieval for Image and Video Databases IV*, 1996.
8. B. Smith. Survey of compressed domain processing techniques. In *Reconnecting Science and Humanities in Digital Libraries, a Digital Library Symposium at the University of Kentucky*, Oct 1995.
9. B. Smith and L. Rowe. Algorithms for manipulating compressed images. *IEEE Computer Graphics and Applications*, 13(5):34–42, Sept 1993.
10. K. Tate and Z.N. Li. Depth map construction from range-guided multiresolution stereo matching. *IEEE Transactions on Systems, Man, and Cybernetics*, 24:134–144, 1994.
11. G. Wallace. The JPEG still-picture compression standard. *Communications of the ACM*, 34(4):31–44, April 1991.

Robust Total Least Squares Based Optic Flow Computation

Alireza Bab-Hadiashar and David Suter
Intelligent Robotics Research Centre,
Department of Electrical & Computer Systems Engineering,
Monash University, Clayton Vic. 3168, AUSTRALIA.
E-mail: [ali, suter]@basil.eng.monash.edu.au

Abstract This paper considers the problem of finding a robust solution to the optic flow problem. The optical flow field is recovered by solving a system of over-determined linear equations with all the data matrices containing both outliers and noise. Here, we present a novel and very effective solution for this problem called weighted total least squares. The weights for this method are computed using a new robust statistical method named least median of squares orthogonal distances. Unlike the total least squares which is only robust to noise, this method is extremely robust to both noise and outliers and can tolerate up to 50% of equations in the system to be contaminated by outliers. The proposed weighting method is fast and the total computation remains inexpensive. To demonstrate the performance of the proposed algorithm, we compare the accuracy of our algorithm for computing optic flow field for a number of synthetic and real image sequences and show that the proposed method, despite being very simple and straightforward, out performs all methods used for comparison.

1 Introduction

In this paper, we consider the problem of finding a robust solution to a differential based optic flow problem. The differential techniques invariably involve some form of what has become known as the Optic Flow Constraint (OFC). The OFC can be written as (Horn & Schunck, 1981):

$$\frac{\partial I}{\partial x}u_x + \frac{\partial I}{\partial y}u_y + \frac{\partial I}{\partial t} = 0 \qquad (1.1)$$

which relates the spatial $\partial I/\partial x$, $\partial I/\partial y$ and temporal $\partial I/\partial t$ derivatives of the image brightness function, at each point, to the optic flow (u_x, u_y) at that point. Since there is only one equation in two unknowns, it cannot be solved for both the x and y components of the optic flow, without additional assumptions or information (the well-known aperture problem). In other words, using just the information we have so far, the problem is ill-posed. Various alternative strategies to make the problem well-posed (regularize the problem) have been suggested but regardless of the strategy for overcoming the aperture problem, one usually arrives at a set of over-determined linear equations that one must solve for the optic flow at each point.

Elements of the data matrices in the final equation for solving the optic flow (which are the spatial and temporal derivatives) have to be numerically estimated, and therefore contain noise. Also, the assumptions made to overcome aperture problem (constant motion, affine motion, etc., e.g. Bergen et al., 1992) are likely to be violated due to multiple motions, transparencies, etc. Therefore, we are in fact faced with the problem of solving a set of over-determined linear equations *where all the data matrices contain both outliers and noise*. This is just one example, from many problems in computer vision, for which the solution requires solving an over-determined set of linear equations with outliers and noise in all the data matrices.

Thus, we consider the problem of finding a robust solution x to a set of over-determined linear equations:
$$Ax \approx b \qquad (1.2)$$
where the data matrices A and b contain both outliers and noise (without A being rank deficient). By *robust solution*, we refer to a solution x exactly satisfying the equation $(A_s + \Delta A_s)x = (b_s + \Delta b_s)$ where the subscription s refers to the largest sub-group of equations consistent with the Least Median of Squares Orthogonal Distances (LMSOD), defined in section 2, while the Frobenius norm[1] of the perturbation matrix $\Delta = [\Delta A_s \Delta b_s]$ is kept minimum (by using the total least squares).

Finding a consistent solution to a set of over-determined linear equations $Ax \approx b$ where both of the data matrices A and b contain noise has been studied for a long time. Total least squares (TLS) is the method of choice for solving this problem (VanHuffel & Vandewalle, 1991). However, the TLS, as well as the ordinary least squares (LS) problem, is extremely sensitive to the influence of any outliers. Indeed, the breakdown point (the smallest number of contaminated data that can cause the estimator to take on values arbitrary far from true estimate, Rousseeuw and Leroy, 1987) of the TLS is only one. This means that even one contaminated element, in either of data matrices, can result in an arbitrary bad solution. It should be noted here that the TLS is often preferred over LS because the TLS solution, unlike the LS solution, which is only consistent where the observation matrix b is error free, remains consistent even when all the data matrices are noisy. Being consistent means that the estimated solution converges to the true solution as the number of equations tends to infinity (based on the assumption that all the elements of Δ are uncorrelated random variables with equal variance).

The TLS method has been frequently employed to solve many different computer vision problems. Providing a comprehensive list of all these attempts is beyond the scope of this paper but we briefly review a few relevant works. It should be noted here that these solutions have a serious limitation associated with the sensitivity of the TLS to contaminated data (outliers).

Chu and Delp (1989) have suggested using TLS for solving the set of over-determined equations resulting from an optic flow formulation. Their study addresses the rank deficient problem (where the data matrix A in the final over-determined set of linear equations is rank deficient) but fails to address the problem of having discontinuities in either the image brightness function or the optic flow itself (which commonly happens in any practical applications).

Wang et al. (1992) used the TLS to recover the smooth flow where the chances of having outliers are limited. It is important to note that, assuming the flow is smooth is not sufficient to ensure there are no outliers in the final equations. Secondly, the assumption of smooth flow, over a predefined area, is too conservative to be useful in practical applications. Weber and Malik (1995) also presented a method for estimating the optic flow based on the TLS method. In this work, the authors allow the outliers to corrupt the results when they solve the linear equations but they reject the final results

---

[1] The Frobenius norm of a m x n matrix M, with entries m_{ij}, is defined as: $\|M\|_F = \sqrt{\sum_{i=1}^{m}\sum_{j=1}^{n} m_{ij}^2}$.

based on some weighting scheme of the singular value decomposition (SVD) of the augmented data matrices. We think that this approach is too conservative. Indeed, this method under utilizes the available information by allowing the outliers to contaminate the estimate in the first place and then attempts to reject the bad results when the damage is irreversible (see section 3 for performance comparison).

Chaudhuri and Chatterjee (1991) presented a performance analysis of the TLS method for 3-D motion estimation. In this study, using synthetic data with additive uncorrelated Gaussian noise, they conclude that the TLS out performs LS method in deriving both the motion of deformable objects from range data, and the motion of a rigid object under perspective projection.

The rest of paper is organized as follows. Section 2 presents our robust TLS based solution for the optic flow problem formulated as the solution of a set of over-determined linear equations. In section 3, we calculate the optic flow for a number of image sequences and compare the results with the performance of other methods that claim high accuracy. A brief summary concludes the paper.

2. Robust TLS Based Technique

The study of robust estimators with high breakdown point, allowing a substantial portion of the data to be contaminated by outliers, has been actively researched for decades among statisticians. In 1984, Rousseeuw proposed the method of Least Median of Squares (LMedS) for the standard regression (SR) problem. LMedS has a breakdown point of 50%. Although this estimator is very robust to outliers, its theoretical performance in the presence of noise, and its computational complexity, are not attractive. Rousseeuw and Leroy (1987) proposed a very powerful method known as reweighted least squares (RLS). In this method first, a fast approximate solution to the LMedS problem is found. Then all the data points are categorized into outliers and inliers, based on their scaled residuals with respect to the LMedS solution. Finally, the regressor for the inliers is calculated using the LS technique.

Meer et al. (1991) presents a comprehensive survey on the applications of robust statistics in computer vision problems. Stewart (1997) provides a comprehensive review of the main robust estimators commonly used in computer vision literature.

The LMedS has been used (independently and concurrently) to recover the optic flow by Ong and Spann (1996a and 1996b) and Bab-Hadiashar and Suter (1996 and 1997). Although the crux of algorithms presented in both set of papers are similar, their algorithms differ in the way that the optic flow fields are calculated. To resolve the issue of multiple motions, Ong and Spann compare the flow computed for a block with the flow computed for a shifted (right and downward) block. The estimate from the block with the greatest number of inliers is retained. Although this technique may prove to be effective for occluding objects, it cannot resolve transparent motion. Bab-Hadiashar and Suter however, use a small patch centered at every pixel and employ a robust LMedS based estimator to reject the OFC belonging to pixels with motions widely different from the majority of pixels in the original patch. Then, they solve all the remaining OFC equations (inliers) using the LS technique. Moreover, the Bab-Hadiashar and Suter approach has an extra stage that performs a reliability check. Although we follow the Bab-Hadiashar and Suter's approach for calculating the optic flow (and we compare our results with their results), the presented work here goes well beyond the common basic ideas developed in both set of papers.

In this paper, we propose the Weighted Total Least Square (WTLS) method, similar to the RLS method proposed by Rousseeuw and Leroy (1987), for solving an overdetermined set of linear equations $Ax \approx b$ where the data matrices A and b contain both outliers and noise without A being rank deficient. The proposed method differs from the RLS method in two ways. Firstly, unlike the RLS method, the outliers (of equation 1.2) in this method are detected using the LMSOD technique. LMSOD seeks an approximate solution x which exactly satisfies $(A+\Delta A)x = b+\Delta b$ while minimizing the median of squares orthogonal distances between solution x and the geometrical entity (line, plane, hyper-plane, etc) represented by every equation in the original set. The second major difference is that in the WTLS method, unlike the RLS method which uses ordinary least square to solve the inlier group, the total least square technique is used to solve the remaining system of over-determined linear equations (after rejecting the outliers).

It is important to note here that this method has all the advantages of the TLS method without being sensitive to the influence of outliers. Comparing LMSOD to the LMedS, it is trivial to show that LMSOD method also has the breakdown point of 50%.

2.1 Proposed algorithm

For the sake of clarity, we describe the proposed algorithm in six steps:

1. Estimate the spatio-temporal derivatives of the image brightness function. We choose to use, for our experiments, convolutions with derivatives of Gaussian functions with equal spatial and temporal standard deviations (Nagel, 1995).

2. Select a patch of the image, over which we are going to assume some motion consistency. The precise form of the motion consistency is not essential: we are simply assuming a single or dominant population (we only recover the dominant population if there is more than one - our method can be elaborated to remove the dominant population and re-solve for any secondary populations). In this paper, we restrict the motion consistency to one of two forms: constant motion and affine motion with in a patch.

3. Use a fast and robust approximate LMSOD solution to obtain a temporary estimate of the solution x (free from the influence of any existing outlier). Here, we propose a method similar to the one presented by Rousseeuw and Leroy (1987). The method starts by randomly choosing a group of sample equations. Each sample must group contain n equations where n is the number of rows in the solution matrix x (equation 1.2). Moreover, all the chosen equations must be independent (to ensure the existence and uniqueness of the solution). By solving every such set of equations and finding the median of the squared orthogonal distances between this solution and the rest of the equations in the original set, one can find the solution, which approximately satisfies the LMSOD.

Similar to the LMedS case, one needs to choose only one sample group of size n that belongs to majority, in order to return the approximate solution associated with the majority. Therefore only a small number of sample groups is required to have the probability of having at least one good sample close to 1 (Rousseeuw & Leroy, 1987).

4. Having found the approximate LMSOD, one can weight the different equations based on the vertical distance between the LMSOD solution and the geometrical entity (line, plane, hyper-plane, etc.) represented by every equation in the original system of equations. Here, we closely follow the recommendations made by Rousseeuw and Leroy (1987) for scaling the residuals in two steps. The detail of the weighting scheme is described in appendix. We then identify the outliers by comparing the scaled residuals with some constant threshold. After identifying the outliers, we eliminate the outlier equations (weight them by zero) to arrive at a new system of over-determined linear equations $A_s \, x = b_s$ in which the number of equations are now less than or equal the original set.

5. The final solution can be obtained by solving the new system of over-determined linear equations using the total least squares technique (VanHuffel & Vandewalle, 1991):

$$x = (A_s^T A_s - \kappa^2 I)^{-1} A_s^T b_s \tag{2.1}$$

where κ is the smallest singular value of the augmented matrix $[A_s \, b_s]$ and I is the identity matrix.

6. This last step in this algorithm is to ensure that the final solution is acceptable. Thus, a measure of reliability is proposed here to examine the validity of the end results. The detail of this step is described in the following section.

2.2 Measure of Reliability

Although the LMSOD techniques have the highest possible breakdown point (50%) of all known robust estimators, it has the potentially fatal flaw in that is still produces an estimate, even if the number of outliers is more than 50%. Moreover, there are extreme cases where an image patch may not contain sufficient data (lack of texture) or data so badly corrupted (aliasing for example) for any estimate to be valid. Thus we still need to validate the estimate produced by our method. A tool for the validation process can be modeled on "the coefficient of determination" (Kvalseth, 1985). The coefficient of determination, denoted R^2, has been defined for the Standard regression problem in at least nine different ways. However, although we are guided by analogy with the SR problem, we are interested in robust forms of TLS. We define our own measure, which is also called R^2, similar to the one presented by Bab-Hadiashar and Suter (1997). For the WTLS technique, we want to ensure that the Frobenius norm of the perturbation matrix $\Delta=[\Delta A_s \, \Delta d_s]$ is small enough for the solution to be acceptable. Since it has been shown that κ (the smallest singular value of the augmented matrix $[A_s \, d_s]$) is equal to the Frobenius norm of the perturbation matrix Δ for the calculated x (VanHuffel & Vandewalle, 1991), we propose the following R^2 statistic:

$$R^2_{WTLS} = 1 - \frac{\kappa^2}{\sum_i (d_{si} - \bar{d}_{si})^2} \tag{2.2}$$

where d_{si} represents the different elements of vector d_s and maximum number of i is set by the number equations regarded as inliers.

3 Optic Flow Computation

To evaluate the performance of the proposed estimator for recovering the optic flow field, we compute the flow field and the error statistics for a few synthetic and real image sequences whose the "ground truth" motion is known. As mentioned in section 1, a common approach to the optic flow problem is to formulate the flow field as a solution to an over-determined system of linear equations (similar to equation 1.2). The number of unknowns in this approach (number of columns in matrix A) depends on the model of motion in every patch of the image. Constant (2 unknowns) and affine (6 unknowns) motions are the most common models of motion proposed in the optic flow literature. To keep the computation minimum, we first solve the LMSOD (step 3) for all the OFC contained in a square window with constant model of motion. Then, we calculate the weights for every OFC based on its residual with respect to LMSOD. We simply reject the constraints, whose scaled residual is above some threshold (step 4). The final steps in estimating the flow field are to solve the new system of over-determined linear equations using total least squares and compute the associated R^2 statistics. To improve the accuracy at the last stage, we solve the weighted set of OFC using an affine motion model (six unknowns). The idea behind this is very simple. The outliers contaminating the OFC (due to multiple motions, transparency, etc) are essentially independent of the motion model and by rejecting the outliers using constant motion model, the computational time is reduced. It is important to note that this argument is only justified for small windows where the chance of disregarding good points at the tail of the affine model by the robust solution calculated using constant model is negligible. One, of course, may achieve slightly better results by using the affine model of motion in both steps.

3.1 Synthetic and Real Image Sequences

To demonstrate the effectiveness of our method for dealing with motion boundaries, we use a number of synthetic and real image sequences as benchmark. These image sequences are: New-Sinusoid1, Yosemite and Otte (see Bab-Hadiashar, 1997 for a detailed description of each image sequence). The error statistics related to each image sequences and their comparison with a number of other methods is shown in tables 3.1 to 3.3.

These results show that our method is very robust to the existing depth and motion discontinuities. It can be seen from these tables that our WTLS based method outperforms other methods used in comparison.

4 Conclusion

This paper presents a novel method for solving a system of over-determined linear equations when the parameters of the equations are contaminated with both outliers and noise. The solution to this type of problem is frequently sought in the study of different computer vision problems. The proposed algorithm uses a new method named the least median of squares orthogonal distances combined with the well-known total least squares for dealing with the outliers and noise, respectively. A fast method for computing an approximate solution to the LMSOD is also proposed which makes the computation inexpensive. The performance of this method has been demonstrated by solving the optic flow problem. Although the presented algorithm is conceptually very straight forward, it out-performs any other (often very sophisticated) optic flow technique.

Table 3.1: Error analysis using New-Sinusoid1 image sequence.

Technique	Avg. Error	Std. Dev.	Density
Fleet and Jepson(σ =2.5,τ =1.25)	7.39°	10.84°	43.4%
Fleet and Jepson(σ =2.5,τ = 2.5)	1.41°	3.65°	46.0%
WLS2(σ =1.0,5x5,m=30,without check)	1.56°	7.12°	100%
WLS2(σ =1.0,5x5,m = 30, R^2 = 0.9999)	0.05°	0.06°	84.6%
WLS6(σ =1.0,5x5,m=30,without check)	1.51°	5.86°	100%
WLS6(σ =1.0,5x5,m = 30, R^2 = 0.9999)	0.05°	0.06°	83.5%
WTLS2(σ =1.0,5x5,m=30,without check)	2.82°	8.82°	100%
WTLS2(σ =1.0,5x5, m = 30, R^2 = 0.9999)	0.05°	0.06°	76.1%
WTLS6(σ =1.0,5x5,m=30, without check)	1.51°	6.23°	100%
WTLS6(σ =1.0,5x5, m = 30, R^2 = 0.9999)	0.08°	0.22°	88.4%

Table 3.2: Error analysis using Otte image sequence (Otte & Nagel, 1994).

Technique	Avg. Error	Std. Dev.	Density
Fleet and Jepson (σ = 2.0, τ = 1.25)	2.08°	3.77°	50.6%
Fleet and Jepson (σ = 2.0, τ = 2.50)	2.56°	4.08°	57.1%
Fleet and Jepson (σ = 2.5, τ = 1.25)	2.05°	3.85°	55.8%
Fleet and Jepson (σ = 2.5, τ = 2.50)	2.53°	4.25°	62.2%
Giachetti and Torre (1996)	5.33°	.°	100(25)%
WLS2(σ =2.0,15x15,m=30,without check)	3.39°	6.55°	100%
WLS2(σ =2.0,15x15,m=30,R^2 = 0.99)	1.50°	2.22°	59.1%
WLS6(σ =2.0,15x15,m=30,without check)	3.51°	6.48°	100%
WLS6(σ =2.0,15x15,m=30,R^2 = 0.99)	1.44°	1.92°	55.9%
WTLS2(σ =2.0,15x15,m=30,without check)	3.74°	8.09°	100%
WTLS2(σ =2.0,15x15,m=30,R^2 = 0.99)	1.61°	2.60°	71.2%
WTLS6(σ =2.0,15x15,m=30,without check)	3.67°	7.37°	100%
WTLS6(σ =2.0,15x15,m=30,R^2 = 0.99)	2.46°	4.71°	82.0%
WTLS6(σ =2.0,15x15,m=30,R^2=0.999)	1.55°	2.34°	51.6%

Table 3.3: Error analysis using Yosemite image sequence.

Technique	Avg. Error	Std. Dev.	Density
Fleet and Jepson (σ = 1.5, τ = 1.25)	4.95°	12.39°	30.6%
Fleet and Jepson (σ = 1.5, τ = 2.5)	4.29°	11.24°	34.1%
Weber and Malik (1993)	3.42°	5.35°	45.2%
Szeliski and Coughlan (1994)	3.06°	7.54°	39.6%
Weber and Malik (1995)	4.31°	8.66°	64.2%
Giachetti and Torre (1996)	2.82°	6.98°	70.79%
WLS2(σ =2.0,15x15,m = 30, without check)	3.17°	6.46°	100%
WLS2(σ =2.0,15x15,m = 30, R^2 = 0.99)	3.13°	7.07°	76.2%
WLS6(σ =2.0,15x15,m = 30, without check)	2.86°	6.76°	100%
In the following results, the cloud region is not included.			
Black (1994)	3.52°	3.25°	100%
Black and Jepson (1994)	2.29°	2.25°	100%
Black and Anandan (1996)	4.46°	4.21°	100%
Ju et al (Skin & Bones, 1996)	2.16°	2.00°	100%
WLS2(σ =2.0,15x15,m=30,without check)	2.51°	2.57°	100%
WLS6(σ =2.0,15x15,m=30,without check)	2.02°	2.05°	100%
WTLS2(σ =2.0,15x15,m=30,without check)	2.56°	2.34°	100%
WTLS6(σ =2.0,15x15,m=30,without check)	1.97°	1.96°	100%

The first column of entries determines the method applied to generate the row of error statistics. In our method (WTLS) the numbers 2 and 6 represent the constant and affine motion models, respectively. The numbers in brackets depict the size of the Gaussian smoothing (σ is the standard deviation of the filter), the size of local patch used (p), the number of pairs of lines used to approximate the LMedS or the LMSOD (m), and the reliability threshold (R^2), in that order.

5 References

Bab-Hadiashar A., Suter D. 1996 "Robust Optic Flow Estimation Using Least Median of Squares" Proceeding of IEEE International Conference on Image Processing ICIP'96, Lausanne, 513-516.

Bab-Hadiashar A., 1997 "Accuracy and Robustness in Visual Motion Analysis" PhD dissertation, Monash University, Australia.

Bab-Hadiashar A., Suter D. 1997 "Optic Flow Calculation Using Robust Statistics" Proceeding of IEEE Conference on Computer Vision and Pattern Recognition CVPR'97, Puerto Rico, 988-993.

Bergen J.R., Anandan P., Hana K.J., Hingorani R. 1992 "Hierarchical model-based motion estimation" Proc. Secd. Europ. Conf. Comp. Vis., ECCV-92, Springer-Verlag, 237-252.

Black M. J. 1994 "Recursive non-linear estimation of discontinous flow field" ECCV'94, 138-145.

Black M.J., Anandan P., 1996 "The robust estimation of multiple motion: parametric and piecewise-smooth flow fields" Computer Vision and Image Understanding, Vol. 63(1), 75-104.

Black M. J. Jepson 1994 "Estimating multiple independent motions in segmented images using parametric models with local deformations" workshop on Motion of Non-rigid and Articulated Objects, 220-227, Austin.

Chaudhuri S., Chatterjee S. 1991 "Performance Analysis of Total Least Squares Method in Three-Dimensional Motion Estimation" IEEE Transactions on Robotics and Automation, 7(5), 707-714.

Chu C. H., Delp E. J. 1989 "Estimating displacement vector form an image sequence" J. Opt. Soc. Am. A 6(6), 871-878.

Fleet D. J., Jepson A. D. 1990 "Computation of component image velocity from local phase information" Intern. J. Comput. Vis. 5: 77-104.

Giachetti A., Torre V. 1996 "Refinement of Optical Flow Estimation and Detection of Motion Edges" Proceedings. ECCV'96, Cambridge, UK, 15-18 April, 151-160.

Horn B.K.P., Schunck B.G., 1981 "Determining optical flow" Artificial Intelligence 17, 185-204.

Ju S. X., Black M. J., Jepson A. D. 1996 "Skin and Bones: Multi-layer, Locally Affine, Optical Flow and Regularaization with Transparency", CVPR'96, San Francisco, 307-314.

Kvalseth T. O. 1985 "Cautionary note about R^2", The American Statistician, 39(4), 279-285.

Meer P. Mintz D. Rosenfeld A. Kim D. Y. 1991 "Robust regression methods for computer vision: A review" Intern. J. Comput. Vis. 6(1): 59-70.

Nagel H.H., 1995 "Optical flow estimation and the interaction between measurement errors at adjacent pixel positions", Intern. J. Comput. Vis. 15, 271-288.

Ong E.P., Spann M., 1996a "Robust multiresolution of optical flow" Procd. IEEE Int. Conf. Acous. Spch. Sigl. Proc., Atlanta, Georgia, 4, 1938-1941.

Ong E.P., Spann M., 1996b "Robust computation of optical flow" British Machine Vision Conf., Edinburgh, 573-582.

Otte M., Nagel H. H. 1994 "Optical Flow Estimation: Advances and Comparisons" Proc. ECCV 94, Stockholm, Sweden, 2-6 May 1994, 51-60.

Rousseeuw P. J. 1984 "Least Median of Squares Regression" Journal of the American Statistical Association, 79, 871-880.

Rousseeuw P. J. Leroy A. M. 1987 "Robust Regression and Outlier Detection", John Wiely, New York.

Shizawa M., Mase K., 1990 "Simultaneous multiple optical flow estimation" Proc. of 10th Int. Conf. on Pattern Recognition, Atlantic City, New Jersey, 274-278.

Stewart C., 1997 "Bias in robust estimation caused by discontinuities and multiple structures", IEEE Transactions on Pattern Analysis and Machine Intelligence, to appear.

Szeliski R., Coughlan J. 1994 "Hierarchical spline-based image registration", In Proceedings CVPR'94, Seattle, 194-201.

VanHuffel S., Vandewalle J. 1991 "The Total Least Squares Problem: Computational Aspects and Analysis" 1st Ed., SIAM, Philadelphia.

Wang S., Markandey V., Reid A. 1992 "Total least squares fitting spatiotemporal derivatives to smooth optical flow field" Proc. of the SPIE: Signal and Data processing of Small Targets, vol 1698, 42-55.

Weber J., Malik J. 1993 "Robust computation of optical flow in a multi-Scale differential framework", Procd. of Int. Conf. on Computer Vision, ICCV-93, Berlin, May, 12-20.

Weber J., Malik J. 1995 "Robust Computation of Optical Flow in a Multi-Scale Differential Framework", Intern. J. Comput. Vis. 14: 67-81.

Appendix: Residual Scale and Outlier Threshold

In our method, having obtained an approximate solution, based on an approximate LMSOD, we wish to assess the reliability of each equation. The following procedure, which is similar to the recipe proposed by Rousseeuw and Leroy (1987), is used for detecting outliers.

We first calculate, for each equation in the original system of linear equations, a residual r_i by finding the distance between the LMSOD solution and the geometrical entity (line, plane, hyper-plane, etc.) represented by that equation. Then we calculate a scale factor s^0 according to:

$$s^0 = 1.4826(1+\frac{5}{p-2})\sqrt{\operatorname*{med}_i r_i^2} \qquad (A.1)$$

where p is the number of equations in the original system of linear equations.

We then associate a binary weight w_i so that the weight is 0 for any constraint whose residual r_i is such that absolute of r_i/s^0 is greater than 2.5 (and the weight is otherwise equal to 1).

Rather than using these weights to directly reformulate the problem now as a (weighted) Total Least Squares problem, we go through one more step of scaling. This is because the original weights were chosen, according to equation (A.1), using the median *involving the outliers*. Since we now have a better idea of which are truly outliers, we calculate:

$$\sigma^* = \sqrt{\frac{\sum_{i=1}^{p} w_i r_i^2}{\sum_{i=1}^{p} w_i - p}} \qquad (A.2)$$

and we, finally, reject those constraints for which the associated absolute value of r_i/s^0 is greater than 2.5.

Image Processing via the Beltrami Operator *

R. Kimmel,[1] R. Malladi,[1] and N. Sochen[2]

[1] Lawrence Berkeley National Laboratory
University of California, Berkeley, CA 94720.
[2] Raymond and Beverly Sackle Faculty of Exact Sciences
Tel-Aviv University, Israel.

Abstract. We present a framework for enhancing images while preserving either the edge or the orientation-dependent texture information present in them. We do this by treating images as manifolds in a feature-space. This geometrical interpretation leads to a natural way for grey level, color, movies, volumetric medical data, and color-texture image enhancement. Following this, we invoke the Polyakov action from high-energy physics, and develop a minimization procedure through a geometric flow. This flow, based on manifold volume minimization yields a natural enhancement procedure. We apply this framework to edge-preserving denoising of grey value and color images, for volumetric medical data, and orientation-preserving flows for grey level and color texture images.

1 Introduction

In this paper, we present a general framework for processing images of various types like grey scale, color, and those that have orientation-dependent information such as textures. We do this by treating images as embedded maps that flow towards minimal surfaces. In other words, our view on images is that they are $2D$ or $3D$ manifolds embedded in higher dimensional space; for example a grey-scale image is a surface in (x, y, I) space and a color image is a surface embedded in a $5D$ space, i.e. the (x, y, I^r, I^g, I^b). We then use the Polyakov action, that is a general way of measuring area for a manifold embedded in a given space. The edge-preserving enhancement procedure is a result of minimizing this "action" and is expressed via a geometric flow. Our framework has the following properties: (1) It is the most general way of writing the geometrical scale-space and enhancement algorithms for grey-scale, color, volumetric, time-varying, and texture images, (2) it unifies many existing partial differential equation based schemes for image processing, and (3) the schemes are edge-preserving and hence suitable for segmentation tasks.

Texture plays an important role in the understanding process of many images, specially those that involve natural scenes. Preserving the orientation information while diffusing a given texture image is important in certain cases, say in

---

* This work is supported in part by the Applied Mathematics Subprogram of the Office of Energy Research under DE-AC03-76SFOOO98, ONR grant under NOOO14-96-1-0381, and in part by the National Science Foundation under grant PHY-90-21139.

denoising a fingerprint image. We imagine a procedure that preserves domains of constant/homogeneous texture, enhances the texture in each domain, and thereby enhances the boundaries between neighboring domains with different textures. Weickert in [23, 24] presents a coherence enhancing flow based on a structure tensor idea. In Section 4 we first link the coherence enhancing flow to the Beltrami geometric framework. Then we extend the method by inverting the diffusion direction across the edge for better enhancement and sharpening results.

2 Images as Embedded Maps that flow toward Harmonic Maps

Our geometric framework finds a seamless link between the L_1 ([19, 3] TV and its variants) and the L_2 norms (used in [13] and its variants) based on the geometry of the image and its interpretation as a surface[2]. The aspect ratio between the gray level and the xy plane, used as a parameter, enables us to switch between the two commonly used norms; see [22] for details.

In this work, we also propose a flow in a rich feature space which is different from the image space. Other flows in similar feature spaces were recently proposed in [20, 18, 5, 21, 25]; see also [23, 24] for orientation preserving flows. All these approaches begin with a flat metric [7] that does not yield a meaningful minimization process when going to more than one channel[3]. The main difference between these schemes and the one we propose is the geometric interpretation of the information as a manifold flowing so as to minimize its volume. Our geometric perspective of a color image as a surface embedded in a higher dimensional space enabled us to define a simple and natural coupling in the multi-channel color space. Other schemes have also considered image as a surface [2, 8, 26, 12], some even used the image information to build a Riemannian metric for segmentation [4]. However, these methods were not generalized to feature space or any co-dimension higher than one. We now describe the details of our framework.

2.1 The Metric

The basic concept of Riemannian differential geometry is distance. Let us start with the map $\mathbf{X} : \Sigma \to \mathbb{R}^3$, where Σ is a $2D$ manifold. We denote the local coordinates on the two dimensional manifold Σ by (σ^1, σ^2). The map X is explicitly given by $(X^1(\sigma^1, \sigma^2), X^2(\sigma^1, \sigma^2), X^3(\sigma^1, \sigma^2))$. Since the local coordinates σ^i are curvilinear, and not orthogonal in general, the distance square between two close points on Σ, $p = (\sigma^1, \sigma^2)$ and $p + (d\sigma^1, d\sigma^2)$ is not $ds^2 = d\sigma_1^2 + d\sigma_2^2$. In fact, the

---

[2] TV (Total Variation) schemes are based on minimizing the L_1 norm, namely $\int |\nabla I|$, the L_2 norm minimizes $\int |\nabla I|^2$, while the area of the gray level image surface is given by $\int \sqrt{1 + |\nabla I|^2}$.

[3] This flat metric is called 'structure tensor' in [23, 24].

squared distance is given by a positive definite symmetric bilinear form called the metric, whose components we denote by $g_{\mu\nu}(\sigma^1, \sigma^2)$, i.e.

$$ds^2 = g_{\mu\nu} d\sigma^\mu d\sigma^\nu = g_{11}(d\sigma^1)^2 + 2g_{12} d\sigma^1 d\sigma^2 + g_{22}(d\sigma^2)^2, \qquad (1)$$

where we used Einstein summation convention in the second equality.

2.2 Polyakov Action

Let us briefly review our framework for non-linear diffusion in computer vision. The equations are derived by a minimization problem from an action functional. The functional in question depends on *both* the image manifold and the embedding space. Denote by (Σ, g) the image manifold and its metric and by (M, h) the space-feature manifold and its metric, then the map $\mathbf{X} : \Sigma \to M$ has the following weight

$$S[X^i, g_{\mu\nu}, h_{ij}] = \int d^m \sigma \sqrt{g} g^{\mu\nu} \partial_\mu X^i \partial_\nu X^j h_{ij}(\mathbf{X}), \qquad (2)$$

where m is the dimension of Σ, g is the determinant of the image metric, $(g^{\mu\nu})$ is the inverse of the image metric, the range of indices is $\mu, \nu = 1, \ldots, \dim \Sigma$, and $i, j = 1, \ldots, \dim M$, and (h_{ij}) is the metric of the embedding space. This functional, for $m = 2$, was first proposed by Polyakov [16] in the context of high energy physics.

Given the above functional, we have to choose the minimization. We may choose for example to minimize with respect to the embedding alone. In this case the metric $(g_{\mu\nu})$ is treated as a parameter and may be fixed by hand. Another choice is to vary only with respect to the feature coordinates of the embedding space, or we may choose to vary the image metric as well. In [22] we show how different choices yield different flows. Some flows are recognized as existing methods, other choices are new and will be described below.

Using standard methods in variational calculus (see [22]), the Euler-Lagrange equations with respect to the embedding are:

$$X^i_t = -\frac{1}{2\sqrt{g}} h^{il} \frac{\delta S}{\delta X^l} = \frac{1}{\sqrt{g}} \partial_\mu (\sqrt{g} g^{\mu\nu} \partial_\nu X^i). \qquad (3)$$

The operator that is acting on X^i is the natural generalization of the Laplacian from flat spaces to manifolds and is called *the second order differential parameter of Beltrami* [10], or for short *Beltrami operator*, and is denoted by Δ_g. For the grey scale image case, the flow $I_t = \Delta_g I$, is *edge-preserving*. The generalization to any manifold embedded with arbitrary co-dimension is given by using Eq. 3 for all the embedding coordinates and the induced metric; see [22] for more details. In what follows we apply this operator to construct an orientation-preserving flow on texture images. But first let us look at the color image case more closely.

3 Color

We apply the Beltrami flow to the 5 dimensional space-feature needed in color images. The embedding space-feature space is taken to be Euclidean with Cartesian coordinate system. The image, thus, is the map $f : \Sigma \to \mathbb{R}^5$ where Σ is a two dimensional manifold. Explicitly the map is

$$f = \left(X^1(\sigma^1, \sigma^2), X^2(\sigma^1, \sigma^2), I^r(\sigma^1, \sigma^2), I^g(\sigma^1, \sigma^2), I^b(\sigma^1, \sigma^2)\right).$$

Note that there are obvious better selections to color space definition rather than the RGB flat space.

We minimize our action (2) with respect to (I^r, I^g, I^b). For convenience we denote below (r, g, b) by $(1, 2, 3)$, or in general notation i. The induced metric is given in this case as follows:

$$\begin{aligned} g_{11} &= 1 + (I_x^1)^2 + (I_x^2)^2 + (I_x^3)^2, \\ g_{12} &= I_x^1 I_y^1 + I_x^2 I_y^2 + I_x^3 I_y^3, \\ g_{22} &= 1 + (I_y^1)^2 + (I_y^2)^2 + (I_y^3)^2. \end{aligned} \quad (4)$$

The action functional under this choice of the metric is the Euler functional $S = \int d^2\sigma \sqrt{g}$. It is simply the area of the image surface. Minimization with respect to I^i gives the Beltrami flow

$$I_t^i = \frac{1}{\sqrt{g}} \partial_\mu (\sqrt{g} g^{\mu\nu} \partial_\nu I^i), \quad (5)$$

which is a flow towards a minimal surface that preserves edges. As an example, we show the result of color denoising in Fig. 1.

4 The Metric as a Structure Tensor

In [9, 11], Gabor considered an image enhancement procedure based on a single small time step along a directional flow. It is based on the anisotropic flow via the <u>inverse</u> second directional derivative in the 'edge' direction (∇I direction) and the geometric heat equation (second derivative in the direction parallel to the edge). The same idea of steering the diffusion direction motivated many recent works[4]. Cottet and Germain [6] used a smoothed version of the image to direct the diffusion, while Weickert [23] smoothed also the structure tensor $\nabla I \nabla I^T$ and then manipulated its eigenvalues to steer the smoothing direction. Eliminating one eigenvalue from a structure tensor $\sum_i \nabla I^i \nabla I^{iT}$, was used in in [21], in which the tensors are not necessarily positive definite. However, in [24], the eigenvalues are manipulated to result in a positive definite tensor.

Motivated by all of these results we will first link the anisotropic orientation diffusion (coherence enhancement) to the geometric framework, and then invert

---

[4] See [17] for many interesting extensions and applications of the locally isotropic flow.

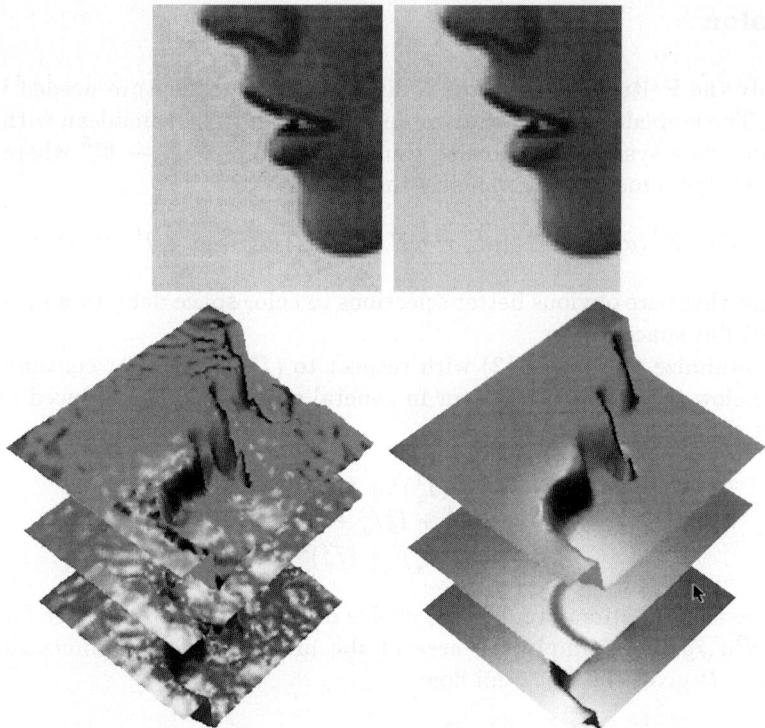

Fig. 1. Color results: The top row shows noisy image on the left and the denoised one on the right. To better depict the edge-preserving property of our method, in the bottom row we render as surfaces the three color channels of both the noisy and the reconstructed image.

the diffusion direction across the edge. Let us first show that the diffusion directions can be deduced from the smoothed metric coefficients $g_{\mu\nu}$ and may thus be included within the Beltrami framework under the right choice of directional diffusion coefficients.

The induced metric $(g_{\mu\nu})$ is a symmetric uniformly positive definite matrix that captures the geometry of the image surface. Let λ_1 and λ_2 be the largest and the smallest eigenvalues of $(g_{\mu\nu})$, respectively. Since $(g_{\mu\nu})$ is a symmetric positive matrix its corresponding eigenvectors u_1 and u_2 can be chosen orthonormal. Let $U \equiv (u_1|u_2)$, and $\Lambda \equiv \begin{pmatrix} \lambda_1 & 0 \\ 0 & \lambda_2 \end{pmatrix}$, then we readily have the equality $(g_{\mu\nu}) = U\Lambda U^T$. Note also that

$$(g^{\mu\nu}) \equiv (g_{\mu\nu})^{-1} = U\Lambda^{-1}U^T = U\begin{pmatrix} 1/\lambda_1 & 0 \\ 0 & 1/\lambda_2 \end{pmatrix} U^T, \qquad (6)$$

and that $g \equiv \det(g_{\mu\nu}) = \lambda_1\lambda_2$.

We will use the image metric in its natural geometric interpretation, i.e. as a structure tensor. The coherence enhancement Beltrami flow $I_t = \Delta_{\hat{g}} I$ for color-texture images is then given as follows:

1. Compute the metric coefficients $g_{\mu\nu}$. For the N channel case (for color $N = 3$) we have (see Eq. (4)) $g_{\mu\nu} = \delta_{\mu\nu} + \sum_{k=1}^{N} I_\mu^k I_\nu^k$.
2. Diffuse the $g_{\mu\nu}$ coefficients by convolving with a Gaussian of variance ρ, thereby $\tilde{g}_{\mu\nu} = G_\rho * g_{\mu\nu}$. For 2D images $G_\rho = e^{-(x^2+y^2)/\rho^2}$.
3. Change the eigenvalues, λ_1, λ_2, $\lambda_1 > \lambda_2$, of $(\tilde{g}_{\mu\nu})$ so that $\lambda_1 = \alpha^{-1}$ and $\lambda_2 = \alpha$, for some given positive scalar $\alpha \ll 1$. This yields a new metric $\hat{g}_{\mu\nu}$ that is given by: $(\hat{g}_{\mu\nu}) = \tilde{U} \begin{pmatrix} \alpha^{-1} & 0 \\ 0 & \alpha \end{pmatrix} \tilde{U}^T = \tilde{U} \Lambda_\alpha \tilde{U}^T$.
4. Evolve the k-th channel via Beltrami flow, that by the selection $\hat{g} \equiv \det(\hat{g}_{\mu\nu}) = \lambda_1 \lambda_2 = \alpha^{-1} \alpha = 1$ now reads

$$I_t^k = \Delta_{\hat{g}} I^k \equiv \frac{1}{\sqrt{\hat{g}}} \partial_\mu \sqrt{\hat{g}} \hat{g}^{\mu\nu} \partial_\nu I^k = \partial_\mu \hat{g}^{\mu\nu} \partial_\nu I^k$$
$$= \mathrm{div}\left(\tilde{U} \begin{pmatrix} \alpha & 0 \\ 0 & \alpha^{-1} \end{pmatrix} \tilde{U}^T \nabla I^k\right) = \mathrm{div}\left(\tilde{U} \Lambda_\alpha \tilde{U}^T \nabla I^k\right). \quad (7)$$

Note again that both for gray level and color images the above flow is similar to the coherence-enhancing anisotropic diffusion with the important property of a uniformly positive definite diffusion tensor. For color images, $(g_{\mu\nu}) = \mathcal{I} + \sum_i \nabla I^i \nabla I^{iT}$, where $\mathcal{I}$ is the identity matrix, and I^i are the color channels $((I^r, I^g, I^b) \equiv (I^1, I^2, I^3))$. In this case all that is done is the identity added to the structure tensors $\nabla I \nabla I^T$ for gray and $\sum_i \nabla I^i \nabla I^{iT}$ for color. This addition does not change the eigenvectors and thus the above flow is equivalent to Weickert schemes [23, 24]. Next, we introduce a new inverse/direct diffusion model.

4.1 Beyond a Metric: Inverse Diffusion Across the Edge

Let us take one step further, and exit our 'metric' framework by defining $(g_{\mu\nu})$ to be a non-singular symmetric matrix with one positive and one negative eigenvalues. That is, instead of a small diffusion we introduce a controlled inverse diffusion across the edge. Here we extend Gabor's idea [9, 11] of inverting the diffusion along the gradient direction.

Inverting the heat equation is an inherently unstable process. However, if we keep smoothing the metric coefficients, and apply the heat operator in the perpendicular direction we get a coherence-enhancing flow with sharper edges that is stable for a short duration of time. The idea is simply to change the sign of one of the modified eigenvalues in the algorithm described in the previous subsection. In other words, in step 3 of the previous scheme we change the eigenvalues of $(\tilde{g}_{\mu\nu})$ such that the largest eigenvalue λ_1 is now $\lambda_1 = -\alpha^{-1}$ and $\lambda_2 = \alpha$, for some given positive scalar $\alpha < 1$.

For the gray level case with $\rho = 0$ it simplifies to highly unstable inverse heat equation. However, as ρ increases the smoothing along the edges becomes fundamental and the scheme is similar in its spirit to that of [9]; also see [14, 1].

4.2 Color Orientation-Enhancing Results

In [23] the coherence enhancement flow was applied on several color masterpieces by van Gogh, which resulted in a 'coherence enhancement of expressionism'. In the next example we attempt to 'enhance and sharpen impressionism'. We apply first the anisotropic oriented diffusion flow and then the new oriented diffusion along/inverse diffusion across the edge on a color painting by Claude Monet, see Fig. 2.

Fig. 2. Top: Original picture "Femme à l'ombrelle tournée vers la gauche," by Claude Monet 1875 ("woman with umbrella turning left") 521 × 784 (left), the result of orientation-preserving diffusion (middle), and the result of inverse/direct diffusion flow ($\rho = 4$) (right).

References

1. L Alvarez and L Mazora. Signal and image restoration using shock filters and anisotropic diffusion. *SIAM J. Numer. Anal*, 31:590-605, 1994.
2. A Blake and A Zisserman. *Visual Reconstruction*. MIT Press, Cambridge, Massachusetts, 1987.
3. P Blomgren and T F Chan. Color TV: Total variation methods for restoration of vector valued images. CAM TR, UCLA 1996.
4. V Caselles, R Kimmel, and G Sapiro. Geodesic active contours. In *Proc. ICCV'95*, pages 694–699, Boston, Massachusetts, June 1995.

5. A Chambolle. Partial differential equations and image processing. In *Proc. IEEE ICIP*, Austin, Texas, November 1994.
6. G H Cottet and L Germain. *Image processing through reaction combined with nonlinear diffusion.* Math. Comp. Vol. 61, 659–673, 1993.
7. S Di Zenzo. A note on the gradient of a multi image. *Computer Vision, Graphics, and Image Processing*, 33:116–125, 1986.
8. A I El-Fallah, G E Ford, V R Algazi, and R R Estes. The invariance of edges and corners under mean curvature diffusions of images. In *Processing III SPIE*, volume 2421, pages 2–14, 1994.
9. D Gabor. Information theory in electron microscopy. *Laboratory Investigation*, 14(6):801–807, 1965.
10. E Kreyszing. *Differential Geometry.* Dover Publications, Inc., New York, 1991.
11. M Lindenbaum, M Fischer, and A M Bruckstein. On Gabor's contribution to image enhancement. *Pattern Recognition*, 27(1):1–8, 1994.
12. R Malladi and J A Sethian. Image processing: Flows under min/max curvature and mean curvature. *Graphical Models and Image Processing*, 58(2):127–141, March 1996.
13. D Mumford and J Shah. Boundary detection by minimizing functionals. In *Proc. of CVPR*, San Francisco, 1985.
14. S J Osher and L I Rudin. Feature–Oriented Image Enhancement Using Shock Filters. *SIAM J. Numer. Analy.*, 27(4):919-940, 1990.
15. P Perona and J Malik. Scale-space and edge detection using anisotropic diffusion. *IEEE-PAMI*, 12:629–639, 1990.
16. A M Polyakov. *Physics Letters*, 103B:207, 1981.
17. M Proesmans, E Pauwels, and L van Gool. Coupled geometry-driven diffusion equations for low level vision. In B M ter Haar Romeny, editor, *Geometric–Driven Diffusion in Computer Vision*. Kluwer Academic Publishers, The Netherlands, 1994.
18. Y Rubner and C Tomasi. Coalescing texture descriptors. In *Proc. of the ARPA Image Understanding Workshop*, Feb. 1996.
19. L Rudin, S Osher, and E Fatemi. Nonlinear total variation based noise removal algorithms. *Physica D*, 60:259–268, 1992.
20. G Sapiro. Vector-valued active contours. In *Proc. IEEE CVPR'96*, pages 680–685, 1996.
21. G Sapiro and D L Ringach. Anisotropic diffusion in color space. *IEEE Trans. Image Proc.*, 5:1582-1586, 1996.
22. N Sochen, R Kimmel, and R Malladi. A general framework for low level vision. in press: *IEEE Tran. on Image Processing*, 1997.
23. J Weickert. *Multiscale texture enhancement.* In *Computer analysis of images and patterns; Lecture Notes in Computer Science*, Vol. 970, Springer, pp. 230-237, 1995.
24. J Weickert. *Coherence-enhancing diffusion of colour images.* In Proc. VII National Symposium on Pattern Rec. and Image Analysis, Barcelona, Vol. 1, pp. 239-244, 1997.
25. R Whitaker and G Gerig. Vector-valued diffusion. In B M ter Haar Romeny, editor, *Geometric–Driven Diffusion in Computer Vision*. Kluwer Academic Publishers, The Netherlands, 1994.
26. S D Yanowitz and A M Bruckstein. A new method for image segmentation. *Computer Vision, Graphics, and Image Processing*, 46:82–95, 1989.
27. A. Yezzi. Modified curvature motion for image smoothing and enhancement. *IEEE Trans. IP*, to appear, 1997.

Efficient Contour Extraction in Color Images

Aldo Cumani

Istituto Elettrotecnico Nazionale "Galileo Ferraris"
Str. delle Cacce 91, I-10135 Torino, Italy

Abstract. An extension of second-order differential edge detection methods to color images was previously proposed. This work shows how, by casting the problem into a subpixel resolution oriented framework, an efficient algorithm for edge detection and contour traversal can be devised for both graylevel and color images. The results from an implementation on a standard PC show that the computational cost is largely dominated by convolution-type operations, whose impact can be substantially reduced by using specialized image processing devices.

1 Introduction

Edge detection methods based upon differential operators are widely used in the early processing of graylevel images. In particular, methods based upon the analysis of zero-crossings (ZC) of some second-order differential operator applied to image data have already been explored extensively (see e.g. [1, 2, 3]).

In contrast, differential methods have received little attention in the case of color images. The difficulty in extending differential methods to multi-band images stems from the fact that in the latter case the image function is vector-valued: after computing the gradients of the individual image components, there remains the problem of how to combine them into one meaningful output.

In [4], the author has proposed an extension of the second-directional derivative approach to color images. The main critical point of that approach is the inherent sign ambiguity in the definition of the direction of maximal contrast (the equivalent of the luminance gradient for graylevel images). That work has been criticised by some authors [5, 6] both for the use of costly computational techniques in the resolution of the sign ambiguity, and for the need to resort to subpixel techniques. In this connection, the aim of this paper is to show that:

- by casting the problem in the framework of subpixel-resolution contour detection, an efficient algorithm can be devised for both locating edge points and traversing connected edge chains;
- by a suitable reformulation, the computational cost of the sign disambiguation step can be kept quite low with respect to the rest of the algorithm.

The algorithm is introduced in the next section by first considering graylevel images. It is then extended to color images following the theory in [4]. Finally, some results from an implementation of the algorithm on a standard PC are shown. Even if the reported timings do not allow for real-time usage, their analysis indicates that the computational cost is largely dominated by convolution operations, whose impact can be substantially reduced when porting the algorithm to a specialized image processing device.

2 Edge Finding: Graylevel Images

Let us first consider the case of a graylevel image. A typical second-order differential contour extraction algorithm can be sketched as follows:

G1 Smooth the data $I(x,y)$, e.g by convolution with a Gaussian: $\mathcal{I} = \mathcal{G} \bullet I$
G2 Compute first and second order derivatives of the smoothed image $\mathcal{I}$ and combine them to obtain the second directional derivative D_I of the smoothed luminance in the direction of the gradient:

$$D_I = (\nabla \mathcal{I})^\top (\nabla \nabla^\top \mathcal{I})(\nabla \mathcal{I})/\|\nabla \mathcal{I}\|^2$$

G3 Find edge points by searching for transversal ZC (i.e. sign changes) of D_I and applying a suitable test for discriminating gradient maxima from minima (a threshold on gradient amplitude can also be applied at this step);
G4 Link the so found edge points to form connected chains (contour lines).

We now show that, even if subpixel resolution is not sought for, an analysis of the problem in view of subpixel-resolution contours can lead to an efficient algorithm for coping with both steps G3 and G4 above. To this end, let us consider the D_I values obtained at each pixel as point samples of the continuous function $D_I(x,y)$ at the nodes of a rectangular grid (Fig. 1). Let us focus on a particular grid mesh, having as corners four adjacent pixels. As D_I is only known at the mesh corners, we assume that its value inside and on the mesh sides is well approximated by a bilinear formula:

$$D_I(x+u, y+v) \simeq (1-u)(1-v)D_I(x,y) + (1-u)vD_I(x, y+1) + \\ u(1-v)D_I(x+1, y) + uvD_I(x+1, y+1) \qquad (1)$$

where (x,y) is the location of corner c_0 and $0 \le u \le 1$, $0 \le v \le 1$.

Although somewhat arbitrary, the bilinear approximation has the distinct advantage that on each side of the mesh it reduces to *linear* interpolation between the values at the side's endpoints. Therefore, when the latter differ in sign, a ZC point is easily found on the considered side, and its location does not depend on which of the two adjacent meshes sharing this side is considered. Let us now consider the signs of D_I at the four mesh corners. Excluding, for the sake of simplicity, that D_I be exactly zero at any of these four pixels, there are 16 distinct possible cases, summarized in Table 1. Note that:

- the "code" (case) number (CD) is simply the 4-bit number formed by the sign bits of D_I at the corners, read in row-major order (LSB first);
- except for codes 0,6,9 and 15, all other codes correspond to a single contour line crossing two sides of the mesh, which cut an edge element (edgel) from that line; with the above numbering, edgels with the same endsides but opposite direction correspond to complementary (one's complement) codes;

- it is easy to see that for any two adjacent meshes, if one mesh has an edgel with its "head" on the common side, the other must have an edgel "tail" on the same side (and vice versa).
- the special cases 0 and 15 correspond to no ZC (hence no edgel). In cases 6 and 9 there are four ZC, implying at least 2 edgels within the mesh; taking for good the bilinear formula, a simple analysis [8] shows that the edgels structure can be disambiguated by a criterion depending only on the D_I values at corners (see Table 2). Note anyway that, under fairly general smoothness assumptions, the two ZC lines within the mesh cannot both be loci of maximal gradient, so that the application of the maximum/nonmaximum discrimination test leads again to one of the single-edgel cases. A rough (but effective) criterion consists in comparing the gradient direction at the mesh top left corner to the "average" gradient direction for that mesh code, i.e. the clockwise normal to the "typical" edgel shown in Table 1; this test only needs the dot product of the gradient with the typical gradient direction obtained from a 16-entry lookup table.

The above observations show that the sign coding scheme of Table 1, after double-edgel reduction and non-maximum suppression, constitutes a natural *chain code* for traversing ZC lines. For a given mesh, the "next mesh" function is easily realized by feeding its code to a 16-entry lookup table yielding the $\Delta x, \Delta y$ steps relative to the current position; a zero code signals the end of the chain.

The contour following algorithm can therefore be specialized as follows:

G4.1 Build a "code image" array C with $C[y, x]$ equal to the 4-bit code of the mesh with top left corner at (x, y);

G4.2 Scan the code array. For every non-zero code, start a new contour:

G4.2.1 Find the contour head by traversing it backwards (with suitable provisions to avoid looping in case the contour is closed);

G4.2.2 Scan the contour by traversing it forwards, outputting edgels, and zeroing traversed code cells.

Finally, notice that subpixel computations (i.e. the location of the ZC on the mesh sides) are confined to the "outputting edgels" substep of step G4.2.2, so they can be avoided if one is not actually interested in subpixel results.

3 Edge Finding: Color Images

We follow, with some modifications, the theory outlined in [4]. An RGB color image is a 3-component vector field $f(x, y)$ over the image plane. We suppose that the components f_i of f (the R, G and B channels) have been smoothed by Gaussian filtering, as the strong regularizing property of Gaussian filters ensures the existence and continuity of the derivatives of $f_i(x, y)$ of any order [1].

The *squared local contrast* of f at a point $P = (x, y)$ in the direction of the unit vector $u = (u_1, u_2)$ is the quantity

$$S(P, u) = u^\top \Gamma u = E u_1^2 + 2 F u_1 u_2 + G u_2^2 \qquad (2)$$

with the shorthand notation

$$\Gamma = \begin{pmatrix} E & F \\ F & G \end{pmatrix} = \nabla f (\nabla f)^T, \quad E = \frac{\partial f}{\partial x} \frac{\partial f}{\partial x}, \quad F = \frac{\partial f}{\partial x} \frac{\partial f}{\partial y}, \quad G = \frac{\partial f}{\partial y} \frac{\partial f}{\partial y}$$

The extrema of (2) are the eigenvalues of Γ and are attained when u is the corresponding eigenvector. In particular, the maximum of (2) is

$$\lambda = (E + G + \sqrt{(E-G)^2 + 4F^2})/2 \qquad (3)$$

and the corresponding eigenvector n can be computed as follows:

$$v = (v_1, v_2) = (E - G, 2F) \qquad (4)$$

$$c = v_1/\sqrt{v_1^2 + v_2^2}, \quad n_1 = \sqrt{\frac{1+c}{2}}, \quad n_2 = \sqrt{\frac{1-c}{2}} \mathrm{sign}(v_2) \qquad (5)$$

The main difference with [4] is in the more efficient computation of n through (5) (the original formulation used trigonometric functions). Now, the contour points are defined as the loci where (3) has a local maximum along the direction n. This leads to considering transversal ZC of the directional derivative of (3):

$$D_S(P) = \nabla \lambda \cdot n = E_x n_1^3 + (2F_x + E_y) n_1^2 n_2 + (G_x + 2F_y) n_1 n_2^2 + G_y n_2^3 \qquad (6)$$

where x and y subscripts indicate derivation with respect to x and y, respectively. The required derivatives of E, F and G can be easily expressed in terms of first and second derivatives of the smoothed image components f_i (see e.g. [6]).

Note that a straightforward application of the algorithm of the preceding section is not possible, due to the intrinsic sign ambiguity of n. Indeed, since (2) is quadratic, the squared contrast defined by (2) is the same in direction n and in the opposite $-n$. But since D_S is *cubic* in n, the choice of sign is not immaterial. Some authors [5] have claimed that this ambiguity can be removed by computing n with a procedure similar to the one proposed above (although their formulas are actually different); in the author's opinion, this claim has no real foundation, as the sign ambiguity is intrinsic in the definition (2) of contrast.

In [4] it is shown that a consistent choice of sign can (theoretically) be effected in any region where the vector v as defined by (4) has no zeros, or along any closed line along which the vector n rotates by an integer multiple of 2π; in particular, this rotation is zero if the region enclosed by the line does not contain zeros of v. The latter suggests to use a "minimal rotation" criterion for fixing the signs of D_S around a given mesh by pretending that the angle between the directions of n at any two adjacent corners be less than $\pi/2$. More specifically:

- let $s_i \epsilon \{-1, 1\}, i = 0..3$ be the chosen sign at corner c_i, so that after correction the maximal contrast directions shall be $s_i n_i$; fix $s_0 = 1$.
- traversing the mesh clockwise, check the rotation between the (sign corrected) n at c_i and the (raw) n at c_{i+1}. This is accomplished by testing the sign of $d = s_i n_i \cdot n_{i+1}$: if it's negative, then $s_{i+1} = -1$, else $s_{i+1} = 1$;

- if at the end we find $s_4 = -1$, this means that a minimal rotation choice of signs is not possible, so we set the mesh code to zero (to indicate that the mesh is not to be processed further).

Note that reversing the sign of n implies reversing that of D_S; but the (re-ordered) sequence of signs s_i makes a 4-bit corrective value whose bitwise XOR with the raw mesh code yields the code for the mesh with the corrected signs for D_S. The choice (5) for the raw definition of n allows further savings: in (5) n_1 is definitely nonnegative, so if on a given mesh v_2 has the same sign on all four corners, n on that mesh is confined to either the 1st or 4th quadrant of the (n_1, n_2) plane. This allows to restrict the rotation test to only those meshes where v_2 exhibits sign changes; the latter are easily found by 4-corner encoding the sign of v_2 in a manner analogous to that used for the sign of D_S.

Finally, notice that the need for a per-mesh correction of D_S signs also affects the chain traversal step. Indeed, let us consider the common side of two adjacent meshes. If both meshes pass the sign test, then the less-than-$\pi/2$ criterion ensures that the existence and location of a ZC on the common side is preserved when passing from one mesh to the other, but there is no guarantee that the corrected signs at the side's endpoints be the same when seen from either mesh. This means that the edgels corresponding to the sign-corrected codes may show head-head or tail-tail collision on the common side. This drawback can be easily overcome by augmenting the traversal algorithm with a little memory: knowing from which side the mesh has been entered, it's a trivial task to check whether the mesh code yields an edgel starting from that side, and in case of failure one only needs to complement the code to obtain the correct edgel direction.

4 Test Results

In order both to test the correctness of the above reasonings, and to gather some insights on the relative impact of each step on the overall computational cost, the above algorithm has been implemented in C on a standard PC equipped with a Pentium™ processor running at 166 MHz. Gaussian smoothing has been implemented by sequential row/column 1-dimensional convolutions, exploiting the separability of the Gaussian; derivatives have been estimated from the filtered data by convolution with 3x3 Beaudet masks [7] as proposed in [4].

Figure 2 shows the contours found on two 256x256 test images (note that no contrast amplitude threshold has been applied, so that most of the edge lines in low-contrast areas are only garbage). Table 3 shows the corresponding timings, split between the various algorithm steps. Even if the timings reported here are still far from allowing real-time usage, some remarks are in order.

First, all computations were done in double precision floating-point. From the standpoint of an implementation on a single general-purpose CPU, the use of limited-precision fixed-point arithmetic may or may not improve things (some tests on the Pentium™ have shown even worse performance with 32 bit integers instead of double floats). This situation changes radically when porting the algorithm to a specialized processing device: note that the timings in Table 3 are largely dominated by convolution operations (Gaussian filtering and computation of derivatives), and the the cost of these operation does not depend upon the image contents, but only on its size (and on the scale of the Gaussian for what concerns smoothing). Such filtering operations can be performed quite efficiently by a suitable pipeline processor such as e.g. the convolution unit on the Imaging Technology ITEX 150/40 series image processors.

The timing relative to the computation of D_S from the smoothed data has been split into the gradient calculations proper and the evaluation of n and D_S, that involves some nonlinear operations (row labeled "SQRT"). In a dedicated implementation, the latter could be sped up by using lookup tables.

The row labeled "encoding" comprises sign disambiguation, and the reported figures show that its contribution to overall cost is not so heavy. Finally, the row labeled "traversal" (which includes subpixel ZC localization) confirms our claim that the proposed schema allows for fast contour tracking; also note that the latter step is the only one that cannot be implemented by a pipeline processor, as a contour line may traverse pixels located anywhere in the image.

Acknowledgments

The author wishes to thank his colleagues of the Gruppo di Visione Artificiale of IEN and CNR for helpful discussions. A particular thank to A. Guiducci and P. Grattoni, who first suggested the use of sign-encoding in [8].

References

1. Torre, V., Poggio, T.: On edge detection. IEEE Trans. Patt. Anal. Mach. Intell. **PAMI-8**, 2 (1986) 147-163.
2. Canny, J.: A computational approach to edge detection. IEEE Trans. Patt. Anal. Mach. Intell. **PAMI-8**, 6 (1986) 679-698.
3. Clark, J.J.: Authenticating edges produced by zero crossing algorithms. IEEE Trans. Patt. Anal. Mach. Intell. **PAMI-11**, 1 (1989) 43-57.
4. Cumani, A.: Edge detection in multispectral images. CVGIP: Graphical Mod. and Image Process. **53** (1991) 40-51.
5. Alshatti, W., Lambert, P.: Using eigenvectors of a vector field for deriving a second directional derivative operator for color images. Proceedings, 5th Int. Conf. Comp. Analysis of Images and Patterns CAIP'93 (1993) 149-156.
6. Koschan, A.: A comparative study on color edge detection. Proceedings, 2nd Asian Conf. on Computer Vision ACCV'95 (1995) 574-578.
7. Beaudet, P.: Rotationally invariant image operators. Proceedings, Int. Joint Conf. on Pattern Recognition (1987) 579-583.
8. Grattoni, P., Guiducci, A.: Contour coding for image description. Pattern Recognition Letters **11** (1990) 95-105.

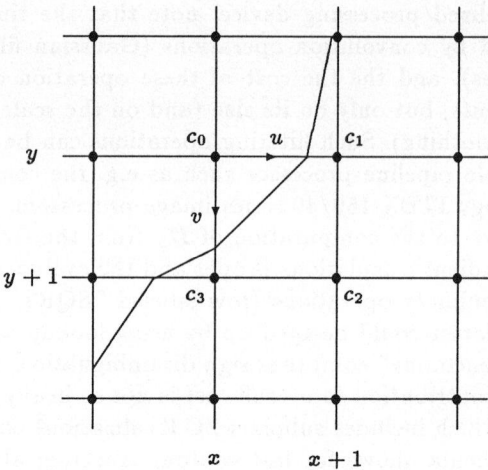

Fig. 1. Subpixel contour lines over the pixel grid.

Fig. 2. Two examples of application of the contour finding algorithm. Left to right: 1) the average $(R+G+B)/3$ of the three smoothed channels; 2) the found contours; 3) the same, encoded by log of contrast amplitude (smoothing scale $\sigma = 1$ for both images).

Table 1. Encoding of edgel structure within a 4-pel mesh from the signs of D_I at the mesh corners. CD: mesh code; SB: sign bits of D_I; ED: edgel structure within the mesh (see text for special marks).

CD	SB	ED	CD	SB	ED	CD	SB	ED	CD	SB	ED
0	00 00	•	1	10 00	↗	2	01 00	↘	3	11 00	→
4	00 10	↙	5	10 10	↑	6	01 10	(II)	7	11 10	↗
8	00 01	↗	9	10 01	(II)	10	01 01	↓	11	11 01	↘
12	00 11	←	13	10 11	↖	14	01 11	↗	15	11 11	•

Table 2. Resolving double-edgel mesh codes. $\Delta = a_0 a_2 - a_1 a_3$, where a_i is the value of D_I at the corner c_i.

code	Δ	edgels	code	Δ	edgels
6	+	2,4	6	−	7,14
9	+	11,13	9	−	1,8

Table 3. Timings of algorithm steps for some examples. The cases WOMAN/1 and SHIPS/1 are those displayed in Fig. 1.

	image/σ of Gaussian		WOMAN/0.5	WOMAN/1	WOMAN/2	SHIPS/1
image complexity	# contours		4638	2232	746	2763
	# points		23172	17438	9605	21664
timings (in ms)	smoothing		346	572	976	572
	D_S	gradients	647	656	662	665
		SQRT	208	213	216	217
	encoding		113	95	75	109
	traversal		91	70	40	84

Color Edge Detection Using Orthogonal Polynomials

R.Krishnamoorthi and P. Bhattacharyya
Deptt. of Computer Science and Engineering
I.I.T, Kharagpur, India

Abstract

An orthogonal polynomials based maximizing signal-to-noise ratio (SNR) scheme for edge detection in 2-D color image is proposed in this paper. The proposed framework takes into account not only the spatial interaction within each of the three color planes (R, G, B) but also the interaction between the color planes. The edge detected output by the proposed scheme is also compared with two existing color edge detection schemes.

1 Introduction

An edge in color image is defined by discontinuity in a three dimensional color space. The well known R, G, B color space has been chosen here in which the discontinuity can be assumed to be in the form of Euclidean distance. Traditionally, color edge detection has been treated as a simple extension of the monochrome edge detection in separable color space. The use of the Hueckel operator in the luminance-chrominance color space has been proposed in [1] and an application of the compass gradient edge operators to color images has been highlighted in [2] . However, the existing color edge detection methods are based on extension of the well known gradient based techniques proposed for monochrome images. In these methods [3], edges are computed in three chosen color components separately and are merged finally by using some specified procedures. However, this approach may be very unsatisfactory in certain cases where the image gradients show the same strength but in opposite direction. Moreover, these edge detection methods are mainly based on a single criterion, namely, maximization of amplitude response of some derivative operators to the image. However, a better criterion, namely, "maximization of the operators' responses towards edges compared to the operators' responses towards noise" that has been proposed in [4] for monochrome images is difficult to extend for composite color images. Consequently, the problem of color edge detection in vector space has been proposed [5]. Here, color images are treated as vector fields and an edge detection method has been suggested. The "entropy operator" as edge detector is proposed in [6] . Here the entropy of brightness is defined in a local region in the picture. For color images, entropy of brightness as well as color is defined in the local region.

In this paper, a computational framework for facilitating color edge detection, in RGB color space is presented. Using a set of orthogonal polynomials difference operators are configured and employed to represent a color image region. A color edge detector based on maximizing SNR is then devised in presence of Guassian noise. The proposed color edge detector is also compared experimentally with two existing color edge detection schemes.

2 Orthogonal Polynomials Based Framework

Since an edge can be detected based on the local properties of the image, a local point-spread operator is required to be devised such that it is a cartezian as well as color coordinate separable and deblurring operator. The three dimensional point-spread function $M(x, y, z)$ can be considered to be a real valued function defined for $(x, y, z) \in X \times Y \times Z$ where X, Y and Z are ordered subsets of real values. In the case of a color image of size $(n \times n \times n)$ where X (rows) consists of a finite set, which for convenience can be labeled as $\{0, 1, \ldots, n-1\}$, the functions $M(x, y, z)$ reduces to a sequence of functions

$$M(i, t) = u_i(t), i = 0, 1, \ldots, n-1 \tag{1}$$

As shown in equation 2 the process of image analysis can be viewed as the linear transformation defined by the point-spread operator $M(x, y)(M(i, t) = u_i(t))$,

$$\beta'(\zeta, \eta, s) = \int_{x \in X} \int_{y \in Y} \int_{z \in Z} M(\zeta, x) M(\eta, y) M(s, z) I(x, y, z) dx dy dz \tag{2}$$

where ζ, η, s are coordinates in the 3-D transformed space and $I(x, y, z)$ is a color image region wherein x and y are two spatial coordinates and z indicates the color space. Considering both X, Y and Z to be finite set of values $\{0, 1, 2 \ldots, n-1\}$ equation 2 can be written in matrix notation as follows

$$|\beta'_{ijk}| = (|M| \otimes |M| \otimes |M|)^t |I| \tag{3}$$

where the point-spread operator $|M|$ is

$$|M| = \begin{vmatrix} u_0(t_1) & u_1(t_1) & \ldots & u_{n-1}(t_1) \\ u_0(t_2) & u_1(t_2) & \ldots & u_{n-1}(t_2) \\ & & \vdots & \\ u_0(t_n) & u_1(t_n) & \ldots & u_{n-1}(t_n) \end{vmatrix} \tag{4}$$

$\otimes$ is the outer product and $|\beta'_{ijk}|$ be the n^3 matrices arranged in the dictionary sequence. $|I|$ is the image and $|\beta'_{ijk}|$ be the coefficients of transformation. We consider a set of orthogonal polynomials $u_o(t), u_1(t), \ldots, u_{n-1}(t)$ of degrees 0, 1, 2, ..., n-1, respectively. The generating formula for the polynomials is as follows.

$$u_{i+1}(t) = (t - \mu) u_i(t) - b_i(n) u_{i-1}(t) \; for \; i \geq 1, \tag{5}$$

$$u_1(t) = t - \mu, \quad \text{and} \quad u_0(t) = 1,$$

where

$$b_i(n) = \frac{<u_i, u_i>}{<u_{i-1}, u_{i-1}>} = \frac{\sum_{t=1}^{n} u_i^2(t)}{\sum_{t=1}^{n} u_{i-1}^2(t)}$$

and

$$\mu = \frac{1}{n}\sum_{j=1}^{n} t_j$$

Considering the range of values of t to be $t_j = j, j = 1, 2, 3, ..., n$, we get

$$b_i(n) = \frac{i^2(n^2 - i^2)}{4(4i^2 - 1)}, \quad \mu = \frac{1}{n}\sum_{j=1}^{n} t_j = \frac{n+1}{2}$$

Next, we construct point-spread operators $|M|$ s of different width from the above orthogonal polynomials using equation 4.

2.1 Complete set of difference operators for color images

In case of R-G-B color space, the elements of finite set X, Y and Z are labeled as $\{1, 2, 3\}$. The point-spread operator in (4) that defines the linear transformation of color images can be obtained as $|M| \otimes |M| \otimes |M|$. The set of 27 three dimensional basis operators $O_{ijk}, (0 \leq i, j, k \leq 2)$ can be computed as follows.

$$O_{ijk} = \hat{u}_i \otimes \hat{u}_j \otimes \hat{u}_k$$

where $\hat{u}_i$ is the (i+1) st column vector of $|M|$. It can be shown easily that O_{ijk}s (except O_{000}, because β_{ooo} is the DC component) are symmetric finite difference operators. β'_{ijk}s are the coefficients of the linear transformations defined as follows.

$$|\beta'_{ijk}| = |\mathcal{M}|^t |I| \tag{6}$$

where $|\mathcal{M}|$ is the 3-D point-spread operator defined as $|\mathcal{M}| = |M| \otimes |M| \otimes |M|$. Now it can be shown easily that the orthogonal transformation (6) defined by the orthogonal system $|\mathcal{M}|$ is complete.

2.2 Responses Towards Color Edge and Noise

The coefficients $|\varsigma|$ of the complete 3-D orthogonal transformation defined by the set of polynomials are mean squared amplitude responses per unit volume of the n^3 (n = 3) basis operators. In general, $|\beta'|$ ($= |\mathcal{M}|^t|I|$) are mean squared amplitude responses of the operators, where

$$|\varsigma| = (|\mathcal{M}|^t|\mathcal{M}|)^{-\frac{1}{2}}|\beta'| \tag{7}$$

Division of each component of $|\beta'|$ by the corresponding element of $(|\mathcal{M}|^t|\mathcal{M}|)^{-\frac{1}{2}}$ results in unity noise gain during convolution with each of the 26 finite difference operators O_{ijk}s. In the presence of random noise only each of the 26 $|\varsigma|$ values which gives an estimate of the standard deviation of a component is the same as the standard deviation of the noise.

3 Separation of Responses towards color edges from responses towards noise

In presence of additive noise the proposed polynomials based model for the color image is

$$I(x,y,z) = \sum_{i=0}^{n-1}\sum_{j=0}^{n-1}\sum_{k=0}^{n-1} \beta_{ijk}\, u_i(x)u_j(y)u_k(z) + \eta(x,y,z) \qquad (8)$$

where the ηs are the additive noise which are normally distributed with zero mean and constant variance σ^2 and are uncorrelated. Since ς_{ijk}^2s are the corresponding mean squares of β_{ijk}s, each ς_{ijk}^2 is a $\chi^2\sigma^2$ variate with one degree of freedom. In other words, due to completeness criterion, the local variance of the color image region is decomposed into the $(n^3 - 1)$ number of non-negative quadratic forms ς_{ijk}^2 as follows.

$$\sum_{i=0}^{n-1}\sum_{j=0}^{n-1}\sum_{k=0}^{n-1}(I_{ijk} - \beta_{000})^2 = \sum_{i=0}^{n-1}\sum_{j=0}^{n-1}\sum_{k=0}^{n-1}\varsigma_{ijk}^2 \; at\; not(i=j=k=0) \qquad (9)$$

where β_{000} is the local mean $\left(\frac{1}{n^3}\sum_{i=0}^{n-1}\sum_{j=0}^{n-1}\sum_{k=0}^{n-1} I_{ijk}\right)$.

Without any ambiguity ς_{ijk}^2 may be termed as variances corresponding to the responses of the last $(n^3 - 1)$ basis, difference operators. Our next task is to group these $(n^3 - 1)$ variances into two disjoint subsets, say ψ_s and ψ_e as the estimates of variances corresponding to the responses towards edges and noise, respectively.

The desired grouping can be accomplished by observing the fact that the polynomial finite difference operators $O_{01}^2, O_{10}^2, O_{01}^3, O_{10}^3$ (the superscript denotes the width of the polynomial basis operator) can be used to represent some of the widely known enhancement/thresholding edge operators for monochrome images. In general, the set of 3-D polynomials based finite difference operators $\{O_{0j0}\}|_{0<j\leq n-1} \cup \{O_{i00}\}|_{0<i\leq n-1} \cup \{O_{00k}\}|_{0<k\leq 2}$ may be considered as the color edge operators. Thus the set of mean squared amplitude responses $\{\varsigma_{0j0}\}_{0<j\leq n-1} \cup \{\varsigma_{i00}\}_{0<i\leq n-1} \cup \{\varsigma_{00k}\}_{0<k\leq 2}$ of these color edge operators are basically the responses towards edges and the responses of the remaining operators are the responses towards noise.

3.1 The statistical Grouping criterion for Color edges

In order to frame the desired grouping criterion, we introduce here some statistical concepts and terminology. In statistics, a linear contrast A in the given input data sample $X(=\{x_1, x_2, ..., x_k\})$ is defined numerically shown as $A = \sum_{i=1}^{k}\lambda_i x_i$ where λ_is are some values such that $\sum_{i=1}^{k}\lambda_i = 0$. Suppose B is another linear contrast in X such that $B = \sum_{i=1}^{k}\mu_i x_i$ where $\sum_{i=1}^{k}\mu_i = 0$. Two contrasts A and B are said to be uncorrelated if $\hat{\lambda}(=\{\lambda_1, \lambda_2, \lambda_3, ...\lambda_k\})$

and $\hat{\mu}(=\{\mu_1,\mu_2,\mu_3,...,\mu_k\})$ are orthogonal. Now, it can be shown easily that uncorrelated linear contrasts are unbiased statistical estimates. The physical significance of which is that their effects are mutually independent to each other. Furthermore, if $A' = A/(\sum_i^k \lambda_i^2)^{\frac{1}{2}}$ then A' may be considered to be a linear contrast per unit length.

In case of the proposed polynomials based transformation the ς_{ijk}s (except ς_{000}), which are the mean squared amplitude responses of the finite difference operators O_{ijk}s (except O_{000}), can be shown as uncorrelated linear contrasts per unit length. ς_{ijk} may be called the linear contrast in $|I|$ due to jointly the ith order finite difference along x axis, j^{th} order finite difference along y axis, and k^{th} order finite difference along the color coordinate. ς_{i00} is termed as the linear contrast due to i^{th} order finite difference along x axis only, whereas ς_{0j0} is the linear contrast due to the jth order finite difference along y axis only and ς_{00k} is the linear contrast due to the kth order finite difference along the color coordinate. As ς_{ijk}s are unbiased statistical estimates and ς_{ijk}^2 are estimates of variances corresponding to the proposed finite difference operators' amplitude responses the latter can be considered to be unbiased statistical estimates of variances. In fact each ς_{ij}^2 is a $\chi^2\sigma^2$ variate with one degree of freedom. Since $\{\varsigma_{i00}\} \cup \{\varsigma_{0j0}\} \cup \{\varsigma_{00k}\}$ are the responses towards color edges, $\psi_s = \{\varsigma_{i00}^2\} \cup \{\varsigma_{0j0}^2\} \cup \{\varsigma_{00k}^2\}$ are the unbiased statistical estimates of edge response and $\psi_e = \{\varsigma_{ijk}^2, 0 < i \leq n-1, 0 < j \leq n-1, 0 < k \leq 2\}$ are the unbiased statistical estimates for the noise present.

In order to ensure that a set of $\chi^2\sigma^2$ variates with known degrees of freedom are basically the estimates of the same noise variance, Nair's test for homogeneity among variances [7] is used. The significance of the responses towards color edges compared to noise has then been measured by performing another statistical test viz. F-ratio test [8], after computing the mean square error variance, $\bar{\eta}_o^2$.

4 Edge detection by maximizing Signal-to-Noise Ratio

Algorithm

Input : Color image with three components R, G and B of size 256 * 256 * 3
Output : Edge detected image of size (256 * 256)

Step 1 If(end of image), go to step 8.
       Else extract a small color image region [I] of size (n*n*3).

Step 2 Compute the mean squared amplitude responses ς_{ijk} as follows:
$$|\varsigma_{ijk}| = (|\mathcal{M}|^t|\mathcal{M}|)^{-\frac{1}{2}}|\mathcal{M}|^t|I_{ijk}|$$

Step 3 Apply Nair's test to determine whether the variance $\varsigma_{ijk}^2 \in \psi_e$ are estimates of the same variance.

If (yes) compute the mean squared error variance $\bar{\eta}_0^2$ from them. Else go to step 1.

Step 4 Repeat the following for the set ψ_s of all the mean squared amplitude responses.

   1. Perform variance ratio test (F test) with the error variance $\bar{\eta}_0^2$ as the denominator.
   2. Add those mean squared variances for which the null hypothesis is rejected.

Step 5 Compute the RMS value of the sum of the mean squares obtained at step 4.

Step 6 If the RMS value $\geq$ a threshold T then the presence of an edge is detected at the center position of $|I|$
Else go to step 1.

Step 7 Go to step 1.

Step 8 Stop.

5 Experiments and Results

The orthogonal polynomials based framework for color edge detection has been experimented with different untextured color images. One such original untextured color image viz. bird image, is shown in figure 1. The presence of color edges is detected based on the fact that out of 26 mean-squared amplitude responses six, namely, $\varsigma_{001}, \varsigma_{002}, \varsigma_{010}, \varsigma_{020}, \varsigma_{100}, \varsigma_{200}$ are responses towards color edges whereas the remaining twenty mean squared amplitude responses $\{\varsigma_{ijk} | \varsigma_{ijk} \in \psi_e\}$ are responses towards noise. This has been verified by conducting experiments with a large number of randomly selected color image samples. The edge detection method using maximization of SNR has been applied on this image as stated in the previous section and the edge detected output is shown in figure 2. The edge detected output image by the proposed orthogonal polynomials based color edge detection scheme is also compared with (i) Color edge detection using vector order statistics [5] and (ii) Edge extraction using entropy operator [6] which are shown in figure 3 and 4 respectively. From these outputs, it is evident that the proposed color edge detection scheme shows either the same or better performance (for example see neck and head portions) than the other two color edge detection schemes.

6 Conclusion

An orthogonal polynomials based framework for color edge detection in RGB color space is presented in this paper. The framework takes into account not

Figure 3: Edge Detection Using Vector Order Statistics Scheme

Figure 1: Original Bird Image

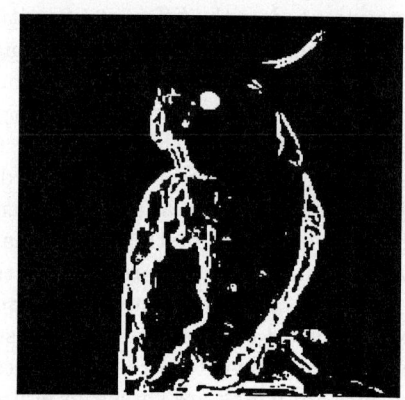

Figure 2: Edge Detection Using Max. SNR Scheme

Figure 4: Edge Detection Using Entropy Scheme

only the spatial interaction within each of the three color planes but also the interaction between different planes. The proposed framework is based on a complete set of difference operators which are easily configurable from a set of orthogonal polynomials. The operators are employed to represent a color image region as a linear combination of the operator's responses towards color edge and noise. A simple statistical design of experiments paradigm is used for separating out the responses towards color edge from that of noise. The framework supports in devising a color edge detector based on maximizing signal-to-noise ratio (SNR). The proposed color edge detector is experimented on various untextured color images and are compared with two existing color edge detecting schemes.

References

[1] R. Nevatia, "A color edge detector and its use in scene segmentation," *IEEE Transactions on Systems, Man and Cybernatics*, vol. 7, no. 11, pp. 820–826, 1977.

[2] G. Robinson, "Color edge detection," *Optical Engineering*, vol. 16, pp. 479 – 484, 1977.

[3] S. D. Zenzo, "A note on the gradient of a multi image," *Computer Graphics and Image Processing*, vol. 33, pp. 116–125, 1986.

[4] J. Canny, "A computational approach to edge detection," *IEEE Transactions on Pattern Analysis and Machine Intelligence*, vol. 8, no. 6, pp. 679 – 698, 1986.

[5] P. E. Trahnias and A. N. Venetsanopoulos, "Color edge detection using vector order statistics," *IEEE Transactions on Image Processing*, vol. 2, no. 2, pp. 259 – 264, 1993.

[6] A. Shiozaki, "Edge extraction using entropy operator," *Computer Vision Graphics and Image Processing*, vol. 36, pp. 1 – 9, 1986.

[7] D. J. Bishop and U. S. Nair, "A note on certain methods of testing for homogeneity of a set of estimated variances," *Biometrika*, vol. 30, pp. 89 – 99, 1939.

[8] R. A. Fisher and F. Yates, *Statistical Tables for Biological, Agricultural and Medical Research*. Oliver and Boyd, London, 1947.

Fast and Robust Segmentation of Natural Color Scenes

Volker Rehrmann and Lutz Priese

Image Recognition Lab, University of Koblenz-Landau,
Rheinau 1, 56075 Koblenz, Germany

Abstract. We present a fast and robust system for color-based segmentation. The system is based on hierarchical region-growing on a special hexagonal topology. In contrast to common region-growing techniques it is independent of the starting point and the order of processing. It is generally applicable in natural color scenes and algorithmically efficient. The use of local and global information and a new color similarity measure contribute to the robust segmentation results. The system is successfully applied in two difficult applications from the field of autonomous vehicle guidance.

1 Introduction

One of the most important tasks of an image analysis system is image segmentation, the identification of homogeneous regions in an image. In the literature several methods for segmentation are distinguished. Common are *edge detection, split and merge, region growing* and *clustering* techniques. Most of the extensive research on image segmentation in the last three decades has been done for gray scale images. However, as the technical equipment for color image acquisition becomes cheaper and more common, color image analysis becomes more and more important. Nearly all techniques for gray scale image segmentation have been transferred to color images. A survey on color image segmentation can be found in [SK94].

Most papers on color segmentation follow the clustering method. Here the pixels are mapped to feature vectors in a feature space. Now statistical methods are applied to find some clusters in this feature space. These clusters, re-mapped to the image, form the color segments. A well-known clustering technique is recursive histogram splitting ([Oht85]), applied by many researchers ([Cel90]). The advantage of clustering methods is the global view of the data often in form of histograms. However, although histograms provide a global view of the feature data, they do not represent the spatial information of the underlying image. The extension of clusters in feature space is often ambiguous and the statistical methods trying to solve this problem are computationally expensive.

Region growing techniques start with initial cells, pixels or small regions and let them grow by sequential merging with neighbored, similar regions. The pure local methods tend to chaining mismatches by merging differently colored segments. The centroid-linkage techniques are sequential methods and are therefore dependent on the choice of starting point and the order in which the pixels are processed.

We tried to develop a new method combining the advantages of local (simplicity and fastness) and global (robustness and accuracy) techniques. It is a hierarchical region growing method that is inherently parallel and therefore independent of the choice of the starting point and the order of processing. It uses local and global information and achieves very robust segmentation results in natural color scenes which is also explained by the use of a newly developed color similarity measure. Our idea was first published in [PR93]. Since then a lot of improvements have been applied. This paper describes our entire color segmentation system, called *CSC* (Color Structure Code). Further details can be found in [PR97]. In section 2 we introduce the hexagonal, hierarchical island structure on which our method is based. Section 3 describes the actual segmentation method. In Section 4 the new color similarity measure is presented. Section 5 discusses the complexity of our approach. Finally we present some results and conclusions in section 6.

2   Hexagonal, Hierarchical Island Structure

Our segmentation method follows a hierarchical region growing on a special hexagonal topology (firstly introduced by Hartmann [Har87]). This hierarchical topology is formed by so–called *islands* of different levels. One island of level *0* consists of seven neighbored pixels in the hexagonal topology. The partition of the image is organised in such a way that the islands are overlapping. One island of level *n+1* consists of seven overlapping islands of level *n* (s. Fig. 1). Repeating this until one island covers the whole image, the number of islands decreases from level to level by a factor 4. Operating on a hexagonal topology

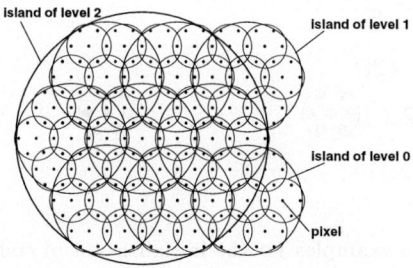

Fig. 1. The hexagonal, hierarchical island structure.

leads to some difficulties in practice. Nearly all imaging devices scan the pixels in an orthogonal scheme. In order to avoid additional efforts we use the hexagonal island hierarchy only as a logical structure on an orthogonal raster (s. [PR97]).

3 Color Structure Code

The generation of the *CSC* operates essentially in four phases. In the *preprocessing phase* noise suppression is accomplished by the use of a nonlinear filter which, in addition, strengthens the sharpness of contours. We investigated different filters with the desired properties and selected the fastest of them: a variant of the symmetric nearest neighbor filter (s. [PR97]). In an *initialization phase* the image is partitioned into small, atomar color regions within an island of level 0. These small color regions are growing in the *linking phase* in a hierarchical manner to complete regions. Within the linking phase it is possible to detect that color regions connected by a chain of smoothly changing colors have to be split again. This is done in the *splitting phase*.

3.1 Initialization Phase

In the initialization phase color homogeneous regions in level 0 islands of seven pixels are detected and mapped to initial code elements (s. Fig. 2). Such an initial code element consists of those pixels of level 0 islands that are neighbored and whose mutual color distance lies below a certain threshold. A code element is a data structure describing color regions within an island. In Figure 2 two examples are shown: A homogeneous island resulting in one code element and another island, where an edge goes through it, resulting in two code elements that describe two differently colored small regions in the island. Hence, a code element of level 0 describes a small colored region within an island of level 0. Note, this operation is a pure local operation within one island, whose processing can be done independently for each island. Instead of starting with one seed pixel, the CSC starts concurrently in all islands of the image. The result of the initialization phase is a set of code elements, each one describing a small color patch. In the following linking phase these small color patches are checked for continuity and grow hierarchically to complete, connected color segments.

Fig. 2. Two examples for the initialization of code elements.

3.2 Linking Phase

In the linking phase code elements of level n are linked to new code elements of level $n+1$ in seven, neighbored overlapping islands of the hexagonal island structure (s. Fig. 3). Code elements will be linked if the regions represented by them are connected and similar in color. The connectivity of code elements can easily be determined within the hexagonal island structure: two code elements are connected if they share a common subregion in their common sub island. On level *1* this simply means that they share a common pixel. The linking operations are repeated for all islands on every level, starting from level 1 and ending on the topmost level, where only one island covers the whole image (level 8 with 512x512 pixel images). By repeated linking those code elements form a code tree. A new code element c on level n is stored together with pointers to the code elements on level $n-1$ from which c was formed (s. Fig. 3), the so called *sub code elements*. Code elements that do not find any partner for linking on some level n form the root of such a code tree. Thus, a connected homogeneous region is represented by a tree in our *CSC* data structure. The larger a region is the higher is its root level in the hierarchical data structure. The root contains raw information about the size, location, and mean color of a region. More details can be obtained by descending the tree.

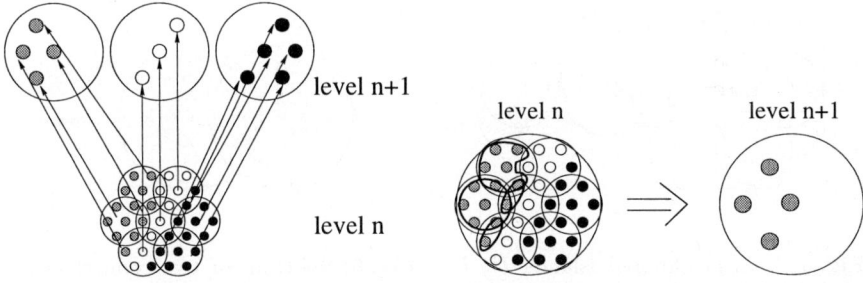

Fig. 3. An example for the linking of code elements. The opposite Figure 4 shows a zoom on part of this example.

Fig. 4. The four marked overlapping code elements of level n form a new code element on level $n+1$.

The linking of code elements within one island is similar to the operation in the initialization phase. Instead of linking single pixels, regions are linked. Again all operations within one island can be done independently of the other islands. The segmentation results are not depending on the order of execution. All small color regions of the initialization phase are growing concurrently within one level. The overlapping of the islands leads to efficient connectivity checks of code elements. The hexagonal island structure assures that the regions are growing in all directions in contrast to quad-tree-like structures.

3.3 Splitting Phase

A typical error in local region growing techniques is the linking of differently colored regions due to a chain of connected pixels with smoothly changing colors. Segmentation algorithms that only use local information are unable to detect region boundaries with low contrast. A good segmentation algorithm has to use local and global information. We solve this problem by additional color similarity checks between connected code elements on every linking level. If the color distance lies above a certain threshold the two code elements won't be linked although they are connected by a chain of color similar pixels. Note, two connected regions in the CSC structure are always connected by a chain of color similar pixels. Consider the example in Figure 5. If the color distance between r_1 and r_2 is too high, they won't be linked, although all their subregions on level $n - 1$ are locally homogeneous. It is the global view at this level that makes it possible to detect the smooth transition of one color to the other. The fact that r_1 and r_2 won't be linked results in two different complete segments. Due to the overlapping structure they possess a common subregion, thus $r_1 \cap r_2 \neq \emptyset$. Therefore r_1 and r_2 have to be explicitly separated, which means the common subregion has to be partitioned between r_1 and r_2.

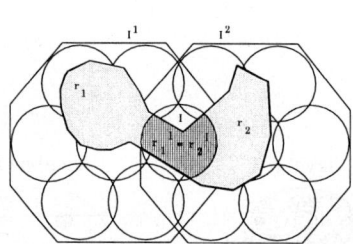
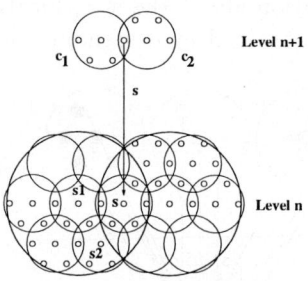

Fig. 5. Two neighbored islands I^1, I^2 with their common sub island I and the common subregion r_1^I of r_1 and r_2.

Fig. 6. Splitting of two code elements c_1 and c_2.

Consider the general case of Figure 6. Two connected code elements c_1 and c_2 are not color similar. They possess the common sub code element s. s has to be partitioned between c_1 and c_2. At first, the color value of s is compared with those of c_1 and c_2. s is assigned to that code element that is closer in color which means s has to be deleted from the other code element, say c_1. This is simply a pointer deletion. This does not mean that the whole region represented by s is assigned to c_2. This may not be an accurate border between c_1 and c_2. Although s has been deleted from c_1 it is still connected with the region represented by c_1 via s_1 and s_2 due to the overlapping structure. Now, s has to be separated from s_1 and s_2. This is done in a recursive procedure in the same way as with c_1 and c_2. The only difference is that the color values to compare with are those of c_1 and c_2 as they represent the most global information. When splitting s and s_1 it is possible that their common subregion is assigned to s_1. That is why not necessarily the entire region of s is assigned to c_2. s can loose some subregions in the recursive descent. The recursion stops at level 0. With this simple and elegant recursive algorithm (main operation is deletion of pointers) very accurate borders can be found. Some care has to be taken that the connectivity of code elements doesn't get lost (s. [PR97]).

4 Color Similarity

The color similarity measure is of particular importance for the quality of the segmentation results. The color similarity measure in the CSC is a color predicate D. Given two colors c_1, c_2 in a three-dimensional color space, D is defined by:

$$D(c_1, c_2) = \begin{cases} true & : \quad c_1 \text{ and } c_2 \text{ are color similar} \\ false & : \quad otherwise \end{cases}$$

Most published color similarity predicates are measures calculating the ratio of the mean colors distances and their variances. The color features are usually RGB, recently CIELAB and CIELUV. The measures are more oriented on statistical properties than on human color sensations. As those measures often don't correspond to human judgement, we developed a new color predicate in the HSV color space. In the HSV color model a color is described by the three attributes hue, saturation and value. A description of HSV and a conversion from and to RGB can be found e.g. in [FvDFH90]. Perez and Koch ([PK94]) discussed the advantages and disadvantages of color spaces using hue, saturation and intensity. The main advantage is that the hue value remains constant if the intensity of illumination changes or if the saturation of a color is decreasing. These advantages make the hue coordinate so valuable in the segmentation of natural scenes, where illumination can't be controlled and is often changing. The drawback of the HSV space is an unremovable singularity at the V axis, where R=G=B (saturation = 0). At low intensities and at low saturations the hue value is very unstable. Because of these problems many researchers recommended other color spaces, but we think it is worth using the HSV space and taking special consideration of the drawbacks. It is clear from this fact that the HSV-space is not suited for Euclidean distance measures. It is impossible to use a constant threshold over the entire HSV color space to decide about the similarity of colors. With well saturated, bright colors the hue value is an excellent discrimination feature, while bigger differences in saturation and intensity can be tolerated to become invariant against the variations in illumination. On the other hand, the hue value is useless with unsaturated and dark colors as it is either undefined or very unstable. In this range of colors the most important feature is intensity. From color metric we know that the sensed hue difference becomes lower as the saturation and intensities are decreasing. To imitate this human color sensation the allowable thresholds for the color similarity predicate have to be chosen dependent on the color location in HSV space. Our realization of this idea is to use a table of color thresholds. The valid thresholds are determined depending on the saturation and intensity of the actually analyzed colors. Therefore we quantized saturation and intensity into 16 steps. For each of the 16x16 different locations in color space we developed thresholds for hue, saturation and value. The thresholds have been empirically determined in a large number of experiments. A graphical representation of the table entries for the three color attributes is shown in Figure 7. The access to the table and the definition of the color predicate using the tables are as follows: D is defined by:

$D(h_1, s_1, v_1, h_2, s_2, v_2) = \text{TRUE} \Leftrightarrow$
$|h_1 - h_2| \leq \text{hue\_thresh}$ with hue_thresh = huetab[min(s_1, s_2)][max(v_1, v_2)] and
$|s_1 - s_2| \leq \text{sat\_thresh}$ with sat_thresh = sattab[min(s_1, s_2)][max(v_1, v_2)] and
$|v_1 - v_2| \leq \text{val\_thresh}$ with val_thresh = valtab[min(s_1, s_2)][max(v_1, v_2)]

Of course, the values for saturation and value have to be shifted into the range 0..15 before accessing the table. In the calculation of the absolute hue difference the special modular arithmetic ($360° \equiv 0°$) must be considered.

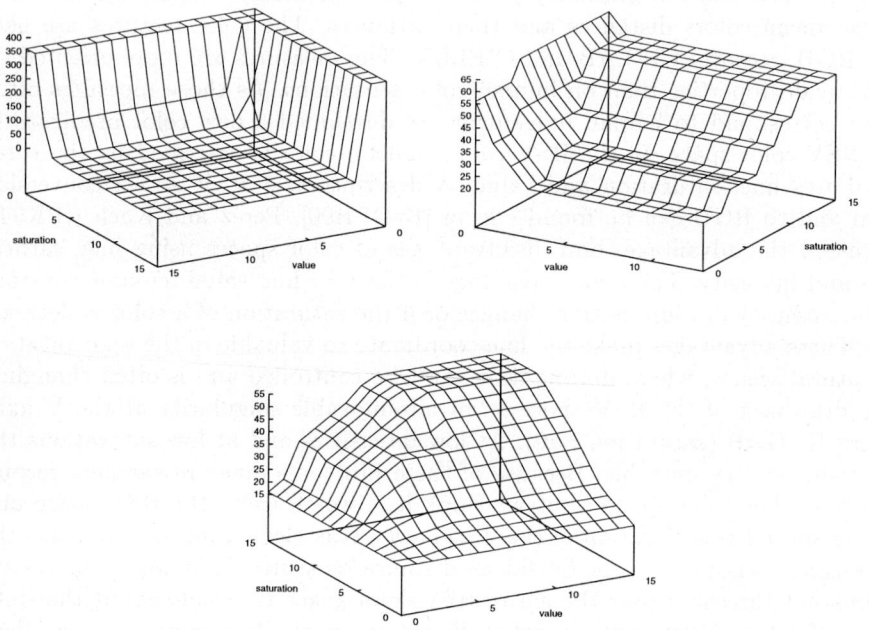

Fig. 7. A graphical representation of the tables of thresholds (hue, sataturation and value from left to right).

Note, that the viewpoints are different for the three tables in Figure 7. They have been changed to better visualize the shape of the particular thresholds. It can be seen that the hue thresholds for the upper quarter of the table (saturation $>$ 7, value $>$ 7) are nearly constant. The thresholds are changing when approaching low saturation and intensity. The extremal value for the hue threshold is reached when s=0 or v=0, resulting in a hue threshold of 360 which simply means ignoring hue. The thresholds for saturation are increasing from low saturated to high saturated colors with the exception of very dark colors (black), where due to noise the saturation can vary a lot. The thresholds for value are low for low saturated and dark colors and increase with increasing saturation and intensity. These tables of thresholds work very well for natural color scenes.

5 Complexity

Let us briefly discuss the complexity of our approach. Let N denote the number of pixels. Then N times the number of operations for the processing of one pixel (6 color distance calculations in snn-filtering) is the complexity for the preprocessing phase (s. [PR97]). The two main operations in the initialization and linking phase are the comparison of color similarity and the check of connectivity of two color regions. In the initialization phase all neighbored pixels have to be compared resulting in $\frac{N \cdot 6}{2} = 3 \cdot N$ color similarity comparisons. The connectivity check is not necessary as the connectivity of pixels is given by the image data structure. The number of color similarity comparisons in the linking phase depends on the number of color regions in each island. As the number of islands decreases from level to level by a factor of 4 the number of color regions (code elements) on average also decreases by a factor 4. This is exactly the case with a totally homogeneous image where in every island on every level exactly one code element appears. The more different color regions appear in the image the more code elements appear on the lower levels. But as the regions are smaller they are not linked to the higher levels and hence there are less code elements on higher levels. These considerations are experimentally verified stating that on average the number of code elements in each level corresponds approximately to the number of islands in each level (plus a small offset). The total number of color similarity comparisons including the initialization phase for an image with less than 512x512 pixels (less than 8 levels) is therefore about

$$3 \cdot N \cdot \sum_{n=0}^{8} (\frac{1}{4})^n \approx 3 \cdot N \cdot \frac{4}{3} = 4 \cdot N$$

which is a very good result for a technique exploiting global information.

Color regions in neighbored islands are not necessarily connected. Therefore their connectivity has to be checked. From the same argumentation as above it follows that the number of connectivity checks in the linking phase is about N. To be connected two code elements must have a common subregion. This can be checked in constant time. In our implementation it is simply a check whether the two code elements have at least one identical pointer to their sub code elements. Note, that it is the hexagonal, overlapping island structure allowing for this efficient connectivity check. In a non overlapping structure the connectivity has to be checked at the pixel level which is an expensive task for large regions.

In general, the time required for the preprocessing, initialization and linking phase is nearly constant and is not much depending on the input image. This is not the case for the splitting phase. The cost of one splitting call (separation of two code elements) depends on the actual linking level and the length of the common border. The number of all splitting calls depends on the nature of the input image. In a typical natural color scene the time of the splitting phase can be neglected (less than 5% of the total processing time) because the number of splitting calls is small.

The overall runtime depends as well on appropriate implementation techniques. As all operations are local operations which are performed many times, it is worthwhile to optimize these operations by standard programming tricks (e.g. use tables instead of calculations). The runtime on a SUN ULTRA SPARC I with 167 MHz including color space conversion is on average 700 msec for 512x512 images, and 180msec for 256x256 images.

6 Conclusions

As it is still too expensive to print color images in the proceedings we refer to [PR97] for examples of segmentation results. Up to now the CSC proved to be a reliable scene segmenter in two larger applications. The first application is the recognition of traffic signs from a moving car. The traffic sign recognition system is based on the CSC and was successfully integrated into a prototype of an autonomous vehicle by Daimler-Benz ([PKL+94]). The system operates close to real-time with excellent recognition rates. The second application is a new research project using color segments as features in the analysis of motion in natural color image sequences ([RR97]). The efficient CSC segmentation allows for the processing of color image sequences and the stability of color segments along with elaborate matching techniques leads to promising results in difficult tasks like tracking of objects in color outdoor scenes and motion segmentation.

References

[Cel90] M. Celenk. A color clustering technique for image segmentation. *Computer Vision, Graphics, and Image Processing*, 52:145–170, 1990.

[FvDFH90] J. D. Foley, A. van Dam, S. K. Feiner, and J. F. Hughes. *Computer Graphics: principles and practice*. Addison Wesley, second edition, 1990.

[Har87] G. Hartmann. Recognition of Hierarchically Encoded Images by Technical and Biological Systems. *Biological Cybernetics*, 57:73–84, 1987.

[Oht85] Y. Ohta. *Knowlege-based Interpretation of Outdoor Natural Color Scenes*. Pitman Advanced Publishing Program, Boston, Massachusetts, 1985.

[PK94] F. Perez and C. Koch. Toward color image segmentation in analog VLSI: Algorithm and Hardware. *Intern. Journal of Computer Vision*, 12(1):17–42, 1994.

[PKL+94] L. Priese, J. Klieber, R. Lakmann, V. Rehrmann, and R. Schian. New Results on Traffic Sign Recognition. In *Proceedings of the Intelligent Vehicles Symposium*, pages 249–254. IEEE, 1994. Paris, Oct. 24-26.

[PR97] V. Rehrmann and L. Priese. Fast and Robust Segmentation of Natural Color Scenes. Technical Report, Computer Science Department, University of Koblenz-Landau, 1997. http://www.uni-koblenz.de/~lb/lb_publications.e.html.

[PR93] L. Priese and V. Rehrmann. A Fast Hybrid Color Segmentation Method. In S. J. Pöppl and H. Handels, editors, *Mustererkennung 1993*, pages 297–304. Springer Verlag, 1993. 15. DAGM-Symposium, Lübeck, 27.-29.Sept. 1993.

[RR97] V. Rehrmann and M. Rothhaar. Detection and Tracking of Moving Objects in Color Outdoor Scenes. In *30th ISATA: Dedicated Conference on Robotics, Motion and Machine Vision in the Automotive Industry*, 1997. Florence, 16-19th June 1997.

[SK94] W. Skarbek and A. Koschan. Color Image Segmentation - A Survey -. Technical report 94-32, Computer Science Department, TU Berlin, 1994.

Segmentation and Tracking Using Colour Mixture Models

Yogesh Raja, Stephen J. McKenna and Shaogang Gong

Dept. Computer Science, Queen Mary and Westfield College, London.
E-mail: jpmetal@dcs.qmw.ac.uk

Abstract. A system is described that provides robust and real-time focus-of-attention for tracking and segmentation of multi-coloured objects. Gaussian mixture models were used to estimate the probability densities of object foreground and scene background colours. Tracking was performed by fitting dynamic bounding boxes to image regions of maximum probability. Two scenarios are presented: (1) real-time face tracking based upon a skin colour model and (2) dynamic body segmentation for virtual studios based upon combined foreground and background models.

1 Introduction

This work was initially motivated by a requirement in the broadcasting industry for an effective method for segmenting moving people from image sequences in order to perform superimposition onto virtual studios. Currently, the state of the art in virtual studio superimposition involves the use of "chroma-keying" techniques which perform colour-based segmentation to replace blue regions in an image with an alternative image. These techniques require painstaking preparation of a studio so that the background is entirely covered in blue material. Care is also taken to ensure that actors have no blue colours about their appearance. A new system for performing the task of segmentation without the need for this preparatory effort is desired and the work presented here offers a contribution to this end. This work also has implications for areas such as teleconferencing, vision-based man-machine interfaces and face recognition.

Colour has been used in machine-based vision systems for tasks such as segmentation [9] and recognition [2, 4, 10]. Colour cues have been shown to offer several significant advantages over geometric information for certain tasks in visual perception, such as robustness under partial occlusion, rotation in depth, scale changes and resolution changes [10]. Furthermore, colour processing can often utilise efficient algorithms yielding real-time performance on standard hardware.

The techniques presented here use colour as a cue for object localisation, segmentation and tracking. Multi-coloured objects are modelled using colour mixtures which estimate probability density functions in colour space. Whilst a single colour, e.g. skin tone, can be adequately modelled as a Gaussian distribution, multiple colours can be modelled using a mixture of Gaussians. Additionally, modelling scene background enables classification of pixels as object or background by computing posterior probabilities.

Two scenarios are described in this work. Firstly, detection and tracking of human faces was performed using a face colour model. A relatively simple colour model was used and real-time performance was obtained on a standard PC platform. Secondly, the more difficult task of segmenting humans from video sequences was considered for a virtual studio application. More complex, scene-specific colour models incorporating both human body foreground and scene background were used. Pixels were classified as foreground (object) or background by computing posterior probabilties from the two mixture densities. Background pixels were subsequently replaced with an alternative background.

The remaining parts of this paper are organised as follows. Colour modelling is discussed in Section 2 with emphasis on the use of Gaussian mixtures. Section 3 desribes a colour tracking framework. Section 4 presents the human face tracking application. Body tracking and segmentation for virtual studios are presented in Section 5. Finally, Section 6 gives conclusions and future work.

2 Statistical Colour Mixture Models

A major difficulty with using colour cues in machine vision is the *colour constancy* problem which arises due to variation in colour values brought about by lighting changes. This is particularly apparent in RGB (red, green, blue) space. Intensity is distributed throughout all three parameters, rendering colour values highly sensitive to scene brightness. A simple approach to colour constancy is to use the HSV colour space which consists of hue angle (H), colour saturation (S) and brightness (V). In order to obtain a limited level of intensity invariance, colours can be modelled in HS-space.

2.1 Modelling the Foreground

Colour histograms [10] are a simple non-parametric method for modelling. In a histogram, the density at a point in a colour space quantised into n bins is approximated by the fraction of pixels which fall into the corresponding bin. If n is too large, the estimated density will be "noisy" and many bins will be empty. If n is too small then the distribution's structure will be "smoothed" away. The use of histograms for estimating colour densities is only possible because n can be kept relatively small and because there are many data points (pixels) available. A potentially more effective "semi-parametric" technique for colour density estimation is the use of Gaussian mixture models. The conditional density for a pixel, ξ, belonging to an object $\mathcal{O}$ is modelled as a mixture with m component densities:

$$p(\xi|\mathcal{O}) = \sum_{j=1}^{m} p(\xi|j) P(j) \qquad (1)$$

where a mixing parameter $P(j)$ corresponds to the prior probability that pixel ξ was generated by component j and where $\sum_{j=1}^{m} P(j) = 1$. Each mixture component is a Gaussian with mean μ and covariance matrix Σ, i.e. in the case of a 2D colour space:

$$p(\xi|j) = \frac{1}{2\pi|\Sigma_j|^{\frac{1}{2}}} e^{-\frac{1}{2}(\xi-\mu_j)^T \Sigma_j^{-1}(\xi-\mu_j)} \qquad (2)$$

Expectation-Maximisation (EM) provides an effective maximum-likelihood algorithm for fitting such a mixture to a data set [1, 8]. Fig. 1 shows an example of a Gaussian mixture model of a multi-coloured object in HS-space.

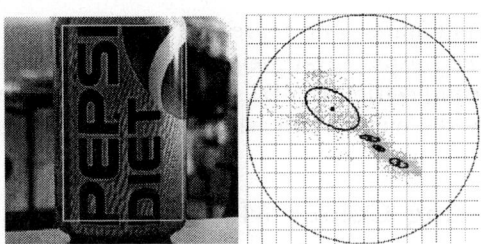

Fig. 1. A multi-coloured object and its Gaussian mixture model in HS-space. The mixture components are shown as elliptical contours of equal probability.

Outlier points, which can be caused by image noise and specular highlights, have little influence upon the mixture model. Once a model has been learned it can be converted into a look-up table for efficient on-line indexing of colour probabilities.

2.2 Modelling the Background

In virtual studios, it is desirable to model the colour distribution of the background scene in addition to the objects to be tracked. Given density estimates for both the object, $\mathcal{O}$, and the background scene, $\mathcal{S}$, the probability that a pixel, ξ, belongs to the object is given by the posterior probability $P(\mathcal{O}|\xi)$:

$$P(\mathcal{O}|\xi) = \frac{p(\xi|\mathcal{O})P(\mathcal{O})}{p(\xi|\mathcal{O})P(\mathcal{O}) + p(\xi|\mathcal{S})P(\mathcal{S})} \qquad (3)$$

The probability of misclassifying a pixel is minimised by classifying it as the class with the greatest posterior probability. A pixel is therefore classified as object foreground if and only if $P(\mathcal{O}|\xi) > 0.5$. The prior probability, $P(\mathcal{O})$, was set to reflect the expected size of the object within the search area of the scene $[P(\mathcal{S}) = 1 - P(\mathcal{O})]$. This approach has the advantage that object and scene models can be acquired independently. In the virtual studio scenario, this means that a single background scene model can be acquired and subsequently used with many different people.

Alternatively, a single combined colour distribution can be estimated using both object and background data. If a Gaussian mixture model is used, the Gaussian components can be treated as basis functions which form the hidden layer in a neural network. The output layer of this network is trained to classify data as either object or background. This is a form of Hyper Basis Function (HyperBF) network [6]. The kth output unit computes a function $f_k(\xi)$:

$$f_k(\xi) = \sum_{j=1}^{m} w_j p(\xi|j) \qquad (4)$$

There are two output units: one representing the object class and the other the background. The output layer weights w_j can be determined by applying Singular Value Decomposition (SVD) [7]. Each input pixel is assigned to the class represented by the output unit with the highest activation.

3 Tracking Using Colour Models

The tracking dynamics involve estimating the position, width and height of the object. This box provides a focus of attention for further processing. The position and size of the box are found by computing the mean $\boldsymbol{\mu}^t = (\mu_x, \mu_y)$ and standard deviation $\boldsymbol{\sigma}^t = (\sigma_x, \sigma_y)$ of the local colour probability distribution within a rectangular search area centred on $\boldsymbol{\mu}^{t-1}$ in the image domain at time t. The dimensions of this search area are determined by scaling the dimensions of the bounding box at time $t - 1$. The experiments presented in this paper were performed with search areas $\frac{3}{2}$ times the height and width of the bounding box.

For a given time frame t, the box position $\boldsymbol{\mu}^t$ is estimated as an offset from the position $\boldsymbol{\mu}^{t-1}$:

$$\boldsymbol{\mu}^t = \boldsymbol{\mu}^{t-1} + \frac{\sum_{\mathbf{x}} p(\boldsymbol{\xi}_{\mathbf{x}})(\mathbf{x} - \boldsymbol{\mu}^{t-1})}{\sum_{\mathbf{x}} p(\boldsymbol{\xi}_{\mathbf{x}})} \qquad (5)$$

where $\mathbf{x}$ ranges over all image coordinates in the region of interest and $\boldsymbol{\xi}_{\mathbf{x}}$ is the HS colour vector at image position $\mathbf{x}$. To improve accuracy, probabilities $p(\boldsymbol{\xi}_{\mathbf{x}})$ are thresholded. Probabilities lower than the threshold are taken to be background and are consequently set to zero in order to nullify their influence on $\boldsymbol{\mu}^t$ and $\boldsymbol{\sigma}^t$. The size of the bounding box is estimated by computing the standard deviation of the image probability density:

$$\sigma^t = \sqrt{\frac{\sum_{\mathbf{x}} p(\boldsymbol{\xi}_{\mathbf{x}})\{(\mathbf{x} - \boldsymbol{\mu}^{t-1}) - \boldsymbol{\mu}^t\}^2}{\sum_{\mathbf{x}} p(\boldsymbol{\xi}_{\mathbf{x}})}} \qquad (6)$$

In the next two sections, the above colour model was applied to both face and human body tracking/segmentation. Face tracking requires only a simple foreground model whereas body tracking/segmentation benefits from the use of combined foreground and background models.

4 Face Tracking Using a Skin Colour Model

Human face tracking has a wealth of possible applications such as security, teleconferencing and human-computer interfaces. A fast and robust method for isolating faces for subsequent recognition is a prerequisite for a fully automated face recognition system. The tracking technique presented here provides a useful component for such systems [5].

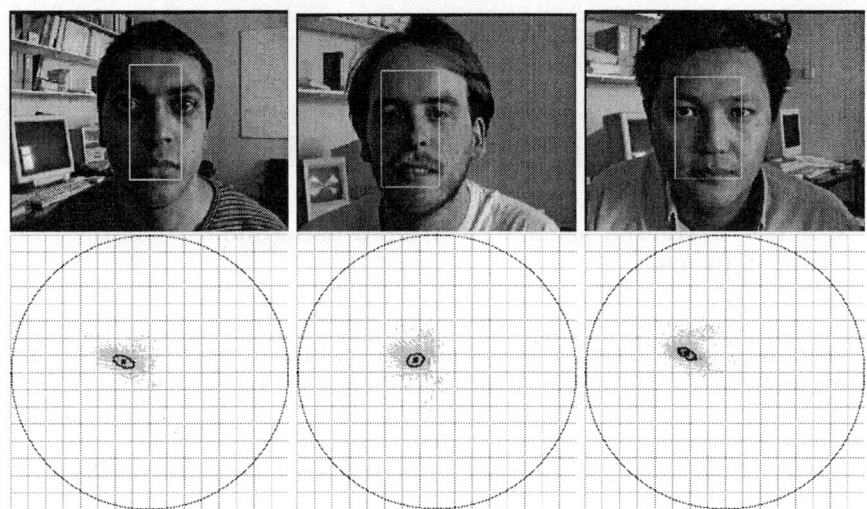

Fig. 2. The tight clustering of skin colour for three types of skin colour is illustrated here. The top row shows the face regions used to build the mixture models. The bottom row illustrates the colour distributions in HS-space.

Skin colour was modelled by collecting colour samples from images of faces. The face colours occupied a relatively compact area of HS-space (e.g. as found by [3]) and, in fact, a single Gaussian with full covariance was often sufficient to model the distribution of skin colour (Fig. 2). Fig. 3 shows a sequence of a face being tracked with a moving camera against a cluttered background. The tracker's ability to deal with changes in scale, large rotations in depth and partial occlusion are all clearly demonstrated.

This tracking system was implemented on a standard PC with a 200MHz Pentium processor, a Matrox Meteor colour frame grabber and a Sony EVI-D31 active camera with pan/tilt actuators and a zoom lens. The active camera can be driven by maintaining the mean position of the image probability distribution at the centre of the image. The tracking process is performed at approximately 15 frames per second. Tracking is robust without the use of temporal prediction. However, a recursive filter such as a Kalman filter might yield some improvement in performance and in particular help prevent the tracker "jumping" from one face to another. Problems are inevitably caused by large changes in the spectral

Fig. 3. A face is tracked against a cluttered background while the camera pans, tilts and zooms.

composition of scene illumination. In particular, it has been found necessary to use two models, one for interior lighting and one for exterior natural daylight.

5 Multi-coloured Object Tracking and Segmentation

In this section, colour mixture models are used to track multi-coloured objects. If a multi-coloured object consists of several distinct and differently coloured patches, then it may be beneficial to decompose the object and model each patch using a separate colour model (see e.g. [4]). Furthermore, if such colour patches are relatively homogenous, each can be directly modelled using a single Gaussian thus avoiding the need for the iterative EM algorithm. However, many objects cannot be thus decomposed and their colours are better modelled using a mixture. [1]

An example of tracking performance is shown in Fig. 4, where a soft drinks can was located and tracked robustly under changing background, scale, rotation in depth and occlusion. Only the colour distribution of the object was modelled.

The distributions in colour space formed by multicoloured objects are multi-modal and can span wide areas of the colour space. Thresholding probabilities generated by a foreground model alone is often ineffective due to severe overlap between background and foreground colour distributions. The third image in Fig. 4 illustrates this problem. The bounding box is overly large as a result of colours similar to the can lying within the search space. In many situations, however, the colour distribution of the background scene can also be modelled. Fig. 5 shows an example in which both the person and the background have been modelled. Pixels were classified as object or background using equation (3) with

---

[1] The number of Gaussian components to use is currently determined empirically.

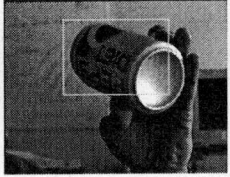

Fig. 4. Using only a foreground Gaussian mixture model, tracking was robust against a cluttered background with changing camera position and zoom, differing orientations of the object and large occlusions. Colours similar to the can on the bookshelf in the third image lie within the search space and consequently contribute to the estimation of the bounding box size.

the prior probabilities set to $P(\mathcal{S}) = P(\mathcal{O}) = 0.5$. A multi-resolution approach was taken in which segmentation was performed in a coarse-to-fine manner. Once the position and size of the bounding box had been estimated, superimposition of the object onto an alternative background sequence was performed. Only pixels inside the search area of the tracker were classified. All pixels outside this area were rendered as background.

6 Conclusions and Further Work

A general framework was presented for modelling the colour distributions of multi-coloured objects using Gaussian mixtures and for using these models to perform tracking and segmentation. Expectation-Maximisation provided an effective algorithm for training the mixtures. The method has been shown to work consistently in two quite different scenarios. Firstly, real-time face tracking was performed using a simple foreground colour model which was robust under changing camera position and zoom. Secondly, body tracking and segmentation were performed by combining foreground and background colour models for use in a virtual studio application.

Pixel-wise classification based only upon colour information is obviously insufficient in order to guarantee perfect segmentation. However, it is often surprisingly effective. In addition, the image of posterior probabilities provides a rich source of information and could be combined with other visual processes. To this end, current work is being done to incorporate shape constraints into the system to exploit the results obtained from the segmentation technique.

References

1. C. Bishop. *Neural Networks for Pattern Recognition.* Oxford University Press, 1995.
2. G.D. Finlayson. Colour object recognition. Master's thesis, Simon Fraser Univ., 1992.

Fig. 5. Segmentation results. The top row shows three images from a sequence. The second row illustrates the segmentation accuracy (performed using a multi-resolution approach). The third row shows the reconstructed (superimposed) sequence.

3. M. Hunke and A. Waibel. Face locating and tracking for human-computer interaction. In *28th Asilomar Conf. on Signals, Systems and Computers*, 1994.
4. J. Matas, R. Marik, and J. Kittler. On representation and matching of multi-coloured objects. In *IEEE ICCV*, pages 726–732, 1995.
5. S. J. McKenna, S. Gong, and Y. Raja. Face recognition in dynamic scenes. In *BMVC*, 1997.
6. T. Poggio and F. Girosi. Networks for approximation and learning. *Proceedings of The IEEE*, 78(9), September 1990.
7. W.H. Press, S.A. Teukolsky, W.T. Vetterling, and B.P. Flannery. *Numerical Recipes in C*. Cambridge University Press, 1992.
8. R. A. Redner and H. F. Walker. Mixture densities, maximum likelihood and the em algorithm. *SIAM Review*, 26(2):195–239, 1984.
9. W. Skarbek and A. Koschan. Colour image segmentation - a survey. Technical report, Technical University of Berlin, 1994.
10. M. J. Swain and D. H. Ballard. Colour indexing. *IJCV*, pages 11–32, 1991.

Object Tracking Using Adaptive Colour Mixture Models*

Stephen J. McKenna, Yogesh Raja and Shaogang Gong

Dept. of Computer Science, Queen Mary and Westfield College, London.
E-mail: {stephen,jpmetal,sgg}@dcs.qmw.ac.uk

Abstract. The use of adaptive Gaussian mixtures to model the colour distributions of objects is described. These models are used to perform robust, real-time tracking under varying illumination, viewing geometry and camera parameters. Observed log-likelihood measurements were used to perform selective adaptation.

1 Introduction

Colour can provide an efficient visual cue for focus of attention, object tracking and recognition allowing real-time performance to be obtained using only modest hardware. However, the apparent colour of an object depends upon the illumination conditions, the viewing geometry and the camera parameters, all of which can vary over time. Approaches to colour constancy attempt to reconstruct the incident light and adjust the observed reflectances accordingly (e.g. [2]). In practice, these methods are only applicable in highly constrained environments. In this paper a statistical approach is adopted in which colour distributions are modelled over time. These stochastic models estimate an object's colour distribution on-line and adapt to accommodate changes in the viewing conditions. They are used to perform robust, real-time object tracking under variations in illumination, viewing geometry and camera parameters.

Swain and Ballard [7] renewed interest in colour-based recognition through their use of colour histograms for real-time matching. Kjeldson used Gaussian kernels to smooth the histograms [3]. These colour histogram methods can be viewed as simple, non-parametric forms of density estimation in colour space. They gave reasonable results only because the number of data points (pixels) was always high and because the colour space was coarsely quantised. In the absence of a sufficiently accurate model for apparent colour, good parametric models for density estimation cannot be obtained. Instead, a semi-parametric approach has been adopted using Gaussian mixture models. Estimation is thus possible in a finely quantised colour space using relatively few data points without imposing an unrealistic parametric form on the colour distribution. The mixture models are adapted on-line using stochastic update equations. It is this adaptation process which is the main focus of this paper.

---

* Supported by an EPSRC/BBC CASE Studentship and EPSRC Grant GR/K44657.

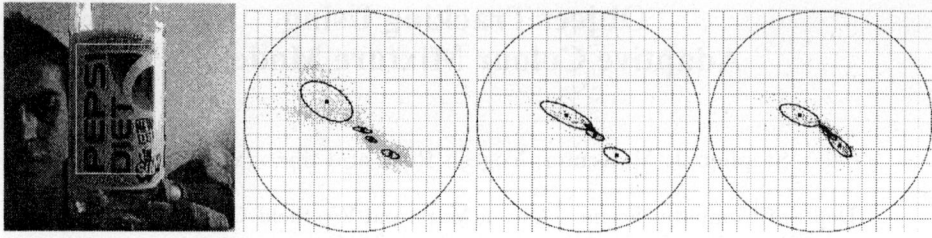

Fig. 1. *A mixture model superimposed onto plots of a bottle's colour distribution. Hue corresponds to angle and saturation to the distance from the centre. Ellipses show the four Gaussian components. The leftmost plot shows the original mixture model. The remaining two plots show the model adapting to the illumination and viewing conditions.*

In order to boot-strap the tracker for object detection and re-initialisation after a tracking failure, a set of predetermined generic object colour models which perform reasonably in a wide range of illumination conditions can be used. Once an object is being tracked, the model adapts and improves tracking performance by becoming specific to the observed conditions.

2 Colour Mixture Models

The conditional density for a pixel, $\mathbf{x}$, belonging to an object, $\mathcal{O}$, is modelled as a Gaussian mixture with m component densities:

$$p(\mathbf{x}|\mathcal{O}) = \sum_{j=1}^{m} p(\mathbf{x}|j)\pi(j)$$

where a mixing parameter $\pi(j)$ corresponds to the prior probability that $\mathbf{x}$ was generated by the jth component, $\sum_{j=1}^{m} \pi(j) = 1$. Each mixture component, $p(\mathbf{x}|j)$, is a Gaussian with mean $\boldsymbol{\mu}$ and covariance matrix $\boldsymbol{\Sigma}$. Expectation-Maximisation (EM) provides an effective maximum-likelihood algorithm for fitting such a mixture to a data set [1, 6].

Colour distributions were modelled in a 2D hue-saturation space. The intensity component was not used. Furthermore, pixels with very low intensity were discarded because the observed hue and saturation became unstable for such pixels. Likewise, pixels with very high intensity were discarded. Gaussian mixture models have been used to perform real-time object tracking given reasonably constrained illumination conditions. The resulting tracking system was surprisingly robust under large rotations in depth, changes of scale and partial occlusions [4, 5]. However, in order to cope with large changes in illumination conditions in particular, an adaptive model is required.

3 Adaptive Colour Mixture Models

A method is presented here for modelling colour dynamically by updating a colour model based on the changing appearance of the object. Fig. 1 illustrates a colour mixture model of a multi-coloured object adapting over time. At each frame, t, a new set of pixels, $X^{(t)}$, is sampled from the object and can be used to update the mixture model[2]. These colour pixel data are assumed to sample a slowly varying non-stationary signal. Let $\psi^{(t)}$ denote the sum of the posterior probabilities of the data in frame t, $\psi^{(t)} = \sum_{\mathbf{x} \in X^{(t)}} p(j|\mathbf{x})$, where by Bayes rule:

$$p(j|\mathbf{x}) = \frac{p(\mathbf{x}|j)\pi(j)}{p(\mathbf{x}|\mathcal{O})}$$

The parameters are first estimated for each mixture component, j, using only the new data, $X^{(t)}$, from frame t:

$$\boldsymbol{\mu}^{(t)} = \frac{\sum p(j|\mathbf{x})\mathbf{x}}{\psi^{(t)}} \qquad \pi^{(t)} = \frac{\psi^{(t)}}{N^{(t)}}$$

$$\boldsymbol{\Sigma}^{(t)} = \frac{\sum p(j|\mathbf{x})(\mathbf{x} - \boldsymbol{\mu}_{t-1})^T(\mathbf{x} - \boldsymbol{\mu}_{t-1})}{\psi^{(t)}}$$

where $N^{(t)}$ denotes the number of pixels in the new data set and all summations are over $\mathbf{x} \in X^{(t)}$. The mixture model components then have their parameters updated using weighted sums of the previous recursive estimates, $(\boldsymbol{\mu}_{t-1}, \boldsymbol{\Sigma}_{t-1}, \pi_{t-1})$, estimates based on the new data, $(\boldsymbol{\mu}^{(t)}, \boldsymbol{\Sigma}^{(t)}, \pi^{(t)})$, and estimates based on the old data, $(\boldsymbol{\mu}^{(t-L-1)}, \boldsymbol{\Sigma}^{(t-L-1)}, \pi^{(t-L-1)})$ (see Appendix):

$$\boldsymbol{\mu}_t = \boldsymbol{\mu}_{t-1} + \frac{\psi^{(t)}}{D_t}(\boldsymbol{\mu}^{(t)} - \boldsymbol{\mu}_{t-1}) - \frac{\psi^{(t-L-1)}}{D_t}(\boldsymbol{\mu}^{(t-L-1)} - \boldsymbol{\mu}_{t-1})$$

$$\boldsymbol{\Sigma}_t = \boldsymbol{\Sigma}_{t-1} + \frac{\psi^{(t)}}{D_t}(\boldsymbol{\Sigma}^{(t)} - \boldsymbol{\Sigma}_{t-1}) - \frac{\psi^{(t-L-1)}}{D_t}(\boldsymbol{\Sigma}^{(t-L-1)} - \boldsymbol{\Sigma}_{t-1})$$

$$\pi_t = \pi_{t-1} + \frac{N^{(t)}}{\sum_{\tau=t-L}^{t} N^{(\tau)}}\left(\pi^{(t)} - \pi_{t-1}\right) - \frac{N^{(t-L-1)}}{\sum_{\tau=t-L}^{t} N^{(\tau)}}\left(\pi^{(t-L-1)} - \pi_{t-1}\right)$$

where $D_t = \sum_{\tau=t-L}^{t} \psi^{(\tau)}$. The following approximations are used for efficiency:

$$\psi^{(t-L-1)} \approx \frac{D_{t-1}}{L+1} \qquad (1)$$

$$D_t \approx (1 - 1/(L+1))D_{t-1} + \psi^{(t)} \qquad (2)$$

The parameter L controls the adaptivity of the model[3].

---

[2] Throughout this paper, superscript $^{(t)}$ denotes a quantity based only on data from frame t. Subscripts denote recursive estimates.

[3] Setting $L = t$ and ignoring terms based on frame $t-L-1$ gives a stochastic algorithm for estimating a Gaussian mixture for a stationary signal [1, 8].

4 Selective Adaptation

An obvious problem with adapting a colour model during tracking is the lack of ground-truth. Any colour-based tracker can lose the object it is tracking due, for example, to occlusion. If such errors go undetected the colour model will adapt to image regions which do not correspond to the object. In order to alleviate this problem, observed log-likelihood measurements were used to detect erroneous frames. Colour data from these frames were not used to adapt the object's colour model.

The adaptive mixture model seeks to maximise the log-likelihood of the colour data over time. The normalised log-likelihood, $\mathcal{L}^{(t)}$, of the data, $X^{(t)}$, observed from the object at time t is given by:

$$\mathcal{L}^{(t)} = \frac{1}{N^{(t)}} \sum_{\mathbf{x} \in X^{(t)}} \log p(\mathbf{x}|\mathcal{O})$$

At each time frame, $\mathcal{L}^{(t)}$ is evaluated. If the tracker loses the object there is often a sudden, large drop in its value. This provides a way to detect tracker failure. Adaptation is then suspended until the object is again tracked with sufficiently high likelihood. A temporal filter was used to compute a threshold, T_t. Adaptation was only performed when $\mathcal{L}^{(t)} > T_t$. The median, ν, and standard deviation, σ, of $\mathcal{L}$ were computed for the n most recent above-threshold frames, where $n \leq L$. The threshold was set to $T = \nu - k\sigma$, where k was a constant. In all the experiments described here, $k = 1.5$, $n = 2f$ and $L = 6f$, where f denotes the frame rate in Hz.

Fig. 2. *Eight frames from a sequence in which a face was tracked using a non-adaptive model. The apparent colour of the face changes due to (i) varying illumination and (ii) the camera's auto-iris mechanism which adjusts to the bright exterior light.*

Fig. 3. *The sequence depicted in Fig. 2 tracked with an adaptive colour model. Here, the model adapts to cope with the change in apparent colour. Only the last four images are shown for conciseness. Performance in previous frames was similar.*

5 Experiments

The adaptive mixture modelling described in the previous two sections was integrated with an existing colour-based tracking system [4, 5] implemented on a standard 200MHz Pentium PC platform with a Matrox Meteor frame-grabber. This system performs tracking at approximately $f = 15$Hz. The tracker estimates the centroid, height and width of the object. New samples of data for adaptation are gathered from a region of appropriate aspect ratio centred on the estimated object centroid. It is assumed that these data form a representative sample of the objects' colours. This will hold for a large class of objects.

Figs. 2 and 3 illustrate the use of the mixture model for face tracking and the advantage of an adaptive model over a non-adaptive one. In this sequence the illumination conditions coupled with the camera's auto-iris mechanism resulted in large changes in the apparent colour of the face as the person approached the window. Towards the end of the sequence the face became very dark, making hue and saturation measurements unreliable. In Fig. 2, a non-adaptive model was trained on the first image of the sequence and used to track throughout. It was unable to cope with the varying conditions and failure eventually occured. In Fig. 3, the model was allowed to adapt and successfully maintained lock on the face.

Fig. 4 illustrates the advantage of selecting when to adapt. The person moved through challenging tracking conditions, before approaching the camera at close range (frames 50-60). Since the camera was placed in the doorway of another room with its own lighting conditions, the person's face underwent a large, sudden and temporary change in apparent colour. When adaptation was performed in every frame, this sudden change had a drastic effect on the model and ultimately led the tracker to fail when the person receded into the corridor. With selective adaptation, these sudden changes were treated as outliers and adaptation was suspended, permitting the tracker to recover.

Fig. 5 depicts the tracking of a multi-coloured item of clothing with adaptation performed in every frame. Although tracking was robust over many frames, erroneous adaptation eventually resulted in failure. Fig. 6 shows the last four frames from the same sequence tracked correctly using selective adaptation.

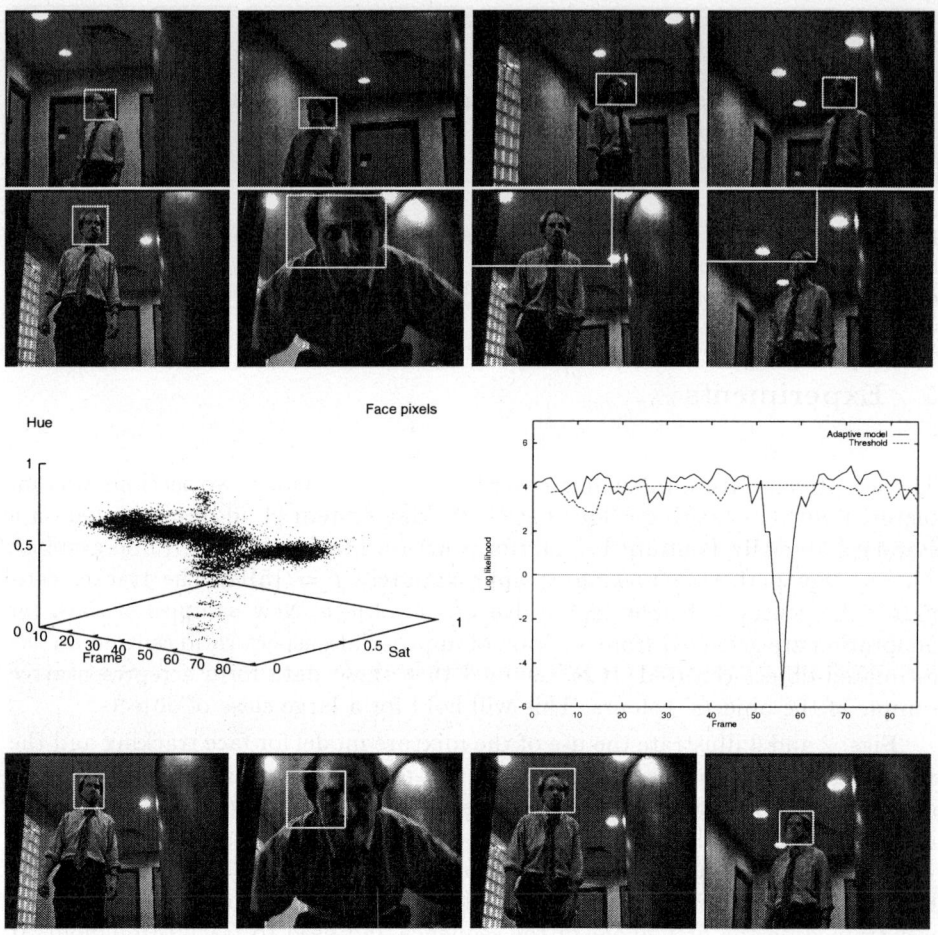

Fig. 4. *At the top are frames 5, 15, 25, 35, 45, 55, 65 and 75 from a sequence. There is strong directional and exterior illumination. The walls have a fleshy tone. At around frame 55, the subject rapidly approaches the camera which is situated in a doorway, resulting in rapid changes in illumination, scale and auto-iris parameters. This can be seen in the 3D plot of the hue-saturation distribution over time. In the top sequence, the model was allowed to adapt in every frame, resulting in failure at around frame 60. The lower sequence illustrates the use of selective adaptation. The right-hand plot shows the normalised log-likelihood measurements and the adaptation threshold.*

6 Conclusions

Objects' colour distributions were modelled using Gaussian mixture models in hue-saturation space. An adaptive learning algorithm was used to update these colour models over time and was found to be stable and efficient. These adaptive models were used to perform colour-based object tracking in real-time under

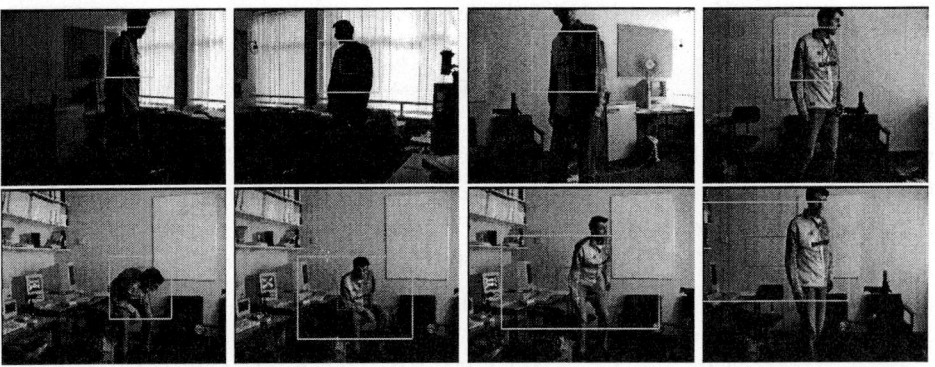

Fig. 5. *A green, yellow and black shirt tracked using the adaptive mechanism. Eventually, tracking inaccuracies cause the model to adapt erroneously and the system fails.*

Fig. 6. *The sequence shown in Fig. 5 tracked using selective adaptation. The shirt was correctly tracked throughout. Only the last four frames are shown for brevity.*

varying illumination, viewing geometry and camera parameters. Outlier detection based on a normalised log-likelihood statistic was used to detect tracking failures. This adaptive scheme outperformed the non-adaptive colour models.

Topics for further work include (i) emphasised co-operation with other visual cues during periods when colour becomes unreliable, (ii) adaptive modelling of background scene colours and (iii) model order selection, i.e. adaptation of the mixture size.

Appendix

Here we derive the update equations for the adaptive mixture model components. For each mixture component, let $\boldsymbol{\mu}_t$ and $\boldsymbol{\Sigma}_t$ be the mean and the covariance matrix estimated from the $L+1$ most recent time-slots:

$$\boldsymbol{\mu}_t = \frac{\sum_{\tau=t-L}^{t} \sum_{\mathbf{x} \in X^{(\tau)}} p(j|\mathbf{x})\mathbf{x}}{\sum_{\tau=t-L}^{t} \psi^{(\tau)}}$$

$$\boldsymbol{\Sigma}_t = \frac{\sum_{\tau=t-L}^{t} \sum_{\mathbf{x} \in X^{(\tau)}} p(j|\mathbf{x})(\mathbf{x} - \boldsymbol{\mu}_{\tau-1})^T (\mathbf{x} - \boldsymbol{\mu}_{\tau-1})}{\sum_{\tau=t-L}^{t} \psi^{(\tau)}}$$

The above expressions are both of the form:

$$\boldsymbol{\theta}_t = \frac{\sum_{\tau=t-L}^{t} \boldsymbol{\theta}^{(\tau)} \psi^{(\tau)}}{D_t}$$

where $\boldsymbol{\theta}_t$ denotes either $\boldsymbol{\mu}_t$ or $\boldsymbol{\Sigma}_t$. A recursive expression for $\boldsymbol{\theta}_t$ is derived as follows:

$$\boldsymbol{\theta}_t = \frac{1}{D_t} \left(\sum_{\tau=t-L-1}^{t-1} \boldsymbol{\theta}^{(\tau)} \psi^{(\tau)} + \boldsymbol{\theta}^{(t)} \psi^{(t)} - \boldsymbol{\theta}^{(t-L-1)} \psi^{(t-L-1)} \right)$$

$$= \frac{1}{D_t} \left(\boldsymbol{\theta}_{t-1} \sum_{\tau=t-L-1}^{t-1} \psi^{(\tau)} + \boldsymbol{\theta}^{(t)} \psi^{(t)} - \boldsymbol{\theta}^{(t-L-1)} \psi^{(t-L-1)} \right)$$

$$= \frac{1}{D_t} \left(\boldsymbol{\theta}_{t-1} \sum_{\tau=t-L}^{t} \psi^{(\tau)} - \boldsymbol{\theta}_{t-1} \psi^{(t)} + \boldsymbol{\theta}_{t-1} \psi^{(t-L-1)} + \boldsymbol{\theta}^{(t)} \psi^{(t)} - \boldsymbol{\theta}^{(t-L-1)} \psi^{(t-L-1)} \right)$$

$$= \boldsymbol{\theta}_{t-1} + \frac{\psi^{(t)}}{D_t} (\boldsymbol{\theta}^{(t)} - \boldsymbol{\theta}_{t-1}) - \frac{\psi^{(t-L-1)}}{D_t} (\boldsymbol{\theta}^{(t-L-1)} - \boldsymbol{\theta}_{t-1}) \qquad (3)$$

Approximation (1) yields (2) which approximates the sum D_t. The update expression for the prior π_t is obtained similarly to (3). If the number of data points is the same in every time frame (i.e. $N^{(\tau)} = N$, for all τ) then we have:

$$\pi_t = \pi_{t-1} + \frac{\pi^{(t)} - \pi^{(t-L-1)}}{L+1}$$

References

1. C. Bishop. *Neural Networks for Pattern Recognition*. Oxford University Press, 1995.
2. D. A. Forsyth. *Colour Constancy and its Applications in Machine Vision*. PhD thesis, University of Oxford, 1988.
3. R. Kjeldsen and J. Kender. Finding skin in color images. In *2nd Int. Conf. on Automatic Face and Gesture Recognition*, 1996.
4. S. McKenna, S. Gong, and Y. Raja. Face recognition in dynamic scenes. In *BMVC*, 1997.
5. Y. Raja, S. McKenna, and S. Gong. Segmentation and tracking using colour mixture models. In *Asian Conference on Computer Vision*, 1998.
6. R. A. Redner and H. F. Walker. Mixture densities, maximum likelihood and the EM algorithm. *SIAM Review*, 26(2):195–239, 1984.
7. M. J. Swain and D. H. Ballard. Colour indexing. *IJCV*, pages 11–32, 1991.
8. H. G. C. Traven. A neural network approach to statistical pattern classification by "semiparametric" estimation of probability density functions. *IEEE Trans. Neural Networks*, 2(3):366–378, 1991.

A Learning Approach to Fixating on 3D Targets with Active Cameras

Narayan Srinivasa and Narendra Ahuja

The Beckman Institute for Advanced Science and Technology
University of Illinois at Urbana-Champaign
405 N. Mathews Avenue, Urbana, IL 61801

Abstract

Fixation of an active camera pair on a given target requires that the pan and tilt angles of the cameras must be set to bring the target to image centers. However, the calibration needed to achieve a specific configuration of real cameras involves tedious estimation of a number of imaging parameters. Fortunately, this excercise is not essential for fixation if images are acquired and used as feedback during the fixation process to continuously direct the cameras to the target. This paper defines a direct mapping from the changes in the direction of target motion in the image plane to changes in camera angles necessary to reduce the disparity between image center and the image plane target location. The mapping captures camera calibration, as well as other effects such as deviations from the assumed imaging model which are difficult to characterize and capture in calibration. The mapping is formulated as a task in nonlinear function approximation and learnt from real data. For computational efficiency, learning is done at multiple resolutions and using a PROBART network. Experimental results are presented using an active vision system.

1 Introduction

For active vision systems to be useful in performing real tasks, it is critical that the camera control and processing be real-time. The increased availability of powerful and cheap computers and real-time image processing capabilities are making such systems feasible. This paper is aimed at the capability of fixation which is an integral part of active visual analysis [2, 3, 4, 5, 9, 14, 15]. To fixate on a three-dimensional (3D) point target, the orientations (joint angles) of the cameras are changed such that the optical axes of the stereo cameras move to intersect at the 3D point. It is possible to obtain an exact analytical expression for the camera joint angles required to fixate on a 3D target point. To use the expression for fixation requires an accurate calibration of the various camera parameters [16] which is usually tedious and time consuming. However, complete calibration is more than what is required to fixate. It suffices to know how to continuously approach and thus converge to the state of fixation from current camera configuration using the images obtained during the fixation as feedback instead of directly transitioning to the state of fixation in one reconfiguration step. This paper uses a Direction-to-Joints (DTJ) mapping which models the relationship between incremental changes in the direction of image plane motion of the scene points and the corresponding incremental changes in joint angles. DTJ exploits the property that for a given image of a 3D target, as camera joint angles are changed by a small amount, the direction in which the image of the target moves is independent of the current joint angles and the 3D target location.

We present an approach to learn the DTJ mapping at multiple resolutions of incremental camera motions. Initially, the coarse resolution DTJ mapping is used to rapidly bring the target roughly to the vicinity of the camera center. Then, increasingly fine resolution DTJ mappings are used to monotonically reduce the residual disparity between image center and target location to accurately fixate on the 3D target. An interesting aspect of the approach is that a single target at a fixed 3D

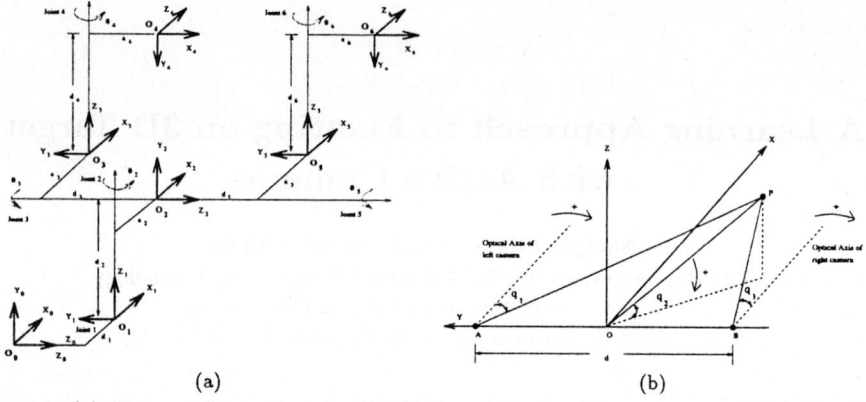

Figure 1: (a) The coordinate frames of UIAVS and the model parameters. (b) A simplified schematic of the UIAVS.

location is sufficient to learn the DTJ mapping over the entire joint space for which the target is visible. Thus, it is easy to implement the learning in an autonomous mode on a real active vision system. The learning process presented is also self-organizing because all the training inputs and outputs are self-generated. Finally, as stated earlier, the learning approach encompasses, and therefore, does not require explicit calibration.

2 Background

The most common form of experimental setup used in active vision research has two motorized cameras mounted as a 'head' [1, 5, 10, 11]. The University of Illinois Active Vision System (UIAVS) [1] is one such system. Typically, each camera is mounted on a separate motor that controls the azimuth (pan) angle of its optical axis. Both motors are supported by a common base mounted on another motor, which controls the common elevation (tilt) angle of both cameras. The two motors allow the cameras to assume arbitrary azimuth and an arbitrary but common elevation angle. In this paper, the UIAVS will be used to perform all the experiments.

The complete kinematic model of the UIAVS to be used in the experiments is shown in Figure 1(a). Each movable joint of the system is assigned a reference frame $X_i - Y_i - Z_i$. The panning motors for the two cameras are assigned the frames O_4 and O_6; these frames can rotate with respect to the frame O_3 attached to the base on which the two cameras are mounted and θ_4 and θ_6 represent the two pan angles. The tilt motor is assigned the frame O_3 and this frame can rotate with respect to frame O_2 attached to the head of the camera by the tilt angle θ_3. If the base on the which the two cameras are mounted is not flat, then the amount of tilt in the right camera may be different from the left camera. To account for this, an additional angle θ_5 is asspciated with frame O_5 as shown in the Figure 1(a). Both the cameras can also be reoriented in unison by a motor attached to frame O_2. This frame can rotate with the respect to the translation unit frame O_1 by the pan angle θ_2 for the head. In our experiments, this angle is fixed. The frame at O_1 can translate with respect to frame O_0 by an amount d_1. This parameter is also fixed in all our experiments. The frame at O_0 represents the world frame. The constants a_i, d_i for $(i = 1, \cdots, 6)$ represent the offsets of the various frames from one another and are assigned using the Denavit-Hartenberg (D-H) convention [8]. In general, the parameters $d_1, \theta_2, \theta_3, \theta_4, \theta_5$ and θ_6 represent the six degrees of freedom of the camera configuartion. Similarly, the focus, zoom and aperture settings represent the six degrees of freedom available for optical configuration.

Using the above kinematic model of the UIAVS, it is possible to obtain an exact analytical expression for the location of each fixated and non-fixated 3D point that is visible to the stereo cameras as follows. To compute the 3D location (x, y, z) of a point with respect to the base coordinate frame O_0, the composite transformation $^0T_4$ from coordinate frame O_4 of the left camera to base coordinates O_0. Using the perspective projection model for a pin hole camera, the world point (x, y, z), defined with respect to frame O_0, can be obtained from its image location $[u^L, v^L]$ at the left camera as:

$$\begin{bmatrix} x \\ y \\ z \end{bmatrix} = {}^0T_4 P_L^{-1} \begin{bmatrix} u^L \\ v^L \end{bmatrix} \tag{1}$$

where the transformation matrix P_L is defined in terms of N_{sx} and N_{sy} which are the number of camera pixels per row and column; d_x and d_y are the width and height of camera pixels; N_{fx} and N_{fy} are the number of camera pixels per row and column that are digitized and stored for processing; u_0^L and v_0^L are the image coordinates of the image plane center for the left camera; and λ_L is the effective focal length for the left camera.

Thus, if the image location $[u^L, v^L]$ of a 3D target is known, its exact 3D location can be computed by using equation 1. However, the T matrices involve extrinsic parameters such as a_2 and d_2 in Figure 1, as well as optical parameters such as λ_L. Therefore, these parameters must be estimated through extensive camera calibration before the 3D estimation can be done [7]. In this paper, we argue that this tedious and time consuming calibration process can be avoided for the purposes of fixating on 3D targets by defining a direct mapping from the change in direction of the image location of an unfixated target to the incremental camera motion required to fixate.

3 Existence of DTJ mapping

To develop the basic terminology for defining the components of the DTJ mapping, consider the simplified kinematic model of the UIAVS shown in Figure 1(b). The angles q_1 (and q_3) represent the pan angles of the left (and right) cameras with respect to a "straight-ahead" direction (X-axis). The tilt is represented by the angle q_2 between the plane defined by the optical axes and the XY plane. The pan and tilt axes of each camera pass through its optic point (A and B in Figure 1(b)). These three angles are independently controlled. The direction conventions for these angles are shown in Figure 1(b). The point of fixation (P in Figure 1(b)) lies on both optical axes and projects onto both image centers.

Let an arbitrary 3D point P be imaged at (u, v) using the pin-hole imaging model and for the configuration (q_1, q_2) of the left camera as shown in Figure 1(b). Then, for an incremental change $(\Delta q_1, \Delta q_2)$ in the camera orientation, the corresponding change in the image position $(\Delta u, \Delta v)$, for the simplified kinematic model of the UIAVS, can be derived using the image Jacobian [12] as

$$\begin{bmatrix} \Delta u \\ \Delta v \end{bmatrix} = \begin{bmatrix} \frac{\lambda^2 + u^2}{\lambda} & \frac{-uv}{\lambda} \\ \frac{uv}{\lambda} & \frac{-\lambda^2 - v^2}{\lambda} \end{bmatrix} \begin{bmatrix} \Delta q_1 \\ \Delta q_2 \end{bmatrix} \tag{2}$$

where λ is the focal length of the camera. In order to establish the existence of the DTJ mapping, let us rewrite the left hand side of equation (2) in terms of the image plane direction components (n_x, n_y) where $n_x = \frac{\Delta u}{C}$ and $n_y = \frac{\Delta v}{C}$ and $C = (\Delta u^2 + \Delta v^2)^{\frac{1}{2}}$. The magnitude C depends upon the depth of the target from the active camera (closer the target, larger the magnitude) as well as the magnitude of $(\Delta q_1, \Delta q_2)$. The direction vector (n_x, n_y) depends on (u, v) and the ratio $\frac{\Delta q_1}{\Delta q_2}$.

For the general kinematic model of the real UIAVS (as described in the previous section), equation 2 will continue to hold provided the intrinsic and extrinsic parameters of the UIAVS are included in the 2X2 matrix. To verify this, a simple experiment was performed using the real UIAVS. A set of random 3D targets (small dark patches on a white background) at various 3D locations were viewed one at a time by the UIAVS using a variety of camera joint angles. For each pair of 3D target location and camera orientation, the camera pan and tilt angles were changed over a wide range at fixed angular increments ($\Delta q_1 = \Delta q_3 = 0.001°$ and $\Delta q_2 = 0.003°$ where the resolution of the joint encoders of the pan and tilt units is $0.001°$). From the entire resulting data, the observed directions of target motion in image were recorded for small neighborhoods of (u, v) values of target location, regardless of 3D target location and camera joint angles. Results for seven such neighborhoods in the left camera image are shown in Figure 2(a). As can be seen in Figure 2(b), the image plane directions are similar for an entire neighborhood, and are different for the different neighborhoods (appearing as seven different curve segments for seven different clusters of (u, v) values). This serves as empirical evidence for the existence of the DTJ mapping.

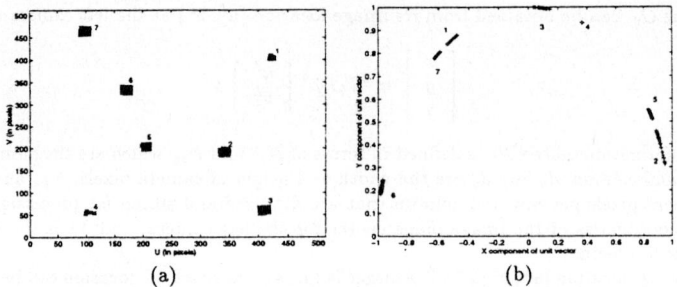

Figure 2: (a) Seven clusters of image coordinates used to record the direction information. (b) Seven clusters of the unit directional vectors recorded over the seven neighborhoods.

4 Estimation of DTJ mapping

Estimating the relationship between $(u, v, \Delta u, \Delta v)$ and $(\Delta q_1, \Delta q_2)$ will be viewed as a nonlinear function approximation task, and performed by a learning algorithm. To obtain the data for learning, (u, v, n_x, n_y) values are observed for a range of $(\Delta q_1, \Delta q_2)$ selections. This data is then used to incrementally learn the DTJ mapping. A single target at a fixed 3D location suffices, provided it is visible to the cameras at all orientations. An additional benefit of using a single target is that the training examples for learning can be generated automatically making the implementation easier.

5 Learning Algorithm and Architecture

We have adopted the algorithm for incremental function approximation using the Probabilistic Adaptive Resonance Theory (PROBART) architecture as described [13] and extended it to learn the multiresolution DTJ mapping. One of the most important properties of this algorithm is that the majority of processing involves simple compare and add operations (as outlined below) which makes learning extremely efficient. Another important property is that the estimation of the mapping, and thus the learning, is accomplished incrementally as inputs are presented.

5.1 PROBART Representation and Notation

In developing an algorithm to perform the above computation, fuzzy set theory methodology [17] is used to represent classes as well as to perform computations [6, 13]. For concreteness, we will explain the notation for the 2-dimensional space; it generalizes to other spaces in a straight forward manner.

Class: A class is represented by specifying two diagonally opposite vertices of its rectangle. This is done by a vector consisting of the coordinates of one vertex followed by the complement (with respect to 1) of the coordinates of the diagonally opposite vertex. Thus, for example, the output class represented by the rectangle defined by vertices (x_1, y_1) and (x_2, y_2) is represented by the vector $(x_1, y_1, 1 - x_1, 1 - y_1)$. For a (class consisting of) a single point (vertex) (x_1, y_1), the representation is the 4-tuple $(x_1, y_1, 1 - x_1, 1 - y_1)$, denoted by its *weight* vector $\mathbf{W}$.

Norm: The norm $|V|$ of a vector (class) is the sum of the city block distances of the class from the points $(0, 0)$ and $(1, 1)$.

Distance: The distance between a sample point and a class rectangle is denoted by the city block distance to the nearest point in the class.

AND Operation: The fuzzy AND (or $\wedge$) between two classes is the vector whose elements are obtained by taking pairwise MIN of the corresponding elements of the operand vectors. AND of two classes (points) denotes the result of adding one to the other, possibly resulting in expansion.

Choice Function: The choice function is used to determine the class defined by its weight $\mathbf{W}$ that is closest to a given point. Given a new point $\mathbf{I}$ and a class $\mathbf{W}$, the choice function is defined as $\frac{|\mathbf{I} \wedge \mathbf{W}|}{|\mathbf{W}|}$, which assumes highest value for that class which is at shortest distance from $\mathbf{I}$. If the choice function is one for a given class, then the class is a *fuzzy subset choice* for input $\mathbf{I}$. This means that the input

I is completely contained within the class W. If more than one class is a fuzzy subset choice, then a small but positive parameter α is added to the denominator to break the tie to identify the class that maximizes $|\mathbf{W}|$ among the fuzzy subset choices.

Vigilance function: The vigilance function is used to enforce the restriction on class size. For example, given a sample I and a class W, W is allowed to (expand and) include I if the value of the vigilance function, defined as $\frac{|\mathbf{I} \wedge \mathbf{W}|}{|\mathbf{I}|}$, is no smaller than a certain *a priori* (user specified) threshold ρ called the vigilance parameter.

5.2 Multi-Resolution PROBART Algorithm

The PROBART algorithm consists of two identical modules based on the Fuzzy Adapative Resonance Theory (fuzzy ART) [6]. The first fuzzy ART module learns classes from inputs while the other fuzzy ART module creates classes for the outputs. These classes are then mapped to each other. The multi-resolution PROBART algorithm is now described using the inputs and outputs for the left camera. The learning algorithm for the right camera is the same as for the left camera. We perform learning at a range of step sizes (resolutions) of $(\Delta q_1, \Delta q_2)$. This is because while using the mapping to determine the $(\Delta q_1, \Delta q_2)$ for a desired image plane motion direction, it may be desirable to perform the $(\Delta q_1, \Delta q_2)$ control in a coarse-to-fine manner for computational efficiency. The large camera motions (or *ballistic mode* of fixation) will bring the target image roughly near the image center, and, the fine resolution mapping will help bring the target into fixation more accurately. In the rest of this paper, we will consider only two levels of resolution, one coarse and one fine, without loss of generality. Initialize all class representations $\mathbf{W}_J$ to be unit vectors. Present the normalized input vector $\mathbf{I} = (u_1, v_1, n_x, n_y)$. This input is classified as a *coarse* or *fine* resolution input depending on the magnitude of the outputs $(\Delta q_1, \Delta q_2)$ that caused the image to move in the (n_x, n_y) direction. Thus, for example, if the magnitude of $(\Delta q_1, \Delta q_2)$ is large (as defined by the user), then it is used by the coarse resolution network. The input and its complement are stored as a single vector. The class J that is closest to the input I is computed using the choice function as $T_J = \frac{|\mathbf{I} \wedge \mathbf{W}_J|}{\alpha + |\mathbf{W}_J|}$. The network then makes a hypothesis that the selected class J is the appropriate classification for the given input. This hypothesis is then tested using the vigilance criterion $\frac{|\mathbf{I} \wedge \mathbf{W}_J|}{|\mathbf{I}|} \geq \rho$ where ρ is called the vigilance parameter set by the user.

If node J satisfies the vigilance criterion, then the input I is in *resonance* (hence the name adaptive resonance) with the class weights W_J. The class weights are then updated as: $\mathbf{W}_J = |\mathbf{I} \wedge \mathbf{W}_J|$ The updated weights store the prototype for the class J. If node J does not satisfy the vigilance criterion, then the class with the next maximum choice function value is selected and the process is repeated. If there are no currently existing classes that satisfy the vigilance criterion, then a new class is created to represent the input. The bound on the size of the hyperrectangle, $|D_J|$, for each class can be defined as $|D_J| \leq M(1 - \rho)$ where M is the number of features in the input. Thus, if the vigilance parameter ρ is small, the size of the hyperrectangles are bigger (or the classification is coarse) and vice versa. The training process continues until the input feature space is covered with hyperrectangles. It should be noted that the number of hyperrectangles created at the end of training depends on the vigilance parameter. Using the previous steps, the corresponding output $(\Delta q_1, \Delta q_2)$ is also classified into some class K. The weights F_{JK} of links between the winning classes J and K of the two fuzzy ART modules is incremented by one. These weights measure the frequency of coactivation of a given pair of input and output classes. Initially, $\mathbf{F}$ is zero for all j and k. If there are more inputs and outputs to be mapped, repeat the entire process.

Once all the inputs and outputs are presented to the algorithm, the outputs $(\Delta q_1, \Delta q_2)$ can be computed for a given input class J as $\Delta q_i = \sum_{k=1}^{R} F_{Jk} Z_{ik} / \sum_{k=1}^{R} F_{Jk}$ $(i = 1, 2)$ where Z_{ik} is zeroth moment of Δq_i for the output class k, R is the total number of output classes during training, and $F_{Jk}/\sum_{k=1}^{R} F_{Jk}$ represents the estimated probability of association between the input class J and the output class k.

5.3 PROBART Architecture

The algorithm is implemented using the PROBART network architecture which allows incremental learning of the mapping. We use one PROBART network for each level of resolution. Each PROBART network consists of two Fuzzy ART networks [6] as shown in Figure 3. The Fuzzy ART network is capable of unsupervised classification of binary or analog inputs in real-time. Fuzzy ART_1 in Figure 3 is used as the input processing module. At each resolution, the input layer L_1 module receives two types of inputs: (1) Image coordinates (u_1, v_1) of the non-fixated 3D point target in the left image and

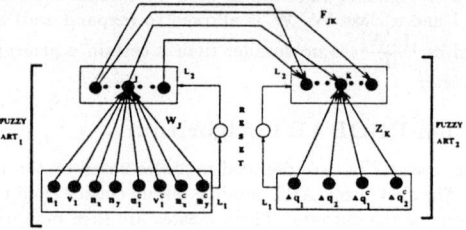

Figure 3: The PROBART network architecture.

the (2) Image direction (n_x, n_y) pointing towards the image center. The L_1 layer of the Fuzzy ART_2 module receives the changes in camera pan and tilt $(\Delta q_1, \Delta q_2)$ that cause the image of a 3D target at (u_1, v_1) to move in the (n_x, n_y) direction.

The complement coded input in the L_1 layer is propagated forward to the L_2 layer. The nodes in L_2 layer correspond to different input classes. Each node computes its own choice function for the given input. The node with the most excitation is hypothesized as the approriate category for the input. This hypothesis is then tested using the vigilance criterion. If the selected node passes this test, then the input features are encoded into the weight vector $\mathbf{W}_j$ that connects the inputs to the winner node. The winner node is then a category that represents the input. If the vigilance criterion is not satisfied, then the selected node is shut off (by a *reset* signal as shown in Figure 3) and a new node is selected in L_2. The entire process is repeated until either a previously selected category is found to satisfy the vigilance criterion or a new node is selected in L_2 to represent the presented input. In this manner the inputs to the L_1 layers are stored into classes in the L_2 layers of the two Fuzzy ART modules. These created classes are then mapped to each other via the weights F_{jk} as shown in Figure 3.

6 Training and Performance Evaluation

The UIAVS is moved to various camera configurations for which a single target at a fixed 3D location is viewed. Only those pan and tilt angles are considered for which the image of the target is visible in both images. The camera configurations are obtained using a random generator of pan and tilt angles within the allowable joint ranges. At each of these non-fixated camera configurations, the pan and tilt of the camera are randomly incremented (using the incremental motion generator) at multiple resolutions to generate the training inputs $V_k = (u_1, v_1)$ and $D = (n_x, n_y)$, and outputs $Q_k = (\Delta q_1, \Delta q_2)$ at each time step k. The direction vector D is computed by using the image coordinates of the fixed target at any two consecutive camera configurations.

There are two modes in which the fixation can occur. The first mode is based on the availability of a *continuous visual feedback*. The role of the incremental motion generator during training is now replaced by the trained network. In this mode, a randomly selected 3D point target is first viewed by the active camera at some non-fixated camera configuration. In order to fixate on the 3D target, the distance between the current location of the target in the image and the image center is computed, and the coarsest level resolution which can perform movement by the distance is identified. The desired direction for each camera is obtained from the vector connecting the current image coordinate of the 3D target and the image plane center. The selected coarse resolution network is used to generate large increments in the joint angles which are used to reconfigure the cameras. Next, the control is transferred to a network at the next higher resolution and the process repeated. In this manner, the incremental changes in joint angles are generated by an appropriately chosen PROBART network until the target is accurately fixated by the highest resolution network of interest.

The second mode of the fixation process is based on the availability of only an *intermittent visual feedback*. At the initial time step, the image coordinates of the target are known. Using this information, the direction to the image center is computed. Using the coarsest resolution network, the change in joint angle is predicted for direction. At this step, if we have no visual feedback, the new location of the target in the image due to a camera motion is unknown. So, we predict a new location of the target in the image by multiplying the direction with a constant magntude M (fixed at different values for different resolutions) and adding it to the previous image location. The predicted location is then used to compute the new direction of the target to the image center and this process is repeated. The transfer of control from coarse to fine resolution is made exactly as in the visual feedback mode. When the image of the target is predicted to be within 0.5 pixels from the image center, a visual feedback is provided to verify if the camera has actually fixated on the target. If not, the actual image location is used to provide a new image location and the fixated process is repeated until the target is fixated.

7 Experiments and Results

The PROBART neural networks at multiple resolutions were interfaced with the camera joint actuators and the image signals from the UIAVS. The active vision setup is mounted on a mobile robot to provide additional mobility. However, the mobile robot was not used during any of our experiments. The camera motion can be controlled by the tilt and pan units. The experimental setup for training the PROBART network is shown in Figure 4. A single dark patch was placed on a wall at a depth of about

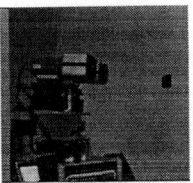

Figure 4: Experimental setup during training.

4 meters from the active vision system. The image of the patch was extracted by thresholding and its centroid was used as the target.

In the experiments reported here, the DTJ mapping was learned at two resolutions. The angle changes at the coarse resolution for pan and tilt angles was randomly selected to be within 0.5° and 1° while those for the fine resolution were selected to be between 0.001° and 0.003°. The resolution of the joint encoders is 0.001°. The allowable joint angle range for both the pan and tilt angles was $[-60°, 60°]$. Since the image contains 512 X 512 pixels, each coordinate was normalized to the range $[0, 512]$. The focal length λ was fixed at $30mm$. The regimen outlined in section 6 was adopted to train the PROBART neural networks. Training data was collected at 60,000 camera orientations were collected. At each of these orientations, the active camera was incrementally moved both by coarse and fine camera motions. For each such motion, the (n_x, n_y) and $(\Delta u, \Delta v)$ were sampled. This training data was clustered by the coarse resolution network (of each camera) into 302 input and 16 output clusters. Similarly, the fine resolution networks created 1231 and 49 clusters for the inputs and outputs respectively. Once the networks were trained on the inputs and outputs generated by observing a single target, the performance of the trained network was evaluated by placing 3D targets one at a time in a 4x4x6 cubic meter volume in front of the active camera. Each of these targets were fixated using the continuous and intermittent visual feedback modes. The image and joint trajectories during fixation in the continuous mode are shown in the first row of Figure 5 for the left camera and for a target at $(X = 3.5, Y = 2.0, Z = 2.0)$. This plot corresponds to the prediction of the multi-resolution PROBART network. The image is rapidly brought to within 6 pixels of the image center by the coarse resolution network. Then, the fine resolution network brings the target accurately into fixation. An accuracy of 0.05 pixels was obtained for all the targets. The average time taken to bring each of these targets was about 6 secs using the multi-resolution networks.

In order to compare the multi-resolution approach to just a single JTD mapping at fine resolution, the above plots were repeated without the coarse resolution JTD mapping as shown in the second row of Figure 5. It can be seen that the time taken to bring the same target into fixation is 20 times more than

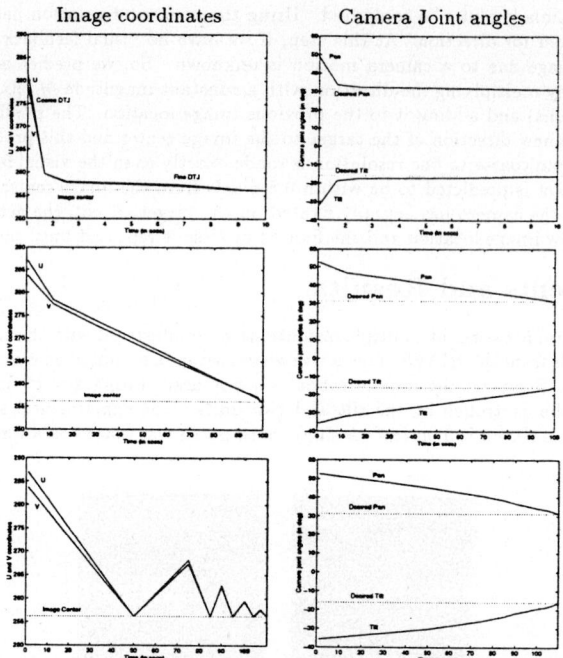

Figure 5: For the left camera: the first row is for fixation using continuous visual feedback; the second row is for continuous visual feedback and only the fine resolution DTJ mapping; the third row is for intermittent visual feedback.

in the multi-resolution case. The accuracy of fixation is however the same for single and multi-resolution cases. For the intermittent mode of visual feedback, the time taken to fixate on a target clearly depends on the accuracy of the mapping. The main parameter that affects this accuracy is the constant M. In the multi-resolution case, M can be large (for the coarse resolution) or small (for the fine resolution).

However, for large values of M, the system can have large overshoots and thus take prohibitively long periods of time to fixate. These overshoots can also be detrimental to the camera control equipment. To prevent this, we used the active camera only for small values of M. For comparison, the image and joint trajectories for the same 3D target are plotted for the intermittent mode as shown in the third row of Figure 5. The value of M was set to 0.05 pixels. This setting was reasonable because it prevented overshoots in the image trajectory for all the 3D targets. The inflection points in the image trajectory correspond to instances of visual feedback during the fixation process as shown in the first graph of the third row in Figure 5. For all the targets in our experiments, the maximum number of feedbacks was less than 10. It can be seen from Figure 5 that the intermittent mode takes longer compared to even the single resolution mode. However, there are no overshoots and the accuracy of fixation is not compromised.

8 Conclusions

We have addressed the problem of learning to fixate on 3D point targets using real active cameras. This is achieved by exploiting a Direction-To-Joints or DTJ mapping that relates camera motion to image motion. The learning does not require the camera to be calibrated. Once the DTJ mapping is learned at multiple resolutions and using a single target, it is possible to rapidly fixate on any other visible 3D target. This is achieved by using the learned DTJ mapping in a control loop with continuous or intermittent visual feedback. Experiments were performed on the UIAVS to verify the feasibility and accuracy of the proposed approach presented. The results obtained suggest the algorithm is easy to implement on a real active vision system and that the fixation achieved is accurate.

References

[1] L. Abbott and N. Ahuja. Surface reconstruction by dynamic integration of focus, camera vergence and stereo. In *Proc. IEEE International Conference on Computer Vision*, pages 532–543, 1988.

[2] N. Ahuja and A. L. Abbott. Active stereo: Integrating disparity, vergence, focus, aperture, and calibration for surface estimation. *IEEE Transactions on Pattern Analysis and Machine Intelligence*, 15(10):1007–1029, 1993.

[3] J. Aloimonos, I. Weiss, and A. Bandyopadhyay. Active vision. *International Journal of Computer Vision*, 1:333–356, 1988.

[4] R. Bajcsy. Active perception. *Proceedings of the IEEE*, 78:996–1005, 1988.

[5] D. Ballard and C. Brown. Principles of animate vision. In Y. Aloimonos, editor, *Active Perception*. Hillsdale, N.J. : Lawrence Erlbaum Associates, 1993.

[6] G. A. Carpenter, S. Grossberg, and D. B. Rosen. Fuzzy ART: Fast stable learning and categorization of analog patterns by an adaptive resonance system. *Neural Networks*, 4:759–771, 1991.

[7] S. Das and N. Ahuja. A comparative study of stereo, vergence, and focus as depth cues for active vision. In *Proc. IEEE Conf. on Computer Vision and Pattern Recognition*, pages 194–199, 1993.

[8] J. Denavit and R. S. Hartenberg. A kinematic notation for lower-pair mechanisms based on matrices. *ASME Journal of Applied Mechanics*, pages 215–221, 1955.

[9] E. D. Dickmanns and W. Graefe. Applications of dynamic monocular machine vision. *Machine Vision and Applications*, 1:241–261, 1988.

[10] N. J. Ferrier. The harvard binocular head. Technical Report 91-8, Harvard Robotics Laboratory, 1991.

[11] F. Fuma, E. P. Krotkov, and J. Summers. The pennsylvania active camera system. Technical Report MS-CIS-86-15, GRASP Laboratory, University of Pennsylvania, 1986.

[12] S. Hutchinson, G. D. Hager, and P. I. Corke. A tutorial on visual servo control. *IEEE Transactions on Robotics and Automation*, 12(5):651–670, 1996.

[13] S. Marriott and R. F. Harrison. A modified fuzzy artmap architecture for the approximation of noisy mappings. *Neural Networks*, 8:619–641, 1995.

[14] M. J. Swain and M. Stricker. Promising directions in active vision. Technical Report CS 91-27, University of Chicago, 1991.

[15] W. B. Thompson and J. K. Kearney. Inexact vision. In *Proc. Workshop on Motion: Representation and Analysis*, pages 15–22, 1986.

[16] G.-Q. Wei and S. D. Ma. Implicit and explicit camera calibration: Theory and experiments. *IEEE Transactions on Pattern Analysis and Machine Intelligence*, 16:469–480, 1994.

[17] L. Zadeh. Fuzzy sets. *Information Control*, 8:338–353, 1965.

Automatic Detection and Tracking of Human Heads Using an Active Stereo Vision System*

Cheng-Yuan Tang[†‡], Yi-Ping Hung[†] and Zen Chen[‡]

[†] Institute of Information Science, Academia Sinica, Taipei, Taiwan
[‡] Institute of Computer Science & Information Engineering,
National Chiao Tung University, Hsinchu, Taiwan
E-mail: hung@iis.sinica.edu.tw

Abstract

A head tracking system for automatically detecting and tracking human heads in complex backgrounds is developed. In this paper, two issues are addressed: the detection of human heads and the development of a head tracking system. First, based on an elliptical model for the human head, we propose a Maximum Likelihood (ML) detector to reliably locate human heads in images having complex backgrounds. This ellipse-based ML head detector is relatively insensitive to illumination and rotation of the human heads, and its computation is similar to template matching. Second, we develop a head tracking system that can monitor the entrance of a person, detect and track the person's head, and then control the stereo cameras to focus their gaze on this person's head. Difference images are used to detect the entrance of a human. The ellipse-based ML head detector and the mutually-supported constraint are used to extract the corresponding ellipses in a stereo image pair. Then, the 3D position computed from the centers of the two corresponding ellipses can be used for fixation. A well-calibrated active stereo head, the IIS-head, is used to perform the experiments and demonstrate that our approach is feasible and promising.

1. Introduction

Due to its potential for surveillance, security and human-computer interface, tracking human motion with computer vision techniques has become a popular research field [1][4][6][9][10][12]. In this field, many researchers are specifically interested in tracking human heads or faces [1][6][9][12]. For the problem of human head tracking, two important issues should be addressed: what to track (*i.e.*, detection of the human head) and how to track. Because it is difficult to automatically detect human heads or faces in images having complex backgrounds, much previous research either bypassed the problem of human head detection by manually locating the head in the image before tracking [1][6] or dealt only with images having simple backgrounds [3][9][15][17]. However, for many practical applications, detection and tracking of human heads have to be automatic (*i.e.*, no manual initialization is allowed) and should not be limited to simple backgrounds.

The approaches to head (or face) detection can be model-based [9][15][17][18], feature-based [2][19], neural network-based [13][14] or color-based [15][17]. In the model-based approach, several researchers modelled the shape of a human head with an ellipse, and then used ellipse fitting methods to locate the human head [9][15][17].

---
*This work was partially supported by National Science Council of Taiwan under grant NSC 86-2745-E-001-007.

However, it is not easy to extract the contour of a human head reliably when the background is complex.

Moghaddam and Pentland [13] adopted a Maximum Likelihood (ML) estimator which used an eigenspace decomposition method to detect human faces and hands. They selected a training set of facial feature templates and estimated the likelihood function. However, the computation of the eigenspace decomposition and of the likelihood distance measure used are quite expensive. Colmenarez and Huang [5] proposed another ML face detector. They presented a visual learning approach that used non-parametric probability functions and entropy analysis to build a probability model. Then, the detection of face was carried out by searching with the model over several scaled versions of the input image. In this paper, we propose a simpler ML head detector by modelling the human heads with an elliptical templates, and then develop a head tracking system based on a well-calibrated active stereo vision system [16].

2. ML Detection of Human Heads

It has been noticed that the contour of a human head can be approximated by an ellipse [9][15][17]. An ellipse can be described by the following equation:

$$\frac{(x-x_0)^2}{S_x^2} + \frac{(y-y_0)^2}{S_y^2} = 1, \tag{1}$$

where $(x_0, y_0)^T$ is the center of the ellipse. Let $r_0 = (x_0, y_0)^T$ and $s \equiv S_x = c_1 S_y$, where c_1 is a constant. Then, r_0 and s are the parameters that describe the ellipse. As shown in Figure 1, v_s^i, $i = 1, 2, ..., N$, denote the points located uniformly on an ellipse of size s centered at r_0. Then, u_s^i such that each $u_s^i = v_s^i - r_0$ represents the displacement of v_s^i from the reference point r_0. In this paper, we define the elliptical template as follows:

$$T_{r_0,s}(r) = \sum_{i=1}^{N} h_i \, \delta(r - u_s^i - r_0), \tag{2}$$

where $r = (x, y)^T$, $\delta(.)$ is the delta function, and $h_i = [h_{xi}, h_{yi}]^T$ are the weighting factors.

Based on the elliptical template, we formulate the human head detection as a Maximum Likelihood (ML) estimation problem [11]. Let G be an image defined as the vector function $G : D \rightarrow R^2$, where $D = \{(x, y) : x, y = 1, ..., M\}$ and R^2 is the set of all

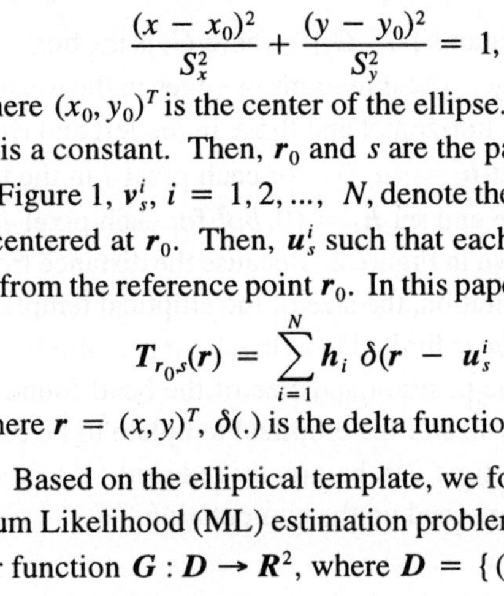

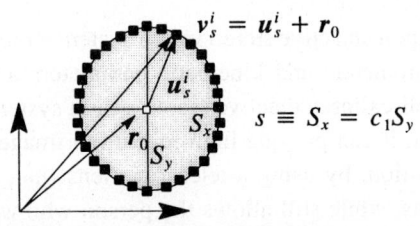

Figure 1. The elliptical template used for modelling human heads (each square on the ellipse represents an image point).

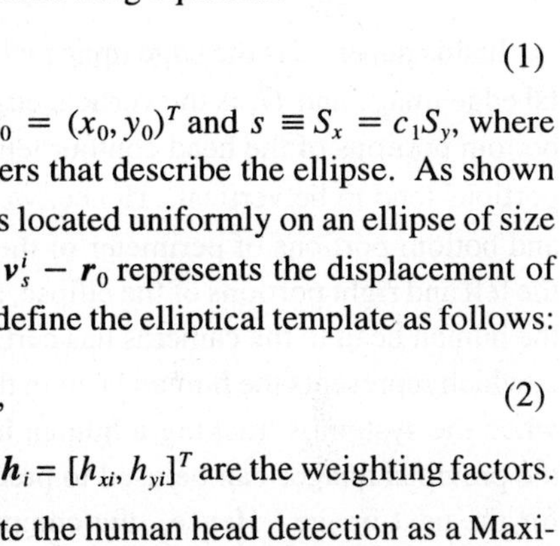

Figure 2. The horizontal edge (G_x) and the vertical edge (G_y) are used for the upper-bottom portion and the left-right portion of the elliptical head template, respectively.

possible intensity gradient vectors. Given r_0 and s, we assume that a noisy edge image G containing an elliptical contour can be modeled as follows:

$$G(r)|r_0, s = T_{r_0, s}(r) + \eta(r), \tag{3}$$

where $\eta(r)$ is assumed to be Gaussian, i.e., $\eta(r) \sim N(0, \sigma_\eta^2 I)$. Therefore, $p(G(r)|r_0, s)$ is also Gaussian, and can be shown to be

$$p(G|r_0, s) = \frac{1}{C} \exp\left\{ \frac{2 \sum_{i=1}^{N} h_i^T G(u_s^i + r_0) - G^T G - const}{2\sigma_\eta^2} \right\}, \tag{4}$$

where $C = (2\pi\sigma_\eta^2)^{N/2}$. The ML estimate of the head position, r_0, and size, s, can be obtained by maximizing $p(G|r_0, s)$, which is equivalent to

$$\max_{r_0, s} \sum_{i=1}^{N} h_i^T G(u_s^i + r_0). \tag{5}$$

In fact, the operation required by equation (5) is similar to that required by template matching. Notice that the template defined in equation (2) only has values on the ellipse.

In this paper, G is the edge image which contains $[G_x, G_y]^T$, where G_x is the horizontal edge image and G_y is the vertical edge image. The directions of edges in the top and bottom portions of the head contour tend to be horizontal and those in the left and right portions tend to be vertical. Hence, we can set $h_i = (a_i, 0)^T$ for each pixel i in the top and bottom portions of perimeter of the ellipse and set $h_i = (0, b_i)^T$ for each pixel i in the left and right portions of the ellipse, as shown in Figure 2. Because the distance from the human head to the cameras has certain limitation, the size of the elliptical template, s, which represents the human head in the image is limited to a search range. Moreover, when the system is tracking a human head, the position and size of the head found in the previous images can be used to predict the size of the elliptical template to be used for the next image. Hence, the computation time can be greatly reduced. Also, we adopt the adaptive EJO technique [8] to further speed up the computation.

3. The Head Tracking System

In this work, the head tracking system is built upon an active stereo vision system which has been calibrated so that all the camera parameters and kinematic parameters are available. The advantages of using such a well-calibrated active stereo vision system for head tracking include the following. First, it can provide high-resolution images for face detection, reconstruction, and recognition, by using a telescopic lens (*i.e.*, a lens having a long focal length) or a zoom lens, while still allows the person who we want to track to move around in a wide area. Second, it can use the epipolar geometry (or more precisely, the mutually-supported constraint described in this section) to improve the correctness of head detection. Third, it can simplify the fixation process by providing all the camera parameters and the kinematic parameters needed by the computation of inverse kinematics.

The head tracking system presented in this paper has four modes: the entrance-detection mode, the tracking mode, the fixation mode and the disappearance mode. The entrance-detection mode is used to determine whether a human head is entering into the field of view, so that the head tracking system can be completely automatic (*i.e.*, without manual initialization). For this purpose, we use the difference image between the background image (containing no human head) and the present image. For each pixel, if the absolute difference is larger than a threshold, we count one. Let N_D be the number of pixels whose absolute difference values are larger than a threshold. If N_D is larger than a pre-specified threshold, which means a target is moving into the field of view, then the system will switch to the tracking mode.

In the tracking mode, we use the ellipse-based ML head detector described in the last section to locate the human heads seen in the images. However, because of noise and the modelling error (*e.g.*, the contour of a human head is usually not an exact ellipse and the noise is usually not Gaussian), the estimate of the head position obtained by using the ML detector may occasionally be incorrect. To improve the correctness of head detection, instead of using only the most likely one, we first keep a few candidates of human heads when maximizing equation (5) and then adopt a mutually-supported constraint to eliminate those incorrect candidates. The mutually-supported constraint used in our head tracking system is briefly described below.

Consider a pair of stereo images shown Figure 3. Let r_l and r_r be the centers of the elliptical candidates found in the left and right images, respectively. Let s_l and s_r be their corresponding elliptical sizes. Since our active stereo vision system is well calibrated, given r_l in the left image, we can determine its corresponding epipolar line, EL_{r_l}, in the right image. Similarly, EL_{r_r}, the epipolar line in the left image corresponding to r_r, can also be determined if r_r is given. Next, let $Dist(r_r, EL_{r_l})$ denote the 2D distance between the point r_r and the epipolar line EL_{r_l}, which can be easily computed if r_l, r_r, and the camera parameters are given. Similar case holds for $Dist(r_l, EL_{r_r})$. In our head tracking system, two elliptical candidates specified by (r_l, s_l) and (r_r, s_r) can only form a valid stereo correspondence if they satisfy the following mutually-supported constraint: $Dist(r_r, EL_{r_l}) < Threshold_1$, $Dist(r_l, EL_{r_r}) < Threshold_1$ and $|s_l - s_r| < Threshold_2$, as illustrated in Figure 3.

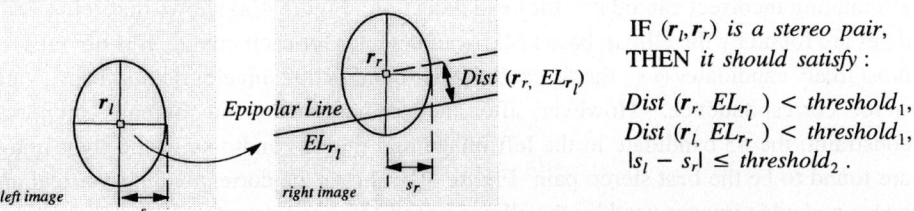

Figure 3. An illustration of the mutually-supported constraint.

Once the best and valid stereo correspondence of the elliptical candidates is obtained, it can be used to infer an ellipsoid in the 3D space since the camera parameters are available in our system. The center of this ellipsoid can be used to control the fixation of the active stereo cameras when the human head is moving toward the borders of the images. It can also be used to predict the 3D motion of the human head using Kalman filtering. This prediction is useful for reducing the search range of applying the ML head detector in the next image frame. Finally, if the human head disappears (due to tracking failure or out of surveillance range) for a few cycles, the system will first enter into the disappearance mode and then switch back to the entrance-detection mode. The algorithm is summarized below:

Entrance-Detection Mode
If N_D > Threshold, then Goto *Tracking Mode*.

Tracking Mode
(1). Search for heads by using the ellipse-based ML detector in both images and then verify them with the mutually-supported constraint.
(2). If the human head disappears for a few cycles, then Goto *Disappearance Mode*.
(3). Use the centers of the ellipses to calculate a 3D position.
(4). Estimate the 3D motion parameters and predict the head motion.
(5). If the prediction goes out of the safety-margin, then Goto *Fixation Mode*.

Fixation Mode
(1). Control the stereo cameras to focus their gaze on the moving human head.
(2). Goto *Tracking Mode*.

Disappearance Mode
(1). Grab a new pair of stereo images which contain only the background.
(2). Goto *Entrance-Detection Mode*.

4. Experimental Results

In the following experiments, the camera system we used for head tracking is the IIS head, a reconfigurable binocular head that employs eight motors to control its stereo cameras. The IIS head has been calibrated with a simple four-stage method and has achieved the accuracy of one pixel prediction error and 0.2 pixel epipolar error, even when all the eight motors are moving simultaneously [16]. Based on this calibration accuracy, our tracking system not only can use the mutually-supported constraint to verify the candidates of heads found by the ellipse-based ML head detector, but also can easily use the inverse kinematics to control the binocular head to focus their gaze on the moving human head.

First, we use an example to show the utility of the mutually-supported constraint in eliminating incorrect candidates for head detection. Figure 4(a) shows that four candidates are found by the ellipse-based ML head detector for each image. It is obvious the most likely candidates (*i.e.*, the #1 candidates) found in both images do not form a valid stereo correspondence. However, after the verification by the mutually-supported constraint, the #3 candidate in the left image and the #3 candidate in the right image are found to be the best stereo pair. Figure 4(b) shows the corresponding vertical and horizontal edge images used by the ellipse-based ML head detector.

Figure 5 shows an example of applying our head tracking system to detect and track a human head seen in an image sequence. Notice that the background of this image sequence changed most of the time because of the camera movement for fixation. Due to the camera movement, difference images can hardly be used for detecting the human head in the tracking mode. However, our ellipse-based ML head detector, together with the mutually-supported constraint, works pretty well. In Figure 5(c), the size of the head was larger than those in other images because the person moved closer to the camera system at that time instant. In Figure 5(g), the person tracked purposely rotated his head to test the performance of the tracker. In Figures 5(e) and 5(f), partial occlusion caused by the hand (or hands) did not prevent our head detector from working. In Figures 5(h) and 5(i), another person walked into the field of view unexpectedly and passed

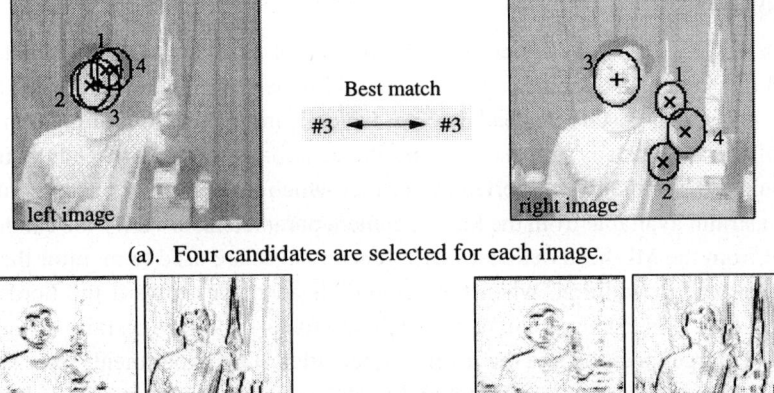

(a). Four candidates are selected for each image.

(b). The horizontal and vertical edge images computed from the above two images are used by the ellipse-based ML head detector.

Figure 4. An illustration of using the mutually-supported constraint to eliminate incorrect candidates found by the ML detector.

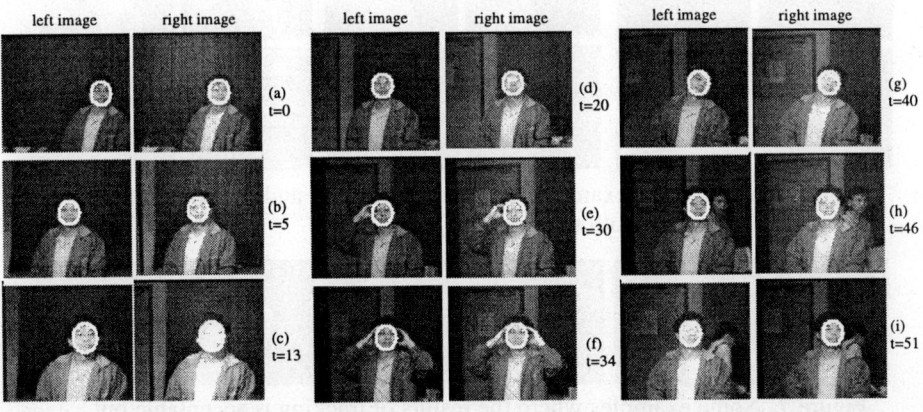

Figure 5. An example of applying our system to detect and track a human head.

through behind the person that the system was tracking, and the system can still keep track of the target.

Finally, to demonstrate the performance of our tracking system, we show a few other examples of tracking different persons in different backgrounds. Figure 6 shows some experimental results where the elliptical templates can properly locate the human heads, which are regarded as successful detection. Sometimes, the head detection may not be very precise due to occlusion, shadowing, and rotation, but the results of tracking may still be acceptable, as shown in Figure 7. Of course, our system might occasionally fail to detect the correct human head for one or two frames. However, as long as the correct human head can be detected in the next few frames, the previous detection failure can be easily recovered.

5. Conclusion

In this work, we develop an automatic head tracking system which can monitor the entrance of a person, track his head and then control the stereo cameras to focus their gaze on it. An ellipse-based ML head detector which is insensitive to illumination and rotation of human heads is proposed to find the candidates of human heads reliably. We then adopt the mutually-supported constraint, which makes use of the epipolar geometry constraint available from the known camera parameters, to verify the candidates obtained from the ML head detector. Difference images are used to monitor the entrance of a person. In addition, when the person's head moves toward the borders of the images, the stereo cameras can be controlled to focus their gaze on it by using the computed 3D target position and the inverse kinematics. Our experiments have shown that this head tracking system can detect and track human heads in complex background reliably. The tracking results (*i.e.*, the human heads) can be used as an input for face recognition systems. Also, we can use our head tracking system to compute 3D head

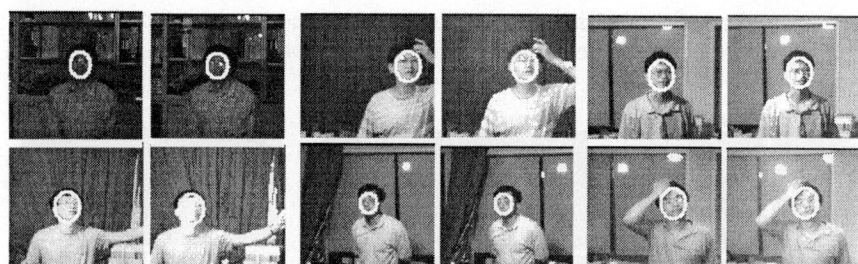

Figure 6. A few other examples where the detection and tracking of head motion are both successful.

Figure 7. Some examples where the results of tracking is acceptable but the estimates of the head position and size are not very precise.

orientation for other applications [7]. In the future, we are planning to perform the tracking and on-line reconstruction of human heads simultaneously.

References

[1] S. Basu, I. Essa and A. Pentland, "Motion Regularization for Model-Based Head Tracking", *Proc. of ICPR'96*, Vienna, Austria, Aug. 1996, pp. 611-616.

[2] M. C. Burl, T. K. Leung and P. Perona, "Face Localization via Shape Statistics", *International Workshop on Automatic Face- and Gesture- Recognition*, Zurich, 1995, pp. 154-159.

[3] G. Chow and X. Li, "Towards A System for Automatic Facial Feature Detection", *Pattern Recognition*, Vol. 26, No. 12, 1993, pp. 1739-1755.

[4] J. H. Chuang and H. Y. Chen, "A Real-Time Visual Tracking System Using FPGA", *Proc. of Conf. CVGIP*, Taiwan, 1996, pp. 143–150.

[5] A. J. Colmenarez and T. S. Huang, "Maximum Likelihood Face Detection", *2nd International Conf. on Automatic Face- and Gesture- Recognition*, Killington, Vermont, 1996, pp. 307-311

[6] A. Gee and R. Cipolla, "Fast Visual Tracking by Temporal Consensus", *Image Vision Computing*, No. 14, 1996, pp. 105-114.

[7] T. Horprasert, Y. Yacoob and L. S. Davis, "Computing 3-D Head Orientation from a Monocular Image Sequence", *2nd International Conf. on Automatic Face- and Gesture- Recognition*, Killington, Vermont, 1996, pp. 242-247.

[8] H. C. Huang, Y. P. Hung and W. L. Hwang, "Adaptive Early-Jump-Out Technique for Fast Motion Estimation in Video Coding", *Proc. of ICPR'96*, Vol. 2, pp. 864-868.

[9] A. Jacquin and A. Eleftheriadis, "Automatic Location Tracking of Faces and Facial Features in Video Sequences", *International Workshop on Automatic Face- and Gesture- Recognition*, Zurich, 1995, pp. 142-147.

[10] Y. Kameda, M. Minoh and K. Ikeda, "Three Dimensional Motion Estimation of a Human Body Using a Difference Image Sequence", *Proc. of ACCV'95*, Vol. 2, pp. 181-185.

[11] K. F. Lai and R. T. Chin, "Deformable Contours: Modeling and Extraction", *IEEE Trans. PAMI*, Vol. 17, No. 11, Nov. 1995, pp. 1084-1090.

[12] S. McKenna and S. Gong, "Tracking Faces", *2nd International Conf. on Automatic Face- and Gesture- Recognition*, Killington, Vermont, 1996, pp. 271-276.

[13] B. Moghaddam and A. Pentland, "Maximum Likelihood Detection of Faces and Hands", *Proc. of Inter. Workshop on Automatic Face– and Gesture– Recognition,* Zurich, 1995, pp. 122–128.

[14] H. A. Rowley, S. Baluja and T. Kanade, "Human Face Detection in Visual Scenes", Tech. Rep., Carneige Mellon Univ., 1995.

[15] E. Saber and A. M. Takalp, "Face Detection and Facial Feature Extraction Using Color, Shape and Symmetry-Based Cost Functions", *Proc. of ICPR'96*, Vol. 3, pp. 654-658.

[16] S. W. Shih, Y. P. Hung and W. S. Lin. "Four Stage Method for Accurate Calibration of an Active Binocular Head", *Proceedings of the Workshop on 3D Computer Vision 97*, 1997, pp. 49-56.

[17] K. Sobottka and I. Pitas, "Extraction of Facial Regions and Features Using Color and Shape Information", *Proc. of ICPR'96*, Vol. 3, pp. 421-425.

[18] G. Yang and T. Huang, "Human Face Detection in a Complex Background", *Pattern Recognition*, Vol. 27, No. 1, 1994, pp. 53-63.

[19] K. C. Yow and R. Cipolla, "A Probabilistic Framework for Perceptual Grouping of Features for Human Face Detection", *2nd International Conf. on Automatic Face- and Gesture- Recognition*, Killington, Vermont, 1996, pp. 16-21.

Front Propagation and Level-Set Approach for Geodesic Active Stereovision

Rachid DERICHE - Christophe BOUVIN - Olivier FAUGERAS

INRIA, 2004 route des Lucioles, BP 93
F-06902 Sophia-Antipolis Cedex, France
Email: der@sophia.inria.fr

Abstract. A framework for matching complex 2D planar curves lying at the intersection of a 3D virtual plane and the 3D scene being observed using a weakly calibrated stereo system, is presented in this article. Using an energy minimization based approach, we reformulate this stereo problem as a front propagation problem. The Euler Lagrange equation of the designed energy functional is first derived and the flow minimizing the energy is obtained. The curves to be matched are modelized as geodesic active contours [2, 8, 7] evolving toward the minimum of the designed energy, under the influence of internal and external correlation image dependent forces. This original scheme may be viewed as a geodesic active stereo model which basically attract the given curves to the bottom of a potential well corresponding to pixels having similar intensities. Using the level set formulation scheme of Osher and Sethian [9], complex curves can be matched and topological changes for the evolving curves are naturally managed. The final result is also relatively independent of the curve initialization. Promising experimental results have been obtained on various real images

1 INTRODUCTION

In this paper, an original approach to deal with the important problem of recovering geometric information from a weakly calibrated stereo pair of images is presented. Given two different views of a 3D scene, and a virtual 3D plane, the method we propose allows to recover the 2D projections, in the two images, of the 3D planar curves corresponding to the intersection of the virtual plane with the different objects in the scene being observed. The 3D plane is specified just by the knowledge of the 2D projections on the two images of 3 points lying in this plane. An arbitrary curve is first initialized in one of the two images. This curve, and its associated homographic curve in the second image are then designed to move under the influence of internal and external image dependent forces while minimizing an energy functional. The Euler-Lagrange equation of this new functional is derived and its associated PDE is then solved using the level set formulation scheme of Osher and Sethian [9] by viewing it as a front propagating with internal and external image correlation dependent speed [2, 8, 7]. With this technique, complex curves can be matched, and the final result is relatively independent of the curve initialization. Topological changes for the evolving curves are also naturally obtained in this setting.

The problem addressed is very important from the point of view of the applications. Matching 2D planar curves provides useful 3D information as the *relative positioning* of any point in the scene with respect to this plane. This is an important application in robotics tasks as *obstacle avoidance*. This technique can also help in *providing all the contours located at the same depth (iso-height)* from a stereo pair of aerial or satellite image. It can also be applied in an *automated surveillance system*, where it can be used to detect any entrance through a virtual specified 3D plane...The ideas presented in this paper can also be applied to develop different computer vision applications such as motion detection and tracking [11, 10], shape from shading,.

The outline of this paper is as follows : In section 2, we first present the approach we have developed to specify interactively the virtual 3D plane. We then formulate our new geodesic active stereo paradigm in section 3, and make use of the Osher-Sethian level set approach to implement the derived flow in section 4. Experimental results obtained on real pair of images are then shown in section 5 before we draw our conclusion in section 6.

2 Setting the 3D virtual plane

In this section, we present the approach we have developed to specify interactively the 3D virtual plane. Let $C(p) : [0,1] \leadsto R^2$ be a parameterized planar curve. Let $I_l : [0,a] * [0,b] \leadsto R^+$ and $I_r : [0,a] * [0,b] \leadsto R^+$ be the left and right images provided by a weakly calibrated stereo system.

Let's suppose that two cameras are looking at a plane Π. Given a 3D point M and its 2D projections m in the left image I_l and m' in the second image I_r, then it is well known that there is a linear relationship in projective space between the points m and m' :

$$\mathbf{m'} = H_\Pi \mathbf{m}$$

. H_Π is a $3 * 3$ matrix defined up to a scale factor. It defines the homography of the plane Π between the two images. Since for each pair of matched point $(m.m')$, we have 2 equations involving the pixel coordinates of m and m', then it is easy to see that four matched points allows to recover H_Π.

For a large class of images, it has been found that it is very difficult for an external user to specify a plane by giving four matched points lying in this plane. However, due to the fact that a plane is completely defined by 3 points in the space, it is much more easier for an user to specify this plane by giving interactively the 3 points. This is possible if we work with a weakly calibrated system, that is to say a stereo system where the epipolar geometry is known through the knowledge of the so-called *Fundamental matrix* **F**. In such case, one possible solution to find H_Π is to solve the associated equation using the set of 4 points including the three pair of points matched by the user, and the pair of epipoles **e** and **e'** determined from the knowledge of **F** and considered as the fourth pair of corresponding points [3] (See Figure (1) for an easy geometrical interpretation of this point). This way to specify the plane Π has been tested and found to be much more satisfactory and much more robust against noise than the first solution.

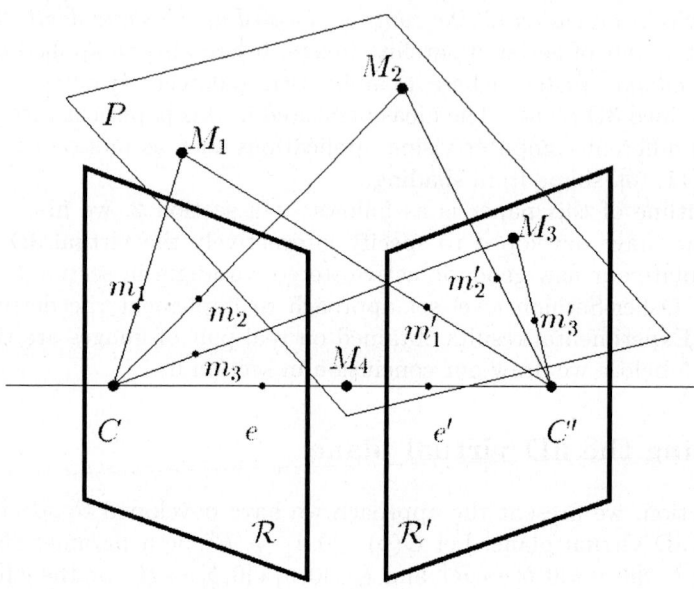

Fig. 1. 3 matched points and the epipoles allow to recover the homography of the plane from where the matched points are issued

3   Geodesic active stereo

In this section, we are interested in solving the following problem : Given two different views of some 3D objects, and a virtual 3D plane set by an external operator, recover the 2D projections, in the two images, of the 3D planar curves corresponding to the intersection of the virtual plane set by the user. Our key idea is to proceed as follows : An arbitrary curve $C(p)$ is first initialized in one of the two images. This curve, and its matched curve in the second image $H_\Pi C(p)$ are then represented by two propagating fronts forced to move towards the 2D projections, in the two images, of the 3D planar curves corresponding to the intersection of the virtual plane Π with the different objects in the scene. This leads us to define a speed function from the image data that can be applied in the propagating front as a stopping criterion that attains zero as the two curves get matched. In order to derive a well adapted stopping criterion, we use the framework of energy minimization and express our problem as a geodesic active stereo problem.

we associate to the given curve an energy functional $E(c)$ and try to find for a given set of parameters α, λ, the curve $C(p)$ and $C'(p) = H_\Pi C(p)$ that minimizes the following energy $E(C)$: [6, 2]

$$\underbrace{\alpha \int_0^1 |C'(p)|^2 \, dp}_{E_s(C)} + \underbrace{\lambda \int_0^1 (I_l(C(p)) - I_r(H_\Pi C(p)))^2 \, dp}_{E_D(C)} \quad (1)$$

where α is a real positive constant related to the *elasticity* of the curve, $E_s(C)$ an energy term that basically control the smoothness of the contours to be detected, $E_D(C)$ stands for the attraction energy term of the curves towards the same contours of the object and λ, a real positif constant weighting the actions of the two energy terms $E_s(C)$ and $E_D(C)$.

Following the work on geodesic active contours presented in [2], we then transform the problem of minimizing this energy functional into a problem of geodesic computation in a Riemannian space, according to a new metric defined by $g_{ij}dx_i dx_j$ with $g_{ij} = (I_l(C(p)) - I_r(H_\Pi C(p)))^2 \delta_{ij}$:

$$E(C) = \int_0^1 |(I_l(C(p)) - I_r(H_\Pi C(p)))| |C'(p)| \, dp \quad (2)$$

Since $|C'(p)dp| = ds$ where ds is the Euclidean arc-length element, this scheme may be viewed as a geodesic active stereo model which basically attract the given curves to the bottom of a potential well corresponding to pixel having similar intensities.

The Euler-Lagrange equation of this new functional is then derived and the following flow that deforms the initial curve $C(p, 0) = C_0(p)$ towards the local minima of (2) is then obtained :

$$\begin{cases} \frac{\partial C}{\partial t} = (|(I_l(C(p)) - I_r(H_\Pi C(p)))|\kappa \\ \quad + \nabla(|I_l(C(p)) - I_r(H_\Pi C(p))|).N)N \\ C(p, 0) = C_0(p) \end{cases} \quad (3)$$

where N denotes the inward Euclidean normal vector to the curve $C(p, t)$ and κ represents the Euclidean curvature. Lets's mention that the problem tackled in this article, and the way we proceed to solve it, using geodesic active contours and the level-set approach, is completely different from some previous work dealing with the application of classical deformable contours to stereo [1, 6].

4 A level Set Approach to Implement the flow

A classical numerical approach to implement the flow (3) is to consider the Lagrangian formulation by producing the associated equations of motions for the position vector p and updating these positions using a simple difference approximation scheme. Due to the fact that this formulation leads to a physical coordinate system that moves with the front, stability problems may occur during the evolution process and topological changes in the moving front are almost impossible to manage.

An alternative technique solving these serious limitations has recently been presented by S.Osher and J.A.Sethian [9]. The flow (3) is adapted to this level set formulation scheme by viewing the propagating curve as the zero level set $u = 0$ of a time dependent surface $u(p.t)$. The one parameter family of moving curves $C(p,t)$ is then associated with a one parameter family of moving surfaces $u(p,t)$ in such a way that the level set $u = 0$ always yields the moving front. The following resulting PDE equation acting on u is derived and solved using techniques borrowed from hyperbolic conservation laws [9]

$$\begin{cases} \frac{\partial u}{\partial t} = |\nabla u|(|(I_l(C(p)) - I_r(H_\Pi C(p)))|div(\frac{\nabla u}{|\nabla u|})) \\ \qquad + \nabla(|I_l(C(p)) - I_r(H_\Pi C(p))|).\nabla u \\ u(x,y,0) = u_0(x,y) \end{cases} \quad (4)$$

Using this technique, complex curves can be matched, and the final result is relatively independent of the curve initialization. Topological changes for the evolving curves are also naturally obtained in this setting, since the level surface in which the evolution is performed, needs not be simply connected. Hence, during their evolution, the matched curves may change connectivity, split and merge, allowing the simultaneous detection and matching of all the planar curves.

On implementing this scheme, various solutions have also been tested in order to obtain the best possible results. Hence, as it it was expected, using just the difference between the grey level values of the matched points in the criterion (1) to be minimized has not been found sufficient to move the curves towards the expected solution. In order to robustify this approach, the following better solution has been considered : Replace the term taking into account the difference in grey level values by a more classical correlation term as the following one [5] :

$$C_6(x,y) = \frac{\sum_{i,j} A_{i,j} B_{i,j}}{C.D} \begin{cases} A_{i,j} = [I_l(x+i,y+j) - \overline{I_l(x+i,y+j)}] \\ B_{i,j} = [I_r(x+i,y+j) - \overline{I_r(x+i,y+j)}] \\ C_{i,j} = \sqrt{\sum_{i,j}[I_l(x+i,y+j) - \overline{I_l(x+i,y+j)}]^2} \\ D_{i,j} = \sqrt{\sum_{i,j}[I_r(x+i,y+j) - \overline{I_r(x+i,y+j)}]^2} \end{cases} \quad (5)$$

where $\overline{I}$ denotes the average of I on a given neighborhood. Another interesting implementation issue that has been shown to improve the quality of the results has been to work on rectified images. The best rectification is the one performed with respect to the plane Π given by the user, leading to a null disparity for all the points of the space lying in the specified plane. Readers interested by the level set methods and the stereo problem may refer to [4] where an extension of this approach for an arbitrary number of images is considered.

5 EXPERIMENTAL RESULTS

We now present some experimental results obtained using this geodesic active stereo algorithm. The numerical implementation is based on the algorithm for surface evolution via level set developed by Osher and Sethian. In all the figures, the initial curve has been given in the right image and correspond to the exterior rectangular curve,

Figure 2 presents the evolution of the geodesic stereo pair on a real textured and deformed cube. Note how far the initialization has been given and how accurate is the final matching. The plane has been specified by the three points that can be seen in white.

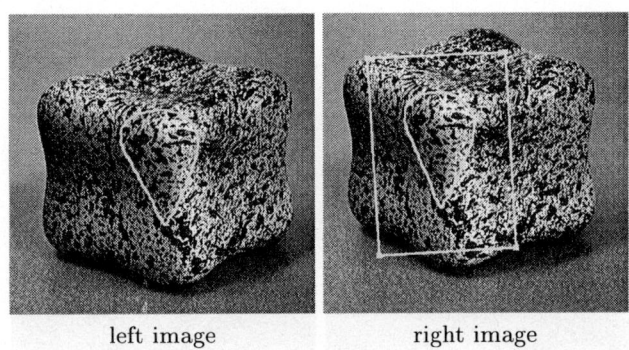

left image　　　　　　　right image

Fig. 2. Geodesic active stereo acting on a textured and deformed cube. The initial curve is the rectangular one in the right image. The 2 other curves are the results of the geodesic stereo matching process

Figure 3, illustrates the capability of the method on an image representing a human face.

Figure 4 presents the evolution of the geodesic active stereo algorithm on Hervè's head. This example is interesting because it illustrates the capability of the method within not sufficiently well textured areas as Hervè's face.

Figure 5 presents the evolution of the geodesic active stereo algorithm on a real textured and deformed cube but with a 3D plane intersecting the cube in two different parts. This example is particularly interesting because it illustrates the capability of the method to deal with different intersections and to effectively manage the change of topology.

An MPEG version of all the examples shown in this article can be found at the first author WEB page http://www.inria.fr/robotvis/personnel/der/demos.html Others experimental results can also be found at this page.

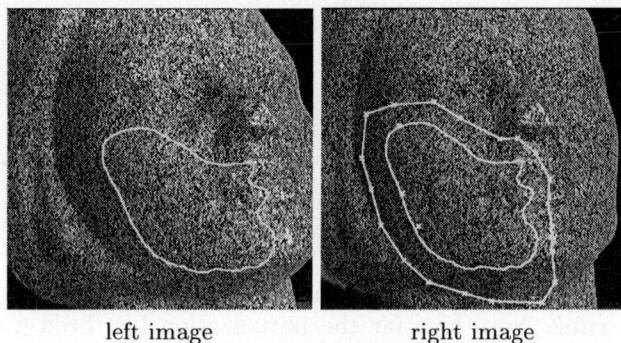

left image　　　　　　　　right image

Fig. 3. Geodesic active stereo acting on an human face : The initial curve is the exterior one in the right image. The 2 other curves are the results of the geodesic stereo matching process.

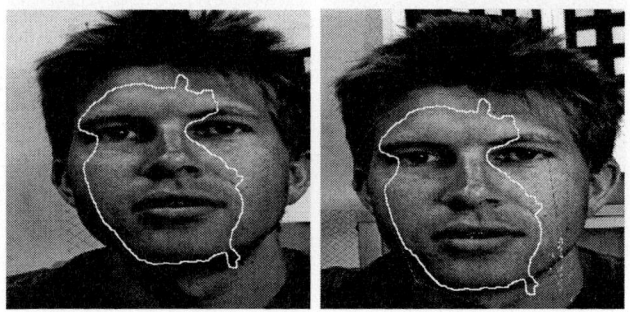

Fig. 4. Geodesic active stereo acting on Herve's face

6 CONCLUSION

Since the pioneering work of S. Osher and J.A. Sethian in developing level set methods for capturing fronts, these techniques have been applied to a wide range of applications, including problems in fluid mechanics, combustion, computer animation, image processing, robotic navigation etc. More recently these techniques started to be applied in the computer vision domain including motion detection and tracking [11, 10], shape from shading, geodesic active contours, skeletons, color scale space, and more. We have shown in this article that the level set method can also be applied effectively for solving 3D vision problems such as stereo and that promising results can be obtained using this interesting tool.

References

1. B. Bascle and R. Deriche. Stereo matching, reconstruction and refinement of 3-D curves using deformable contours. In *Proceedings of the 4th International Conference On Computer Vision*, Berlin, Germany, 1993.

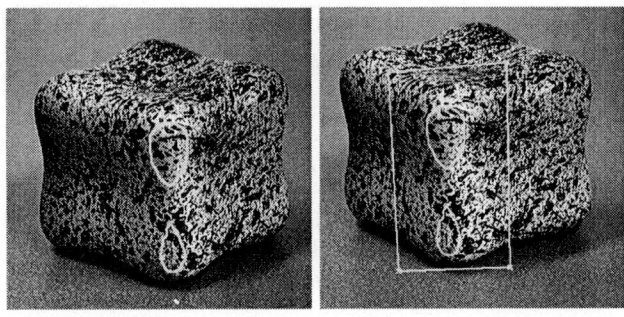

left image   right image

Fig. 5. Geodesic active stereo acting on a textured and deformed cube (II). The initial curve is the rectangular one in the right image. The 4 other curves are the results of the geodesic stereo matching process. Note how the initial rectangular curve has been split into 2 curves during its evolution

2. V. Caselles, R. Kimmel, and G. Sapiro. Geodesic active contours. Technical report, HP Labs, September 1994. A shorter version appeared at 5th ICCV'95 - Boston.
3. Gabriella Csurka, Cyril Zeller, Zhengyou Zhang, and Olivier Faugeras. Characterizing the uncertainty of the fundamental matrix. *CVGIP: Image Understanding*, 1997. To appear.
4. O. Faugeras and R. Keriven. Variational principles, surface evolution, pde's, level set methods and the stereo problem. *IEEE Trans. on Image Processing*, 1997. To appear in a Special Issue on Partial Differential Equations and Geometry-Driven Diffusion in Image Processing and Analysis.
5. Olivier Faugeras, Bernard Hotz, Hervé Mathieu, Thierry Viéville, Zhengyou Zhang, Pascal Fua, Eric Théron, Laurent Moll, Gérard Berry, Jean Vuillemin, Patrice Bertin, and Catherine Proy. Real time correlation based stereo: algorithm implementations and applications. Technical Report 2013, INRIA Sophia-Antipolis, France, 1993. Submitted to *The International Journal of Computer Vision*.
6. M. Kass, A. Witkin, and D. Terzopoulos. Snakes: Active contour models. In *First International Conference on Computer Vision*, pages 259–268, London, June 1987.
7. R. Malladi, J. A. Sethian, and B.C. Vemuri. Shape modeling with front propagation: A level set approach. *PAMI*, 17(2):158–175, February 1995.
8. R. Malladi, J.A. Sethian, and B.C. Vemuri. A topology independent shape modeling scheme. *SPIE*, 2031:246, 1993.
9. S. Osher and J. Sethian. Fronts propagating with curvature dependent speed : algorithms based on the Hamilton-Jacobi formulation. *Journal of Computational Physics*, 79:12–49, 1988.
10. N. Paragios and R. Deriche. A PDE-based Level Set Approach for Detection and Tracking of Moving Objects. Technical Report 3173, INRIA, France, May 1997. http://www.inria.fr/rapports/sophia/RR-3173.html.
11. Nikolaos Paragios and Rachid Deriche. Detecting multiple moving targets using deformable contours. In *International Conference on Image Processing*, Santa Barbara, CA, USA, October 1997.

A Bayes Nets-Based Prediction/Verification Scheme for Active Visual Reconstruction

Éric Marchand, François Chaumette

IRISA / INRIA Rennes
Campus de Beaulieu
35042 Rennes-cedex, France

Abstract. *We propose in this paper an active vision approach for performing the 3D reconstruction of polyhedral scenes. To perform the reconstruction we use a structure from controlled motion method which allows an accurate estimation of primitive parameters. As this method is based on particular camera motions, perceptual strategies able to appropriately perform a succession of such individual primitive reconstructions are proposed in order to recover the complete spatial structure of complex scenes. The algorithm described in this paper is based on the use of a prediction/verification scheme managed using decision theory and Bayes nets. It allows the visual system to get a more complete high level description of the scene.*

1 Overview

Our goal is to obtain a complete and precise description of a scene using the visual data provided by a controlled camera mounted on the end effector of a robot arm. The idea of using active schemes to address vision issues has been introduced a few years ago. Since the major shortcomings which limit the performance of vision systems are their sensitivity to noise, their low accuracy and their lack of reactivity, the aim of active vision is generally to elaborate control strategies for adaptively setting camera parameters (position, velocity,...) in order to improve the knowledge of the environment. In our application, the purpose of active vision is handled at two levels: a **local aspect** where active vision is used to constrain the camera motion in order to improve the quality of the reconstruction results, and a **global aspect** which is used to ensure full scene reconstruction.

The method we have used to estimate the 3D structure of the primitives assumed to be present in the scene is fully described in [3]. It is based on the measure of the camera velocity and the corresponding motion of the primitive in the image. More precisely, we use a *"structure from controlled motion"* method which consists of constraining the camera motion in order to obtain a precise and robust estimation of 3D geometrical primitives. When no particular strategy concerning camera motion is defined, important errors in the 3D structure estimation can be observed. This is due to the fact that the quality of the estimation is very sensitive to the nature of the successive camera motions. An active vision paradigm is thus necessary to improve the accuracy of the estimation results by generating adequate camera motions. Indeed, it has been shown that two vision-based tasks (called *fixation* and *gazing* tasks) have to be realized in order to obtain a robust and non-biased estimation. In this paper, we restrict ourselves to polyhedral objects. The only considered primitives are thus 3D segments, which must appear centered and vertical (or horizontal) in the image during the camera motion.

Since the proposed structure estimation method involves fixating at and gazing on the different primitives in the scene, this can be done on only one primitive at a time, and reconstructions have to be performed in sequence for each primitive of the scene [7]. Our incremental strategy leads to an exploration process which is handled at two levels:

- When a new primitive appears in the camera field of view, it is estimated. In that case, we do not need to compute explicitly new viewpoints. This level is called **local exploration**. It allows to split the observed areas into free-space and reconstructed objects and to bridge the gap between a local model of the scene in terms of isolated 3D segments and a global model in terms of objects (segments, junctions, and polygons).
- When a local exploration ends, a more complex strategy is used in order to gaze on parts of the 3D space which have not been already observed. This level is called **global exploration** and is described in [7].

This paper deals with the local exploration strategies. They are based on a prediction/verification scheme which is described in Section 2. Finally, we present real-time experimental results carried out on a robotic cell in Section 3.

2 A Bayes Nets-Based Prediction / Verification Scheme

Let us consider the scene depicted on Figure 1.a. The model obtained using a simple incremental reconstruction algorithm is given in Figure 1.b. We can notice that the 3D model is composed of five segment which *a priori* came apart, a few segments have not been taken into account because of their small size, and a long segment has not been estimated (because it was always occulted).

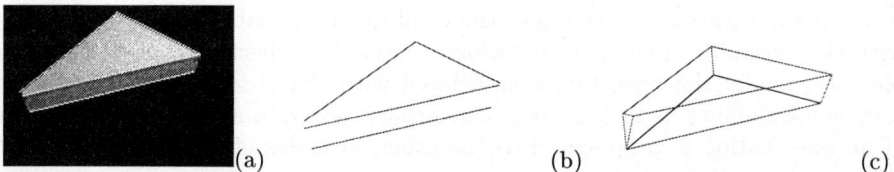

Fig. 1. Polyhedral scene: (a) view of the scene, (b) model of the "polyhedron" scene acquired using an incremental algorithm, (c) model of the same scene acquired using the prediction/verification scheme

To cope with these problems, we propose a prediction/verification scheme based on the use of Bayesian networks which allows us to obtain a high level description of the scene. As an example, the method proposed afterwards allows us to complete the model of the object depicted in Figure 1.a as shown on Figure 1.c.

2.1 Prediction/Verification Scheme and Bayes Nets

Prediction/verification approaches intend to solve the problem of the agreement between models and data. In our application, uncertainty appears either in the 3D data acquired using the structure from motion approach or in the extraction of segments in images. Mainly, the consequence of this uncertainty is the confrontation of different possible alternatives for guiding the reconstruction and the exploration of the scene. The goal of decision theory is to provide well defined and mathematical approaches for making a decision in presence of uncertainty. Bayes nets [8] seem to be well adapted to our problem. They allow us to model "expert" reasoning. They are adapted to the automatic generation of action while performing this reasoning. Thus we can directly introduce perception strategies within the scene interpretation process. Bayes nets have been already used in computer vision (e.g. [1], [10]). Using Bayes nets in active vision is more recent. Most discriminant works have been proposed by Rimey and Brown [9] with the TEA-1 system (selective perception for visual search), Buxton and Gong [2] (traffic analysis), and Djian and Rives [4] (for object recognition). The goal of these systems, including ours, are obviously different ; however, the realization of the task always requires the execution of perception actions such as sensor motions.

Bayes nets allows us to represent join probabilities distributions of a set of variables using a set of *a priori* knowledge on the relations between these variables. A Bayes net is a directed acyclic graph where nodes represent the discrete random variables and where links between nodes represent the causality between the variables. Such a net can be used to represent the knowledge available on a particular domain. The graph structure and the *a priori* knowledge introduced in the graph (as conditional probability tables) must be defined in function of the application. The advantages of Bayes nets lies in the ability to reflect the *a priori* knowledge available on the application. The knowledge is reflected at two levels: first in the structure of the net through the nature and the number of nodes (variables), the different states of these variables and the relations (links) between these variables; second in the conditional probability tables associated with the variables of the net and which reflect the expert reasoning. These tables also model the uncertainty associated with the observations. Finally, the propagation allows us to take each new observation into account. The influence of an observation is propagated to the other variables of the net according to the causality relations.

2.2 Our approach

The information available on the scene is composed by a set $\mathcal{S}(\mathcal{T}_0^{t-1})$ of 3D segments. $\mathcal{S}(\mathcal{T}_0^{t-1})$ is a subset of $\mathcal{O}(\mathcal{T}_0^{t-1})$ which represents all the known objects of the scene (*i.e.* 3D segments but also junctions, polygons, etc.). The goal is to determine the relations between segments and to infer either the presence of new segments either the existence of more complex objects. As our reconstruction is incremental, we have to determine the consequence of the introduction of a new segment S_t in $\mathcal{S}$. Therefore, this module is used each time that a new segment is introduced in $\mathcal{S}$.

Our approach can be decomposed into three steps. For each couple of segments $(S_{t'}, S_t), t' \in [0, t-1]$, we propose hypotheses on the relation between these two segments. Then, we verify if these hypotheses match the observations. Finally, the system propose a new model of the scene resulting from the integration of the new segment.

Prediction Dealing with two segments $S_{t'}$ and S_t, the possible actions are the followings: fuse the segments, create a junction, or add a link (a new segment) between $S_{t'}$ and S_t. Therefore the aim of the prediction step is to create some hypotheses leading to the realization of one (or more) of these actions. The hypotheses are directly linked to the actions:

- H_1: there is a junction between $S_{t'}$ and S_t ;
- H_2: there are one or two segments between $S_{t'}$ and S_t.
- H_3: $S_{t'}$ and S_t are identical ;
- H_4: there are no (or some other) relation between $S_{t'}$ and S_t.

We have a multi-step strategy. First, we compute the belief we have in simple topological relations between $S_{t'}$ and S_t (proximity ($p(N)$), coplanarity ($p(C)$), and collinearity ($p(P)$)). Then, according to these beliefs, it is possible to classify the pair of segments into five classes (see the first raw of the Table in the Fig. 2). Classes are C_1: CNP (coplanar, neighbor and parallel) , C_2: $CN\neg P$, C_3: $C\neg NP$, C_4: $C\neg N\neg P$, and C_5: $\neg C\neg N\neg P$.

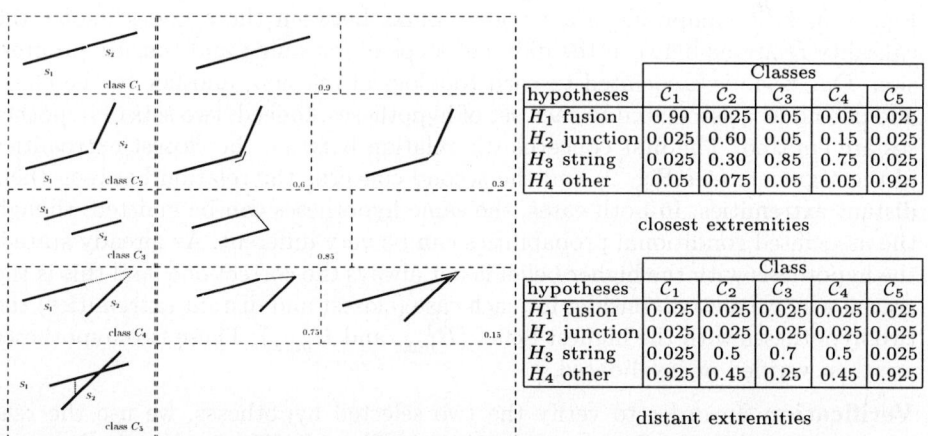

Fig. 2. Elementary classes and associated hypothesis (closest extremities), and conditional probabilities table P(hypotheses | classes) for the closest and distant extremities

Using the belief we have in the belonging of the couple of segments to each class, the system can infer the belief in each possible hypothesis. We have defined decision strategies which are able to determine the best hypothesis according to the available knowledge. These strategies are coded in conditional probability tables $P(\mathcal{H}|C)$ where $\mathcal{H}$ is the hypothesis and C the class (see Fig. 2). These tables are defined in an empirical way from a set of elementary considerations about topological relationship that we usually find in a group of segments. These considerations often reflect the truth, though they provide no guarantee. However, extreme precision is not required. Rather, they must reflect the knowledge we want to transmit to the system.

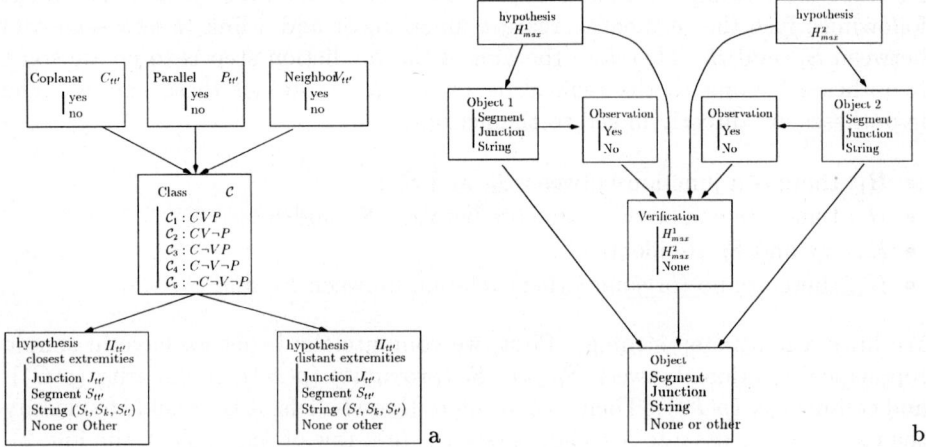

Fig. 3. (a) Prediction net, (b) Verification net

The prediction step reasoning can be encoded in a simple Bayes net (see Fig. 3.a). It is composed of six nodes. Links between these nodes depict the causality relations between the different steps of reasoning and thus its progression. One node is associated to each topological relation, another to the class, and one node is associated to each set of hypotheses. Indeed, two sets of hypotheses are emitted. The first concerns the relation between the closest extremities of the segments (see Fig. 2) and the second concerns the relation between their distant extremities. In both cases, the same hypotheses can be emitted, though the associated conditional probabilities can be very different. As already stated, the hypothesis with the higher belief is not always the correct one, and this is the reason why we always consider for each case (closest and distant extremities) the two hypotheses with the highest belief (H^1_{max} and H^2_{max}). These two hypotheses are then verified or invalidated.

Verification In order to verify the two selected hypotheses, we use the reasoning encoded in the Bayes net depicted in Fig. 3.b. We use two similar nets, each associated with one of the two sets of hypotheses (*i.e* close and distant extremities). Considering the two hypotheses, we first define the nature (segment, junction, string) and the position of the created object associated with each hypothesis. Then, we compute the belief in the existence of this object using the observation node. Finally, knowing the belief in each hypothesis and the belief in the related observation, it is possible to determine the most probable hypothesis (or to reject both).

The most important node in the verification net is the observation node. Sometimes, the hypotheses can be verified (or invalidated) using direct observation in the images previously acquired. In such cases, the validation is performed using the 3D information associated with the hypotheses and the 2D observation. We perform a back-projection of the 3D objects in each image previously acquired by the camera and we try to associate this projection to the observed data in more than one image (to avoid false matching). For each possible matching, we compute the belief granted to this matching. The case of a single segment or of a junction is simple. If this junction exists, it has already been observed (because the presence of the two segments has been already verified). Thus, the verification is performed as described above. In the case of a string, with three segments, the presence of two of them is certain (they have been used to predict the presence of the third). However, the last one has not been yet reconstructed (most of the time), and its presence is not validated. When no matching is found in the images previously acquired, it is necessary to know why. The first possibility is that the segment under consideration does not exist, the second is that it is occluded by another object. In the latter case, it is necessary to move the camera to a new viewpoint from which the segment can be observed. Rather than computing explicitly a viewpoint (*e.g.* [11]) and researching *off-line* the considered segment, we prefer to turn the camera around a segment which belongs either to the occluding polygon or to a plane to which the considered segment belongs. During this motion, automatically generated by visual servoing [5], an image processing is performed *on-line* to detect the appearance of the researched segment.

Modeling. At this step of the reconstruction process, we have a model of the scene composed of 3D segments, 3D junctions, or even a coplanar string of segments. It is finally quite easy to use this information in order to get 3D polygons. To this end, we use the junction information and the coplanarity information already used in the hypotheses generation (see [6] for further details). This three-step approach allows us to get a high level and more complete representation of the scene.

3 Experimental results

We present in this section the reconstruction results obtained for a polyhedral object (see Fig. 4.a). This scene allows us to illustrate the interests of the proposed method. As already stated, as they are too small, some of the vertices of the polyhedron may be not reconstructed using a simple incremental reconstruction process. Furthermore, due to the local approach used in that process, others remain occluded and thus non reconstructed. We now focus on two aspects of the Bayes nets prediction verification scheme.

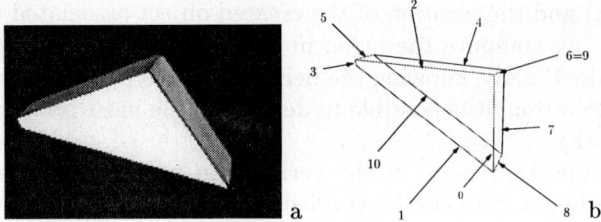

Fig. 4. Polyhedral scene: (a) view of the scene, (b) model of the same scene acquired using the prediction/verification scheme and numbering of the reconstructed segments in the order of their introduction in the 3D map

Fig. 5. Polyhedral scene: arrows point at the next primitive to be estimated

Consider that segments S_0 and S_1 have been already estimated and that S_2 has just been reconstructed (see Fig. 5.abc), the system considers the relation between S_2 and S_0 and between S_2 and S_1. Dealing with S_2 and S_0, the system concludes easily to the presence of a junction between them. Dealing with the couple (S_1, S_2), there is around 1cm between their closest extremities. The belief for S_2 and S_1 to be neighbor is 61% and to be coplanar is 99% ; thus they are likely to belong to the class C_2. According to the strategies encoded in the hypotheses Bayes net, it is likely that there exists a junction with a 46% belief and a a segment between them with a 41% belief. The remaining 13% are shared between the two other hypotheses. After the verification process, and according to the observations, the former hypothesis (junction) is verified with a 60% belief. This high value (even if this hypothesis is false, see Fig. 4.a) results from the fact that these two segments are very close in the different images (around 5 pixels). Thus the observations reinforce this hypothesis. However, the latter hypothesis is verified with a 95% belief. Indeed, a 2D segment is observed at the predicted position in many images. Finally, according to the belief in each hypothesis, to the belief in the observations, a new segment S_3 is added to the model of the scene (with a confidence of 53%, while the confidence in a junction creation is only 37%). This underlines the interest to consider a multi-hypotheses approach. A classical approach might have chosen the first (and wrong) hypothesis.

Let us now consider a second interesting case. When segment S_7 is reconstructed, within relations with other segments, the system proposes the creation of a junction with S_4 and the creation of a segment between their two distant extremities. Such a segment has never been observed (and could not have been observed according to the current knowledge on the scene and on the camera trajectory). Therefore, as described in the previous section, the camera gazes on S_7, and turns around it (see Fig. 6). During this motion, automatically generated by visual servoing, observers are looking for a moving segment located at its expected position in the images. The discovered segment is then reconstructed and introduced in the scene model (see Fig. 6.c).

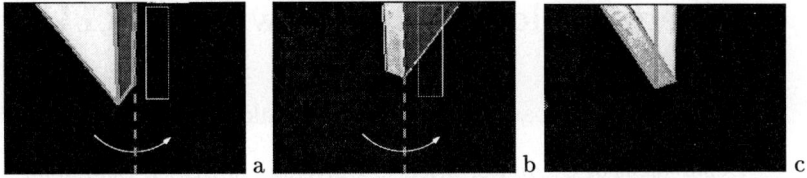

Fig. 6. Verification of a hypothesis: (a) rotation around S_7, (b) S_{10} is discovered, (c) and reconstructed

4 Conclusion

The main goal of the prediction/verification approach was to bridge the gap between a representation in terms of isolated primitives and a representation in terms of objects. Sub-goals were to deal with small segments and to propose a partial solution to the occlusion problem. Although the considered scene and images are quite simple, the resulting system is able to perform a complete and accurate reconstruction of these scenes using images acquired and processed at nearly video rate. Finally, experiments carried out on a robotic cell have proved the validity of our approach.

References

1. J.M. Agosta. The structure of Bayes nets for vision recognition. In *Proc. of 4th Workshop on Uncertainty in Artificial Intelligence*, Minneapolis, Aug. 1988.
2. H. Buxton, S. Gong. Visual surveillance in a dynamic and uncertain world. *Artificial Intelligence*, 78(1-2):431–459, Oct. 1995.
3. F. Chaumette, S. Boukir, P. Bouthemy, D. Juvin. Structure from controlled motion. *IEEE PAMI*, 18(5):492–504, May 1996.
4. D. Djian, P. Probert, P. Rives. Active sensing using Bayes nets. In *Proc. of Int. Conf. on Advanced Robotics, ICAR'95*, pages 895–902, Sant Feliu de Guixols, Spain, Sep. 1995.
5. B. Espiau, F. Chaumette, P. Rives. A new approach to visual servoing in robotics. *IEEE Trans. on Robotics and Automation*, 8(3):313–326, June 1992.
6. E. Marchand. *Stratégies de perception par vision active pour la reconstruction et l'exploration de scènes statiques*. PhD thesis, Univ. of Rennes 1, IRISA, June 1996.
7. E. Marchand, F. Chaumette. Controlled camera motions for scene reconstruction and exploration. In *CVPR'96*, pages 169–176, San Francisco, June 1996.
8. J. Pearl. *Probabilistic reasoning in intelligent systems : Networks of plausible inference*. Morgan Kaufmann Publisher Inc., 1988.
9. R.D. Rimey, C. Brown. Control of selective perception using Bayes nets and decision theory. *IJCV*, 12(2/3):173–207, Apr. 1994.
10. S. Sarkar, K. Boyer. Integration, inference, and management of spatial information using Bayesian networks: perceptual organization. *IEEE PAMI*, 15(3):256–274, Mar. 1993.
11. K. Tarabanis, P.K. Allen, R. Tsai. A survey of sensor planning in computer vision. *IEEE Trans. on Robotics and Automation*, 11(1):86–104, Feb. 1995.

Actively Building Models with VIRTUE

J. Lang[1] and M. R. M. Jenkin[2]

[1] Department of Computer Science, University of British Columbia,
Vancouver, BC, Canada, V6T 1Z4 (jlang@cs.ubc.ca)
[2] Department of Computer Science, York University,
North York, Ontario, Canada, M3J 1P3 (jenkin@cs.yorku.ca)

Abstract. This paper presents the sensing plan of VIRTUE; an active vision system based around a <u>VIR</u>tual <u>T</u>rinoc<u>U</u>lar st<u>E</u>reo-head. VIRTUE is used to build polyhedral volumetric models of unknown objects based on recovered 3-D line segments. Partial models and a viewpoint enumeration scheme are used to guide the image acquisition process and to determine 'where to look next'. Results of the active vision recovery of a number of objects are provided as well as volumetric and surface errors associated with the resulting models.

1 Introduction

The complete sensing of a 3-D object with vision necessarily involves multiple views. In passive approaches to object recovery, views may be chosen based upon some pre-set resolution and sampled exhaustively, e.g. a fixed angle step-size for an object rotating in front of a stationary sensor [8, 6]. If an object is sampled exhaustively then resources are spent collecting redundant information. More efficient sensing is possible if a *sensing plan* [7, 11] is generated. When no *a priori* model of the object is available, this sensing plan must be created dynamically as only information gathered during the sensing process is available. A sensing plan must address two principle tasks; (i) what views are necessary to scan an object completely and (ii) where and how should the sensor be positioned in order to view the object?

Any view selection mechanism must consider the capabilities of the sensor. VIRTUE's active modeling system relies upon a trinocular stereopsis algorithm to recover 3D line-segments from the surface structure of the object in view. Thus before considering possible view selection mechanisms, it is worthwhile to examine how VIRTUE obtains local scene structure.

VIRTUE is a virtual trinocular stereo head implemented using a single camera mounted on the end-effector of a robotic manipulator. VIRTUE's three virtual cameras form the vertices of an equilateral triangle and fixate points which lie on the normal which projects to the triangle's center (see Figure 1). VIRTUE is calibrated in order to use the stereo system to recover metric information concerning its environment. Figure 2 shows VIRTUE fixating a point on the object used to calibrate the head. The three configurations which make up one 'virtual' trinocular view are superimposed in Figure 2. Full details of VIRTUE's design, and its forward and inverse kinematics are described elsewhere (see [5]).

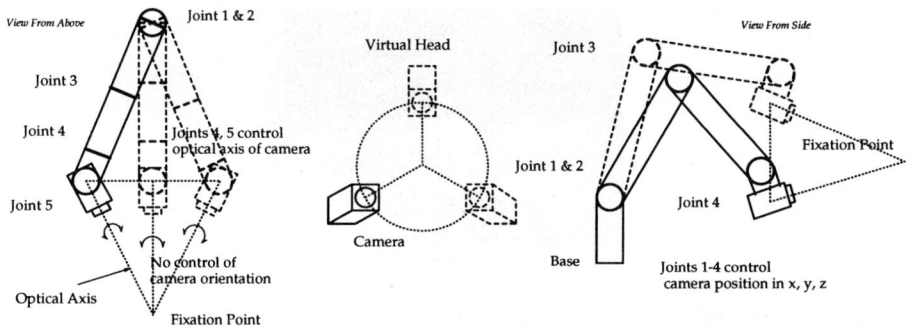

Fig. 1. Single Eye Virtual Trinocular Stereo-Head (VIRTUE)

VIRTUE recovers 'raw' 3D line segments in 'world' coordinates using a trinocular stereo algorithm based on the algorithm of Ayache [1, pp. 129-149] modified so as to omit Ayache's global matching constraint. Ayache's global matching constraint is not required here as any global matching errors are dealt with by a multiple view integration process which has more information concerning the structure of the object.

1.1 Static Scene Exploration

Fundamental to any active vision system is a mechanism for determining 'where to look next'. One potential approach to this problem for a known object would be to take the current object model and the known properties of the sensor and representation, and to determine the next view so as to maximize the additional information available from that view. Although this *synthesis* approach would appear to be an attractive approach and has had success in a number of applications (see [9, 10, 2] for example), we do not attempt to find an analytical solution to the problem of "Where to Look Next?" for a number of reasons.

1. No *a priori* model exists for the object under study.
2. The dimensionality of VIRTUE's parameter space is very large and the space is very complex.
   (This parameter space includes parameters of the object model, VIRTUE's kinematic parameters and its manual camera settings, as well as uncontrolled lighting parameters.)
3. It is difficult to capture analytically the performance of the stereo matching process given the unknown photometric properties of the features.

Given these difficulties, an analytic approach to the problem of determining 'where to look next' seems problematic and an enumeration scheme is more promising. In an enumeration scheme a finite set of sensing configurations are constructed, and an evaluation function is established which allows the best remaining 'view' of the object to be selected (see e.g. the system in [11]). An enumeration scheme may reduce the parameter space by selecting *a priori* an appropriate set of sensor configurations, as well as allowing simple heuristics and approximate tests to help choose among feasible sensing configurations.

Fig. 2. VIRTUE Fixating. Superposition of the three camera positions required for a single fixation of VIRTUE.

2 An Enumeration of VIRTUE's Configurations

VIRTUE and its stereo-algorithm are designed for depth estimation near the fixation point and matches are limited by the viewing pyramid of each camera. Thus the scene volume over which depth recovery is possible for a single view is limited. The total scene or working volume over which VIRTUE can recover objects must be divided into individual viewing volumes.

The location and size of VIRTUE's working volume was chosen to be identical with the upper part of the cube of the calibration object; a parallelepiped of $14.98cm \; x \; 14.98cm \; x \; 11.73cm$. The workspace is digitized into 30 parallelepiped viewing volumes grouped into two overlapping three-dimensional grids. The fixation point for each viewing volume is the center of the subspace's parallelepiped. Not only are the fixation points discretized but the possible fixation directions are as well. In the current implementation 12 possible viewing directions from the geodesic tessellation of the viewing sphere are generated for each fixation point. For a given fixation point and direction, a single 'optimal' fixation distance is pre-computed based on VIRTUE's known kinematics and the minimum distance which ensures that the complete subspace is within VIRTUE's sensing capabilities. For each combination of fixation point and viewing direction, the inverse kinematics of VIRTUE need to be solved. The inverse kinematics can be solved for 122 of the 360 configurations. 238 configurations are excluded because of obstacle avoidance (i.e. table, workspace and robot base) and limitations of the kinematics of the manipulator. VIRTUE's reachable configurations are shown in Figure 3.

2.1 Evaluating the Enumeration

The process of determining where to look next involves considering each of the unviewed viewing configurations and determining which view is the most likely to provide additional salient information concerning the object. The enumeration of potential sensing configurations is based on the following tests for visibility and concavities.

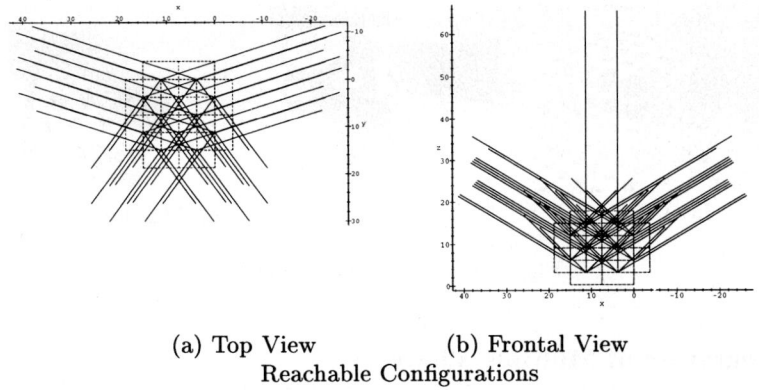

(a) Top View　　　(b) Frontal View
Reachable Configurations

Fig. 3. A line of sight is shown between the fixation point and the center of the trinocular head for each reachable configuration.

Visibility Test It is obviously undesirable to select a viewing configuration that views a region which is known to be occluded. Unfortunately an accurate measure of the 'visibility' of a workspace would be very expensive to calculate. As this value is likely to only be approximate due to the dynamic nature of the enumeration task, VIRTUE approximates the 'visibility' measure of a subspace by evaluating it with respect to a one-voxel-thin surface model of the object. This surface model is represented as an octree. The octree is constructed from the more detailed and accurate polyhedral model representing the intermediate results of the modeling process.

In the octree model 'visibility' is evaluated using the 'line-of-sight' between a fixation point and a viewing position computed as the product of the transparency of all voxels on the line-of-sight. A transparency of 1 is assigned to non-surface voxels, a surface voxel associated with a 3-D line segment is assigned a transparency of 0 and other surface voxels are assigned a transparency of 0.3. The total visibility of a subspace is the sum of the visibilities of each line of sight starting from a voxel in the subspace.

Dealing with Concavities Concavities require special treatment in order to fully explore the space within them. Within the framework of our work, concavities can be dealt with using multiple similar views. Here, a similar view is defined as a view with the same fixation point as the current view and which is connected to the current view by an edge in the geodesic tessellation of viewing directions.

(a) Cube   (b) Ball   (c) 'blue-box'

Fig. 4. Objects to be modeled

2.2 Integration of Multiple Views

Occlusions, measurement errors and matching outliers complicate the integration of multiple views. Thus before integrating line segments into an object model the complete set of segments is examined in order to identify potential outliers and to reduce the measurement error associated with correct matches. In order to perform this gating and to reduce the measurement error, line segments are grouped according to their similarity before being combined into the object model. Recovered line segments which are not grouped are rejected.

2.3 The Enumeration Scheme

The enumeration algorithm operates by iteratively selecting fixation points and views until all visible subspaces have been evaluated. The set of possible views is divided into two sets; external views, which cannot be obstructed by structure in other subspaces, and internal views. Each of these two sets is processed separately, with the set of external views being processed first. From the remaining viewpoints in the current set, the viewpoint is chosen which is the most 'visible' (based on the visibility test described above), and VIRTUE obtains matches from this subspace. These matches are then integrated into the object model. The object model is based on a polyhedral representation which contains the edges obtained from different fixations. As each view is processed a concavity test is performed to see if the associated subspace may contain a concavity. If this test detects the possibility of a concavity then additional similar views are selected of this workspace in order to more fully explore the concave region. The process of selecting viewpoints continues until no workspace exists with a sufficiently high 'visibility'. Once both sets have been processed, the complete model is returned. The enumeration algorithm can be summarized as;

1. Do for all boundary views, and then for all interior views:
   (a) Select view with maximum visibility value based on the current octree model.
   (b) Keep acquiring neighboring views if new concave structure was discovered in the previous acquired view.
   (c) Continue with next fixation point and step 1(a).
2. Return complete tetrahedral model.

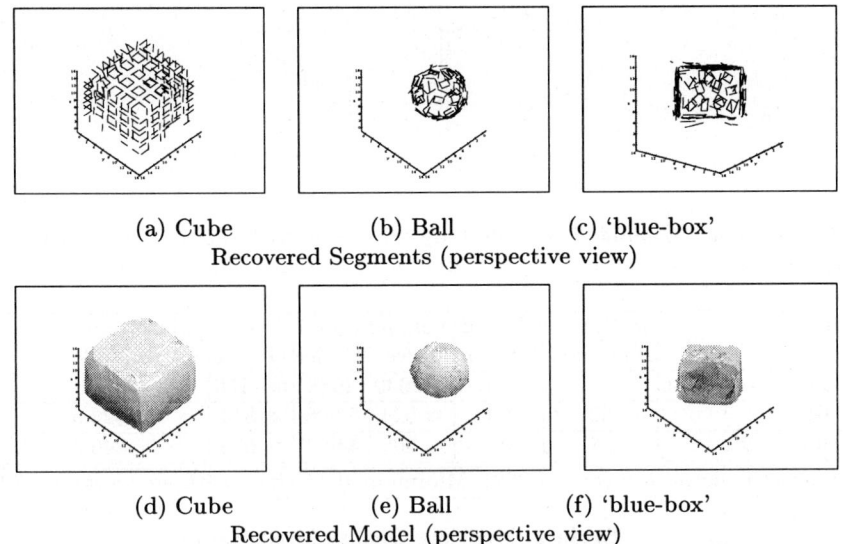

(a) Cube   (b) Ball   (c) 'blue-box'
Recovered Segments (perspective view)

(d) Cube   (e) Ball   (f) 'blue-box'
Recovered Model (perspective view)

Fig. 5. Modeling results

The algorithm terminates when all fixation points have been processed, i.e. the complete visible workspace has been explored, or when all of the remaining fixation points fail the visibility test.

3 Experimental Results

The VITRUE active object recovery system has been employed to recover a number of objects including the calibration cube, a ball and an open box (a miniature recycling 'blue-box'). Views of these objects are included in Figure 4. The line structure of the ball and the box have been enriched by the addition of surface texture (through label stickers). Also notice the ramp mounted in the opening of the 'blue-box'. The resulting global models are shown in Figure 5.

Table 1 summarizes the algorithm's performance on these three objects. It shows the number of boundary and interior views required to explore each object as well as the volume and surface area of the recovered model. The surface area and volume of each object was measured manually and this is included in Table 1 for comparison.

Calibration Object The volume of the model is close to the volume of the calibration cube (see Table 1). However, some deformations of the model are visible; the sides are not completely planar and some corners are missing. Figure 6(a) shows contour lines (at height $11.5cm$) of the calibration cube and the recovered model in comparison. When applying the algorithm to the calibration object, the set of interior views were not included in the algorithm. Although the object

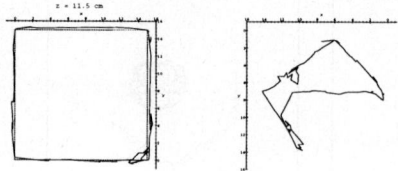

Fig. 6. (a) right: *Ground Truth* (dotted) plus recovered surface (solid) for a slice through the Calibration Cube (b) left: Boundary of a slice through the 'blue-box'

Object	# Views Selected		Volume (cm^3)		Surface (cm^2)	
	Boundary	Interior	Object	Model	Object	Model
Calibration Cube	27	n/a	≈ 2630	2688	≈ 1150	1520
Ball	29	20	≈ 524	575.8	≈ 314	526.5
"Blue-Box"	38	46	≈ 309	444.3	≈ 655	846.3
Reachable views: 122 (out of 360). Minimum # of views with enumeration: 22.						

Table 1. Modeling Quality Measures

is convex and fills the workspace volume, the enumeration scheme would not rule out all interior views since surface triangles are only considered 'partially opaque'.

Ball The model obtained for the ball (see Figure 5(b) and (e)) underlines some additional properties of the modeling process. The best the modeling can achieve is a polyhedron bounded by the line surface texture; here the labels and hexagonal patches of the ball. Outliers and measurement errors cause the model of the ball to be more voluminous and possess more surface area than the ball itself (see Table 1).

Blue-box Figure 5(c) and (f) shows the model and line segments recovered for the 'blue-box'. Again, label stickers are used to augment the surface texture because of the lack of original line structure on the box. The ramp inside the box shows how concavities are dealt with by the enumeration scheme. Figure 6(b) shows a slice through the recovered 'blue-box' object. The enumeration scheme selected a total of 84 boundary and interior views for the "blue-box" (see Table 1). This is reasonable due to the large concavity of the box visible from numerous views. The algorithm verifies concave structure at a fixation point by selecting additional neighboring views to the initially chosen view.

4 Discussion and Future Work

This paper describes the modeling and view selection components of a system capable of acquiring geometric models from intensity images.

The approach has been shown to be a valid option for objects with simple geometry and with sufficiently rich line structure. The use of a sensing plan reduces the number of views required while still allowing reasonably accurate models to be recovered. The major limitation of the existing system is the use of a line-based stereopsis algorithm which requires significant line structure in order to encode the surface of the object.

Currently, the enumeration scheme evaluates views only based on occlusion of subspace and previously sensed concave structure in a subspace. However, it would also be desirable to evaluate subspaces according to the expected success and precision of a sensing operation from a certain view.

Views of an object contain much more information than the shape of an object, i.e. color, shading or texture. A vision based modeling system has the potential to acquire not only geometric models but also model the complete visual appearance of an object.

Acknowledgments

The financial support of the Federal Networks of Centres of Excellence IRIS project and NSERC Canada is gratefully acknowledged.

References

1. N. Ayache. *Artificial Vision for Mobile Robots*. MIT Press, Cambridge, MA, 1991.
2. C. K. Cowan and P. D. Kovesi. Automatic sensor placement from vision task requirements. *IEEE PAMI*, 10(3):407–416, 1988.
3. O. Faugeras. *Three-Dimensional Computer Vision*. MIT Press, Cambridge, MA, 1993.
4. O. Faugeras, E. Lebras-Mehlmann, and J.-D. Boissonnat. Representing stereo-data with the delaunay triangulation. *A. I.*, 44:41–87, 1990.
5. J. Lang. Where to look next: An active vision approach to object modeling. M.Sc. Thesis, York University, 1996.
6. Y. E. Li and M. R. M. Jenkin. Shape from rotation using stereo. In *Prov. VI'92*, pages 119–124, Vancouver, BC, 1992.
7. S.O. Mason and A. Grün. Automatic sensor placement for accurate dimensional inspection. *CVIP*, 61(3):454–467, 1995.
8. R. Szeliski. Shape from rotation. Technical Report 90/13, Digital Equipment Corporation,Cambridge Research Lab, December 1990.
9. K. A. Tarabanis and R. Y. Tsai. Computing occlusion-free viewpoints. In *Proc. IEEE CVPR*, 1992.
10. K. A. Tarabanis, R. Y. Tsai, and A. Kaul. Computing occlusion-free viewpoints. *IEEE PAMI*, 18(3):279–292, 1996.
11. G. H. Tarbox and S. N. Gottschlich. Planning for complete sensor coverage. *CVIU*, 61(1):84–111, 1995.

Using RBF Networks to Map GWT Ridge Images to Pose

Alexandra Psarrou and Jonathan Tanner

School of Computer Science, University of Westminster, Harrow Campus
Watford Road, Northwick Park, Harrow HA1 3TP, London, U.K.
E-mail: {psarroa,tannerj1}@wmin.ac.uk

Abstract. A Pose estimation system is proposed that uses an RBF network to map Gabor Wavelet Transformations (GWT) of faces to a pose angle. In particular we show (a) how the functional description of the GWT face images can be used for their parameterisation and reduction of their dimensionality and (b) how the dimensionality reduced GWT images can lead to the definition of a ridge pose space V where the face representations are sparsified and the distance measure in V correlates well with the perceived image similarity.

1 Introduction

Face recognition is one of the areas in machine vision that has attracted immense attention in the last decade due to the numerous commercial and law enforcement applications. The methods used can be divided into (a) feature-based and (b) view-based [19, 13, 10, 9]. A survey of techniques recently employed in the recognition of faces can be found in [4].

Independent of the method used, the main challenge in the creation of face models and their recognition is to capture the non-linear appearance of faces under varying viewpoints and lighting conditions. This has led to the development of dynamic face models where temporal face signatures are used to track and subsequently identify faces in images [10, 18]. One of the main problems in the use of dynamic face models is the accurate and fast estimation of pose for bootstrapping and tracking faces. Fast pose estimation requires dimensionality reduction to remove redundant information in the original images. A well known example is the "eigenface" approach [17, 19] based on the Principal Component Analysis (PCA). However, PCA can only be used to create linear models for face recognition and therefore cannot provide an accurate pose estimation [1].

This paper is concerned with the estimation of poses using a view-based representation of faces based on *ridges* that are derived from parameterised Gabor wavelet transformed (GWT) [5] face images. The advantage of using GWT to filter face images is that it allows us to obtain representations that are invariant under scaling as well as changes in illumination conditions, skin tone, and hair colour [9]. In addition it permits investigation into the role of locally oriented features with regard to pose changes [15].

The properties of GWT have been investigated by a number of researchers in tracking [14, 16], recognition of faces [11, 13, 20, 21] and pose estimation

[15, 12]. In particular, Lades *et al* have used the response of a set of 2D Gabor filters tuned to different orientations and scales called "jets" [13]. However, norms of the difference between feature vectors based either on the magnitude or on the phase of the Gabor filters response are proved inadequate to discriminate an imposter against the authentic person [7]. Accordingly, a local discriminant criterion is proposed to overcome the sensitivity of the method [6]. More recently, McKenna and Gong [15] use composite Gabor wavelet transformed face images of four orientations to create pose templates. Following Brunelli [3] these templates are then used to track similar processed face images and determine their pose.

In our approach instead of using a statistical analysis to reduce the dimensionality of the original image, we remove redundant information by taking advantage of the functional description of the Gabor wavelet transformed images. A pose space is then defined such that image representations are sparsified. RBF networks [2, 8] are used to map image data to pose, as they are well suited to handling sparse high-dimensional data.

In the next section we first describe the Gabor wavelets used in filtering our face images followed by the functional analysis and dimensionality reduction of the GWT face images and their parameterisation in terms of *GWT ridges*. In section 3 we address the sparcifying of the *ridge images* and the creation of a *Ridge Pose Space* (RPS). RBF networks and their use in our framework is discussed in Section 4 before we conclude in Section 5.

2 The 2D Gabor Wavelet Transform

Gabor wavelets (GW) are biologically motivated convolution kernels which are derived by sinusoidally modulating Gaussian functions of different spatial frequencies and orientations [5]. Each Gabor wavelet $\psi_\mathbf{k}$ is a complex function composed of a Gaussian enveloped sine functions, in which the width σ of the Gaussian functions is related to the frequency and orientation $\mathbf{k}$ of the sinusoids, and is given by,

$$\psi_\mathbf{k}(\mathbf{x}) = \exp\left(-\frac{\mathbf{k}^2\mathbf{x}^2}{2\sigma^2}\right)[\exp(-i\mathbf{k}\cdot\mathbf{x}) - \exp(\sigma^{-2})] \qquad (1)$$

where $\mathbf{x}$ is a position in the wavelet. When convoluted with an image, Gabor wavelets respond to variations in image intensity and therefore Gabor wavelet transformed images are invariant to illumination, skin tone and hair colour. Gong *et al* have exploited this property of GW and have shown that Gabor wavelets of different orientations play some role in "regularising" pose distribution [9].

The complex formulation of the Gabor wavelets has the advantage that when used with a sequence of face images GW responses can be described in terms of phase and amplitude. Phase gives a measure of how the GW responses change in successive images and provides an accurate localisation of the response. Amplitude responses give a measure of the similarity between the sinusoidal modulated Gaussian functions and the image. The real and imaginary parts of the GW response is primarily used to feature matching between image sequences.

In our work we are concerned with the estimation of poses from static images. We therefore convolve the face images with only the real part of the Gabor wavelets as we do not estimate feature translation.

Figure 1 shows the magnitude of a face sequence that has been filtered with the real part of Gabor wavelets of four orientations. The GW frequency is chosen

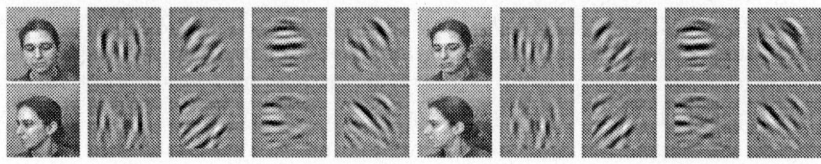

Fig. 1. Responses of the real part of GW of a face sequence in four orientations.

such that common facial features between individuals are responded to. In the images above the vertical orientated filters respond to the eyes, and mouth and the horizontal orientation respond to the eyes, and nose. Note that head rotation in the horizontal direction is most strongly detected by horizontal GW, as shown in columns two and seven above.

2.1 Parameterised GWT Images

The dimensionality of the Gabor wavelet transformed images can be reduced if we take into account the way the magnitude of the real part of the GW response varies in an image. In particular, using a functional description of GWT images we show that the convolution of any image with the real part of a Gabor wavelet results in a transformed image whose magnitude varies locally according to a bell shaped function.

Functional description of GWT images Given that convolution of a function f, with an image I is described in vector notation by the the dot product of f and I

$$f \cdot I = \int f(x) I(x) dx \qquad (2)$$

then if there is a point in the image in which $f \cdot I \neq 0$ this denotes that the image approximates locally the function f.

In our case the real part of a Gabor wavelet function is of the form

$$f = exp(-kx^2/\sigma^2) \cos x \qquad (3)$$

Therefore, convoluting an image I with Equation (3) gives maximum magnitude response at a point x_0 in the image, where the image I is locally described by

$$I = \alpha exp(-kx_0^2/\sigma^2) \cos x_0 + E(x_0) \qquad (4)$$

where the error $|E(x)|$ is at a minimum.

Using Equations (3) and (4) the convolution of the real part of the GW with the image I at a distance δx from x_0 is given by

$$f(x) \cdot I(x_0 + \delta x) = f(x) \cdot f(x_0 + \delta x) + f \cdot E(x) \qquad (5)$$

As $f \cdot E(x) \approx 0$ within some range $\pm \delta x$

From Equations (3) and (5) we can derive that

$$f(x) \cdot I(x + \delta x) \approx \alpha (exp(-2kx^2/\sigma^2)) \int \cos x \cos(x + \delta x) dx \qquad (6)$$

Equation (6) shows that the magnitude of the Gabor wavelet transformed images near any local maxima can be described by a well-defined bell-shaped function.

It is therefore clear that the Gabor wavelet transformed images still contain a significant amount of redundant information and that their dimensionality can be reduced without involving any statistical analysis but rather by recording the maximum and minimum responses of the Gabor wavelets. This process of local nonmaxima (or nonminimal) suppression results in the derivation of thin lines in the GWT image, called *ridges*.

GWT ridges Figure 2 shows the ridges that have been derived from the local nonmaximal (and nonminimal) suppression of the GW responses in a GWT face image. Without loss of generality, and ease of discussion we consider only the

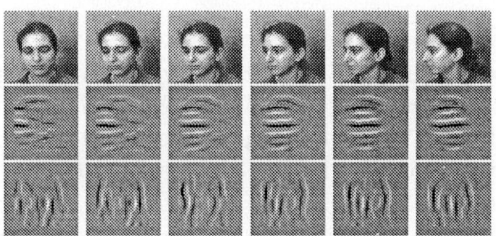

Fig. 2. Ridges derived from the local nonmaximal and non-minimal suppression of horizontal the vertical GW responses.

responses of the horizontal GW. To parameterise the GWT images we divide them into a set of approximately 10 horizontal bands. Figure 3 shows that each band consists of a set of approximately vertical ridges. Each ridge, r, is param-

Fig. 3. Ridges found in each band of a GWT image are approximately vertical.

eterised in terms of its position in the GWT image and the magnitude of the GW responses along its length, here referred to as the *amplitudes* of a ridge.

$r = (x, y, \mathbf{a})$

where x is the x coordinate of a ridge

y is the band number

$\mathbf{a}$ contains the GW amplitudes along a line

In subsequent parts of this paper a *ridge image* will refer to the composite of such parameterised GWT images for all GW orientations.

3 Ridge Pose Space

In order to determine a face pose a suitable vector space is required to represent the ridge images. This space V, called the Ridge Pose Space (RPS), is constructed so that image representations are sparsified and points in V are close, if and only if they are close in pose angle.

3.1 Building Ridge Pose Space

Building the Ridge Pose Space requires three stages:

1. The creation of *ridge exemplars* that denote the most likely ridge image form of each pose.
2. The matching of ridge exemplars in order to determine the commonality of ridges between poses and determine the new ridges that are introduced in each pose.
3. The description of the ridges that appear in a set of poses in an integrated representation.

Creation of ridge exemplars The set of ridge images of a group of individuals at various poses is used to construct the set $S = \{\{S_1\}, \{S_2\}, ..., \{S_N\}\}$, where N is the number of poses the system can estimate and S_i is a set of ridge images that have been derived from face images of some range of pose angles $d\theta_i$ centered at θ_i. Where $\theta_i > \theta_j$ iff $i > j$.

The most likely form of a ridge image from S_i is denoted by R_i. The set $R = \{R_1, R_2, ..., R_N\}$ is called the *ridge exemplar set*. Associated with the ridge exemplar of each pose is a covariance matrix M_i that contains the covariances between positions and amplitudes of the ridges in R_i, and therefore captures the variations in face.

Matching ridges in the exemplar set R In order to create an integrated representation of the ridges found in R we need to examine their position and amplitude variance and determine their likely commonality.

Ridges are matched only between successive ridge images R_i, R_{i+1}. Matching of ridges starts with the ridge in R_i with the lowest variance, and ends with a match of the highest variance ridge. A ridge is matched only once and only to its closest neighbours. The likelihood that two ridges in R_i, R_{i+1} match is given by,

$$L(pose) = \delta \begin{pmatrix} \sigma_{xx} & 0 \\ 0 & \sigma_{aa} \end{pmatrix}^{-1} \delta^T$$

where δ represents the difference in position and amplitude of the ridges in R_i, R_{i+1} and $\sigma_{xx}^2, \sigma_{aa}$ are the position and amplitude variances respectively of a ridge in R_i. If $L > L_{threshold}$ the ridges are considered not to match.

Integrated ridge representation To generate Ridge Pose Space V we need an integrated representation that describes uniquely all ridges found in the ridge exemplar set. The degrees of freedom of V can be described by a set U.

Initially $U = \emptyset$. Let R_1 be the first set of members in U. By matching successive ridge images R_i, R_{i+1} we can calculate a difference set $d_{i,i+1}$

$$d_{i,i+1} = \begin{cases} \emptyset & \text{if } R_{i+1} \subset R_i \\ R_i - (R_i \cap R_{i+1}) & \text{if } R_{i+1} \supseteq R_i \end{cases}$$

that contains the set of ridges that differ between two ridge images R_i and R_{i+1}.

If $d_{i,i+1} \neq \emptyset$ the dimension of U is increased. The new dimensions are then used to represent the new ridges in $d_{i,i+1}$.

Note that if l is the number of ridge vectors used to define a point in V, then l is also the total number of distinct ridges in the set R. Hence in general a ridge image, p, in V has, ignoring indices, the following form

$$p = \{r_1, r_2, ..., r_k\} \cup \{v_1, v_2, ..., v_m\} \quad (7)$$

where r_i is a ridge vector
v_i is a ridge vector of the form $(x, y, \mathbf{a} = \mathbf{0})$
$(k + m)$ is the dimension of V

This space V in effect, provides a representation of ridge images in a higher dimensional space that takes into account the temporal relation of ridges in successive poses.

3.2 Templates

Templates are used to describe exemplar points in the Ridge Pose Space V. Each template T_i is generated from a exemplar ridge image R_i and has associated with it the covariance matrix that contains the covariances between positions and amplitudes of the ridges in R_i.

3.3 Distance Measure

The distance measure in V has the property that images that appear to be similar are close to one another in V.

Let $p = \mathbf{x_1}, \mathbf{x_2}, ..., \mathbf{x_l}$ and $q = \mathbf{y_1}, \mathbf{y_2}, ..., \mathbf{y_l}$ be points in V. Where $x_i = (u, \mathbf{a})$, and $y_i = (v, \mathbf{b})$. Then the distance between points p and q is given by

$$|p - q|^2 = \sum_1^l |\mathbf{x_i} - \mathbf{y_i}|^2 \quad (8)$$

$$|\mathbf{x_i} - \mathbf{y_i}|^2 = (u - v)^2 + (a_1 - b_1)^2 + ... + (a_k - b_k)^2 \quad (9)$$

where k is the dimension of the ridge amplitudes $\mathbf{a}, \mathbf{b}$.

4 RBF Network

Networks of Radial Basis Functions (RBF) are well suited to handling sparse data of high dimensionality [2, 8]. In our work we use RBF networks to map a point in the Ridge Pose Space V to a pose angle. The RBF network consists of a set of N Gaussian functions, where N is the number of ridge exemplar images available in the pose estimation system. The centre of each Gaussian function i is given by the template T_i. Each Gaussian function measures the distance from a *ridge measurement* X to its template value. Before a face image I is input to the RBF network we need to represent it in the Ridge Pose Space V. This is achieved by first deriving its ridge image, R_I, representation and then computing the likelihood L of matching templates $T_1, ..., T_N$, as given by

$$L = \delta M^{-1} \delta^T \tag{10}$$

where δ=distance between ridges
M is the template covariance matrix

The template T with the highest likelihood is then used to generate the *ridge measurement* X from the its ridge image R_I. The distance measure is the RBF network is then given by,

$$|T_i - X|^2 = \sum^k (u-v)^2 + |(\mathbf{a}-\mathbf{b})|^2 + ... + (\mathbf{u_k}-\mathbf{v_k})^2 + |(\mathbf{a_k}-\mathbf{b_k})|^2 \tag{11}$$
$$+ \sum |\mathbf{a_{k+1}}|^2 + |\mathbf{b_{k+1}}|^2 + ... + |\mathbf{a_n}|^2 + |\mathbf{b_n}|^2 \tag{12}$$

where k is the number of matching ridges
(n-k) is the number of ridges that do not match a template

Using the weighted sum of the RBF unit responses the RBF network then calculates a pose estimate.

5 Conclusion

In this paper we discussed the needs of a pose estimation system and proposed one that tackles the problem of both dimensionality reduction and having a sparsified pose representation. In particular we looked at face images filtered with the real part of a Gabor wavelet. Functional analysis of such images led to the non-maximal suppression of Gabor wavelet transformed responses and their parameterisation in terms of ridges. Ridges were then described using an integrated representation that takes into account the temporal relationship of ridges in successive poses. This resulted in the construction of a Ridge Pose Space V where face images are sparsified and points in V are close, if and only if they are close in pose angle. Finally we propose that networks of Radial Basis Functions are used to map the parameterised GWT face images to pose, as RBF networks are well suited for handling sparse data of high dimensionality.

References

1. P. Belhumeur and et al. "Eigenfaces vs fisherfaces: Recognition using class specific linear projection". In *ECCV'96*.
2. C.M. Bishop. *Networks for pattern recognition*. Oxford University Press, 1995.
3. R. Brunelli and T. Poggio. "Face recognition: Features versus templates". *IEEE PAMI*, 15(10):1042–1052, October 1993.
4. R. Chellapa, C.L. Wilson, and S. Sirohey. "Human and machine recognition of faces: A survey". *Proceedings of the IEEE*, 83:705–740, 1995.
5. J. G. Daugman. "Uncertainty relation for resolution in space, spatial frequency and orientation optimized by two-dimensional visual cortical filters". *J. Opt. Soc. Am.*, pages 1160–1169, 1985.
6. B. Duc, S. Fischer, and J. Bigün. "Face authentification with sparse grid Gabor information". In *Proc. IEEE Int. Conf. ASSP*, 1997.
7. S. Fischer, B. Duc, and J. Bigün. "Face recognition with Gabor phase and dynamic link matching for multi-modal identification". Technical Report LTS 96.04, Signal Proc. Lab., Swiss Federal Inst. of Technology, 1996.
8. F. Girosi. "Some extensions of radial basis functions and their applications in artificial intelligence". *Computers and Mathematics with Applications*, 24(12):61–80, 1992.
9. S. Gong, S.J. McKenna, and J.J. Collins. "An investigation into face pose distributions". In *Proc. 2nd ICAFGR*, Vermont, 1996.
10. S. Gong, A. Psarrou, and et al. "Head tracking and dynamic face recognition". In *European Workshop on Combined Real and Synthetic Image Processing for Broadcast and Video Production*, Hamburg, Germany, 1994.
11. A.J. Howell and H. Buxton. "Towards unconstraint face recognition from image sequences". In *Proc of the 2nd ICAFGR*, Vermont, October 1996.
12. N. Krüger and et al. "Determination of face position and pose with a learned representation based on labeled graphs". Technical Report 96-03, Institut für Neuroinformatik, Ruhr-Universität Bochum, January 1996.
13. M. Lades and et al. Distortion invariant object recognition in the dynamic link architecture. *IEEE Transactions on Computers*, 42(3):300–311, 1993.
14. T. Maurer and C. von der Malsburg. "Tracking and learning graphs on image sequences of faces". In *Proc. ICANN*, Bochum, 1996.
15. S.J. McKenna and S. Gong. "Real-time face pose estimation". Submitted to Real Time Imaging Journal March 1997.
16. S.J. McKenna, S. Gong, R. Würtz, J. Tanner, and D. Banin. "Tracking facial feature points with Gabor wavelets and shape models". In *Proc. Audio- and video-based biometric person authentification*, pages 35–42, 1997.
17. A. Pentland, B. Moghaddam, and T. Starner. "View-based and modular eigenspaces for face recognition". In *IEEE proc. on CVPR*, 1994.
18. A. Psarrou, S. Gong, and H. Buxton. "Modelling Spatio-temporal Trajectories and Face Signatures on Partially Recurrent Neural Networks". In *IEEE ICNN*, Perth, Australia, 1995.
19. M. Turk and A. Pentland. "Eigenfaces for recognition". *Journal of Cognitive Neuroscience*, 3(1), 1991.
20. L. Wiskott, J.M. Fellous, N. Krüger, and C. von der Malburg. "Face recognition and gender determination". In *IWAFGR*, Zurich, 1995.
21. R.P. Würtz. *"Multilayer dynamic link networks for establishing image point correspondences and visual object recognition"*. Verlag Harri Deutsch, 1995.

3-D Pose Estimation and Model Refinement of an Articulated Object from a Monocular Image Sequence*

Nobutaka SHIMADA, Yoshiaki SHIRAI, Yoshinori KUNO and Jun MIURA

Dept. of Computer-Controlled Mechanical Systems, Osaka University,
Yamadaoka 2-1, Suita, Osaka, 565 Japan.
E-mail: shimada@mech.eng.osaka-u.ac.jp
URL: http://www-cv.mech.eng.osaka-u.ac.jp/~shimada/

Abstract. This paper proposes a method to precisely estimate both the shape (link lengths) and pose (joint angles) of a articulated object from a monocular image sequence. Normal model-fitting method often leads to wrong estimates due to depth ambiguity in monocular views. The paper proposes a filtering method with constraint knowledge of the object represented as inequalities. The method calculates the probability distribution satisfying both the observation and the constraints. When multiple solutions are possible, they are preserved until a unique solution is determined. Experimental results show that the depth ambiguity is incrementally reduced if the informative observations are obtained.
Keywords: 3-D reconstruction, articulated object, depth ambiguity, inequality constraints

1 Introduction

Automatic estimation of the shape and pose of articulated objects like human bodies or animals is useful for man-machine interface, 3-D modeling in computer graphics, virtual reality, etc. Many visual methods have been developed for that purposes [1][2][3]. Such a methods utilizes a 3-D shape model and fits the model to corresponding image features. In case the shape of model (lengths and widths of parts) is initially approximated and fixed, the estimated pose (joint angles) is not precise due to the approximation error of the shape.

Thus it is convenient to refine the shape of the initial model while observing an image sequence. Once the feature correspondences are resolved, the model can be fitted to the image features by least squares method [4][5]. In monocular cases, however, these methods don't always lead to the correct solution because of *depth ambiguity*. Fig.1 shows an example for it. In the figure, supposing the 1-D edge points are observed as features for the 2-D link system, any link longer than the correct length can explain the observed features. That is the ambiguity between the length and the joint angle.

In order to resolve the ambiguity, O'Rourke [6] and Brooks [7] utilize constraint knowledge of the objects. If the ambiguity of one parameter is limited by a constraint, the depth ambiguity is also limited. If we know the joint angle is constrained within a certain range, shown in Fig.1, the candidates 'A' and 'B' in Fig.1 are found to be impossible. The maximal link length is achieved at 'C'

---

* This work is supported in part by Grant-in-Aid for Scientific Research from Ministry of Education, Science, Sports, and Culture, Japanese Government.

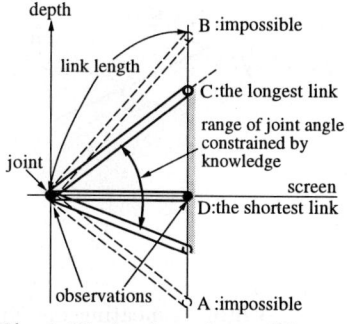
Fig.1 Limitation of possible parameter ranges with constraints

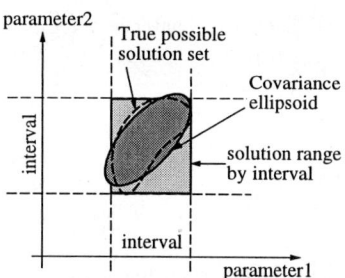
Fig.2 Interval description and possible solution set

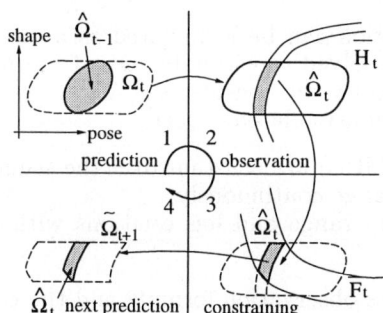
Fig.3 Incremental update of the solution set : $\widetilde{\Omega}$: predictions, $\widehat{\Omega}$: estimates, $\bar{H}$: observations and $\bar{F}$: constraints

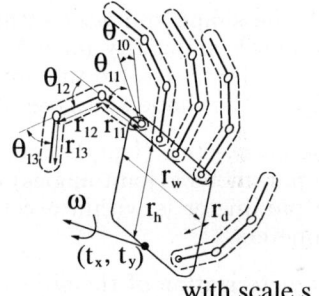
Fig.4 3-D hand model

in Fig.1 and the minimum at 'D'. The ambiguity can be reduced more and more by other constraints or various observations over the image sequence.

The methods [6][7] handle the ambiguity of the shape and pose parameters as *intervals* represented as a maximal and a minimal value deduced from temporal observations and the constraints. However, the ambiguity isn't sufficiently limited by the interval of each parameter because the correlations between each parameter is not considered. (see the square region in Fig.2)

We propose to represent the depth ambiguity as *a possible set* in the multi-dimensional parameter space using a filtering method. The ambiguity is sufficiently limited because of the relationship of the parameters represented as a correlation (see the broken contour in Fig.2). As shown in Fig.3, the 1st phase of filtering is *prediction* from the previous estimate. The 2nd phase is *update* of the present estimate with an current observation. The 3rd phase is *constraining* to resolve the depth ambiguity. Because the constraints are not equations but inequalities like $-20 \leq \theta \leq 40 deg$, the special method is needed to compute the constraining.

Here, we introduce the modified extended Kalman filter (modified EKF) with inequality constraints. The standard EKF solution is modified so as to satisfy the inequalities by *truncation* of the probability distribution in the part not satisfying them. Then possible solution set, which are represented by the covariance ellipsoid in Fig.2, is incrementally reduced (i.e. the shape and pose get precise) by various observations.

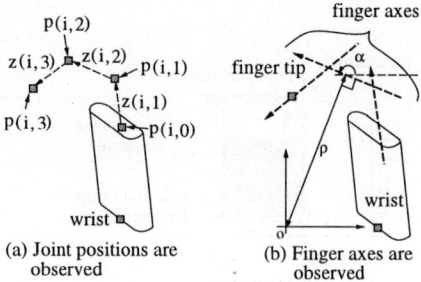
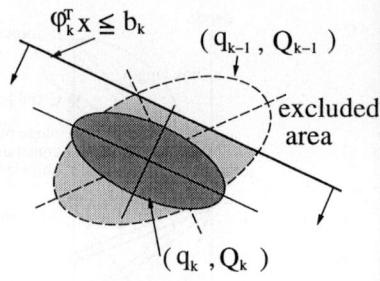

(a) Joint positions are observed (b) Finger axes are observed

Fig.5 Observed features   **Fig.6** Truncating the EKF solution

2 Refinement of Shape and Pose Parameters

Although the shape and pose estimation can be formulated as a model-fitting problem, depth ambiguity must be resolved to estimate them from monocular images. In order to resolve the ambiguity, we consider the following constraint knowledge of the shape and pose of an articulated object.

(a) shape parameters (lengths and widths) are constant over the sequence.
(b) pose parameters (joint angles) change continuously.
(c) each parameter is within a certain range and has relations with the other parameters.

The parameterization of the object, its observation formula and the constraints are first modeled in Sec.2.1. The modified EKF algorithm with these constraints is proposed in Sec.2.2. Additionally, in order to consider the symmetrical ambiguity to the screen, we propose generation and preservation of multiple solutions in Sec.2.3.

2.1 Modeling of Articulated Object

Here, we consider the object (Fig.4) is observed by scaled orthogonal projection. We define a m-dimensional state vector of the shape and pose as

$$\boldsymbol{x} = (\boldsymbol{t}^T, s, \boldsymbol{\omega}^T, \theta_{10}, \cdots, \theta_{53}, \dot{\theta}_{10}, \cdots, \dot{\theta}_{53}, r_{11}, \cdots, r_{53})^T \qquad (1)$$

where $\boldsymbol{t}$, s, $\boldsymbol{\omega}$, θ, $\dot{\theta}$, and r respectively denote a wrist position (t_x, t_y), scale of the projection, 3-D direction of palm, joint angles, its velocities and lengths of links. Supposing the constancy of the shape ((a) in Sec.2), $\dot{r}$ is not included. The transition and observation formulas are represented as

$$\boldsymbol{x}_{t+1} = \boldsymbol{A}\boldsymbol{x}_t + \boldsymbol{u}_t \qquad (2)$$
$$\boldsymbol{y}_t = \boldsymbol{h}(\boldsymbol{x}_t) + \boldsymbol{w}_t \qquad (3)$$

where $\boldsymbol{y}_t$ is a n-dimensional observation vector. $\boldsymbol{u}_t$, $\boldsymbol{w}_t$ are white noises with zero mean and variances $\boldsymbol{U}$, $\boldsymbol{W}$. Supposing linear prediction, $\boldsymbol{A}$ is represented as the following $(2m+n) \times (2m+n)$ matrix:

$$\boldsymbol{A} = \begin{bmatrix} \boldsymbol{I}_m & \boldsymbol{I}_m & \boldsymbol{O} \\ \boldsymbol{O} & \boldsymbol{I}_m & \boldsymbol{O} \\ \boldsymbol{O} & \boldsymbol{O} & \boldsymbol{I}_n \end{bmatrix} \qquad (4)$$

where $\boldsymbol{I}_m$ denotes $m \times m$ identity matrix. $\boldsymbol{U}$ is determined by considering the continuity of the pose changes ((b) in Sec.2).

The observation formula of the wrist and joint position and the finger axes is next modeled in detail. In Fig.5, the 2-D projection of the jth joint position of ith finger is described as

$$p(i,j) = (p_x(i,j), p_y(i,j))$$

$$= L \cdot R(\omega)R(\theta_{i0}) \cdot \sum_{k=1}^{j} r_{ik} \left(\cos(\sum_{l=1}^{k} \theta_{il}), \sin(\sum_{l=1}^{k} \theta_{il}), 0 \right)^T + t \quad (5)$$

where L and R represent projection and rotation matrices. We suppose a straight line (α_{ij}, ρ_{ij}) is extracted as the jth axis of ith finger. α_{ij} and ρ_{ij} respectively denote the direction of the line and the distance from the origin. They are formulated as

$$\alpha_{ij} = \arctan(z_y(i,j)/z_x(i,j)) \quad (6)$$

$$\rho_{ij} = (z_y(i,j)p_x(i,j) - z_x(i,j)p_y(i,j))/|z(i,j)| \quad (7)$$

$$z(i,j) = (z_x(i,j), z_y(i,j)) = \begin{cases} p(i,j) - p(i,j-1) & \cdots j \neq 0 \\ p(i,j) & \cdots j = 0 \end{cases} \quad (8)$$

The observation function h consists of the $p(i,j)$, α_{ij}, ρ_{ij} observed in one frame. In case of occlusion, $p(i,j)$, α_{ij}, ρ_{ij} of the occluded part are not contained in h. For example, if we obtain the wrist, finger tips and the all of the finger axes except the most proximal one, h is described as

$$h(x) = (t^T, \alpha_{12}, \rho_{12}, \alpha_{13}, \rho_{13}, p(1,3)^T, \cdots, \alpha_{52}, \rho_{52}, \alpha_{53}, \rho_{53}, p(5,3)^T)^T. \quad (9)$$

In addition to above, the following constraints of the object is considered ((c) in Sec.2). They are formulated as

$$\theta_{min,i} \leq \theta_i \leq \theta_{max,i} \quad (10)$$

$$|\theta_i - \theta_j| \leq \Delta\theta_{ij} \quad (11)$$

$$r_{min,i} \leq r_i \leq r_{max,i} \quad (12)$$

$$|r_i - r_j| \leq \Delta r_{ij}. \quad (13)$$

The above inequalities must be simultaneously satisfied.

2.2 Modified EKF with Inequality Constraints

Because the observation function h is non-linear, the current state $\hat{x}_t$ and variance P_t are approximately estimated by EKF as

$$\hat{x}_t = \tilde{x}_t + K_t \{y_t - h(\tilde{x}_t)\} \quad (14)$$

$$P_t = (I - K_t \frac{\partial h}{\partial x_t}\bigg|_{\tilde{x}_t})(AP_{t-1}A^T + U) \quad (15)$$

where K_t is the Kalman gain matrix, $\tilde{x}_t = A\hat{x}_{t-1}$. In the calculation of $\partial h/\partial x_t$, $\partial p/\partial x_t$ is directly obtained from the derivatives of Eq.(5). $\partial \alpha/\partial x_t$ and $\partial \rho/\partial x_t$ are also obtained from the derivatives of Eq.(6) and (7) which are reduced to $\partial p/\partial x_t$. However, the solution of Eq.(14) still includes errors due to the depth ambiguity.

In order to resolve the ambiguity, we modify the EKF solution with the constraints Eq.(10)...(13). Note that they are *inequalities*, not equations, which can't be introduced into EKF as observations. A method is proposed in which inequality constraints are modified to equations with slack variables [8]. This method linearizes the modified constraints with respect to the slack variables even if the original constraints are linear. The linearized constraints are inaccurate. Another way is to introduce the constraints as an initial distribution. But it is also inappropriate because the effect of the initial distribution decreases by filtering at every frame.

In our method, the distribution of EKF solution $(\hat{x}_t^*, P_t^*)$ outside the constraints is truncated as shown in Fig.6 and then the ambiguity can be reduced. Eq.(10)...(13) are generally represented as

$$\varphi_k^T x \le b_k \qquad (k=1\cdots K). \tag{16}$$

Because it is difficult to exactly compute the distribution truncated with all constraints, it is approximated by sequential truncation with an each single constraint.

Suppose the distribution with a mean q_{k-1} and a variance Q_{k-1} is truncated by the constraint $\varphi_k^T x \le b_k$, where $q_0 = \hat{x}_t^*$ and $Q_0 = P_t^*$. This computation is reduced to the case where the mean is o, the variance is identity matrix I and the constraint is $(1, 0, \cdots, 0)^T x' \le c_k$, by applying the following transformation:

$$x' = RW^{-\frac{1}{2}}T^T(x - q_{k-1}) \tag{17}$$

where R, T is orthogonal, W is diagonal and

$$TWT^T = Q_{k-1} \tag{18}$$

$$RW^{\frac{1}{2}}T^T \varphi_k = (1, 0, \cdots, 0)^T \tag{19}$$

$$c_k = (b_k - \varphi_k^T q_{k-1})/\{(\varphi_k^T Q_{k-1} \varphi_k)^{\frac{1}{2}}\}. \tag{20}$$

In this case, the truncated mean μ_k and variance S_k is computed as

$$\mu_k = (\nu_k, 0, \cdots, 0)^T \tag{21}$$

$$S_k = \text{diag}\{1 + c_k \nu_k - \nu_k^2, 1, \cdots, 1\} \tag{22}$$

$$\nu_k = -\sqrt{\frac{2}{\pi}} \exp(-\frac{c_k^2}{2})/(1 + \text{erf}(\frac{c_k}{\sqrt{2}})) \tag{23}$$

where $\text{erf}(\cdot)$ represents the error function and $\text{diag}\{a, b, \cdots\}$ represents a diagonal matrix whose diagonal elements are $a, b, \cdots$. Then the truncated mean and variance are expressed as

$$q_k = TW^{\frac{1}{2}}R^T \mu_k + q_{k-1} \tag{24}$$

$$Q_k = TW^{\frac{1}{2}}R^T S_k RW^{\frac{1}{2}}T^T. \tag{25}$$

Finally, the fully truncated mean and variance are obtained by recursive computation: $\hat{x}_t = q_K$ and $P_t = Q_K$. Because the computation of Eq.(18) takes much time, the kth truncation is skipped for efficiency if c_k, mahalanobis distance from q_0 to the plane $\varphi_k^T x = b_k$, is greater than a threshold.

2.3 Multiple Estimation

Estimation by EKF may fail because the distribution becomes multimodal due to the depth ambiguity. Fig.7 shows a example of a 2-D link system in which 1-D joint positions are observed. At the 18th frame(Fig.7(b)), In spite that there

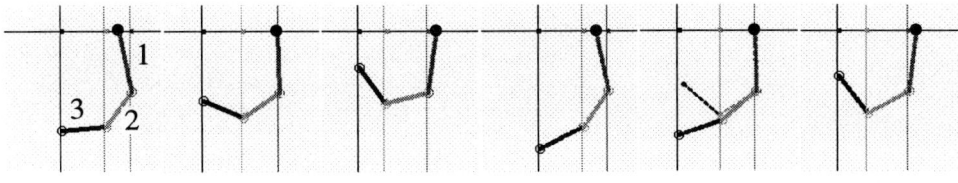

(a) t=16/cor. (b) t=18/cor. (c) t=20/cor. (d) t=16/est. (e) t=18/est. (f) t=20/est.

Fig. 7. Multiple estimation: (a)-(c) are correct pose and (d)-(f) are estimates. In (e), solid line shows the wrong estimate obtained by normal EKF and broken line shows the alternative estimate.

are two possible solution (see (e)), normal EKF only produces either. Although sampling method [9] can treat the multimodal case, it is hard to compute when the state dimension is large.

In our method, we generate and preserve multiple estimates. This means that the multimodal distribution is approximated by sum of multiple gaussian distributions. For the ith link, the multimodal problem arises when the link is nearly parallel to the screen, namely when $\partial h_i/\partial \theta_i|_{\tilde{x}_t} \simeq 0$, where h_i and θ_i respectively denote observations of the ith link and the proximal joint angle. If the prediction $\tilde{x}_t$ satisfies above equation, the following processes are activated.

1. Generate the $\tilde{x}_t^{sym}$ which is identical to $\tilde{x}_t$, except that the ith link is symmetrical to $\tilde{x}_t$ with respect to the screen.
2. At both of $\tilde{x}_t$ and $\tilde{x}_t^{sym}$, calculate $\partial h/\partial x_t$ in Eq.(15)
3. With each $\partial h/\partial x_t$, calculate the estimate by modified EKF respectively with the original prediction $\tilde{x}_t$ and its variance.

At most, 2^n (n:the number of links) estimates are possible. They are preserved until unique solution is determined. In case that the area truncated by the constraints is more than a threshold, such a estimates is eliminated as illegal. The rest are also preserved for robustness by use of beam-search[?].

In Fig.7, two estimates are simultaneously generated at the 18th frame (e) and the wrong estimate is eliminated at the 20th frame (f) and the following estimation is successfully continued.

3 Experimental Results

We show a estimation result by a simulation with a synthesized image sequence (Fig.8). In the sequence, a hand-like object moves to right and left, rotates, and folds its fingers. We utilize the constraints shown in Tab.1. The initial estimate and variance is set so that the correct value is included in 99% confidence region.

Table 1. The constraints used in the simulation

pose constraints	$\|\theta_{i2} - \theta_{i3}\| \leq 10deg,$ $-20deg \leq \theta_{10} \leq 0deg,$ $0deg \leq \theta_{20} \leq 20deg,$ $0 \leq \theta_{ij} \leq 90deg$ $\|\theta_{11} - \theta_{21}\| \leq 35deg$ for $i = \{1,2\}, j = \{1,2,3\}$	shape constraints	$0 \leq r_{i1} - r_{i2} \leq 30,$ $\|r_{i2} - r_{i3}\| \leq 10,$ $55 \leq r_{i1} \leq 70,$ $30 \leq r_{i2} \leq 50,$ $25 \leq r_{i3} \leq 50$ for $i = \{1,2\}$

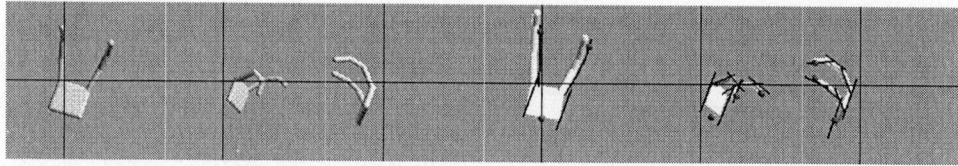

(a) t=1/cor.   (b) t=32/cor.   (c) t=55/cor.   (d) t=1/est.   (e) t=32/est.   (f) t=55/est.

Fig. 8. Simulational estimation results of 3-D articulated object: (a)-(c) are correct pose and (d)-(f) are estimates.

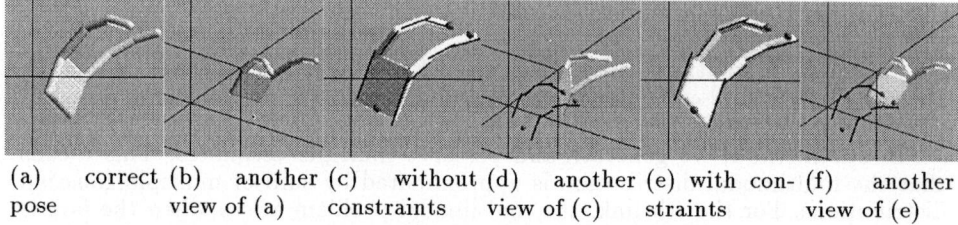

(a) correct pose   (b) another view of (a)   (c) without constraints   (d) another view of (c)   (e) with constraints   (f) another view of (e)

Fig.9 Comparison of estimates with constraints to no constraints (t=23)

In Fig.8, (a)-(c) show correct shapes and poses and (d)-(f) show the estimates. In (d)-(f), the black spheres and straight lines are observed wrist positions, finger tips and finger axes which are perturbed by Gaussian noise. Although the estimation without constraints makes a wrong estimate at the 23th frame due to the depth ambiguity (Fig.9(c),(d)), our method correctly estimates the pose with the constraints in the Tab.1 (Fig.9(e),(f)). Fig.10 shows that two different shapes (b) and (c) are correctly identified using the same initial shape model shown in (a) and the same constraints.

4 Conclusion and Discussion

In this paper, we propose a method to simultaneously estimate the shape and pose of articulated objects from a monocular image sequence. We estimate the pose at each frame and simultaneously refine the initial shape model using the modified EKF with inequality constraints by truncating the probability distribution. Then the depth ambiguity is incrementally reduced with informative observations over the sequence. In addition, we resolve the ambiguity of symmetrical poses by generating and preserving multiple solutions until unique solution is determined. We show the effectiveness of our method applying to simulated images. In case of real images, feature correspondance is important problem and we proposed a silhouette-matching method [10] as its solution. Now we are trying to combine that with this method and estimate both shape and poses of a real human hand.

However, we still have a problem. In some cases, the variance of the estimate improperly decreases. This is caused by the error of linearization of the observation. One way to solve this problem is to use a boundary description instead of the probability distribution. In general, however, the computation of the boundary in multi-dimensional space is almost impossible. An approximation method was proposed [11] which approximate the boundary by an ellipse in a multi-

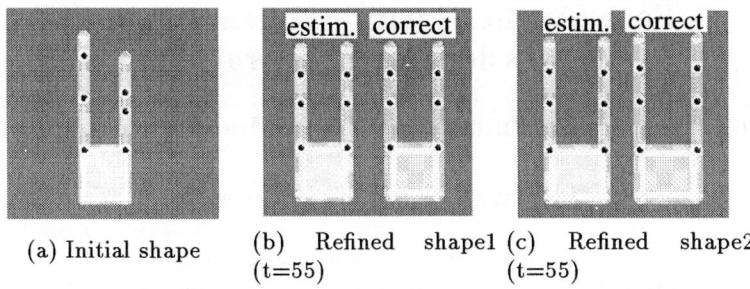

(a) Initial shape (b) Refined shape1 (t=55) (c) Refined shape2 (t=55)

Fig.10 The result of shape refinement (simulation)

dimensional space and updates the ellipse with each observation iteratively. It is the future works to apply that method to our problem.

References

1. M. Mochimaru and N. Yamazaki. "The Three-dimensional Measurement of Unconstrained Motion Using a Model-matching Method". *ERGONOMICS, vol.37, No.3*, pp. 493–510, 1994.
2. J. Davis and M. Shah. "Recognizing Hand Gestures". *ECCV'94.*, pp. 331–340, 1994.
3. M. Yamamoto and K. Koshikawa. "Human Motion Analysis Based on A Robot Arm Model". In *CVPR'91*, pp. 664–665. IEEE, 1991.
4. J. M. Rehg and T. Kanade. "Visual Tracking of High DOF Articulated Structures: an Application to Human Hand Tracking". *ECCV'94*, pp. 35–46, 1994.
5. D. Lowe. "Fitting Parameterized Three-Dimentional Models to Images". *IEEE trans., Pattern Anal. Machine Intell.,vol.13,No.5*, pp. 441–450, 1991.
6. J. O'Rourke and N. I. Badler. "Model-Based Image Analysis of Human Motion Using Constraint Propagation". *IEEE Trans. of Pattern Anal. and Machine Intell.,PAMI-2, No.6*, pp. 522–536, 1980.
7. R. A. Brooks. "Symbolic Reasoning Among 3-D Models 2-D Images". *Artificial Intelligence, Vol.17, No.1-3*, pp. 285–348., 1981.
8. Y. Hel-Or and M. Werman. "Recognition and Localization of Articulated Objects". In *Proc. of Workshop on Motion of Non-Rigid and Ariticulated Objects '94*, pp. 116–123. IEEE, 1994.
9. M. Isard and A. Blake. "Contour Tracking by Stochastic Propagation of Conditional Density". *ECCV'96.*, pp. 343–356, 1996.
10. N. Shimada, Y. Shirai, and Y. Kuno. "Hand Gestrue Recognition Using Computer Vision Based on Model-matching Method". In *Proc.of 6th International Conference on HCI*, pp. 11–16. Elsevier, 1995.
11. E. Fogel and Y. F. Huang. "On the Value of Information in System Identification – Bounded Noise Case". *Automatica, vol.18, No.2*, pp. 229–238, 1982.

Face Synthesis with Arbitrary Pose and Expression from Several Images

– An Integration of Image-Based and Model-Based Approaches–

Yasuhiro MUKAIGAWA[1], Yuichi NAKAMURA[2] and Yuichi OHTA[2]

[1] Department of Information Technology, Faculty of Engineering,
Okayama University, Tsushima-naka 3-1-1, Okayama 700, JAPAN
[2] Institute of Information Sciences and Electronics,
University of Tsukuba, Tennodai 1-1-1, Tsukuba, 305 JAPAN

Abstract. We propose a method for synthesizing face views with arbitrary poses and expressions combining multiple face images. In this method, arbitrary views of a 3-D face can be generated without explicit reconstruction of its 3-D shape. The 2-D coordinate values of a set of feature points in an arbitrary facial pose and expression can be represented as a linear combination of those in the input images. Face images are synthesized by mapping the blended texture from multiple input images. By using the input images which have the actual facial expressions, realistic face views can be synthesized.

1 Introduction

A human face includes various information such as individuality and emotion. Techniques for generating a facial image has been studied for many applications. However, a face is the most difficult objects for image synthesis, because we are extremely sensitive to differences between real face images and synthesized face images. We deal with both pose and expression, and aim to synthesize realistic facial images which is almost indistinguishable from real images.

In order to synthesize a facial image with arbitrary poses and expressions, some CG (Computer Graphics) techniques are used. A 3-D shape model of a human head is often used and the model is deformed according to the expressions. CV (Computer Vision) is often considered as a useful model acquisition method for CG. The rendering techniques, however, require accurate models which we cannot expect without special devices such as a high precision range finder[1].

It is still intractable with ordinary efforts to measure a precise 3-D face structure with various expressions. As for modeling, FACS (Facial Action Coding System)[2] is often used. A facial expression is described as a combination of the AU (Action Unit). A facial image with an expression is synthesized by deformation defined for each AU, but it is difficult to reproduce the slight changes such as a wrinkle.

Thus, synthesized images are still far from a real face appearance as you may see in many applications. Even small modeling errors cause undesirable effect to the synthesized images.

On the other hand, there is another paradigm called *image-based rendering*. It aims to synthesize realistic images by using the textures from real images. For

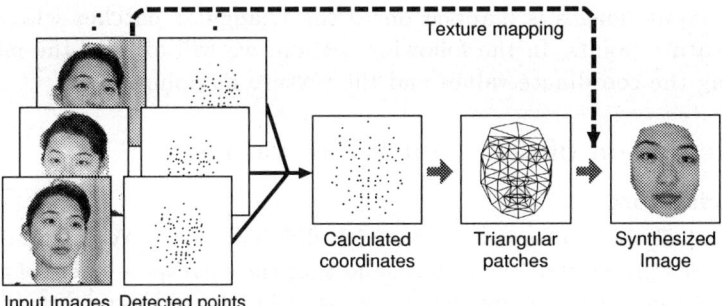

Fig. 1. Flow of the process

example, view morphing[3] method generates a new image easily, though this method generates only intermediate poses between two views.

In order to change the facial poses and expressions of an input image, Poggio, et al. proposed some methods. The *linear classes* method[4] assumes that one facial image can be represented as a linear combination of multiple facial images of other persons, and synthesizes a new image which has another pose and expression of the target person. However, this method requires a large number of images of other persons for reproducing the individuality. Also, the peculiarities of textures such as moles or birth-marks are not preserved. The *parallel deformation* method[5] synthesizes intermediate views of two expression images with two poses, but synthesized images are limited to intermediate views between the two views.

Therefore, to reproduce a true 3-D rotation with human individuality using *image-based rendering* is still an open problem. For this purpose, we propose a new method to deal with both rotation and expression in a same framework. We use linear combination to deal with 3-D appearance changes, and texture blending to deal with texture changes. These successfully avoid the difficult problems in 3-D reconstruction and image synthesis. In other words, our method realizes stable image-to-image conversion, that is from the input images to the output synthesized image, by combining *image-based* and *model-based* approach.

The principle of our framework is based on the structure-from-motion theory. Ullman, et al.[6] proved that an arbitrarily oriented object can be recognized by a linear combination of the 2-D coordinate values of feature points. We have applied this principle to image synthesis and showed that a facial image with arbitrary pose can be synthesized from only two images without explicit 3-D reconstruction[7]. This method can be applied to facial expression synthesis.

2 Basic scheme

Fig.1 shows the flow of the process. A set of images with different facial poses and expressions is used as input. First, 2-D coordinate values of the feature points in the synthesized image are calculated by a linear combination of the feature points detected from the input images. Then, the blended texture which is taken from the input images is mapped on to the triangular patches whose vertices are the feature points. In the following section, we will explain the method for calculating the coordinate values and the texture mapping.

3 Calculation of 2-D coordinate values

3.1 Facial poses

Let B_1 and B_2 be two input images with different poses. We assume that the feature points are located on the face, and that the correspondences of all feature points between two input images are known. Let (x_k^1, y_k^1) and (x_k^2, y_k^2) be 2-D coordinate values of the k-th feature point on images B_1 and B_2, respectively. These 2-D coordinate values are the result of rotating and projecting the points (X_k, Y_k, Z_k) of the 3-D space. The vectors and matrices which indicate these coordinate values are defined in the followings:

$$\boldsymbol{x}^1 = [x_1^1, x_2^1, \cdots, x_n^1] \tag{1}$$

$$\boldsymbol{y}^1 = [y_1^1, y_2^1, \cdots, y_n^1] \tag{2}$$

$$\boldsymbol{x}^2 = [x_1^2, x_2^2, \cdots, x_n^2] \tag{3}$$

$$\boldsymbol{y}^2 = [y_1^2, y_2^2, \cdots, y_n^2] \tag{4}$$

$$\boldsymbol{P} = \begin{bmatrix} X_1 & X_2 & \cdots & X_n \\ Y_1 & Y_2 & \cdots & Y_n \\ Z_1 & Z_2 & \cdots & Z_n \end{bmatrix} \tag{5}$$

For simplification, we assume rigidity, orthographic projection, and no translation. As shown in equations (6) and (7), the vectors which indicate the coordinate values of the feature points of each input image are represented as a multiplication of the 2 × 3 transformation matrix and the 3-D coordinate values of the feature points.

$$\begin{bmatrix} \boldsymbol{x}^1 \\ \boldsymbol{y}^1 \end{bmatrix} = \begin{bmatrix} \boldsymbol{r}_x^1 \\ \boldsymbol{r}_y^1 \end{bmatrix} \boldsymbol{P} \tag{6}$$

$$\begin{bmatrix} \boldsymbol{x}^2 \\ \boldsymbol{y}^2 \end{bmatrix} = \begin{bmatrix} \boldsymbol{r}_x^2 \\ \boldsymbol{r}_y^2 \end{bmatrix} \boldsymbol{P} \tag{7}$$

Let $\hat{\boldsymbol{x}}$ and $\hat{\boldsymbol{y}}$ be sets of 2-D coordinate values of the feature points in another view $\hat{B}$. Let $\hat{\boldsymbol{r}}_x$ and $\hat{\boldsymbol{r}}_y$ be the first row vector and the second row vector of the transformation matrix corresponding to $\hat{B}$. If $\boldsymbol{r}_x^1, \boldsymbol{r}_x^2, \boldsymbol{r}_y^1$ are linearly independent, a set of coefficients a_{x1}, a_{x2}, a_{x3} which satisfy equation(8) should exists, because the rank of $\hat{\boldsymbol{r}}_x$ is 3. This means that the X-coordinate values of the all feature points on $\hat{B}$ can be represented as a linear combination of 2-D coordinate values on B_1 and B_2, as shown in equation(9).

$$\hat{\boldsymbol{r}}_x = a_{x1}\boldsymbol{r}_x^1 + a_{x2}\boldsymbol{r}_x^2 + a_{x3}\boldsymbol{r}_y^1 \tag{8}$$

$$\hat{\boldsymbol{x}} = a_{x1}\boldsymbol{x}^1 + a_{x2}\boldsymbol{x}^2 + a_{x3}\boldsymbol{y}^1 \tag{9}$$

The base vectors of the linear combination are not always linearly independent. In order to get sufficient base vectors stably, we apply the principal component analysis of the four base vectors (x^1, y^1, x^2, y^2), and we use the first three eigen vectors (p_1, p_2, p_3) as the base vectors. As shown in equations (10) and (11), X- and Y-coordinate values can be stably represented as linear combinations of the linearly independent vectors.

$$\hat{x} = a_{x1}p_1 + a_{x2}p_2 + a_{x3}p_3 \qquad (10)$$

$$\hat{y} = a_{y1}p_1 + a_{y2}p_2 + a_{y3}p_3 \qquad (11)$$

As shown above, the 2-D coordinate values of the feature points can be easily calculated once the coefficients of the linear combination are obtained. In order to determine the coefficients corresponding to the synthesized image, we have two ways; specifying by a sample image and directly specifying the orientation.

Specifying by a sample image : A sample image is used for specifying the pose. We synthesize a new image whose pose is the same as this sample image. On the sample images, at least four representative points need to be detected. The coefficients are calculated so that these points of the sample image coincide with the corresponding points of the input images by using a least square method.

Directly specifying the orientation : For specifying the pose with an angle, we use representative points whose 3-D coordinates are known. We call these points control points. The control points are rotated to the requested pose, and projected on to the 2-D coordinates. By using these 2-D coordinate values, the coefficients are determined in the same way using sample images.

The coefficients of the linear combination are determined from representative points, and the 2-D coordinate values of all feature points on the facial image with arbitrary poses can be calculated.

3.2 Facial expressions

We try to synthesize a new image with arbitrary expression. Let $B_j(1 \leq j \leq m)$ be input images which have the same pose but different facial expressions. If these input images include a sufficient variety of facial expressions, the coordinate values of the feature points with any expressions can be approximated by the linear combination of those in the input images.

Let x_j and y_j be sets of coordinate values of the feature points on the input image B_j. As shown in equation (12), the vectors $\hat{x}$ and $\hat{y}$, which indicate sets of coordinate values on a new facial expression $\hat{B}$, are represented as a linear combination. The 2-D coordinate values of the feature points are calculated with the appropriate coefficients $b_1, \cdots, b_m$.

$$\hat{x} = \sum_{j=1}^{m} b_j x_j, \qquad \hat{y} = \sum_{j=1}^{m} b_j y_j \qquad (12)$$

It is convenient to specify the facial expression of the synthesized image by the interpolation of the input images, because no heuristic knowledge is necessary. In this method, a variety of synthesized facial expression depends on the kind of the input images. In other words, if we can prepare a sufficient variety of facial expressions as input images, we can obtain more realistic images.

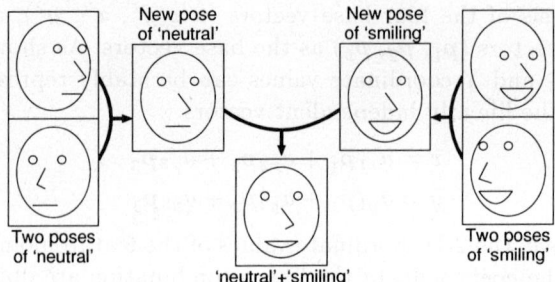

Fig. 2. Integration of pose and expression generation

3.3 Integration of the poses and expressions

Since both pose and expression are represented as a linear combination of the coordinate values of the feature points, these can be easily integrated. We assume that a set of input images includes some facial expressions and has at least two different poses for each expression. As shown in Fig.2, we try to synthesize a new image with arbitrary pose and expression from the input images. First, the facial pose in the input images is adjusted to a specified pose for each expression by the method explained in section 3.1. Then, the facial expression is adjusted to a specified expression by the method explained in the section 3.2.

The 2-D coordinate values corresponding to a facial pose with an expression can be represented as the linear combination of base vectors shown in equations (13) and (14). Note that a_{xi}^j, a_{yi}^j, and $\boldsymbol{p}_i^j$ ($i = 1, 2, 3$) are the coefficients and the base vectors of the j-th expression in the equations (10) and (11), respectively.

$$\hat{\boldsymbol{x}} = \sum_{j=1}^{m} b_j (\sum_{i=1}^{3} a_{xi}^j \boldsymbol{p}_i^j) = \sum_{j=1}^{m} \sum_{i=1}^{3} b_j a_{xi}^j \boldsymbol{p}_i^j \tag{13}$$

$$\hat{\boldsymbol{y}} = \sum_{j=1}^{m} b_j (\sum_{i=1}^{3} a_{yi}^j \boldsymbol{p}_i^j) = \sum_{j=1}^{m} \sum_{i=1}^{3} b_j a_{yi}^j \boldsymbol{p}_i^j \tag{14}$$

4 Texture mapping

4.1 Texture blending

Basically, the textures taken from input images which have similar poses and expressions are mapped onto the synthesized image. However, if we take all the textures from one image with the closest facial pose, the synthesized image will be warped unnaturally as the facial pose changes. This undesirable warping is caused by a drastic deformation of texture. The same can be said to the facial expression. It is obvious that the natural expression can be synthesized by directly using textures of the similar expression rather than by deforming the textures of different expressions. For example, the wrinkles that appear when smiling cannot be synthesized by warping the texture taken from a neutral expression.

We can solve this problem by texture blending. In our method, facial images are synthesized by mapping the blended texture taken from multiple input images. The weight for blending is set larger for the similar pose and expression.

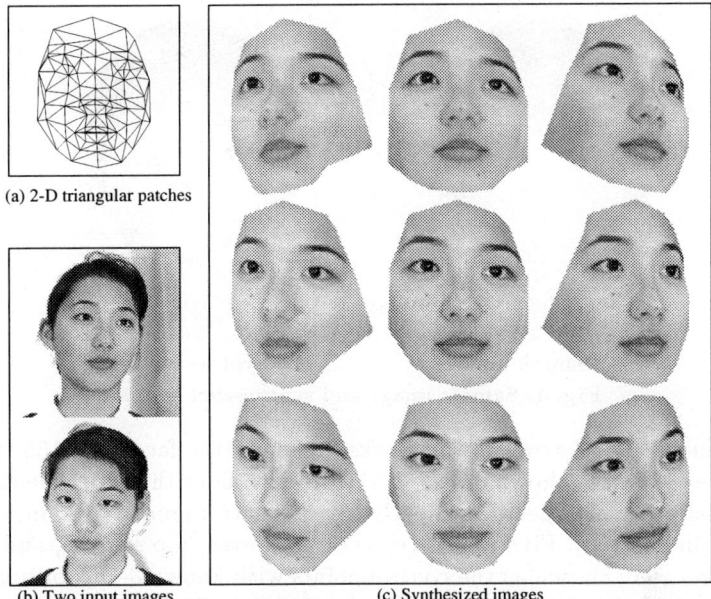

Fig. 3. Synthesized images of various poses using two input images

Although the facial orientation of the input image is unknown, a rough orientation can be estimated by the factorization method[8]. First, several feature points which do not move by the facial expression changes, such as the top of nose, are selected. Then, the relative facial orientation of the sample image is estimated. Since this value is only used for determining the texture blending weights, small errors are not critical. The weights of the texture blending are determined to be inversely proportional to the square of the angle difference of facial orientations. The weights of the blending for facial expression are determined in proportion to the coefficients b_j in the equation (12).

4.2 Two dimensional texture mapping

The texture is mapped by using triangular patches whose vertices are the feature points. For each triangular patch, the texture is clipped from multiple input images, and deformed by affine transformation according to the 2-D coordinate values of the feature points. Then, the textures are blended by using the weights and mapped. In this step, it is checked whether each patch is obverse or reverse. The texture is not mapped on to the reverse patch. This simple judgment is enough for facial image synthesis, because complicated occlusion never occurs.

5 Experiments

We chose 86 feature points on a face. The number is relatively small compared to the ordinary facial models used in other applications. These feature points are located not only on the facial components such as eyes and mouth, but also on the movable parts such as cheeks. The 2-D coordinate values of the points are

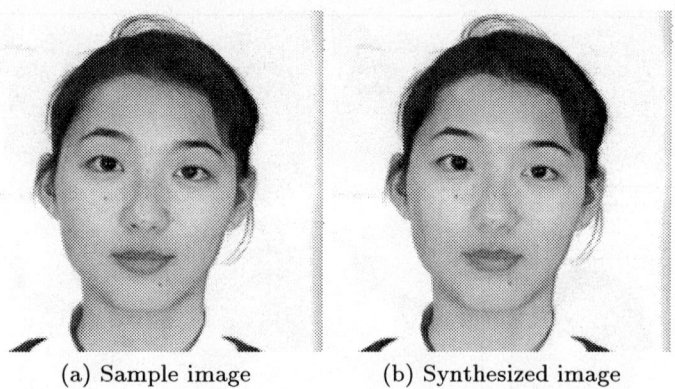

(a) Sample image (b) Synthesized image

Fig. 4. Sample image and synthesized image

given manually by referring to the marks drawn on the face. The 156 triangular patches are created as shown in Fig. 3(a). First, we show the experimental results whose facial poses are specified directly by the control points. Two input images are shown in Fig.3(b). Five points (ends of both eyes, top of nose, and bottoms of both ears) were chosen as the control points with known 3-D coordinates. The synthesized images with various poses are shown in Fig. 3(c). We can see that true 3-D rotations are realized.

Next, in order to compare the synthesized image and the real image, we synthesize a new image which has the same pose as the sample image from two input images shown in Fig.3(b). A sample image was prepared as shown in Fig.4(a). On the sample image, five feature points (ends of both eyes, top of nose, and bottoms of both ears) were manually detected. A new facial image was synthesized and overwritten onto the sample image as shown in Fig.4(b). The outer region of the face such as hair, boundary, and background is copied from the sample image and the inner region of the face is overlaid by the synthesized image. As we can see in this example, the synthesized images are often indistinguishable from real images.

Last, we show the results of changing facial expressions of a sample image. The input images include two different poses for each three different expressions as shown in Fig.5(a). Fig.5(b) shows two sample images which are the same person as the input images. New images with different expressions were synthesized and overwritten onto each sample image as shown in Fig.5(c).

6 Conclusion

We have proposed a new method for synthesizing facial views with arbitrary poses and expressions without explicitly modeling the 3-D shape and the facial expression. We have shown that both of the poses and the expressions are treated in a unified way. We have implemented this method and demonstrated that the natural facial images can be synthesized.

The number of facial expressions that can be synthesized in the current system is limited, because we used only few kinds of facial expressions. In order to

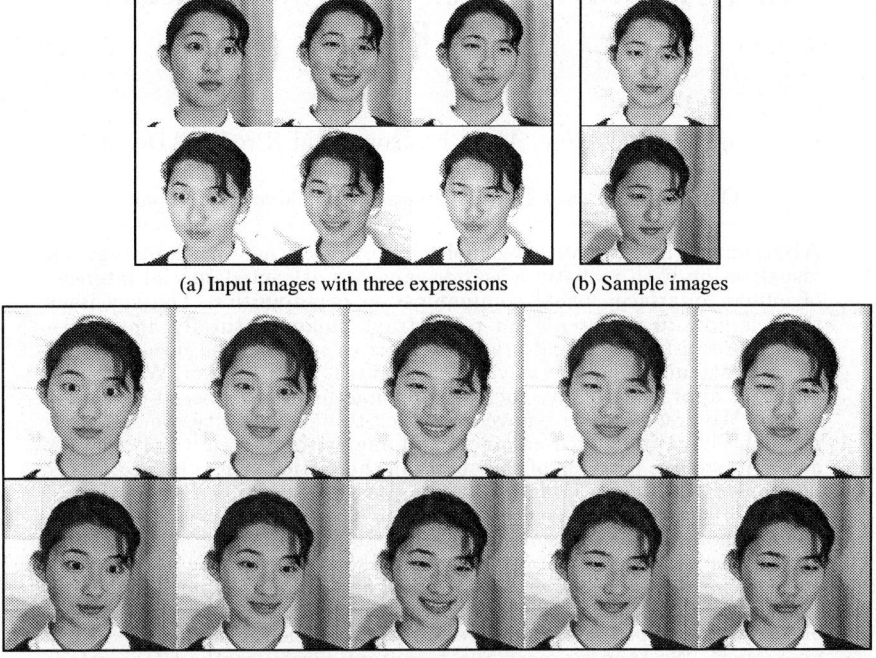

Fig. 5. The results of changing facial expressions of sample images

synthesize *arbitrary* facial expression, we are now working on a new framework which utilizes a set of typical expressions obtained from a sequence of images.

References

1. T.Akimoto and Y.Suenaga, "Automatic Creation of 3D Facial Models", IEEE Computer Graphics and Applications, September 1993, pp.16–22, 1993.
2. P.Ekman, and W.V.Friesen, "Facial action coding system", Consulting Psychologists Press, 1977.
3. S.M.Seitz, and C.R.Dyer, "View Morphing", Proc. SIGGRAPH'96, pp.21–30, 1996.
4. T.Vetter and T.Poggio, "Linear object classes and image synthesis from a single example image," A.I.Memo No.1531, Artificial Intelligence Laboratory, Massachusetts Institute of Technology, 1995.
5. D.Beymer, A.Shashua, and T.Poggio, "Example based image analysis and synthesis," A.I.Memo No.1431, Artificial Intelligence Laboratory, Massachusetts Institute of Technology, 1993.
6. S.Ullman, and R.Basri, "Recognition by linear combinations of models", IEEE Trans. PAMI, vol.13, no.10, pp. 992–1006, 1991.
7. Y.Mukaigawa, Y.Nakamura, and Y.Ohta, "Synthesis of Arbitrarily Oriented Face Views from Two Images", Proc. ACCV'95, Vol.3, pp.718-722, 1995.
8. C.Tomasi, and T.Kanade, "The factorization method for the recovery of shape and motion from image streams", In Proceedings of Image Understanding Workshop, pp. 459–472, 1992.

Live Facial Expression Generation Based on Mixed Reality

Hiromi T. TANAKA, Akira Ishizawa, and Hiroaki ADACHI

Computer Science Dept., Ritsumeikan University, Japan

Abstract. Virtual reality technology provides a new methodology for visualization with realistic sensation, and has attracted special interests of human interface, visual communication communities. The key issue there is how to represent and reconstruct *human* naturally and realistically. Accordingly, recent study of facial expression has been received growing attention and intensively investigated. In this paper, We propose a hybrid approach to Live facial expression generation based on mixed reality. We first propose a novel approach to mixed reality, we call Augmented Virtuality, which enhances and augments the reality of complex and delicate live motions of the object in the virtual space, by projecting portions of live video images observing the deformation and motion of the object onto the surface of its static CG model. We also propose a new method of adapting color properties for smooth merging of real and virtual spaces, and also propose a new method of extraction of the region effective for merging, based on the optical flow analysis of both range and color images in which in the virtual space, shape and texture changes are observed. We apply this technique to real time generation of realistic eye expression. We then propose the homotopy sweep method for surface deformation using 3D control vectors, and apply this technique to the animation of mouth/lips expression. Our approach has the advantages of describing the geometric shapes and the deformation of circular muscle simply, and of reconstructing realistic deformation efficiently. Experimental results demonstrate the effectiveness of the proposed hybrid approach in representive and visualizing live facial expression in real time.

1 Introduction

Virtual reality technology provides a new methodology for visualization with realistic sensation, and has attracted special interests of human interface, visual communication communities. The key issue there is how to represent and reconstruct *human* naturally and realistically. Accordingly, recent study of facial expression has been received growing attention and intensively investigated in the field of human interface, computer vision, and psychology.

Generally, facial expression are generated by deformation of facial muscles such orbicularis oris, orbicularis oculll, and venter frontalls. Many previously proposed approaches [7, 8, 9] for the facial expression problem, have been based on the analysis of facial feature points such as Facial Action Coding System(FACS). And facial expressions have been generated by deforming the face wireframe model according to the displacement of such feature points. However, individual structure of every different facial muscle and its delicate and complex motions have not been described correctly and efficiently with such feature points.

In this paper, we propose a novel approach to live facial expression generation based on mixed reality, which merges live video images in a 3D face CG model to represent and reconstruct the fine details of live facial expression.

First, we propose a novel approach to mixed reality, we call Augmented Virtuality(AV), which enhances and augments the reality of complex and delicate live motions in the virtual space, by projecting portions of live video images which observes the deformation and motion of the object, onto the surface of its static CG (Wireframe) model.

We apply this technique to real time eye expression generation. Consequently, live eye expressions are realized and visualized in real time in the virtual space by continuously projecting video images onto the 3D face model.

Next, we propose a smooth surface deformation method based on the homotopy sweep technique, which deforms a surface by interpolating continuous transition between coarsely sampled time-varying cross-section contours with three dimensional velocity vectors.

The homotopy sweep method was originally proposed for solid surface generation using the continuous transition among a set of two cross-section contours, where the three dimensional shape and volume are described as a "swept volume" of two dimensional cross-section contours along a space curve called a trajectory. In this work, we extend the homotopy sweep method to surface deformation by replacing "the space-trajectory" by "time-trajectory". We then successfully apply the proposed method to real time animation of mouth/lips expression. Our approach to adapt the homotopy sweep method, is based on that mouth/lips expression is composed of deformation of circular muscle around the mouth and these muscles can be modeled as a set of closed contours.

Experimental results show the effectiveness and feasibility of the proposed approach to live facial expression generation in real time.

2 Image (Projection)-based Eye Expression Generation

2.1 Representing and Visualizing Live Motions with Augmented Virtuality

In this section, we introduce a novel approach to mixed reality, we call Augmented Virtuality(AV), which enhances and augments the reality of complex and delicate live motions in the virtual space, by projecting portions of live video images which are observing the deformation and motion of the object, onto the surface of its static CG (Wireframe) model.

This technique extends the use of real image to the *dynamic* region, and releases the complexity of 3D modeling to represent live motion of the object, and also provides to present the stereoscopic images from arbitrary directions. First, we extract merging regions from the video images which observe live motions, that is, effective for merging, based on the optical flow analysis of both range and color real images. We then adapt properties of color texture image of the CG model to that of the merging region, according to the changes in the color values of the corresponding regions. This process decreases the contrast along the boundary of the merging region and achieves smooth merging of real and virtual spaces in real time. The novelty of our approach is that merging regions from real video images are extracted based on the optical flow analysis of images observing both shape and texture changes of the objects.

We apply this technique to real time generation of live eye expression. Consequently, live facial expressions are realized and visualized in real time in the virtual space by continuously projecting video images onto the 3D face models.

2.2 Region Effective for Merging

First, we define a *region effective for merging (REM)* in which our technique works effectively to enhance the reality of 3D CG model in representing and visualizing live motions.

When we evaluate cost and quality in generating dynamic views, live video images are not sufficient to represent dynamic changes, if the 3D shape deformation of the region occurs in large scale, moreover, if such deformation can be generated easily. On the other hand, video images is efficient, as a texture image on the region in which complex and delicate motions occur over solid and static surface structure. In short, we define REMs as regions where the 2D texture changes are much lager and more complex than 3D shape deformation. In the following, we describe REM in case of a face surface.

2.2.1 3D shape Deformation Analysis

As the analysis of 3D shape deformation, we refer to the result of face surface segmentation based on the motion analysis reported by Ueno [6], where a face surface was divided into 4 parts as shown in Figure 1.

As Figure 1(a) shows, the candidates regions for REMs are extracted as regions with sliding motion or stable regions(Figure 1(a) and Figure 1(d))

2.2.2 2D texture Change Analysis

The apparent velocity of the object is computed as optical flows from spatio-temporal gradients of the image intensities, with the condition that the object moves keeping the distribution of the image intensities unchanged. The optical flow $\mathbf{v} = (u, v)$ is defined from the following equation.

$$\frac{\partial \rho}{\partial x} \cdot \frac{dx}{dt} + \frac{\partial \rho}{\partial y} \cdot \frac{dy}{dt} + \frac{\partial \rho}{\partial t} = 0 \qquad (1)$$

where ρ is image intensity, $u = dx/dt$ and $v = dy/dt$ denote the x and y elements of the velocity respectively.

Figure 2 shows the result of the optical flow analysis computation from video images of facial expressions. The candidates for REMs are extracted over eye region, as shown in Figure 3(b).

2.2.3 Extraction of REMs

From the result shown in Figure 3(a) and Figure 3(b), REM is extracted over the eye region, and shown in Figure 3(c). Then REM is manually selected on the wireframe model by considering symmetry, as shown in Figure 3(d)

2.3 Adapting Color Properties

The difference in color tone is outstanding along the boundaries of the merging region, because the different lighting conditions cause significant changes in color properties between the texture image of 3D CG model and the live images.

We describe the method of adapting color properties for smooth merging of virtual and real spaces. In order to reflect *live states* of the objects to the virtual space, we adapt properties of color texture image of the CG models to that of the live video image, according to the changes in the color values of the corresponding regions.

2.3.1 Color Adaptation Algorithm

We define the reference color of REM to measure the difference in the color tone which is dominant in quantity and also in quality in representing live states of the object (e.g. a skin color for human). We extract the reference color based on the HSV (H:hue, S:saturation, V:value) color histogram analysis. The algorithm is shown below.
1. RGB to HSV conversion of input images.
2. Computing the color histograms on each H, S, V value in the REM
3. Finding a peak value of H, and average values of both S and V
4. Adapting color properties based on the equations as shown below,

$$\begin{aligned} H_{-CG} &= H_{-CG} + \Delta H, & \Delta H &= Hpeak_{-video} - Hpeak_{-CG} \\ S_{-CG} &= S_{-CG} \cdot \Delta S, & \Delta S &= Saverage_{-video}/Saverage_{-CG} \\ V_{-CG} &= V_{-CG} \cdot \Delta V, & \Delta V &= Vaverage_{-video}/Vaverage_{-CG} \end{aligned} \qquad (2)$$

5. HSV to RGB conversion of adapted texture image of the CG model

2.4 Computation Steps

Figure 12 and Figure 14 shows two input images: a color texture image of the CG model; and a live video image by a CCD camera shown in Fig.14(b).

step1. Extract REMs from live video images
step2. Extract the dominant color of REM
step3. Adapt color properties of the whole color texture image of the CG model to that of the video image
step4. Merge a portion of video images in the color texture image of the 3D CG model
step5. Display the 3D CG model, i.e, the static wireframe mapped with dynamic changing texture

Figure 14(a) shows our experimental video camera mounted on the helmet to observe a fixed region independent of the object's motion, as shown in Figure 14(b). It results in avoiding feature corresponding problems in real time.

In step 1, the complexity of the texture changes in different facial expressions is analyzed from optical flow analysis. The texture change of around eyes region is significant as observed from the optical flow in Figure 2 where Figure(a) and Figure(b) show the directions and magnitude of the optical flow respectively. In addition to that, the deformation of 3D shape is small in the region around eyes as shown in Figure 1. Compared with other regions that are important in generating facial expression, the shape deformation in eyes region is smaller than the texture change. Therefore, the region of spectacles shape around eyes is extracted for merging to the virtual space.

3 Deformation-based Mouth/Lips Expression Generation

3.1 Homotopy Sweep-based Surface Deformation

The homotopy sweep method was originally proposed for solid surface generation using the continuous transition among a set of two cross-section contours, where the three dimensional shape and volume are described as a "swept volume" of two dimensional cross-section contours along a space curve called a trajectory [10] (cf. Figure 5).

The technique provides a convenient and intuitive control of surface generation, and allows the generalized cylinders to be modeled with a smaller set of input parameters. The problem is, however, how to determine the path connecting two adjacent contours, out of infinitely possible ones. In this work, we extend the homotopy sweep method to surface deformation by replacing "the space-trajectory" by "time-trajectory". In this section, we describe the method to control and find the path with the use of a scaling function of 3D control vectors along the contours. We then successfully apply the proposed method to real time generation of mouth/lips expression. Our approach based on that mouth/lips expression is composed of deformation of circular muscles around the mouth and these muscles can be modeled as a set of closed contours.

3.2 Definition

Let O be the origin of the world coordinate system, A be a point on a contour C_1 and B be a corresponding point of A on the contour C_2 after a unit sampling period Δt(cf. Figure 6). The points A,B are defined in the local cylindrial coordinate system $(O, \mathbf{Ur}(\theta), \mathbf{U}_\theta(\theta), \mathbf{Uz}(\theta))$ at A, where the x, y, and z axises are defined by $\mathbf{Ur}(\theta) = \frac{\mathbf{OA'}}{\|OA'\|}, \mathbf{U}_\theta(\theta)$ and $\mathbf{Uz}(\theta) = \mathbf{Ur}(\theta) \times \mathbf{U}_\theta(\theta)$ respectively. Then A on C_1 and B on C_2 are given by

$$A = C_1(\theta) = [C_{1r}(\theta), 0, C_{1z}(\theta)], \quad B = C_2(\theta) = [C_{2r}(\theta), C_{2\theta}(\theta), C_{2z}(\theta)]$$

We define the homotopy in the local system to give the 3D deformation. Then a path $M(\theta,t)$ between A and B is given by (cf. Figure 6),

$$\mathbf{OM}(\boldsymbol{\theta},\mathbf{t}) = r(\theta,t)\mathbf{U_r}(\boldsymbol{\theta}) + \theta(\theta,t)\mathbf{U_\theta}(\boldsymbol{\theta}) + z(\theta,t)\mathbf{U_z}(\boldsymbol{\theta}) \qquad (3)$$

where $r(\theta,t)$ is the homotopy value in $\mathbf{Ur}(\boldsymbol{\theta})$ direction, $\theta(\theta,t)$ is the homotopy value in $\mathbf{U_\theta}(\boldsymbol{\theta})$ direction, and $z(\theta,t)$ is the homotopy value in $\mathbf{Uz}(\boldsymbol{\theta})$ direction.

3.3 Surface Deformation by Homotopy Sweep Method

The interpolation between the two values C_{1r} and C_{2r} in the $\mathbf{Ur}(\boldsymbol{\theta})$ direction is computed from Eq.(4).

$$r(\theta,t) = (1 - R_n(t))(1 + Sc_1(\theta,t))C_{1r}(\theta) + R_n(t)(1 + Sc_2(\theta,t))C_{2r}(\theta) \qquad (4)$$
$$(5)$$

where $0 \leq t \leq 1$ R_n is a blending function given by Eq.(10), and Sc_1 and Sc_2 are scaling functions for the contours C_1 and C_2 are used in our proposed scaling function Eq.(11), which allow to control the contour deformation in arbitrary 3D directions.

The interpolation between the two values 0 and $C_{2\theta}$ in the $\mathbf{U_\theta}(\boldsymbol{\theta})$ direction is computed from Eq.(6).

$$\theta(\theta,t) = (1 - R_m(t))(1 + Sc_1(\theta,t))N + R_m(t)(1 + Sc_2(\theta,t))(C_{2\theta}(\theta) + N) - N \qquad (6)$$
$$(7)$$

where $0 \leq t \leq 1$ N is an offset value which make the scaling function effective at t=0, and is used in the interpolation from the interval $[0, C_{2\theta}(\theta)]$ to $[N, C_{2\theta}(\theta) + N]$, a blending function R_m is given by Eq.(10) and the scaling functions Sc_1 and Sc_2 are computed from Eq.(11).

The interpolation between the two values C_{1z} and C_{2z} in the $\mathbf{Uz}(\boldsymbol{\theta})$ direction is computed from Eq.(8).

$$z(\theta,t) = (1 - R_k(t))(1 + Sc_1(\theta,t))C_{1z}(\theta) + R_k(t)(1 + Sc_2(\theta,t))C_{2z}(\theta) \qquad (8)$$
$$(9)$$

where $0 \leq t \leq 1$ a blending function R_k is given by Eq.(10) and Sc_1 and Sc_2 are the scaling functions defined in Eq.(11).

3.4 A Blending Function

A Blending function $R_n(v)$ which arbitrarily controls surface deformation between the contours C_1 and C_2 with parameter n, with G^1 geometric continuity, is given by[10].

$$R_n(v) = \frac{(1+n)v^2}{(1+n)v^2 + (1-n)^2} \qquad (10)$$

where $0 \leq v \leq 1$, $-1 \leq n$

Figure 7 shows the graph of a blending function when the parameter n changes, Figure 8(a) shows the generalized cylinder of n = -0.9, and Figure 8(b) shows the generalized cylinder of n = 9.

3.5 A Scaling Function with 3D Control Vectors

With the proposed scaling function, however the control is not allowed to arbitrary directions along the contour. We solve this limitation by adding another weight parameter k to specify a valid range where the control vectors are effective. Then scaling functions with 3D control vectors for C_1 and C_2 are given by

$$Sc_1(\theta,t) = t(t-1)(\frac{t}{k_1} - 1)\mathbf{p_1}(\boldsymbol{\theta}), \quad Sc_2(\theta,t) = -t(t-1)(\frac{t}{1-k_2} - 1)\mathbf{p_2}(\boldsymbol{\theta}) \quad (11)$$

(12)

where $0 \le k_1$, $0 \le k_2$ Sc_1 and Sc_2 are the scaling functions for the contours C_1 and C_2, the parameter k is a weight parameter($0 \le k$), $\{\mathbf{p_1}(\boldsymbol{\theta})\} = \{(p_{1r}(\theta), p_{1\theta}(\theta), p_{1z}(\theta))\}$, and $\{\mathbf{p_2}(\boldsymbol{\theta})\} = \{(p_{2r}(\theta), p_{2\theta}(\theta), p_{2z}(\theta))\}$ are 3D control vectors along C_1 and C_2 respectively.

Figure 9 demonstrates the property of a scaling function applied to free-formed surface generation, where tangent vectors along cross-section contours are used as a set of 3D control vectors.

3.6 3D Control Vectors

The following form of 3D control vectors $\mathbf{p_1}$, $\mathbf{p_2}$ are provided along contours C_1 and C_2 in the local coordinate system, as shown in Figure 6.

$$\mathbf{p_1}(\boldsymbol{\theta}) = \begin{bmatrix} v_{1r}(\theta)/C_{1r}(\theta) \\ v_{1\theta}(\theta)/N \\ v_{1z}(\theta)/C_{1z}(\theta) \end{bmatrix}, \quad \mathbf{p_2}(\boldsymbol{\theta}) = \begin{bmatrix} v_{2r}(\theta)/C_{2r}(\theta) \\ v_{2\theta}(\theta)/(C_{2\theta}(\theta) + N) \\ v_{2z}(\theta)/C_{2z}(\theta) \end{bmatrix} \quad (13)$$

3.7 Algorithm

We have applied the proposed homotopy sweep method to the real time animation of mouth/lips expression.

step1 Sample key frames at every Δt from input video images according to the given sampling rate.
step2 Obtain the mouth/lips contour using 3D coordinates of marker points
step3 Compute the homotopy sweep using 3D control vectors computed along the contour
step4 Deform the 3D mouth wireframe model of mouth/lips region based on the interpolation in step3
step5 Display the 3D deformed wireframe model

3.7.1 The Mouth Model of Circular Muscle

Figure 10(b) shows a wireframe model of a mouth/lips region considering the circular muscle structure. This model consists of four nested contours along circular muscle fibers. P_0 corresponds to marker points on the outer and inner contours (cf. marker points in Figure 10(a)). Our homotopy sweep method deforms the surface according to displacement of four marker points at every Δt.

The inbetween points between the outer and inner contours (A1,A2) is correspond to $\frac{A_0 A_1}{A_0 A_3} = \frac{1}{3}$ and $\frac{A_0 A_2}{A_0 A_3} = \frac{2}{3}$ because the the ratio of displacements of circular muscle is constant along the muscle fibers.

The inbetween points on the same contour (P_1-P_9) is correspond to a point on an ellipse approximated from the four markers. In addition, the 3D shape of the muscle fibers is estimated using a cosine curve.

We compute the inbetween points on the same contour and different contours with displaced the homotopy sweep method.

4 Experiment

Figure 13 shows our hybrid system of live facial expression generation based on mixed reality technique.

We have implemented the proposed algorithm in C language on a SGI Indy R4600, and applied the algorithm to the real time generation of live facial expression. We used a CCD camera Panasonic WV-KS102 mounted on the helmet for video image acquisition. The acquired image was 628 x 492. We have obtained both range and color texture images of size 512x512 using the Cyberware Digitizer to construct a 3D CG face model.

We first analyzed the color change of the facial expression image as color texture analysis as shown in Figure 2. We nominated the region (a) in Figure 1 as the REM based on definition in Section 2. In current stage we extracted the REM manually, as shown in Figure 3, based on the definition in Section 2.

Secondly, we analyzed color in the REM and extracted the skin color as the reference color by the color histogram. Then, we adapted the color properties of (texture image of) 3D CG model for smooth merging. The result of this adaptation is shown in Table 1.

Finally, we presented the model merged with video images in Figure 4.

We also aquired images in which an examinee pronounces a set of single vowels "a", "i", "u", "e", "o". We selected each image as a key frame and we gave $\frac{1}{4}$ of displacement between the present frame and the previous frame.

Nextly, we computed the homotopy sweep between the two key frames. We have used the following parameters to compute the homotopy sweep.

- The parameter n of blending function is 0
- The parameter k of scaling function is 0.5
- The number of generated frames between two key frames is 20

Figure 11 shows original images and synthesis images.

Figure 15 shows the final result of the hybrid system in the receiver side.

5 Conclusion

We proposed a hybrid approach to Live facial expression generation based on mixed reality. We first proposed a novel approach to mixed reality, we call Augmented Virtuality, which enhances and augments the reality of complex and delicate live motions of the object in the virtual space, by projecting portions of live video images observing the deformation and motion of the object onto the surface of its static CG model.

We also proposed a new method of adapting color properties for smooth merging of real and virtual spaces, and also proposed a new method of extraction of the region effective for merging, based on the optical flow analysis of both range and color images in which in the virtual space, shape and texture changes are observed. We applied this technique to real time generation of realistic eye expression.

We then proposed the homotopy sweep method for surface deformation using 3D control vectors, and applied this technique to the animation of mouth/lips expression. Our approach has the advantages of descripting the geometric shapes and the deformation of circular muscle simply, and of reconstructing realistic deformation efficiently.

Experimental results demonstrated the effectiveness of the proposed hybrid approach in representive and visualizing live facial expression in real time.

References

1. H.D.Foley, A.V.Dam: Fundamentals of Interactive Computer Graphics. Addison-Wesley (1982)
2. M.Bajura, H.Fuchs: Merging Virtual Objects with the Real World. Proc.'92 SIGGRAPH (July 1992) 203-210
3. P.Milgram: Applications of Augmented Reality of Human-Robot Communication. Proc.Int.conf.'93 IEEE/RSJ
4. P.Milgram: A class of displays on the reality-virtuality continuum. SPIE Vol.2351 Telemanipulator and Telepresence Techonologies (1994)
5. P.Milgram: A Taxonomy of Mixed Reality Visual Display. IEICE TRANS. INF. &SYST. Vol.E77-D No.12 (1994) 1321- 1329
6. M.Ueno et al.: A Construction of High Definition Wire Frame Model of Head and Its Hierarchical Control for Natural Expression Synthesis. Tech. Rep. IEICE PRU92-77 (1992-12) 9-16
7. C.S.Choi, H.Harashima, T.Takebe: Analysis of Facial Expression Using Three-Dimensional Facial Model. IEICE TRANS. INF. &SYST. D-II Vol.J74 No.6 Jun 1991 766-777
8. K.Aizawa, H.Harashima, T.Saito: A Model-Based Analysis Synthesis Image Coding Scheme. IEICE TRANS. INF. &SYST. B-I Vol.J72 No.3 Mar. 1989 200-207
9. M.Kaneko, A.Koike, Y.Hatori: Synthesis of Moving Facial Images with Mouth Shape Controlled by Text Information. IEICE TRANS. INF. &SYST. D-II Vol.J75 No.2 Feb. 1992 203-215
10. C.Tai, K.Loe, T.Kunii: Integrated Homotopy Sweep Technique for Computer-Aided Geometric Design. Computer Vision Springer-Verlag 1991 583-595
11. L.Moubaraki, H.Tanaka, Y.Kitamura, J.Ohya, F.Kishino: Homotopy-Based 3D Animation of Facial Expression. Tech.Rep.IEICE IE94-07Jul.1994 9-16
12. M.Kaneko, Y.Hatori, A.Koike: Coding of Facial Images Based on 3-D Model of Head and Analysis of Shape Changes in Input Image Sequence. IEICE TRANS. INF. &SYST. Vol.J71-B No.12 Dec. 1988 1554-1563
13. K.Mase, A.Pentland: Automatic Lipreading by Optical-Flow Analysis. IEICE TRANS. INF. &SYST. D-II No.6 Jun 1990 796-803

Table 1. Result of Adapting Color Properties

	H peak	S average	V average
Video Image	10.00	0.579	0.775
3D CG Model	15.00	0.344	0.345
Result	10.00	0.582	0.710

(H: hue, S: saturation, V: value)

Table 2. Processing Speed

Process	Process Type	FPS
Sender	Video Image Input	30
	REM Color Change Detection	17
	Mouth/Lips Motion Detection	1.2
	Change Parameters Transmission	20
Receiver	Change Parameters Input	20
	Texture Color Adaptation	0.3-30
	Mouth/Lips WF Deformation	15
	Facial Expression Generation	7

(FPS: Frames / Second)

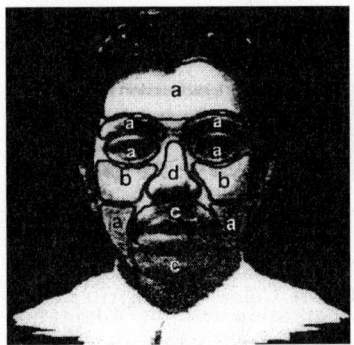

(a). region mainly sliding along the surface
(b). region bulging or sinking against the surface
(c). smoothly deforming region
(d). stable region

Fig. 1. Face Segmentation (Ueno p15: Figure 8)

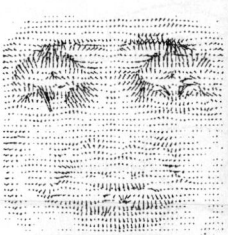 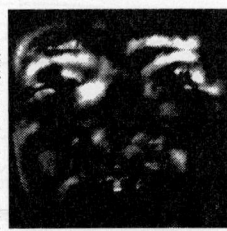

(a) Optical Flow   (b) Magnitude of Optical Flow

Fig. 2. Optical Flow Analysis

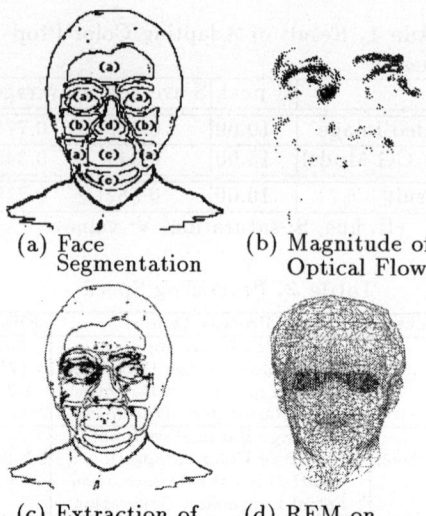

(a) Face Segmentation
(b) Magnitude of Optical Flow
(c) Extraction of REM
(d) REM on Wireframe

Fig. 3. Region Effective for Merging (REM)

(a) Before Color Adaptation
(b) After Color Adaptation

Fig. 4. Color Adaptation Result

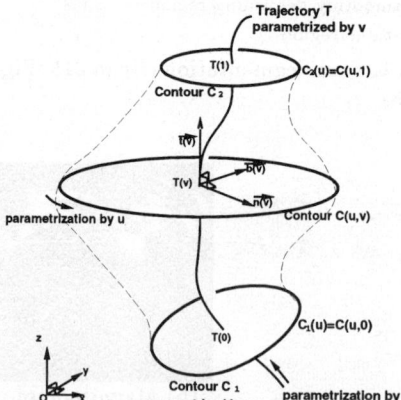

Fig. 5. Definition of Homotopy Sweep Method

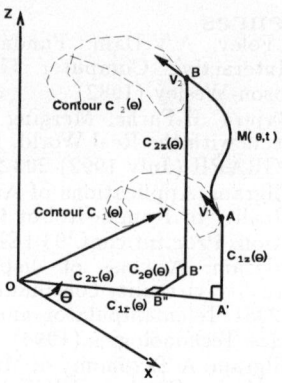

Fig. 6. Cylinder Frame for Interpolation

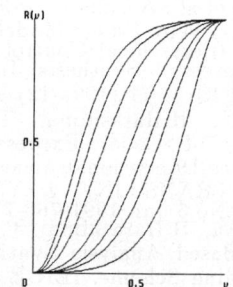

Fig. 7. Blending Function R_n for Interpolation $n=-0.9,-0.7,-0.5,0,2,5,9$

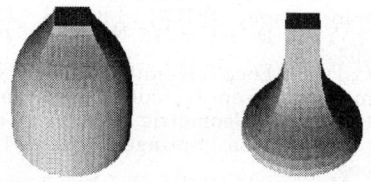

(a) $n=-0.9, S_1=S_2=0$
(b) $n=9, S_1=S_2=0$

Fig. 8. Generalized Cylinders based Homotopy Sweep Method

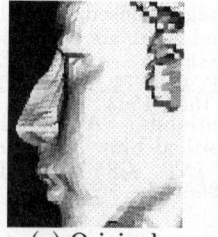

(a) Original
(b) Interpolated

Fig. 9. Face Surface Modeling with Homotopy Sweep

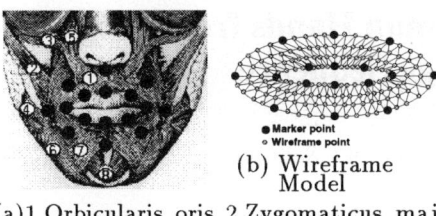

(a) 1.Orbicularis oris 2.Zygomaticus major 3.Zygomaticus minor 4.Risorius 5.Levator labii superioris 6.Depressor anguli oris 7.Deprssor labii inferioris 8.Mentalis
! | Marker points

Fig. 10. Marker Points and CG Model

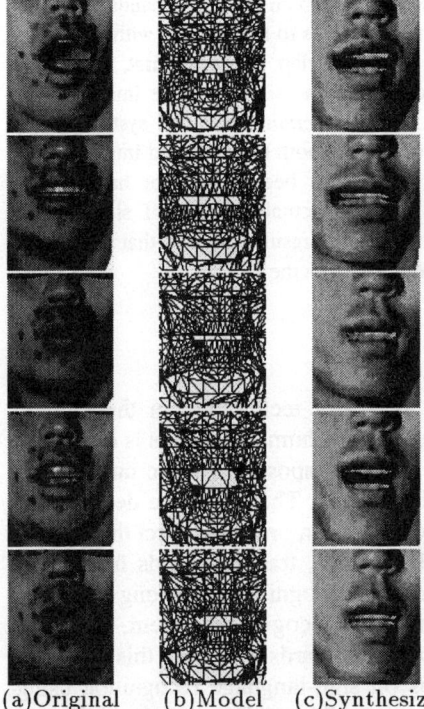

(a)Original (b)Model (c)Synthesized

Fig. 11. Mouth/Lips Expression Animation With Homotopy Sweep

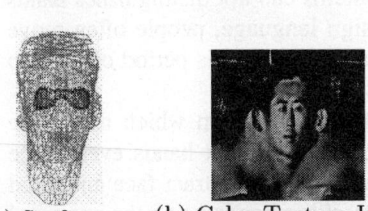

(a) Surface Model (b) Color Texture Image

Fig. 12. 3D CG Model

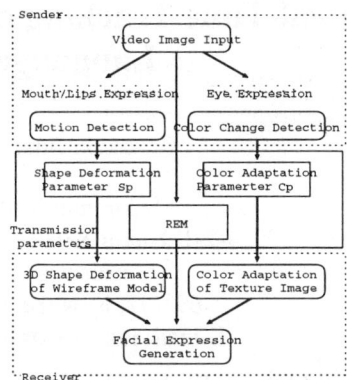

Fig. 13. Hybrid System for 3D Facial Animation

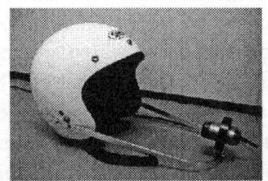

(a) Helmet Mounted Camera (b) Video Image

Fig. 14. Helmet Mounted Camera and Video Image

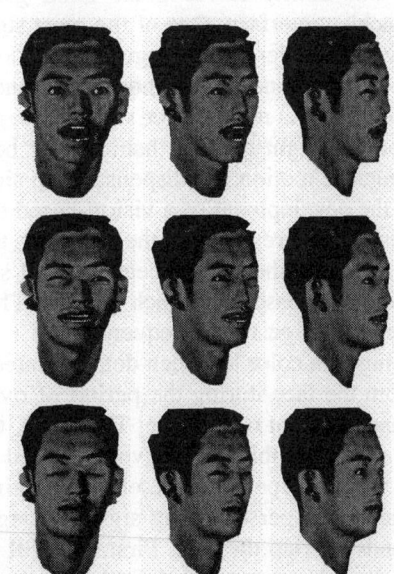

Fig. 15. Facial Expression Animation Results

Real-Time Tracking of Human Hands from a Sign-Language Image Sequence

Kazuyuki Imagawa [1,2], Shan Lu [1], Seiji Igi [1]

[1] Communications Research Laboratory,
4-2-1, Nukui-Kitamachi, Koganei, Tokyo, 184,JAPAN
[2] Matsushita Electric Industrial Co., LTD.,
693-47, Kawazu, Iizuka, Fukuoka, 820, JAPAN
E-mail: imagawa@qrl.mei.co.jp, {lu, igi}@crl.go.jp

Abstract: We have developed a real-time system which tracks the hands of a person doing sign language. The system enables us to track hands without markers or colored gloves even if the hands overlap the face. First, the system extracts the hand and face regions from the sign-language image sequence using an improved *histogram backprojection*. Next, the system tracks hands from blobs which are computed from both the extracted image and the time differential image. The system has been tested for hand tracking using both primitive motions and the actual motions of sign-language used by native signers. The experimental results indicate that the system is able to track hands while the hand overlaps the face.

1. Introduction

The ability to recognize hand gestures is a key technology in the human-machine interface. One of the most structured sets of human gestures is evident in sign languages. Sign languages are known to be composed of basic units which consist of hand shape, movement, and position [1,2]. Therefore, the detection of these units is required for sign-language recognition. In order to detect these units, both the right and left hands must be tracked. Thus, tracking hands from sign-language motion is indispensable in sign-language recognition. Our long term goal is the development of a vision-based sign-language recognition system. The topic described here is one of the problems to be solved towards achieving this goal.

To date, there have been several systems on sign-language recognition using the vision-based approach [3,4,5]. These systems find and track a hand from a sign-language image sequence and interpret its motion. Although hand tracking using skin color has been demonstrated, these systems can not distinguishes hands from the face during the period of overlap. In sign language, people often move hands in front of the face. Therefore, tracking hands during this period of overlap is a major problem in the vision-based approach.

In this paper, we present a real-time hand tracking system which tracks the hands of a person doing sign language. The system can track hands even if the hands overlap the face. First, we will describe a method to extract face and hand regions. Next, we will describe a procedure for tracking hands using the extracted images. Finally, we will present experimental results.

2. Extraction of Hand and Face Regions

Our system extracts hand and face regions using the color distribution of sign-language images. For real-time processing, we prepared a three-dimensional look-up table (3-D LUT) which was used to set the color distribution. The system then sets the 3-D LUT to extract hand and face regions before tracking hands, and it extracts these from all frames of the sequence using the 3-D LUT. In order to detect hand motion when the hand overlaps the face, the contrast in the extracted images needs to be sufficiently increased. To do this, we developed a new method, which is based on *histogram backprojection* [6]. We then introduced this method into the 3-D LUT by using the first frame of the sequence.

What follows is an overview of our method. First, all pixels in the first frame of the sequence (see Fig. 1a) are classified into a target region or a background region. For classification, the contours of the targets, which are hand and face regions, are manually traced (see Fig. 1b).

Next, based on a multi-dimensional color histogram of both the target regions M and the background region B, a combined histogram C is defined as

$$C_j = \max\left(\frac{w_M M_j - w_B B_j}{C_{max}}, 0\right) \times D,$$

where j is the index of each histogram bin, C_{max} is the maximum value of the combined histogram, D is the range of the output image, e.g., 255 for an 8-bit image, and w_M and w_B are weighted values representing the sensitivity of C in each histogram, which have been implemented as $w_M = w_B = 1$ in our system. The combined histogram C is set to the 3-D LUT.

In order to use the method in our system, a small 3-D LUT memory for an 8 bits × 3 RGB color space was prepared. Each axis was divided into widths of d, which was defined as 32 in our system. Here, the 3-D LUT has 512 ($2^3 \times 2^3 \times 2^3$) unit cubes. All pixels in the image were converted using the 3-D LUT as follows. If the color value of the pixel is the same as the mean color value of a unit cube, the value of the pixel is converted to the value of the unit cube. If the color pixel value is a value besides the mean color value of a unit cube, the pixel value is converted to a value which is interpolated by the *PRISM* algorithm [7] using the values of 6 neighboring cubes. The extracted image is shown in Fig. 1c. The extraction stage was implemented on a standard PC-AT with a custom daughterboard which provided the color extraction through the above method at full frame rate (30Hz).

3. Tracking Hands

After the system has extracted hand and face regions from a sign-language image sequence, the system tracks right and left hands using the extracted image sequence. We will first describe the scene and hand representation used in our system. We will then describe the initialization and hand tracking algorithms.

(a) Original image (b) Target regions are determined by tracing these contours. (c) Extracted image

Fig. 1. Extracting face and hand regions

3.1. Defining the Scene

We have assumed that a scene for recognizing sign language is relatively static and sign-language motion can be acquired with a stationary camera. The requirements for a person who does sign language can be defined as follows.
- The person should sit on a chair in front of the camera.
- The person should wear long-sleeved clothes which are not skin colored.

3.2. Representing the Hand and Face

In our system, the hand and face of the person doing sign language are represented as a set of two groups of blobs which are computed from the extracted image sequence. One group is made up of *static blobs* computed from an image of a single frame in the sequence. The other group is *motion blobs* computed from a time differential image between continuous frames. The size and position of each blob are recorded during each time step. Then, each hand is assigned to one of these blobs.

In order to compute blobs, the images are thresholded because they are noisy, caused either by the camera or the color extraction board. Then, connected regions, which have more pixels than the value determined by the system to reduce the influence of noise, are found and their bounding boxes are computed. A fast bounding box merging algorithm [8] is used to cluster small groups of blobs. Clustered blobs are then regarded as one blob. Figure 2 shows a typical example of static blobs and motion blobs.

3.3. Initialization

The initialization process detects each location of the face and hands using various heuristic rules. We assume that the face region and the hand regions do not touch or overlap one another in the first frame. Then, the biggest static blob located at the approximate center of the image is detected as a "face blob", and then two of the biggest blobs in the other blobs are detected as a left or right "hand blob". The initial location of the hand is computed using these blobs. The system repeats initialization until a hand blob is detected. The initialization process is also used for recovery from tracking errors.

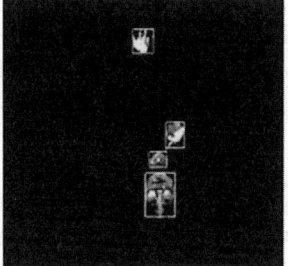

Static blob   Motion blob

Fig. 2. Static blobs and motion blobs. By using a fast bounding box merging algorithm, clustered blobs can be regarded as one blob, as marked by the white bounding boxes.

3.4. Tracking

After initialization, the system tracks the locations of both hands. During each time step, the following steps are computed:

1. The location of each hand for the current frame is estimated and the next location for each hand is predicted.

2. Static blobs and motion blobs for the next frame are computed. Then, the hand blobs are determined using the predicted location for each hand.

3. The location of each hand is updated using the hand blob.

Each of these steps will be described later in more detail.

Predicting the Location of Hands

In the first step, we apply Kalman filters [9] to estimate and predict the location of each hand using the current location of the hand blob. Kalman filter approaches have previously used to track humans or face motions [10, 11, 12]. For example, Rohr [10] estimated the model parameters of a pedestrian by using a Kalman filter. Goncalves et al. [11] applied a Kalman filter to estimate the position and motion of an arm. Wren et al. [12] use Kalman filters to predict the parameters for their human model.

In our system, Kalman filters are applied to estimate and predict each location of the hands. We assumed that the movement of the hand would be sufficiently small during step interval ΔT. Then, a dynamic process could be used to describe the x or y coordinate at the center of the hand on the image plane with state vector $\mathbf{x}$ which includes the position and velocity of each coordinate. The dynamic process is defined as

$$\mathbf{x}_{k+1} = \mathbf{F}_k \mathbf{x}_k + \mathbf{G}_k w_k, \text{ where } \mathbf{F}_k = \begin{bmatrix} 1 & \Delta T \\ 0 & 1 \end{bmatrix}, \mathbf{G}_k = \begin{bmatrix} \Delta T^2/2 \\ \Delta T \end{bmatrix}$$

The system noise is modeled by w_k, an unknown scalar acceleration, whose statistical characteristics are Gaussian and white.

Then, an observation model can be given by

$$z_{k+1} = \mathbf{H}\mathbf{x}_{k+1} + v,$$

where $\mathbf{x}_{k+1}$ is the actual state vector at time $k+1$, v is measurement noise, and z_{k+1} is the location of the hand blob at time $k+1$.

The noise covariances were determined by experiments so that the system can achieve optimal tracking. The system estimates and predicts the location of the hand from the location of the hand blob through using these models.

Labeling blobs

After the system predicts the location of the hand, the static blobs and motion blobs are computed for the next frame. Then, the system assigns one of the static blobs to the face, and labels the blob the 'face blob'. Also, the system assigns one of all blobs to each hand using the predicted location, and labels the blob the 'hand blob'.

First, a face blob is determined from the static blobs using the past location of the face. In the sign-language image sequence, the location of the face does not change very much. For each static blob, the Euclidean distance between the location of the blob and the past location of the face is calculated. Then, the nearest static blob is labeled a 'face blob'. Thus, the face blob is eliminated from candidate blobs for the hands.

A hand blob is determined from both the static blobs and the motion blobs. For each blob, the Euclidean distance between the location of the blob and the predicted location of each hand is calculated. Then, the nearest blob for the predicted location within a threshold value is labeled a 'hand blob'. The threshold value is determined by the length of the palm because a motion blob may be generated by the edge. Then, blobs, which are expected to be assigned to the same hand, are eliminated from candidate blobs for the other hand. If the labeled blob is a static blob, motion blobs inside the area of the labeled blob are eliminated. If the labeled blob is a motion blob, the static blob which includes the labeled blob is eliminated. Thus, both hand blobs are labeled 'hand blobs'.

When the hand overlaps the face, these regions appear as one static blob. This static blob is labeled a 'face blob'. Therefore, a hand blob can only be determined from motion blobs. If a person moves the head while the hand stops in front of the face, a motion blob originating from the head may appear, and the blob may be labeled a 'hand blob'. However, we have assumed that the head does not move as much as the hands during the motions of sign-language when the hand stops in front of the face.

Updating the Observed Locations of Hands

After each blob is labeled, the system updates the observed location for each hand using the hand blobs. The observed location is established by the center of the hand blob. If a hand blob exists in the current frame, the observed location is updated using the current hand blob. However, if the hand stops in front of the face, no hand blobs originating from the hand appear. Here, the observed location is updated by the location of the last hand blob because this situation is caused by stopping the hand in front of the face.

4. Experiments

We tested our system through in two different experiments. First, a basic experiment on hand tracking was carried out. In this experiment, the image sequence of primitive motions using one hand was used in order to investigate whether the system could track the hand when hands overlapped the face. Next, we conducted a practical hand tracking experiment in which sign-language image sequences by native signers were used.

4.1. Basic Hand Tracking Experiment

In the basic experiment, the system was tested on simple movements by moving one hand across the front of the face as shown in Fig. 3. Each movement was done using two types of motions as follows.
- Continuous motion: The subject moved the hand continuously.
- "Halfway pause" motion: The subject paused the hand in front of the face for a few seconds.

The hand was moved either continuously or in a "halfway pause" motion along each trajectory. In order to measure the actual location of the hand, the subject grabbed the magnetic sensor which measured its location while the subject moved the hand.

(a) straight motion　　(b) straight motion with right-　　(c) spiral motion
　　　　　　　　　　　　angle direction change

Fig. 3. Primary trajectories used in the basic experiment

First, we will present typical results obtained in this experiment. Figure 4(a) shows the tracking results of continuous motion. The smoothing properties of the Kalman filter can clearly be seen. Figure 4(b) shows the results of spiral "halfway pause" motion. This is the worst result in this experiment. In the spiral "halfway pause" motion, the hand blob was trapped twice by a blob generated by other factors apart from the motion of the hand while it overlapped the face. First, the hand blob was trapped by a motion blob generated by the wrist while the hand moved. Second, the hand blob was trapped by a motion blob generated by noise from the edge of the hand while it stopped in front of the face. This indicates that the system is sensitive to other blobs when the hand stops in front of the face.

Next, we measured both RMS error and maximum deviation for all movements. The mean value of the RMS error for all movements was 1.8 cm horizontally (X) and 2.2 cm vertically (Y) when the hand overlapped the face, whereas they were 1.4 cm at X and 1.3 cm at Y when the hand did not overlap the face. We also obtained a maximum deviation which was 6 cm except for spiral "halfway pause"

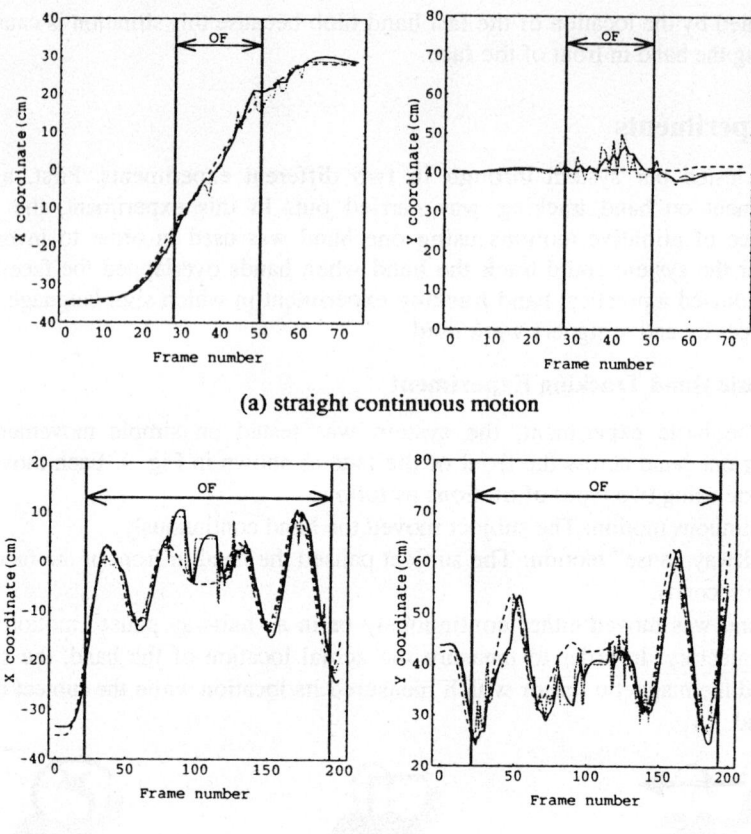

(a) straight continuous motion

(b) spiral "halfway pause" motion

Fig. 4. Typical results of basic experiment: trajectory for the location of the hand blob (dotted), tracked trajectory of the system (solid), and trajectory of the magnetic sensor (dashed). Interval "OF" represents the frames the hand overlaps the face. In order to classify these frames, we checked whether the hand region and the face region appeared as one static blob or not.

motion. These results indicate that the system is capable of tracking the hand even when the hand overlaps the face.

4.2. Practical Hand Tracking Experiment

In the practical experiments, the system was tested on the sign-language image sequences of three native signers. In each image sequence, the experiment used sign language motion for approximately 1 minute, (1800 frames). In order to estimate the tracking results, the number of frames, in which the tracking results were within the hand regions, were counted, and we then classified whether these frames were overlapping periods or not. As a results of this experiment, we obtained a success rate of 91% in tracking all sequences, 85% during the overlapping period, and 95% when the hand did not cross the face.

5. Conclusion and Future Work

The problem of hand tracking when the hand overlaps the face has been addressed in this paper. This is a prerequisite for most methods of tracking hands in sign language. This paper presented a new hand tracking system which enables us to track hands without markers or colored gloves even when the hand overlaps the face. Experimental results indicate that the system is able to track a hand overlapping the face accurately in terms of basic movements, and we obtained smooth estimates by applying a Kalman filter. Moreover, the experimental results indicate that the system is capable of tracking hands during real sign-language motions, although it cannot track hands completely.

In the near future, our system will be integrated with a feature extractor in sign-language or hand-gesture recognition systems.

Acknowledgements

We would like to thank the members of the Deaf Association of Tokorozawa in Japan for their cooperation in the experiment.

References

[1] W. Stokoe, D. C. Casterline, and C. G. Groneberg: "A Dictionary of American Sign Language on Linguistic Principles.", *Linstok Press, London*, (1976).
[2] K. Kanda: "A Computer Dictionary of Japanese Sign Language", *The 5th International Symposium of Sign Language Research*, pp.409-419 (1992).
[3] S. Tamura, and S. Kawasaki: "Recognition of Sign-language Motion Images", *Pattern Recognition*, Vol. 21, No. 4, pp. 343-353 (1988).
[4] C. Charayaphan, and A. E. Marble: "Image-processing system for interpreting motion in American Sign Language", *Journal of Biomedical Engineering*, Vol. 14, pp. 419-425 (1992).
[5] T. Starner, and A. Pentland: "Real-Time American Sign Language Recognition from Video using Hidden Markov Models", In Proc. *International Symposium on Computer Vision*, pp. 265-270 (1995).
[6] M. J. Swain, and D. H. Ballard: "Color Indexing", *Int. J. Comput. Vision*, Vol. 7, No. 1, pp. 11-32 (1991).
[7] K. Kanamori, H. Kotera, O. Yamada, H. Motomura, R. Iikawa, and T. Fumoto: "Fast color processor with programmable interpolation by small memory (PRISM)", *J. of Electronic Imaging*, Vol. 2, No. 3, pp. 213-224 (1993).
[8] S. S. Intille, J. W. Davis, and A. F. Bobick: "Real-time Closed-World Tracking", *MIT Media Laboratory, Perceptual Computing Technical Report*, No. 403, (1996).
[9] R. E. Kalman: "A New Approach to Linear Filtering and Prediction Problems", *Trans. ASME, J. Basic Eng.*, Vol. 82D, No. 1, pp. 35-45, (1960).
[10] K. Rohr: "Towards Model-Based Recognition of Human Movements in Image Sequences", *CVGIP: Image Understanding*, Vol. 59, No. 1, pp. 94-115 (1994).
[11] L. Goncalves, E. D. Bernardo, E. Ursella, and P. Perona: "Monocular Tracking of the Human Arm in 3D", *ICCV95*, pp. 764-770, (1995).
[12] C. Wren, A. Azarbayejani, T. Darrel, and A. Pentland: "Pfinder: Real-Time Tracking of the Human Body", *IEEE PAMI*, Vol. 19, No. 7, pp. 780-785, (1997)

The Model-Based Dynamic Hand Posture Identification Using Genetic Algorithm

Cheng-Chang Lien[1] and Chung-Lin Huang[2]
[1]Institute of Electronic Engineering
Chin Min Technology College
Tou-Fen, Miaoli, Taiwan, R.O.C.
e-mail: cclien@ravel.ee.nthu.edu.tw

[2]Institute of Electrical Engineering
National Tsing-Hua University
Hsin-Chu, Taiwan, R.O.C.
e-mail: clhuang@ee.nthu.edu.tw

Abstract

This paper proposes a genetic-based hand posture identification system which consists of an efficient model fitting method and a labeling technique. The model fitting method consists of (1) finding the closed form inverse kinematics solution, (2) defining the alignment measure, and (3) developing a genetic-based dynamic posture fitting process. Different from the conventional computation intensive hand model fitting methods, we develop (1) an off-line training process which uses the inverse-kinematics to find the closed-form solution function, and (2) a fast on-line model-based hand posture identification process. In the experiments, we will illustrate that our genetic-based hand posture identification system is effective and real-time implementable.

1. Introduction

The hand gesture recognition can be classified into the gloves-based methods[1] and vision-based methods[2]. All the gloves-based methods are designed to detect the hand shape and finger motion in real time. However, the mechanical gloves are expensive, uncomfortable to wear, and finger-movement-limited. The vision-based methods provide a more humane and user-friendly input environment. The vision-based methods can be further categorized into the contour-based methods[3-6] and model-based methods[7-14]. The contour-based methods have problems in the edges extraction and the occlusion between the fingers. The model-based methods can be categorized into hand without markers[7-10] and hand with markers[12-14] methods. Most of the current gesture recognition methods belong to the first category, such as DigitEyes proposed by Rehg et al.[7-8], Virtual Gun developed by Kuch et al.[9], and 3-D deformable hand model developed by Heap et al.[10]. All these systems have applied the projection information of 3-D hand model and image features(edge data) to track the hand motion with 27 degree of freedom (DOF). However, these tracking systems require the hand to be placed in certain pose and position, and the projection of 3-D model will lose the depth information so that it cannot guarantee the accuracy of the estimated joint angles. With the 3-D positions of markers, the second category methods[12-14] track the hand motion with 27 DOF precisely.

To solve the problem with large degrees of freedom, the system requires complex computation. Therefore, constraint-based kinematics methods [12,13] have been proposed to reduce the search space and the number of variables. In the constraint-based hand model fitting algorithms, one of the most time-consuming steps is the fingers-fitting process which is developed to solve the inverse kinematics problem. However, the inverse kinematics proposed by Hans et al.[15] does not generate closed form solutions. The joint angles are computed by the binary search method. In this paper, we propose a genetic-based hand model fitting method to improve the fitting efficiency. The model fitting method consists of (1) finding the closed form inverse kinematics solution for the finger fitting process, (2) defining the method of alignment measure for the wrist fitting process, and (3) applying a genetic-based searching method for the dynamic gesture fitting process.

2. The Training Process for Inverse Kinematics Solution Function

In this section, we will review the hand model, and then develop a training process (using two separable steps) to find the finger inverse kinematics solution functions.

2.1 The Kinematics of The Hand Model

According to the D-H (Denavit-Hartenberg) rules[16], we define the local coordinate system for each finger. For a joint i of a finger, it has four structural kinematics parameters (1) a_i : the length of the link i, (2) α_i : the angle of rotation about positive x_{i-1} axis measured from the positive z_{i-1} axis to the positive z_i axis ,i.e., the twist angle of link i, (3) θ_i : the angle of rotation about positive z_{i-1} axis measured from the positive x_{i-1} axis to the positive x_i axis, and (4) d_i : the perpendicular distance between the x_i axis and the x_{i-1} axis. The four structural kinematics parameters are used to define the local transformation matrix A as

$$A_{i-1}^i = \begin{bmatrix} \cos\theta_i & -\cos\alpha_i \sin\theta_i & \sin\alpha_i \sin\theta_i & a_i \cos\theta_i \\ \sin\theta_i & \cos\alpha_i \cos\theta_i & -\sin\alpha_i \cos\theta_i & a_i \sin\theta_i \\ 0 & \sin\alpha_i & \cos\alpha_i & d_i \\ 0 & 0 & 0 & 1 \end{bmatrix}. \quad (1)$$

If a vector p_i is known in the ith coordinate frame, it can be expressed in the $(i-1)$th coordinate system as p_{i-1}, i.e.,

$$p_{i-1} = A_{i-1}^i p_i. \quad (2)$$

From[12], we summarize some important constraints for the inter-joint-angle relationship and the movements of fingers: (1) The joints on four fingers are planar manipulators except the metacarpophalangeal joints(MP joints). (2) The relationship between joint angles 3 and 4 of finger II-V is represented by $\theta_4 = 2/3 \theta_3$. (3) The joint angle 2 around the y axis of middle finger is negligible, i.e., $\theta_2^y(III) = 0$.

2.2 The Generation of the Solution Function for Finger Inverse Kinematics

In the solution function generating process, we apply two steps to make the fingertip reach the goal position (r, η):

(1) Bend the finger along a certain ϕ angle to the desired r distance (see Figure 1(a)),
(2) Rotate the link 1(MP joint) of the finger from angle ϕ to the angle η (see Figure 1(c)).

Since the movement of fingers can be decomposed into two separable steps, we may define two solution functions to describe the relationship between the goal position of finger tip, i.e., (r, η), and the joint angles, i.e., $(\theta_2, \theta_3, \theta_4)$. Because $\theta_4 = 2/3 \theta_3$, the DOF of each finger can be reduced to 2. In the training process, we apply an exhaust search method to find as many solutions as we can for all possible goal positions of the finger tips. Then we assume that (1) all the possible r, θ_2, and θ_3 values are located on an analytic curve (i.e., Figure 1(b)), and (2) all the possible η, r, and θ_2 values are located on a conical surface (i.e., Figure 1(d)). Therefore, we can apply the regression method to identify the coefficients of these two analytical (or solution) functions. The first function which can be used to analyze the relationship among r, θ_2 and θ_3 for a certain elevation angle ϕ is

$$r = f(\theta_2, \theta_3), \ 0 \leq r \leq \sum_{i=1}^{n} l_i, \quad (3)$$

where l_i is the length of link i, n is the number of the links for each finger. The second function which relates the slant angle η with the variables r and θ_2 is

$$\eta = g(r, \theta_2), \ -45° \leq \eta \leq 90°. \quad (4)$$

Once the above two equations have been developed, they can be applied to find the closed form solution of the finger inverse kinematics.

From the observation of all possible solutions in (r, θ_2, θ_3) space, we assume that the first solution function is an analytical curve which can be expressed by intersecting the conical surface with a vertical plane as

$$r = f(\theta_2, \theta_3) = a\theta_2^2 + b\theta_2\theta_3 + c\theta_3^2 + d\theta_2 + e\theta_3 + f, \text{ and } \theta_2 = m\theta_3 + k \quad (5)$$

Given the constraint, $\theta_2 = m\theta_3 + k$, the first solution function can be simplified as

$$r = A\theta_3^2 + B\theta_3 + C \quad (6)$$

where $A = am^2 + bm + c$, $B = 2amk + bk + dm + e$, and $C = ak^2 + dk + f$. Given r, θ_3 can be easily derived. The similar assumption can be made for the second solution function $\eta = g(r, \theta_2)$. The solution function can be described by another conical surface in (η, r, θ_2) space as

$$\eta = a\theta_2^2 + b\theta_2 r + cr^2 + d\theta_2 + er + f. \quad (7)$$

Given r, the solution function can be simplified as

$$\eta = l\theta_2^2 + m\theta_2 + n \qquad (8)$$

where $l = a$, $m = br + d$, $n = cr^2 + er + f$. Given the destination slant angel η and radius r of the goal position of finger tip, we may derive the joint angle θ_3 and then the joint angle θ_2. Given as many solutions (r, θ_2, θ_3) and (η, r, θ_2) as possible, we can apply the curve fitting and surface fitting methods to generate the solution functions. Here, we apply the total least-square regression method[17] to find the coefficients of equations (5) and (7). Although the regression method is only an approximation, the estimated coefficients can still make equations (5) and (7) acceptable for the following hand model fitting process.

3. The Hand Model Fitting

Here, we propose a hand model fitting method to fit the hand model with less computation time than the Lee and Kunii's method[12] with similar accurate results. To increase the speed of model fitting, we propose a three-phase fitting method: the wrist fitting, the finger fitting(for fingers II ~V), and the thumb fitting.

3.1 The Wrist Fitting Process

The wrist fitting process takes advantage of the following three hand motion constraints to determine the angles ϕ_x, ϕ_y and ϕ_z. The first constraint indicates that the fingers II~V are planar manipulators. The second constraint mentions the restriction of abduction motion or adduction motion for the middle finger. The third constraint assumes that the motion of the palm is a rigid body motion. The parameters ϕ_x and ϕ_y are calculated by using the first two constraints, and the third parameter ϕ_z is computed by using the third constraint. The transformation between the space coordinate and the wrist coordinate is described as

$$p_o = W_0^1 W_1^2 W_2^3 p_w = W_o^3 p_w, \qquad (9)$$

where p_w is the vector p_o represented in the wrist coordinate, W_0^1, W_1^2, and W_2^3 are the rotation transform matrices about the x, y, and z axes.

3.1.1 The calculation of ϕ_x and ϕ_y

In order to estimate the parameters ϕ_x and ϕ_y, we apply the coplanar constraint on the movement of all fingers and the constraint of limited abduction and adduction on the movement of the middle finger. The coplanar constraint for the bending movement of the middle finger can be depicted in Figure 2. There are two vectors ($\vec{v}_1$ and $\vec{v}_2$) located on the hand model which originate from the position of joint 1 of the middle finger to the position of finger tip and the position of MP joint, respectively. The third vector $\vec{v}_3$ originates from the position of joint 1 of middle finger of the hand model to the tip position of middle finger of real hand. The normalized unit vectors $\{\vec{u}_1, \vec{u}_2, \vec{u}_3\}$ (the unit vectors of $\vec{v}_1$, $\vec{v}_2$, and $\vec{v}_3$) form a parallelepiped whose volume can be used to determine how closely the three vectors are located on the same plane. Here, the cubical volume is also called the alignment measure, which can be used to determine the fitness between the finger tip position and the finger bending plane of the hand model. The alignment measure is defined as

$$V = |(\vec{u}_1 \times \vec{u}_2) \cdot \vec{u}_3|. \qquad (10)$$

Similarly, we can apply the alignment measurements for the finger 2(index), finger 4(ring), and finger 5(little) fitting respectively.

Because the movements of abduction and adduction of the index, ring and little fingers may generate inaccurate rotation angles(ϕ_x and ϕ_y) of the wrist, we apply different weighting to compute the rotation angles ϕ_x and ϕ_y of the wrist. We define the alignment measure for the rotation angles ϕ_x and ϕ_y of the wrist as

$$V(\phi_x, \phi_y) = w_1 V_1 + w_2 V_2 + w_3 V_3 + w_4 V_4 \qquad (11)$$

where (1) w_i is the weighting for finger i, (2) V_1, V_2, V_3, and V_4 are the alignment measures for the index finger, middle finger, the ring finger, and the little finger, respectively. The highest weighting value is assigned to the middle finger, i.e., $w_2 > w_i$, $i=1, 3, 4$. The direction of the wrist can be adjusted precisely by applying the constraint that limits the abduction or adduction motion for the middle finger. Using the algorithm mentioned in rough searching process, we can rotate the hand model around the direction of ϕ_x and ϕ_y with a small deviation $\Delta\phi_x$ and $\Delta\phi_y$ until the alignment measure for the middle fingers of hand model and real hand is below certain lower threshold.

3.1.2 The calculation of ϕ_z

Similar to the previous calculation of ϕ_x and ϕ_y, we define three vectors and apply the volume estimation method to acquire the angle ϕ_z. In Figure 3, there are two vectors originating from the wrist position to the MP positions of index and little fingers of the hand model. The third vector originates from the wrist position to the position of marker b. Similarly, the alignment measure for the rotation angle ϕ_z of the wrist can be defined as

$$(\phi_z) = V_5 \tag{12}$$

where V_5 is the alignment measure(see equation (10)) for the palm. If the minimum value of V is found to be smaller than certain threshold then these three vectors are coplanar and the palm of the hand model is precisely aligned with the markers on the real hand.

3.2 The finger fitting process

The finger fitting process can be further decomposed into two phases. The former is the MP joint fitting phase which calculates the angles of adduction and abduction for fingers II~V. The latter is the inverse/forward-kinematics fitting phase.

a) **The abduction or adduction fitting phase of the MP joints.** Here, we assume that the limited abduction motion or adduction motion for the middle finger is negligible. Therefore, we do not need to calculate the angle of abduction motion or adduction motion for middle finger. By representing the tip positions of fingers in the coordinate of MP joint, the angle of abduction motion or adduction motion can be calculated by rotating the finger about the y-axis until the alignment measure is minimized.

b) **Inverse kinematics fitting phase.** Given each marker's 3-D position in space coordinate, we apply the method of coordinates transformation to acquire the distance r and slant angle η in the MP joint coordinate. By applying the closed form solution functions, we can find the joint angles for the five fingers.

3.3 The thumb fitting process

Because the M joint (directive joint) of the thumb has two DOF with large rotation angle and the bending direction of the thumb is different form the other fingers, we use two local transformation matrices A_0^1 and A_1^2 to describe the these movements of the M joint. The transformation matrix A_0^1 indicates that the M joint rotates θ_0 about the z-axis and then rotates α_0 about the x-axis such that the links 1, 2, and 3 form a planar manipulator bending with angles θ_1, θ_2, and θ_3. The value of the parameter α_0 is determined by the hand structure and in our 3-D hand model, $\alpha_0 \approx 90°$. Similar to the middle finger, the MP joint of the thumb has limited adduction or abduction movement. Therefore, we do not estimate the adduction or abduction angle of MP joint.

When observing the movement of the M joint of the thumb, we may find that there are many ways for the joint angles, (θ_0, θ_1), to vary from $(\theta_{0start}, \theta_{1start})$ to $(\theta_{0stop}, \theta_{1stop})$. From our experimental observation of normal movement of the thumb, we assume that the interdependence of joint angles θ_0 and θ_1 is linear and can be approximated by

$$\theta_1 = m\theta_0 + b. \tag{13}$$

Such an approximation may cause a little inaccurate estimation for the joint angle θ_1, but it still be used to find the angle θ_1 efficiently after the value of θ_0 has been found. In the experiments, we find that m = -0.643 and b = 50. The value of θ_0 is found by applying the method of alignment measure to the thumb. The fitting process of the thumb is implemented by the following steps :
1. Apply the coplanar constraint to find the joint angle of θ_0.
2. Apply the interdependency of θ_0 and θ_1 (equation (13)) to find the joint angle of θ_1.
3. Calculate the joint angles θ_2 and θ_3 by using the closed form solution function.

4. Optimization using the Genetic Algorithm

To increase the accuracy and speed of the wrist fitting process, here, we apply the genetic algorithm (GA) to search the optimum solution of wrist angles ϕ_x, ϕ_y and ϕ_z.

4.1 The Genetic Optimization Method

The GA consists of three processes: the initialization process, the chromosome sampling process, and the new population generating process.

1) The initialization process. To initialize a population, we select *pop_size* number of chromosomes randomly. Each chromosome vector was coded as a vector of floating numbers representing ϕ_x, ϕ_y, and ϕ_z, i.e., the rotation angles for the wrist. The ranges of the three parameters are described as

$$-70° \leq \phi_x \leq 70°, -45° \leq \phi_y \leq 30°, -30° \leq \phi_z \leq 45°. \tag{14}$$

2) The chromosome sampling process. When the GA is used to optimize the alignment measure function, the superior chromosomes will grow up exponential. However, the phenomenon of rapid (premature) convergence is often caused by the presence of super individuals which are better than the average fitness of the population. In this paper, we apply the following two techniques to improve the premature phenomenon: the modified sampling method and the power law scaling method.

a) The modified sampling method. Zbigniew[18] proposed a modified genetic algorithm (modGA) to improve the premature disadvantage. The modGA select independently *r* (not necessarily distinct) chromosomes from population P(t-1) for reproduction and *r* (distinct) chromosomes to die. The selections are based on the fitness values of the chromosomes (stochastic universal sampling method).

b) Non-uniform power law scaling. For each generation, the non-uniform power law scaling technique[21] changes the characteristics of the function in order to improve the premature phenomenon and increase the chance of finding the global minimum. The power law scaling method changes the characteristics of the original function as:

$$F_i' = (F_i)^k. \tag{15}$$

where $k \approx 0$ forces a random search, while $k > 1$ makes the sampling allocating only the most suitable chromosomes, and *i* is the generation number.

3) The population generating process. Here, we apply the following operations proposed by Zbigniew[18] for population generation.

a) The mutation operator. The non-uniform mutation can provide the fine tuning capabilities to the system. If $\Phi^t = <\phi_1, \phi_2, ..., \phi_k, ..., \phi_m>$ is a chromosome and the element ϕ_k, whose domain is $[l_k, u_k]$, was selected for mutation, then the mutation result can be described as a vector $\Phi^t = <\phi_1, \phi_2, ..., \phi'_k, ..., \phi_m>$ with $k \in \{1,...m\}$, and

$$\phi_k' = \begin{cases} \phi_k + \Delta(t, u_k - v_k) & \text{if a random digit is 0,} \\ \phi_k - \Delta(t, v_k - l_k) & \text{if a random digit is 1,} \end{cases} \tag{16}$$

and the $\Delta(t, y)$ is defined as :

$$\Delta(t, y) = y \cdot (1 - r^{(1-\frac{t}{T})^b}), \tag{17}$$

where *r* is a random number from [0..1], *T* is the maximal generation number and *b* is a system parameter determining the degree of non-uniformity.

b) The crossover operator. Here, we apply the arithmetical crossover which is a linear combination of two vectors to perform the crossover operation. If two vectors Φ_v^t and Φ_w^t are to be crossed, the resulting offspring are $\Phi_v^{t+1} = a\Phi_w^t + (1-a)\Phi_v^t$ and $\Phi_w^{t+1} = a\Phi_v^t + (1-a)\Phi_w^t$.

Figure 3 shows the fitness value for each generation during the modGA optimization process, and the generation number is 35. The fitness value shown in Figure 4 indicates that it can converge to a neighborhood region of the global minimum.

4.2 The Dynamic Tracking of the Hand Postures

The rough searching process of the wrist fitting is implemented by using the modGA optimization method. It can find the optimal rotation angle of the wrist. Because of the abduction and adduction movements of the fingers II, IV, and V, the rotation angles ϕ_x, ϕ_y and ϕ_z are found with small error. The

fine tuning process (section 3.1.1) needs to be applied on the middle finger to determine the exact solution for the wrist angles ϕ_x, ϕ_y and ϕ_z. The fine tuning process is illustrated in the following steps:
1. Search for the optimal rotation angles, ϕ'_x and ϕ'_y, by minimizing the alignment measure $f_{\pi}^{-}(\phi_x,\phi_y)$ of the middle finger in small range $|\phi'_x - \phi_x| < \delta_{\phi_x}$ and $|\phi'_y - \phi_y| < \delta_{\phi_y}$.
2. Search for the optimal rotation angles, ϕ'_z, by minimizing the alignment measure $f_{alig}^{z}(\phi_z)$ of the middle finger in small range $|\phi'_z - \phi_z| < \delta_{\phi_z}$.

After the wrist fitting process, the finger fitting process will be operated by using the inverse kinematics solutions. In the first image frame, the modGA searches in a larger range which is defined by equation (14) for the wrist fitting process. After the first image frame, the search range for the modGA will be limited to a much smaller range which is defined as $-5° \le \phi_x \le 5°$, $-5° \le \phi_y \le 5°$, $-5° \le \phi_z \le 5°$.

5. Experimental Results

(a) The accuracy analysis of the closed form solution for the five fingers. The accuracy analysis of the closed form solution is performed by simulating all the possible combinations of the θ_2, and θ_3 and then calculating the value r and η defined in section 2. The criterion for the error analysis is defined by the mean absolute error defined as

$$e = \sum_{i=1}^{n} abs(e(\theta_i))/n, \qquad (18)$$

where $e(\theta_i) = \theta_i - \theta'_i$, θ_i is the simulated angle for the joint i which is used to generate the finger tip position as the input and θ'_i is the joint angle acquired from the closed form solutions. In this experiment, the maximum error is about 3° and the average error is about 1.5°.

(b) The experimental results of the genetic-based dynamic hand posture identification. In the experiment, the image sequence of the hand posture is sampled with a rate 6Hz. The lower frame rate leads to a larger motion between adjacent frames, however, we can apply our genetic-based model fitting method to these images to verify the robustness of the algorithm. Figure 5 shows the image sequence of gestures. Figure 6 shows the 3-D hand model of the recognized gestures in Figure 5. Figure 7 shows the error for 3-D model fitting to the postures shown in Figure 18. The fitting error is analyzed by

$$E(j) = \sum_{i=1}^{5} d_i(j) / \sum_{i=1}^{5} l_i \qquad (19)$$

where $E(j)$ is the error rate for jth adjustment, d_i is the distance from the tip position of finger i of the fitted hand model to the tip position of the same finger i of the real hand, and l_i is the length of the finger i. The fitting process running on a pentium-90 PC requires 5 seconds for the first frame and 1.5 seconds for the following frames.

6. Conclusion

This paper has proposed a genetic-based model-fitting method to fit the hand model very efficiently. The model-fitting method consists of three techniques: (1) generating closed form solutions for the inverse-kinematics process; (2) applying the alignment measure for the wrist fitting process; (3) developing a genetic-based wrist fitting methods. Our hand model fitting can be completed in 1.5 seconds and the error rate is below 0.013. It has greatly improved the fitting efficiency and accuracy.

References
[1] D. J. Sturman and D. Zeltzer, "A survey of glove-based input," *IEEE computer graphics & application*, Jan. 1994, pp.30-39.
[2] P. Wellner, "Digitaldesk," *Comm of the ACM*, No. 7, vol. 36, July 1993, pp. 87-96.
[3] J. Davis and M. Shah, "Visual gesture recognition," *IEE Proc.-Vis image signal process.*, Vol. 141, No. 2, April 1994.
[4] T. J. Darrell and A. P. Pentland, "Recognition of space-time gestures using a distributed representation," *M.I.T. media laboratory vision and modeling group*, technical report No. 197.
[5] S. Tamura and S. Kawasaki, "Recognition of sign language motion images," *Pattern Recognition*, Vol. 21, No. 4, pp.343-353, 1988.

[6] Y. Cui and J. J. Weng, "Hand sign recognition from intensity image sequences with complex backgrounds", *IEEE Proc.- Automatic Face and Gesture Recognition*, pp.259-264, 1996.
[7] M. Rehg and T. Kanade, "Digiteyes : vision-based hand tracking for human-computer interaction," *IEEE workshop on motion of non-rigid and articulated objects*, Nov. 1994, pp. 16-22.
[8] M. Rehg and T. Kanade, "Visual tracking of high DOF articulated structures: an application to human tracking," *Proc. ECCV*, Vol. 2, pp. 35-46, 1994.
[9] J. J. Kuch and T. S. Huang, "Vision based hand modeling and tracking for virtual teleconference and telecollaboration, " *ICCV*, June 1995, pp. 666-671.
[10] T. Heap and D. Hogg, "Towards 3D tracking using deformable model", *IEEE Proc.- Automatic Face and Gesture Recognition*, pp. 140-145, 1996
[11] I. J. Mulligan, A. K. Mackworth, and Peter D. Lawrence, "*A model-based vision system for manipulator position sensing,*" *IEEE Proc.- Workshop on interpretation of 3D Scenes 1989*, pp. 186-193.
[12] J. Lee and T. L. Knuii, "Model-based analysis of hand posture," *IEEE computer graphics & application*, Sep. 1995, pp.77-86.
[13] J. Lee and T. L. Knuii, "Constraint-based hand animation," *Models and techniques in computer animation, Tokyo: Springer-Verlan*, pp. 110-127, 1993.
[14] B. Dorner, "Hand shape identification and tracking for sign language interpretation", Looking at people workshop, Chambery, France, 1993.
[15] H. Rijpkema and M. Girard, "Computer animation of knowledge-based human grasping," *Computer graphics, Proc. Siggraph*, Vol. 25, No. 4, July 1991, pp.339-347.
[16] A. J. Koivo, Fundamentals for control of robotics manipulators, John Wiley & Sons, INC. 1989.
[17] P. Lancaster and K. Salkauskas, Curve and surface fitting, Academic press, 1986.
[18] Zbigniew Michalewicz, Genetic algorithms + data structures = evolution programs, Springer-Verlag. Berlin Heidelberg New York 1992.

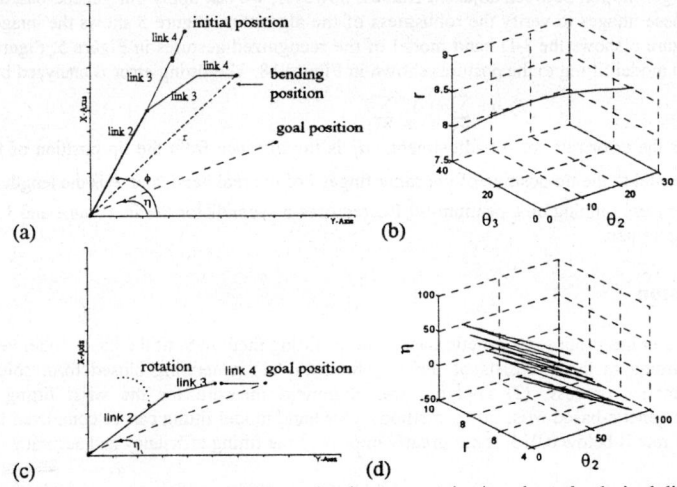

Figure 1. (a) The first step is to move the finger tip along a certain ϕ angle to the desired distance r. (b) All the possible r, θ_2, and θ_3 are illustrated. (c) The second step is to rotate the link 1(MP joint) of the finger from angle ϕ to the angle θ. (d) All the possible η, r, and θ_2 are illustrated.

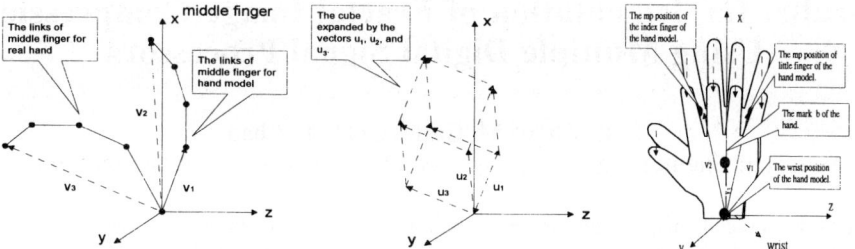

Figure 2. The middle finger is a coplanar manipulator.

Figure 3. The three vectors that are applied to calculate the angle ϕ_z.

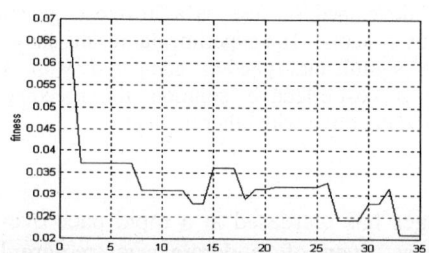

Figure 4. The best fitness value for each generation.

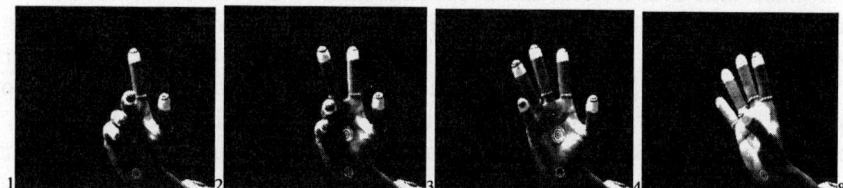

Figure 5. The images of various gestures with sampling rate 6 frames/sec.

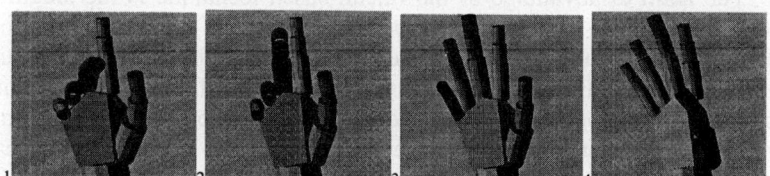

Figure 6. The 3-D hand models of the recognized gestures in Figure 5.

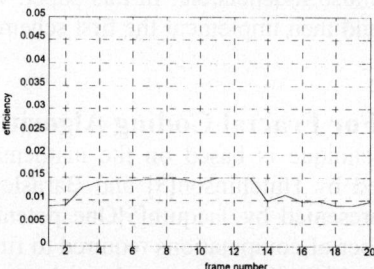

Figure 7. The fitting efficiency for frame 1 to 20 for frame rate 6Hz.

Parallel Implementation of Fractal Image Compression Using Multiple Digital Signal Processors

S. K. Chow, M. Gillies and S. L. Chan
School of Engineering and Technology, Deakin University, Australia 3168

Abstract: Fractal image compression technique provides very high compression ratios for natural scenes and has the advantage of being resolution independent. However, the encoding process based on the self-similarity search between range and domain blocks is very computationally intensive. This prohibits their real-time application. In this paper, we first propose two parallel schemes and then present one of the parallel implementations on multiple DSP cards. This card is developed for use as a low-cost, general-purpose digital signal-processing card for ISA bus systems. The experimental results show that the proposed parallel implementation yields a significant speedup compared with serial computation. This implementation provides a cost-effective solution for speeding up the fractal image compression and makes it competitive with other methods.

1 Introduction

The use of digital images has increased at a rapid pace over the past decade. It is noted that digital images, even of moderate size, require large storage space or transmission time over a telecommunication channel. As a result, the need for image compression is likely to increase. Recently fractal compression of digital images has drawn much attention. Fractal compression methods have some interesting properties. The most noteworthy is the potentially achievable high compression ratio. Another interesting feature of fractal compression is that unlike the JPEG/DCT technique, it is independent of the resolution of the image file. Fractal images contains infinitely small details and do not suppress those high frequency data associated with sharp edges.

The main disadvantage of the fractal-based technique is the length of time required for encoding. During the encoding, each range block requires to search through the whole image to find the best-matched domain block under an affine transform. These self-similarity searches are computationally intensive. A number of solutions such as block classification have been proposed to overcome this drawback. However, all these methods concentrate on reducing the complexity of searching. Unlike other lossy image compression techniques, fractal compression is intrinsically well suited to parallel implementation. Each block is encoded independently and the domain block searches are also independent. In this paper, we propose two possible parallel encoding scheme and then implement the first scheme on developed multiple ISA bus based DSP cards.

2 Parallel Solutions For Fractal Coding Algorithm

The idea of the fractal technique is based on the mathematical theory of iterated function systems developed by Hutchinson[3] and Barnsley[1]. Its application for image compression was presented by Jacquin[5]One potential drawback of fractal technique is the large number of computations required to find the domain block and the set of transformations that will map to a desired range block. A number of solutions have been proposed to overcome this problem. Jacquin[5] classified blocks into flat, edge, and texture region. Later Fisher, et at.[4] preclassified the blocks into 72 classes according to intensity variance. The other method proposed by Munro and Dudbridge [6] used a richer set of transformation to limit the search in a smaller pool of domain blocks. However, the rapidly growing number of pixels per image often do not allow practical implementations of fractal compression algorithm on conventional computers owing to their limited processing capabilities. Therefore, parallel computing may provide promising solution to the problem.

In order to exploit parallelisation of an algorithm, one needs precise information about all processing requirements of the complete computational task. For this reason, the encoding algorithm, including all its operations, data and control structures as well as flow of information, is investigated. It is found that the intensive computational demands come from the large volume of data that requires processing, rather than from the complexity of the algorithms themselves. This means that the use of data parallelism becomes important.

There are various ways of performing the parallel encoding process. The most straightforward method is to partition the whole image into segments of equal size and allocate each to a processor. An alternative approach may be found by incorporating some form of pipelining. In this a master processor is used to determine the ranges, while each of several slave processors compares a range to some subset of the domain pool. Each slave processor determines the best domain in its subset, and returns it to the master processor. The master processor then collects the optimal domain and sends the next range for matching. Fig. 1. depicts the operation of the algorithm. The advantage of this approach is that the communication load involved will not impose a performance penalty as the number of processor increases. An analysis of the ring architecture for fractal image encoding process was described by Chow[2].

3 Parallel Implementation
3.1 C25 DSP Card For The ISA Bus
Our implementation uses multiple ISA bus based DSP cards. This card is developed for use as a low cost, general-purpose digital signal processing (DSP) card for ISA bus systems. The main components of the card are as follows:

- TMS320C25 50MHz DSP
- 32k × 16 15ns Static RAM – arranged as 16k × 16 program & 16k × 16 data
- GAL20V8B – I/O address decoder
- GAL20V8B – glue logic
- 74LS245 Octal bus tranceiver
- 74LS279 Quad Set-Reset Latch
- 74LS393 Dual 4-bit Binary Counter

Since there is no program ROM on the card, user defined program are loaded into the program half of the SRAM. Data transfers to and from the card are performed using 16-bit direct memory access (DMA). In this case, "data" refers to the DSP program and data information. DSP data can also be transferred via the C25 serial port. The serial port can be connected to another C25 card or other serial devices, such as audio codecs. Three serial clock speeds are available, 1/4, 1/8, and 1/16 of the

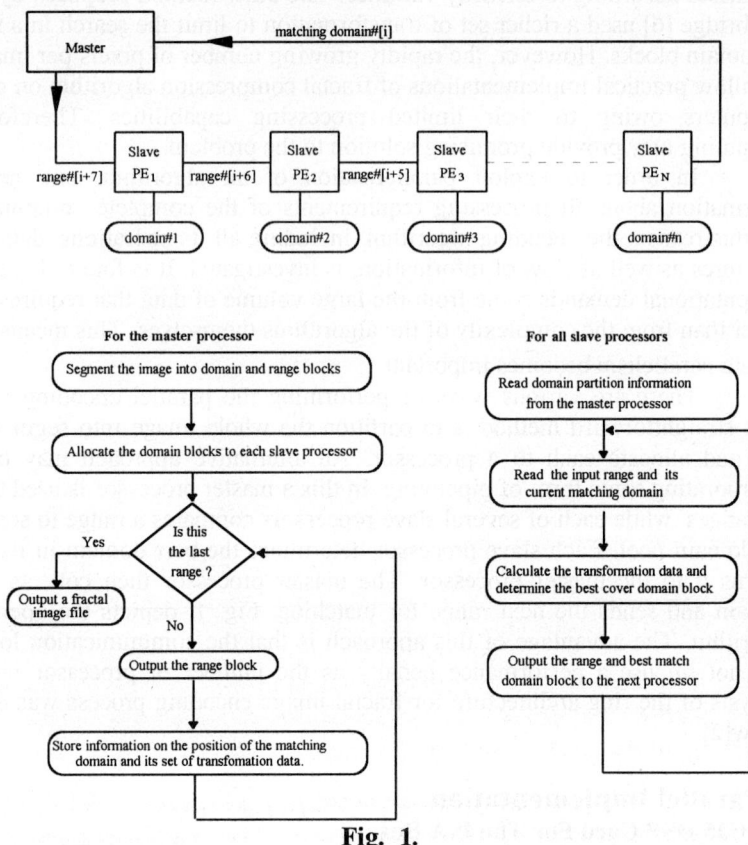

Fig. 1.

C25 CLKOUT frequency (12.5MHz). Hardware jumpers are used to define the card's I/O port range, DMA channel, and IRQ line. The DSP card memory is "mapped" into the boundary of a single 64K segment within the PC system memory (see Fig. 2.).

3.2 Description Of The Encoding Procedure
Since the computational load for pipelining scheme is more likely to impose a performance penalty of the serial communication port in the DSP card, partitioning the whole image into segments of equal size is employed (see Fig. 3.).

The hardware configuration for this experiment consisted of a master processor (the CPU on the PC motherboard) and two slave processors (C25-ISA cards). A simple management program is used to load sections of the uncompressed image into the local memory of each C25-ISA card. When the compression of an image section is complete, the management program stores the compressed section. Each card has a separate file for its compressed data. After both cards have finished processing, new image data is sent to each card. Both cards are using the same compression program, so the time taken to process the data is identical. Therefore, the first card will only have to pause for a short time while waiting for the second card to

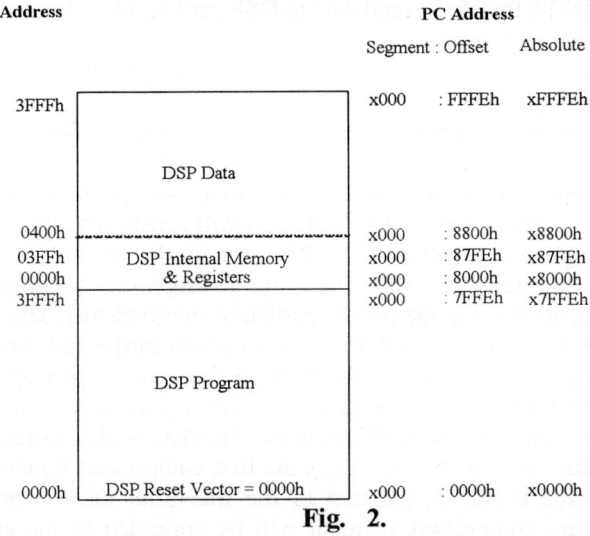

Fig. 2.

finish. The steps involved during the execution of the management program are as follows:

Step 1. Check for valid input arguments Each DSP-ISA card has its own I/O base address and IRQ line. All cards share a common DMA channel. These values must be checked (validated) before continuing. If these values are not valid, the program ends.

Step 2: Allocate/reserve 64k bytes of PC memory The local memory of the DSP-ISA card is aligned within the boundaries of one 64k-byte segment. A memory allocation function is employed to reserve an entire (unused) segment in PC memory.

Step 3: Open files The management program uses three different file "groups". These are the DSP program, the uncompressed image, and one or more compressed image data files.

Step 4: Read DSP program into PC memory. This step reads the DSP program binary file and stores the data at the start of the memory segment (x000:0000). This start address corresponds to the start (reset vector) of the C25 program memory.

Step 5. Select first segment of image & store in PC memory The image data is stored at start location (PC memory offset) C000h. This corresponds to DSP local data memory location 2000h

Step 6: Invoke DMA transfer – send data to DSP card (card 1) The management program performs a dummy I/O port read, which asserts and latches the DSP HOLD line. After the DSP enters HOLD mode, the HOLDA line is asserted, which starts the DMA transfer.

Step 7: Reset & "start" DSP card 1 After completion of the DMA transfer, the management program resets the DSP and un-latches the HOLD line. The DSP compression program is now running.
Step 8: Select second segment of image & store in PC memory
Step 9: Invoke DMA transfer – send data to DSP card (card 2) Same as step 6. In this case, the data is transferred to card 2.
Step 10: Reset & "start" DSP card 2 Same as step 7. DSP compression program is now running on card 2.
Step 11: Wait for end of compression – IRQ (either card can finish first) Both cards have image segments of identical height and width. Therefore, the image compression time (program execution time) will be the same for both cards. DSP-ISA card 1 was started first, so it will finish first. When it has finished the compression routine, the C25 external flag is strobed, which in turn strobes the relevant IRQ line. The management program executes the interrupt service routine for the particular interrupt line. This service routine sets a flag (variable) to tell the program which card requires reading.
Step 12: Read data from card (1) and store in file (data appended to end of file) Another DMA transfer is used to transfer the compressed data from the DSP card local memory to the PC memory. The data is then stored in the relevant output file for that card. If this is the first compressed segment (segment 1), file storage begins at the start of the file (after the 4-byte header). Each subsequent compressed segment will be appended to the end of this first section.
Step 13: Wait for end of compression – IRQ (second card) Same as step 11. This time, a different interrupt line and interrupt service routine are used.
Step 14: Read data from card (2) and store in file (data appended to end of file) Same as step 12. The first block of data stored in this case will be the compressed data for image segment 2.
Step 15: If more segments to compress, goto step 5. For this particular exercise, 4 segments were used, so the loop was executed twice. When repeating steps 5 through 10, image segment 3 is transferred to card 1, while image segment 4 is sent to card 2.
Step 16: Close all files and restore original interrupt vectors. When all image segments have been processed, close all opened files and restore interrupt vectors to their original settings.

Two output files were produced in this case. The first file contained the compressed data for image segments 1 and 3 – processed by DSP-ISA card 1. The second file contained the data from card 2, that is, image segments 2 and 4. After decompression, two individual images were produced. These images could be combined to form one image and then displayed, or displayed separately, one as the left half, and one as the right half of the original image.

The first step in the process is to open the necessary files and load the DSP program into PC memory. The image compression management (loader) program is "hardwired" for this exercise to work with uncompressed images measuring 160 pixels wide by 96 pixels high and 256 gray levels. Images are stored in RAW file format, with no headers. Each pixel is represented by an 8-bit value. Image dimensions and palette information is not included (not required). Four equal sized segments measuring 80 × 48 are selected, the first segment located at the bottom left of the main image. The origin (0,0) for this image type is at the bottom left side.

The first segment (1) is loaded into the PC memory, and along with the DSP program, is written to DSP-ISA card number 1 using a DMA transfer. After the transfer, the management program resets the DSP and allows the program to run. The next segment (2) is now loaded into memory, then transferred to card number 2. This card is reset and also allowed to run. If a shared DMA channel is used, each card must be loaded and "set running" before the next card can be loaded. This is due to the nature of the DMA control hardware on the DSP-ISA card. To invoke the DMA transfer, the DSP is placed in the HOLD mode. When HOLDA (HOLD Acknowledge) is asserted, the DMA transfer begins. At the end of the transfer, the management program must disable HOLD, which in turn causes the DSP to release the DMA Request (DRQ) control line. This line can now be used by the other DSP-ISA card(s).

After loading both cards, the management program waits in an endless loop for either card to finish processing its data. Since both cards contain the same program (same number of loops) and the same size image data, card 1 will finish first. The compression time for the particular algorithm used does not depend on the "content" of the image segment.

When card one has finished, it will generate a hardware interrupt (IRQ). The compressed image data is read from the DSP-ISA card local memory using another DMA transfer. The compressed data is then stored in an output file. Each card has its own ouptut data file. The management program returns to the loop and waits for card 2 to finish, if it has not already done so. The compressed data is read from card 2 and stored using the same procedure as card 1. In this case, a different output file is used. Each card now receives a new image segment. The procedure above is repeated, with the new compressed data appended to the end of the previous data. This procedure can be repeated with 3 or 4 cards and smaller image segments. The timing diagram of the whole process is shown in Fig. 4.

4 Experimental Results

The parallel fractal-based encoding algorithm, implemented in two DSP cards, was tested for a castle image. The castle image is shown in Fig. 5., which 160x96 pixels digitized at 8 bits The result of the parallel implementation is listed in Table 1. Image 1 shows the encoding result of Hurd & Barnsley algorithm. The computing time requires 477 CPU second. We have modified Hurd & Barnsley algorithm and programmed in the C and assembly language. The compressed images are presented in Fig. 7 and 8. An implementation on a TMSC25 DSP card yields an encoding time of 513 and 5484 CPU second. As can be observed, the processing time, as compared to Hurd & Barnsley technique, is degraded significantly. However, an improvement in PSNR is obtained when using our modified algorithm. Fig. 9. shows the decomposition image of our parallel algorithm. Parallel execution of the domain searches using two DSP cards took 591 CPU seconds. From this data, it is observed that, doubling the number of processor gives a processing time approximately divided by 4 and significant enhancement of PSNR. The minimal speedup attained is in order of 4. It is believe that real-time processing can be archived if more DSP cards are permitted.

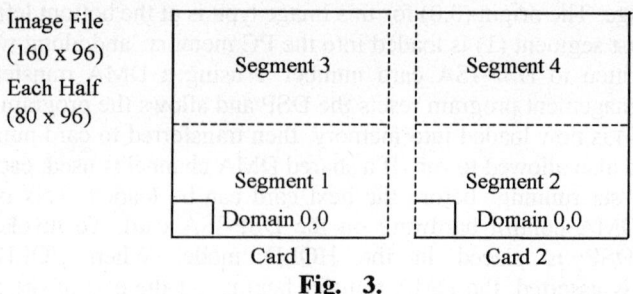

Fig. 3.

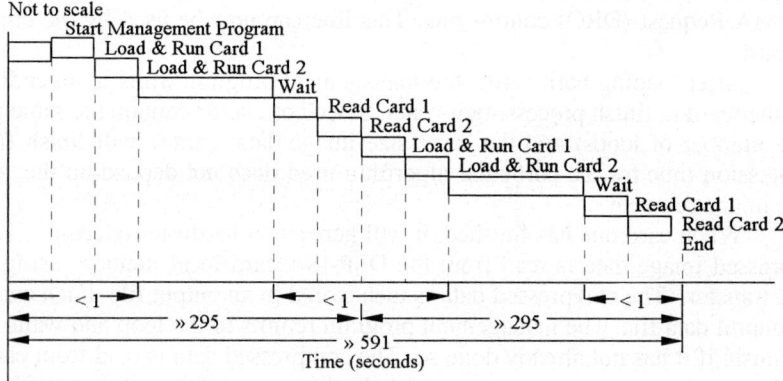

Fig. 4.

5 Conclusions

Fractal image compression exploits natural affine redundancy presents in most images to achieve high compression ratio. However, the search for the redundancy, represented as contractive transformations, has high computational demands. In this paper, we have presented an efficient parallel implementation of the fractal image coding on multiple DSP cards. Results show that the proposed solution provides significant improvement in domain block search. We have reduced the encoding time significantly by an order of four. It is believed that the approach offer a cost-effective solution for speeding up the fractal image compression and make it competitive with other methods from an execution speed perspective.

Image	Compression Time		NMSE	PSNR 1
1	477 sec	7.95 min	0.000015420592	51.8098
2	513 sec	8.55 min	0.000015420592	51.8098
3	5484 sec	91.4 min	0.000046943336	46.9751
4A	591 sec	9.85 min	-	-
4B	591 sec	9.85 min	-	-
5	(591 sec)	(9.85 min)	0.000554234895	36.2539

Table 1

References

[1] M.F. Barnsley, "Fractals Everywhere", Academic Press, 1988, New York.
[2] S.K.Chow and S.L.Chan "A Design for Fractal Image Compression using Multiple Digital Signal Processors", Proc. International Picture Coding Symposium, Melbourne, Australia, Vol.1, 1996, p.303-308.
[3] J.E. Hutchinson, "Fractals and Self-similarity", Indiana Univ. Math. J., Vol.35, 1981, pp. 713-747.
[4] E.W. Jacobs, Y.Fisher, R.D. Boss, "Image compression : a study of the iterated transform method", Signal Process. Vol.29, 1992, pp. 251-263.
[5] A.E. Jacquin, "Fractal image coding: a review", Proceedings of the IEEE, Vol.81, 1993, pp.1451-1465.
[6] D.M. Munro and F. Dudbridge, "Fractal block coding of images", Electron. Lett. Vol.28, 1992, pp.1053-1055.
[7] TMS320C2x User's Guide, Texas Instruments, 1993.

Fig. 5. (Original Image)

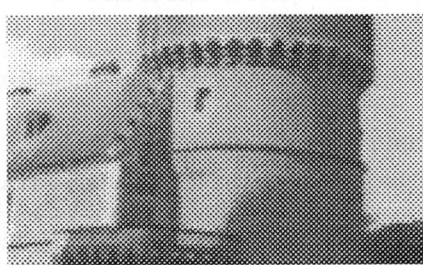

Fig. 6. (Image 1)

Fig. 7. (Image 2)

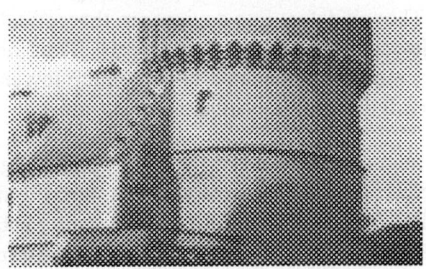

Fig. 8. (Image 3)

Fig. 9. (Image 4A)   (Image 4B)

Fig. 10. (Image 5)

Comparison of Mean Field Annealing and Multiresolution Analysis in Missing Data Estimation

Hairong Qi, Wesley E. Snyder, Griff L. Bilbro
Electrical and Computer Engineering Department, Box 7911
North Carolina State University, Raleigh, NC 27695-7911, U.S.A

Abstract

The project we are working on is to help develop and test a low cost, large area, high resolution X-ray detection system with a high dynamic range. The large area is achieved by butting two or more scintillator/fiber/CCD combinations together. An algorithm is thus required to compensate the defects come from the detector induced errors including missing single pixel due to individual defective detectors or missing column(s)/row(s) due to misalignment of adjacent CCD's in the blurred and noise corrupted images. Mean field annealing, as a global optimization technique, is the proposed algorithm. This paper, however, proposes a new approach based on multiresolution analysis where the defect compensation is implemented by removing the characteristics created by the missing columns/rows from the *detail images* of lower resolution. Experiments will be carried out to compare the performance of these two approaches. Future research directions are discussed at last.

1 Introduction

The requirement for a missing data estimation algorithm is originated by the project where two or more scintillator/fiber/CCD will be butted together to implement a low cost, large area, high resolution X-ray detection system with a high dynamic range. Here, fiber optics is used to develop the light-guide system such that light from an X-ray scintillator may be guided down to a CCD array preserving the alignment of all the pixels in a 1100 x 1200 array. Though defect free devices are theoretically available, they are much more expensive than devices with only a few defects. Thus an algorithm is required to compensate the defects and misalignment, allowing much lower cost detectors to be used, thereby reducing the overall cost of the system.

Mean field annealing (MFA) is the proposed algorithm for this project. It combines annealing technique and mean field approximation to find the global minimum. By minimizing an objective function according to different problem definitions, MFA can achieve noise removal with sharp edges preserved [1], restoration of locally-homogeneous [4] images, optimal image interpolation of missing data [8], etc. As we can see in Sec. 5, MFA can obtain good results in missing data estimation as well.

Multiresolution analysis (MRA), on the other hand, is proposed here in order to compare with the performance of MFA. Since its formulation in 1986, MRA has found its applications in many areas of computer vision, such as image compression, edge detection, texture analysis, image restoration, etc. Having a stable mathematical foundation from wavelet theory [2][6][7], MRA provides us a new tool to analyze image contents. In this paper, we try to implement the missing data estimation by analyzing the detail images of lower resolution, removing the characteristics that generated by the missing columns/rows, and estimate the missing data through reconstruction process.

The organization of this paper is as follows: after problem definition in Sec. 2, we introduce the MFA approach in Sec. 3, and MRA approach in Sec. 4; experiments and results will be presented in Sec. 5; conclusions and future research directions are discussed in Sec. 6.

2 Problem Definition

The project we are working on is to help develop and test a low cost, large area, high resolution X-ray detection system with a high dynamic range. The large area is achieved by butting two or more scintillator/fiber/CCD combinations together (Fig. 1). In developing such a kind of system, the trade-off between cost and image quality must be considered. Defect free devices are much more expensive than devices with only a few defects. Therefore, an algorithm is required to compensate such defect so to reduce the cost of the overall system.

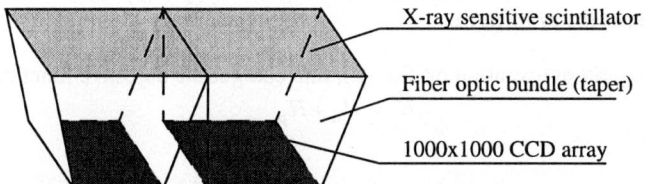

Figure 1. Two scintillator/fiber/CCD combinations may be butted together.

The defect of the image mainly come from three sources: 1) *detector induced errors* including missing single pixel due to individual defective detectors or missing column(s) or row(s) due to misalignment of adjacent CCD's; 2) *blur* caused by "point source" x-ray systems; and 3) *spatially-varying noise*.

$$g_i = (f \otimes H)_i + n_i + d_i \tag{1}$$

Given the corrupted image g, generated from Eq. (1), the proposed algorithm should be able to find the estimated $\hat{f}$ which is a perfectly-aligned, defect-free detector based on the information of blur, noise, and detector induced errors. The whole process is like the model in Fig. 2.

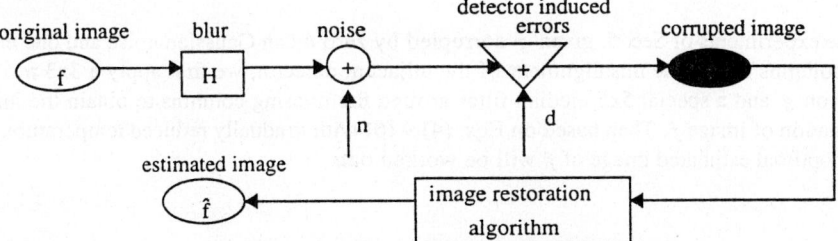

Figure 2. A model for image corruption and reconstruction process.

3 Mean Field Annealing (MFA) Approach

MFA is the proposed algorithm in our project because it has good performance in both noise removal and image interpolation. Though the determination of the missing data is (in general) impossible, MFA is still optimal in its ability to estimate those values, given the suitability of the blur, noise, and detector induced errors. Even in the case that the blur is not accurately modeled, the prior term allows edge-preserving interpolation very well.

The problem defined in Sec. 2 can be modified into a global optimization alternative, i.e. to seek image f, which will maximize the *a-posteriori* conditional probability $P(f|g)$. To solve this problem, Bayes' rule is used to derive Eq. (2),

$$Max(P(f|g)) = Max\left(\frac{P(g|f)P(f)}{P(g)}\right) \quad (2)$$

where the denominator $P(g)$ is independent of f and therefore does not affect the maximization process. We denote the conditional probability $P(g|f)$, which depends on the corruption process, as the *noise term*, and the *a-priori* probability $P(f)$ which is independent of observed image, as the *prior term*.

By taking the logarithm operation on Eq. (2), one can get the objective function as Eq. (3),

$$H = H_n + H_p \quad (3)$$

where,

$$H_n = \sum_{i,j} \frac{(f_{i,j} - g_{i,j})^2}{2\sigma^2} \quad (4)$$

$$H_p = -\sum_{i,j} \left(\frac{\beta}{T}\right) exp\left(-\frac{(\nabla f)_{i,j}^2}{2T^2}\right) \quad (5)$$

The algorithm carries out the minimization by combining conventional gradient descent (Eq. (6)) with *annealing*. The annealing minimization process starts with large T and gradually reduces the temperature over time. This process avoids most local minima and produces an optimal restored image.

$$f_{i,j}^{k+1} = f_{i,j}^k - \alpha \frac{\partial H}{\partial f_{i,}} \quad (6)$$

In the experiments of Sec. 5, given g corrupted by zero mean Gaussian noise and one missing columns due to the misalignment of the adjacent detector, we first apply a 3x3 median filter on g and a special 5x5 median filter around the missing columns to obtain the initial estimation of image f. Then based on Eqs. (4) ~ (6) with gradually reduced temperature, the final optimal estimated image of f will be worked out.

4 Multiresolution Analysis (MRA) Approach

Multiresolution analysis, since its formulation in the fall of 1986 by Mallat and Meyer, has been found important in analyzing the content of images. Natural applications of MRA are recognized as the image compression, edge detection and texture analysis. Recently, researchers found some more applications and got good results, such as image restoration and noise removal.

MRA tries to understand the content of the image at different resolutions, because the details of an image generally characterize different physical structures of the scene.

The 2-D MRA first convolves the rows with a one-dimensional filter, retain every other row, then convolves the columns of the resulting signals with another one-dimensional filter and retain every other column. The framework is drawn as Fig. 3 where $\phi(x)$ is the scaling function (like a low-pass filter) and $\psi(x)$ is the wavelet function (like a high-pass filter.

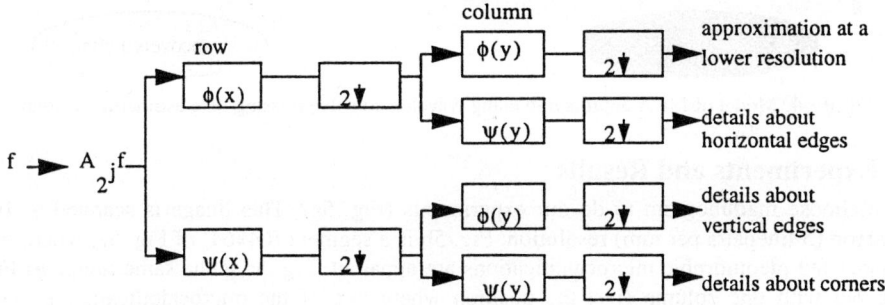

Figure 3. Decompose an image from a higher resolution to a lower resolution.

The application of wavelets in the area of image interpolation is not as popular as that in denoising, especially for missing data estimation. Only few papers try this area. Chang et. al. [3] is one of them where they propose a wavelet based method which estimates the higher resolution information needed to sharpen the image, i.e. it can extrapolate the wavelet transform of the higher resolution based on the evolution of the wavelet transform extrema across the scales.

For the missing data estimation problem, we propose the multiresolution decomposition based method, where in each resolution, the sharp edges caused by the missing data is detected and deleted, the information missed there is then interpolated (estimated) based on the vicinity of each pixel. On the other hand, since the details in the original image will disappear while the resolution goes smaller and smaller, theoretically, we can reconstruct the optimal image by using the approximated image in the smallest resolution and the processed detail images in the same resolution to obtain the approximation image in the adjacent larger resolution, and repeat this process until we get the final optimal image which is in the same resolution as the original image (Fig. 4). Though theoretically correct, we can't reduce the image resolution to infinity, thus some defect might occur depends upon the property of the original image.

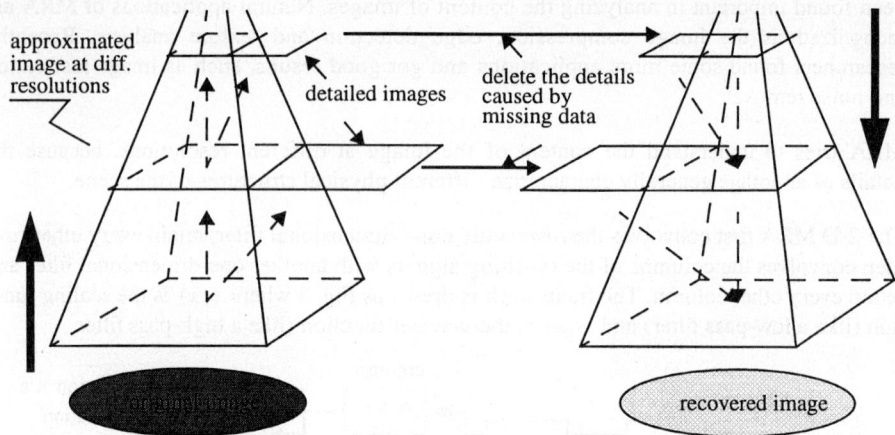

Figure 4. Model of MRA approach in using wavelets to solve missing data estimation problem.

5 Experiments and Results

We choose mammogram to do our experiments (Fig. 5a). This image is scanned at 100 micron (5 linepairs per mm) resolution. Fig. 5b is a segment (64x64) of Fig. 5a, where two associated pleomorphic microcalcifications are apparent. Fig. 5c is the same image as Fig. 5b but with one column (just the position where one of the microcalcification locates) missing. Five methods are used to recover the missing column and the results are shown in Fig. 6. Fig. 7 is the decomposition of the missing column image with db2 as the mother wavelet. The vertical edge caused by the missing column is clear.

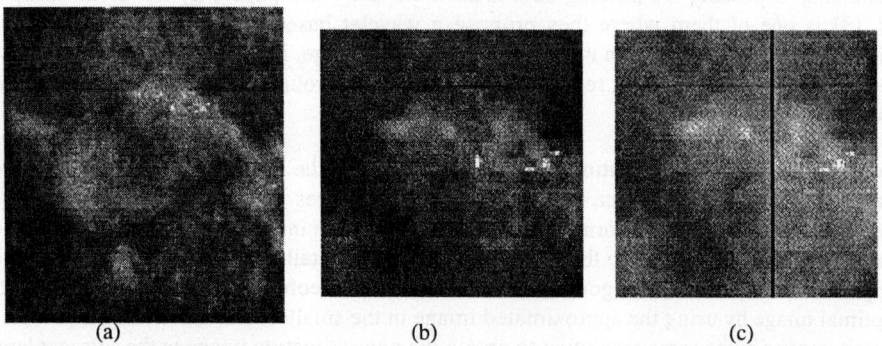

Figure 5. (a) original image; (b) testing image; (c) testing image with one column missing.

We have mentioned two recovering methods in previous sections (MFA and MRA). The three others are much simpler.
- *random copy method*: randomly copy the value from the left or right neighbor of the missing pixel;
- *average substitution method*: use the average value of the left and right neighbors as the estimation of the missing pixel; and
- *median substitution* method: use the median value of the local neighbors as the estimation of the missing pixel.

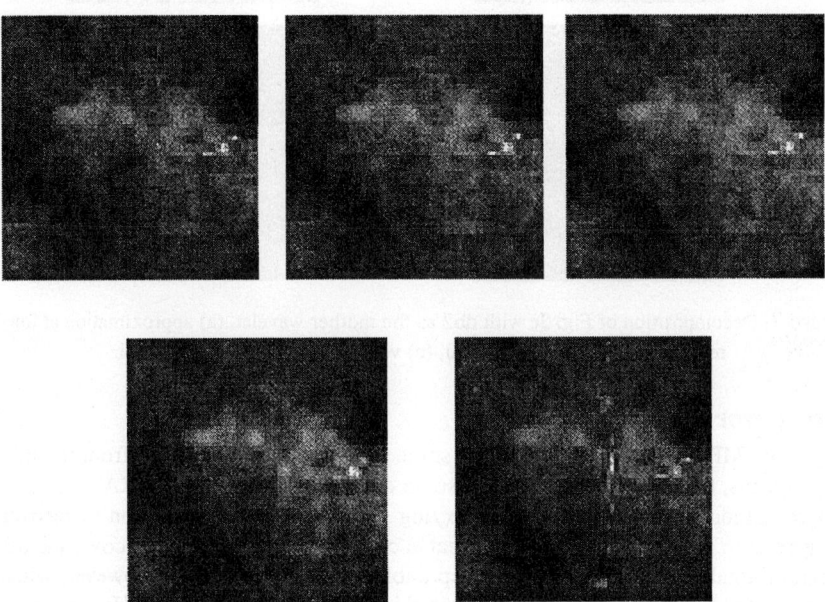

Figure 6. Recovered image. (from upperleft to lower bottom: by random copy, by average substitution, by median substitution, by MFA and by MRA).

From Fig. 6, we can see that the upper three images didn't recover the missed microcalcification at all, but the bottom two can both recover it to some extent. On the other hand, result from MFA is more blurred than that from MRA, and MRA also generates some apparent artifacts the same time it preserves the characteristics. The oversmoothing from MFA mainly comes from the contribution of prior term where a globally smooth image is supposed to be reached.

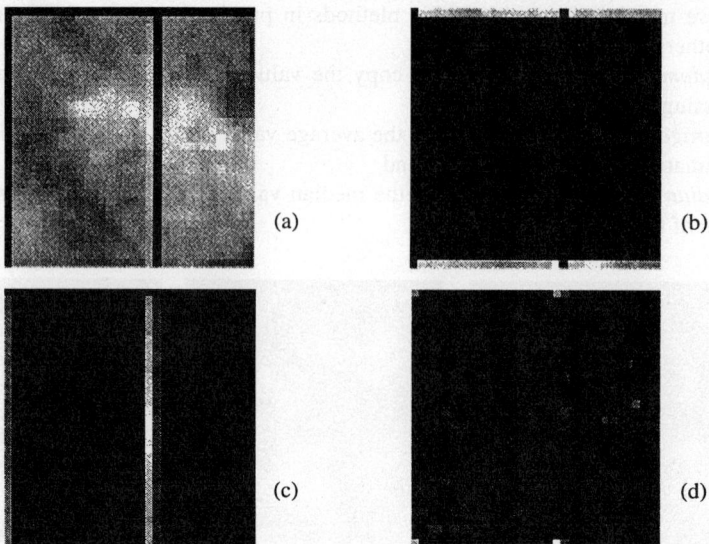

Figure 7. Decomposition of Fig. 5c with db2 as the mother wavelet. (a) approximation at lower resolution, (b) horizontal detail, (c) vertical detail, (d) corner detail.

6 Future work

Though both MFA and MRA can recover some important characteristics from the missing column images, they both, however, have some disadvantages, such as MRA's artifacts and MFA's oversmoothing. Recently, we are trying to use the blur information to recover the missing column exactly or at least with great accuracy. This method can recover the missed information totally if the blur kernel is separable and exactly known. However, when the image size goes larger and larger, the recovered image turns very unstable. How to solve this ill-conditioning problem is the next research topic.

7 Reference

1. Bilbro GL, Snyder WE: Applying mean field annealing to image noise removal. *J. of Neural Network Computing*, pp5-17, Fall, 1990.
2. Daubechies I: Ten lectures on wavelets. Capital City Press, Montpelier, Vermont, 1992.
3. Chang GS, Cvetkovic Z, Vetterli M: Resolution enhancement of images using wavelet transform extrema extrapolation. *Proceedings ICASSP*, v4, pp2379-2382, 1995.
4. Hiriyannaiah H, Bilbro GL, Snyder WE, Mann R: Restoration of locally homogeneous images using mean filed annealing. *J. of the Optical Society of America A*, pp1901-1912, December, 1989.

5. Mallat S: A theory for multiresolution signal decomposition: the wavelet representation. *IEEE Trans. Pattern Anal. Machine Intell.* 11 (7): 674-693, 1989.
6. Mallat S: Wavelets for a vision. *Proceedings of the IEEE* 84 (4): 604-614, April 1996.
7. Rioul O, Vetterli M: Wavelets and signal processing. *IEEE Signal Processing Magazine*, pp14-38, October 1991.
8. Wang CX: Optimal image interpolation using optimal method. *Ph.D. thesis*, North Carolina State University, 1996.

Acknowledge

This project is supported by US Army Research Office (ARO) through Contract No. DAAH04-93-D-0003D. We acknowledge the support of ARO and thank Dr. William A. Sander for his encouragement.

Segmentation of MRF Based Image Using Hierarchical Genetic Algorithm*

Jin Wook Kim, Eun Yi Kim, Se Hyun Park, Hang Joon Kim
Department of Computer Engineering, KyungPook National University
Taegu, 702-701, South Korea
kimhj@bh.kyungpook.ac.kr

Abstract - In this paper, a segmentation of an Markov Random Field based image is proposed. We use a hierarchical image model consists of color, blurring and noise parameters and define an energy function for proper image segmentation criterion. In general, it is not easy to search optimal parameter values which optimize the given energy function. To search optimal parameter values effectively, Hierarchical Genetic Algorithm is used and the experimental results show the effectiveness of the proposed method.

1. Introduction

Image segmentation is a process of segmenting an image into a group of homogeneous regions according to whose characteristics such as color and texture. It is the front-end processing stage in object recognition and image understanding systems. The accuracy of this task is a crucial factor in determining overall system performance.

So far many segmentation methods have been proposed using various features[1-3]. Hansen F.R. and H. Elliott used an Markov Random Field (MRF) based image segmentation using intensity feature vector [1], but if two regions are of different color but the intensity of them are similar then the two regions are merged into one. F. Cohen and Z. Fan used texture feature [2]. However, in a syndicated image it is easy to extract texture feature, but in a real scene it is hard to extract reliable texture. Il Yo Kim and Hyun S. Yang proposed a split-and-merge method using color feature spatial relationship between each region [3]. However, this method has a drawback that the final segmented image is much influenced by the initially segmented image.

For reliable segmentation it is necessary a proper image modeling to consider degraded image. Stuart Geman and Donald Geman proposed an MRF based image restoration method considering intensity, blurring and noise [8].

In this paper, an MRF based image segmentation method is proposed. We transform RGB color space into HSI color space [4] to segment image more stably. An image is modeled hierarchically with color, blurring and noise parameters. Our goal is to find appropriate parameter values which guarantee proper segmentation by minimizing the given energy function [1,6]. Hence, Hierarchical Genetic Algorithm (HGA) is used to search optimal parameter values [5]. The algorithm is a heuristic optimization technique of focusing on a set of promising solutions which eventually lead to convergence on a globally optimal solution. It is robust in dealing with our image model since it can perform parameter optimization hierarchically. The method we propose here is robust, unsupervised and highly parallel.

The rest of this paper is organized as follows. In section 2, we explain an image modeling based on Gibbs and Markov random field. Section 3 describes segmentation using HGA. Experimental results in section 4 are followed by conclusions in section 5.

---

* This research was supported by KOSEF under core research grant #971-0908-050-2

2. Image modeling based on MRF

Since all of our work deals with image we first consider an image observation model then we derive an energy function for segmentation criterion.

2.1 Image characteristics

Usually an input image is degraded. Typical degradation model includes noise, blurring and nonlinear transformation [1] as follows.

$$G = \phi(I) \odot N \qquad (2.1)$$

where G is an observed image, I is an image to be estimated along with its attributes, ϕ, in general, is a nonlinear operator, N is a corrupting noise, and $\odot$ is an invertible nonlinear operation.

We regard that nonlinear transformation ϕ is linear transformation occur in a degraded image and $\odot$ is an additive operator. So an input image G can be described as follows.

$$G = H(I) + N \qquad (2.2)$$

where H is a blurring matrix of 3×3 and N, composed of mean μ, is noise field independent of an input image.

2.2 Gibbs and Markov random field

Let $I = \{ F_{i,j} : 1 \leq i \leq N, 1 \leq j \leq N \}$ be a set of pixels. Suppose that there exists a neighborhood system denoted by $\Gamma = \{ n(F_{i,j}) : 1 \leq I \leq N, 1 \leq j \leq N \}$, where $n(F_{i,j})$ is a set of pixels in I that are neighbors of $F_{i,j}$. Let $X = \{ X_{i,j} : 1 \leq I \leq N, 1 \leq j \leq N \}$ denotes any family of random variables each $X_{i,j}$ of which is associated with pixels $F_{i,j}$ and $\Lambda = \{ \lambda_i, \ldots, \lambda_m \}$ be a set of possible labels so that $X_{i,j} \in \Lambda$ for all i and j. Let Ω be the set of all possible configurations: $\Omega = \{ \omega = (x_{1,1}, \ldots, x_{N,N}) : x_{i,j} \in \Lambda, 1 \leq i,j \leq N \}$ then X is an MRF with respect to Γ if

$$P(X = \omega) > 0 \quad \forall \omega \in \Omega \qquad (2.3)$$

$$P(X_{i,j} = x_{i,j} | X_{k,l} = x_{k,l}, F_{i,j} \neq F_{k,l}) = P(X_{i,j} = x_{i,j} | X_{k,l} = x_{k,l}, F_{k,l} \in n(F_{i,j})) \qquad (2.4)$$

for every $F_{i,j} \in I$ and $(x_{1,1}, \ldots, x_{N,N}) \in \Omega$. $P(\cdot)$ and $P(\cdot|\cdot)$ are the joint and conditional probability density functions, respectively. By Hammersley-Clifford theorem, a priori and a posteriori probabilities of image labeling ω can be represented like (2.5) and (2.6) respectively.

$$P(X = \omega) = \frac{e^{-U(\omega)/T}}{Z} \qquad (2.5) \qquad P(X = \omega | G = g) = \frac{e^{-U^P(\omega)}}{Z^P} \qquad (2.6)$$

where T is the temperature, Z a normalizing constant, and U an energy term. U is obtained by summing over applicable clique potentials $V_c(\omega)$. $U(\omega)$ and $U^P(\omega)$ is a priori and a posteriori energy function respectively represented as (2.7) and (2.8).

$$U(\omega) = \sum_{c \in C} V_c(\omega) \qquad (2.7)$$

In (2.7), C is a possible clique set in neighbor, V is a potential function defined as follows.

$$V_c(\omega) = \begin{cases} -1 : L(ch_i) = L(ch_j) \, and \, F(ch_i) = F(ch_j) \\ 1 : L(ch_i) = L(ch_j) \, and \, F(ch_i) \neq F(ch_j) \\ 0 : L(ch_i) \neq L(ch_j) \end{cases} \tag{2.8}$$

where L represents the label number of a chromosome ch and F represents the color feature of a chromosome. We define the energy function as follows

$$U^p(\omega) = U(\omega) + \alpha \, \| I - (\phi(B) \odot N) \|^2 \tag{2.9}$$

where $\alpha = 1 / $ (mean value of HIS of image), B is an estimated image and at pixel level,

$$U^p_{i,j} = U_{i,j} + \alpha \, \| I_{i,j} - (\phi(B_{i,j}) \odot N) \|^2 \tag{2.10}$$

3. Image segmentation

We employ the maximum a posteriori (MAP) estimation approach such that $P(X=\omega|G=g)$ is maximized, that is, U^p should be minimized. To minimize U^p, we have to consider color feature, blurring matrix and noise. Searching of all of these values which optimize the given energy function requires much computation time. So we use HGA to search each parameter set hierarchically.

3.1 Hierarchical genetic algorithm

The HGA uses a multi-level search strategy illustrated in Figure 1, in which a higher level cluster investigates a wide search space[5]. For a problem which contains many independent parameters to consider, the HGA facilitates an expanding search parameter approach as shown in figure 2. The lower level searches with a few parameters. The solutions found at the lower level are sent to the higher level, which appends extra parameters. The higher level GA module exploits the previously searched parameter information from the lower level so that it needs not search the entire parameter set.

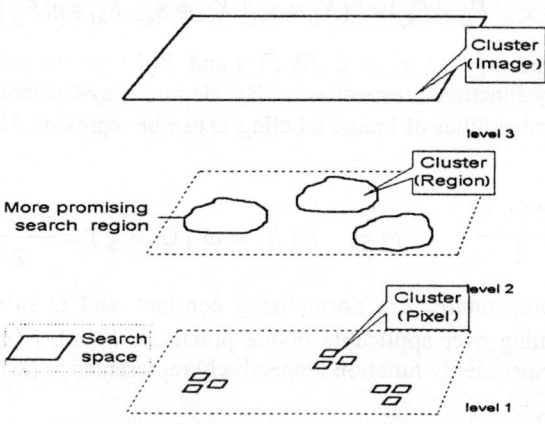

Figure1. The structure of HGA.

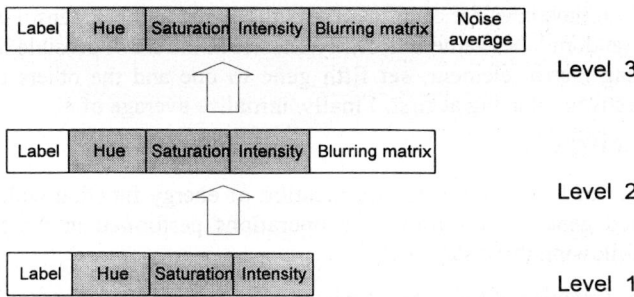

Figure 2. Parameter structure.

A chromosome for each pixel of first layer consists of two parts as shown in figure 3(a). Figure 3(b) and 3(c) show the structure of a chromosome in layer 2 and 3.

The first level only searches color features then these features are sent to second level where optimal blurring matrices of each region are searched. Note that in this level only blurring matrix are manipulated. Finally, these parameters are sent to the higher level where optimal noise parameter values are searched.

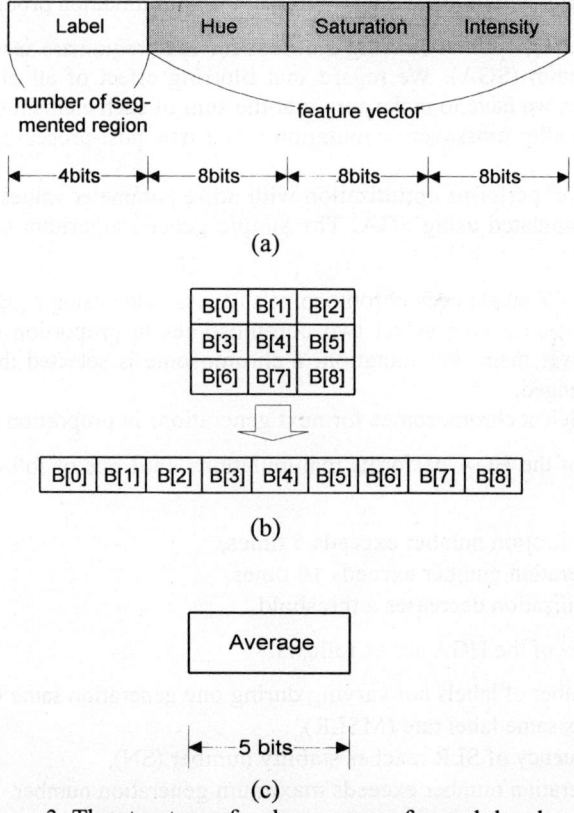

Figure 3. The structure of a chromosome for each levels.

The values of chromosomes for each layer are initialized in three steps. First, each label part is given a random label value uniformly chosen in the set of possible labels. Second, for each blurring matrix element, set fifth gene to one and the others to zero which means that there is no blurring at first. Finally, initialize average of N.

3.2 Behavior of HGA

The first layer of HGA performs optimization of energy function with color feature using distributed genetic algorithm. The operations performed in the first layer are consist of the following three steps [7].

(1) *Evaluation.* Calculate fitness for each chromosome. the energy function(2.10) is evaluated as fitness function.
(2) *Selection.* Replace each chromosome by a neighbor according to the fitness of the neighboring chromosomes, where the neighborhood includes the point under selection and the eight contiguous points. We use an elitist selection scheme for the selection mechanism.
(3) *Mating.* In this stage crossover and mutation is occurred. For each chromosome, a neighbor which has the same label with the current chromosome is chosen randomly. Recombine the current chromosome with this neighbor, and replace the current chromosome by one of two offspring that has the better fitness value. We also randomly alter any or all of the feature values with mutation probability.

In the second layer, optimal blurring matrices for each region are searched using Simple Genetic Algorithm (SGA). We regard that blurring effect of all pixels in a region is equal. However, we have to make sure that the sum of each element of a blurring matrix should be 1 so after crossover or mutation a heuristic post-processing is performed for integrity.

The third layer performs optimization with noise parameter values of the image. This layer also manipulated using SGA. The simple genetic algorithm can be described as follow [9].

(1) *Evaluation.* Calculate each chromosome's fitness value using equation (2.9)
(2) *Mating.* For crossover, select two chromosomes in proportion to its fitness value then crossover them. For mutation, a chromosome is selected then one of its gene value is changed.
(3) *Selection.* Select chromosomes for next generations in proportion to fitness value

Each layer of the HGA performs manipulations until one of following conditions is satisfied.

- When stabilization number exceeds 5 times,
- When generation number exceeds 10 times,
- When stabilization decreases a threshold.

The stop criteria of the HGA are as follows.

When the number of labels not varying during one generation same label rate (SLR).
- SLR ≥ Max same label rate (MSLR),
- When frequency of SLR reaches stablity number (SN),
- When generation number exceeds maximum generation number.

4. Experimental results

To evaluate the practical performance of the proposed method, experiments were performed using 50 images. The images were taken from a video camera located at a fixed position. We used a SAMSUNG SV-33 video camera as an input device, and implemented the method on a Pentium PC with Visual C++ language in Windows NT environment. Maximum label number is 16 and the size of blurring matrix is 3×3. Maximum generation number is 500 and the neighbor is surrounding eight points. SR is 99.95% and SN is set to 5. Finally, MSLR is set to 99.99%. Figure 4(b) shows the result of optimizing only with feature vectors.

Figure 4(c) shows the result of optimization with both of feature vector and blurring matrix. Figure 4(d) shows the result of optimizing with all parameters such as feature vector, blurring matrix and noise.

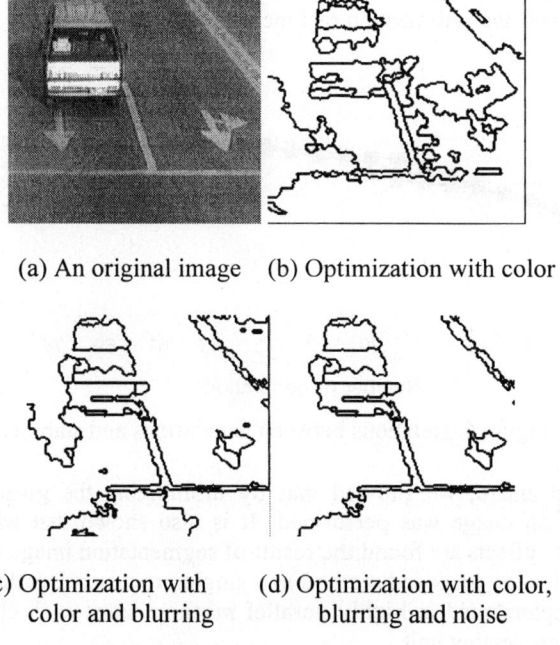

(a) An original image   (b) Optimization with color

(c) Optimization with color and blurring   (d) Optimization with color, blurring and noise

Figure 4. Results of segmentations.

Figure 5 shows the relation between generation number and total sum of energy functions of figure 4. As shown in this figure, total sum of energy functions decreases as generations pass.

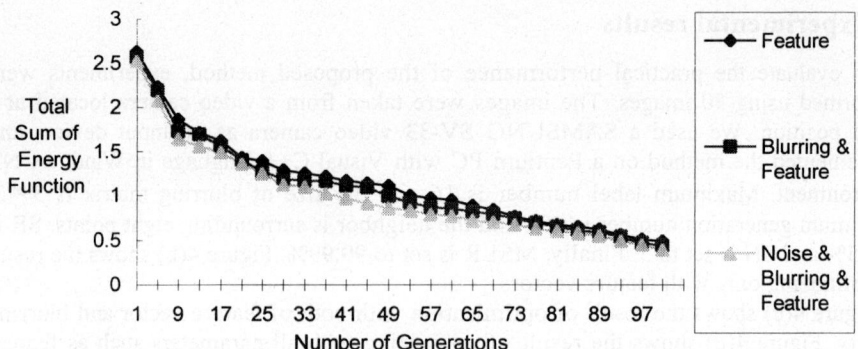

Figure 5. Relation between generations and total sum of energy functions.

Figure 6 shows the relationship between stabilization rates and generations of figure 4. As generations pass, the stabilization rate increases.

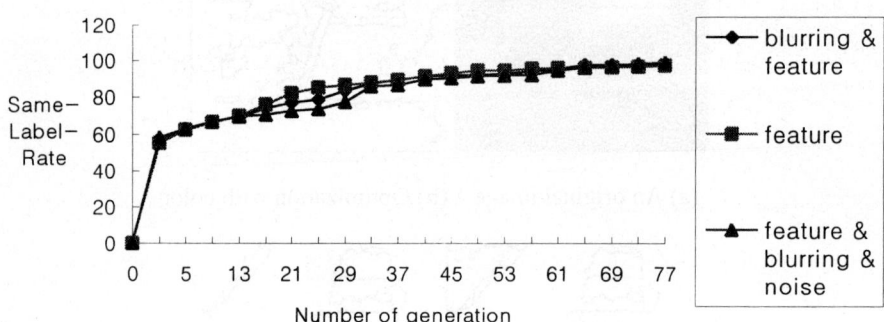

Figure 6. Relations between generations and stability.

From figure 5 and 6, we showed that by minimizing the given energy function, segmentation of an image was performed. It is also shown that when optimal color, blurring and noise effects are found the result of segmentation image was most clear.

We implemented the proposed method in a single processor. But the proposed method can be easily implemented in highly parallel machine since each chromosome can be allocated in one processing unit.

5. Conclusions

In this paper, we have proposed segmentation method of MRF based image using HGA. An input image was modeled hierarchically with color, blurring and noise parameters. The energy function for segmentation criterion was defined considering all parameters. For effective optimization of given energy function, we used HGA to find adequate parameter values.

Experimental results showed that the proposed method has the potential to work well in segmenting real world images. The advantages of the method are that: (1) it provides robustness in dealing with degraded and corrupted images; (2) HGA has robustness in optimizing the energy function which has many parameters hierarchically represented.

Although the performance of the proposed method is good, it can be improved further by using a more efficient color model. Future works also include an implementation of the method with parallel facilities.

References

[1] F. R. Hansen and H. Elliott, "Image Segmentation using simple Markov field models," *Computer Graphics Image Processing,* vol. 20, pp. 101-132, 1982
[2] F. Cohen and Z. Fan, "Maximum likelihood unsupervised texture image segmentation", *CVGIP*, vol. 54, no. 3, pp. 239-251, 1992.
[3] I. Y. Kim and H. S. Yang, "Efficient Image Labeling based on Markov Random Field and Error Backpropagation Network", *Pattern Recognition*, vol. 26, no. 11, pp. 1695-1707, 1993.
[4] R. G. Gonzalez and R. E. Woods, *Digital image processing*, Addison-Wesley Publish Co. 1992.
[5] J. W. Kim and B. P. Zeiger, "Hierarchical Distributed Genetic Algorithms: A Fuzzy Logic Controller Design Application", *IEEE EXPERT*, vol. 11, no. 3, 1996.
[6] N. R. Pal and S. k. Pal, "A Revies On Image Segmentation Techniques," *Pattern Recognition,* vol.26, no.9, pp 1277-1294, 1983
[7] P. Andrey and P. Tarroux, "Unsupervised Image Segmentation using A Distributed Genetic Algorithm", *Pattern Recognition*, vol. 27, no. 5, pp. 659-673, 1994.
[8] S. Gemam and D. Geman, "Stochastic relaxation, Gibbs distributions, and the baysian restoration of images", *IEEE Trans. PAMI*, vol PAMI-6, no. 6, pp. 721-741, 1984.
[9] D. E. Golberg, *Genetic Algorithm in Search, Optimization and Machine Learning.* Addison Wesley, Reading, Massachusetts 1989.

Motion Compensated Color Video Classification Using Markov Random Fields*

Zoltan Kato, Ting-Chuen Pong, John Chung-Mong Lee

Hong Kong University of Science and Technology, Computer Science Dept.,
Clear Water Bay, Kowloon, Hong Kong, Tel: +852 2358 7000 — Fax:+852 2358 1477,
email: kato@cwi.nl, tcpong@cs.ust.hk, cmlee@cs.ust.hk

Abstract. This paper deals with the classification of color video sequences using Markov Random Fields (MRF) taking into account motion information. The theoretical framework relies on Bayesian estimation associated with MRF modelization and combinatorial optimization (Simulated Annealing). In the MRF model, we use the CIE-**luv** color metric because it is close to human perception when computing color differences. In addition, intensity and chroma information is separated in this space. The sequence is regarded as a stack of frames and both intra- and inter-frame cliques are defined in the label field. Without motion compensation, an inter-frame clique would contain the corresponding pixel in the previous and next frame. In the motion compensated model, we add a displacement field and it is taken into account in inter-frame interactions. The displacement field is also a MRF but there are no inter-frame cliques. The Maximum A Posteriori (MAP) estimate of the label and displacement field is obtained through Simulated Annealing. Parameter estimation is also considered in the paper and results are shown on color video sequences using both the simple and motion compensated models.

1 Introduction

Image classification is an important early vision task where pixels with similar features are grouped into homogeneous regions. Many high level processing tasks (surface description, object recognition, for example) are based on such a preprocessed image. Using color information can considerably improve capabilities of image classification algorithms compared to purely intensity-based approaches. However, we need a good color space in order to use color information in the same way as humans perceive color differences. There are several metrics proposed for computer vision [5]. We use the CIE-**luv** [5] color space here because it separates luminance and chroma information and it is easy to compute color differences in this metric.

The visual motion derived from a sequence of time-varying images [11, 4] is also a valuable source of information. Basically, it can be used to detect motion in the scene but it is also possible to derive more detailed information such as

---

* This research was supported by Hong Kong Research Grants Council under grants *HKUST616/94E* and *HKUST661/95E*, and ITDC grant *AF/66/95*

position, orientation of a visible surface or 3D reconstruction of a scene. Herein, we are interested in computing displacement vectors [11, 13, 4] in order to build a motion compensated Markov Random Field (MRF) image classification model.

When we have a sequence of color images, still image MRF models [2, 9, 12, 7] can be easily extended to take into account the information in the previous and next frames [13, 11] (see Section 2.1). Instead of a 2D neighborhood system, we can use a 3D one with inter-frame cliques. If the camera or the objects in the scene are not moving then this model yields good segmentations. In the case of moving objects, however, this static model can fail.

To overcome the problem caused by moving objects, we introduce a displacement field (DF) [13] in Section 2.2 in order to take into account motion information in the label field. For simplicity, the DF is defined over the same lattice as the label field by placing a new lattice between two neighboring frames. DF is a vector-valued MRF giving the displacement vector at each site between two frames in the sequence. The estimation of the DF is done in parallel with the label field and no external algorithm or initialization is needed. The energy function of the so-defined system is minimized by the Metropolis algorithm [10]. The result is the classification of the input frames *and* the displacement vectors between frames.

Usually, MRF-based segmentation methods suffer from a lack of parameter estimation. The majority of the proposed methods are supervised, which limits their practical use because a human intervention is needed to compute the model parameters. Herein, we are interested in completely *data driven* algorithms since in real-life applications, these parameters are usually unknown and one has to estimate them without human intervention. In Section 2.3, we consider parameter estimation of the proposed model. Finally, some results are presented in Section 3.

2 MRF model

In this section, we describe a spatio-temporal MRF model for color video classification. First, we define a model which uses only color information and then we extend our model to take into account motion information.

2.1 Color video sequence classification (Label Field)

Let us suppose that the observed images consist of three spectral component values (**luv**) at each pixel denoted by the vector $\mathbf{f}_s^t$, where $s \in \mathcal{S}$ is the spatial index and $t \in \mathcal{T}$ is the temporal index. We are looking for the labeling $\hat{\omega}$, which maximizes the a posteriori probability $P(\omega \mid \mathcal{F})$, that is the *maximum a posteriori* (MAP) estimate. Bayes theorem tells us that:

$$P(\omega \mid \mathcal{F}) = \frac{1}{P(\mathcal{F})} P(\mathcal{F} \mid \omega) P(\omega). \tag{1}$$

Actually $P(\mathcal{F})$ does not depend on the labeling ω and we make the assumption that:

$$P(\mathcal{F} \mid \omega) = \prod_{t \in \mathcal{T}} \prod_{s \in \mathcal{S}} P(\mathbf{f}_s^t \mid \omega_s^t). \tag{2}$$

It is then easy to see that the global labeling, which we are trying to find, is given by:

$$\hat{\omega} = \arg\max_{\omega \in \Omega} \prod_{t \in \mathcal{T}} \prod_{s \in \mathcal{S}} P(\mathbf{f}_s^t \mid \omega_s^t) \prod_{C \in \mathcal{C}_\mathcal{S}} \exp(-V_C(\omega_C)) \prod_{C \in \mathcal{C}_\mathcal{T}} \exp(-V_C(\omega_C)) , \tag{3}$$

where $\mathcal{C}_\mathcal{S}$ is the set of spatial (or intra-frame) cliques and $\mathcal{C}_\mathcal{T}$ is the set of temporal (or inter-frame) cliques. It is obvious from this expression that the a *posteriori* probability also derives from a MRF. The energies of cliques of order 1 directly reflect the probabilistic modeling of labels without context, which could be used for labeling the pixels independently. This item ties the resulting segmentation to the original input.

A natural assumption is that $P(\mathbf{f}_s^t \mid \omega_s^t)$ is Gaussian, the classes $\lambda \in \Lambda = \{0, 1, \ldots, L-1\}$ are represented by the mean vectors $\boldsymbol{\mu}_\lambda$ and the covariance matrices Σ_λ. It is then clear that

$$P(\mathbf{f}_s^t \mid \omega_s^t) = \frac{1}{\sqrt{(2\pi)^3 \mid \Sigma_{\omega_s^t} \mid}} \exp\left(-\frac{1}{2}(\mathbf{f}_s^t - \boldsymbol{\mu}_{\omega_s^t}) \Sigma_{\omega_s^t}^{-1} (\mathbf{f}_s^t - \boldsymbol{\mu}_{\omega_s^t})^T\right). \tag{4}$$

We get the following energy function:

$$U(\omega, \mathcal{F}) = U_1(\omega, \mathcal{F}) + U_2(\omega) + U_3(\omega) , \tag{5}$$

$$U_1(\omega, \mathcal{F}) = \sum_{t \in \mathcal{T}} \sum_{s \in \mathcal{S}} \left(\ln(\sqrt{(2\pi)^3 \mid \Sigma_{\omega_s^t} \mid}) + \frac{1}{2}(\mathbf{f}_s - \boldsymbol{\mu}_{\omega_s^t}) \Sigma_{\omega_s^t}^{-1} (\mathbf{f}_s^t - \boldsymbol{\mu}_{\omega_s^t})^T \right) \tag{6}$$

$$U_2(\omega) = \sum_{C \in \mathcal{C}_\mathcal{S}} V_2(\omega_C) \tag{7}$$

where $V_2(\omega_C) = V_{\{s,r\}}(\omega_s^t, \omega_r^t) = \begin{cases} 0 & \text{if } \omega_s^t = \omega_r^t \\ \beta & \text{if } \omega_s^t \neq \omega_r^t \end{cases} \tag{8}$

$$U_3(\omega) = \sum_{C \in \mathcal{C}_\mathcal{T}} V_3(\omega_C) \tag{9}$$

where $V_3(\omega_C) = V_{\{t,t+1\}}(\omega_s^t, \omega_s^{t+1}) = \begin{cases} 0 & \text{if } \omega_s^t = \omega_s^{t+1} \\ \gamma & \text{if } \omega_s^t \neq \omega_s^{t+1} \end{cases} \tag{10}$

where $\beta > 0$ and $\gamma > 0$ are model parameters controlling the homogeneity of the regions and the importance of spatial and temporal interactions. As they increase, the resulting regions become more homogeneous.

2.2 Using motion information (Displacement Field)

To further elaborate our model, we introduce a new field called the *displacement field* (DF). In this way, we can take into account motion information when doing classification of a video sequence. The DF could be defined over a different lattice than the label field (one could use a lower resolution, for instance) but, for simplicity, we define it over the same lattice, placing a new lattice between each neighboring frames $(t, t+1)$. DF is a vector-valued field, $\phi_s^t \in \Phi$ denotes the displacement vector at site s between frames t and $t+1$.

The energy function of the DF is defined as follows:

$$U^{DF} = U_1^{DF}(\omega, \phi) + U_2^{DF}(\phi)$$

$$U_1^{DF}(\omega, \phi) = \sum_{t \in \mathcal{T}} \sum_{s \in \mathcal{S}} V_1^{DF}(\omega_s^t, \phi_s^t) \quad (11)$$

$$\text{where } V_1^{DF}(\omega_s^t, \phi_s^t) = \begin{cases} 0 \text{ if } \omega_s^t = \omega_{s+\phi_s^t}^{t+1} \\ \alpha \text{ if } \omega_s^t \neq \omega_{s+\phi_s^t}^{t+1} \end{cases} \quad (12)$$

$$U_2^{DF}(\phi) = \sum_{C \in \mathcal{C}_{DF}} V_2^{DF}(\phi_C) \quad (13)$$

$$\text{where } V_2^{DF}(\phi_C) = \sum_{r \in C} \|\phi_s^t, \phi_r^t\|_2. \quad (14)$$

Unlike conventional approaches, herein we use the label field ω instead of the color value in the first order potential (Equation (12)). The second order potential (Equation (14)) is a smoothing constraint favoring similar displacement vectors in neighboring sites. Note that we have only intra-frame cliques here. In our tests, we have used a first order neighborhood system.

For motion compensated classification, we have to take into account the DF in the energy function of the label field. For this purpose, we will redefine $V_3(\omega_C)$ (see Equation (10)) in the following way:

$$V_3'(\omega_C) = V_{\{t,t+1\}}'(\omega_s^t, \omega_s^{t+1}) = \begin{cases} 0 \text{ if } \omega_s^t = \omega_{s+\phi_s^t}^{t+1} \\ \gamma \text{ if } \omega_s^t \neq \omega_{s+\phi_s^t}^{t+1} \end{cases} \quad (15)$$

The energy function of the motion compensated model is then given by the following equation:

$$U(\omega, \phi, \mathcal{F}) = U_1(\omega, \mathcal{F}) + U_2(\omega) + U_3'(\omega) + U_1^{DF}(\omega, \phi) + U_2^{DF}(\phi) \quad (16)$$

where $U_3'(\omega)$ is the motion compensated energy function of inter-level cliques (see Equation (10) and Equation (15)). The MAP estimate of the label and displacement field is obtained trough the minimization of $U(\omega, \phi, \mathcal{F})$:

$$(\hat{\omega}, \hat{\phi}) = \min_{\omega, \phi} U(\omega, \phi, \mathcal{F}). \quad (17)$$

Since the energy function has many local minima, we use the Metropolis algorithm [10] to find the global minima. At each iteration, the label field is updated first followed by the DF.

2.3 Parameter Estimation

Our goal is to propose a completely data-driven, unsupervised classification algorithm. Thus, we have to estimate the mean vector μ_λ and the covariance matrix Σ_λ for each class, and the hyper-parameters α, β and γ. The mean vectors and covariance matrices can be obtained from the first frame using an unsupervised classification algorithm (for more details, see [7]). The hyper-parameters are less sensitive. We have found in practice that $\alpha = 15.0$, $\beta = 2.5$ and $\gamma = 2.0$ give good results. Of course, one could also use an estimation algorithm (see [3, 8, 6]) to obtain the right values depending on the input video sequence. However, we found that these algorithms need a huge computing power and the obtained values were very close to our *ad hoc* estimates. The mean vectors and covariance matrices could also be re-estimated during the classification using an adaptive classification algorithm similar to [7] but experiments show that the one frame estimates are good enough to obtain a reasonably good classification.

3 Experiments

The proposed algorithm has been tested on a variety of color video sequences. Herein, we present a few of our results obtained on a variety of color video sequences and also compare the motion compensated and static models. In all cases, the optimization algorithm has been stopped when the number of changed sites was less than 0.01% of the sites.

In Table 1, we give the computing times for the presented video sequences. One can see, that motion compensated classification needs more iterations and more computing time because of the additional displacement field. However, the quality of these results is also better. The computing time depends also on the optimization method. ICM [1] is a deterministic algorithm which converges in a few iterations but it finds only a local minima. This may not be as good as the one given by a stochastic method, like the Metropolis algorithm [10].

In Figure 1 and Figure 2, we compare the results obtained by the static and motion compensated model on two color video sequences. The results in the second (resp. third) column has been obtained by the static (resp. motion compensated) model. One can see that the results are better in the case of motion compensation.

The proposed model can be easily applied to gray-level images, only the first order clique-potentials have to be changed in Equation (6): Instead of a 3-variate Gaussian distribution, we use here a univariate one. In Figure 3, we show the results obtained on the "tennis" sequence using only gray-values. The result clearly shows that color information can improve considerably the final results. The computing time is only slightly lower than in the case of color images (see Table 1).

In Figure 4, we give the classification and displacements obtained on the "tennis" sequence using the Metropolis algorithm. The displacements are displayed over 16×16 blocks. The displacement field is noisy inside homogeneous regions but it is reasonably good over region boundaries. This is good enough for the purpose of motion compensated classification. More accurate displacement field could be obtained through a more elaborated homogeneity constraint in Equation (14).

4 Conclusion

We have proposed an unsupervised, motion compensated color video classification algorithm. The classification model is defined in a Markovian framework and uses a first order potential derived from a three-variate Gaussian distribution in order to tie the final classification to the observed images. The label field has spatio-temporal cliques and the displacement vectors are taken into account by inter-frame (or temporal) cliques. In the DF's energy function we use the label field instead of computing the color differences of corresponding pixels in order to reduce computing time. The energy function is minimized through a Metropolis algorithm [10] and we obtain the classification of the frames *and* the displacement vectors at the same time. The algorithm is unsupervised; only the number of classes is supplied by the user. The method has been tested on a variety of color video sequences and the results are encouraging.

References

1. J. Besag. On the statistical analysis of dirty pictures. *Jl. Roy. Statis. Soc. B.*, 1986.
2. M. J. Daily. Color Image Segmentation Using Markov Random Fields. In *Proc. DARPA Image Understanding*, 1989.
3. D. Geman. Bayesian Image Analysis by Adaptive Annealing. In *Proc. IGARSS'85*, pages 269–277, Amherst, USA, Oct. 1985.
4. F. Heitz and P. Bouthemy. Multimodal Estimation of Discontinuous Optical Flow Using Markov Random Fields. *IEEE-PAMI*, 15(12):1217–1232, Dec. 1993.
5. A. K. Jain. *Fundamentals of Digital Image Processing*. Prentice Hall, 1989.
6. Z. Kato, M. Berthod, J. Zerubia, and W. Pieczynski. Unsupervised Adaptive Image Segmentation. In *ICASSP'95*, Detroit, USA, May 1995.
7. Z. Kato, T. C. Pong, and J. C. M. Lee. Motion Compensated Color Image Classification and Parameter Estimation in a Markovian Framework. Technical Report HKUST-CS97-04, The Hong Kong University of Science and Technology, July 1997.
8. S. Lakshmanan and H. Derin. Simultaneous Parameter Estimation and Segmentation of Gibbs Random Fields Using Simulated Annealing. *IEEE–PAMI*, 11(8):799–813, Aug. 1989.
9. J. Liu and Y. H. Yang. Multiresolution Color Image Segmentation. *IEEE-PAMI*, 16(7):689–700, July 1994.
10. N. Metropolis, A. Rosenbluth, M. Rosenbluth, A. Teller, and E. Teller. Equation of state calculations by fast computing machines. *J. of Chem. Physics, Vol. 21, pp 1087-1092*, 1953.
11. D. W. Murray and B. F. Buxton. Scene Segmentation from Visual Motion Using Global Optimiziation. *IEEE-PAMI*, 9(2):220–228, Mar. 1987.

12. D. K. Panjwani and G. Healey. Markov Random Field Models for Unsupervised Segmentation of Textured Color Images. *IEEE - PAMI*, 17(10):939–954, Oct. 1995.
13. M. I. Sezan and R. L. Lagendijk, editors. *Motion Analysis and Image Sequence Processing*, chapter E. Dubois and J. Konrad: Estimation of 2-D Motion Fields from Image Sequences with Application to Motion-Compensated Processing, pages 53–88. Kluwer Academic Publishers, 1993.

Model	Method	Num. of iterations	CPU time
color "tennis" sequence (23 frames, 6 classes)			
Static	ICM	9	0.61 hours
Static	Metropolis	58	1.94 hours
Motion compensated	Metropolis	400	26.4 hours
gray-level "tennis" sequence (23 frames, 6 classes)			
Motion compensated	Metropolis	400	19.6 hours
color "car" sequence (21 frames, 4 classes)			
Static	ICM	12	0.53 hours
Static	Metropolis	67	2.02 hours
Motion compensated	Metropolis	400	21.8 hours

Table 1. Computing times on a SPARC station 1000.

Original frames   Static   Motion compensated

Fig. 1. Results obtained by the static and motion compensated model (21 frames, 4 classes) using the Metropolis algorithm.

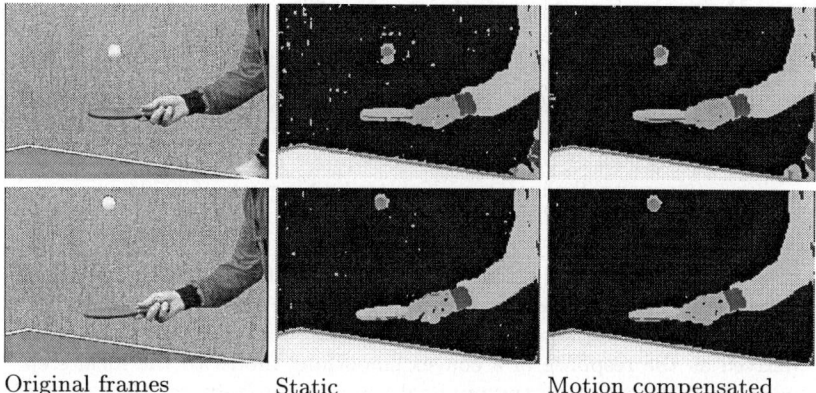

Original frames   Static   Motion compensated

Fig. 2. Results obtained by the static and motion compensated model using the Metropolis algorithm on the "tennis" sequence (23 frames, 6 classes).

Original gray-level frame   Intensity-based   Color-based

Fig. 3. Comparison of intensity- and color-based classification results on the "tennis" sequence (23 frames, 6 classes) using the motion compensated model with the Metropolis algorithm.

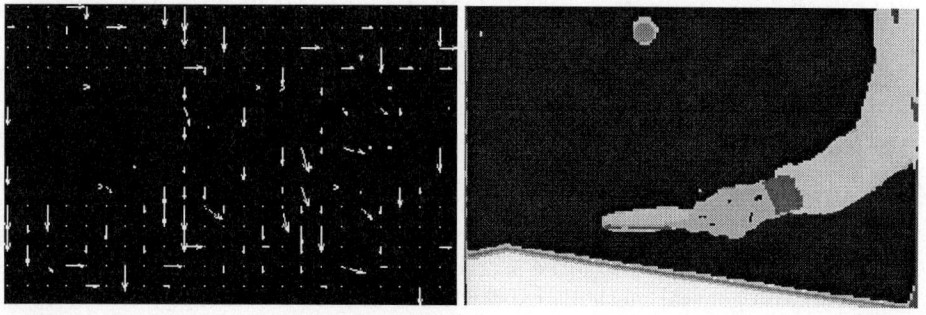

Fig. 4. Classification and displacements obtained on the "tennis" sequence (23 frames, 6 classes) using the Metropolis algorithm.

Edge-Preserving Smoothing by Convex Minimization

S.Z. Li, Y.H. Huang, J.S. Fu, K.L. Chan

School of Electrical and Electronic Engineering
Nanyang Technological University, Singapore 639798

Abstract. This work presents a new approach for analyzing the problem of edge-preserving image smoothing using convex minimization and for selecting smoothing parameters. The close-form (global) solution is derived as the response of a convex smoothing model to the ideal step edge. Insights into how the minimal solution responds to edges in the data and how the parameter values affect resultant edges in the solution are drawn from the analytic expression of the close-form solution. Based on this, a scheme is proposed for selecting parameters to achieve desirable response at edges.

1 Introduction

Edges correspond to discontinuities in the image function and in other image and shape properties over the space and time. Edge-preserving image smoothing provides basis for later processing such as shape recovery, flow and motion estimation, segmentation, active contours, feature detection. It has been one of the most active research areas in image processing and vision.

Edge-preserving image smoothing can be formulated in the regularization or Bayesian framework as minimizing an energy function. The role of an energy function is twofold: (i) as the quantitative measure of the global quality of the solution and (ii) as a guide to the search for a minimal solution. Regard the solution quality, the best result that can be achieved is the maximum *a posteriori* (MAP) estimate when information in both the prior and likelihood distributions is available. The energy function in a posterior typically consists of two terms: one due to the likelihood and the other due to the prior. The truly MAP solution is the *global* minimum of the energy function.

The role of an energy function as a guide to the search may or may not be fully played when a global minimum is the objective. When an energy function is smooth and convex, global minimization is equivalent to local minimization for which the energy descent is sufficient. In this case, the energy function can be fully utilized as a guide to the search. However, when the energy function is nonconvex, no general methods exist which can utilize the energy efficiently to guide the search, and an unfavorable local minimum may result.

Nonconvex energy models for edge-preserving smoothing such as the well-known line-process model and alike have this problem in the minimization process. In recent years, there has been considerable interest in convex models for edge-preserving smoothing with the edge-preserving ability [12, 3, 5, 2, 13, 7].

This class of models overcome the above mentioned disadvantages. The convexity guarantees the unique solution and its stability. Thus local minimization techniques can be efficiently utilized, making continuation or annealing unnecessary.

An important issue in edge-preserving smoothing and segmentation is parameter selection. Different parameters gives different models. They affect edge-preserving smoothing behavior of a model at edges. To overcome the needs for *ad hoc* heuristics, methods have been proposed, *e.g.* using learning parameters from edge-labeled image data based on maximum likelihood [8], or choosing proper filter scales using minimization [4].

This paper contains the following ingredients. Firstly, it gives conditions for an MAP smoothing model to be both edge-preserving and convex. A class of convex models is defined in terms of a system of nonlinear equations constrained by the class of interaction functions capable of preserving edges. Secondly, the response of the convex energy model with Huber function to the ideal step edge is derived in close-form. The analytic, close-form solution provides insights into how the solution responds to edges. It tells us how to predicate the minimum size of edges in the data that can be detected, and the size of detected edges in the solution.

2 Convex Models

A prior potential function g defines a model in terms of minimizing an energy

$$E(f) = U(d \mid f) + U(f) = \sum_i (f_i - d_i)^2 + \frac{\lambda}{2} \sum_i \sum_{i' \in \mathcal{N}_i} g(f_i - f_{i'}) \qquad (1)$$

A g function is usually chosen to be some function of η^2 and so its can be expressed in the following form:

$$g'(\eta) = 2\eta h(\eta) \qquad (2)$$

where h, called an *interaction function*, determines model behaviors. To minimize $E(f)$, it is necessary that the gradient vanish

$$\frac{1}{2} \frac{\partial E(f)}{\partial f_i} = (f_i - d_i) + \lambda \sum_{i' \in \mathcal{N}_i} (f_i - f_{i'}) h(f_i - f_{i'}) = 0 \qquad (3)$$

The above is the *model* defined in terms of a system of equations. Its 2D analogue is $[f_{i,j} - d_{i,j}] + \lambda \sum_{(i',j') \in \mathcal{N}_{i,j}} (f_{i,j} - f_{i',j'}) h(f_{i,j} - f_{i',j'}) = 0$. We call the magnitude $|g'(f_i - f_{i'})| = |2(f_i - f_{i'}) h(f_i - f_{i'})|$ the *smoothing strength* with which $f_{i'}$ influences f_i.

The smoothing strength at discontinuities must be bounded in order for edges to be preserved. A discontinuity between two neighboring pixels is manifested by a large derivative, $f_i - f_{i'} \to \infty$. In a model without the edge-preserving ability, f_i can have an infinite influence on $f_{i'}$, and vice versa, when a discontinuity exists between them. In this case, there must be $\lim_{(f_i-f_{i'})\to\infty}(f_i-f_{i'})h(f_i-f_{i'}) = \infty$. For example, in the quadratic model where $g(\eta) = \eta^2$, the interaction is constant everywhere, $h(\eta) = 1$, and the smoothing strength is $|2\eta| \to \infty$ at discontinuities, causing oversmoothing. We have the following necessary condition for a model to be *adaptive* to discontinuities.

Axiom 1 *For edges to be preserved in the energy minimum, it is necessary that the following condition is satisfied [6]:*

$$\lim_{\eta\to\infty} |g'(\eta)| = \lim_{\eta\to\infty} |2\eta h(\eta)| = C \qquad (4)$$

where $C \geq 0$ is a constant.

Smoothing at discontinuities is entirely prohibited when $C = 0$, whereas limited smoothing is allowed when $C > 0$. In the line process model with $g(\eta) = \min\{\eta^2, \alpha\}$, the interaction is $h(\eta) = 1$ if $|\eta| < \sqrt{\alpha}$ or 0 otherwise, and the smoothing strength is $C = 0$ when $|\eta|$ exceeds $\sqrt{\alpha}$. For Huber's robust function $g_1(\eta) = \min\{\eta^2, 2\gamma|\eta|-\gamma^2\}$, the interaction is $h(\eta) = 1$ for $|\eta| \leq \gamma$ or $h(\eta) = \gamma/|\eta|$ otherwise, and the smoothing strength is $C = 2\gamma$. Both edge-preserving models satisfy (4).

To satisfy (4), it is necessary that $h(\eta)$ should approach 0 as $\eta \to \infty$. But how fast? For $C = 0$, $h(\eta)$ must be a higher order infinitesimal than $\frac{1}{\eta}$ as $\eta \to \infty$, that is, $\lim_{\eta\to\infty} h(\eta) = O(\frac{1}{\eta})$. For $C > 0$, they must be infinitesimal of the same order, that is, $\lim_{\eta\to\infty} h(\eta) \sim \frac{1}{\eta}$. The higher order infinitesimal $h(\eta)$ is, the stronger it inhibits smoothing over edges, the sharper the edges. However, if $h(\eta)$ approaches 0 too fast, the corresponding $g(\eta)$ will be nonconvex. So, compromises have to be made between the sharpness of edges and the convexity of the model. For the convexity, we require that $h(\eta)$ satisfy (4) with $C > 0$ for reasons to be explained.

Definition 1. A function h_γ is said to be a *convex adaptive interaction function* (CAIF) parameterized by γ (> 0) if it satisfies: (i) $h_\gamma \in C^0$, (ii) $h_\gamma(\eta) = h_\gamma(-\eta)$, (iii) $h_\gamma(\eta) > 0$, (iv) $h'_\gamma(\eta) \leq 0$ ($\forall \eta > 0$), (v) $(\eta h_\gamma(\eta))' \geq 0$, (vi) $\lim_{\eta\to\infty} |2\eta h_\gamma(\eta)| = C > 0$. The corresponding potential function is $g_\gamma = 2\int_0^\eta \eta' h_\gamma(\eta')d\eta'$. The class of CAIFs, denoted $\mathbb{H}_\gamma$, is the family of all such h_γ. □

The continuity requirement (i) avoids instability of the solution of Eq.(3). The evenness of (ii) is usually assumed for spatially unbiased smoothing. The positive definiteness of (iii) keeps the interaction positive such that the sign of $\eta h_\gamma(\eta)$ will not be altered by $h_\gamma(\eta)$. The monotony of (iv) leads to decreasing interaction (weight) for increasing difference $|\eta| = |f_i - f_{i'}|$. The convexity is guaranteed by (v) which is equivalent to $g''(\eta) \geq 0$. The bounded asymptote of (vi) is for the edge-preserving purpose as stated earlier. Because $h_\gamma(\eta) > 0$, there must be $\lim_{\eta\to\infty} |2\eta h_\gamma(\eta)| = C > 0$. Other properties can also be derived, such as $h_\gamma(0) = \sup_\eta h_\gamma(\eta)$ and $h'_\gamma(0) = 0$.

Theorem 2. *An energy function $E(f)$ with $g = g_\gamma$ is convex.*

Proof: The sum of two convex functions are convex. The first term $U(f \mid d)$ on the RHS of (1) is convex. Hence the convexity of the second term $U(f)$ is sufficient for the convexity of $E(f)$. From property (v) of Definition 1, we have $g_\gamma''(\eta) \geq 0$. Thus $g_\gamma(\eta)$ is convex and so $g_\gamma(\mu u + (1-\mu)v) \leq \mu g_\gamma(u) + (1-\mu)g_\gamma(v)$. Let $x = (x_1, \ldots, x_m)$ and $y = (y_1, \ldots, y_m)$. Because $\frac{1}{\lambda}U(\mu x + (1-\mu)y) = \sum_i \sum_{i' \in \mathcal{N}_i} g_\gamma(\mu(x_i - x_{i'}) + (1-\mu)g_\gamma(y_i - y_{i'})) \leq \sum_i \sum_{i' \in \mathcal{N}_i} [\mu g_\gamma(x_i - x_{i'}) + (1-\mu)g_\gamma(y_i - y_{i'})] = \frac{1}{\lambda}[\mu U(x) + (1-\mu)U(y)]$, $U(f)$ is convex. Therefore, $E(f) = U(f \mid d) + U(f)$ is convex. □

Definition 3. *The system of equations (3) is said to be a convex adaptive model (CAM) if it constrained by $h \in \mathbb{H}_\gamma$.* □

The above defines a class of models which are both convex and edge-preserving. The definition is a broad one. It states the properties that a model should possess, rather than just gives a functional instance. It also includes those models whose corresponding g_γ (and hence the energy) cannot be written in closed-form. This is an advantage of defining the CAM in terms of (3) constrained by $h \in \mathbb{H}_\gamma$ as compared to that in terms of $E(f)$. In practice, it is not necessary to get the explicit expression of g_γ for the purpose of defining the solution. Nonetheless, knowing g_γ is useful for analyzing the convexity of $E(f)$.

3 The Global Solution and Parameters

The model with Huber function g_1 is chosen as the subject of study because it is simple in form and produces good results. Its response (the minimal energy solution) to the ideal step edge is derived in close-form. An analysis of the close-form solution gives insights into how it responds to the edge and this provides guidelines for selecting the involved parameters for desirable, predictable response to edges.

The model of ideal step edge has been used for analyzing image smoothing and edge detection problems (see *e.g.* [11, 1, 9, 10, 4]). In the following, the data $d = (\ldots, d_1, d_2, \ldots, d_m, \ldots)$ is also modeled as an ideal step edge as illustrated in Fig. 1: $d_i = L$ for $i \leq k-1$, $d_i = H$ for $i \geq k$, where $L < H$. The step size in the data is $D_d = H - L$, and the step size in the solution is $D_f = f_k - f_{k-1}$. The global solution is given in the following theorem of which the proof also serves as an analysis.

3.1 The Global Minimum in Close-Form

Theorem 4. *Let $E(f)$ be defined with g_1 and λ, and d as the above. Then the global energy minimum $f = (\ldots, f_1, \ldots, f_m, \ldots)$ in close-form is given as follows:*

At $k-1$ and k,

$$f_{k-1} = \begin{cases} L + (\frac{\epsilon}{1+\epsilon})D_d & \text{if } D_d \leq T_d \\ L + (\frac{\epsilon}{1-\epsilon})\gamma & \text{if } D_d > T_d \end{cases} \quad (5)$$

$$f_k = \begin{cases} H - (\frac{\epsilon}{1+\epsilon})D_d & \text{if } D_d \leq T_d \\ H - (\frac{\epsilon}{1-\epsilon})\gamma & \text{if } D_d > T_d \end{cases} \quad (6)$$

At other points,

$$f_{k-j} = L - \epsilon^{j-1}(L - f_{k-1}) \qquad j = 2, 3, \ldots \quad (7)$$
$$f_{k+j} = H - \epsilon^j(H - f_k) \qquad j = 1, 2, \ldots \quad (8)$$

where

$$\epsilon = \epsilon(\lambda) = \frac{2\lambda + 1 - \sqrt{4\lambda + 1}}{2\lambda} \quad (9)$$

and

$$T_d = (\frac{1+\epsilon}{1-\epsilon})\gamma \quad (10)$$

Proof: For the given data, Huber's function g_1 works locally in one of two "modes": the quadratic mode and the linear mode. The quadratic mode is effective in the flat regions of the data, and at the step of a small size for which $D_f \leq \gamma$. The linear mode is effective only at the step of a large size for which $D_f > \gamma$. The corresponding interaction function is $h(\eta) = \min\{1, \frac{\gamma}{|\eta|}\}$.

In the quadratic mode, the solution is uniquely determined by the following system of linear equations

$$f_i - d_i - \lambda(f_{i+1} - 2f_i + f_{i-1}) = 0 \quad \forall i \quad (11)$$

Letting $\nabla_i = d_i - f_i$, the above system of equations can be rewritten as

$$\nabla_i + \lambda(2\nabla_i - \nabla_{i+1} - \nabla_{i-1}) = 0 \quad \forall i \quad (12)$$

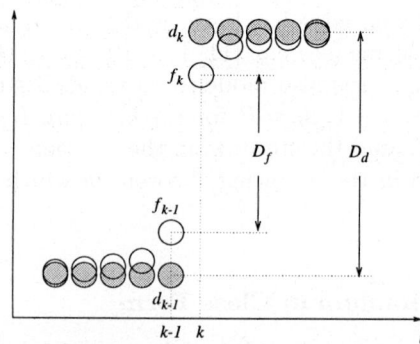

Fig. 1. The step edge data d (in shaded dots) and the solution f (in open dots). D_d is the step size in d. D_f is the step size in f.

Assuming that f_{k-1} is fixed, then the upper part of the solution $(f_{k+1}, f_{k+2}, \ldots)$ are given by
$$\nabla_i = \epsilon \nabla_{i-1} \quad \text{for } i \geq k+1 \tag{13}$$
with ϵ of (9). This can be easily verified to satisfy (12). From (13), we have $\nabla_{k+j} = \epsilon^j \nabla_k$ for $j \geq 1$, and hence (8). Similarly, assuming that f_k is fixed, it can be verified that the system (12) is satisfied by the lower part of the solution $(\ldots, f_{k-3}, f_{k-2})$ given by
$$\nabla_i = \epsilon \nabla_{i+1} \quad \text{for } i < k-1 \tag{14}$$
From (14), we have $\nabla_{k-j} = \epsilon^{j-1} \nabla_{k-1}$ for $j \geq 2$, and hence (7).

In the quadratic mode, f_{k-1} and f_k are the boundary values that satisfy the following two equations
$$\begin{cases} \nabla_k = d_k - f_k = \epsilon(H - f_{k-1}) \\ \nabla_{k-1} = d_{k-1} - f_{k-1} = \epsilon(L - f_k) \end{cases} \tag{15}$$
the first one being related to (13) and the second one to (14). Solving these gives $f_{k-1} = L + (\frac{\epsilon}{1+\epsilon}) D_d$ and $f_k = H - (\frac{\epsilon}{1+\epsilon}) D_d$, and hence the first lines of (5) and (6).

In (5) and (6), T_d is the minimum step size in d that enables Huber function to work in the linear mode locally. In a convex model, the step size in f, $D_f = f_k - f_{k-1}$, varies continuously with respect to D_d, and therefore the value of D_f at transition between the two modes swap $D_f = \gamma$ or $D_d = T_d$. From the f_k and f_{k-1} derived for the quadratic mode, we have $D_f = \frac{1-\epsilon}{1+\epsilon} D_d$. Equating it to γ, we obtain the transitional value of D_d as $T_d = \frac{1+\epsilon}{1-\epsilon} \gamma$ given in (10).

The linear mode is turned on between $k-1$ and k when $D_f > \gamma$ or $D_d > T_d$. According to (3), the local solution in this case is determined by the following:
$$f_{k-1} - d_{k-1} + \lambda[-\gamma + (f_{k-1} - f_{k-2})] = 0 \tag{16}$$
$$f_k - d_k + \lambda[\gamma + (f_k - f_{k+1})] = 0 \tag{17}$$
Substituting $f_{k-2} = L - \epsilon(L - f_{k-1})$, $d_{k-1} = L$, $f_{k+1} = H - \epsilon(H - f_k)$ and $d_k = H$, and re-arranging the equations, we obtain
$$f_{k-1} = L + \frac{2\lambda\gamma}{1+\sqrt{4\lambda+1}} \tag{18}$$
$$f_k = H - \frac{2\lambda\gamma}{1+\sqrt{4\lambda+1}} \tag{19}$$
These can be rewritten as the second lines of (5) and (6). To see this, we need only to prove $\frac{\epsilon}{1-\epsilon} = \frac{2\lambda}{1+\sqrt{4\lambda+1}}$. We can do this by showing $\epsilon(1+\sqrt{4\lambda+1}) = 2\lambda(1-\epsilon)$. We find both sides of this equation equal to $2\lambda(1 - \sqrt{4\lambda+1}$ after expanding $\epsilon = \epsilon(\lambda)$ by (9). Hence the second lines of (5) and (6) are proven. □

The above is for $f = \{f_i\}$ and $d = \{d_i\}$ of infinite length extending from $i = -\infty$ to $i = \infty$. The solution at the boundaries can be found as $f_{k-\infty} = L$ and $f_{k+\infty} = H$ because ϵ takes a value in $(0, 1)$ for $\lambda > 0$. When $f = (f_1, \ldots, f_m)$

is of finite length, the above theorem provides a good approximation. We can see this by examining ϵ^j in (7) and (8). Assuming $\lambda = 10$, then $\epsilon = 0.73$ and $\epsilon^j = 0.042$ for $j = 10$. This means that f_i are quite close to the boundary values when i is 10 points apart from the edge.

Theorem 5. *For the solution f given in Theorem 4, the resultant step size $D_f = f_k - f_{k-1}$ is*

$$D_f = \begin{cases} D_f^q = (\frac{1-\epsilon}{1+\epsilon})D_d & \text{if } D_d \leq T_d \\ D_f^l = D_d - \Delta & \text{if } D_d > T_d \end{cases} \quad (20)$$

where

$$\Delta = \frac{2\epsilon\gamma}{1-\epsilon} = \frac{2\epsilon T_d}{1+\epsilon} \quad (21)$$

Proof: Subtracting (5) from (6) yields D_f. □

D_f^q explains several observations made in smoothing using the quadratic regularization: (i) when $\lambda = 0$, $\epsilon(\lambda) = 0$ and $R(\lambda) \triangleq \frac{D_f^q}{D_d} = 1$, meaning the step sizes in f and d are the same; (ii) when $\lambda > 0$, $\epsilon'(\lambda) > 0$ and $R(\lambda) < 1$, implying D_f^q is smaller than D_d; and (iii) $R'(\lambda) < 0$, implying that D_f^q decreases with λ. The step size change rate $\frac{dD_f^l}{dD_d} = 1$ in the linear mode, meaning D_f^l changes as much as D_d does, whereas in the quadratic mode, $\frac{dD_f^q}{dD_d} < 1$.

Because the edge positions k are not known in advance and the data are always subject to noise, in pract.ce, the solution cannot be really calculated analytically from the data. However, the theoretic results give insights into how the solution is related to edges and also how to determine suitable parameter values for detecting edges of certain sizes.

4 Conclusions

Convex models have the advantages of the uniqueness of solution and the computational efficiency. We have defined a class of convex models for edge-preserving smoothing in terms of a system of nonlinear equations constrained by a class of interaction functions given in Definition 1. The close-form solution of the model with Huber function in response to the ideal step edge gives insights into how the parameters affect the edge-preserving smoothing. The critical value T_d in (10), or equivalently the loss in step size Δ in (21), is identified to be of critical important in the edge-preserving property. The T_d or Δ is determined precisely by λ and γ, by which the minimal size of detectable edges can be predicated. Similar response at the edges can be maintained by fixing T_d even if λ and γ are changed to a large extent. The influence of λ on the solution is global whereas that of γ is limited near the edges.

Acknowledgement This work was supported by NTU project RG 43/95.

References

1. A. Blake and A. Zisserman. *Visual Reconstruction.* MIT Press, Cambridge, MA, 1987.
2. C. Bouman and K. Sauer. "A generalized Gaussian image model for edge preserving MAP estimation". *IEEE Transactions on Image Processing,* 2(3):296–310, July 1993.
3. P. Green. "Bayesian reconstructions from emission tomography data using a modified EM algorithm". *IEEE Transactions on Medical Imaging,* 9(1):84–93, March 1990.
4. H. Jeong and C. I. Kim. "Adaptive determination of filter scales for edge detection". *IEEE Transactions on Pattern Analysis and Machine Intelligence,* 14:579–585, 1992.
5. K. Lange. "Convergence of EM image reconstruction algorithm with Gibbs smoothing". *IEEE Transactions on Medical Imaging,* 9(4):439–446, December 1990.
6. S. Z. Li. "On discontinuity-adaptive smoothness priors in computer vision". *IEEE Transactions on Pattern Analysis and Machine Intelligence,* 17(6):576–586, June 1995.
7. S. Z. Li, Y. H. Huang, and J. Fu. "Convex MRF potential functions". In *Proceedings of IEEE International Conference on Image Processing,* volume 2, pages 296-299, Washington, D.C., 23-26 October 1995.
8. S. G. Nadabar and A. K. Jain. "Parameter estimation in Markov random field contextual models using geometric models of objects". *IEEE Transactions on Pattern Analysis and Machine Intelligence,* 18:326–329, 1996.
9. P. Perona and J. Malik. "Scale-space and edge detection using anisotropic diffusion". *IEEE Transactions on Pattern Analysis and Machine Intelligence,* 12(7):629–639, July 1990.
10. M. Petrou and J. Kittler. "Optimal edge detection for ramp edges". *IEEE Transactions on Pattern Analysis and Machine Intelligence,* 13(5):483–490, 1991.
11. I. Pitas and A. N. Venetsanopoulos. "Edge detectors based on nonlinear filters". *IEEE Transactions on Pattern Analysis and Machine Intelligence,* 8:538–550, 1986.
12. D. Shulman and J. Herve. "Regularization of discontinuous flow fields". In *Proc. Workshop on Visual Motion,* pages 81–86, 1989.
13. R. L. Stevenson, B. E. Schmitz, and E. J. Delp. "Discontinuity preserving regularization of inverse visual problems". *IEEE Transactions on Systems, Man and Cybernetics,* 24(3):455–469, March 1994.

Author Index

A

Abdallah, Samer M. I-386
Abdellatif, Mohamed.........I-208
Abe, Keiichi..........................I-450
Abe, Norihiro.....................II-487
Achermann, B. II-726
Adachi, Hiroaki I-688
Adán, Antonio....................I-482
Aggarwal, J.K. II-275
Ahuja, Narendra I-623
    II-33
    II-291
    II-323
Akutsu, Takashi I-321
Alexandrov, V.V. I-426
Alferez, Ronald-Bryan....I-542
Aloimonos, Yiannis.........II-283
Anderson, J.A.D.W. II-177
Arakawa, Kenichi................I-321
Ariki, Yasuo II-695
Åström, Kalle.....................II-169

B

Bab-Hadiashar, Alireza..I-566
    II-599
Bai, Xuesheng......................I-240
Bajcsy, P. II-291
Beigi, Homayoon S.M. I-531
Bennet, Karen....................II-575
Berger, Marie-Odile II-360
Bhattacharyya, P.............I-590
Bhuiyan, Md. Shoaib.........II-25
Bilbro, Griff L. I-722
Binkert, M. II-726
Black, Michael S. II-267
Blum, Stefan A. I-128
Bolle, Ruud............................I-2
    II-283
Booth, William I-176
    II-623
Bottino, A. II-416
Bouvin, Christophe..........I-640

Brady, Michael.....................I-120
    I-152
    II-41
Brown, Michael S. I-558
Bryll, Robert K. II-591
Bunke, Horst II-299
    II-726
Burschka, Darius................I-128
Buxton, Hilary I-523
Byne, J.H.M. II-177

C

Caelli, Terry II-551
Califano, Andrea I-32
Cerrada, Carlos................I-482
Cha, Sung-Hyuk....................I-370
Chaen, Atsushi I-288
Chai, Jinxiang I-272
Chan, Chorkin II-121
Chan, Kap Luk I-746
    II-225
Chan, S.L. I-714
Chan, Syin I-402
Chang, Chun-Ming...............I-192
Chaumette, François.........I-648
Chen, Dongwei I-192
Chen, Fang..........................II-49
Chen, Liang-Hua II-742
Chen, Meng Chang II-742
Chen, Ming I-168
Chen, Tie Qi II-631
Chen, Youbin......................I-160
    I-168
Chen, Yunqiang II-9
Chen, Zen I-632
Chow, S.K. I-714
Chuang, Jen-Hui II-535
Chung, Jae-Moon I-346
Cipolla, Roberto................I-515
Collina, Costantino.........I-338
Colville, Scott I-32
Cui, Yuntao..........................I-418

Cumani, Aldo I-582

D

Daum, M. I-72
Davis, Larry S. II-201
　　　　　　　　　　　　　　　II-267
Deguchi, Koichiro..............I-48
　　　　　　　　　　　　　　　I-56
Dennis, Tim J. I-264
Deriche, Rachid I-640
Ding, Xiaoqing I-160
　　　　　　　　　　　　　　　I-168
　　　　　　　　　　　　　　　II-145
Dudek, G. I-72
Duric, Zoran I-305

E

Eason, Richard O. I-112
Endoh, Toshio II-1

F

Faugeras, Olivier..............I-640
Fejes, Sandor II-267
Feliu, Vicente I-482
Fermüller, Cornelia II-283
Fernandes, João L. I-64
Ferri, Massimo I-329
　　　　　　　　　　　　　　　I-338
Fountain, S.R. II-57
Frenkel, B.E. I-426
Frosini, Patrizio I-329
　　　　　　　　　　　　　　　I-338
Fu, J.S. I-746
Funahashi, Tatsushi............I-474
Funt, Brian II-257

G

Germain, Bob...................I-32
Gillies, M. I-714
Gofuku, Akio I-208
Gong, Shaogang I-507
　　　　　　　　　　　　　　　I-607
　　　　　　　　　　　　　　　I-615
　　　　　　　　　　　　　　　II-679
Goss, Simon II-551

Greiffenhagen, Michael..I-418
Grimson, Eric I-498
Gross, Ari I-184
Gu, Jin II-455
Gu, Weikang I-144
　　　　　　　　　　　　　　　II-511
Guibas, Leonidas J. I-104
Guo, Fanxia II-145

H

Han, Chin-Chuan................II-742
Hanmandlu, M. I-458
　　　　　　　　　　　　　　　I-466
Hariatoglu, Ismail............II-267
Harley, Eric II-575
Harwood, David II-267
Hasegawa, Tsutomu..............I-442
Hata, Tadashi II-639
Hauta-Kasari, M. I-248
Hayashi, Jun-ichiro II-1
Hedley, Mark II-376
　　　　　　　　　　　　　　　II-392
Heyden, Anders II-169
Hiura, Shinsaku II-495
Hong, Lin I-16
Hoshide, Tsuyoshi..............I-442
Hotta, Kazuhiro II-89
Howell, A. Jonathan..........I-523
Hu, Wei I-362
Hu, Zhencheng..................II-519
Huang, Chung-Lin...............I-706
Huang, Qian....................I-418
Huang, Thomas S................II-455
　　　　　　　　　　　　　　　II-463
Huang, Y.H. I-746
Hügli, H. I-490
Hung, Yi-Ping..................I-632
Hwang, Wey-Shiuan.............II-503

I

Igi, Seiji I-698
Ikeuchi, Katsushi II-209
　　　　　　　　　　　　　　　II-350
　　　　　　　　　　　　　　　II-408
　　　　　　　　　　　　　　　II-424

Imagawa, Kazuyuki I-698
Inokuchi, Seiji................II-495
Ipson, Stanley S..............I-176
Ishii, Naohiro....................I-40
Ishikawa, Noriyuki..........II-695
Ishikawa, Seiji I-354
Ishikawa, Takahiro..........II-671
Ishizawa, Akira I-688
Itoh, Hidenori....................I-474
Iwahori, Yuji I-40
                               II-25
Iwai, Yoshio.....................II-639
                              II-647
Iwata, Akira II-25

J

Jain, Anil I-16
Jenkin, Michael R.M. I-656
Jenkinson, Mark................II-41
Jia, Jin I-450
Jiang, Xiaoyi....................II-299
                              II-726
Jo, Kang-Hyun II-368
Jojic, Nebojsa II-455
                              II-463
Jost, T. I-490
Jurie, Frederic.................II-440

K

Katafuchi, Norifumi.........I-200
Kato, Hirokazu.................II-495
Kato, Kunihito II-1
Kato, Zoltan I-738
Katsuyama, Yutaka..........II-137
Kaveti, Satish...................I-378
Kawaguchi, Eiji I-112
Kim, Chang-Hun II-241
Kim, Eun Yi......................I-730
Kim, Hang Joon...............I-730
Kim, Hyoung Seop...........I-354
Kim, Jin Wook..................I-730
Kimmel, R. I-88
                               I-574
King, Irwin.......................I-410
                              II-559

Kirihara, Satoshi II-448
Kittler, J...........................II-307
Koshimizu, Hiroyasu II-1
                              II-663
Krishnamoorthi, R. I-590
Kuno, Yoshinori I-672
                              II-368
Kurita, Takio.....................II-89

L

Lai, Kok F. I-402
Laikov, E.V. I-426
Laine, Andrew..................I-192
Lan, Zhong-Dan I-313
Lang, J. I-656
Lang, S. I-136
Latecki, Longin I-184
Lau, Tak Kan....................I-410
Laurentini, A. II-416
Lee, John Chung-Mong I-738
Lee, Mi-Suen II-315
Lee, Sang Uk....................I-96
Lenz, R. I-248
Leow, Wee Kheng............II-17
Li, Fuxing I-120
Li, Stan Ziqing I-746
                              II-225
Li, Yi...............................II-719
Liang, Ping.....................II-400
Liao, H.Y. Mark...............II-742
Liao, Simon X. I-394
Lien, Cheng-Chang............I-706
Lin, Xiaofan......................I-160
Liu, Jinhui........................I-160
Liu, Peilin II-209
Liu, Tianrong..................II-225
Liu, Yongmei II-153
Loft, Peter J. II-679
Lourakis, Manolis............II-527
Lourens, Tino II-193
Lovato, Alberto I-329
Lu, G.Z. I-136
Lu, Hanqing....................II-719
Lu, Shan I-698
Lu, Weier II-687

Lu, Yi II-631
Luk, W.S. II-559

M

Ma, SongDe I-136
　　　　　　　　　　　　I-272
　　　　　　　　　　　　II-9
　　　　　　　　　　　　II-607
　　　　　　　　　　　　II-719
Maes, Stéphane H. I-531
Malladi, R. I-574
Marchand, Éric I-648
Maru, Noriaki I-256
Masai, Hiroyuki II-105
Masegawa, Tsutomu II-185
McIvor, Alan M. I-434
McKenna, Stephen J. I-507
　　　　　　　　　　　　I-607
　　　　　　　　　　　　I-615
Medioni, Gérard II-315
Melhi, Muhammed I-176
Menard, Christian I-550
Michalski, Ryszard S. I-305
Miller, Jason II-631
Minoh, Michihiko II-332
Mirmehdi, M. II-307
Mishima, Taketoshi II-89
Mita, Takeshi II-495
Miura, Jun I-672
Miyazaki, Fumio I-256
Miyazaki, Tsuyoshi I-474
Mohr, Roger I-313
Mokhtarian, Farzin II-73
Mori, Hiroki II-655
Mori, Takeaki II-471
Morishima, Shigeo II-671
Morooka, Ken'ichi II-185
Motomura, Nachi I-354
Mukai, Toshiharu II-583
　　　　　　　　　　　　II-734
Mukaigawa, Yasuhiro I-680
Murakami, Kazuhito II-1
　　　　　　　　　　　　II-663
Murakami, Masamitsu I-40
Murase, Hiroshi I-321

Murata, Akio II-487

N

Nagai, Isaku I-208
Nakamura, Yuichi I-680
Nakatani, Hiromasa II-711
Nalwa, Vishvjit I-10
Naoi, Satoshi II-137
Nathan, Krishna I-1
Nebot, Eduardo M. I-386
Nevatia, Ram II-259
Ngan, Phillip M. II-615
Niimi, Michiharu I-112
Nishijima, Masakazu II-217
Nishikawa, Atsushi I-256
Nishimura, T. II-734
Noras, James M. II-623
Noronha, Sanjay II-259
Nozaki, Koichi I-112

O

Ohara, Shuichi I-200
Ohba, Kohtaro II-424
Ohnishi, Noboru I-232
　　　　　　　　　　　　I-346
　　　　　　　　　　　　II-153
　　　　　　　　　　　　II-583
Ohta, Hiroshi II-711
Ohta, Yuichi I-680
Ohya, Jun II-655
　　　　　　　　　　　　II-703
Oka, R. II-734
Okatani, Takayuki I-48
　　　　　　　　　　　　I-56
Okudaira, Masashi I-200
Ong, Eng-Jon II-679
Or, S.H. II-559
Orphanoudakis, Stelios ... II-527
Oshiro, Naoki I-256
Osumi, Noriyoshi II-471
Otsuka, Takahiro II-703

P

Pagliari, Carla L. I-264
Palmer, P.L. II-307

Pankanti, S. I-2
Park, In Kyu I-96
Park, Se Hyun I-730
Parkkinen, J. I-248
Pawlak, Miroslaw................I-394
Peake, G.S. II-97
Pearce, Adrian R. II-551
Perrin, Benoit II-323
Pong, Ting-Chuen................I-738
Porcellini, Eleonora I-338
Pramadihanto, Dadet II-647
Priese, Lutz I-598
Psarrou, Alexandra...........I-664

Q

Qi, Hairong I-722
Qiu, G. II-81
Quan, Long I-313
Quek, Francis K.H. II-591

R

Raja, Yogesh I-607
　　　　　　　　　　　　　　I-615
Ramakrishna, R.S. I-296
Ratakonda, Krishna............II-33
Ratha, Nalini I-2
Rehrmann, Volker................I-598
Rubner, Yossi I-104
Rye, David C. I-386

S

Saito, Hideo II-448
Saito, Takafumi II-471
Saji, Hitoshi II-711
Sakauchi, Masao II-209
　　　　　　　　　　　　　　II-408
Salden, Alfons II-65
Samarasekera, Supun I-418
Sano, Mutsuo I-200
Sato, Yoichi II-350
　　　　　　　　　　　　　　II-424
Schütz, C. I-490
Se, Stephen............................I-152

Seales, W. Brent...............I-362
　　　　　　　　　　　　　　I-558
　　　　　　　　　　　　　　II-233
Seki, Hirohisa I-474
Sera, Hajime......................II-671
Shah, Shishir II-275
Shantaram, V. I-458
　　　　　　　　　　　　　　I-466
Shen, Helen C. II-455
Shen, Weicheng I-24
Shimada, Nobutaka I-672
Shinya, Mikio II-471
Shirai, Yoshiaki................I-672
　　　　　　　　　　　　　　II-368
　　　　　　　　　　　　　　II-432
Simon, Gilles II-360
Snyder, Wesley E...............I-722
Sochen, N. I-574
Srinivasa, Narayan..........I-623
　　　　　　　　　　　　　　II-323
Sugie, Noboru......................I-232
　　　　　　　　　　　　　　II-153
Sugimoto, Akihiro............II-161
Sugimoto, Noriko...............II-543
Sugiyama, Yoshiaki..........II-695
Suh, Tae-Jung II-241
Sumi, Yasushi.....................II-249
Suter, David I-566
　　　　　　　　　　　　　　II-49
　　　　　　　　　　　　　　II-599

T

Takai, Mikiko II-663
Takebe, Hiroaki.................II-137
Takemura, Haruo I-288
Tan, T.N. II-57
　　　　　　　　　　　　　　II-97
Tanaka, Hiromi T..............I-688
Tanaka, Yutaka I-208
Tang, Cheng-Yuan I-632
Taniguchi, Rin-ichiro...II-479
Taniguchi, Yasuhiro II-432
Tanner, Jonathan................I-664
Teoh, Eam Khwang...............I-378

Terzopoulos, Demetri II-671
Tian, Ying-li I-216
           I-224
Tomasi, Carlo I-104
Tominaga, Shoji I-80
           II-258
Tomita, Fumiaki II-249
Tong, W.B. I-136
Toriu, Takashi II-1
Torreão, José R.A. I-64
Toyooka, S. I-248
Trajković, Miroslav II-376
           II-392
Tsai, Chi-Hao II-535
Tsai, Wei-Hsin II-535
Tso, S.K. I-136
Tsotsos, John K. II-575
Tsui, Hung-Tat I-216
           I-224
           II-384
Tsuruta, Naoyuki II-479
Tyan, Hsiao-Rong II-742

U

Uchimura, Keiichi II-519
Utsumi, Akira II-655

V

Vaidyanathan, B. I-296

W

Waibel, Alex II-687
Waltenberg, Peter T. I-434
Wang, Han I-378
Wang, Pingtao II-408
Wang, W. I-248
Wang, Xinli II-129
Wang, Yuan-Fang I-542
           II-400
Watanabe, Toyohide II-105
           II-113
           II-217

Weng, John (Juyang) II-503
Westling, Mark F. II-201
Wheeler, Mark D. II-350
Wirtz, Brigitte I-499
Wong, K.H. II-559
Wong, Pak-Kwong II-121
Woodham, Robert J. I-40
Wu, Haiyuan II-647
Wu, Wei II-209
Wu, Youshou I-160
           I-168
           II-145
Würtz, Rolf P. II-193

X

Xie, M. I-280
Xu, Donglai II-623
Xu, Gang II-543
Xu, Guangyou I-240

Y

Yachida, Masahiko II-639
           II-647
           II-655
Yacoob, Yaser II-267
Yamada, Masashi I-474
Yamamura, Tsuyoshi II-153
Yamazawa, Kazumasa I-288
Yang, Chuei-Yaw II-535
Yang, Jie II-687
Yang, Jun I-232
Yang, Qing II-607
Ye, Xiuqing II-511
Ye, Yimimg II-575
Yeung, S.Y. I-216
           I-224
Yokokura, Naoko II-113
Yokoya, Naokazu I-288
Yonemoto, Satoshi II-479
Yow, Kin Choong I-515
Yu, Gwo-Jong II-742
Yuan, Cheng Jiun II-233
Yun, Il Dong I-96